Molecular Solar Fuels

RSC Energy and Environment Series

Editor-in-Chief:
Professor Laurence Peter, *University of Bath, UK*

Series Editors:
Professor Heinz Frei, *Lawrence Berkeley National Laboratory, USA*
Professor Ferdi Schüth, *Max Planck Institute for Coal Research, Germany*
Professor Tim S. Zhao, *The Hong Kong University of Science and Technology, Hong Kong*

Titles in the Series:
1: Thermochemical Conversion of Biomass to Liquid Fuels and Chemicals
2: Innovations in Fuel Cell Technologies
3: Energy Crops
4: Chemical and Biochemical Catalysis for Next Generation Biofuels
5: Molecular Solar Fuels

How to obtain future titles on publication:
A standing order plan is available for this series. A standing order will bring delivery of each new volume immediately on publication.

For further information please contact:
Book Sales Department, Royal Society of Chemistry, Thomas Graham House, Science Park, Milton Road, Cambridge, CB4 0WF, UK
Telephone: +44 (0)1223 420066, Fax: +44 (0)1223 420247, Email: books@rsc.org
Visit our website at http://www.rsc.org/Shop/Books/

Molecular Solar Fuels

Edited by

Thomas J. Wydrzynski and Warwick Hillier
The Australian National University, Canberra, Australia

RSC Publishing

RSC Energy and Environment Series No. 5

ISBN: 978-1-84973-034-1
ISSN: 2044-0774

A catalogue record for this book is available from the British Library

Published by The Royal Society of Chemistry,
Thomas Graham House, Science Park, Milton Road,
Cambridge CB4 0WF, UK

Registered Charity Number 207890

For further information see our web site at www.rsc.org

Foreword

It has been increasingly recognized that one of the major problems facing our generation and later ones is the reliance on fossil fuels, their effect on climate, their gradual exhaustion, and the great need for alternative energy sources. Since the most conspicuous and abundant source of energy is the sun, solar energy conversion has become the magical concept to conjure the development of organic and inorganic photovoltaics and, even better, the direct production of chemical energy carriers.

The focus of the present volume is on the many facets of new energy technologies where functional components reflect single features of the photosynthetic apparatus in nature. These features relate to the structure of the photosynthetic apparatus, its reaction kinetics and mechanisms, its underlying design and realization of analogs that mimic, in particular, photosystem II and its catalytic unit that is responsible for the splitting of water into oxygen and (bound) hydrogen. Related to the connected uptake and release of molecular hydrogen a number of chapters are also devoted to hydrogenase enzymes and mimics thereof.

The organization of the individual contributions into sections for *light harvesting*, *photochemical conversion* and *energy storage* in chemical bonds underlines both the complexity of natural photosynthesis and the problems associated with the transferability of biological strategies to large-scale solar technology. This wide ranging state-of-the-art presentation with its many vignettes of past history is also of interest to science history buffs. The present understanding of almost all fundamental aspects of photosynthesis rests on detailed knowledge of structure and organization of the photosynthetic apparatus as well as of its spectroscopic and kinetic fingerprints. The evolving features of Photosystem II over the past eighty years has seen, particularly in the last decade, a steady progress in atomic resolution of the crystal structure with foremost interest in the Mn_4CaO_5 catalytic cluster, the role of its

RSC Energy and Environment Series No. 5
Molecular Solar Fuels
Edited by Thomas J. Wydrzynski and Warwick Hillier

Published by the Royal Society of Chemistry, www.rsc.org

environment and the many water molecules that may constitute the various ion and other delivery channels in the protein. In many cases synthetic analogs of this oxygen evolving apparatus are being sought, with and without electrodes, accompanied by structural studies and theoretical calculations and discussions of detailed mechanisms of the various reactions in these complex systems.

These studies of photosystems that generate energy-rich chemicals that can later be used to generate fuels inspire and complement a parallel approach involving photovoltaic cells where perhaps two may act in concert for water oxidation and hydrogen production, in the optimal version as tandem cells equipped with different light absorbers covering most of the solar spectrum. In many instances, theory plays a role in unifying the structural and spectroscopic information, assisting the understanding of the many chemical reactions, including the frequent transfer of electrons between different sites, proton and proton coupled-electron transfers, and various bond breaking bond forming reaction steps. Readers interested in pursuing or understanding the phenomena involved and assessing future perspectives reaching out to artificial enzymes, protein maquettes and synthetic biology will find in this book sound information on the basics of photosynthesis relevant for the production of solar fuels.

Rudolph A. Marcus
California Institute of Technology
Pasadena CA
Maria-Elisabeth Michel-Beyerle
Technical University of Munich
Nanyang Technological University of Singapore

Preface

World demand for energy is rapidly increasing and projections suggest energy consumption will more than double to some 30 TW by the year 2050. The challenge we face today is to quickly identify carbon-neutral energy sources, and then to develop the infrastructure for implementing and storing these energy sources. One energy source with sufficient scope to meet the growing needs is solar energy. Sunlight accounts for the largest energy input into the earth's surface, providing more energy in one hour (4.3×10^{20} J) than all of the energy consumed by the entire planet in one year. If the sunlight could be harnessed and efficiently stored—such as in the chemical bonds of a molecular fuel, *e.g.* H_2, C–C, C–H carbon skeleton—then a near unlimited supply of clean energy would become available.

About 2.5 billion years ago, the ancestral parent to present-day cyanobacteria evolved a unique method for capturing solar energy, by catalysing the light-driven oxidation of water into molecular O_2, protons and electrons. The biochemical paradigm that evolved was a thermodynamic success, exploiting the chemistry of abundant first-row transition metals. If the efficiency of this reaction were replicated industrially and the released protons were coupled to molecular H_2 production, then an highly efficient fuel cycle could be achieved based on water, where only water forms from the combustion of H_2 with O_2. Alternatively, if the water oxidation reaction could be coupled to CO_2 reduction, then molecular fuels with low carbon credits, such as methanol, could be generated. In the quest for clean, energy-rich fuels, we must take advantage of the natural photosynthetic and hydrogenase systems, by adapting the energy conserving principles that nature has developed through evolution of combinatorial chemistry. These natural systems can provide important markers to guide the development of synthetic photo-oxidative and reductive catalysts that can be used in solar-driven fuel production. To achieve this goal, cutting-edge

RSC Energy and Environment Series No. 5
Molecular Solar Fuels
Edited by Thomas J. Wydrzynski and Warwick Hillier

Published by the Royal Society of Chemistry, www.rsc.org

photochemical conversion technologies and breakthroughs in the understanding of the natural photosynthetic and hydrogen producing systems are essential. An in-depth knowledge of the natural electron cofactors, mechanisms of electron and proton transfer and role of the supporting protein matrices in bio-catalysis must be attained before practical use of energy-rich solar fuels can be developed.

The present volume in the RSC Energy Series is designed to present the recent progress in this area of research by bringing together those researchers working on the natural systems with the researchers developing synthetic chemical catalysts. The book can be divided into four themes. The first part, *Perspectives of Solar Energy and Solar Fuels,* introduces the book with three chapters that justify the development of solar-based fuels. In Chapter 1 practical inputs of solar energy capture and the conversion efficiencies in plants are discussed with an overview of photosynthesis. Chapter 2 provides a summary of solar energy conversion *via* thermal and photochemical means. Chapter 3 addresses photochemistry and how this can be coupled to chemistry of electron and proton transfer mechanisms.

The next two chapters summarize our current knowledge with respect to, *Photochemical Light Capture* and *Solar Energy Conversion to Molecular Fuels.* These sections embrace the details of the natural system, emphasizing several biological aspects that optimize the efficiency. The following chapters summarize the latest developments in the synthetic chemistry of photo-oxidative and reductive catalysts.

For the section on *Photochemical Light Capture*, Chapter 4 summarizes the structure and function of the light-harvesting complexes in photosynthetic organisms. Chapter 5 then details structural aspects of the remarkable machinery in a photosynthetic reaction centre. Chapter 6 covers the physics of light capture in pigment-protein complexes. The next section on *Solar Energy Conversion to Molecular Fuels* deals with transformation of the light energy into a chemical bond—the storage of energy. Chapter 7 in this section provides the latest information on the mechanism of photosynthetic O_2 evolution, which is thermodynamically the most difficult step in the overall water splitting process. Chapter 8 describes the current views on the movement of water through the protein matrix. The controlled access of the substrate water to the catalytic site and release of the O_2 by-product may be critical for the reaction path to avoid producing damaging reactive oxygen intermediates. Chapter 9 describes various inorganic and bio-inorganic complexes of manganese synthesized to model the photosynthetic water-oxidation reaction in plants and cyanobacteria. Chapter 10 explores similar water oxidation chemistry based on different transition metals such as Ruthenium. Chapter 11 summarizes the broad range of hydrogenase enzymes present today as a result of convergent evolution and their various reductive activities, while Chapter 12 reveals the first synthetic molecular model of a hydrogenase catalytic site that produces H_2. The final chapter in this section, Chapter 13 covers the thermodynamics of O–O and H–H bond activation in natural systems and how biology has optimised the reaction pathways.

The last section of the book, *Bio-Engineering Strategies for Molecular Solar Fuels*, covers possible future scenarios for solar fuels. Chapter 14 describes the design properties of synthetic proteins and protein maquettes and how to assemble redox-active cofactors. Chapter 15 describes the synthesis and development of light-activated bio-conjugate molecules. Chapter 16 describes the properties of the light-induced, redox-active cofactors that are bioengineered into non-photosynthetic proteins from bacteria. Chapter 17 describes the conjugation of a photosynthetic reaction centre and a hydrogenase enzyme capable of photolytic H_2 generation. The final Chapter 18 outlines a future research initiative and imperative to develop research momentum in renewable solar based energy conversion technology.

The book is thus designed to bring together researchers from diverse fields in chemistry and biology to work together on creating clean, high energy fuels driven by sunlight.

Contents

RSC Energy and Environment Series No. 5
Molecular Solar Fuels
Edited by Thomas J. Wydrzynski and Warwick Hillier

Published by the Royal Society of Chemistry, www.rsc.org

CHAPTER 1

Harvesting Solar Energy through Natural or Artificial Photosynthesis: Scientific, Social, Political and Economic Implications

A. W. D. LARKUM*

School of Biological Sciences, University of Sydney, NSW 2006, Australia

1.1 Introduction

Humans are facing some difficult choices in the near future. It seems highly probable that greenhouse gas emissions will lead to significant global warming over the next 50 years.[1] This will in turn lead to an increasing constraint on use of carbon-based fuels, especially fossil fuels, for transportation, heating, *etc.* On the other hand solar energy provides our planet with a plenitude of energy, which at the present time is barely utilised by humans. This is a source of energy that is being heralded as the future supply of energy on our planet. And indeed it is clearly a possible source. On the other hand it is important to grasp that this source is not without its own problems, at a number of levels.

Solar energy can be captured and turned into electrical energy by photovoltaic solar panels or by thermal heating systems, with an efficiency of 10% or more.[2] This would displace areas of land or sea used for other purposes or not used at all. Land designated for this purpose would also compete with attempts to use cyanobacteria, algae and plants, ancient photosynthetic organisms on the

RSC Energy and Environment Series No. 5
Molecular Solar Fuels
Edited by Thomas J. Wydrzynski and Warwick Hillier

Published by the Royal Society of Chemistry, www.rsc.org

Earth's surface, to capture solar energy and convert it into bioenergy in the form of stores of organic carbon, albeit by a mechanism with only 0.01–1% efficiency. Nevertheless, this ancient process of photosynthesis is the major driving force for life on our planet. Fortunately, the land or sea needed by photovoltaic/solar thermal systems would be quite small compared with the overall size of our planet: an area of land the size of Egypt or Colombia and slightly larger than that of France would be needed.[3] Much larger areas would be required (>10 fold) if natural photosynthesis were used in the place of photovoltaic solar panels. Furthermore, the use of land for solar harvesting poses ethical problems, which will be discussed below. For example, the use of cropland to produce biofuels has already raised the price of food around the world,[4–6] even at the moderate levels now applied; so without proper planning very serious ethical problems could arise. Finally, the use of natural photosynthesis to sequester stores of organic carbon for bioenergy is less attractive than it might seem in terms of a carbon and solar footprints,[3] as discussed below.

Use of marginal land to produce lignocellulose products that could form the basis for biofuel production has been an unrealised goal for many years. How realistic is this goal and what are the alternatives? The use of algae in ponds and photobioreactors has also been proposed over recent years. Such systems have a rather large solar footprint and their carbon footprint is not as attractive as it might appear at first sight. Thus it is important to recognise these limitations and plan accordingly.

Of course, another strategy is to re-engineer natural photosynthesis by completely artificial means (see later chapters in this book). As discussed in this chapter, natural photosynthesis is hampered by the fact that the basic mechanisms have evolved over the last 2.5–3 billion years and the ability for retro-engineering has been very limited – and will probably continue to be so even under selection by humans. Therefore, artificial photosynthesis or some form of hydrogen generation, involving photosynthesis, has future potential, despite being in an early stage of development. Such developments have the attraction that they hold out the promise of both small carbon and solar footprints.

1.2 Solar Energy Input to the Earth, and Current and Future Energy Usage by Society

The amount of energy reaching the surface of the Earth has been known for many years[7,8] and, despite refinement to show decadal variation,[9] the values have remained much the same. The annual value for the amount of energy reaching the Earth's surface is ~2 500 000 ExaJoules (EJ, where $EJ = 5 \times 10^{20}$ Joules yr^{-1}).[1] Discarding the near infra-red wavelengths that are not available to photosynthetic pigments this leaves ~2 200 000 EJ of the available energy per year.

Values for the amount of solar energy at many points on the Earth's surface are available for many sites (*e.g.*, ref. 10). It is also now possible to make good models for any point at the Earth's surface at any time[11] and to compute daily,

Table 1.1 Budget for solar energy at various sites on the Earth's surface and theoretical primary productivity at various latitudes.

Latitude	*Growing season*	*Daily carbon fixation/ gC m^{-2} d^{-1}*	*Total irradiance/ E m^{-2}*	*Total carbon fixation/ gC m^{-2}*
Equator	All year (365 d)	19.8 17.6	20 238	6823
Darwin (12°28′S)	All year (365 d)	20.3 19.2 13.9	19 710	6602
Tropic of Cancer* 23°30′N	All year (365 d)	21.8 17.6 9.90	18 615	6136
37°N*	7 Months (214 d)	22.3 14.2 4.98	12 208	4031
55°N	5 Months (153 d)	20.8 8.09 0.606	8130	2533

*Values for the Tropic of Capricorn and 37°S are similar. The values for daily carbon are for the summer solstice, the equinox and the winter solstice, except for the Equator where there is only one solstice. Taken from Ritchie, 2010.[11]

seasonal, and annual irradiation levels, both for a horizontal surface and for a surface perpendicular to the sun's rays. Perhaps surprisingly, on a cloudless day, the temperate zones offer good levels of solar energy, even though the season is shorter (Table 1.1).

World energy use is approximately 500 EJ per year (based on values for 2008). Of this total, 80–90% is derived from fossil fuels.[10] In equivalent units, this is an average consumption of 16 TeraWatts (1.6×10^{13} W). In comparison with 1980, this represented an approximate doubling of energy use over the previous 25 years. Further, it is calculated to double again by 2040, with China and India accounting for the greatest increases.

On the basis of the Earth's surface area of ~510 000 00 km^2 and the value of annual solar radiation (above) one can calculate that currently the world's energy needs would be satisfied with an area of 1 150 925 km^2 if 10% of the solar energy could be converted to useable energy, *i.e.* ~0.02% of the total surface area or 0.07% of the land surface, *i.e.* equivalent to a country the size of Ethiopia, Egypt or Colombia.

Assuming that natural photosynthesis can convert solar energy at 1% efficiency (a value in the upper range of current plant productivities – see Section 1.5, Figure 1.2 and Table 1.3) the area would be 10 fold the previously mentioned land area, *i.e.* 11 509 250 km^2, or an area a little bigger than Canada and somewhat smaller than the largest country in the world, Russia.

These then are the realistic figures that we need to work with in considering solar energy conversion into energy or organic products by solar panels or by cyanobacteria, algae and plants and/or by artificial photosynthetic systems.

1.3 Photosynthesis on the Earth is an Ancient Process

Solar energy has sustained life by photosynthesis on the Earth from soon after the emergence of the earliest organisms, which appeared about 3.9 Ga (billion years ago). The earliest form of photosynthesis, which may have evolved 3.5 Ga, occurred in an environment of low atmospheric oxygen, high levels of reducing molecules, such as methane, H_2S (and other sulphur compounds) and H_2 gas, which were used as the oxidising agents necessary for photosynthesis. However, as the reducing atmosphere became exhausted by (i) the activities of the earliest organisms and (ii) the out-gassing to space of gaseous hydrogen, the need for using water as a source of hydrogen placed a strong selection pressure on the evolution of photosynthesis where oxidation occurred by extracting hydrogen from water. This occurred by the evolution of the water splitting mechanism of **oxygenic photosynthesis** – so called because in the process of using the hydrogen from water, oxygen is released as a by-product. The first organisms to perform this revolutionary process were undoubtedly cyanobacteria (or proto-cyanobacteria), which were derived from primitive anoxygenic photosynthetic bacteria.[12]

It is oxygenic photosynthesis, performed by cyanobacteria and their eukaryotic successors, algae and land plants, which is the major process of solar energy conversion on the Earth today (Figures 1.1 and 1.2). And this is also by far the

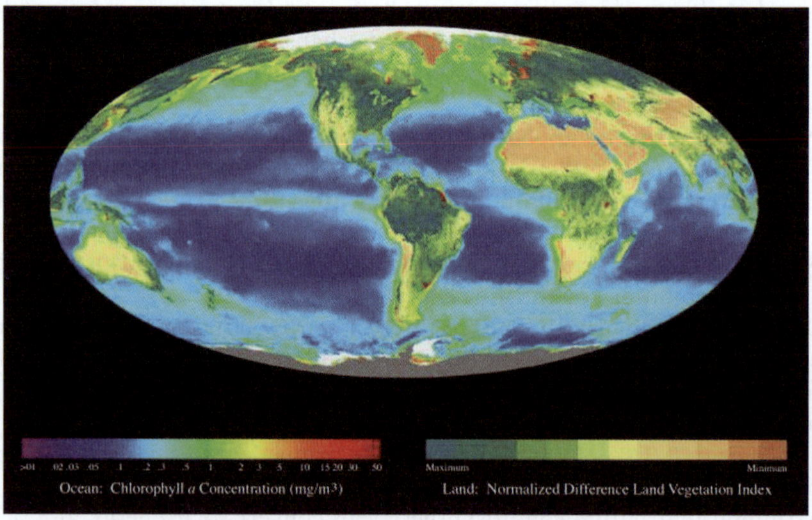

Figure 1.1 Global production of organic matter showing values for land plants and ocean algal and cyanobacterial communities. Note that the majority of primary producers on land are plants. In the sea, the algae are the major producers, but recent estimates have shown that cyanobacteria are very significant primary producers: *Prochlorococcus* spp. having been shown to account for up to 20% or more of oceanic production. Provided with permission by the SeaWiFS Project, NASA/Goddard Space Flight Center and ORBIMAGE.

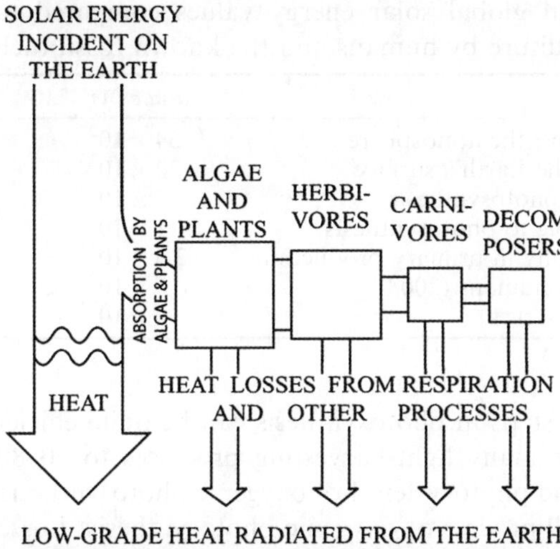

Figure 1.2 Solar energy and the flow of energy in the biosphere. For convenience the small contributions of solar energy to non-oxygenic photosynthesis by photosynthetic bacteria, the contributions of chemosynthetic bacteria and the contributions from geothermal activity have been ignored.

most important process for the input of organic energy into the biosphere: chemical, thermal and radioactive inputs being by comparison negligible.

Since the evolution of the basic mechanisms of photosynthesis took place over 2 billion years ago it is perhaps not surprising that cyanobacteria, algae and land plants all employ a rather uniform chemical apparatus to transform solar energy into the energy stored in organic molecules: the basic mechanisms must have evolved rapidly with the early evolution of cyanobacteria, perhaps 2.7 to 2.4 Ga. Once cyanobacteria came into symbiotic association with eukaryotes (in algae and land plants) it seems that the opportunities for evolution of the basic mechanisms were very restricted.[12] As a result it is not difficult to compute the efficiencies of this rather uniform process in all the different organisms that use oxygenic photosynthesis; the presence of a wide variety of photosynthetically-active pigments in algae make this already diverse group of protists even more diverse in terms of light-harvesting strategies, but nevertheless their absorption properties can be computed quite simply.

The basic equation of photosynthesis is

$$CO_2 + H_2O + \text{energy} \rightarrow [CH_2O] + O_2$$

where $[CH_2O]$ represents a portion of an organic molecule. This equation allows us to compute the organic production of oxygenic photosynthetic organisms and, from a knowledge of the input of solar energy to our planet, to then calculate realistic rates of solar energy conversion into organic molecules (Table 1.2).

Table 1.2 Annual global solar energy values compared to annual energy expenditure by humans and the known fossil fuel reserves.

	Joules yr^{-1}	*%incident energy*
Energy incident above the atmosphere	54×10^{23}	286
Energy incident at the Earth's surface	22×10^{23}	100
Energy available to photosynthesis	7×10^{23}	35
Max. energy available to photosynthesis	1×10^{23}	5
Energy stored annually in primary production	2.5×10^{21}	0.1
Energy expended by humans (2008)	5×10^{20}	0.00025
Known fossil fuel reserves	4×10^{23}	2.5

While the basic steps in photosynthesis can be quite efficient, ranging from 99% efficiency for many light-harvesting processes to 70–80% for bacterial reaction centres and photosystem I of oxygenic photosynthesis,[13] this is not the case for the overall process of oxygenic photosynthesis. Losses in efficiency in the primary processes, *i.e.*, up to the stable conversion of CO_2 into organic matter, occur at a number of levels and are discussed in the next section (Section 1.4, Figure 1.3).

Finally, it is clear that photosynthesis by land plants is more efficient than that of photosynthetic organisms in the sea (cyanobacteria, algae and a few higher plants – mainly seagrasses). This is clearly shown by the fact that the surface of the globe that is covered by land is only about one third of the total, yet the primary production on the land accounts for about half of the total primary production,[14–16] despite the relative large areas of desert on the land area (Figure 1.1).

1.4 Inefficiencies in Oxygenic Photosynthesis

1.4.1 Losses in Absorption

The major flows of energy through the biosphere are set out in Figure 1.2. In considering losses of light (excitation energy) before it enters the photosynthetic process, one must consider, principally, (i) losses in the medium, (ii) losses at the plant/algal interface and (iii) losses in the photosynthetic absorption process.[17]

For land plants, losses in the medium (air) are negligible (Figure 1.3A), but for aquatic algae they may be significant; and if the algae live at any depth they may be considerable (Figure 1.3B). Thus, in developing schemes for algae the use of shallow ponds with thin layers of algae is the preferable strategy – and this necessarily rules out considerations of utilising the ocean or large lakes.

For losses at the plant/algal interface, there will inevitably be scattering or reflection of light (Figure 1.3A and B). However, what is more important is how the photosynthetic light-harvesting system is arranged. In land plants, a major consideration is the absorption of light through the plant canopy. In aquatic systems it is the arrangement of algal cells (unicellular, usually) in the

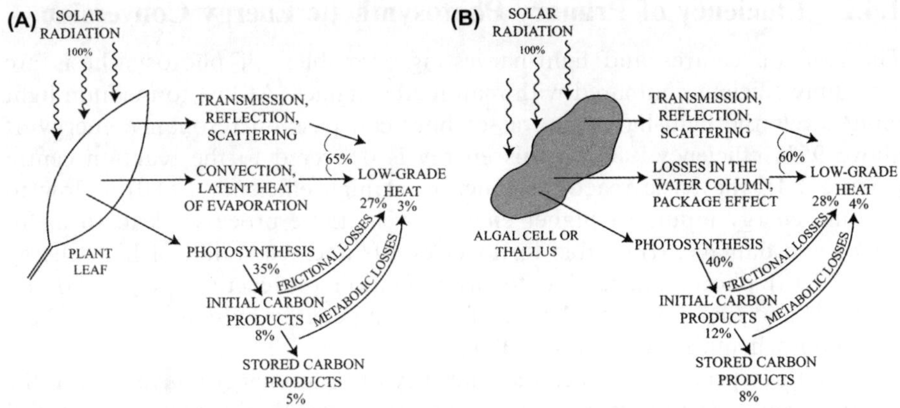

Figure 1.3 Energy conversion and storage in typical oxygen-producing plants and algae or cyanobacteria. Much of the incident energy is lost directly or indirectly as low-grade heat. Only small amounts are stored as chemical free energy in new organic material. The value of 5–6% in efficiency of energy storage is an upper limit obtained for plants and algae over short periods. Values for crops and natural vegetation are listed in Table 2. (A) Plant systems on land. (B) Algal and cyanobacterial systems.

water column: here the most efficient arrangement would be to form a thin algal film or stirred algal suspension: but this is impossible in almost all natural situations. The problem is the rapid reduction in light as it penetrates the plant canopy or algal suspension, and light-harvesting over a range of light intensities that may range from inhibitory to light-limiting; and the fact that the inability of the organism to match this changing situation quickly results in large inefficiencies. Since light in the sea either encounters a cell with high pigment content or penetrates a rather transparent medium, an inefficiency known as the *package effect* occurs, and this further lowers absorption effects in the ocean. Furthermore, harm can be done to the photosynthetic apparatus at high light-intensities, and as a result plants, algae and cyanobacteria put into place a system whereby light energy is wasted by being converted to heat at high intensities (a process called down-regulation) and this further reduces efficiency (see below).

Losses due to plant pigmentation will vary depending on the type of photosynthetic organism. For land plants with their fairly uniform pigmentation and lack of efficient pigments in the green region of the spectrum (*Green Window*), the losses are quite significant. For algae, the extent of losses are more variable; the absorption spectra of typical algae (*e.g.* ref. 16) reveal how variable this range may be. The losses range, typically, from 50% for a cyanobacterium, with phycoerythrin, to 75% for a green alga. Clearly, a cyanobacterium with phycoerythrin would be the best photosynthetic organism to use in harvesting solar energy: however, there would be hidden costs in the extra energy needed to synthesise the additional light-harvesting machinery and in the extra nitrogen needed for the extra protein demand of the phycoerythrin.

1.4.2 Efficiency of Primary Photosynthetic Energy Conversion

The reaction centres and light-harvesting assemblies of photosynthesis are incredibly efficient compared with man-made engines. At low to medium light input levels the light-harvesting assemblies can pass on resonance energy at above 95% efficiency.[13] Once this energy is delivered to the reaction centre (Figure 1.4), the basic reactions function at high efficiency (>80) at low to medium energy input. At higher energy levels, three processes lead to inefficiency and damage: (i) double turnover events and relaxation of high-energy states,[18] (ii) photodamage due to accumulation of reactive oxygen species (ROS),[19] and (iii) down-regulation of incoming energy by conversion to heat (non-photochemical quenching: NPQ).[20]

These three processes reduce the efficiency of solar energy conversion in the reaction centres (RCs) from ~80% at low to medium light intensities, to well below 30% at the highest intensities (see below). The situation is also made worse by the fact that photodamage, which occurs at high light intensities during the middle hours of the day, is a long-term event and recovery may take several hours, so that even during the afternoon, when light intensities are lower, the efficiency of photosynthesis is significantly impaired (and in cold climates the effect may linger to the next day). Thus, the quantum yield of land plant and microalgal photosynthesis drops from ~80% under ideal (very low

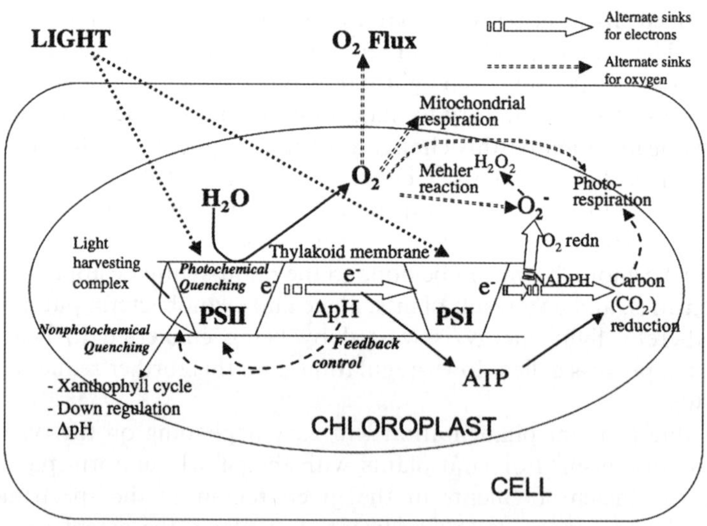

Figure 1.4 Schematic drawing showing the major processes of light absorption, energy conversion and electron transport in chloroplasts of algae and land plants. Photosystem I (PSI) and Photosystem II (PSII) are arranged in thylakoid membranes of chloroplasts, the photosynthetic organelles of algae and land plants. Note that in cyanobacteria there are no chloroplasts and the thylakoid membranes lie within the general cell contents. ATP: adenosine triphosophate; e^-: electron; H_2O: water; O_2: oxygen: ΔpH: delta pH, an acidification of the thylakoid lumen driven by light.

light) conditions to <20% under high light conditions where photoinhibition and NPQ occur.[21]

In addition, oxygenic photosynthetic electron transport is a two-step process with two photosystems (PSII and PSI) in series. Theoretically, there is enough energy in any region of the visible spectrum to drive a single photosystem that would deliver the same result: *i.e.* split water and produce NAPDH $+H^+$. Thus the present two-step process, which evolved over 2 billion years ago, is inherently inefficient. In effect, the two-step process would halve the efficiency of the primary processes. The situation is somewhat ameliorated because energy is recouped in the electron transport pathway between the two photosystems, and the proton gradient, which is generated, can be used to synthesise ATP; however, the recouping of energy is relatively small.

Any recouping of energy is offset by the fact that maximal efficiencies are only obtained in the red region of the spectrum. Shorter wavelengths towards the blue region of the visible spectrum carry more energy per quantum, but can do only the same photosynthetic work, so that a quantum of violet light is only 60% as efficient as a quantum of red light; and the remainder of the absorbed energy is lost as heat. Thus, in a green alga, which absorbs the majority of its light in the violet to blue-green region of the spectrum, the overall efficiency of photosynthesis is cut, due to this effect, by a further factor of *ca.* 25%.

There are a number of enzymic reactions involved in coupling the high-energy state of the reaction centres into the semi-stable chemical products, ATP and NADP $+ H^+$. These intermediate reactions involve the electron transfer pathway from water to NADP $+ H^+$. Much of the time, these processes do not work at maximum efficiency with the result that further losses in efficiency are incurred.

All these inefficiencies mean that the overall processes involved in initial energy capture, *i.e.* primary photochemistry and the production of ATP and NADPH $+ H^+$, are only 20–25% efficient (and perhaps much less) at low light intensities, and considerably less at high light intensities.[21–23] Further losses in secondary processing, growth, disease and carbon storage mean that the ultimate efficiency of photosynthesis is further reduced (Figure 1.3 A and B).

1.4.3 Losses in Secondary Processing

Losses in efficiency at the secondary level begin soon after the primary events of photosynthesis. The process of carbon dioxide fixation occurs as soon as ATP and NADPH $+H^+$ are produced in photosynthetic electron transfer. The sugars produced are used in metabolism to form all the organic compounds that are required by the organism, but this incurs metabolic energy losses. With unicellular cyanobacteria and unicellular algae this metabolism necessarily occurs in the same cell; in multicellular algae and land plants the sugars and other simple organic molecules are translocated to other parts of the organism. The latter process necessarily requires energy and thus the more complex the body plan the more energy is expended in maintenance and the more inefficient is overall photosynthesis (see Table 6.5 in ref. 25).

However, even the most efficient single cells must exist during periods of darkness (or near darkness in bad weather and in the early and late parts of the day). As even simple cells must respire and metabolise during such periods, there is no avoiding considerable losses of newly acquired carbon. The result is a loss of efficiency to the overall photosynthetic process of significant proportions: in massive organisms such as trees the losses can be well over 50%,[25] and in unicellular algae, which cannot control their position in the water column, the losses may also be high.

1.5 Photosynthesis on the Earth Today

1.5.1 Primary Production

In addition to the losses of efficiency outlined above, other losses are inevitable over the life-time of an organism, which must reproduce itself, combat attacks by pathogens and sustain itself during adverse periods of weather, such as storms, and seasonal changes, such as excessive cold, snow and ice. For this reason, storage of fixed carbon will always be smaller than that computed from theoretical calculations. When all this has been factored in we arrive at an integrated rate of carbon storage for any photosynthetic organism and this is called *primary production.*[14–16] For crops and natural communities, many attempts have been made to calculate primary production – usually represented in units of amount of carbon or organic matter fixed per unit area per year or per day (averaged over a year). Table 1.3 presents some typical values (other terminology is used such as net primary production, and different methods may be used to compute the results, but the values are much same). Whichever way the calculations are done, however, the indisputable result is that in terms of efficiency of solar energy conversion, the rates of the even the best crop plants are less than 1% (see below and Table 1.3).

Table 1.3 Primary production of various types of natural and cultivated vegetation. Estimated for dry matter production of crop or vegetation averaged over a year. Values taken from Larkum, 1981[14] (see also ref. 15).

	Rate of dry matter production (tonnes ha^{-1} yr^{-1})	*Efficiency of solar energy conversion (%)*
Tropical		
Napier grass	88	1.6
Sugar cane	66	1.2
Reed swamp	59	1.1
Temperate		
Perennial crops	29	0.5
Annual crops	22	0.4
Grassland	22	0.4
Evergreen forest	22	0.4
Deciduous forest	15	0.3
Savanna	11	0.2
Desert	1	0.02

1.5.2 Crop Yields: Land Plants and Algae

From the previous section is important to note that even the best agricultural crops, grown under average field conditions, have yields of only about 1% at the very highest, and usually much less.[7,15] To achieve even these efficiencies for crop plants, very intensive agricultural practices have to be applied, requiring watering, fertilisers, weed control, tillage and crop rotation. Some new crops, such as switchgrass, jatropha and *Miscanthus* have recently come to the fore.[26–28] These increase the yield for biofuel production by utilising more of the plant, but they do not increase yields by a great extent and in practice many of the new crops have been found wanting.[29–31]

It is also important to realise that land plants have a hidden advantage over algae or plants that live in water: land plants live and photosynthesise in air, not water. Their leaves have evolved to utilise the rapid diffusion of gases in air, which is of the order of 10 000 times faster than that in water. In land plants, an air to water interface is eventually met but leaves are constructed with stomata and air spaces so that this barrier occurs near the final photosynthetic cells, in the palisade parenchyma, where the resistance to carbon dioxide uptake/bicarbonate movement is minimal.[32] For algal cells (and aquatic plants) water presents a much greater resistance to diffusion.[33,34] Even for single cells, the unstirred layers increase resistance to diffusion of carbon dioxide and bicarbonate ions so that without stirring photosynthesis is restricted. Thus, for optimal photosynthesis in algae, efficient stirring must occur. This can be provided either by natural events such as wave action on coasts and tidal streams, *i.e.* in special situations, or by artificial stirring. For algal ponds and photobioreactors, there is only one solution and that is artificial stirring and this is expensive in terms of energy input.

When considering primary production of algae, especially unicellular algae, a wide range of values is available, from very low[35] to very high.[36] The high rates tend to come from theoretical calculations or from small scale, short-term cultures. For example, one can calculate theoretically a rate in terms of net daily turnover and on this basis high values can be obtained,[35] in the range of $8\,\text{gC g}^{-1}\,\text{h}^{-1}$, which for a dense algal suspension of $10\,\text{gC m}^{-2}$ would yield a production of $80\,\text{gC m}^{-2}\,\text{d}^{-1}$ or approximately 777 tonnes $\text{ha}^{-1}\,\text{yr}^{-1}$. This is a rate that is 8.8 times that produced from the most productive agricultural crop, under average growing conditions (Napier grass; Table 1.3). However, it seems likely that this $80\,\text{gC m}^{-2}\,\text{d}^{-1}$ is higher than could be sustained from even a simple bioreactor and does not include the high energy input costs of stirring, *etc.* As any system is scaled up, losses are encountered from many sources. Furthermore, it is important to realise that such values are theoretical and do not represent a realistic production rate, *i.e.* it is not a *cropping rate*. The value that would be required to compare this type of system with any agricultural system, would be the number of crops a year. While this theoretically may be several times what an agricultural crop might produce, it is important to obtain realistic figures for enterprises that would harvest over areas of the size of 1000 ha or so and not use values from what is possible from a small-scale photobioreactor. At present there are no successful systems operating at this scale and so caution should be applied when scaling up from pilot experiments. It is safer for the time being to assume that primary production at agro-industrial scales will be the same or somewhat higher than for land-based crops.

Finally, even if it is possible to produce high rates of primary production from algae, it has to be realised that this will only come about at the expense of high inputs of materials and energy and with the need of considerable human activity. Thus, it may be possible to produce, for example, jet fuel from algae, but this will come at a great cost in terms of input of energy and materials and is unlikely to be sustainable except on a very limited scale.

1.6 Greenhouse Gas Emissions and Global Climate Change

At present there is a very great need to cut the ever-increasing rate of greenhouse gas emissions to keep the planet from rising to unacceptable temperatures. 2 °C has been suggested as a reasonable goal but climate talks at Copenhagen in late 2009 brought into focus the difficulty of doing this. Nevertheless, there will be increasing concerns by international bodies such as the Intergovernmental Panel on Climate Change (IPCC), and national governments, to drastically cut emissions. On the credit side, there will be attempts to ameliorate the climate by (i) halting deforestation and (ii) sequestering greenhouse gases. The principle greenhouse gas, CO_2, could be sequestered by large-scale attempts to produce biofuels. This would have the double advantage of replacing fossil fuels, which produce CO_2, with biofuels that would result in a stable (neutral) CO_2 turnover. However, it is not clear at present how this might be done. Almost any attempt to produce significant quantities of biofuels comes at the cost of production (see previous section) and other negative effects, such as on food production, on biodiversity, on living space, and, paradoxically, on increases in greenhouse gas emissions.[3] These matters are discussed further in the next five sections.

1.7 Future Use of Bioenergy through Natural Photosynthesis

Land plants have a hidden advantage over aquatic plants in that they have evolved homoiohydry[33] to enable them to photosynthesise in air using leaves that have evolved to utilise the rapid diffusion of gases in air, which is of the order of 10,000 times faster than that in water.[32–34] Thus algae or cyanobacteria (in water) must either be limited by stirring and grow slowly – giving rise to a low rate of natural primary production – or be stirred and grow faster, which for algal ponds and photobioreactors means employing some form of artificial stirring needing a high input of energy (see Section 1.6).

Primary production by phytoplankton in lakes or oceans is very low.[14,15] However, in shallow lagoons and ponds, especially those that are polluted, opportunities do exist.[36] Nevertheless the difficulties of adapting new practices to shallow water bodies will certainly delay their use for a number of years.

Because of these advantages of land plants over algae, land plant bioenergy will be most used in the near future. Agricultural and forestry production will sustain many approaches on land in the short to medium term. Two approaches have been promoted recently, (i) a search for new crops that can produce biofuels,[38,39] and (ii) the use of trees and other lignocellulose rich resources for biofuels.[40–43] There is also the possibility of using organic wastes from towns and industrial processes.

1.8 Bioenergy from Algal Ponds and Photobioreactors

As noted in Section 1.5.2, the major disadvantage to overcome with either algal ponds or photobioreactors is stirring, with a second major disadvantage being the provision of nutrients at a sufficient level to maintain optimum growth. Many other problems also exist.[3] Thus, while these systems present novel advantages over land-based agriculture and water-based mariculture, their potential must be thoroughly assessed over this decade, and the disadvantages acknowledged and addressed.

Theoretical yields are discussed in Sections 1.3 and 1.4. There will be much debate for some time to come over the actual yield potentials from algae. This is because in single-celled algae it should be possible to redeploy a greater fraction of absorbed solar energy into fixed carbon compared with land plants, with their stems and root systems. However, the culturing of algae brings with it another set of problems, such as stirring, nutrient supply, optimisation of the light field, optimisation of the growth conditions and protection from pathogens and nuisance "weed" algae.

In the short-term (this decade), the prospect for the use of algae to provide significant amounts of organic carbon, to sustain the human need for bioenergy, is limited. In the longer term the possibilities should be explored, but they come at the cost of a high solar footprint (Section 1.9), a high carbon footprint (Section 1.10) and many unknown factors.[29–31] And it should be cautioned that over the recent past there have been some notable failures in this challenging field.[43]

1.9 Land Use and the Solar Footprint

In the future, land (and water surface) will be at a premium – for food production, for conservation of biodiversity, for solar energy harvesting, for cities and transportation and for many other purposes. As humans begin to look for "free" land to site their solar panels and bioenergy plants, there is great interest in the less-used areas of our planet, such as deserts. However, all land no matter how barren is the patrimony of one group or another and it will not be easy in future to designate these lands for energy harvesting. For example, the Sahara Desert would seem to be an opportune place for solar harvesting but it may not be an easy task to persuade the countries bordering the Sahara, which are poor and with large undernourished populations, to turn over their land for this purpose. Furthermore, security in the Saharan Desert would also pose an additional problem. Thus, in the future, land will be at a premium and any attempt to harvest solar energy will need to consider the solar footprint of the proposed installation.

Solar footprint is the amount of land required by each person (in a country or in the world) to supply the current energy need of a given population. Denholm and Margolis[44] concluded that the solar electric footprint from solar photovoltaic panels is about 181 m^2 per person in the United States. However, when the energy cost of manufacturing the devices is included the area will be substantially larger. There have been no equivalent calculations for bioenergy from land plants or from algae. Since both of the latter are much less efficient than solar panels, the solar footprint must be considerably larger and the total land required would be much larger than the size of Ethiopia, Egypt or Colombia, as calculated for solar panels in Section 1.2. If natural vegetation or natural algal lagoons are harvested, then, to some extent, it may be argued that the footprint is low. Yet in all such proposals there are hidden costs.[45–48] For algal ponds or photobioreactors, the solar footprint would be considerably increased by the need to supply a large amount of energy for stirring and to provide for the hidden energy costs of fertilisers, harvesting and transport.

1.10 Carbon Footprint

The carbon footprint is perhaps best expressed as the amount of greenhouse gases, principally CO_2, that is produced per head of a population (alternatively, it can be expressed as some other criterion such as the output of an industrial complex or country). Carbon footprints, defined in this way, can range from high values for the most advanced industrial societies, such as 19 tonnes per head for the USA, and even higher for a few small Arab and European nations, down to 0.1 or less for non-industrialised nations (*e.g.*, Ethiopia, Uganda, Somalia and Afghanistan).[10]

In future it will be necessary to factor in the contribution of various sources to our per capita carbon footprint. This especially applies to the fuels we use. Petroleum and jet fuel, which are burnt to produce CO_2 add a high contribution to the (high) carbon footprint of industrialised countries, as does electrical usage, which at present comes largely from burning fossil fuels.

The contribution to our carbon footprint of bioenergy crops might seem to be low, since algae and plants take up CO_2 in the production of biofuels. However, biofuels and bioenergy are a high-end product and require much infrastructure, both in production, harvesting of the product and delivery. As a result the production of biofuels from crops, such as bioethanol from sugar cane/sugar beet and oil from palm oil, or from forests, have been found to make for a rather high contribution to the carbon footprint.[30,45–48] Nevertheless, this approach has reached commercial levels for terrestrial plants and the economics and carbon accounting can be done with reasonable confidence.

For algae, despite the prospect for higher yields and higher crop turnover, no commercial scale production has yet been achieved and only crude economic and carbon assessments can be made. Currently, predictions are not heartening.[30,43] Any production of bioenergy crops from algae will be energy and carbon intensive. For obvious reasons, it may be possible to site algal projects adjacent to coal-fired power stations and use waste heat and CO_2 to offset the energy and CO_2 requirements of the algal installations. Overall, though, this is not a viable alternative if thousands of hectares are to be utilised. Furthermore, if solar photovoltaic panels are contemplated to offset the high-energy needs of algal farms, the solar and carbon footprints greatly increase.

1.11 Societal Problems

In the future there will be increasing competition for land (and sea). This in turn will exacerbate societal problems, which will come about through food production, biodiversity, overcrowding, quality of life and waste disposal, to mention just some of the most important issues. The designation of land, previously used to produce food, has clear ethical challenges, but nevertheless is going ahead in a number of countries, such as the USA, Brazil and many East Asian countries.[47–57] There are reliable reports that this move has already driven up the price of food[49,50,54,55,57] and is not sustainable.[58] As a result, several analyses have looked at alternative sources of land. Grazing land could have potential, but this would conflict with the production of meat.[54,55] Marginal land may have potential for new crops; it could also double up as a site for tree production.[59] Some estimates suggest that lignocellulose conversion alone could provide the majority of fuels in the future, although the solar and carbon footprints of such a proposal has yet to be costed and early analyses suggest that this approach is not sustainable.[60] Other disadvantages may be in the form of land designation, increased production of greenhouse gases, pollution and threats to biodiversity.[55,56,58,61]

Finally, it has to be realised that, despite the need to plant more trees and other remedial measures, at present deforestation is a much more serious world-wide issue that is likely to become a chronic feature of future changes.

1.12 Future Use of Bioenergy through Artificial Photosynthesis and Hydrogen Production

It should be clear that a large a part of the problem with natural production on Earth is due to the inefficiencies of the natural photosynthetic organisms (cyanobacteria, algae and land plants), which use biochemical, physiological and ecological systems that were evolved many millions of years ago, and in terms of the basic mechanisms of photosynthesis over 2 Ga. To genetically re-engineer these organisms to be significantly more efficient will be a daunting task and one that seems likely to extend hundreds of years into the future.

As a result, many scientists have begun to look at the possibility of producing bioenergy through the use of artificial photosynthetic systems and chimaerical systems based on the partial reactions of photosynthesis, improved through judicious use of chemical and genetic engineering techniques (see other chapters in this volume).

Here the goal would be to obtain efficiencies that compete with or do better than photovoltaic solar cells.

The chapters of this book contain many ideas on this subject. Artificial reaction centres have already been produced with heartening efficiencies. Here there is the possibility of using photopigments that can harvest over a more extensive range of wavelengths of light than chlorophylls and which are more stable. The ability of photosystem II reaction centres to split water, produce reducing equivalents and release oxygen is a process that will surely be mimicked in artificial systems in the future. Alternatively, photovoltaic cells might be utilised to photolyse water and to produce either hydrogen or reducing equivalents, as well as oxygen. Then units based on photosystem I could be used either to produce hydrogen or to reduce enzymic systems such as $NADP^+$.

Another challenge ahead is to try to utilise the hydrogenase properties of a number of organism (not all photosynthetic) to produce hydrogen. Early results hold out promise for the future.

The challenge here is that by incorporating higher efficiencies much smaller areas of land would be required than for natural bioenergy. Furthermore, the use of hydrogen as a fuel has obvious advantages over carbon fuels for reducing carbon footprints. The net result could be clean photosynthetic systems with both low carbon and dollar footprints.

1.13 Conclusions

Oxygenic photosynthesis evolved over 2.4 billion years ago under conditions where maximum efficiency was not a major selection pressure. Today, the most efficient photosynthetic organisms under natural conditions are the land plants. Algae and cyanobacteria, living under light-limited conditions and reliant on high stirring rates for maximum carbon fixation rates, are generally less efficient but offer possibilities in the future: perhaps upwards of 5% efficiency after molecular improvement. Solar photovoltaic panels and solar thermal collectors, however, offer an improved means of collecting solar energy compared with the maximum 1–2% efficiency of plants, algae and cyanobacteria at the present time. Nevertheless, in a world where petroleum products become scarce, natural photosynthesis will become increasingly used for the production of biofuels and organic products, but it seems unlikely that algae and cyanobacteria will be used as a source of bioenergy. In the longer-term, photobiohydrogen production and artificial photosynthesis offer much greater hope for very efficient conversion of solar energy. However, before this can happen a number of problems must be addressed, such as high solar and carbon footprints, land sequestration and loss of biodiversity.

Acknowledgements

This work has greatly benefited over the years by inspiration and advice from many colleagues. Of these I should particularly like to thank Dr Ray Ritchie and Dr David Kramer for help with this manuscript. The Australian Research Council is thanked for grants, which have supported this work.

References

1. International Energy Agency, 2008. http://www.worldenergyoutlook.org.
2. M. A. Green, K. Emery, Y. Hishikawa and W. Warta, Solar cell efficiency tables (version 33), *Prog. Photovoltaics: Res. Appl.*, 2009, **17**, 85.
3. A. W. D. Larkum, *Curr. Opin. Biotechnol.*, 2010, **21**, 271.
4. R. Doornbosch and R. Steenblick, Biofuels: is the cure worse than the disease?, *Round table on sustainable development*, OECD, Paris, France, 2007.
5. D. J. A. Johansson and C. Azar, *Climatic Change*, 2007, **82**, 267–291.
6. C. F. Runge and B. Senauer, How biofuels could starve the poor. *Foreign Affairs*, 2007, May/June.
7. N. K. Boardman and A. W. D. Larkum, Energy exchange in the living world, in *Solar Energy*, ed. H. Messell and S. T. Butler, Shakespeare Head Press, Sydney, Australia, 1976, p. 123.
8. N. K. Boardman and A. W. D. Larkum, Solar energy conversion in photosynthesis, in *Solar Energy*, ed. H. Messell and S. T. Butler, Shakespeare Head Press, Sydney, Australia, 1976, p. 145.
9. M. Wild, H. Gilgen, H. Roesch, A. Ohmura, C. N. Long, E. G. Dutton, B. Forgan, A. Kallis, V. Russak and A. Tsvetkov, *Science*, 2005, **308**, 847.
10. International Energy Agency, 2009, http://www.worldenergyoutlook.org.
11. R. J. Ritchie, *Photosynthetica*, 2010, **48**, 596.
12. A. W. D. Larkum, Evolution of the reaction centres and photosystems, in *Primary Processes of Photosynthesis: Principles and Apparatus*, ed. G. Renger, The Royal Society of Chemistry, Cambridge, UK, vol. 2, 2008, p. 489.
13. A. W. D. Larkum, Light-harvesting systems in algae, in *Photosynthesis of Algae*, ed. A. W. D. Larkum, S. E. Douglas and J. A. Raven, Kluwer Academic Publishing, Dordrecht, The Netherlands 2003, p. 277.
14. A. W. D. Larkum, Marine primary productivity, in *Marine Botany: an Australasian Perspective*, ed. M. C. Clayton and R. J. King, Longman Cheshire, Melbourne, Australia, 1981, p. 369.
15. C. B. Field, M. J. Behrenfeld, J. T. Randerson and P. J. Falkowski, *Science*, 1998, **281**, 237.

16. P. G. Falkowski and J. A. Raven. *Aquatic Photosynthesis*, 2nd edn, Princeton University Press, 2007.
17. A. W. D. Larkum and J. Barrett, Light harvesting processes in algae, *Adv. Bot. Res.*, 1983, **10**, 1.
18. V. P. Shinkarev, *Biophys. J.*, 2005, **88**, 412.
19. A. W. D. Larkum and C. J. Howe, *Adv. Bot. Res.*, 1997, **27**, 257.
20. U. Schreiber, W. Schliwa and U. Bilger, *Photosynthesis Res.*, 1986, **10**, 51.
21. A. Vonshak and G. Torzillo, Environmental stress physiology, in *Handbook of Microalgal Culture*, ed. A. Richmond, Blackwell Publishing, Oxford, UK, 2004, p. 57.
22. R. Radmer and B. Kok, *BioScience*, 1977, **27**, 599.
23. J. R. Bolton and D. A. Hall, *Photochem. Photobiol.*, 1991, **53**, 545.
24. D. A. Walker, *J. Appl. Phycol.*, 2009, **21**, 509.
25. J. A. Raven, *Energetics and transport in aquatic plants*, A.R. Liss, New York, USA, 1984.
26. C. Somerville, *Curr. Biol.*, 2007, **17**, 115.
27. W. E. Tyner, *BioScience*, 2008, **58**, 646.
28. E. A. Heaton, F. G. Dohleman and S. P. Long, *Global Change Biol.*, 2008, **14**, 2000.
29. R. Doornbosch and R. Steenblick, Biofuels: is the cure worse than the disease?, *Round table on sustainable development*, OECD, Paris, 2007.
30. D. C. Holzman, *Environ. Health Persp.*, 2008, **116**, 247.
31. K. Sanderson, *Nature*, 2009, **461**, 710.
32. P. S. Nobel, *Physicochemical and Environmental Plant Physiology*, 2nd edn, Academic Press, New York, USA, 1999.
33. J. A. Raven, *Adv. Bot. Res.*, 1977, **5**, 153.
34. A. W. D. Larkum, G. Roberts, J. Kuo and S. Strother, Gaseous movement in seagrasses, in *Biology of Seagrasses*, ed. A. W. D. Larkum, A. J. McComb, S. A. Shepherd, Elsevier, Amsterdam, The Netherlands, 1989, p. 686.
35. J. Masojidek, M. Koblizek and G. Torzillo, Photosynthesis in microalgae, in *Handbook of Microalgal Culture*, ed. A. Richmond, Blackwell Publishing, Oxford, UK, 2004, p. 20.
36. S. Bastianoni, F. Coppola, E. Tiezzi, A. Colacevich, F. Borghini and S. Focardi, *Biomass Bioenergy*, 2008, **32**, 619.
37. E. A. Heaton, F. G. Dohleman and S. P. Long, *Global Change Biol.*, 2008, **14**, 2000.
38. B. Antizar-Ladislao and J. Turrion-Gomez, *Biofuels, Bioprod. Biorefin.*, 2009, **2**, 455.
39. M. J. Aylott, E. Casella, I. Tubby, N. R. Street, P. Smith and G. Taylor, *New Phytol.*, 2008, **178**, 358.
40. R. Kumar, S. Singh and O. V. Singh, *J. Indust. Microbiol. Biotechnol.*, 2008, **35**, 377.
41. E. M. Rubin, *Nature*, 2008, **454**, 841.
42. E. M. W. Smeets and A. P. C. Faaij, *Climatic Change*, 2007, **81**, 353.
43. A. L. Mascarelli, *Nature*, 2009, **461**, 460.

44. P. Denholm and R. Margolis, The regional per-capita solar electric footprint for the United States, *National Renewable Energy Laboratory; Technical Report*, 2007, NREL/TP-670-42463.
45. M. E. D. de Oliveira, B. E. Vaughan and E. J. Rykiel Jr, *BioScience*, 2005, **55**, 593.
46. R. Righelato and D. V. Spracklen, *Science*, 2007, **317**, 902.
47. J. Fargione, J. Hill, D. Tilman, S. Polasky and P. Hawthorne, *Science*, 2008, **319**, 1235.
48. T. Searchinger, R. Heimlich, R. A. Houghton, F. Dong, A. Elobeid, J. Fabiosa, S. Tokgoz, D. Hayes and T.-H. Yu, *Science*, 2008, **319**, 1238.
49. F. R. Callé, P. de Groot, S. L. Hemstock and J. Woods, *The Biomass Assessment Handbook: Bioenergy for a Sustainable Environment*, Earthscan, UK and USA, 2007.
50. C. F. Runge and B. Senauer, *Foreign Affairs*, 2007, May/June.
51. T. Buchholz, E. Rametsteiner, T. A. Volk and V. A. Luzadis, *Energy Policy*, 2009, **37**, 484.
52. V. Dornburg, I. Lewandowski and M. Patel, *J. Industr. Ecol.*, 2008, **7**, 93.
53. H. K. Gibbs, M. Johnston, J. A. Foley, T. Holloway, C. Monfreda, N. Ramankutty and D. Zaks, *Environ. Res. Lett.*, 2008, **3**, 1.
54. D. J. A. Johansson and C. Azar, *Climatic Change*, 2007, **82**, 267.
55. J. Reilly and S. Paltsev, Biomass energy and competition for land, *MIT Joint Program on the Science and Policy of Global Change*, 2007, Report No. 145.
56. J. P. W. Scharlemann and W. F. Laurance, *Science*, 2008, **319**, 43.
57. D. Mitchell, A note on rising food prices, *World Bank – Development Economics Group (DEC)*, 2008, Policy Research Working Paper No. 4682.
58. F. Danielsen, H. Beukema, N. D. Burgess, F. Parish, C. A. Bruehl, P. F. Donald, D. Murdiyarso, B. Phalan, L. Reijnders, M. Struebig and E. B. Fitzherbert, *Conserv. Biol.*, 2008, **23**, 348.
59. J. E. Campbell, D. B. Lobell, R. C. Genova and C. B. Field, *Environ. Sci. Technol.*, 2008, **42**, 5791.
60. T. W. Patzek and D. Pimentel, *Crit. Rev. Plant Sci.*, 2006, **24**, 327.
61. R. L. Rowe, N. R. Street and G. Taylor, *Renewable Sustainable Energy Rev.*, 2008, **13**, 271.

CHAPTER 2
Solar Energy Utilisation

RON PACE AND ELMARS KRAUSZ

Research School of Chemistry, Australian National University, Canberra, Australia

2.1 Properties of Sunlight

To an excellent approximation, the sun emits radiation as a black body radiator with a surface temperature of ~5800 K (Figure 2.1). The energy intensity, immediately above the earth's atmosphere, is over 1300 W m^{-2}, *i.e.* more than a full kW m^{-2}. When considering the total *useful* sunlight that reaches the Earth's surface, however, a number of 'corrections' have to be allowed for.

Firstly, atmospheric components (aerosols and gases – 'greenhouse components') absorb in the infra-red (IR) and near infra-red (NIR) spectral regions (Figure 2.1). These absorptions correspond to the excitation of molecular vibrational transitions of gaseous molecules, as both fundamental vibrations and overtones. Most atmospheric absorption arises from oxygen-containing molecules such as water, carbon dioxide, nitrogen and sulfur oxides. Nearly all of the absorption loss occurs at photon energies below 1.5 eV (*i.e.* at wavelengths longer than ~750 nm). Ozone is a significant component of the upper atmosphere and absorbs at wavelengths shorter than ~300 nm. The total energy content of this spectral region for sunlight is low, but the protection provided from ultraviolet light *via* ozone absorption is biologically important for terrestrial life.

A second correction is that due to the earth's curvature, which means that sunlight intensity per unit *land surface* area varies with the cosine of the angle of

RSC Energy and Environment Series No. 5
Molecular Solar Fuels
Edited by Thomas J. Wydrzynski and Warwick Hillier

Published by the Royal Society of Chemistry, www.rsc.org

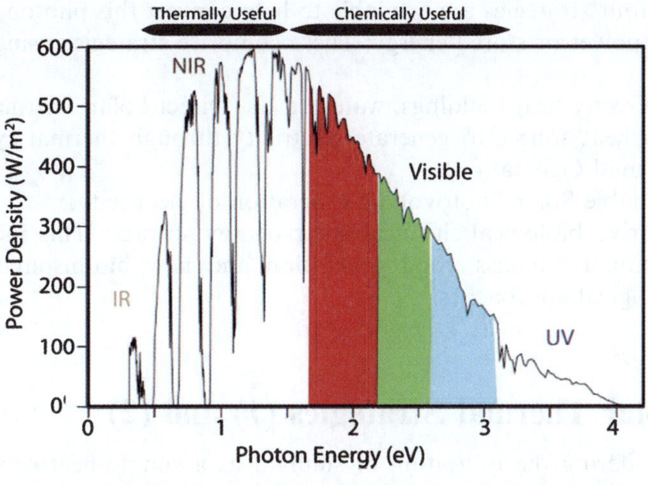

Figure 2.1 Solar irradiance as a function of photon energy in sunlight. The visible region is indicated in colour. Thermally and chemically useful regions (see text) are indicated. The main absorption bands in the IR and NIR regions appear as 'notches' (adapted from ref. 1).

latitude, compounded by the tilt of the earth's axis relative to the ecliptic. For the populated areas of Europe and North America, a ~50% reduction of intensity is apparent, compared to sunlight at normal incidence.

The final important correction is associated with cloud cover. This factor is of course highly variable, being often of little importance in dry inland areas at low latitudes, but very significant in Europe and North America. There are of course large seasonal variations but on average there is one cloud-free day in three. These factors combine to reduce the raw average (over 24 hours) surface power in sunlight to ~10–15% of the solar incidence, for those terrestrial regions in which most advanced economies are located. By contrast, in desert areas at low latitudes (North Africa, Australia), the reduction is to ~25% of solar incidence.

The photon energy spectrum (Figure 2.1) in sunlight may be, for convenience, divided into two bands. These are the 'chemically useful' energies above 1.5 eV (essentially visible light) and 'thermally useful' energies below 1.5 eV (the near infra-red region). The photovoltaic region (see below) actually extends down to ~1 eV, but the distinction is still useful. Chemically useful photons have energies sufficient to break chemical bonds, usually by release of photo-activated electrons. These 'chemically useful' photons comprise ~60% of the available solar energy spectrum at the Earth's surface. The 'thermally useful' photons (~40% of energy spectrum) are in general able to excite only molecular vibrations, which lead to the absorber being heated but not chemically transformed. The energy from the entire solar spectrum may be simply thermalised following capture by a non-reactive absorbant surface but then the sun's energy is available only as heat.

At least four strategies are available to help convert this photon energy flux into useful power or stored energy: These comprise strategies using sunlight:

(1) to directly heat buildings, water or air (Direct Solar Thermal);
(2) as a heat source to generate electricity through thermal cycles (Solar Thermal Generation);
(3) to enable Solar Photovoltaic generation of electricity;
(4) to drive biological/chemical solar energy storage. This includes conventional biomass, food generation and new bio-inspired chemical/biological approaches.

2.2 Solar Thermal Strategies (1) and (2)

When considering the utilisation of sunlight as a simple heat source, fundamental thermodynamic factors come in to play. If the thermal energy is to be used *directly for heating*, then conversion efficiencies can be quite high (~70%) and indeed are limited mainly by practical, mechanical and materials limitations. There is no fundamental limitation to this efficiency.

If, however, the heat is to be converted to mechanical energy or electricity, then a thermal engine or thermoelectric generation cycle must be used. The absolute limit of attainable efficiency (as a fraction), E_{ff}, for such systems is set by the thermodynamic relationship,

$$E_{ff} \leq [T_H - T_L]/T_H \tag{1}$$

In this equation, T_H and T_L are the temperatures (measured in degrees Kelvin, K) of the 'high' and 'low' temperature stages, respectively, in thermal cycles. In almost all cases of practical terrestrial interest, T_L is between ambient (20 °C) and 100 °C (boiling point of water), *i.e.* in the range ~300–400 K.

For a cycle to have an E_{ff} value > 0.4, *i.e.* an efficiency over 40%, after inevitable losses, T_H must be at least 500 °C (~800 K). These are typically the conditions under which fossil or nuclear-fuelled thermal power stations operate, although nuclear-fuelled generators may operate at a somewhat lower T_H and thus have a lower E_{ff}. Systems with effective T_H values of 100 °C or less are generally suitable only for Direct Solar Thermal conversion, unless construction/operating costs are suitably low or other favourable circumstances permit.

2.2.1 Direct Solar Thermal

The use of solar energy for heating, particularly water heating in domestic or larger scale applications is a well-established technology. Conversion efficiencies of ~60% or above are achievable with modern systems.[1] Water is circulated through (typically) roof mounted concentrating collector panels and heated to less than 100 °C. An insulated hot water storage vessel is normally included. However, because of *seasonal*, rather than diurnal variations, such systems are not able to supply even domestic requirements throughout the full year in the geographical areas where most advanced economies are located.

The hot water energy usage of a typical European household corresponds to ~8 kWh per day. Thus, a 3 m^2 south facing solar panel could (only just) provide this amount of heat in mid-summer under normal conditions. In mid-winter, the panel area would need to be ~5–10 times larger, and furthermore would be inoperable in many circumstances (snow, rain *etc.*). Closer to the equator (*e.g.* Australia, southern USA, Central America) the situation is much more favourable and performance is adequate *on average* for most of the year. Backup (electrical-immersion or gas heating) is almost always required for a fully reliable service.

2.2.2 Solar Thermal Generation

One technique for large-scale electricity generation by Direct Solar Thermal means has been demonstrated. This is the solar updraft tower or 'solar chimney', and is based on long-known principles. It is comprised of a large (hundreds of meters) high chimney built in the middle of a collection area covered by a transparent roof (glass/plastic). Air beneath the gently sloping roof is heated and funnelled through the base of the chimney with the rising air turning turbine blades coupled to electric generators. The temperature of the hot air is typically well less than 100 °C. Because the roof materials are simple, with low engineering requirements, this is by far the cheapest solar energy conversion system known, *per unit collection area.*

Using the latest construction approaches, it is proposed[2] that a system with a 7 km diameter collector and a 1000 m tower, operating in high average sunlight (~250 W m^{-2}), could generate ~2 W m^{-2} electrical power. This is less than 1% conversion efficiency, unsurprising given the small $T_H - T_L$ value for eqn (1)), but the output is on average significant at ~75 MW. There is scope to improve these figures by a factor of ~2 by using construction materials which store heat (metals) and operate at somewhat larger T_H. Heat storage and improved construction materials would then increase costs. At present, the power output per unit land area of solar chimneys in desert areas is similar to that of wind farms in suitable coastal regions.[3]

Virtually all the currently operating and proposed systems envisaged to convert sunlight into electric power through thermal means incorporate a system to concentrate the sunlight so as to achieve higher collector temperatures. Sunlight is then 'harvested' over a large effective area and focussed onto collector(s), much smaller in size, which are heated to T_H and serve as the input stage of a thermal cycle. Thermal cycles employed involve steam (Rankine cycles, saturated and superheated steam) and Stirling cycles (with hydrogen as working fluid) (*e.g.* see ref. 4 and 5). Although very high T_H values could in principle be envisaged, re-radiation losses from the hot stage collector surfaces become increasingly significant as T_H rises. These losses per unit area are proportional to T_H^4 (Steffan Boltzmann Law)[6] and thus rise very rapidly with temperature. While various engineering approaches may reduce these losses, they cannot fundamentally be avoided (if radiation is received, it may be re-emitted) and so a practical upper limit of $\sim$350 °C for T_H appears to operate currently, although advanced systems operating at higher values (450–500 °C) have been proposed.

Because of the apparent movement of the sun across the sky, all solar concentrating systems use some form of active heliostat sun tracking arrangement on the collector surfaces to maximise radiation incidence/unit collector area. The mechanical requirements of this and the necessity to avoid self-shadowing of the collectors means that the collector area is only a fraction (typically 30–40%) of the total land area on which the solar capture system operates. Furthermore, due to the high number of individual tracking collectors for large systems (typically hundreds to thousands for > 100 MW installations), the cost, manufacturability and reliability of the reflector-tracker components become important factors in the overall operability of the system. Generally for concentrating solar thermal conversion, the array costs totally dominate the total installation cost (*e.g.* see ref. 4).

At present three generic approaches to sunlight concentration for Solar Thermal power generation have been demonstrated (Figure 2.2). All have operated at a sufficient scale to allow a realistic assessment of their large-scale potential. These are as in Figure 2.2:

(a) Linear arrays (Linear Parabolic or Linear Fresnel Reflectors)
(b) Two dimensional (flat) reflector arrays focussing on central tower(s)
(c) Parabolic dishes with individual generators (*e.g.* Stirling Cycle).

We briefly examine representative examples of each.

Figure 2.2 Examples of concentration systems for Solar Thermal power generation: (a) Solar Heat and Power (SHP) linear Fresnel array system installed in NSW, Australia. A linear tube receiver is illuminated by one axis steerable reflector panels (viewed from below). (b) eSolar SunTower Heliostat system installed in Southern California, USA. Tower collectors are seen illuminated on four sides by multiple two axis steerable reflector panels. (c) Stirling Energy Systems (Sandia National Laboratories) solar dish array installed in Albuquerque, NM, USA. Each two axis steerable dish has an integrated Stirling engine, electrical power generator at the dish focus.

(a) Linear Arrays

The latest technology uses linear Fresnel reflectors, which appear to be more efficient than simple parabolic troughs.[4] As they are nearly 'continuous' along one dimension these systems make efficient use of the site space, with a ratio of collector/total area near 50%. The long ($\sim$50 m) flat reflector panels have fixed orientation (north/south) and tracking of these reflector strips in the segmented Fresnel surface occurs by motion around single horizontal axes parallel to the array direction. With this geometry, the solar energy is concentrated onto receivers, which are also parallel to and fixed above the array surfaces. These thermal receivers are absorbent 'tubes' through which pressurised water is pumped and heated (*i.e.* 'line' focus system).

Mills *et al.*[4] describe a proposed system for a 240 MW demonstration plant, based on a smaller functioning prototype operating in Eastern Australia (Figure 2.2a). This would operate at $\sim$300 °C maximum, using saturated steam Rankine cycle turbines, without superheating. Such operation is similar to that in 'pressurised water' nuclear power plants, for which huge investments have been made in developing turbine technologies and in optimising thermal cycle efficiency. In addition, the heated, pressurised water can be stored in deep, metal-lined caverns so as to allow power generation overnight. Based on published performance data, this system has a conversion efficiency (per unit collector area) in the range 15–20%. Using this technology, all current US electricity usage (3600 TW h yr^{-1}) could be supplied by a desert sited collection area of $\sim$23 000 km^2 (*i.e.* about 150 km by 150 km).

(b) Towers with Heliostats

Several large-scale demonstration systems are operating or under construction in Europe (Spain) and the US. On the basis of published performance data, the most efficient (in power output/unit land area) overall appears to be the current eSolar technology,[7] which delivers a peak electrical output (strong sunlight) of over 55 W m^{-2} (land area). This system comprises a field of many small ($\sim$1.5 m^2) individually steered (two axis) plane reflectors, which focus on tall receiver towers with thermal collectors on each (Figure 2.2b). There are 12 000 mirrors per tower and each tower heats water sufficient to generate 2.5 MW in Rankine cycle steam turbine generators. This array/tower combination comprises a 'solar module' and multiple solar modules may feed a single turbine generator plant. The total mirror area is $\sim$45% of total site area. One can estimate a solar to electrical conversion efficiency, per unit collector surface area of at least 25% for this system. This is somewhat higher than the linear Fresnel system described above, probably due to the use of large and compact (60 ton) water heating thermal receivers operating at higher temperatures (not quoted). The 'point focus' design substantially reduces re-radiation losses mentioned above. At present the eSolar system does not incorporate any significant energy storage facility for off-peak operation. Some of the systems planned or in operation in Europe have thermal storage capabilities, generally in the form of a molten salt. Additionally, supplementary heating from natural gas is incorporated in some of these designs. An ambitious program, from the DESERTEC consortium,[8] proposes to locate many (in the order of 100) large thermal-powered generating fields in high sunlight locations near the Mediterranean coast, with each field producing 10 GW on average. It is envisaged that the electrical power from these fields to population centres, principally in Europe, will be transmitted *via* high voltage DC transmission lines.

(c) Parabolic Dishes

In this approach, relatively large (~16 m diameter, 200 m^2 area) steerable (two axis) parabolic reflector dishes are each equipped with a thermal collector at the focal point of the dish ('point focus'). A typical example is that from Stirling Energy Systems (Figure 2.2c).[5] In this technology, the thermal receiver is in fact a compact, integrated Stirling cycle engine and generator, capable of 25 kW peak output. The 'parabolic' surface is composed of 82 flat mirror segments, rigidly mounted on the steerable frame. Satellite dish technology is quoted as underlying the design/construction of the system. The dishes may be sited at one per 500 m^2 land area, giving a mirror/site area ratio of ~40%. The Stirling cycle engines operate with H_2 as the working fluid, for maximum efficiency.

These units are the result of a long process of technological refinement, with the initial dish design principally intended for other uses. The overall conversion efficiency of sunlight to electric power is just over 30%, which is the record for any solar conversion system up to 2008.[9] Because each dish generator is totally self-contained, the system is indefinitely scalable. However the overall mechanical complexity of this type of solar thermal generation technology is probably greater than that of any other large-scale system at present. The system does not lend itself easily to thermal energy storage for off peak operation, in contrast to other competing solar thermal technologies. These systems physically separate the thermal collectors and electricity generating plant.

In general, the solar thermal conversion technologies are comparable and reasonably competitive in overall performance. Because of its simple construction, ease of operation and high land coverage ratio (almost 50%), the linear Fresnel system is very attractive. This option can produce, on average, over 20 W m^{-2} land area, which is quoted as being the best such performance of any solar thermal generating system.[4]

2.2.3 Solar Photovoltaic

The photovoltaic effect (direct conversion of light into electrical output by certain solid materials) has been known since 1839. Semiconductor-based photovoltaic electricity generation as a practical technology, emerged in the 1950s, soon after the invention of the p-n junction transistor. For a long time this approach was not cost competitive for large-scale electric power generation, but that is now rapidly changing with numerous recent developments. The literature on the field is vast and what is given here is only a short overview. A good, more detailed recent summary is found in ref. 10.

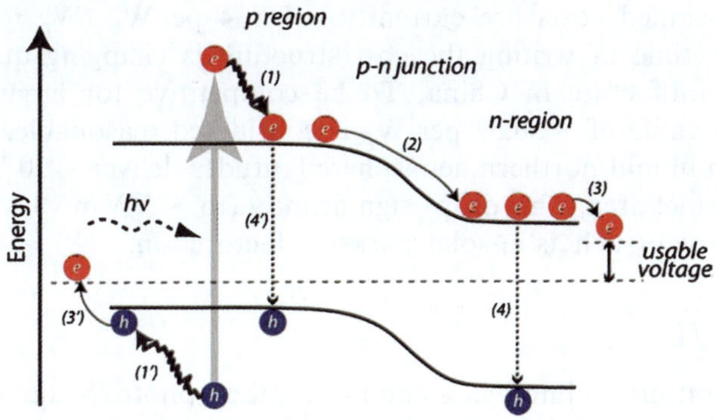

Figure 2.3 The basic operating principle of a single junction semiconductor (silicon) photovoltaic device is shown. (4′) identifies the band gap separation, which most solar photons (*hv*, grey arrow) exceed in energy. This excess energy is lost in thermalisation back to the band gap values of the electrons (red dots) and holes (blue dots) *via* processes (1) and (1′). Electron-hole recombination processes (4) and (4′) are further important loss mechanisms, particularly (4′). Un-recombined electrons and holes separate across the p-n junction and diffuse to the contact electrodes, which connect to the external circuit. Further losses (3′) and (3) occur at these points, resulting in a usable output with is typically much less than the original photon energy. (Adapted from ref. 11.)

The basic principle of a semiconductor photovoltaic device is illustrated in Figure 2.3. Photons whose energy exceed that of the semiconductor band gap are absorbed, creating an electron–hole pair which become separated across the p-n junction boundary and diffuse to surface electrodes, where they are collected and drive the external circuit. Photon energy in excess of the band gap value is normally lost through thermalisation of the photo-excited electron energy back to the upper band gap edge, and recombination of electron holes pairs before separation across the p-n junction is an important loss mechanism in practical devices. The nomenclature of photovoltaics is somewhat imprecise, but broadly, three ‘generations’ of the technology are recognised:

Generation I

This constitutes the overwhelming bulk of systems now in production. The devices are constructed from single junction silicon wafers, of single crystalline or polycrystalline form. Thermodynamically, these devices are limited to ~31 % efficiency under standard direct sunlight, rising to ~40% under highly concentrated sunlight. This limit is being approached with high quality single-crystal based materials achieving a ~25% efficiency in direct sunlight. Most common devices, particularly polycrystalline substrate units, which are far cheaper to manufacture, have efficiencies somewhat less than 20%. Modern systems employ a number of strategies (surface passivation, textured anti-reflection coatings *etc.*) to improve the efficiency of light capture and cell operation. Installed costs are currently ~US$4 per W_p (W_p refers to peak watt). At the time of writing the cost structure is changing quickly due to increasing manufacture in China. To be competitive for large-scale power generation, a value of ~$0.33 per W_p is considered reasonable. Current systems installed in mid-northern hemisphere latitudes deliver ~20 W m^{-2} (daily average) of panel area. This drops significantly (to ~5 W m^{-2}) when the total land area utilised (such as in solar parks) is factored in.

Generation II

Second generation technologies comprise other photovoltaic systems with useful operating efficiencies and with the potential, at least, to be manufactured more cheaply than crystalline Si. At present they are in early stages of commercial application. They fall broadly into two classes:

(a) Thin Film systems, which employ various strategies to reduce cost/unit area below *Generation I* technologies, while retaining acceptable efficiencies.
(b) Solar Concentrators with advanced multi-junction PV systems, based on GaAs and related low abundance group III–V materials.

At present, amorphous Si is the most widely employed thin film technology, comprising, in its several forms, ~6% of the total installed PV market. The most advanced examples are hetero-junction systems, employing thin layers of both crystalline and amorphous Si, with two junctions in a sandwich *n*–*I*(intrinsic)–*p* arrangement.[11] These have a conversion efficiency exceeding 20%.[12]

Another approach, as yet somewhat further from commercialisation, involves CdTe and Cu(InGa)Se_2 (called CIGS, or CIS if no Ga present) type systems. These have laboratory-demonstrated efficiencies of ~20% and hold promise of relatively cheap manufacture, although they utilise materials that are in much lower abundance than Si. At present the technologies also face challenges in fabrication, particularly in the interfaces with the transparent conducting oxide contacts. This is a significant consideration, as these are devices intended for large area applications and all material layers must be as thin as possible (of order 1–5 μm thick layers).

The highest efficiency PV cells known are multi-junction devices based on GaAs and related materials. These require sophisticated fabrication techniques, such as molecular beam epitaxy and chemical vapour deposition to achieve their high conversion performance (currently ~40%, in principle ~60%).[13] Their cost per unit area is too great for conventional large area applications, but they become competitive when used in conjunction with solar concentrators, when operating with concentration factors of 100–2000. Under these conditions (concentration factor > 500), the economics are dominated by the concentrator technology used, so costs become closer to those discussed above for Solar Thermal systems.

Overall conversion efficiencies/unit capture area of assembled concentrator systems ($\sim$10%) are less than those for high performance *Generation I* technologies, in part because they operate at higher temperatures, which always lowers PV performance. However, heat from the concentrators (obtained from the necessary cooling of the PV assemblies) may be separately collected and utilised (*e.g.* for water desalination), which improves the attractiveness of the approach in specific high solar incidence locations. For example, Heliodynamics[14] supply units that generate $\sim$14 W m^{-2} electrical and $\sim$80 W m^{-2} thermal (daily average) in sunbelt latitudes.

Generation III

This is a rather loose term, embracing what are regarded as the ultimate technical aims of PV technology. These incorporate thin film construction from cheap, abundant relatively non-toxic materials, combined with sunlight conversion efficiencies approaching the theoretical, multi-junction limits (>75%). In practice these aims have become somewhat uncoupled and we focus here on one important class of devices, that which addresses the first two of these ideal requirements but has still far to go on the third.

The class we initially consider is organic PV technology (Figure 2.4). These devices employ non-crystalline, organic polymer substrates in which dye molecules are dispersed. Photo conversion occurs by strong photon

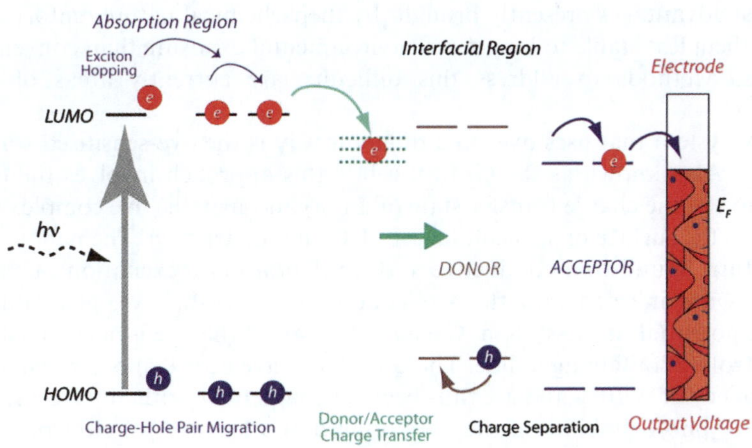

Figure 2.4 Organic PV schematic. Light absorbed in donor layer dye molecules produces excited states. These excitations are able to hop *via* an exciton mechanism until they reach the hetero junction region, where the charge separation occurs. E_F, the output voltage, has an upper limit of the HOMO (donor)–LUMO (acceptor) energy difference. This energy gap can be chemically engineered to be better utilise the photon energy of the sun than can be achieved for Si PV devices systems. Figure adapted from ref. 15.

absorption in a dye molecule to form an excited state. Excitation is transferred between dye molecules by an exciton hopping mechanism until a hetero junction is reached. Charge separation occurs across the hetero junction, where there is access to donor molecules on one side and acceptor dye molecules on the other. Critically, these differ in their HOMO and LUMO levels.

Diffusion of the hole and electron to separate collector surfaces of holes in the donor region and electrons in the acceptor region generates the output voltage. Because carrier diffusion and electron-hole recombination rates are unfavourable by comparison with semiconductor systems, efficiencies for organic PV systems are currently low ($\sim$5%), although in principle they can capture a greater fraction of the photon energy in charge separation than Si PV's, with appropriate selection of dye molecules. However, this is a rapidly emerging PV technology and due to its potential for low-cost, non-vacuum based fabrication techniques based on high volume, low-temperature manufacturing methods used for plastic sheeting, it remains attractive.

At present, the efficiency 'learning curve' for this technology is rising rapidly. As a similar fundamental limit on conversion efficiency ($\sim$30%) generally holds for these devices as for semiconductor single junction systems, there are significant opportunities for improvement. Because of the versatility, mechanical flexibility and the range of well-established chemical technologies that can be exploited in their manufacture, organic PV systems may find application and penetrate markets hitherto thought unavailable to conventional photovoltaics (active fashion elements, decorating *etc.*). The fabrication and cost advantages presently brought by their chemical nature, unfortunately makes them less stable to long-term environmental exposure than conventional systems. Methods to address this difficulty are currently areas of active research.

A PV system that uses dyes in a different way is the dye-sensitised solar cell (DSSC). Also known as the Grätzel cell,[15] this approach involves the optical excitation of the charge transfer state of an organo-metallic dye complex that is bound to the surface of semiconductor (TiO_2 nanoparticles). The nanoparticles are in turn bound to a conductive surface. Upon photoexcitation of the dye, electrons are injected into the semiconductor from the dye, generating the driving potential of the system. Cyclic reduction of the dye is accomplished by an electrolyte containing iodide. These systems have been demonstrated to have an efficiency of $\sim$10% and are thus become competitive with Si PV cells. Issues related to photo-stability of the dyes utilised and leakages of electrolyte from the cells have been addressed and this technology is now undergoing commercialisation.

Figure 2.5 reproduces an informative diagram from Green *et al.*,[10] summarising a cost comparison for the various PV technologies, including (particularly for Generation III) plausible projections beyond the current state of the technologies. Generation III clearly holds promise for attaining the large scale cost competitive limit of $\sim$\$0.33 per W_p for PV technology.

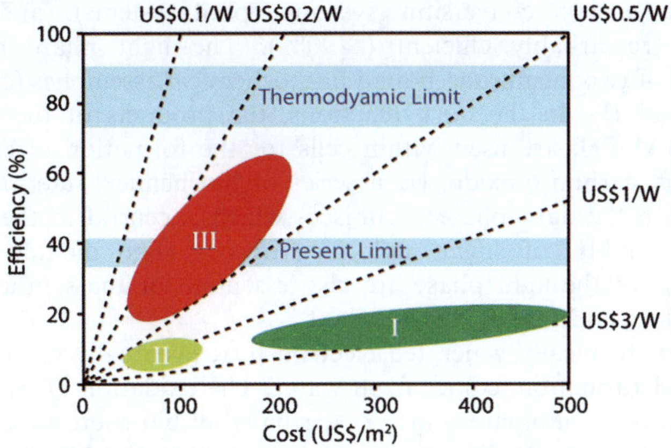

Figure 2.5 Cost *versus* efficiency analysis for Generation I, II and III PV technologies. Adapted from ref. 10.

2.2.4 Bio-solar

Nature's method of solar energy conversion is of course photosynthesis. This process occurs in all biological habitats on earth having access to sufficient sunlight. The bulk of photosynthesis ($\sim$90%) occurs in the oceans, within tens of meters of the water surface (euphotic zone). The species involved are principally chlorophyll-containing unicellular organisms (cyanobacteria) and small (multicellular) phytoplankton.

From an energy utilisation perspective, bio-solar processes offer at present two practically reasonable alternatives:

(1) *Generation of biomass* by photosynthetic growth (principally as carbohydrate, cellulose *etc.*) which may be dried and then combusted to yield thermal energy. This process could utilise any convenient photosynthetic species.
(2) *Generation of biofuels* (high-energy chemicals) during photosynthetic growth, *via* natural or specifically modified natural enzymatic pathways. These fuels would include alcohols (ethanol, methanol) or bio-diesel, from high yielding terrestrial (C4 type) plants, and gaseous hydrogen, principally from unicellular species (*e.g.* cyanobacteria), grown in aquaculture or fermentation reactors.

The overall process of photosynthesis consists of two main phases, the so-called light and dark reactions.[17] In the first, light energy is absorbed by 'antenna' chlorophyll molecules and drives electrochemical reactions producing 'energy rich' biological compounds: ATP (adenosine triphosphate) and reduced pyridine nucleotide (NADPH).

Molecular oxygen is a by-product in this process and is released to the atmosphere. The early steps in this chain are actually nature's own photovoltaic energy conversion systems (photosystems), and these are individually remarkably efficient (~90%). The light phase requires the cooperation of two membrane-bound photochemical assemblies (called Photosystems I and II). In the dark reactions, the products of the light phase, ATP and NADPH, are used within cells for the formation of carbohydrate (sugars) from carbon dioxide, *via* a series of biochemical intermediates. The key enzyme in the dark phase is rubisco, which is central to the progressive chemical 'assembly' of sugar molecules from carbon dioxide and water. The products of the light phase are also essential for the synthesis of lipids and other plant constituents by the plant.

The photo-chemically generated electron flux, eventually resulting in ATP and NADPH formation, comes from water. The oxidation of water to molecular oxygen is the ultimate source of virtually all bio-energetic electrons utilised by living creatures. The process occurs within the Mn_4/Ca containing oxygen evolving complex (OEC) located in Photosystem II. The reaction:

$$2H_2O \xrightarrow{4h\nu} 4H^+ + O_2 + 4e^- \qquad (2)$$

achieves almost all the elements of direct, efficient conversion of light into a clean, high energy fuel *i.e.* hydrogen (generated from protons). However, because natural photosynthetic organisms have not evolved to specifically optimise this process, beyond less-utilised secondary biochemical pathways, much of the potential conversion efficiency is in fact lost (see below).

Mellis[18] has analysed the maximal efficiencies possible from solar energy conversion in photosynthesis occurring within microalgae, these being likely organisms type for practical applications. The efficiencies are summarised in Figure 2.6.

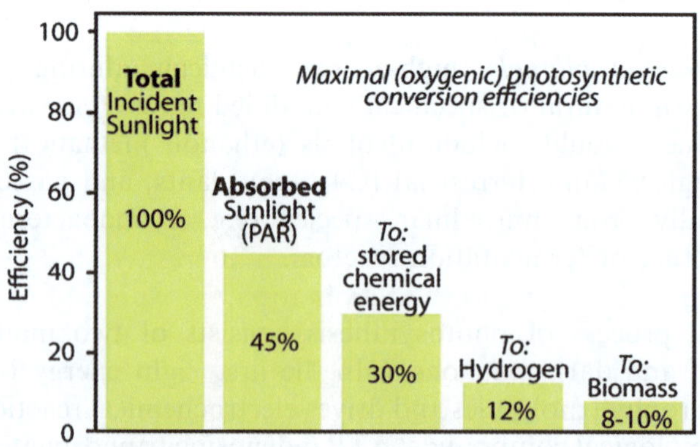

Figure 2.6 Bar diagram of theoretically maximum solar energy conversion efficiencies in oxygenic photosynthesis. This assumes an average incident sunlight energy (full solar spectrum) of ~5 kW h m^{-2}. Adapted from ref. 18.

Many factors combine to progressively reduce the overall photosynthetic conversion efficiency from sunlight in a typical daily cycle. Initially, two losses occur because only that part of the 'chemically useful' solar spectrum between ~400 and 700 nm is utilisable for photosynthesis and more importantly, photosynthetic organisms have evolved to be most efficient at *low* light levels and essentially waste a great deal of light. At 'high' light levels, the excess light energy is simply dissipated (through fluorescence or heat).

The precise light intensity at which the process of photosynthesis saturates is organism dependent, but is typically at ~20–25% of the peak light levels on sunny days in mid-latitude locations (*i.e.* ~200 W m^{-2}). This limitation applies only to those parts of photosynthetic organisms (cells, chloroplasts *etc.*) directly facing the sun. However, there is normally a substantial component of the overall photosynthesising biomass lying deeper within high-density foliage, tissue or culture, where the saturating effects are mitigated by the diminishing intensity of penetrating light. Integrating over such typical effective optical density distributions, Mellis[18] estimates an overall capture efficiency of ~45%. Through intermediate stages, this results in potential yields of H_2 and biomass formation in the range of 10% (Figure 2.6). In practice however, achievable yields are at present much lower. The two bio-solar approaches currently receiving significant attention are discussed below.

Terrestrial Crops

Several high-yielding crop species have been proposed for bio-solar energy conversion/storage. These include sugarcane, tropical plantations, short rotation coppice trees, *Miscanthus* grass, switch grass and others. A useful summary is found in ref. 19. However, even the highest-yielding tropically grown species have an average effective yield of only ~1–2 W m^{-2} of growing area (most much less). This is quite low by comparison with other technologies discussed above. Given the (likely increasing) importance of arable land for food production it seems unlikely to these authors that specifically grown plant species will ever make a significant contribution, generally, to large scale solar energy conversion.

H_2 from algal/cyanobacterial Sources

This area is currently receiving much research interest, with promising results emerging. However, the fundamental limitation, as expressed in Figure 2.6, is unlikely to be exceeded in the foreseeable future, due to the nature of the basic, biochemical processes in photosynthesis. The electrons exiting Photosystem II in eqn (2) are effectively available for recombination with protons, within endogenous hydrogenases forming H_2, only after they receive a second photo-energisation through Photosystem I. They emerge from this with a redox potential of ~ −1.2 V. Thus, over 2 eV of energy is lost per molecule when these electrons form H_2 (redox potential ~0 V) in hydrogenases fed by ferredoxin-mediated electron transfer, directly or indirectly, from the acceptor side of Photosystem I.

The most direct pathway for H_2 production from hydrogenases would employ the electrons directly flowing from the highly negative potential iron–sulfur complexes of Photosystem I.

$$H_2O \rightarrow H_2 + \tfrac{1}{2}O_2 \qquad (3)$$

This process requires 4 photosynthetic photons per H_2 molecule produced. To operate at full Photosystem II capacity (*i.e.* at saturating light levels), would require a fully O_2 tolerant hydrogenase, as well as a promoter enabling gene expression at elevated O_2 levels and subsequent hydrogenase assembly under such conditions. At present this capacity does not exist; the natural pathways generally require anaerobic conditions for hydrogenase expression and operation (*e.g.* concomitant with nitrogen fixation). Although a capacity for H_2 formation is widespread in nature, this ability occurring within various photosynthetic organisms, many of these organisms have yet to be examined in detail (reviewed in ref. 19). Active programs are underway to create hydrogenase enzymes with heightened O_2 tolerance through genetic means.[20]

A second possible pathway for H_2 formation involves electron and proton flow to the plastoquinone pool, from anaerobic metabolism of previously stored carbon, mediated by a dehydrogenase, through Photosystem I, thence to ferredoxin and hydrogenase (as above). The pathway is summarised in (4a) and (4b), (after ref. 21).

$$H_2O + CO_2 \xrightarrow{h\nu} [CH_2O] \text{ (biomass)} \qquad (4a)$$

$$\text{dark} + \text{anaerobiosis} \xrightarrow{-O_2} \text{hydrogenase induction}$$

$$\xrightarrow{h\nu} H_2 + CO_2 \qquad (4b)$$

This process requires 6 light quanta per H_2. Such a pathway operates most effectively in two temporally separated stages, a photosynthetic growth stage followed by an anaerobic stage in which hydrogenase is expressed. Photosystem I then uses light to pump electrons from previously stored reductants *via* the plastoquinone pool into ferredoxin, while the protons are used for ATP synthesis. The anaerobic stage requires O_2 production be substantially depleted (in cyanobacteria) or virtually eliminated (in green algae). Separation of the gaseous H_2 and O_2 evolving steps is advantageous for gas separation and indeed safety. The current strategy to lower intracellular O_2 production is to reversibly inhibit Photosystem II activity by sulfur deprivation of green algae. This suppresses the biosynthetic repair of Photosystem II, which is inevitably damaged during photoinhibition. Replenishment of the sulfur nutrient permits regeneration of Photosystem II and replenishment of stored carbon during a subsequent aerobic light cycle, when the hydrogenase does not function.

Such a system has been developed by the US DOE (NREL, ref. 20) and has advanced to pilot scale level using mutants derived from cyanobacteria and algae. A further refinement is to genetically engineer the photosynthetic species to have significantly reduced chlorophyll antennae sizes.[18] This substantially raises the light level at which photosynthetic rates saturate, allowing the use of more dense bacterial suspensions in the circulating reactor streams and more efficient light penetration of the reaction medium.

Figure 2.7 shows a diagram of the type of system proposed, which has been subjected to detailed engineering analysis. This operates with a modified *C. reinhardtii*, algal species, immobilized in a thin-bed reactor that may be fed alternately by nutrient solutions appropriate for operation in the aerobic or anaerobic, H_2 generating phases, whilst being illuminated. Currently, this system has a demonstrated overall light energy to H_2 conversion efficiency of $\sim 1\%$. Near term improvements in the biochemical performance of the algal species is expected to at least double this figure. This appears to set the performance benchmark for this type of bio-solar technology at present and into the reasonably foreseeable future.

In considering this outcome, it is useful to compare this process with other currently realizable approaches enabling the direct conversion of solar energy into H_2 *via* water splitting. The most straightforward method, which has been demonstrated on laboratory scale (*e.g.* ref. 22 and 23), is to use available

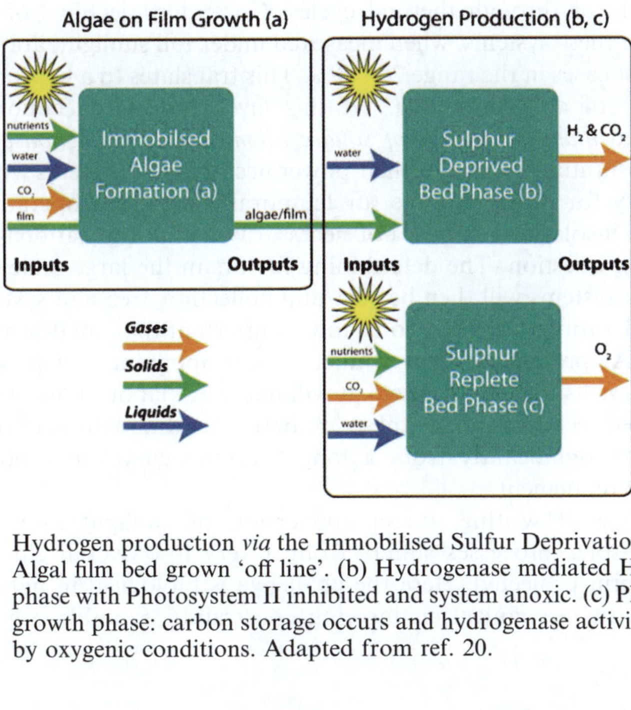

Figure 2.7 Hydrogen production *via* the Immobilised Sulfur Deprivation Process. (a) Algal film bed grown 'off line'. (b) Hydrogenase mediated H_2 production phase with Photosystem II inhibited and system anoxic. (c) Photosynthetic growth phase: carbon storage occurs and hydrogenase activity suppressed by oxygenic conditions. Adapted from ref. 20.

photovoltaic technology coupled to water electrolysis using a range of newly developed, abundant transition metal based water oxidizing (cathodic reaction) catalysts. Those based on Mn (as in Photosystem II) and Co seem to be most promising.[20] A plausible figure for the electrolysis efficiency, based on current examples and using water at approximately neutral pH, is at least 50%. In the plant OEC enzyme centre, which employs a specific Mn cluster as the catalytic species and executes the most energetically demanding reaction in nature, the effective figure approaches 90%.

With currently achievable *Generation I* PV efficiencies near 20%, such a combined system would appear to outperform even optimistic projections for bio-solar H_2 production. At present, the structure and mechanism of the water-oxidising site in Photosystem II is under intense study by many researchers, including the authors, with the intent of using this remarkably efficient enzyme as a basis to develop 'bio-inspired' catalysts, using abundant elements, for H_2 production from water using electrical energy.[24]

2.3 Conclusion

Over the last decade, substantial advances have occurred in a range of technologies enabling the direct conversion sunlight to directly usable energy, either as heat, electricity or chemical fuels. The efficiency of these conversions drops progressively as one advances through the three above outcomes. At present, the most important technologies, in terms of their overall significance for industrial societies, is that class which produces electric power as the output, either directly or through thermal cycles. Current projections of the overall efficiencies of these systems, when measured under full sunlight, are surprisingly close, in most cases in the range 20–30%. This translates to a figure in the order of 50 $W\,m^{-2}$ of collection area. At this level, advanced economies would typically need *many thousands of square kilometres of collection area* to contribute a substantial fraction of their power needs. These systems would need to have capacity for power storage (or temporally staggered operation) and be sited in high insolation regions, not necessarily within the national borders of the state(s) in question. The determining factors in the large-scale use of such solar-electric systems will then be cost/unit collection area and system lifetime. These considerations are likely to be more important than modest variations in efficiencies. At present, a 'clear winner' is not apparent, but if photovoltaic systems emerge as the frontrunners, it will likely be elaborations of Generation I silicon based technologies, as these are based on abundant materials and are now profiting significantly from a long 'learning curve' in production and performance refinement.

At the time of writing, direct conversion of sunlight into usable fuel through bio-solar processes suffers from a very low overall efficiency, even when near-term projected. Here the challenge is biological in nature and not associated with mechanical engineering constraints. Suitable organisms that can sustain acceptable efficiencies ($\sim$10% overall) do not yet exist and it is unclear if they ever will without major biological/genetic engineering breakthroughs. One starts with the fundamental difficulty that oxygenic photosynthetic organisms, which are the only ones presently feasible to use, have been evolutionarily adapted to operate under those precise conditions that limit enzymatic bio-hydrogen production.

The future of solar fuel generation may lie in our ability to 'extract' from biology a key enzymatic processes in, for example, H_2 generation by water electrolysis, and to then implement this bio-mimetic catalysis on an industrial scale.

References

1. An excellent technical overview and summary of these and other energy systems is to be found in: J. C. Mackay, *Sustainable Energy – Without the Hot Air*, 2009, UIT Cambridge Ltd., Cambridge, UK, ch. 6 and 7. Available free on line from www.withouthotair.com.
2. Schlaich Bergermann und Partner, Structural Consulting Engineers, www.sbp.de.
3. J. C. Mackay, *Sustainable Energy – Without the Hot Air*, 2009, UIT Cambridge Ltd., Cambridge, UK, p. 183.
4. D. Mills and P. Le Lièvre, Solar Heat and Power Pty. Ltd, www.ausra.com/pdfs/Paper_CompetitiveSolarElectricity.pdf.
5. Stirling Energy Systems, www.stirlingenergy.com.
6. http://en.wikipedia.org/wiki/Stefan%E2%80%93Boltzmann_law.
7. eSolar Utility Scale Solar Power, www.esolar.com.
8. DESERTEC Consortium, www.desertec.org.
9. https://share.sandia.gov/news/resources/releases/2008/mechanics.html.
10. D. Ginley, M. A. Green and R. Collins, *MRS Bull.*, 2008, **33**, 355–364.
11. M. A. Green, *Phys. E (Amsterdam, Neth.)*, 2002, **14**, 65–70.
12. Sanyo Corporation White paper: http://us.sanyo.com/Dynamic/customPages/docs/solarPower_22_3_Cell_Efficiency_White_Paper_Dec_07.pdf.
13. P. Pérez-Higueras, E. Muñoz, G. Almonacid and P. G. Vidal, *Renewable Sustainable Energy Rev.*, 2011, **15**, 1810–1815.
14. Heliodynamics, www.hdsolar.com.
15. J. L. Brédas, J. E. Norton, J. Cornil and V. Coropceanu, *Acc. Chem. Res.*, 2009, **42**, 1691–1699.
16. M. Grätzel, *Inorg. Chem.*, 2005, **44**, 6841–6851.
17. *Advances in Photosynthesis and Respiration*, ed. Govindjee and T. D. Sharkey, Springer, Dordrecht, The Netherlands, vol. 22.
18. A. Mellis, *Plant Science*, 2009, **177**, 272–280.
19. J. C. Mackay, *Sustainable Energy – Without the Hot Air*, 2009, UIT Cambridge Ltd., Cambridge, UK, ch. 6.

20. National Renewable Energy Laboratory, Subcontract Report NREL/SR-560-46674, September 2009.
21. O. Kruse, J. Rupprecht, J. H. Mussgnug, G. C. Dismukes and B. Hankamer, *Photochem. Photobiol. Sci.*, 2005, **4**, 957–969.
22. M. W. Kanan and D. G. Nocera, *Science*, 2008, **321**, 1072–1075.
23. R. Brimblecombe, G. F. Swiegers, G. C. Dismukes and L. Spiccia, *Angew. Chem.*, 2008, **120**, 7445–7448.
24. P. Gatt, R. Stranger and R. J. Pace, *J. Photochem. Photobiol., B*, 2011, **104**, 80–93.

CHAPTER 3

Converting Photons to Electron and Proton Shifts from Water for Fuel Production

CHRISTIAN HERRERO,[a] CYRILLE COSTENTIN[b] AND ALLY AUKAULOO[a,c]

[a] CEA, iBiTec-S, Service de Bioénenergétique Biologie Structurale et Mécanismes (SB^{2}SM), F-91191 Gif-sur-Yvette, France, CNRS, URA 2096, F-91191 Gif-sur-Yvette, France; [b] Laboratoire d'Electrochimie Moléculaire UMR CNRS - P7 7591 Université Paris Diderot - Paris, 15 rue Jean-Antoine de Baïf 75205 Paris Cedex 13, France; [c] Institut de Chimie Moléculaire et des Matériaux d'Orsay, UMR-CNRS 8182, Université de Paris-Sud XI, F-91405 Orsay, France

3.1 Introduction

Artificial photosynthesis aims to extract the main principles of natural photosynthesis and put them in a man-made device to transform solar energy into chemical energy and store it in the form of chemical bonds. While photosynthesis is responsible for energy input into the biosphere, it uses only a small fraction of the solar energy that reaches the Earth's surface. Indeed the amount of solar energy reaching the planet is so great that many scientists consider that its capture and usage provides the best long-term solution to the energy problem and the associated environmental constraints brought about by our

RSC Energy and Environment Series No. 5
Molecular Solar Fuels
Edited by Thomas J. Wydrzynski and Warwick Hillier

Published by the Royal Society of Chemistry, www.rsc.org

current dependence on fossil fuels. Intriguingly, solar-based electricity accounts for only *ca.* 0.1% of the world's electricity with an additional 1% of the world's energy coming from solar-based biomass. Our derisory use of solar energy when viewed in respect to its unmatched potential must define the contours of solar energy research. Can a sustainable solar energy-driven cycle (without net CO_2 production), like that of photosynthesis (energy input)/respiration (energy usage), be mimicked on a scale sufficient to support human activities? Different approaches based on solar energy conversion are currently ongoing:

1. Solar electricity, where the challenge is to convert light energy to electricity using photovoltaic devices. An intense effort is now being devoted to find new cost effective materials to implement this technology in all facets of our energy consumption.
2. Solar thermal energy can help to supply thermal energy and also perform high temperature catalytic processes for the synthesis of a fuel from renewable sources.
3. Solar fuel in the form of biomass. However, it is argued that that its capacity to meet the human energetic needs is limited due to incompatibilities with our need for arable land.
4. Another more challenging task resides in designing robust highly efficient man-made device to capture and drive the uphill reactions for the production of a fuel. Artificial photosynthesis seeks to capture and use solar energy to perform multi-electron bond breaking and bond making catalytic reactions in order to produce a fuel starting with low energy content renewable sources like water and carbon dioxide. Research in this area is highly active and spans fields from solid-state chemistry to molecular approach.

In this chapter, we will focus on the molecular approach that chemists are seeking in order to capture sunlight resulting in the electron and proton removal from water, and consequently the use of these in order to perform multi-redox catalysis to the desired products (eqn (1)).

$$2H_2O + 4h\nu \rightarrow O_2 + 4H^+ + 4e^- \quad (1)$$

Although this research field is moving at a fast pace, it is still at a fundamental research level. Basic discoveries still need to be realised at several levels before technology can be deployed. We start this chapter with a short overview on Photosystem II, the enzyme responsible for the use of light energy to remove electrons and protons from water, which are then transported further in the photosynthetic machinery for the reduction of CO_2. Further on, we will discuss the main units of the molecular assemblies that chemists must design for the capture of visible light and its storage in a chemical form. These include a photoactive chromophore, electron relay and molecular catalyst for water oxidation. We also give the basic principles underlying the photo-induced electron transfer processes and the proton coupled electron transfer processes, two important concepts that are inherent in the framework of designing and studying the transformation of light into a charge separated state that will trigger multi-electron/proton catalytic processes. The topic of this chapter introduces the steps taken towards using water and light to provide the reactants needed for the production of clean fuels. However, in this chapter we do not discuss the prolific research in the design of non-noble metal-based catalysis for the reduction of protons or reduction of carbon dioxide to reduced forms of carbon.

3.2 Lessons from Photosystem II

Photosystem II is the enzyme responsible for the light driven oxidation of water, thereby providing electrons and protons to fix carbon dioxide. From the viewpoint of the chemist, its structure/function is composed of two main two parts: (i) a photochemical charge separating system which produces a reductant and a strong oxidant species on opposite sides of the membrane, and (ii) a catalytic module acting as the reservoir for accumulating sufficient oxidising power at a cluster of high valence manganese ions to oxidise water.

Several reviews deal with the structure and mechanism of the natural photocatalytic machinery at great length and thus these aspects will be presented in this article only briefly. A view of the main cofactors is given in Figure 3.1 and the sequence of electron transfer processes after a photonic excitation is indicated.[1,2]

1. Charge separation is triggered between pigment molecules very rapidly after excitation of the chlorophylls in the enzyme (step 1 in Figure 3.1). A radical pair is detected which appears to be made up of a chlorophyll (P_{D1} in Figure 3.1) cation and a pheophytin (Ph_{D1} in Figure 3.1) anion.
2. The pheophytin anion radical then donates an electron to a quinone (Q_A) thereby stabilizing the charge separated state by shifting the charges away from each other.
3. The highly oxidising chlorophyll cation ($P_{680}{}^{+} \sim +1.2\,V$ *vs*. NHE) returns to its initial state by abstracting an electron from a tyrosine residue (Tyr_Z). The tyrosine's phenolic carbonyl group loses a proton as the tyrosyl radical forms and the proton is captured by a nearby imidazole group (see Scheme 3.1) presumably in a concerted manner. This process further increases charge separation by shifting the charges away from each other.
4. The neutral tyrosyl radical ($\sim +0.97\,V$ *vs*. NHE) then goes on to oxidise the Mn ions of the Mn complex (designated OEC, Oxygen Evolving Complex, in Figure 3.1).
5. The semiquinone anion formed is further stabilized by a lateral electron transfer step to a second quinone, Q_B.

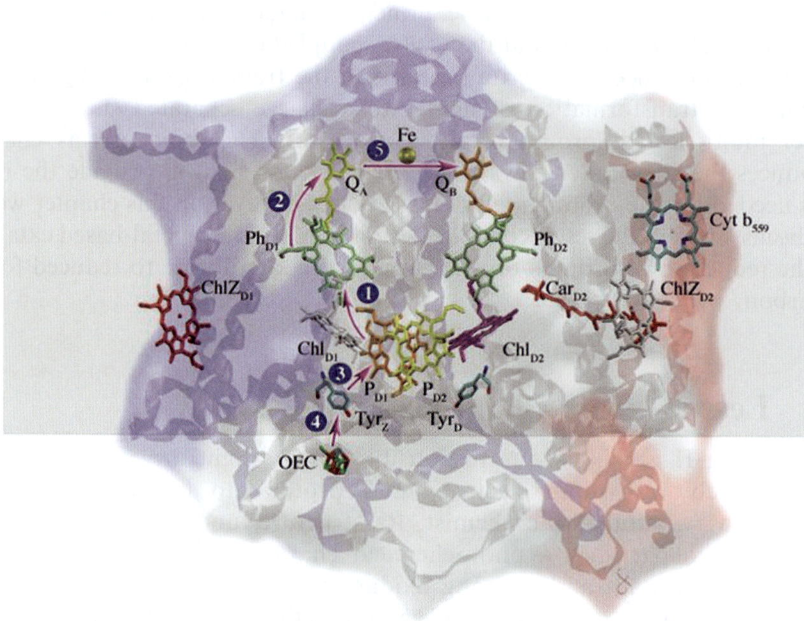

Figure 3.1 Map of the main cofactors of Photosystem II with the protein matrix (thanks to Dr A. W. Rutherford and Dr C. Fufezan).

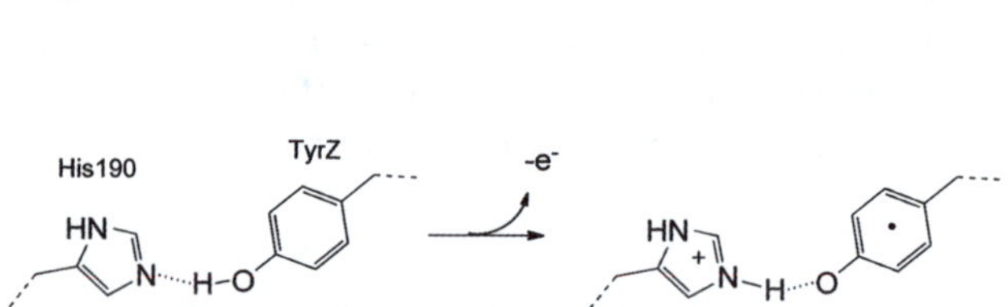

Scheme 3.1 The concerted proton electron transfer mechanism at the TyrZ/His190 pair in Photosystem II.

The final state formed after the excitation of the enzyme by one photon contains a semiquinone (Q_B^-) and a high valence form of the Mn cluster. This state contains more than half of the energy (approx. 1 eV) of the absorbed photon. In order to complete the reduction of plastoquinone, a second turnover must occur. In order to oxidise water, a total of 4 photochemical events must take place, with each one resulting in an increase in the manganese cluster's valence. It is important to note that we often look at the electron transfer process uniquely as a stepwise increase in the oxidation state of the manganese ions. We should bear in mind that this is a formal view for electron counting purposes. However, the water molecule is also being activated by deprotonation and loss of electron density at the oxygen atom.

The oxygen evolving centre (OEC) is a Mn_4Ca cluster which is the only biological catalytic site able to perform the four electron oxidation of water to molecular oxygen. It cycles through four different redox states (S_0–S_4) after stepwise oxidation by P_{680}, and re-reduction by two molecules of water.

Different proposals for the mechanistic pathways of the OEC prompted from X-ray crystallography,[3–5] spectroscopy, and theoretical calculations[6–8] assume a highly oxidised manganese oxo motif as an intermediate in the water oxidation reaction.

These oxo functionalities can be either terminal, with the oxygen atom bound by a multiple bond to the metal ion, or singly bound to multiple metallic centres. Both models have been hypothesized as working mechanisms in the oxygen evolving complex (OEC) of photosystem II.

One model, an acid–base mechanism (Scheme 3.2) assumes that all the oxidising equivalents are taken in by the Mn ions [$Mn^{IV}{}_3Mn^{V}$], with the oxo motif forming at the Mn^{V} centre which is the lone Mn outside the Mn_3Ca cube. The highly oxidised Mn ion polarizes the oxygen atom and renders it electrophilic. This moiety is then susceptible to nucleophilic attack by a water molecule as depicted in Scheme 3.2.[6,9,10]

A second proposal involves radical character of two adjacent oxo groups of a [$Mn^{III}Mn^{IV}{}_3$, $O^{\bullet}$ $O^{\bullet}$] system.[11] In this case, coupling of the two radicals *via* peroxo and hydroperoxo-like intermediates yields formation of the O–O bond (RC, radical centre mechanism).

A tentative schematic mechanism for the four electron, four proton oxidation process is shown in Figure 3.2. This mechanism proposed by Batista *et al.*[2] favours the so called nucleophilic attack and is based on structural and theoretical data. Therefore, at this stage of our knowledge on the OEC, it is probable that this mechanistic proposal will be revised based on upcoming new breakthroughs in the mechanistic studies of the OEC and from new crystal structures of PSII.[5]

In this proposal, three of the manganese ions (inside the cubane structure) share the oxidising power to drive the oxidation process while the chemistry for the water activation and the oxygen–oxygen bond formation takes place on one manganese ion.

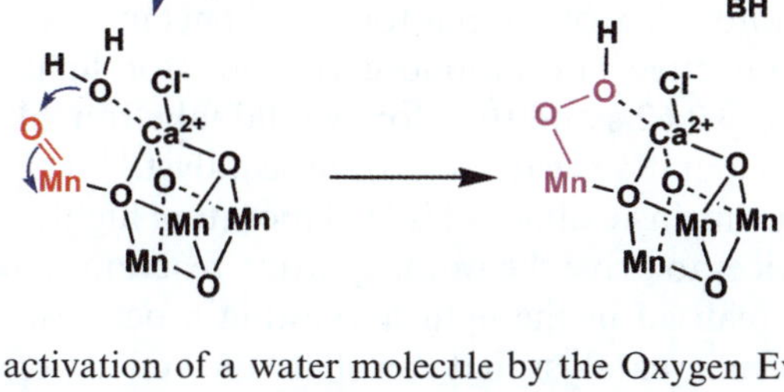

Scheme 3.2 AB activation of a water molecule by the Oxygen Evolving Complex.

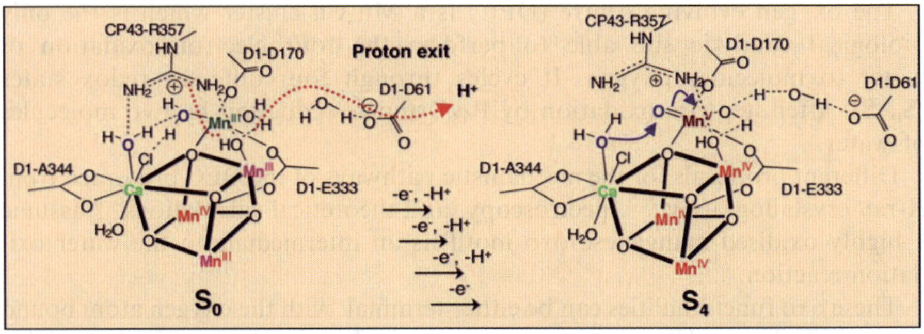

Figure 3.2 Mechanistic proposal for the functioning of the OEC by Batista *et al.* We point out to the reader that recent studies on the valence states of the manganese ions in the S_0 state have been revised to the high valent state, *i.e.* 4Mn(III) and 1Mn(IV) in the S_0 state.[12]

3.3 Synchronising Light Absorption and Charge Shifts

A fascinating feature of the functioning of Photosystem II is the coupling of the one-photon and the one charge-shift process that result in the accumulation of four oxidising equivalents at the manganese catalytic unit, the OEC. A rough estimate considering a normal photon flux density of 1×10^{15} photons cm^{-2} s^{-1} indicates that the hole transfer proceeds on the millisecond timescale and is modulated by the number of chromophores in the Light Harvesting Complex; (typically around 80 chlorophylls) leading to a value close to 30 ms for the conversion of four photons to four holes at the OEC. This time lapse is smaller for systems with larger amounts of light capturing chlorophylls, thereby leading to better efficiency while diminishing the delay to holding the "hot" catalytic intermediates. Nature has another trick to preserve the reactivity pattern of such highly oxidised species by providing a unique microenvironment through the protein matrix to inhibit deleterious quenching reactions. In the molecular world, synchronising the photon capture to performing multi-electron catalytic processes is probably the most challenging task, and breakthroughs in this field will probably bring artificial photosynthesis to technological deployment. In the absence of an antenna like Light Harvesting System, a simplified calculation for a molecular photosentiser with a cross-section for light absorption of 1.6×10^{15} cm^{-2} and with the higher limit of extinction coefficient of absorption $= 10^9$ mol^{-1} cm^{-2} (corresponding to those of porphyrin derivatives) leads a photon flux of 1.6 photons s^{-1}. This means that under normal sunlight each photoactive chromophore is ready for light induced electron transfer around every 0.62 s and therefore would take roughly 2.5 s to build up the four holes at a putative water oxidising catalyst.[13] These estimated data already point to the intrinsic difficulties of elaborating highly efficient molecular photocatalytic devices to drive the water splitting reactions. On one side, special efforts need to be realised in the light harvesting processes, while at the same time a robust molecular catalyst that can share the activity of the OEC must be elaborated. Above all, new ways to couple photodriven electron transfer processes to multi-electron catalytic processes must be unravelled.

3.4 Molecular Construction of an Artificial Photoanode

A molecular-based system intended to perform the reactions needed to mimic the reactions leading to water oxidation should be capable of, firstly, efficient absorption of visible light, and secondly, converting this energy into a chemical potential capable of triggering the accumulation of oxidising equivalents at a catalytic centre through a series of consecutive electron and proton removal steps, *via* activation of H_2O. The collected electrons and protons can then be used for fuel production, such as production of hydrogen, or for reductive fixation of carbon dioxide at another catalytic site. A minimalist system should therefore consist of a photoactive unit (PU) and a catalytic unit (CU) that, in principle, should be connected through an electron relay (ER) moiety that could help in the electron transfer pathway and inhibiting charge recombination reactions. To date, most of the work accomplished in this area has been done by using mono- or bi-molecular systems (PU–CU or PU + CU) in solution and in the presence of sacrificial electron acceptors. This approach, although interesting from a proof of concept standpoint, suffers the downfalls of any system which is limited by diffusion controlled reaction rates, instability of charge accumulated states and, more importantly, the loss of the reaction products for the production of a fuel. The way to overcome these issues is to graft the designed molecular assembly PU–CU on to semi-conductor nanoparticules of TiO_2 fixed on an indium tin oxide (ITO) electrode (Figure 3.3). This strategy would allow fast photo-induced electron injection into the conduction band of the material, relatively slow charge recombination reactions and, more importantly, the recovery of the reaction products of water splitting, H^+ and e^- ultimately allowing production of fuel.

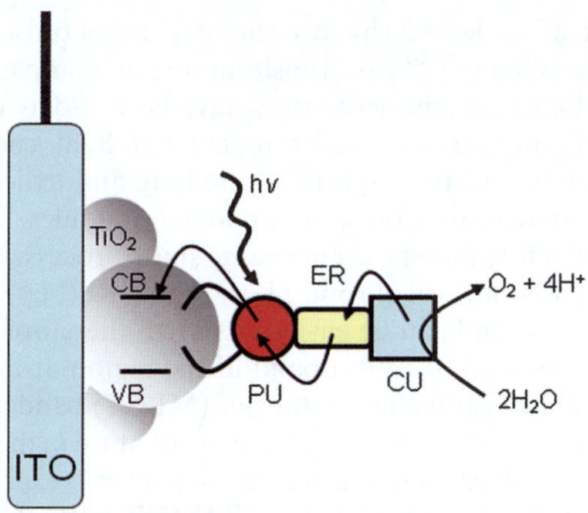

Figure 3.3 Representation of a chromophore–catalyst assembly attached to TiO_2 nanoparticles on an indium tin oxide electrode.

3.5 Antennas

As previously mentioned in section 3.2, one of the main issues is to harvest light energy efficiently. The role of an antenna is to increase the absorption cross-section area and, through a series of energy transfer steps, funnel that energy to a reaction centre where a charge separated state is formed. The solar spectrum has a maximum intensity just below 500 nm and extends well into the infrared region, where a photon with $\lambda = 1000$ nm still has 1.2 eV energy.[14] In nature, the efficient geometrical arrangement of the pigments is related to the tertiary structure of the PSII antenna subunits (Light Harvesting Complexes).[15] In an artificial photosynthetic device, this sophisticated spatial arrangement is far from reach on a synthetic level. The distance between the pigments to be used, their respective angle, and electronic coupling, must be engineered through the covalent bonds that unite them. Two mechanisms are responsible for energy transfer processes. Förster energy transfer occurs in cases where the distance between the chromophores is larger than their transition dipole strength, the dipoles have sufficient strength and the lifetime of the emitting chromophore is sufficiently long. Dexter energy transfer takes place when the chromophores are brought into van der Waals contact and there is orbital overlap between them. Systems such as carotenoid-tetrapyrrole triad and dyads,[16–18] and (phenylethynyl)anthracene–porphyrin–fullerene heptads[19] have been shown to combine large absorption spectral range (350–700 nm), high extinction coefficients (10^5 M^{-1} cm^{-1}), and quantum yields for energy transfer of close to 100%. Different approaches, such as the use of dendrimers[20] or self assembled π-stacked arrays[21,22] have also been reported in the literature. However, the connection of antenna pigments to a photochemically active chromophore to trigger a catalytic reaction has not been realised yet.

3.6 Photoactive Chromophore

Different families of molecules have been used as photosensitisers, ranging from tetrapyrrole systems[23,24] to transition metal complexes (Ru, Re, Ir, Os).[20,25,26] Synthetic porphyrin derivatives have been widely used to mimic the natural chlorophyll pigments in order to convert light energy into charge-separated states. However, their syntheses are long and tedious, therefore preventing their common utilisation as photoactive modules. Still, tetrapyrollic macrocycles are currently being used in charge shifts processes in photocatalytic reactions. Another well-known chromophore is the $[Ru(bpy)_3]^{2+}$ complex.[27–29] This complex is robust and can be easily prepared, therefore making it a good candidate. It presents an absorption band in the region around 450 nm corresponding to a metal to ligand charge transfer (MLCT) band with an extinction coefficient of about 13 000 M^{-1} cm^{-1}.[30] Upon irradiation in the MLCT band, the input light energy is converted into a $1(d\pi6) \rightarrow 1(d\pi5\pi^*)$ excited state, which in turn relaxes to form the lowest triplet state (^{3}MLCT) in less than a picosecond.[31] This state is sufficiently long-lived to allow reductive or oxidative electron transfer processes to occur before regeneration of the ground state. The oxidation potential of $[Ru(bpy)_3]^{3+}$ species is around 1.2 V *vs.* NHE which is close to that of the primary donor of PSII, $P_{680}{}^{+}$, therefore making it a suitable candidate to reproduce the oxidation reactions performed by the natural system.[29,31,32]

Two different proposals describe the nature of the triplet state of the parent ruthenium(II) trisbipyridine complex.[20] On one hand, a localised description where the metal promoted electron is present on only one bipyridine moiety, on the other, a delocalised view where the excited electrons are extended over all three bipyridines. A recent DFT study described the electronic properties of the ground and excited states of the $[Ru(bpy)_3]^{2+}$ complex. The lowest triplet state calculated $^3A_2$, as represented in Figure 3.4, was the symmetric $[Ru^{III}(bpy^{[\cdot-]1/3})_3]^{2+}$.[33]

It is important to notice that in the computational methodology used, no asymmetry was introduced in the bipyridine ligands that could alter the transition moments. This calculated electronic snapshot may result from the statistical mean of three equivalent and equiprobable situations in which the electron is localised on one of the ligands. However, the computed data for the singlet state $^2A_1$ of the oxidised species after relaxation are in agreement with experimental data for a metal centred oxidised form.

Synthetic methods are available for the modification of the bipyridine ligands surrounding the ruthenium atom.[27,34] In this manner it is possible to obtain mono- and di-substituded ligands with either electron donating or withdrawing groups, as well as synthetic handles either for further synthesis or for grafting onto surfaces or formation of triads.[35–38] Oxidation potentials for this group of molecules range from 1.0 to 1.30 V *vs.* NHE, depending on the nature and number of substitutions.

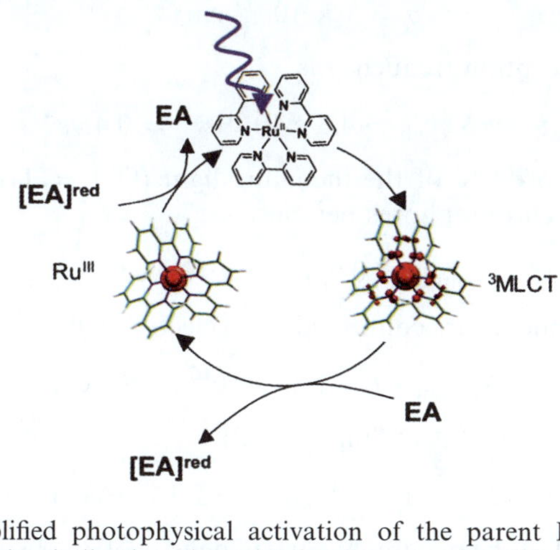

Figure 3.4 Simplified photophysical activation of the parent Ru(II) trisbipyridine complex indicating the spin density map describing the ^{3}MLCT state $[Ru^{III}(bpy^{[\cdot-]1/3})_3]^{2+}$ and the one electron oxidised Ru(III) formed after light absorption and oxidative quenching with an electron acceptor (EA).

These structural changes on the ligand allow the modulation of the photophysical properties of the complex. For instance, direct changes in the relative energies of the dπ and π^* orbital levels observed as bathochromic shifts of up to 50 nm are achieved by the addition of ester or carboxylate groups to the bipyridine ligands.[34,39,40] The emission lifetimes of these compounds are also affected by substitution patterns on the bipyridines. Electron withdrawing substituents diminish the fluorescence lifetimes and intensities by up to 70% compared to the parent $[Ru(bpy)_3]^{2+}$. For all the above-mentioned reasons, we will focus in this review on the derivatives of the $[Ru(bpy)_3]^{2+}$ complex as the photoactive chromophore unit. On a practical basis, for a Ru(II)-trisbipyridine type chromophore with absorption bands around 400–500 nm, the average photon flux in this spectral region is about 1.5×10^{14} cm^{-2} nm^{-1} s^{-1} integrated over the 400–500 nm window.

$$\Phi_{ph} \approx 1.5 \times 10^{16}\ \text{cm}^{-2}\ \text{s}^{-1}.$$

An approximation for mean molar extinction coefficient of a Ru chromophore in the blue region is:

$$\varepsilon \approx 8000\ \text{M}^{-1}\ \text{cm}^{-1} = 8000\ \text{L mol}^{-1}\ \text{cm}^{-1}$$

The molar extinction coefficient ε (if expressed in units of L mol^{-1} cm^{-1}) is directly related to the absorption cross-section σ (in units of cm^2) *via* the Avogadro constant, N_A:

$$\sigma = 1000\ln(10)\frac{\varepsilon}{N_A} = 3.82 \times 10^{-21}\varepsilon$$

Thus, the mean absorption cross-section of a Ru complex in the blue region can be estimated to be

$$\sigma = 3 \times 10^{-17}\ \text{cm}^{-2}$$

The photon absorption frequency is

$$f = \sigma \times \Phi_{ph} = 4.5 \times 10^{-1}\ \text{s}^{-1} = 0.45\ \text{s}^{-1}$$

For absorption of 90% of the incoming light (OD = 1) we need the following amount of chromophores per cm^2

$$N_{chrom} = 0.9/\sigma = 3 \times 10^{16}\ \text{cm}^{-2}$$

If these chromophores are connected to an electrode, the maximum current is

$$i = N_{chrom} \times f = 1.35 \times 10^{16}\ \text{e}^-\ \text{s}^{-1}\ \text{cm}^{-2}$$

or with 1e$^-$ = 1.6×10^{-19} As (Amp second)

$$i = 2.16 \times 10^{-3}\ \text{A cm}^{-2} = 2.16\ \text{mA cm}^{-2}$$

which is about half of the maximum current density of an absorber with 500 nm bandgap energy. Therefore, photon absorption frequencies of more than about 10 per second are not realistic for any chromophore. An alternative way an increase in charge flow to a catalytic site can be achieved is by connecting more than one chromophore (antenna) to the catalytic unit.

3.7 Water Activation at Metal Complexes

As mentioned in Section 3.2, the highly oxidised manganese oxo motif is an inherent motif in the water oxidation mechanism. In the so-called dangler model, the OEC is set as a 3 + 1 topology for the manganese ions and the one manganese outside the cube is the actual site for the chemical reaction, while the three Mn ions inside the cube can be viewed as a reservoir of oxidising power. The oxygen–oxygen bond formation is formed through an attack of an incoming water molecule on an electron deficient oxygen Mn bound atom. Another possible mechanism is based upon the radical character of the two oxos of a $[Mn^{III}Mn^{IV}{}_3, O^{\bullet}\ O^{\bullet}]$. In this case, coupling of the two radicals *via* d^3-metal peroxo-like intermediates yields formation of the O–O bond (radical coupling mechanism).

These mechanisms hinge on the ability to form a metal oxo complex *i.e.* activation of a water molecule by a high valence metal ion by successive electron and proton transfer reactions from one or two water molecules, followed by reaction with a substrate.

Higher oxidation states of middle transition metal oxo functionalities are more easily accessible only when electron loss is compensated with proton loss (proton coupled electron transfer, PCET). Otherwise, the price that has to be paid by accumulating positive charges at the metal centre renders these reactions out of reach for photodriven processes. PCET is definitely an important piece of the puzzle for chemists to understand and master in the water oxidation process and is discussed below.

3.8 Mechanistic Aspects: Electron Transfer Coupled to Proton Transfer in Simple Models and Their Implications in Electron Relay and Water Oxidation Reactions

As introduced above, among several challenges in artificial photosynthesis and/or energy storage from water splitting, deciphering the mechanism is probably one of the most difficult ones. On both sides, water oxidation or proton reduction (this last reaction can be replaced, for example, by carbon dioxide reduction in photoelectrochemical devices), the mechanism taking place is complex and requires several electron and proton transfers. We focus in the following on the case of water oxidation. As detailed in the lessons from Photosystem II, formation of dioxygen formally requires "accumulation" of four oxidative equivalents, *i.e.* loss of four electrons. However, the loss of four electrons is "accompanied" by the loss of four protons so that there is no charge variation on the whole process.

From a thermodynamic point of view, formation of dioxygen from two water molecules requires the formal release of four $H^•$ corresponding to a standard free enthalpy of 344 kcal mol^{-1}, indicating that the average energy required for the release of each $H^•$ corresponds to a standard free enthalpy of 86 kcal mol^{-1}. It is interesting to note that this energy corresponds to the energy available in PSII. Indeed, as shown in lessons from Photosystem II, the ultimate oxidant produced by the light capture and charge separation process is a tyrosyl radical ($TyrO^•$) and the standard free enthalpy for $H^•$ uptake by a tyrosyl radical is -87.8 kcal mol^{-1}.[41] The global formal reaction (2) for oxygen production is thus slightly downhill.

$$2H_2O + 4Tyr{-}O^• \leftrightarrows O_2 + 4Tyr{-}OH \quad (2)$$

While useful to indicate that the oxidative power of the tyrosyl radical is in principle enough for the reaction to proceed, this thermodynamic description is somehow misleading for at least two reasons. First of all, in PSII, it is known that both protons and electrons go in different directions. In other words, the tyrosyl radical does not act as an H-atom abstractor but as an electron acceptor. It will also be the case in the light driven mimics: electron and proton uptake from water will not occur as H atom transfer. Secondly, a tyrosyl radical, if used as an oxidant in mimics, is formally not able to abstract an H-atom from a water molecule based on thermodynamic grounds because the first O–H bond in water is a strong bond (BDFE = 122.7 kcal mol^{-1}).[41] The tyrosyl radical is also not able to oxidise a water molecule through a simple outersphere electron transfer leading to a transient cation radical $H_2O^{•+}$ and finally to dioxygen by successive electron transfers (using three other oxidant equivalents) and proton transfers. In fact, only very strong photo-oxidants such as 1,2,3,4,5,6,7,8-octafluorophenazine have been shown to be able to oxidise water to produce $HO^•$.[42,43] Therefore, the catalytic oxidation of water will not proceed as a redox catalysis type reaction but as a chemical catalysis. A discussion on the distinction between chemical catalysis *vs.* redox catalysis is probably beyond the scope of this chapter but can be found elsewhere.[44] In a few words, in redox catalysis, the catalyst plays the role of outersphere electron donor or acceptor and "catalysis" corresponds to a higher efficiency of the reaction if compared with the same reaction taking place at an electrode surface and it is due to a physical process, *i.e.* dispersion of the electrons (or holes) in a 3D space rather than being confined in a 2D space. Chemical catalysis involves intimate interactions between the active form of the catalyst and the substrate; here, water. In other words, a molecular catalyst, involving a transition metal complex will be used to "activate" water molecules and "not only" to oxidise them.

The reason why nature uses the Tyr_Z/His190 pair as electron relay may be interpreted as follows. First of all, as recalled above, the tyrosyl radical is thermodynamically a strong enough oxidant to perform water oxidation. Secondly, the proton coupled electron transfer between the highly oxidising chlorophyll cation and the Tyr_Z/His (Scheme 3.1) is thermodynamically favourable (*ca.* -0.36 eV). It has to be emphasized that, by contrast, a simple electron transfer between $P_{680}{}^{+}$ and Tyr_Z would lead to an uphill reaction of *ca.* $+0.08$ eV.[45] Therefore, despite energy loss, it allows an efficient process preventing charge recombination between $P_{680}{}^{+}$ and $Q_A{}^{-}$, which occurs on a timescale of 200 μs. Moreover, we can note that the presence of a relay prevents reverse electron transfer, *i.e.* oxidative quenching of $P_{680}{}^{*}$ by the highly oxidized OEC, that could compete with Q_A reduction.

At this point, it may be useful to discuss the mechanistic aspect of this Proton Coupled Electron Transfer (PCET). Whatever the intimate pathway followed, the PCET process between $P_{680}{}^{+}$ and Tyr_Z/His190 has the same favourable driving force. However, the competition between charge separation and charge recombination is driven by kinetic characteristics. Consequently, the kinetic features of the PCET are of crucial importance in the charge separation efficiency. In order to tackle this issue, it has to be recalled that three different pathways are available for a PCET involving different sites for electron transfer and proton transfer (Scheme 3.3).

Stepwise pathways correspond to the transfer of an electron followed by the transfer of a proton (EPT) and/or the reverse sequence, *viz.* PET pathway, and a third possibility in which the two steps are concerted giving rise to a CPET (concerted proton–electron transfer) reaction. The CPET pathway is the only one whose rate determining step benefits the whole driving force of the reaction. In other words, the concerted pathway has a thermodynamic advantage in that it allows skipping of high energy intermediates such as $Tyr_Z{}^{\bullet+}$ or the $Tyr_Z{}^{-}$/$HisH^{+}$ pair. This is the main reason why it was originally suggested by Babcock and co-workers[46] that this PCET process is concerted in PSII. One has to bear in mind, however, that there might be kinetic penalties for the concerted reaction. A theoretical description of available theories[47] for concerted electron

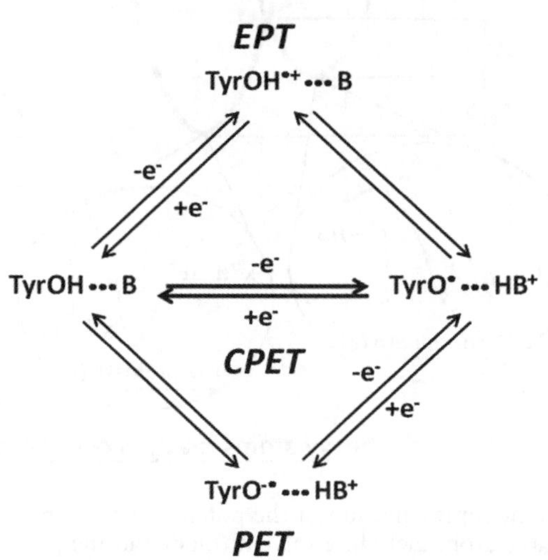

Scheme 3.3 Stepwise and concerted pathways for PCET reactions.

and proton transfers is beyond the scope of this chapter but the main features are summarized in Figure 3.5. The description derives from a double application of the Born–Oppenheimer approximation. Considering only the proton vibrational ground states, the transition state is located at the crossing of the potential energy profiles of the reactant and product electronic states toward the heavy-atom coordinate (blue parabola in Figure 3.5). Both an electron and a proton are transferred at the transition state (upper inset in Figure 3.5). A Marcus-type expression of the rate constant exhibiting a quadratic activation-driving force relationship is thus obtained, as in the case of a simple outersphere electron transfer. For this reason, the concerted pathway has a thermodynamic advantage toward competitive stepwise pathways. Nonetheless, there might be a kinetic penalty due to the transfer of the proton at the transition state. This transfer takes place by proton tunneling through a barrier leading to a decrease of the transition probability as compared to a simple outersphere electron transfer. In this sense, the CPET reaction is nonadiabatic.

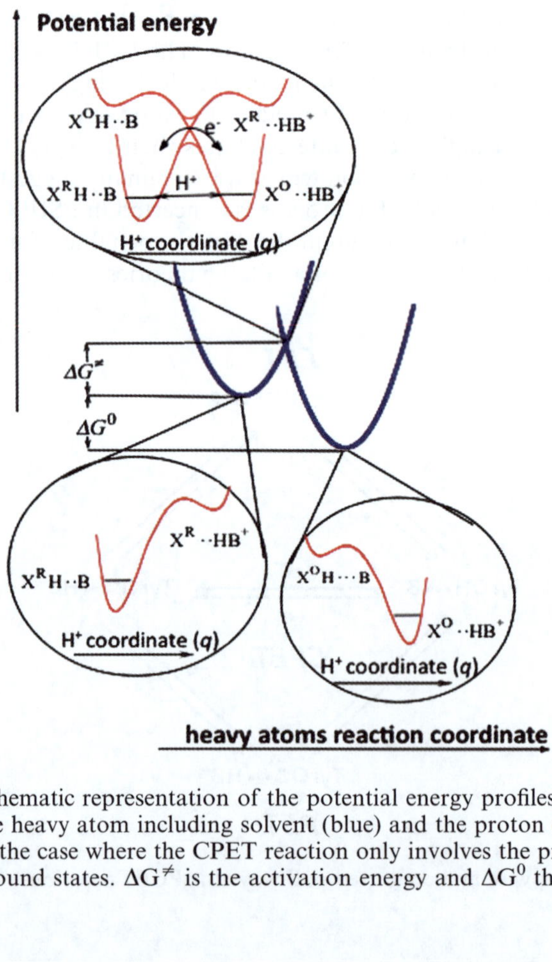

Figure 3.5 Schematic representation of the potential energy profiles as a function of the heavy atom including solvent (blue) and the proton (red) coordinates in the case where the CPET reaction only involves the proton vibrational ground states. $\Delta G^{\neq}$ is the activation energy and ΔG^0 the driving force.

In order to test the possibility of concerted mechanism in Tyr/His-like assemblies, and to understand the main factors governing the dynamics of such a pathway, chemists have designed mimics for the Tyr_Z/His190 pair such as the amino-phenols derivatives as shown in Figure 3.6. Their electrochemical[48–50] and homogeneous[51] oxidation show the occurrence of the CPET pathway.

Those reactions have been shown to be nonadiabatic and very sensitive to the proton donor–acceptor distance. Therefore, the Tyr_Z/His190 pair can be qualified as being designed for an efficient CPET.

Coming now to models mimicking PSII it is worth asking the question whether or not such an electron relay is necessary. Considering a model system where a ruthenium complex is used as photoactive chromophore and TiO_2 is used to collect electron, as shown on Scheme 3.4, the answer depends on the competition between charge recombination (between TiO_2^- and Ru^{III}) and Ru^{III} reduction by the nearby electron donor (either a relay or the water oxidizing catalyst). Because all four oxidation steps corresponding water activation/O_2 formation are expected to be slow processes as compared to charge recombination occurring on a time scale of 0.37 ms,[52] it seems reasonable to design a model with an electron relay as intermediate redox active centre between the chromophore and the catalytic unit.

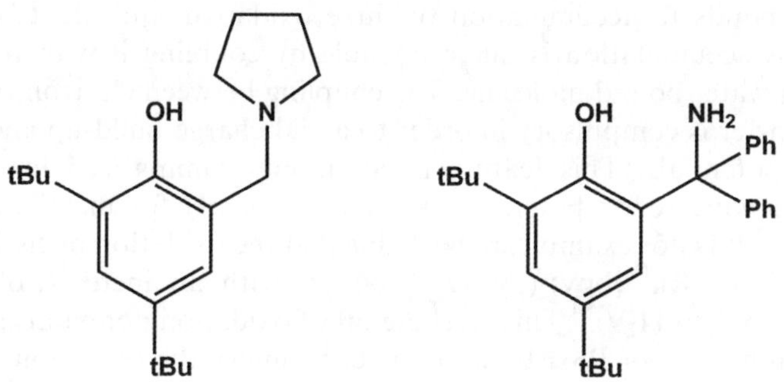

Figure 3.6 Hydrogen-bonded tyrosine models.

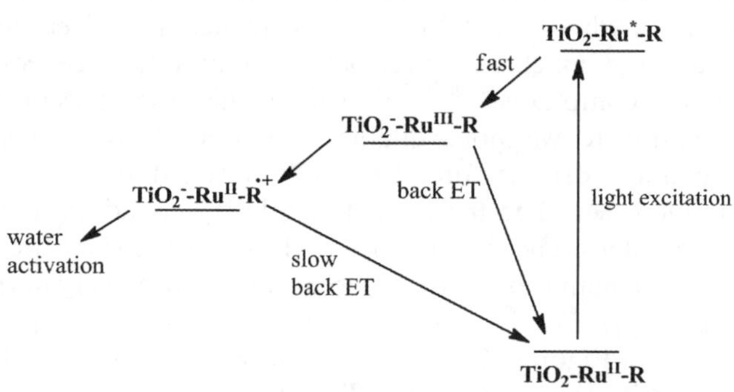

Scheme 3.4 Competing electron transfer pathways from a ruthenium photoactive chromophore on a TiO_2 surface.

The resulting oxidised redox relay is thus the oxidising species for water oxidation. As mentioned previously, only very strong photo-oxidants are able to oxidise water through a simple outersphere electron transfer coupled to deprotonation to produce $HO^{\bullet}$. Therefore, the catalytic oxidation of water will not proceed as a redox catalysis but as a chemical catalysis. Chemical catalysis involves intimate interactions between the active form of the catalyst and the substrate (water). In other words, a molecular catalyst, involving a transition metal complex will be used to activate water molecules. This activation will be performed by a series of electron and proton removal steps from a water bound transition metal catalyst leading to an oxo functionality. Formation of the O–O bond is expected to occur in a subsequent reaction of the oxo thus formed (Figure 3.2). As stated before, this schematic description emphasizes that the water oxidation mechanism can be considered to occur in at least three separated sequences. First of all, activation of water molecules by high valent metal ions through proton-coupled electron transfer processes. Secondly, reaction of the activated oxygen atom to form the O–O bond. This reactivity has been proposed to be either an acid–base reaction or a radical–radical coupling. This will obviously depend on the metal oxygen bond character. Finally, dioxygen is released.

The first challenge is thus to build up an intermediary metal oxo complex that corresponds to accumulation of three oxidative equivalents on a metal centre. This accumulation is made possible by coupling it with the deprotonation of a water bound molecule. The coupling between electron transfer and proton transfer is compulsory in order to avoid charge build-up and very high oxidation potentials. This feature is commonly summarized in a Pourbaix diagram showing redox potentials as function of pH for metal water bound complexes.[53] It is, for example, remarkable that the oxidation of cis-$Ru^{II}(bpy)_2(py)(H_2O)^{2+}$ to $Ru^{IV}(bpy)_2(py)(O)^{2+}$ occurs with an increase of oxidation potential of only 0.11 V.[54] This levelling off of oxidation potentials is useful for light driven models of PSII because in such mimics the oxidation potential is limited by the initial light absorption–charge separation process. Efficiency of PCET reactions allowing formation of oxo species is also of interest and requires kinetics analysis. The mechanism of the PCET oxidation of metal(II)aquo–metal(III)hydroxo–metal(IV) oxo sequence has been addressed in pioneering, albeit not conclusive, electrochemical and homogeneous studies of similar ruthenium complexes,[48,55–58] at a time when the concept of concerted proton–electron transfer was not available or, later on, at a time where the notion of the pH-dependent driving force for a CPET reaction was not yet clearly recognized as incorrect. The mechanistic analysis was additionally obscured by the possible interference of Ru(II) + Ru(IV) comproportionation. The domain of thermodynamic existence of the Ru(III) is indeed very narrow over the accessible range of pH. This is not the case with more recent examples of osmium complexes shown in Figure 3.7, which also show the two chemically reversible cyclic voltammetric waves corresponding to the $Os^{II}OH_2/Os^{III}OH/Os^{IV}O$ sequence.[59,60] Electrochemical approaches to these reactions possess several advantages. Separation of the electron transfer (the electrode) and proton

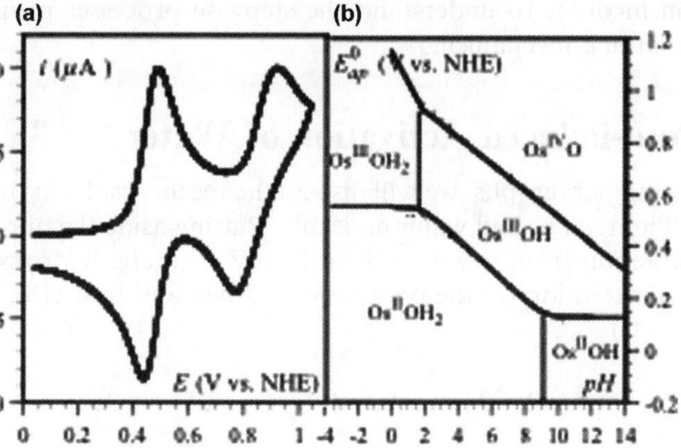

Figure 3.7 (a) Voltammogram for the oxidation of 0.4 mM $[bpy_2pyOs^{II}–OH_2]^{2+}$ in a 0.1 M Britton–Robinson buffer at pH = 3.1 at 0.2 V s^{-1}. (b) Pourbaix diagram defining the zones of thermodynamic stability of the various intervening species.

transfer sites, required to distinguished CPET reactions from H-atom transfers, is readily achieved. Changing the electrode potential is an easy way of varying the driving force of the reaction and the current is an on-line measure of the reaction kinetics. An analysis of the apparent standard rate constant as a function of pH and buffer concentration in water leads to firm conclusions concerning the oxidation mechanisms, and underlines the general factors controlling the reactivity.

It has been shown that the Os^{II}-aquo complex is oxidised along sequential pathways.[54] In contrast, the conversion of Os^{III}-hydroxo into Os^{IV}-oxo species follows a concerted pathway because of the large pK_a gap between the reduced and oxidised species, thus favouring a pathway that avoids intermediates.[55] Such a reaction is made possible by the presence of a large concentration of buffer. Indeed, in the presence of a buffer, the metal complex is associated to the base within a cluster pair, thus forming weakly bonding adducts, as it may be encountered in the active site of proteins. Such assistance could be achieved through weak bonds (*e.g.* hydrogen bonds) with some residue(s), in particular basic residues, and other molecule(s) (*e.g.* water molecules). Those results emphasize the crucial role of proton removal to efficiently reach oxo-transition metal complexes allowing the highly activated water molecules to be able to undergo radical coupling or nucleophilic attack to form a strong O–O bond. Efficiency is determined by the proximity of the proton donor and acceptor sites but it has been recently shown that proton relays may be inserted in order to achieve long distance proton transport through a concerted electron and Grotthuss like double proton transfers.[56] In that respect, more sophisticated metal complexes including ligand scaffolds bearing a coordinating cavity capable of holding high valent metal-oxo motif together with pendant groups that can participate as proton relays in the PCET processes are currently under investigation in order to understand the stepwise processes proposed for the nucleophilic attack mechanism.

3.9 Photo-induced Activation of Water

In the following paragraphs, we will discuss the recent results in the light driven activation of a metal bound water molecule. Putting aside the one electron, one proton abstraction from water to form highly energetic hydroxyl radical, we will focus our attention on the photo-induced two and four electron processes (eqn (3) and (4))

$$H_2O + h\nu + \text{catalyst} \rightarrow [O] + 2H^+ + 2e^- \quad (3)$$

$$2H_2O + h\nu + \text{catalyst} \rightarrow O_2 + 4H^+ + 4e^- \quad (4)$$

3.9.1 Using Water as Oxygen Source in the Oxygen Atom Transfer Reactions

Oxygen atom transfer reactions in mild conditions are an important challenge for chemists. The majority of catalytic oxidation processes known use metal-oxo derivatives that are generated using powerful oxygen-containing oxidants, such as hypochlorite. Another greener approach is to use hydrogen peroxide and metal catalysts to generate the same reactive high valent oxo species. In nature, oxygenation reactions of this kind are carried out by a large class of enzymes (Cytochrome P450), where the active oxidising species, an iron(IV)-oxo porphyrin π-cation radical, is generated through a reductive two electron/proton process of dioxygen following eqn (5).

$$RH + O_2 + 2e^- + 2H^+ \rightarrow ROH + H_2O \quad (5)$$

The reaction depicted in eqn (5) is a particularly interesting pathway that uses water as both an active oxygen atom donor and as a two electron, two proton donor as for instance, the oxidation of organic substrates. An alternative way to write the same reaction is given in eqn (6), where the organic substrate (S) is being oxidised *via* an oxygen atom transfer. As we can see, the net outcome of this reaction is the oxidised substrate and formally an H_2 molecule, *i.e.* a fuel.

$$H_2O + S + h\nu + \text{catalyst} \rightarrow SO + 2H^+ + 2e^- \quad (6)$$

An elegant alternative route therefore would be to photogenerate the corresponding metal-oxo active species using a water molecule bound to a metal through the abstraction of two-electrons and two-protons through PCET processes, as discussed in Section 3.8. The resulting photo-generated reactive metal-oxo would then transfer an oxygen atom to organic substrates yielding at the same time two electrons and two protons. Three different molecular systems have been used in order to perform these reactions.

i. Chromophore/catalyst systems: mono-metallic ruthenium complexes that act as both the chromophore and the catalyst.
ii. Bimolecular chromophore–catalyst systems: where ruthenium(II) tris-pyridine complexes are used to sensitize or activate a metal catalyst for oxygenation reactions
iii. Chromophore–catalyst assemblies: di-nuclear, covalently bound ruthenium complexes where one ruthenium moiety works as a chromophore (Ru_{chrom}) and the other one as the catalyst (Ru_{cat}).

3.9.1.1 Chromophore/Catalyst

Pioneering work in this field comes from the Inoue lab. They were the first to report on the photocatalytic epoxidation of alkenes using water (hydroxide ion) as the oxygen atom donor using tetra (2,4,6-trimethyl)phenylporphyrinatoruthenium(II) [Ru^{II}TMP(CO)] (Scheme 3.5).[61–63] Typical photocatalytic reactions were run under light excitation at 420 nm and in the presence of a highly oxidised platinum salt $(Pt^{IV}Cl_2)^{2-}$ as a sacrificial electron acceptor. The best reported results (quantum yield of 60% and product selectivity of ~95%) were obtained in alkaline conditions (pH ~ 11).

The exact mechanistic pathway explaining how the hydroxide ion is activated through light irradiation is not known but the proposed mechanism is depicted in Scheme 3.6. Illumination in the 400 nm region (Soret band) of the Ru(II)-porphyrin

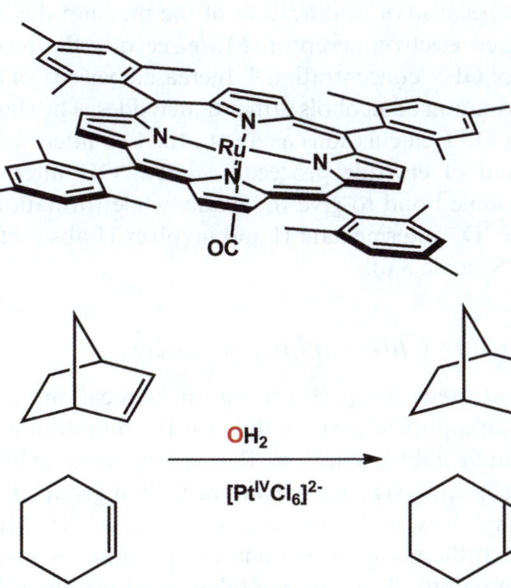

Scheme 3.5 Inoue's light driven oxidation of alkenes using water as both oxygen atom and electron donor.

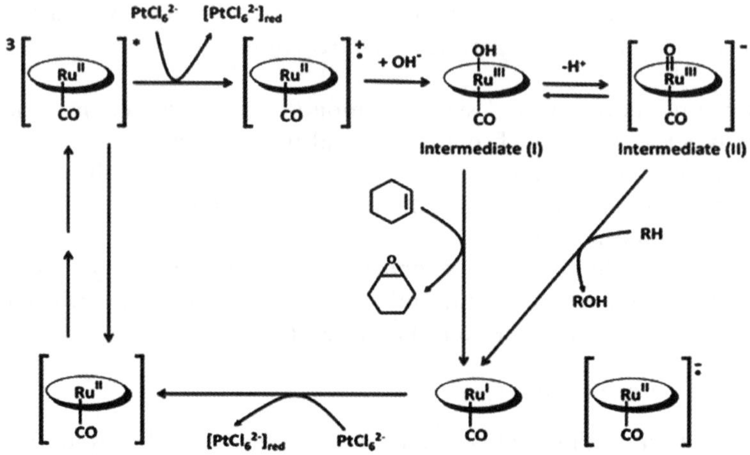

Scheme 3.6 Mechanistic proposal for the photo-oxidation of organic substrates for Inoue's Ru(II)-porphyrin using water as electron and oxygen atom donor.

triggers electron transfer from the triplet excited state of the metalloporphyrin to the sacrificial acceptor that decomposes, generating HCl. The resulting radical cation on the porphyrin reacts with OH^- to give $Ru^{III}OH$ (intermediate I), which can alternatively deprotonate to generate the $Ru^{III}O$ anion (intermediate II). These intermediates are capable of reacting with alkenes to yield solely epoxidation products with quantitative insertion of ^{18}O when $H_2{}^{18}O$ was used. At neutral pH, the epoxidation products were detected after 20 s but rapidly converted into the corresponding diols because of acidification of the medium due to the release of HCl from the reduced electron acceptor. More recently the reactivity has been studied as function of OH^- concentration.[61] Increased basicity of the solution (pH ~12.7) results in formation of alcohols rather than oxides. The change in reactivity was explained using DFT calculations as due to the two intermediates previously postulated: formation of epoxide proceeds *via* $Ru^{III}OH$ intermediate I which attacks the C=C double bond to give the oxide, while formation of the alcohol proceeds *via* the $Ru^{III}O^-$ intermediate II and involves H abstraction followed by hydroxyl insertion (Scheme 3.6).

3.9.1.2 Bimolecular Chromophore–Catalyst

A more common strategy to perform photochemical multi-electron redox processes is the use of a photoactive module for the bimolecular activation of a metal catalyst through light excitation. Pioneering work in the activation of manganese porphyrin catalysts comes from the Calvin group. The excited states of the tetra(N-methyl-4-pyridyl)porphyrin manganese(III) [Mn^{III}TMPyP] are too short lived so that the use of these complexes as photosensitisers to achieve their oxidation is impaired. The strategy taken is to use the [Mn^{III}TMPyP] as an electron donor in a three component system consisting of the Ru(II)-trisbipyridine as sensitiser and a sacrificial electron acceptor in this case the

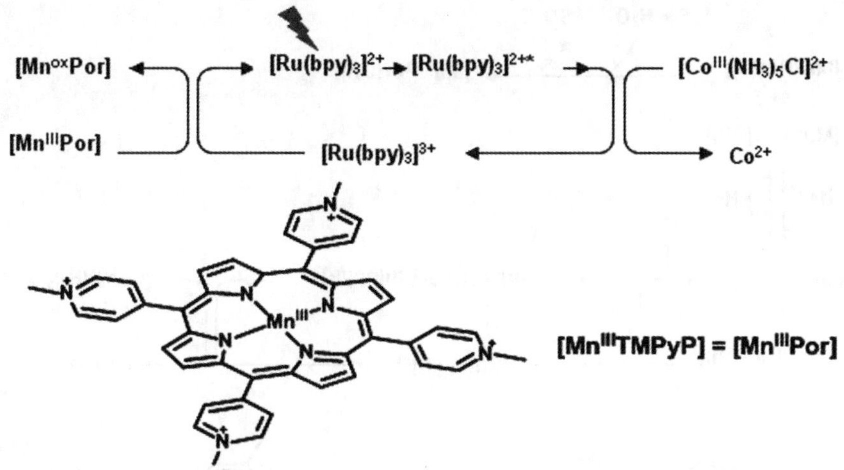

Scheme 3.7 Photo-activation of [Mn^{III}TMPyP(H_2O)] to [Mn^{V}TMPyP(O)].

cobalt(III) chloropentammine complex. A schematic arrangement of this system is represented in Scheme 3.7.

Upon continuous illumination of a degassed solution containing the three components $[Ru(bpy)_3]^{2+}$ (0.1 mM), [Mn^{III}TMPyP] (0.01 mM) and $[Co(NH_3)_5Cl]^{2+}$ (1 mM) at pH 7, the differential spectral changes in the electronic absorption spectra were associated with the oxidation of the [Mn^{III}TMPyP], and matched those observed in the same complex upon chemical oxidation with NaClO at pH 7 which was attributed to the formation of a high valent manganese(V)-oxo porphyrin derivative. The oxidised [Mn^{ox}TMPyP] is unstable and returned quantitatively and reversibly to the initial [Mn^{III}TMPyP], while concomitantly 100% of the $[Ru(bpy)_3]^{2+}$ was recovered. No reactivity of the high valent manganese derivative was reported. Recently, Fukuzumi *et al.*[64] reported on a similar three component arrangement and found that their system behaves as a highly efficient photocatalytic oxygenation catalyst using water as an oxygen source. In this study, $[Ru(bpy)_3]^{2+}$ was used as the molecular sensitiser and $[Co(NH_3)_5Cl]^{2+}$ as the electron acceptor, while two different water soluble sulfonato manganese(III) porphyrin derivatives were used as oxygenation catalysts (see Scheme 3.8). Laser flash photolysis techniques were used to analyse the dynamics of the photo-oxidation of (TMPS)Mn^{III}(OH). Transient absorption spectra of the one electron oxidised species, (TMPS)-Mn^{IV}(O), were obtained by difference spectra after laser pulse excitation of the $[Ru(bpy)_3]^{2+}$ sensitiser at 450 nm in the presence of (TMPS)Mn^{III}(OH), $[Co(NH_3)_5Cl]^{2+}$ and sodium *para*-styrene sulfonate (NaSS) as the substrate in phosphate buffer at pH 7.4. The observed spectral changes match those of an independent experiment where the (TMPS)Mn^{IV}(O) was generated by treating (TMPS)Mn^{III}(OH) with an oxygen containing oxidant such as *m*-CPBA (*meta*-chloroperbenzoic acid) at pH 8. Further observations showed that during the reaction with NaSS, the (TMPS)Mn^{IV}(O) was directly converted to

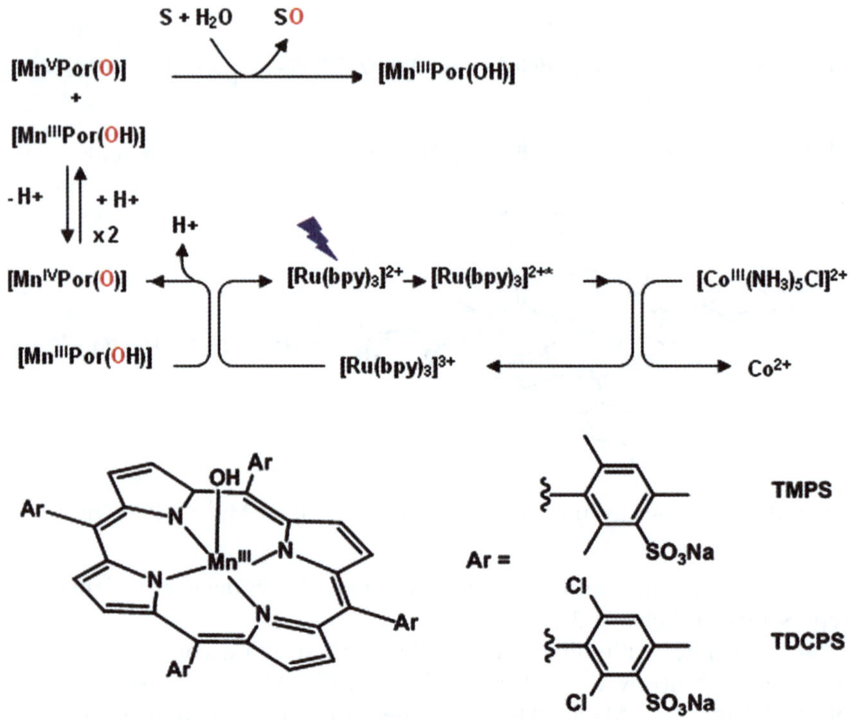

Scheme 3.8 Proposed mechanistic pathway for the photocatalytic oxygenation in the Fukuzumi system.

Table 3.1 Photocatalytic oxygenation of different substrates by $(Por)Mn^{III}(OH)$ (0.01 mM) in the presence of $[Ru(bpy)_3]^{2+}$ (0.1 mM), $[Co(NH_3)_5Cl]^{2+}$ (10 mM). Photoirradiation with a Xe lamp, $\lambda > 450$ nm. TON = mol of product/mol of catalyst. Yield (%) = mol of product/mol of initial substrate × 100. Note the maximum yield is 50% because the product yield is one-half of the conversion of the oxidant.

Catalyst	*Substrate*	*Product*	*TON*	*Yield*
$(TMPS)Mn^{III}(OH)$	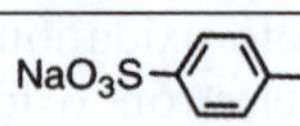	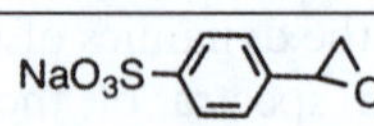	450	45
$(TDCPS)Mn^{III}(OH)$	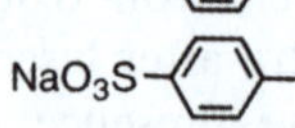	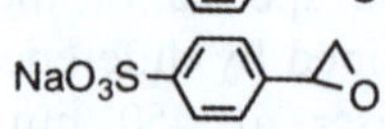	460	46
$(TMPS)Mn^{III}(OH)$	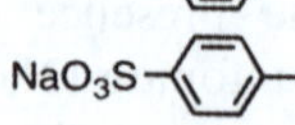	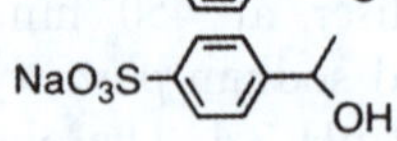	60	6
$(TDCPS)Mn^{III}(OH)$	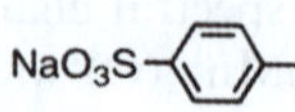	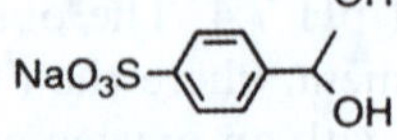	70	7
$(TMPS)Mn^{III}(OH)$	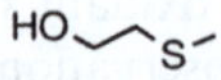	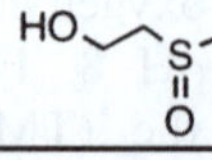	30	3

(TMPS)Mn^{III}(OH), with the reaction obeying second-order rate kinetics with increased rates as the proton concentration is increased. The collection of data suggests that the reaction of (TMPS)Mn^{IV}(O) with NaSS occurs *via* the disproportionation of the photo-oxidised (TMPS)Mn^{IV}(O) complex to give (TMPS)Mn^{III}(OH) and (TMPS)Mn^{V}(O), with the latter being the reactive intermediate in the catalytic oxygenation reaction. Table 3.1 summarises the observed photo catalytic activity towards oxygenation of different organic substrates.

3.9.1.3 Chromophore–Catalyst Assembly

In these kinds of systems both molecular units, chromophore and catalyst, are covalently bound. The advantage of this approach is that the individual properties of the linked chromophore–catalyst couple can be more easily tuned and optimized through design of the ligand.

Work by Meyer *et al.*,[65] as well as more recent work by Rocha *et al.*,[66] has shown that upon light excitation, ruthenium sensitisers are capable, in the presence of electron acceptors (TiO_2, $[Co(NH_3)_5(Cl)]Cl_2$), of sequential electron transfer and activation of covalently bound catalytic sites for the 2 electron oxidation of alcohols as shown in eqn (7).

$$RCH_2OH + h\nu + \text{catalyst} \rightarrow RCHO + 2e^- + 2H^+ \tag{7}$$

Rocha and co-workers synthesized a bis-terpyridine complex that acts as a chromophore covalently attached to a terpyridine–ruthenium–bipyridine moiety that acts as a catalyst (Ru_{cat}). Irradiation of this complex with light > 390 nm in the presence of $[Co(NH_3)_5(Cl)]^+$ leads to the oxidation of benzylic alcohols into their corresponding benzaldehydes, as well as the oxidation of secondary alcohols into ketones with turnover numbers (TON) ranging from 170 to 110 and turnover frequencies (TOF) around 20 h^{-1} (Scheme 3.9). The mechanism postulated is derived from electrochemical experiments, where $Ru^{II}{}_{cat}$ undergoes a two electron/proton oxidation from the original Ru^{II}–OH_2 to form a Ru^{IV}=O complex which is responsible for the oxidation reaction.

These two oxidation processes (Ru^{II}–OH_2/Ru^{III}–OH) and (Ru^{III}–OH/Ru^{IV}=O) are reported to occur at 0.54 and 0.59 V *vs.* SCE, which are both less than the 1.28 V *vs.* SCE that is available upon light-driven formation of the $[Ru(terpy)_2]^{3+}$ sensitiser. Such chromophores have been studied and, while

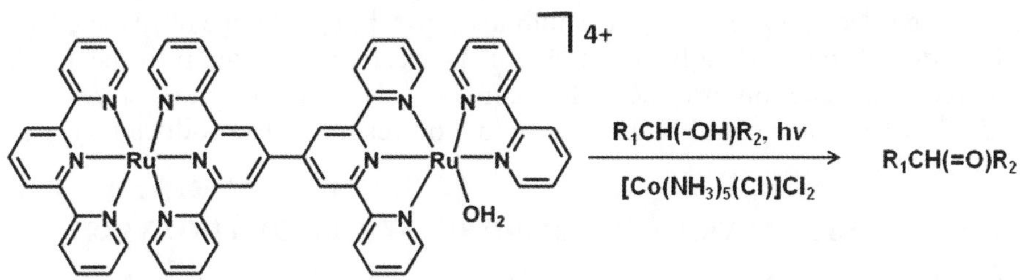

Scheme 3.9 Rocha's sensitiser–catalyst assembly for the two electron/proton photo-oxidation of organic substrates.

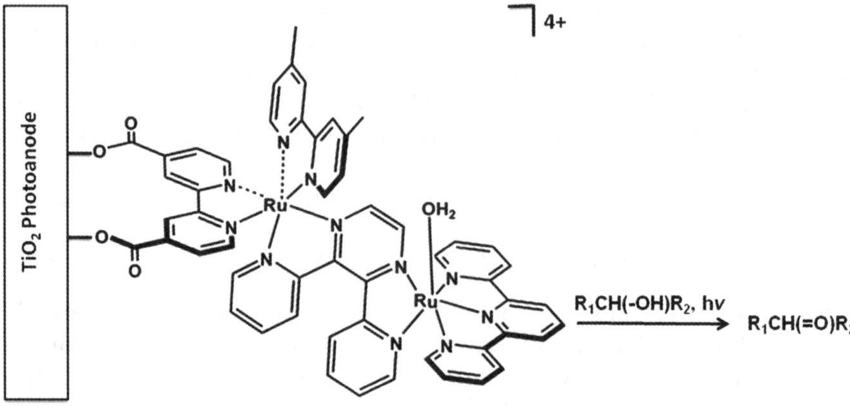

Scheme 3.10 Meyer's heterogeneous photocatalyst for oxidation of alcohols.

having a high oxidation potential and offering the optimal geometry to place acceptor–donor substituents at 180°, they do suffer from a shorter lived excited triplet states (ns) when compared to the parent $[Ru(bpy)_3]^{2+}$ complex (μs), which limits the electron transfer efficiency in solution.[67–69] Nevertheless, this molecule shows photocatalytic activity and is a good example of a covalently bound, chromophore–catalyst dyad.

Earlier work by Meyer on the oxidation of alcohol based on the same (tpy)(bpy)Ru^{II}–OH_2 catalyst covalently bound to a (bpy-COOH)(bpy-Me)Ru^{II} unit which acts as a chromophore.[65] The presence of carboxylic groups allows an extended absorption as well as binding to TiO_2 surfaces (Scheme 3.10). The oxidation potential of the (Ru^{II}–OH_2/Ru^{III}–OH) and (Ru^{III}–OH/Ru^{IV}=O) couples in this system are superimposed at 0.69 V *vs.* SCE, while the oxidation potential for the oxidised chromophore is 1.6 V. The resulting (tpy)(bpy) $Ru^{IV}(O)^{2+}$ oxidises 2-propanol into acetone through a hydride transfer mechanism.[70,71] Although no turnover numbers were reported, production of H_2 at the cathode was observed. The advantage of heterogeneous systems of this kind is two-fold. On one hand, the covalently bound chromophore should be capable of injecting an electron in the TiO_2 surface very rapidly after excitation. This decreases the losses that might arise from having short excited state lifetimes as in the previous case. Consequently, the drawbacks of using solution-based electron acceptors (*i.e.* solubility, absorption, change of pH, *etc.*) are eliminated. The second advantage is that the electrons injected into the anode become true reaction products that can be collected and transferred to a cathode where they could be used in reduction reactions to produce fuel.

3.10 Light Driven Oxidation of Water to Dioxygen

The efficient four-electron oxidation of water using visible light is by far the most challenging task in the field of artificial photosynthesis. Meyer's group was the first to report on chemically driven molecular water oxidation catalysts

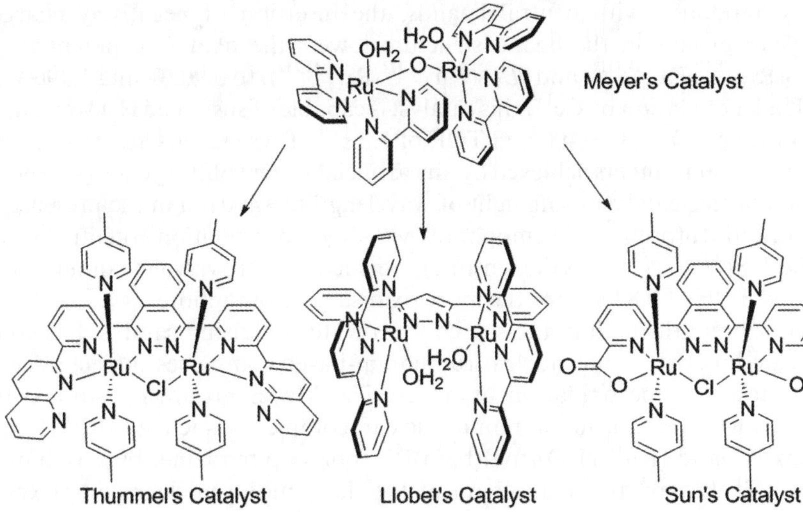

Figure 3.8 Dinuclear ruthenium-based catalysts related to the blue dimer.

in 1985, and since then only few advances in the field of photo-activation of these reactions had been reached until very recently.[72] Meyer's catalyst also known as the blue dimer is a *cis,cis*-$[(bpy)_2(H_2O)Ru(\mu\text{-}O)Ru(H_2O)\text{-}(bpy)_2]^{4+}$ complex (Figure 3.8). This complex holds two Ru–aqua groups which are bound *via* a μ-oxo bond allowing free rotation around the Ru–O–Ru motif. This complex is capable of chemically promoting the reaction $H_2O–Ru^{III}–O–Ru^{III}–OH_2$ to $O{=}Ru^{V}–O–Ru^{V}{=}O + 4e^- + 4H^+$. The resulting Ru-oxo complex undergoes nucleophilic attack from a water molecule to form the intermediate $HO–Ru^{IV}–O–Ru^{IV}–OOH$, which goes on to release oxygen with a TON = 11–13. Visible light-driven oxidation of this complex using a modified ruthenium trisbipyridine $Ru(bpy)_3{}^{2+}$ chromophore and $S_2O_8{}^{2-}$ as the electron acceptor generated oxygen at a rate of 330 μL h^{-1} during the first 15 minutes of the reaction.[73] After this time, the reaction slowed down due to degradation of the ruthenium chromophore.

The Llobet and Thummel groups chemically modified the blue dimer complex in order to lock the geometry of the dinuclear ruthenium complex as well as to increase its stability by replacing the oxygen atom bridging ligand with a two nitrogen donor ligand as the backbone for the two ruthenium ions. In the Llobet system, the *in,in*-$\{[Ru^{II}(tpy)\text{-}(H_2O)]_2(\mu\text{-}bpp)\}^{3+}$ (bpp = 3,5-bis-(2-pyridyl) pyrazolate) has both ruthenium atoms in close proximity resulting in a 2.48 Å distance between the oxygen atoms of their aqua ligands. The $H_2O–Ru^{II}–bpp–Ru^{II}–OH_2$ species can be chemically oxidised to $O{=}Ru^{IV}–O–Ru^{IV}{=}O + 4e^- + 4H^+$. This species evolves into an intramolecular μ-1,2-peroxo followed by a hydroperoxidic intermediate that evolves oxygen (TON = 19).[74–76] These two examples show how ligand design can dictate the mechanism of catalytic reactions. More recently,[77] a dinuclear ruthenium complex with a *cis* structure bearing the ligand 1,4-bis(6′-COOH-pyrid-2′-yl)phthalazine has been reported. Compared to similar compounds with neutral ligands, the insertion of negatively charged carboxylato groups in the ligand structure lowers the oxidation potentials of the ($E_{1/2}$($Ru_2^{II,II}/Ru_2^{II,III}$)) and ($E_{1/2}$($Ru_2^{II,III}/Ru_2^{III,III}$)) to 0.903 V and 1.396 V *vs.* NHE. Under oxidation by Ce^{IV}, this catalyst is capable of sustained H_2O oxidation for 20 h with a TON of 10 400 and TOF of 1.2 s^{-1}. These results are, to date, the highest turnover numbers achieved by any artificial water splitting catalyst. Photo-activation of this catalyst using light of wavelengths >400 nm in the presence of $Na_2S_2O_8$, and $Ru(bpy)_3^{2+}$ chromophores with different oxidation potentials (1.26, 1.40, and 1.54 V *vs.* NHE) as the sensitisers, yielded oxygen with a maximum TON of 580 and TOF of 0.83 s^{-1} for the more oxidising chromophore.

A major breakthrough in the search for a water-oxidising catalyst has come with the finding that mono-nuclear ruthenium-based complexes can act as good catalysts towards the oxidation of water. The Meyer group has studied the reactivity and mechanism of mono-nuclear complexes such as $[Ru^{II}(H_2O)(tpy)(bpz)]^{2+}$ and $[Ru^{II}(H_2O)(tpy)(bpm)]^{2+}$ (bpz = pipyrazine, bpm = bipyrimidine).[78–81] The proposed mechanism for these mono-nuclear complexes is similar to that of the blue dimer molecule and implies nucleophilic attack of an incoming water molecule on an electrophilic oxygen atom at a high valent ruthenium centre. Loss of two electrons and two protons from the $[Ru^{II}–OH_2]$ leads to formation of $[Ru^{IV}=O]$, at this stage, the oxygen atom is not sufficiently activated for the nucleophilic attack to proceed. A third one electron oxidation yields the $[Ru^{V}=O]$ intermediate is formed which then undergoes nucleophilic attack from a water molecule to form a $[Ru^{III}–OOH]$ intermediate with the loss of a proton from the nucleophile water. Therefore, the rather simplistic nucleophilic attack mechanism does not explain the actual bonding pattern for the O_2 molecule. A final one electron oxidation with the concomitant loss of a proton leads formally to the ruthenium(IV) peroxo species $[Ru^{IV}–OO]$ that can also be seen as a valence tautomer of the ruthenium(II) dioxygen adduct $[Ru^{II}–(OO)]$ which then releases one molecule of dioxygen in the triplet state. Llobet and colleagues[82] made a significant contribution based on a mono-nuclear ruthenium-based catalyst using both theoretical calculations and isotopic analysis of the product O_2 in a water splitting reaction using the complex *cis*-$[Ru^{II}(bpy)_2(H_2O)_2]^{2+}$. Experimental data favours a water nucleophilic attack (WNA) mechanism and theoretical data shows evidence that this mechanism is 32 kcal mol^{-1} more favourable than the intramolecular O–O bond formation. This pathway can be explained by a molecular orbital mesomeric representation of the complex, which, upon removal of 4 electrons and 4 protons, can be described as *cis*-$[Ru^{III}(bpy)_2(O^{-0.5})_2]^{2+}$ where the presence of unpaired spin density on the metal and the O atoms renders it electrophilic, thus favouring the WNA mechanism. These examples show that formation of high-valent metal oxos with low overpotentials is key to the water oxidation reaction. The small potential separation between redox states is a consequence of the charge compensation achieved through coupling proton and electron transfer reactions, while the higher oxidation states of ruthenium are stabilized by the metal-oxo bond formation. These thermodynamic limitations become more important when trying to drive these reactions with light, as the oxidising power of the excited chromophores is more limited than that provided by the chemical oxidants used in the previous examples.

Another example of an active mono-nuclear ruthenium catalyst has been presented by the Sun group where they were able to isolate a high oxidation state of a hepta-valent Ru complex bearing negatively charged carboxylato ligands [μ-(HOHOH)[Ru^{IV}L(pic)$_2$(PF_6)$_3$·2H_2O].[83] According to the authors, these observations support the idea of water coordination into a hexa-valent Ru^{IV} centre without the need for ligand exchange. The resulting Ru^{IV} species does not undergo any large structural changes, and a slight opening of the O–Ru–O angle is sufficient to accept a water molecule as the seventh ligand. This system is capable of water oxidation in presence of Ce^{IV} at pH 1 (TON = 120).

Activation of the above mentioned catalyst by light was performed using $[Ru(bpy)_3]^{2+}$ analogues as sensitisers and $[Co(NH_3)_5(Cl)]Cl_2$ and $S_2O_8^{2-}$ as irreversible electron acceptors.[84,85] Depending on the electron acceptor used, formation of oxygen was observed with impressive TOF's of 550 ($Co(NH_3)_5(Cl)]Cl_2$) and 1250 h^{-1} ($S_2O_8^{2-}$) corresponding to reaction times of 50 s and 7 min, respectively. The short duration of the reaction seems to be inherently linked to the electron acceptor used. Cobalt pentamine releases ammonia, which increases the pH of the reaction medium, and decomposition of both the chromophore and catalyst molecules is observed after 7 minutes. Reduction of persulfate salts is done by an oxidative quenching of $[Ru(bpy)_3]^{2+*}$ to $[Ru(bpy)_3]^{3+}$ followed by a thermal reaction, according to eqn (8) and (9):

$$[Ru(bpy)_3]^{2+*} + S_2O_8^{2-} \rightarrow [Ru(bpy)_3]^{3+} + SO_4^{\bullet-} + SO_4^{2-} \quad (8)$$

$$[Ru(bpy)_3]^{2+} + SO_4^{\bullet-} \rightarrow [Ru(bpy)_3]^{3+} + SO_4^{2-} \quad (9)$$

Catalyst deactivation is, in this case, due to the lowering of the pH in the resulting solution. During a 50 s run, the pH of the phosphate buffer used decreased from 7.1 to 2.9 rendering the catalyst inactive. The reaction was also studied by changing the oxidative potential of the chromophores. As expected, $[Ru(bpy)_3]^{2+}$ ($E_{1/2}(Ru^{II}/Ru^{III})$ = 1.26 V *vs.* NHE) gave higher turnover frequencies (591 h^{-1}) for the oxidation reaction than $[Ru(dmbpy)_3]^{2+}$ ($E_{1/2}(Ru^{II}/Ru^{III})$ = 1.10 V *vs.* NHE (dmbpy = dimethylbipyridine) (360 h^{-1}). These results are explained by the decreased driving force available to carry out the reaction for the more electron rich chromophore.

The Bonchio and Hill groups have proposed a molecular catalyst based on a tetranuclear ruthenium-oxo wrapped by polyoxometallate (POM) ligands as water oxidising catalyst.[86–88] The electron withdrawing nature of the POM ligand stabilizes the high valent intermediates and it assists the deprotonation steps. The complex $Cs_{10}[Ru_4(\mu\text{-}O)_4(\mu\text{-}OH)_2(H_2O)_4(\gamma\text{-}SiW_{10}O_{36})_2]$ (Ru_4-POM; POM is the polyoxometalate ligand γ-$SiW_{10}O_{36}$) consists of four Ru^{IV}–OH_2, the adjacent ruthenium centres within each γ-[$SiW_{10}Ru_2$] are bridged by hydroxo ligands, while the ruthenium centres of different monomeric units are

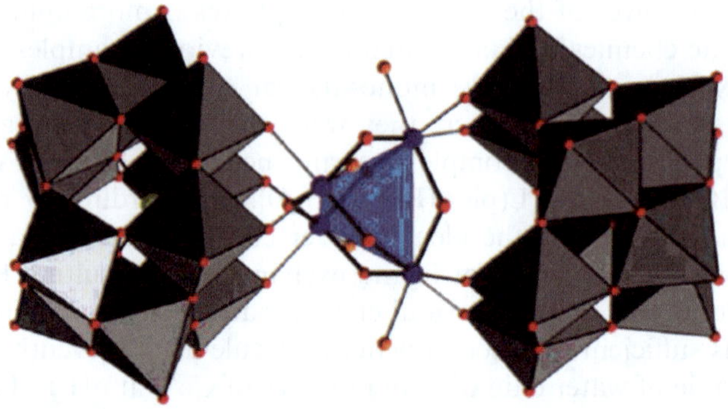

Figure 3.9 Structure of the catalytic core $\{Ru_4(\mu\text{-}O)_4(\mu\text{-}OH)_4(H_2O)_4\}^{6+}$ sandwiched between polytungstate fragments. Polytungstate fragments are shown as gray polyhedra and the Ru_4 as blue tetrahedron, with Ru blue, μ-O red and $O(H_2)$ orange.

bridged by oxo ligands (Figure 3.9). Bonchio *et al.*[89,90] reported that the four Ru^{IV}–OH_2 are oxidised to $(Ru^{V}\text{–}OH)_4$ with potentials of $E_{1/2} = +1.12$, $+0.70$, $+0.53$ and $+0.29$ V *vs.* Ag/AgCl.

Upon oxidation by Ce^{IV} at pH 0.6, the $(Ru^{V}\text{–}OH)_4$ intermediate was reported to undergo nucleophilic attack by a water molecule to produce a hydroperoxo intermediate that evolves oxygen with TON around 500 and initial TOF around 450 h^{-1}. Isotopic experiments using ^{18}O-labelled water demonstrated that it was the origin of the O atom in dioxygen. Hill *et al.*[86,87] reported oxygen evolution with TON of 18, at neutral pH using electrochemically generated $[Ru^{III}(bpy)_3]^{3+}$ as oxidant. Based on the oxidation potentials for the Ru^{IV}–OH_2/Ru^{V}–OH couple, these oxidation reactions should be, in principle, accessible by photoactivation with $Ru(bpy)_3{}^{2+}$ analogues. The use of $S_2O_8{}^{2-}$ salts as acceptor renders these multi-electron reactions easier since two photons are sufficient to oxidise four $Ru(bpy)_3{}^{2+}$ molecules into four $Ru(bpy)_3{}^{3+}$. Hill's group reported the reaction of a mixture of Ru-POM, $Ru(bpy)_3{}^{2+}$ and $S_2O_8{}^{2-}$, under neutral pH conditions, yielding oxygen with a turnover number of 180, an initial turnover frequency of 0.08 s^{-1} and a photoreaction quantum yield (defined as number of O_2 molecules per two absorbed photons) $\Phi = 0.09$ (*i.e.* $\Phi = 0.045$, according to the classical definition of quantum yield). It has been argued that the major limiting factors in the light-induced process are the quantum efficiency of generating $[Ru^{III}(bpy)_3]^{3+}$ and the rate of the bimolecular reaction between the chromophore and the catalyst. Scandola *et al.*[91] reported on a hole transfer within the limit of diffusion rate from the photoproduced oxidant to the catalyst, which was further enhanced when the Ru-POM was adsorbed on nanocrystalline TiO_2. The homogeneous system was highly improved by the use of a tetranuclear Ru^{II} first generation dendrimer, $[Ru\{(\mu\text{-dpp})Ru(bpy)_2\}_3](PF_6)_8$, (bpy = bipyridine; dpp = 2,3-bis(2′-pyridyl)pyrazine) as the sensitiser.[89] Dendrimers, often used as antenna mimics, have enhanced visible absorption throughout a large portion of the visible spectrum, working in a spectral window that is inaccessible to $Ru(bpy)_3^{2+}$.[20] Furthermore, the oxidation potential of the Ru^{II}/Ru^{III} couple in these systems (1.55 *vs.* SCE) is 290 mV higher than that of $Ru(bpy)_3^{2+}$. In these conditions, excitation with 550 nm light gave a quantum yield of $\Phi = 0.30$, which is 6.7 times higher than the previous example.

Although the theme of this chapter relates to molecular-based complexes for the photodriven oxidation of water, there is considerable effort being directed on the field of metal oxides as water oxidising catalyst, such as RuO_2, IrO_2, and Co_3O_4. A recent hybrid system consisting of a colloid of nanoparticles of IrO_2/molecular system containing a chromophore to catalyse oxidation of water is of particular relevance to this chapter. The first experiments reported by Harriman *et al.*[92–94] reported the catalytic oxidation of water using light, $Ru(bpy)_3^{2+}$ and iridium colloids in the presence of an electron acceptor. Using optimal conditions, irradiation of these solutions with light >420 nm yielded a TOF of 160 s^{-1} (corresponding to 40 O_2 molecules per second). The mechanism postulated involves oxidative quenching of photogenerated $Ru(bpy)_3^{3+}$ by Ir centres. This transfer occurs in a stepwise fashion from Ir^{III} to Ir^{IV}, and a subsequent oxidation beyond the Ir^{IV} state leading to oxygen evolution.

No H/D kinetic isotope effect (KIE = 1.0)[95] was evidenced, while large isotope effects had been observed for molecular μ-oxo-bridged ruthenium (H/D = 1.4) or manganese (H/D = 1.7) complexes,[96] where the formation of a metal bound hydroperoxo intermediates is the rate determining step. If similar Ir^{V}–OOH species are involved in the oxidation of water, the lack of an isotope effect is taken as proof that the rate determining step is the electron transfer between $Ru(bpy)_3^{2+}$ and surface Ir centres in the colloid. In order to obtain a more efficient electron transfer between sensitiser/acceptor and sensitiser/catalyst the Mallouk group investigated a tris-heteroleptic ruthenium complex bearing phosphonate and malonate groups (Figure 3.10) designed in order to both be grafted to TiO_2 surfaces and covalently bind small (1–5 nm) diameter IrO_2 iridium nanoparticles,[97,98] thus enhancing the electronic coupling between the two metallic oxide components.

Irradiation with light >420 nm of a photoelectrochemical cell with the ruthenium–iridium complexes at the TiO_2 anode and a Pt wire as the cathode for proton reduction gave an internal quantum yield of 0.9% for overall water splitting when an electrical bias of >330 mV was applied. Formation of oxygen and hydrogen products was observed with a TON for water splitting of 16 per dye molecule. The decrease in TON for this system is explained by the fact that electron transfer from iridium to the oxidised chromophore is very slow (2.2 ms) compared with the fast recombination reaction (0.37 ms) between the photo-injected electron and Ru^{III}. Although to date the efficiency of this cell is far from ideal, it stands as a good example of a water-oxidising, proton-reducing photoelectrochemical device. Further work is aiming towards optimisation of the rate of electron transfer from $IrO_2 \cdot nH_2O$ nanoparticles to the sensitiser, ways of slowing down the back electron transfer, for example by increasing the distance between the complex and the metal oxide surface, and

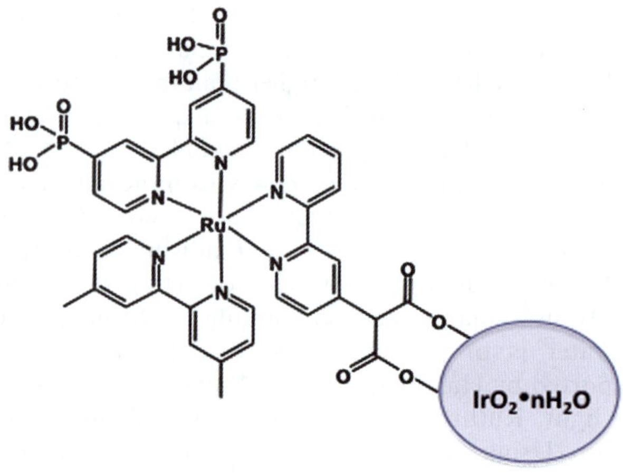

Figure 3.10 Ru–Ir nanoparticle.

ultimately to create a one to one ratio between the catalyst particle and the chromophore.

It is worth noting the majority of known molecular-based catalysts for water oxidation are ruthenium-based, while genuine biomimetic manganese-based molecular catalysts are scarce and their reactivity pattern is still a matter of debate. The intrinsic chemical lability of manganese complexes and their tendency towards formation of non-reactive polynuclear μ-oxo complexes have hampered the preparation of manganese-based catalysts. This theme is discussed in another chapter and therefore we just touch upon one example in the form of the Brudvig–Crabtree catalyst. The dinuclear $[H_2O(terpy)Mn(O)_2Mn(terpy)OH_2]^{3+}$ (Figure 3.11) has been reported to oxidise water using oxygen containing oxidants such as sodium hypochlorite or oxone.[99,100] A mechanism involving the nucleophilic attack of a hydroxide ion onto an activated oxygen atom at a high valent manganese motif was proposed. This proposal has been challenged after electrochemical studies and attempts to drive this catalyst *via* photogenerated $[Ru(bpy)_3]^{3+}$ in solution were unsuccessful.[101]

However, contributions from Yagi and colleagues[102] have shown that the di-μ-oxo manganese complex was able to catalyse the water oxidation reaction when adsorbed on clay surfaces and when treated with powerful one electron oxidant, Ce^{4+}. Such an activity in the heterogeneous phase was attributed to the intermolecular reaction of neighbouring $Mn^{V}{=}O$ units that originated from terminal water bound molecules between two catalytic units. The same group, has recently succeeded in driving the di-μ-oxo manganese catalyst intercalated in a mica matrix (Figure 3.12), using visible light in a bimolecular system where $[Ru(bpy)_3]^{2+}$ was used as a photosensitiser and $S_2O_8^{2-}$ as the electron acceptor.[103] They showed that the excited state of the sensitiser is quenched by the manganese catalyst in mica, this is translated into a one electron oxidation of a manganese centre with the concomitant formation of $[Ru(bpy)_3]^{+}$. Although

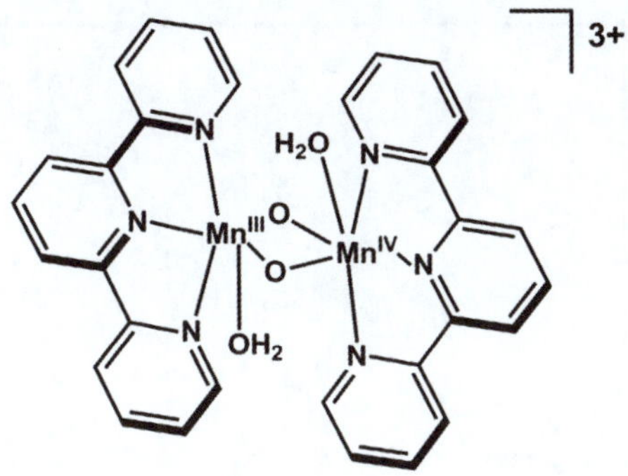

Figure 3.11 Brudvig–Crabtree catalyst.

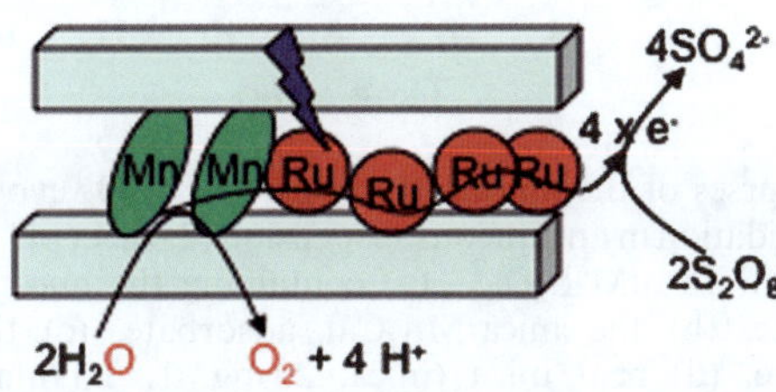

Figure 3.12 Proposed spatial arrangement of the manganese catalyst and the ruthenium sensitiser in between the mica layers. The evacuation of electrons is supposed to follow a vectorial electron transfer through electron hoping from the manganese catalyst and the sensitiser units.

no spectroscopic evidence was observed for the formation of $[Ru(bpy)_3]^+$, the authors proposed successive self exchange electron transfer from the photogenerated $[Ru(bpy)_3]^+$ to the contiguous $[Ru(bpy)_3]^{2+}$ and finally to the sacrificial electron acceptor, the $S_2O_8{}^{2-}$ ions. Isotopic labelling experiments corroborate the idea that the oxygen atoms originate solely from water molecules. Results for the photocatalytic activity in the presence and absence of catalyst are collected in Figure 3.13, showing that oxygen evolution is observed only in the presence of all three components in the heterogeneous system (manganese catalyst, $[Ru(bpy)_3]^{2+}$, mica and $S_2O_8{}^{2-}$).

3.10.1 Perspectives in Molecular Chemistry

The target for molecular chemists is to develop an assembly of sensitiser–catalyst tightly attached to a transparent semiconductor, whereupon light activation would trigger the oxidation of water molecules generating electrons and protons, which are then conveyed to a cathodic half cell for the production of a fuel. As presented in Figure 3.3, the minimal chemical modules are a photoactive chromophore and a catalytic unit held together through a chemical link that could participate as an electron transfer relay. In the following section,

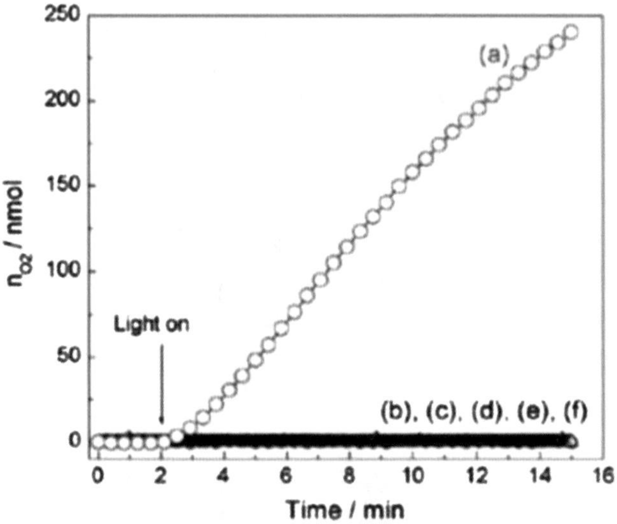

Figure 3.13 Time courses of the amount (nO_2 mol^{-1}) of O_2 evolved in photochemical water oxidation in an aqueous suspension (2.0 mL, pH = 6.2) of 0.2 M acetate buffer and 15 mM $S_2O_8^{2-}$ (a) containing the mica/Mn_2Cat/$[Ru(bpy)_3]^{2+}$ adsorbate, (b) the mica/Mn_2Cat adsorbate, (c) the mica/$[Ru(bpy)_3]^{2+}$ adsorbate, (d) neat mica (mica, 20 mg; 1, 3.3 mmol (164 mmol g^{-1}); $[Ru(bpy)_3]^{2+}$, 0.4 mmol (20 mmol g^{-1})), (e) in the absence of $S_2O_8^{2-}$ under the conditions of (a), (f) in a homogenous solution (pH = 6.2, 2.0 mL) containing 3.3 mmol (1.6 mM) Mn_2Cat and 0.4 mmol (0.2 mM) $[Ru(bpy)_3]^{2+}$ and 15 mM $S_2O_8^{2-}$. (Reproduced with permission from the RSC.)

we will discuss the advances needed in order to develop such a molecular-based photoanode.

3.10.1.1 The Photoactive Chromophore

Capturing and converting sunlight efficiently in the form of a chemical potential are the primary events that will ensure storage of sunlight in a chemical form. Because of its intermittency and diffused form, new ways must be developed to harvest sunlight. In this field, molecular-based sensitisers are expected to palliate the cost effective issues of silicon-based photovoltaic cells. The major leap in this field was realised by a coupling large band gap semiconducting, transparent metal oxides, for instance TiO_2, with redox active light absorbing chromophores.[104,105] Under light irradiation, the excited chromophore injects an electron into the conduction band of the semiconducting material, creating a charge-separated state that initiates electron flow in the presence of a reversible redox mediator. Ruthenium polypyridine complexes are the most commonly photoactive units in this field. Reasons behind this are their unique photophysical properties and their chemical robustness.[27,29] The synthetic versatility to insert differently substituted bipyridine ligands in the coordination sphere of the ruthenium centre is also an important asset for this chromophore. This is in contrast with porphyrin-based photoactive chromophores that are prepared following multistep synthetic reaction schemes. A considerable amount of work has been realised in developing heteroleptic ruthenium complexes whereby changes in the chemical nature of the bipyridine ligand lead to the tuning of the electronic and redox properties of the metal centre. An intrinsic topological issue in using the ruthenium trisbipyridine core in the development of molecular-based photocatalyst is the existence of isomers that may interfere in the studies of electron transfer reactions.[26] Also, photocatalysts based on the $[Ru(bpy)_3]^{2+}$ architecture may lead to different arrangements when anchored on a semiconducting surface. In order to avoid this, $[Ru(terpy)_2]^{2+}$-like complexes have been developed whose substitution sites in the 4-position lead to a linear disposition of the molecular assembly. However, this chomophore suffers from a short lived excited state (20 ns) and is therefore not the ideal candidate for photo-induced electron transfer studies. This issue should be alleviated if this chromophore is tethered on to semiconducting surfaces, where injection of electron in the conduction band from the excited state is much faster (fs). Recently, Abrahamsson and colleagues have reported a novel visible light absorbing ruthenium bistridentate complex (Figure 3.14) with an excited state lifetime of 3 μs. Such behaviour is due to the less constrained structure in this ligand when compared to the terpyridine complex. The novel tridentate ligand provides a structural relaxation around the ruthenium centre which modifies the frontier molecular orbital scheme by shifting the ^{3}MC to higher energy thus short-circuiting the deactivation pathway of the ^{3}MLCT through the ^{3}MC state.[68]

Besides the net gain in the time available to perform charge shift reactions with this new chromophore, it should be noted that the complex exhibits helical isomers and that the synthesis of extended assemblies with this tridentate ligand is not trivial.

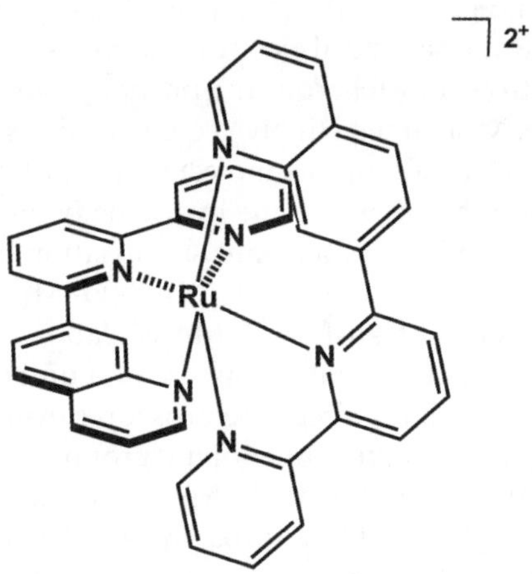

Figure 3.14 Visible-light absorbing ruthenium chromophore exhibiting long lived excited state lifetimes.

To sum up, there is still plenty of room for improvement for future generations of molecular-based chromophores. These range from better absorbing cross-section areas to broader absorbing spectral windows in the visible region. These materials should also satisfy cost effectiveness criteria in order to expect a large-scale deployment.

3.10.1.2 Electron Relays

Nature makes use of a series of cofactors properly displaced in order to drive the several photo-induced vectorial electron transfer reactions, which lead to charge separation and charge accumulation. An important element in the electron transfer pathway is the presence of a redox active electron relay (TyrZ) between the P680 chromophore and the OEC (Scheme 3.1).[106] Its role in PCET reactions has been discussed in section 3.8. Much of the work in the field of light induced electron transfer processes and photocatalytic reactions has been realised using a mixture of the photoactive and the catalytic units. It is less likely that such primary mixtures will be used in the development of a molecular device for the photocatalytic oxidation of water. Following such an approach, different groups have reported on the incorporation of electron relay groups to photoactive chromophores. The presence of such a unit may lead to sufficient decoupling between the photoactive module and the catalyst to preserve the basic properties of each subunit. It can also participate in enhancing the rate of electron transfer through an electron hopping mechanism while at the same time inhibiting the back electron transfer to a highly oxidised catalytic centre. An example of such a system has been reported where a biomimetic model for the TyrZ/His 190 has been bound covalently to a ruthenium(II) polypyridine complex (Figure 3.15).[107–109]

Both crystallographic data and NMR spectroscopy in solution attest for the presence of a hydrogen bonding interaction between the phenol group and the nitrogen of the imino group of the imidazole. Flash photolysis studies in presence of both reversible and sacrificial electron acceptors have shown the presence of intramolecular electron transfer from the phenol group to the photo-induced oxidised chromophore. More interestingly, a comparative study clearly evidenced that the rate of electron transfer in this molecule was orders of magnitude faster than that observed in complexes where the hydrogen bond was not present.

Phenoxyl radicals formed by visible light sensitisation have been found to be sufficiently oxidising so as to activate a Mn^{III}–O–Mn^{III} dimer through an intermolecular one electron transfer.[110] More elaborate systems where the phenol is implicated in the coordination sphere of a dinuclear Mn^{II}–Mn^{II} have also been reported. Although no direct spectroscopic evidence was reported for the role of the manganese linked phenolate group in the electron transfer process, it was shown that the initial Mn^{II}–Mn^{II} states were lifted to a bis-μ-oxo dinuclear Mn^{III}–Mn^{IV}.[111–113] These three intramolecular electron transfers occurred within a potential range of 0.2 V and were followed by ligand exchange and water deprotonation in the coordination sphere of the manganese ion (Scheme 3.11).

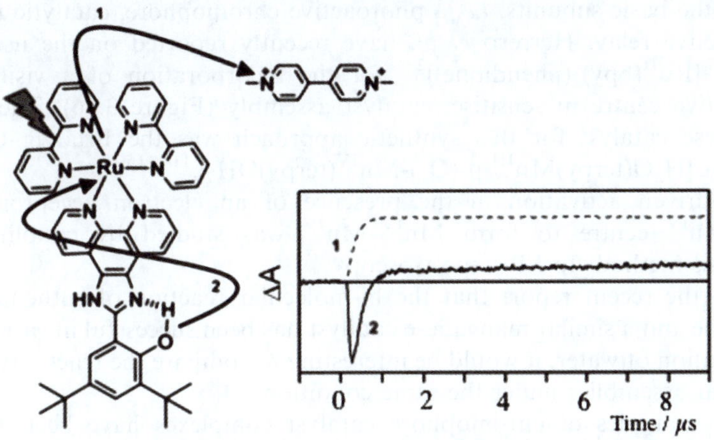

Figure 3.15 Tyrosine–histidine couple mimic with its transient absorption spectrum. (1) Formation of methylviologen radical (605 nm) and (2) recovery of Ru(II) MLCT (450 nm).

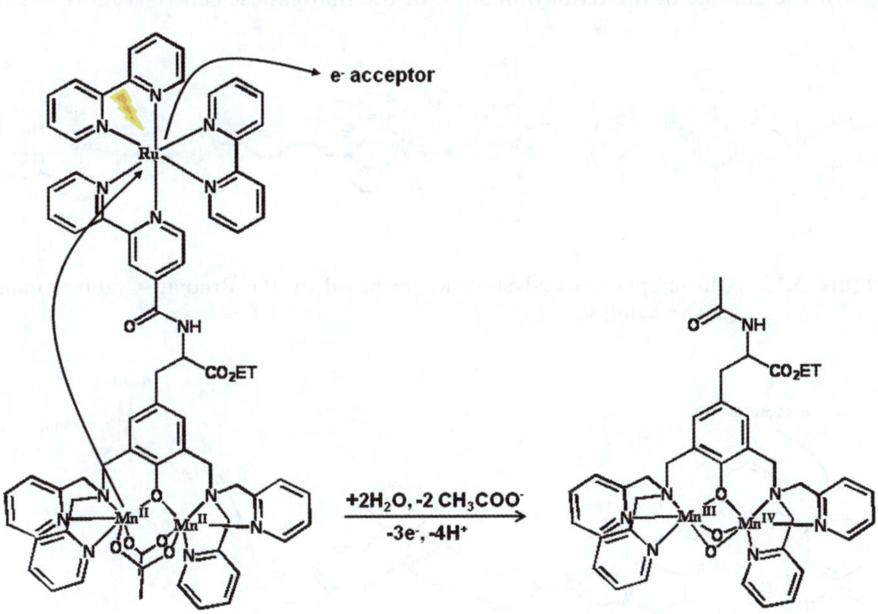

Scheme 3.11 Example of light driven charge accumulation process.

3.10.1.3 Chromophore–Catalyst Assemblies

It should be noted that the development of such molecular assemblies will imply multi-step synthetic procedures. It is therefore crucial to develop new synthetic pathways that lead to the preparation of modular assemblies which contain the basic subunits, *i.e.* a photoactive chromophore, catalytic unit and redox active relay. Herrero *et al.* have recently reported on the use of the synthon $[Ru^{II}(bpy)_2(phendione)]^{2+}$ for the incorporation of a visible light photoactive centre in sensitiser/catalyst assembly (Figure 3.16). The chosen manganese catalyst for this synthetic approach was the Brudvig–Crabtree dinuclear $[H_2O(terpy)Mn^{III}\text{-}\mu\text{-}(O_2)\text{-}Mn^{IV}(terpy)OH_2]$.[114,115]

Light-driven activation in the presence of an electron acceptor of the Mn^{III}–Mn^{IV} centre to form Mn^{IV}–Mn^{IV} was studied by coupling flash photolysis methods to EPR spectroscopy.

Given the recent report that the bi-molecular reaction of ruthenium tris-bipyridine and a similar manganese catalyst has been successful in carrying out the oxidation of water, it would be interesting to compare the reactivity of such molecular assemblies under the same conditions.

Other examples of chromophore–catalyst complexes have been reported based on the synthesis of a $[Ru(bpy)_3]^{2+}$-like complex covalent bound to a manganese(III) salen unit.[116,117] Flash laser photolysis experiments show the activation of the Mn^{III}Salen to a high spin Mn^{IV}Salen *via* intramolecular electron transfer. This reaction was monitored using the EPR techniques to follow the change in the oxidation state of the manganese centre (Figure 3.17).

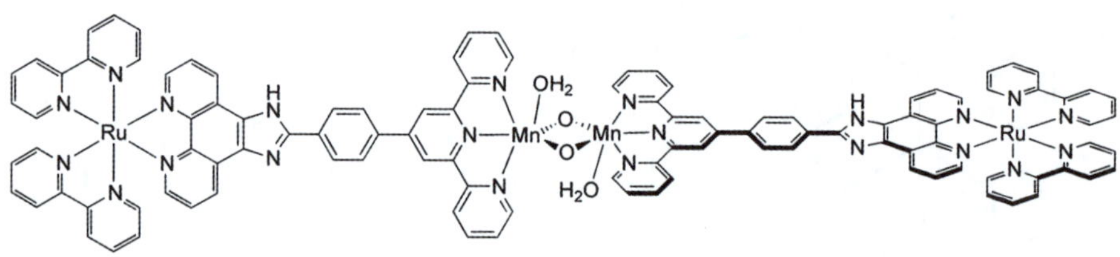

Figure 3.16 Chromophore/Catalyst structure based on the Brudvig–Crabtree manganese catalyst.

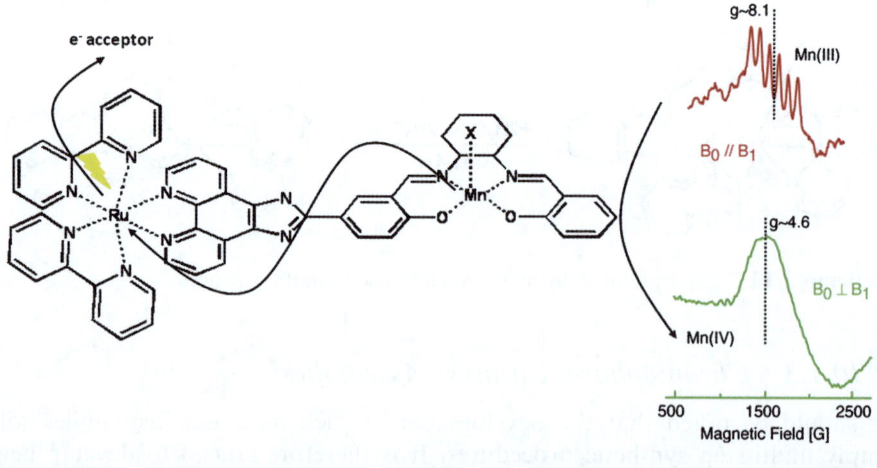

Figure 3.17 Mn^{III} to Mn^{IV} photo-oxidation process in a chromophore–salen complex.

3.10.1.4 Surfaces

The collection of the electrons to be used in fuel forming reactions is a crucial step in this field. Throughout the examples presented in previous sections the use of oxidants in chemically-driven reactions is substituted by irreversible electron acceptors in light-driven homogenous systems. The most promising way to efficiently remove the electrons obtained from the photo-oxidation reactions – and to use them for fuel production in a reductive system – is the transfer of these solution based reactions to the surface of an electrode in a photoelectrochemical cell. The use of n-type semiconductive, optically transparent layers of TiO_2 (but also ZnO or SnO_2) helps avoid many of the drawbacks found in solution-based systems.[104] The most obvious advantage comes from the fact that, in these systems, the photosensitiser is covalently bound to the metal oxide surface. This allows for very efficient and fast electron transfer reactions between the excited state of the chromophore and the conduction band of the semiconductor. Secondly, immobilization of the photocatalysts essentially suppresses potential intermolecular reactions between the highly oxidising species, which might lead to decomposition. This is likely to lead to much higher stability of the system during long-term operation. Lastly, low potential electrons resulting from these photocatalytic processes at a photoanode can be directed to a second half-cell where they can be used at a (photo) cathode for reduction reactions leading to the formation of fuels (H_2, CO, CH_4, *etc.*). With the very high specific surface area of nanocrystalline, mesoporous TiO_2 layers, sufficient light absorption can be obtained even with layer thicknesses of less than 10 μm. Back electron transfer reactions from TiO_2 to the oxidised sensitiser are a recurrent problem in these kind of system which can be circumvented by enhancing the rate of electron transfer from the catalyst to the oxidised dye.

3.10.1.5 Molecular-based Catalysts

The construction of a molecular catalyst for water oxidation is a particularly challenging task. From a synthetic chemistry point of view, the structure of the biological catalyst (Mn_3Ca) and its tertiary structure (amino acids) are out of reach. Additionally, the mechanism of the water oxidation reaction remains the subject of speculation and thus hinders the design of molecules bearing "active sites". Manganese complexes reporting water oxidation have been published, but the catalysis activity has been disappointing and difficult to demonstrate unambiguously. In contrast, ruthenium complexes have shown *bona fide* water oxidation with impressive turnover number numbers. Recent breakthroughs in the field of water oxidising catalysts are drawing the attention of chemists towards carbon-free inorganic type catalysts.[118–120] Is there still room for research of efficient molecular-based catalysts as water oxidising catalysts? At this stage, predicting the future of molecular-based science in the field is quite far fetched. It will be the role of inspirational young chemists to understand the water activation processes, the O–O bond making and the release of O_2 in the triplet state.

3.10.1.6 Catalyst Design

Photosystem II uses a Mn cluster in the OEC that cycles through +3, +4 and +5 oxidation states in order to split water. Besides a great range of oxidation states, the high oxidation state of Mn binds water to form hydroxo and oxo moeities with potentials in the ranges needed for water splitting. The design of molecular catalyst hinges on the different mechanistic proposals for the four electron oxidation process of water. It is worth noticing that this structure–reactivity relationship has only been validated with ruthenium-based catalysts. Below, we present some basic rationale in the design of molecular catalysts.

Radical Coupling

Formation of the O–O bond could result from activation of two proximal, well oriented Mn-ligated water molecules following a two-electron, two-proton abstraction process for each water (Scheme 3.12). Such a mechanism has been postulated many times in the biological literature. An artificial complex aimed at doing this requires a ligand that can hold the two metal ions at the right relative geometry and distance. Each of the metal centres must be surrounded by an oxidatively robust ligand capable of withstanding the formation of high-valent Mn states leaving an exchangeable ligating site for substrate water binding. The co-facial bisporphyrin complexes developed by Naruta and co-workers is an example of such a geometry.[121] Oxidation of the aqua ligands into their oxo equivalents followed by O–O coupling yielded formation of O_2 by the reaction of two proximal HO–Mn^V=O units, although whether O_2 formation is *via* oxo coupling or nucleophilic attack by OH^- is not certain. Future work following this research line will demand the tuning of the electronic properties of the cofacial catalytic units and regulating the distance between them for an optimal interaction.

Nucleophilic Attack

This type of mechanism has been under consideration for several years for the water oxidising enzyme. Salen- or porphyrin-based manganese complexes are known to generate high-valent manganese-oxo species that can oxidise alkenes into epoxides in a two-electron process. If the C–C double bond of the alkene were to be replaced by an activated form of water (*e.g.* OH^-), then it may be possible to use this chemistry to create the O–O bond. However, a 4-electron

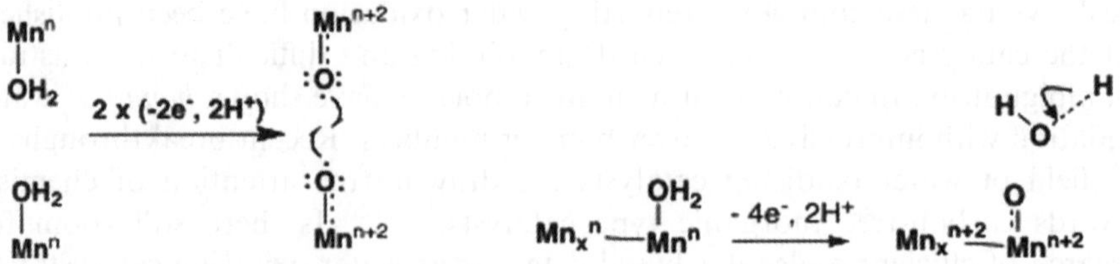

Scheme 3.12 Proposed mechanisms for O_2 evolution. (Left) Radical mechanism. (Right) Nucleophilic attack.

process is required for the liberation of O_2. An elegant synthetic option to do this would be to share the oxidising power on a multi-centred complex that can provide the necessary two additional oxidising equivalents for the nucleophilic attack. Recent reports have evidenced the formation of the O–O through the nucleophilic attack of hydroxy groups on metal oxo complexes include the Mn corroles, where O_2 evolution was detected after nucleophilic attack of OH^- on a Mn^V corrole. Reversible O–O bond cleavage/formation has been observed in a Mn corrole system *via* an acid base mechanism. This finding is of particular interest as it shows the relationship between the O–O bond formation reaction in Photosystem II with the O_2 bond cleavage in mono-oxygenases.

Collins has pioneered the work on oxidatively robust metal complexes based on tetraamidimacrocyclic ligands (TAML) (Scheme 3.13).[122,123] This family of molecules can form highly reactive metal-oxo species that can drive oxidation of recalcitrant aromatic pollutants. Recently, he has shown that an iron catalyst (safe and cheap) was able to drive the oxidation of water using Ce^{4+} as the oxidant. This result constitutes a breakthrough while allowing for large synthetic modifications in the ligands in order to tune the electronic properties of the molecule.

A crucial step in the long running development of molecular-based catalysts will be the intimate understanding of the sequential electron and proton activation of a water molecule sitting in the coordination sphere of a manganese ion for instance. Molecular manganese-based chemistry has repeatedly been faced with the formation of undesired μ-oxo polynuclear manganese derivatives upon subsequent oxidation and condensation reactions. Therefore, the initial activation steps of a bound water molecule at a manganese centre have often been difficult to identify. In a recent study, El Ghachtouli *et al.*[124] have reported on some original aspects of the electron transfer processes at a Mn(III)OH complex (Figure 3.18). They have found that exogenous water molecules can steer the reversibility of the $Mn^{II/III}$ and $Mn^{III/IV}$ redox processes, thereby inhibiting the formation of undesired μ-oxo compounds. Cyclic voltammetry in conjunction with H/D and $^{18}O/^{16}O$ kinetic isotopic effects analysis suggest a H_2O/HO^- ligand exchange process preceding electron transfer for the $Mn^{III/II}$

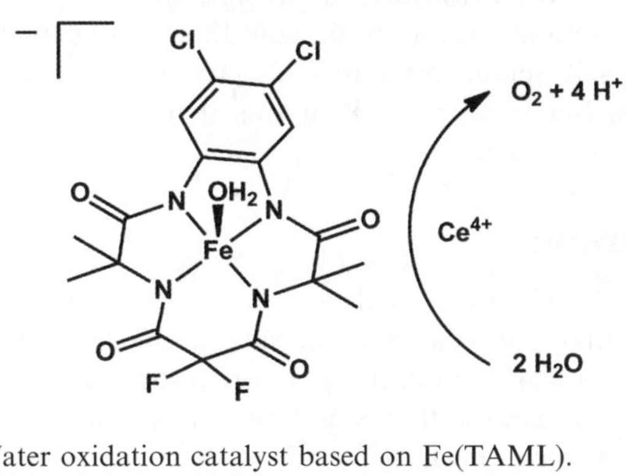

Scheme 3.13 Water oxidation catalyst based on Fe(TAML).

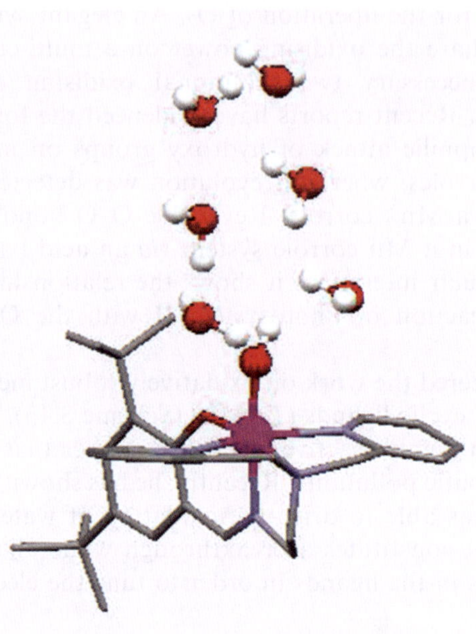

Figure 3.18 Crystallographic structure of a Mn(III)OH complex with secondary sphere water molecules.

couple. No such KIEs are observed in the case of the $Mn^{IV/III}$ couple, for which the electron transfer occurs without proton loss, thus allowing the formation of a genuine Mn(IV)OH species in the presence of water.

The present discussion demonstrates the role of H_2O/HO^- ligand exchange dynamics coupled with the change of oxidation state in manganese chemistry. Such basic research will be invaluable in the understanding of water activation at a manganese centre.

The challenging issue in the design of molecular-based catalysts is to realise the water oxidation reaction has to do with the stability and robustness of the catalytic systems. These molecules will work in very harsh conditions and have to be resistant to strong oxidative processes. Important advances in the development and understanding of oxidising complexes have been made, and self repair mechanisms have been introduced in heterogeneous catalysts, but further work in this field with respect to molecular catalysts needs to be done.

3.11 Conclusion

This chapter is a brief and limited survey of the biomimetic approach for making catalysts and photocatalysts based on the key enzyme involved in energy conversion processes in biology, Photosystem II. Our aim is provide a flavour of the kind of research that is underway in this area and in particular of the strategies that have been adopted recently in different laboratories, including ours. Molecular biomimetics is an exciting interdisciplinary field where synthetic chemistry meets both molecular enzymology and a wide range of physical approaches. The ultimate aim of molecular biomimetics in this area is to obtain devices that will be not only be of commercial value but will contribute to one of the greatest scientific challenges of our time: to improve the efficiency of energy use and to achieve sustainable energy production.

References

1. J. P. McEvoy and G. W. Brudvig, *Chem. Rev.*, 2006, **106**, 4455–4483.
2. J. P. McEvoy, J. A. Gascon, V. S. Batista and G. W. Brudvig, *Photochem. Photobiol. Sci.*, 2005, **4**, 940–949.
3. K. N. Ferreira, T. M. Iverson, K. Maghlaoui, J. Barber and S. Iwata, *Science*, 2004, **303**, 1831–1838.
4. B. Loll, J. Kern, W. Saenger, A. Zouni and J. Biesiadka, *Nature*, 2005, **438**, 1040–1044.
5. Y. Umena, K. Kawakami, J. R. Shen and N. Kamiya, *Nature*, 2011, **473**, 55–65.
6. P. E. M. Siegbahn, *Acc. Chem. Res.*, 2009, **42**, 1871–1880.
7. E. M. Sproviero, J. P. McEvoy, J. A. Gascon, G. W. Brudvig and V. S. Batista, *Photosynth. Res.*, 2008, **97**, 91–114.
8. E. M. Sproviero, J. A. Gascon, J. P. McEvoy, G. W. Brudvig and V. S. Batista, *J. Am. Chem. Soc.*, 2008, **130**, 6728–6730.
9. C. W. Cady, R. H. Crabtree and G. W. Brudvig, *Coord. Chem. Rev.*, 2007, **252**, 444–455.
10. J. M. Peloquin, K. A. Campbell, D. W. Randall, M. A. Evanchik, V. L. Pecoraro, W. H. Armstrong and R. D. Britt, *J. Am. Chem. Soc.*, 2000, **122**, 10926–10942.
11. W. Ruttinger and G. C. Dismukes, *Chem. Rev.*, 1997, **97**, 1–24.
12. L. V. Kulik, B. Epel, W. Lubitz and J. Messinger, *J. Am. Chem. Soc.*, 2007, **129**, 13421–13435.
13. H. Inoue, T. Shimada, Y. Kou, Y. Nabetani, D. Masui, S. Takagi and H. Tachibana, *ChemSusChem*, 2011, **4**, 173–179.
14. H. Neckel and D. Labs, *Solar Phys.*, 1984, **90**, 205–258.
15. J. L. Herek, W. Wohlleben, R. J. Cogdell, D. Zeidler and M. Motzkus, *Nature*, 2002, **417**, 533–535.
16. D. Kuciauskas, S. Lin, G. R. Seely, A. L. Moore, T. A. Moore, D. Gust, T. Drovetskaya, C. A. Reed and P. D. W. Boyd, *J. Phys. Chem.*, 1996, **100**, 15926–15932.
17. S. L. Gould, G. Kodis, P. A. Liddell, R. E. Palacios, A. Brune, D. Gust, T. A. Moore and A. L. Moore, *Tetrahedron*, 2006, **62**, 2074–2096.
18. D. I. Schuster, P. Cheng, P. D. Jarowski, D. M. Guldi, C. Luo, L. Echegoyen, S. Pyo, A. R. Holzwarth, S. E. Braslavsky, R. M. Williams and G. Klihm, *J. Am. Chem. Soc.*, 2004, **126**, 7257–7270.
19. G. Kodis, Y. Terazono, P. A. Liddell, J. Andreasson, V. Garg, M. Hambourger, T. A. Moore, A. L. Moore and D. Gust, *J. Am. Chem. Soc.*, 2006, **128**, 1818–1827.

20. V. Balzani, A. Juris, M. Venturi, S. Campagna and S. Serroni, *Chem. Rev.*, 1996, **96**, 759–833.
21. M. R. Wasielewski, *J. Org. Chem.*, 2006, **71**, 5051–5066.
22. C. Roger, M. G. Muller, M. Lysetska, Y. Miloslavina, A. R. Holzwarth and F. Wurthner, *J. Am. Chem. Soc.*, 2006, **128**, 6542–6543.
23. D. M. Guldi, H. Imahori, K. Tamaki, Y. Kashiwagi, H. Yamada, Y. Sakata and S. Fukuzumi, *J. Phys. Chem. A*, 2004, **108**, 541–548.
24. H. Imahori, *J. Phys. Chem. B*, 2004, **108**, 6130–6143.
25. V. Balzani and A. Juris, *Coord. Chem. Rev.*, 2001, **211**, 97–115.
26. F. R. Keene, *Coord. Chem. Rev.*, 1997, **166**, 121–159.
27. L. Spiccia, G. B. Deacon and C. M. Kepert, *Coord. Chem. Rev.*, 2004, **248**, 1329–1341.
28. J. H. Alstrum–Acevedo, M. K. Brennaman and T. J. Meyer, *Inorg. Chem.*, 2005, **44**, 6802–6827.
29. M. Hang, V. Huynh, D. M. Dattelbaum and T. J. Meyer, *Coord. Chem. Rev.*, 2005, **249**, 457–483.
30. P. A. Anderson, F. R. Keene, T. J. Meyer, J. A. Moss, G. F. Strouse and J. A. Treadway, *J. Chem. Soc., Dalton Trans.*, 2002, 3820–3831.
31. D. P. Rillema, C. B. Blanton, R. J. Shaver, D. C. Jackman, M. Boldaji, S. Bundy, L. A. Worl and T. J. Meyer, *Inorg. Chem.*, 1992, **31**, 1600–1606.
32. F. Rappaport, M. Guergova-Kuras, P. J. Nixon, B. A. Diner and J. Lavergne, *Biochemistry*, 2002, **41**, 8518–8527.
33. M. F. Charlot, Y. Pellegrin, A. Quaranta, W. Leibl and A. Aukauloo, *Chem.–Eur. J.*, 2006, **12**, 796–812.
34. B.-Z. Shan, Q. Zhao, N. Goswami, D. M. Eichhorn and D. P. Rillema, *Coord. Chem. Rev.*, 2001, **211**, 117–144.
35. D. G. McCafferty, B. M. Bishop, C. G. Wall, S. G. Hughes, S. L. Mecklenberg, T. J. Meyer and B. W. Erickson, *Tetrahedron*, 1995, **51**, 1093–1106.
36. N. Garelli and P. Vierling, *J. Org. Chem.*, 1992, **57**, 3046–3051.
37. C. G. Garcia, J. F. d. Lima and N. Y. M. Iha, *Coord. Chem. Rev.*, 2000, **196**, 219–247.
38. G. Sprintschnik, H. W. Sprintschnik, P. P. Kirsch and D. G. Whitten, *J. Am. Chem. Soc.*, 1977, **99**, 4947–4954.
39. P. A. Andenon, G. F. Strouse, J. A. Treadway, F. R. Keene and T. J. Meyer, *Inorg. Chem.*, 1994, **33**, 3863–3864.
40. J.-P. Collin, R. Kayhanian, J.-P. Sauvage, G. Calogero, F. Barigelletti, A. D. Ciancand and J. Fischer, *Chem. Commun.*, 1997, 775–776.
41. J. J. Warren, T. A. Tronic and J. M. Mayer, *Chem. Rev.*, 2011, **110**, 6961–7001.
42. T. Kitamura, Y. Wada, K. Murakoshi, M. Kusaba, N. Nakashima, A. Ishida, T. Majima, S. Takamuku, T. Akano and S. Yanagida, *J. Chem. Soc., Faraday Trans.*, 1996, **92**, 3491–3495.
43. T. Kitamura, H. Fudemoto, Y. Wada, K. Murakoshi, M. Kusaba, N. Nakashima, T. Majima and S. Yanagida, *J. Chem. Soc., Faraday Trans.*, 1997, **93**, 221–229.

44. J. M. Saveant, *Chem. Rev.*, 2008, **108**, 2348–2378.
45. T. J. Meyer, M. H. V. Huynh and H. H. Thorp, *Angew. Chem., Int. Ed.*, 2007, **46**, 5284–5304.
46. C. Tommos and G. T. Babcock, *Acc. Chem. Res.*, 1998, **31**, 18–25.
47. S. Hammes-Schiffer, *Chem. Rev.*, 2011, **110**, 6937–6938.
48. C. Costentin, M. Robert and J. M. Saveant, *Phys. Chem. Chem. Phys.*, 2010, **12**, 13061–13069.
49. C. Costentin, M. Robert and J. M. Saveant, *J. Am. Chem. Soc.*, 2007, **129**, 9953–9963.
50. C. Costentin, M. Robert and J. M. Saveant, *J. Am. Chem. Soc.*, 2006, **128**, 4552–4553.
51. I. J. Rhile, T. F. Markle, H. Nagao, A. G. DiPasquale, O. P. Lam, M. A. Lockwood, K. Rotter and J. M. Mayer, *J. Am. Chem. Soc.*, 2006, **128**, 6075–6088.
52. D. Gust, T. A. Moore and A. L. Moore, *Acc. Chem. Res.*, 2009, **42**, 1890–1898.
53. C. Costentin, *Chem. Rev.*, 2008, **108**, 2145–2179.
54. B. A. Moyer and T. J. Meyer, *Inorg. Chem.*, 1981, **20**, 436–444.
55. K. J. Takeuchi, M. S. Thompson, D. W. Pipes and T. J. Meyer, *Inorg. Chem.*, 1984, **23**, 1845–1851.
56. R. C. McHatton and F. C. Anson, *Inorg. Chem.*, 1984, **23**, 3935–3942.
57. R. A. Binstead and T. J. Meyer, *J. Am. Chem. Soc.*, 1987, **109**, 3287–3297.
58. E. L. Lebeau, R. A. Binstead and T. J. Meyer, *J. Am. Chem. Soc.*, 2001, **123**, 10535–10544.
59. C. Costentin, M. Robert, J. M. Saveant and A. L. Teillout, *Proc. Natl. Acad. Sci., U. S. A.*, 2009, **106**, 11829–11836.
60. C. Costentin, M. Robert, J. M. Saveant and A. L. Teillout, *ChemPhysChem*, 2009, **10**, 191–198.
61. S. Funyu, M. Kinai, D. Masui, S. Takagi, T. Shimada, H. Tachibana and H. Inoue, *Photochem. Photobiol. Sci.*, 2010, **9**, 931–936.
62. S. Funyu, T. Isobe, S. Takagi, D. A. Tryk and H. Inoue, *J. Am. Chem. Soc.*, 2003, **125**, 5734–5740.
63. H. Inoue, S. Funyu, Y. Shimada and S. Takagi, *Pure Appl. Chem.*, 2005, **77**, 1019–1033.
64. S. Fukuzumi, T. Kishi, H. Kotani, Y. M. Lee and W. Nam, *Nature Chem.*, 2011, **3**, 38–41.
65. J. A. Treadway, J. A. Moss and T. J. Meyer, *Inorg. Chem.*, 1999, **38**, 4386–4387.
66. W. Z. Chen, F. N. Rein and R. C. Rocha, *Angew. Chem., Int. Ed.*, 2009, **48**, 9672–9675.
67. M. Maestri, N. Armaroli, V. Balzani, E. C. Constable and A. Thompson, *Inorg. Chem.*, 1995, **34**, 2759–2767.
68. M. Abrahamsson, M. Jager, R. J. Kumar, T. Osterman, P. Persson, H. C. Becker, O. Johansson and L. Hammarstrom, *J. Am. Chem. Soc.*, 2008, **130**, 15533–15542.

69. J. P. Sauvage, J. P. Collin, J. C. Chambron, S. Guillerez, C. Coudret, V. Balzani, F. Barigelletti, L. Decola and L. Flamigni, *Chem. Rev.*, 1994, **94**, 993–1019.
70. M. S. Thompson and T. J. Meyer, *J. Am. Chem. Soc.*, 1982, **104**, 4106–4115.
71. M. S. Thompson and T. J. Meyer, *J. Am. Chem. Soc.*, 1982, **104**, 5070–5076.
72. J. A. Gilbert, D. S. Eggleston, W. R. Murphy, D. A. Geselowitz, S. W. Gersten, D. J. Hodgson and T. J. Meyer, *J. Am. Chem. Soc.*, 1985, **107**, 3855–3864.
73. F. P. Rotzinger, S. Munavalli, P. Comte, J. K. Hurst, M. Gratzel, F. J. Pern and A. J. Frank, *J. Am. Chem. Soc.*, 1987, **109**, 6619–6626.
74. F. Bozoglian, S. Romain, M. Z. Ertem, T. K. Todorova, C. Sens, J. Mola, M. Rodriguez, I. Romero, J. Benet-Buchholz, X. Fontrodona, C. J. Cramer, L. Gagliardi and A. Llobet, *J. Am. Chem. Soc.*, 2009, **131**, 15176–15187.
75. X. Sala, I. Romero, M. Rodriguez, L. Escriche and A. Llobet, *Angew. Chem. Int. Ed.*, 2009, **48**, 2842–2852.
76. I. Romero, M. Rodriguez, C. Sens, J. Mola, M. R. Kollipara, L. Francas, E. Mas-Marza, L. Escriche and A. Llobet, *Inorg. Chem.*, 2008, **47**, 1824–1834.
77. Y. Xu, A. Fischer, L. Duan, L. Tong, E. Gabrielsson, B. Åkermark and L. Sun, *Angew. Chem., Int. Ed.*, 2010, **49**, 8934–8937.
78. J. J. Concepcion, J. W. Jurss, J. L. Templeton and T. J. Meyer, *J. Am. Chem. Soc.*, 2008, **130**, 16462–16463.
79. J. J. Concepcion, J. W. Jurss, M. R. Norris, Z. F. Chen, J. L. Templeton and T. J. Meyer, *Inorg. Chem.*, 2010, **49**, 1277–1279.
80. J. J. Concepcion, M. K. Tsai, J. T. Muckerman and T. J. Meyer, *J. Am. Chem. Soc.*, 2010, **132**, 1545–1557.
81. J. J. Concepcion, J. W. Jurss, M. K. Brennaman, P. G. Hoertz, A. O. T. Patrocinio, N. Y. M. Iha, J. L. Templeton and T. J. Meyer, *Acc. Chem. Res.*, 2009, **42**, 1954–1965.
82. X. Sala, M. Z. Ertem, L. Vigara, T. K. Todorova, W. Z. Chen, R. C. Rocha, F. Aquilante, C. J. Cramer, L. Gagliardi and A. Llobet, *Angew. Chem., Int. Ed.*, 2010, **49**, 7745–7747.
83. J. Nyhlen, L. L. Duan, B. Akermark, L. C. Sun and T. Privalov, *Angew. Chem., Int. Ed.*, 2010, **49**, 1773–1777.
84. Y. H. Xu, T. Akermark, V. Gyollai, D. P. Zou, L. Eriksson, L. L. Duan, R. Zhang, B. Akermark and L. C. Sun, *Inorg. Chem.*, 2009, **48**, 2717–2719.
85. Y. H. Xu, G. Eilers, M. Borgstrom, J. X. Pan, M. Abrahamsson, A. Magnuson, R. Lomoth, J. Bergquist, T. Polivka, L. C. Sun, V. Sundstrom, S. Styring, L. Hammarstrom and B. Akermark, *Chem.–Eur. J.*, 2005, **11**, 7305–7314.
86. Y. V. Geletii, B. Botar, P. Koegerler, D. A. Hillesheim, D. G. Musaev and C. L. Hill, *Angew. Chem., Int. Ed.*, 2008, **47**, 3896–3899.

87. Y. V. Geletii, Z. Q. Huang, Y. Hou, D. G. Musaev, T. Q. Lian and C. L. Hill, *J. Am. Chem. Soc.*, 2009, **131**, 7522–7523.
88. A. Sartorel, M. Carraro, G. Scorrano, R. De Zorzi, S. Geremia, N. D. McDaniel, S. Bernhard and M. Bonchio, *J. Am. Chem. Soc.*, 2008, **130**, 5006–5007.
89. F. Puntoriero, G. La Ganga, A. Sartorel, M. Carraro, G. Scorrano, M. Bonchio and S. Campagna, *Chem. Commun.*, 2010, **46**, 4725–4727.
90. A. Sartorel, P. Miro, E. Salvadori, S. Romain, M. Carraro, G. Scorrano, M. Di Valentin, A. Llobet, C. Bo and M. Bonchio, *J. Am. Chem. Soc.*, 2009, **131**, 16051–16053.
91. M. Orlandi, R. Argazzi, A. Sartorel, M. Carraro, G. Scorrano, M. Bonchio and F. Scandola, *Chem. Commun.*, 2010, **46**, 3152–3154.
92. A. Harriman, J. M. Thomas and G. R. Millward, *New J. Chem.*, 1987, **11**, 757–762.
93. A. Harriman, M. C. Richoux, P. A. Christensen, S. Mosseri and P. Neta, *J. Chem. Soc., Faraday Trans. I*, 1987, **83**, 3001–3014.
94. A. Harriman, I. J. Pickering, J. M. Thomas and P. A. Christensen, *J. Chem. Soc., Faraday Trans. I*, 1988, **84**, 2795–2806.
95. N. D. Morris, M. Suzuki and T. E. Mallouk, *J. Phys. Chem. A*, 2004, **108**, 9115–9119.
96. J. Limburg, J. S. Vrettos, H. Chen, J. C. de Paula, R. H. Crabtree and G. W. Brudvig, *J. Am. Chem. Soc.*, 2001, **123**, 423–430.
97. W. J. Youngblood, S. H. A. Lee, Y. Kobayashi, E. A. Hernandez-Pagan, P. G. Hoertz, T. A. Moore, A. L. Moore, D. Gust and T. E. Mallouk, *J. Am. Chem. Soc.*, 2009, **131**, 926–927.
98. W. J. Youngblood, S. H. A. Lee, K. Maeda and T. E. Mallouk, *Acc. Chem. Res.*, 2009, **42**, 1966–1973.
99. J. Limburg, J. S. Vrettos, L. M. Liable-Sands, A. L. Rheingold, R. H. Crabtree and G. W. Brudvig, *Science*, 1999, 283.
100. C. Baffert, S. Romain, A. Richardot, J.-C. Lepretre, B. Lefebvre, A. Deronzier and M.-N. Collomb, *J. Am. Chem. Soc.*, 2005, **127**, 13694–13704.
101. P. Kurz, *Dalton Trans*, 2009, 6103–6108.
102. M. Yagi and K. Narita, *J. Am. Chem. Soc.*, 2004, **126**, 8084–8085.
103. M. Yagi, M. Toda, S. Yamada and H. Yamazaki, *Chem. Commun.*, 2011, **46**, 8594–8596.
104. G. J. Meyer, *Inorg. Chem.*, 2005, **44**, 6852–6864.
105. K. Kalyanasundaram and M. Gratzel, *Coord. Chem. Rev.*, 1998, **77**, 347–414.
106. M. Sjodin, S. Styring, B. Akermark, L. Sun and L. Hammarstrom, *Philos. Trans. R. Soc., B*, 2002, **357**, 1471–1479.
107. T. Lachaud, A. Quaranta, Y. Pellegrin, P. Dorlet, M. F. Charlot, S. Un, W. Leibl and A. Aukauloo, *Angew. Chem., Int. Ed.*, 2005, **44**, 1536–1540.
108. A. Quaranta, F. Lachaud, C. Herrero, R. Guillot, M. F. Charlot, W. Leibl and A. Aukauloo, *Chem.–Eur. J.*, 2007, **13**, 8201–8211.
109. A. Magnuson, H. Berglund, P. Korall, L. Hammarstrom, B. Akermark, S. Styring and L. Sun, *J. Am. Chem. Soc.*, 1997, **119**(44), 10720–10725.

110. A. Magnuson, Y. Frapart, M. Abrahamsson, O. Horner, B. Akermark, L. C. Sun, J. J. Girerd, L. Hammarstrom and S. Styring, *J. Am. Chem. Soc.*, 1999, **121**, 89–96.
111. P. Huang, A. Magnuson, R. Lomoth, M. Abrahamsson, M. Tamm, L. Sun, B. van Rotterdam, J. Park, L. Hammarstrom, B. Akermark and S. Styring, *J. Inorg. Biochem.*, 2002, **91**, 159–172.
112. P. Huang, J. Högblom, M. F. Anderlund, L. Sun, A. Magnuson and S. Styring, *J. Inorg. Biochem.*, 2004, **98**, 733–745.
113. L. Sun, M. K. Raymond, A. Magnuson, D. LeGourriérec, M. Tamm, M. Abrahamsson, P. H. Kenez, J. Mårtensson, G. Stenhagen, L. Hammarström, S. Styring and B. Åkermark, *J. Inorg. Biochem.*, 2000, **78**, 15–22.
114. C. Herrero, A. Quaranta, S. Protti, W. Leibl, A. W. Rutherford, R. Fallahpour, M. F. Charlot and A. Aukauloo, *Chem.–Asian J.*, 2011, **6**, 1335–1339.
115. H. Wolpher, P. Huang, M. Borgstrom, J. Bergquist, S. Styring, L. Sun and B. Akermark, *Catal. Today*, 2004, **98**, 529–536.
116. C. Herrero, J. L. Hughes, A. Quaranta, N. Cox, A. W. Rutherford, W. Leibl and A. Aukauloo, *Chem. Commun.*, 2010, **46**, 7605–7607.
117. Y. Chen, M. Wang, K. Jin, D. Wang, Y. Na and L. Sun, *Inorg. Chem. Commun.*, 2005, **8**, 606–609.
118. M. Dinca, Y. Surendranath and D. G. Nocera, *Proc. Natl. Acad. Sci., U. S. A.*, 2010, **107**, 10337–10341.
119. M. W. Kanan, J. Yano, Y. Surendranath, M. Dinca, V. K. Yachandra and D. G. Nocera, *J. Am. Chem. Soc.*, 2010, **132**, 13692–13701.
120. M. M. Najafpour, T. Ehrenberg, M. Wiechen and P. Kurz, *Angew. Chem., Int. Ed.*, 2010, **49**, 2233–2237.
121. Y. Shimazaki, T. Nagano, H. Takesue, B.-H. Ye, F. Tani and Y. Naruta, *Angew. Chem. Int. Ed.*, 2004, **43**, 98–100.
122. W. C. Ellis, N. D. McDaniel, S. Bernhard and T. J. Collins, *J. Am. Chem. Soc.*, 2010, **132**, 10990–10991.
123. T. J. Collins, *Acco. Chem. Res.*, 1994, **27**, 279–285.
124. S. E. Ghachtouli, B. Lassalle-Kaiser, P. Dorlet, R. Guillot, E. Anxolabehere-Mallart, C. Costentin and A. Aukauloo, *Energy Environ. Sci.*, 2011, **4**, 2041–2044.

CHAPTER 4

Photosynthetic Light-Harvesting Complexes

AARON M. COLLINS, JIANZHONG WEN AND ROBERT E. BLANKENSHIP

Departments of Biology and Chemistry, Washington University in St. Louis, St. Louis, MO 63130, USA

4.1 Introduction

Photosynthesis is the conversion of photon energy into chemical energy.[1] For the energy of the photon to be stored it first has to be absorbed by a pigment. All photosynthetic pigments have an intense electronic absorption band. The peak of the solar output is in the visible and near infrared region of the spectrum, so it is not surprising that the molecular structures of the pigments are such that they have absorption bands in this spectral region. There are a remarkable number of different molecules that can serve as photosynthetic pigments, but they fall into three distinct classes of types of molecules. The library of pigment molecules that nature has produced includes chlorophyll-type pigments, which are based on cyclic tetrapyrroles called porphyrins or the related compounds chlorins or bacteriochlorins.[2] Other photosynthetically active pigments include bilins, which are open-chain tetrapyrroles, and carotenoids, which are linear polyenes.[3] All of these molecules contain extensive conjugated π-electron systems, which results in low-lying electronic absorption bands. Slight modifications of the chemical structure of the pigments tune the absorption wavelengths so that they cover almost the entire UV/Vis/NIR region of the solar spectrum and adapt different organisms to live in a

RSC Energy and Environment Series No. 5
Molecular Solar Fuels
Edited by Thomas J. Wydrzynski and Warwick Hillier

Published by the Royal Society of Chemistry, www.rsc.org

remarkable variety of photic environments. These environments differ markedly in terms of light intensity (photon flux), light quality (spectral distribution), and light constancy (temporal intensity fluctuation).

In addition to the variety of different chemical structures of pigments found in various photosynthetic organisms, pigment absorption band energies and intensities are modulated *via* coupling to other nearby pigments and the surrounding protein environments. In order to do this, cells build very specific pigment–protein complexes called light-harvesting (LH) complexes. The function of these complexes is to absorb photons and transfer the energy efficiently to another pigment–protein complex, called a reaction center, where the light energy is converted into chemical energy, which in turn can be utilized by other cellular processes. This division of labor is illustrated in Figure 4.1, in which the majority of pigments function as a light-harvesting antenna, absorbing photons and transferring them to the reaction center (RC), where electron transfer processes lead to long-term energy storage in chemical bonds. This chapter focuses on the antenna systems in various types of photosynthetic organisms.

A very diverse group of light-harvesting antenna complexes have been identified, and almost every group of photosynthetic species has a unique collection of light-harvesting complexes.[4] One type of antenna system may be effective for a higher plant that lives in an intense but fluctuating light environment on the surface of the Earth, while a very different arrangement may be optimal for an organism that lives deep in a water column and receives a low but relatively constant flux of light of a significantly different spectral composition. This remarkable diversity of different types of antenna complexes strongly suggests that they have been invented multiple times during the course of evolution.

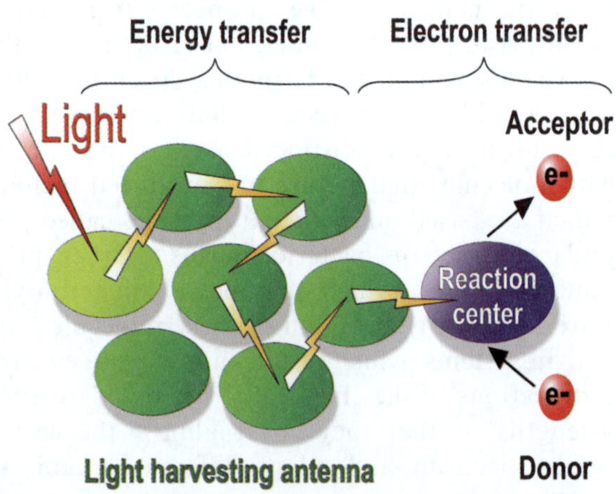

Figure 4.1 Schematic representation of antenna function in photosynthetic organisms. Antenna pigments are represented by the green ovals. Excitation transfer is represented by the yellow arrows. The blue oval represents the reaction center, and the black arrows represent electron transfer processes. Figure courtesy of Neil Hunter.

Achieving a deep understanding of the energy transfer processes in photosynthetic systems is an interdisciplinary research effort. Biologists isolate and identify organisms; biochemists purify natural or genetically modified protein samples; spectroscopists develop and adapt very sophisticated techniques to probe the energy transfer process experimentally; crystallographers determine atomic resolution structures of these protein complexes; and theoreticians extract physical models to connect the structural information with the experimental results.

4.2 Historical Context

The concept of the photosynthetic antenna system grows out of the classic flashing light experiments of Emerson and Arnold in 1932.[5] They showed that a green algal sample produced oxygen when illuminated with a series of short flashes of duration $\sim 10\,\mu s$, now called single-turnover flashes. When the intensity of the flashes was increased, the amount of oxygen produced increased up to a certain amount and then higher intensity flashes produced no additional oxygen. The system was saturated, with a light response curve that is in some ways similar to that exhibited by a typical enzyme (although the saturation curve has exponential rather than hyperbolic shape). When the maximum amount of oxygen produced per flash was compared to the amount of chlorophyll (Chl) present in the sample, it was found that one oxygen molecule was produced per several thousand chlorophylls, so most of the chlorophylls do not directly lead to the production of oxygen.

We now know that the vast majority of the pigments in any photosynthetic organism are part of what is called the antenna system. However, Emerson and Arnold did not use that term for these pigments. The term Photosynthetic Unit was coined by Gaffron and Wohl in 1936.[6] The modern use of that term refers to the collection of antenna pigments, plus their associated reaction centers and electron transport chain. Surprisingly, the first recorded use of the word antenna to describe the pigments that absorb photons and transfer their energy to the reaction center was in 1965 by Roderick Clayton[7] (Douglas Kahn, personal communication). Prior to that time, the concept was described using terms like light-gathering or light-harvesting, providing a more agricultural image compared to the high-tech term antenna, with the context of radio and radar.

The physical basis for the energy transfer process in photosynthesis was first formulated by Arnold and Oppenheimer in the 1940's[8] (see the review by Knox[9] for details). An earlier negative critique on the possibility of energy transfer in photosynthesis by Franck and Teller[10] turned out to be wrong, and held back the field for a time (see the review by Robinson[11] for details). The classic work by Förster, beginning in the 1940's provided the quantitative framework for interpretation of photosynthetic energy transfer processes.[12] A recent introduction to the modern interpretation of Förster theory is given by Sener *et al.*[13]

4.3 Diversity in Light-Harvesting Antennas

Light-harvesting (LH) antennas represent a markedly diverse set of pigment–protein complexes that all function to capture light energy and funnel the excitation to the photochemical reaction center. The absorption properties of LH antennas can be fine-tuned by altering the pigment identity as well as changing the local environment of the pigment within the protein. LH antennas can be broadly categorized as either *integral membrane* or *peripheral membrane* antenna complexes. Integral membrane complexes contain protein regions that span the biological membrane. Peripheral membrane antennas are associated with the membrane but do not span the membrane and are usually found in association with integral membrane proteins. Another categorization is core *vs.* accessory antenna complexes. Core complexes are integral membrane proteins that are always found in physical contact with reaction centers. Accessory complexes can be either integral or peripheral membrane associated. They are somewhat removed spatially from the reaction center, can be mobile, exist in variable stoichiometry and are often involved in regulatory processes. In the following sections we will explore the various antenna complexes that have been described from the different classes of photosynthetic organisms.

Figure 4.2 contains a gallery of molecular structures and absorption spectra of several types of light-harvesting antenna complexes. Their structures and properties are described in more detail below.

4.3.1 Higher Plants and Green Algae

Plants and green algae are photosynthetic eukaryotes that utilize two photosystems connected in series. These are Photosystems (PS) I and II. PSI is a supercomplex that is comprised of a core-complex that is itself the combination of a photochemical RC and its light-harvesting antenna in addition to an accessory membrane antenna, light-harvesting complex I (LHCI), that serves to increases the overall absorption cross-sectional area. Combined, the PSI supercomplex is comprised of at least 19 protein subunits and about 200 cofactors.[14] The current, highest resolution (3.4 Å) X-ray structure of plant PSI supercomplex exposed 17 of these subunits and 178 of the cofactors (with 168 being Chl).[15] The PSI core contains about 100 Chl molecules and at least 20 carotenoids.

PSI has been shown to be monomeric and LHCI binds only to one side of the core,[16] with high specificity and fixed stoichiometry.[17,18] LHCI is comprised of four unique proteins that assemble into heterodimers (Lhca1-4 and Lhca2-3). The addition of LHCI to the core complex increases the number of Chl molecules in the supercomplex by 58.

PSII is also very complex and contains at least 25 different protein subunits. PSII forms a dimeric supercomplex, with two entire PSII complexes, plus core and accessory antenna complexes. Unlike the plant PSI-core, the PSII-core complex has an intrinsically smaller antenna. This necessitates the inclusion of

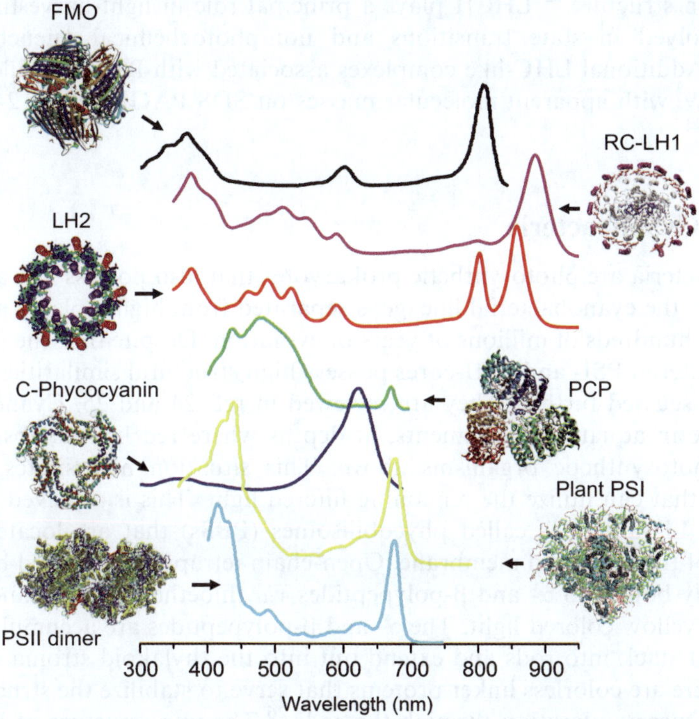

Figure 4.2 Structures of several representative LH antennas as determined by x-ray crystallography along with their respective absorbance spectra. Starting from the top of the figure; Fenna-Matthews-Olsen (FMO) Protein from *Prosthecochloris aestuarii* (PDB accession 3EOJ). The RC-LH1 complex from *Rhodopseudomonas palustris* (PDB accession 1PYH). LH2 complex from *Rhodopseudomonas acidophila* strain 7050 (PDB accession 1NKZ). The peridinin-chlorophyll-protein (PCP) from *Amphidinium carterae* (PDB accession 1PPR). C-phycocyanin, a phycobiliprotein, from *Cyanidium caldarium* (PDB accession 1PHN). Plant PSI (from *Pisum sativum*) (PDB accession 2WSC) and the PSII dimer (PDB accession 3A0B) from *Thermosynechococcus vulcanus*.

distal and extrinsic antenna complexes to ensure linear electron flow between PSI and PSII. In addition to the RC proteins, D1 and D2, the core antenna components of PSII include two Chl *a*-binding proteins, CP43 and CP47 (where CP denotes "chlorophyll protein", and the number indicates the apparent molecular weight as determined by SDS-PAGE analysis). CP43 and CP47 each contain six transmembrane helices that bundle together to form a dimer of trimers motif that serves to coordinate about 15 Chl *a* and 2-3 carotenoids molecules each.[19–21]

The accessory complex called LHCII is the most abundant integral membrane protein found in chloroplasts and is distal to CP43 and CP47. LHCII exists as a trimer that binds a total 42 Chl *a* and *b* and 12 carotenoids (1 neoxanthin, 1 xanthophyll and 2 luteins per monomer) according to the highest resolution structure.[22] LHCII plays a principal role in light-harvesting but is also involved in state transitions and non-photochemical quenching (see below). Additional LHC-like complexes associated with PSII are called CP24, 26 and 29, with apparent molecular masses on SDS PAGE gels of 24, 26 and 29 kDa.[23]

4.3.2 Cyanobacteria

Cyanobacteria are photosynthetic prokaryotes that also possess PSI and PSII. However, the cyanobacterial lineage is separated from higher plants and algae by many hundreds of millions of years of evolution. Despite this, the plant and cyanobacterial PSI- and PSII-cores possess high structural similarities and will not be discussed further. They are reviewed in ref. 24 and 25. Cyanobacteria often live in aquatic environments, at depths where red light is absorbed by other photosynthetic organisms above. This situation necessitates an LH-antenna that can utilize the remaining filtered light. This is achieved by large, extrinsic LH-antennas called phycobilisomes (PBSs) that are located at the surface of the thylakoid membrane. Open-chain tetrapyrroles, called bilins, are covalently bound to α- and β-polypeptides *via* thioether linkages and absorb green to yellow colored light. The α- and β-polypeptides are accumulated into discs that stack into rods and extend out into the thylakoid stroma. Between discs, there are colorless linker proteins that serve to stabilize the structure and regulate energy migration through the rods.[26] The ultrastructure of PBSs can vary among species and even under different growth conditions. PBSs can be described as hemiellipsoidal, hemidiscoidal or simple rods.

PBSs serve as an excellent example of a photosynthetic excitation funnel. The outermost rod biliprotein, phycoerytherin, absorbs the shortest wavelength light, while phycocyanin absorbs and occupies an intermediate position in the rod, while allophycocyanin, which absorbs at the longest wavelengths, is closest to the membrane. Light energy absorbed by PBSs is primarily transferred to PSII, although in some conditions PBSs can migrate and associate with PSI.

Because of their large size, PBSs are quite amenable to electron microscopy.[26–28] Additionally, many biliproteins have been successfully crystallized and their structures are known to high resolution, however the detailed structure of an intact phycobilisome remains to be determined.

4.3.3 Cryptophyte Algae

Cryptophyte algae do not posses PBSs but utilize biliproteins packed into the thylakoid lumen. Two different forms of an α-polypeptides and two copies of the same β-polypeptide combine to form an $\alpha_1\alpha_2\beta\beta$-heterodimer that covalently bind 8 bilin pigments (3 to each β-subunits and 1 to each of the α-subunits).[29] Attachment of various bilins to the apoprotein is not achieved using a conserved amino acid sequence and the energy transfer pathways within the antenna might be unique among various species.[30] Several spectroscopic investigations revealed that cryptophyte biliproteins can transfer excitations to both PSI and PSII. Recently, two different crytophyte biliproteins have been investigated at room temperature, using non-linear ultrafast spectroscopic methods that revealed a long-lasting sharing of excitation over the entire protein that was described *via* a quantum-coherence mechanism.[31]

4.3.4 Dinoflagellates

Dinoflagellates have a novel extrinsic antenna complex that contains a greater number of carotenoids than Chl molecules. This complex is called the peridinin-chlorophyll *a*-protein (PCP). PCP is water soluble and occupies the thylakoid lumen space. The structure has been determined to better than 2.0 Å resolution and is a trimer of mostly alpha helical structure and absorbs light between 400–550 nm.[32] It is interesting to note that the PCP protein has very low structural homology to any other known protein. Each monomer binds 8 peridinin and 2 Chl molecules and this is the highest observed ratio of carotenoid to chlorophyll known for a natural antenna complex. The peridinin molecules appear to be important in light-harvesting and these antenna complexes serve as a good model system to study carotenoid-to-Chl energy transfer processes. Energy absorbed by PCP is preferentially transferred to PSII, although the mechanism of how energy is transferred from the soluble PCP to the membrane-embedded photosystem is not understood.

4.3.5 Purple Bacteria

The purple bacteria are anoxyphototrophs that are often found in anaerobic layers of aquatic environments or sometimes in aerobic marine environments. The purple bacteria contain a rich collection of antenna complexes.[33] All purple bacteria possess an integral membrane core-complex that is the combination of its RC and the so-called LH1 complex. LH1 is composed of an inner and outer ring of α- and β-subunits that encompass the RC. Each subunit binds one bacteriochlorophyll (BChl) *a* (or BChl *b* in some species) so that the pigments are arranged perpendicular to the plane of the membrane, in a ring between the protein subunits. The combination of BChl *a* or *b* as well as interactions with the protein push the wavelengths of light utilized by these organisms from ~800–900 nm in organisms that contain BChl *a* to over 1000 nm in organisms that contain BChl *b*. The long wavelength absorption maximum in LH1 with BChl *a* is typically about 875 nm.

Intact photosynthetic membranes from several species of purple bacteria have been visualized with atomic force microscopy. These remarkable images revealed RCs that can be completely or partially surrounded by the LH1 ring or even form "S-shaped" core-complex dimers.[34] Additionally, the core complex structure from *Rhodopseudomonas palustris* was determined to 4.8 Å resolution using X-ray crystallography.[35]

An accessory integral membrane antenna associated with many but not all purple bacteria is the LH2 complex. The structures of LH2 from *Rhodoblastus acidophilus* (formerly *Rhodopseudomonas acidophila*) and *Phaeospirillum* (formerly *Rhodospirillum*) *molischianum* are known at atomic resolution.[36–38] LH2 is a ring-shaped complex of either eight or nine subunits, depending on species, and like LH1 is composed of an inner ring of α-subunits and outer rings of β-subunits. The diameter of LH2 is about 6.5 nm. Each α/β pair binds three BChl *a*, one BChl absorbing around 800 nm (so called B800) towards the cytoplasmic side of the membrane and two strongly coupled BChl absorbing around 850 nm (B850) buried towards the interior of the membrane. On the exterior of the ring, an extended carotenoid molecule forms a half helix, wrapping around from one subunit to the adjacent one. The carotenoid molecule is considered structurally important and deletion of the carotenoid through mutations leads to LH2s that fail to assemble *in vivo*. Some organisms contain spectral variants of LH2 that have the 850 nm band shifted to shorter wavelengths due to somewhat different interactions of the pigment and protein. The structures of these LH3 and LH4 complexes are not yet known to high resolution.

4.3.6 Green Sulfur Bacteria

Green sulfur bacteria (GSB) are strict anaerobes and possess a complex light-harvesting antenna. Chlorosomes are the principal light-harvesting apparatus of GSB, but are also found in two other phyla; filamentous anoxygenic phototrohps (FAPs) and chloroacidobacteria. Chlorosomes are the largest antenna structure known, being roughly 100–200 nm, 20–50 nm, 10–40 nm (length × width × height) ellipsoidal bodies that contain BChl *c, d*, or *e* and BChl *a* as well as carotenoids housed in a lipid monolayer and can contain between $5 \times 10^4 - 2 \times 10^5$ BChls.[39–41] The large absorption cross-sectional area of chlorosomes permits photon capture and energy transfer in extremely low light intensities. The organization of pigments within the chlorosome is unique in that it results from pigment–pigment interactions as opposed to a protein scaffold that serves to coordinate pigments as observed in all other antenna complexes. This is further corroborated by the fact that the interior of the chlorosome lack proteins. Recently, it has been determined that the pigments in the chlorosome from *Chlorobaculum* (formerly *Chlorobium*) *tepidum* are arranged in a *syn-anti* stacking mode where the 3^1-OH of one BChl coordinates to the central Mg of the adjacent BChl and the farnesyl tails alternate in and out of the plane of BChl molecules. These structures build-up to form concentric tubes or lamellae that extend the long axis of the chlorosome.[42]

Excited BChl in the chlorosome transfers the excitation towards the RC through two BChl *a* containing complexes, the baseplate and the Fenna–Matthews–Olson (FMO) protein. The baseplate complex is thought to reside on the side of the chlorosome that faces the cytoplasmic membrane and serve as an intermediary in energy transfer between BChl *c, d*, or *e* in the chlorosomes and BChl *a* of the FMO protein.[43]

The FMO protein is a water-soluble antenna complex that is sandwiched between the chlorosomes and the cytoplasmic membrane and serves as the immediate energy donor to the reaction center. The FMO protein structure has been determined to high resolution (1.3 Å) and the structure is revealed as a trimer of mostly beta-sheet structure.[44] Each monomer binds at least 7 BChl *a* and recent high-resolution structural refinement revealed partial occupancy of an 8th BChl *a* between each of the monomers. Lastly, the orientation of FMO on the membrane has recently been determined using mass spectrometry and chemical labeling.[45] These results reveal that the C-3 rotational symmetry axis is aligned perpendicular to the membrane with the BChl #3 of each monomer closest to the membrane. This confirms predictions based on theoretical calculations. The FMO protein exhibits quantum coherence effects that suggest that there may not be a simple pathway of excitation energy hopping that takes place in this system.[46]

4.3.7 Chloroacidobacteria

A newly discovered phototroph, *Chloracidobacterium thermophilum*, belonging to the phyla Acidobacteria has been isolated.[47] Not much has been done to describe its photosynthetic apparatus, however this organism contains chlorosomes and a variant of the FMO protein, along with a Type I reaction center. Interestingly, unlike green sulfur bacteria, *C. thermophilum* is an obligate aerobe.

4.3.8 Heliobacteria

The heliobacteria contain the simplest known photosystem and the smallest known antenna of any photosynthetic organism. A single pigment–protein complex serves as both antenna and reaction center. Light-harvesting is achieved by a modest pool of about 22 BChl *g* that are bound to the reaction center as a core complex.[48] Low-temperature absorption and linear dichroism measurements indicate that there are at least three distinct pools of BChl *g* with the Q_Y transition of the longest wavelength pigments being oriented essentially parallel to the plane of the membrane and the same transition of the shorter wavelength pigments being oriented at a large angle.[49] Work on heliobacteria has been hindered because the extreme sensitivity of BChl *g* to convert irreversibly into a form of Chl *a* in the presence of oxygen and light.[50]

4.3.9 Filamentous Anoxygenic Phototrophs (FAPs)

FAPs (also previously known as Green Nonsulfur Bacteria or Green Filamentous Bacteria) have an unusual collection of properties and share LH antenna features with other distantly related phototrophs. The bulk of work has been done on the thermophilic organism *Chloroflexus aurantiacus*. For example, some species of FAPs contain chlorosomes that are similar to those of GSB and *Cab. thermophilus* but are slightly smaller. FAPs contain no analog to

FMO and excitation from the chlorosome is funneled though the baseplate complex (see above) to a membrane-bound LH antenna. The membrane bound antenna has a sequence, spectroscopic and structural homology that shows similarities to both LH1 and LH2 from purple bacteria.[51–54]

4.4 Mechanisms of Energy Transfer in Photosynthetic Antenna Complexes

Although there is not yet a generalized mechanism to describe all the excitation energy transfer processes that take place in different photosynthetic systems, our understanding of these processes has deepened in the past years.

The appropriate theory for a given system depends on the relative strength of pigment–pigment and pigment–protein coupling.[55,56] An important parameter in understanding the spectroscopic properties of light-harvesting complexes is the site energy of a pigment (transition energy). For the same kind of pigment in a protein complex, it will have different site energies due to interactions with different local protein environments. Such pigment–protein interactions include: electrostatic interactions with surrounding charged amino acids; hydrogen bonding between the pigment and the protein; the axial ligation to the central magnesium; electrostatic interaction with the protein backbone, *etc.* Such tuning can affect the site energy so that the absorption maximum shifts by up to tens of nanometres. Besides the above static shift of the site energy, there is also a dynamic regulation of the site energy induced by the coupling of the electronic transition with the protein vibrations.

The premier theory of excitation energy transfer between molecular electronic transition dipoles is that of Förster Resonance Energy Transfer, often called FRET.[12,13] This mechanism applies to the condition in which the electronic coupling between two pigments is small compared to the difference of their site energies. Normally, this requires that the two pigments be well separated spatially. Excitation energy is basically localized on an individual pigment and non-radiatively transfers between them through an incoherent hopping process. Under this situation, the coupling energy between the two pigments p and q can be described using a point-dipole approximation.

$$V_{pq} = \frac{1}{4\pi\varepsilon_0\varepsilon_r}\left[\frac{\vec{\mu}_p \cdot \vec{\mu}_q}{R_{pq}^3} - \frac{3(\vec{\mu}_p \cdot \vec{R}_{pq})(\vec{\mu}_q \cdot \vec{R}_{pq})}{R_{pq}^5}\right] \tag{1}$$

All the protein environment effects are taken into account by the dielectric constant ε_r. $\vec{\mu}_p$ and $\vec{\mu}_q$ are the optical transition dipoles of pigments p and q, and $\vec{R}_{pq}$ is the center-to-center spatial vector from p to q.

The energy transfer rate from p to q can be described by:

$$k = \frac{8.79 \times 10^{-5} k_f \kappa^2}{(R_{pq}^6 n^4)} \int \varepsilon(\lambda) F_D(\lambda) \lambda^4 d\lambda \tag{2}$$

where k_f is the fluorescence decay rate constant of the energy donor; $\kappa^2 = (\cos\alpha - 3\cos\beta_p\cos\beta_q)^2$ is an orientation factor where α is the angle between transition dipoles; and $\vec{\mu}_p$ and $\vec{\mu}_q$ are β the angles between $\vec{\mu}_p$ and $\vec{R}_{pq}$, $\vec{\mu}_q$ and $\vec{R}_{pq}$; n is the medium refractive index; the rate k depends on the sixth-order distance between the donor and acceptor; R_{pq}^6; $\int \varepsilon(\lambda)F_D(\lambda)\lambda^4 d\lambda$ is the spectral overlap of the donor emission spectrum and the acceptor absorption spectrum where $\varepsilon(\lambda)$ is the molar extinction function of the acceptor on a wavelength scale and $F_D(\lambda)$ is the normalized emission spectrum of the donor.

This mechanism is used to describe the energy transfer process between chlorophylls in the PCP complex successfully[57,58] but rarely gives fully satisfactory results in elucidating the observed energy transfer phenomena in most other photosynthetic complexes.[55,59] The Chl molecules in the PCP complex are separated by around 18 nm, while the pigment molecules in all the other known light-harvesting complexes are densely packed and the excitonic coupling between them is moderate or strong. The excitation is instead delocalized on multiple pigments simultaneously and the overall excitation wave function can be viewed as coherent superpositions of the monomeric excited states. Nevertheless, the classical description of the Förster energy transfer is easy to understand and can provide a qualitative description of the light-harvesting process in the photosystems. The basic concept used to describe the electronic–vibrational coupling provides the foundation of more advanced theories developed using quantum mechanics to describe the energy transfer mechanism in the strong coupling cases.

For the strong excitonic coupling but weak exciton–vibrational coupling case, the wave function of the exciton states is viewed as the coherent superposition of localized excited states. The relaxation of the delocalized excitation states (under this condition, the traditional point-dipole approximation in the Förster mechanism is not valid due to the short distance between the coupling pigments) can be described by several theories, among which the Redfield relaxation theory is widely applied.[60–62] In this theory, the weak exciton–vibrational interaction is treated as a perturbation in contrast to the Förster mechanism, where the weak excitonic coupling is treated as a perturbation. The rate of the exciton relaxation derived is dependent on the energy difference between the two exciton states; the overlap of the two exciton wavefunctions; the correlation of the protein vibrations (uncorrelated vibrations lead to fast exciton relaxation) and how well the vibrational environment can dissipate the excess energy during exciton relaxation.[63,64]

Versions of modified Redfield theory[65–68] were also developed to treat the strong excitonic coupling and strong exciton–vibrational coupling systems in which the nuclear reorganization was also taken into account non-perturbatively. In the Redfield theory, the nuclear relaxation is assumed to be fast compared to the energy transfer between different exciton states and only a single phonon is considered in the exciton–nuclear interaction. This limitation breaks down in the modified Redfield equation where the nuclei relax into different equilibrium states in the strong exciton–vibrational coupling conditions.[62,69]

In the photosynthetic light-harvesting process, an energy gradient or funnel is almost always formed in the individual light-harvesting complexes by different excitonic coupling or environment tuning of the site energies of different pigments. This energy gradient on one hand guides the energy flow to the reaction center and provides the energetic driving force. On the other hand, the excess energy generated has to be dissipated. This dissipation is achieved by exciton–vibrational coupling. The uncontrollable environmental fluctuations are usually assumed to be deteriorating in the energy transfer process. However, it was recently found that certain fluctuations can dynamically manipulate the site energies, which can increase the overlap of different energy levels and may actually aid energy transport through a dissipative network by opening up additional energy transfer pathways. This noise-assisted transport[70–72] has been proposed to enhance both the rate and yield of energy transfer, especially in the quantum coherence energy evolution that will be mentioned later.

As discussed before, the Förster theory has proved successful to predict the energy transfer rate between two weakly coupled donor and acceptor molecules, but can not be employed to describe the energy transfer in a confined geometry with multi-chromophores in which the exciton relaxation theory should be considered. In the past decade, a generalized Förster energy transfer mechanism[73–76] has been developed to explore and quantify the energy transfer dynamics in large light-harvesting complexes in which not all the pigments are coupled strongly with each other but they form certain domains.[77–79] Within each domain, the pigments are tightly coupled and share the excitation. The interaction between different domains is weak. Redfield or modified Redfield theory can be used to describe the exciton relaxation within the domains. The generalized Förster theory can be used to describe the energy transfer between the domains. Here each domain is viewed as a supermolecule with collective exciton states.

It is exciting to see the great breakthroughs of understanding the energy transfer process in the individual light-harvesting complexes. However, on a large scale, it is still not well understood how the efficient long-range energy transfer process in a native photosynthetic membrane is achieved; under that condition, the excitation energy needs to migrate a long distance from the peripheral light-harvesting complex to the reaction center. On the microscopic scale, the details of the exciton–bath interaction, such as the tuning of the site energy by the protein environment, the effect of environmental fluctuation on the transfer efficiency, is still not clearly understood.

Both free (B)Chl in organic solvents and the isolated light-harvesting complexes in most cases have a lifetime around several nanoseconds. In order to be utilized for photosynthesis, light energy captured by these antenna complexes must migrate rapidly by non-radiative processes. This typically occurs in sub-picosecond to a few hundreds of picosecond timescales to compete with the fluorescence and other decay processes. The developments of advanced laser and optical spectroscopy techniques in the past decades have made observing these ultrafast processes possible.[80] Techniques such as pump–probe,[81–83] hole burning,[84,85] single molecule spectroscopy,[86,87] photon echo,[88] *etc.*, have been developed or applied to study the light-harvesting complexes and have already generated tremendous amounts of experimental results, which are used to test or advance the development of theories.

Recently, a new ultrafast nonlinear spectroscopy technique called two-dimensional electronic spectroscopy[89–93] has been used to probe energy transfer dynamics in several light-harvesting complexes.[46,94–96] This new technique has the ability to record both the population and phase information of the wave function. The exciton coupling and dynamics can be directly viewed in the off-diagonal peaks. These experiments have demonstrated the existence of strong and long-lived quantum coherences in the photosynthetic complexes at liquid nitrogen temperature[46,97] and also at physiological temperature.[98–100] Such long-lived coherence has been proposed to improve energy transfer efficiency in photosynthetic systems by allowing an excitation to follow a quantum random walk as it approaches the reaction center. These observations have generated considerable interest in understanding the possible role of quantum coherence effects in the remarkably efficient excitation energy transfer in the FMO protein and other pigment-binding protein complexes.

4.5 Regulation of Antenna Systems

The light absorbed by the antenna system is finally utilized to generate a charge separation at the reaction center, where electrons are extracted to join the metabolic pathways to fix carbon or be involved in other cellular processes. Compared to the ultrafast energy transfer in the light reactions, the rates of the following dark reactions are slow, which often forms the bottleneck of the whole light-driven energy storage process. It is extremely dangerous for the cells if too much reducing power is generated by transferring too much photon energy to the reaction center, which can easily happen under full sunlight. In addition, the quantity and quality of the light in natural environments can vary over several orders of magnitude on the time scales of milliseconds to seasons. To avoid these deadly situations and adapt organisms to environmental fluctuations, cells have generated the ability to dynamically control the energy transfer process to prevent photodamage, although different groups of photosynthetic species often use different regulation methods.

In plants, algae and cyanobacteria, PSI and PSII work together to split water and evolve oxygen. Electrons extracted from water by PSII are transferred to PSI, and their serial connection means that the rate of electron transport between the two photosystems must be similar. Thus, the light energy received by both photosystems should also be similar to achieve maximum efficiency.[101] "State transitions" is a phenomenon generally adapted by oxygenic species to balance the energy input to PSI and PSII under low light conditions.

State transitions are a short-term adaption mediated mainly by the reversible phosphorylation of the LHCII in plants and green algae and its migration between PSI and PSII, which is mainly controlled by the redox state of plastoquinone pool and cytochrome b_6f complex.[102,103] When the illumination is favouring PSII, LHCII is phosphorylated, detached from PSII and migrates to PSI. The absorption cross-section of PSI is thus increased. This is called State II. If PSI is preferentially excited, LHCII is dephosophorylated and migrates back to PSII to redistribute the excitation energy. This is called State I. Through state transitions, the system distributes the excitation such that the light-limited photosystem receives more energy.

Under high light or saturated light conditions, plants and algae have evolved ways to get rid of excess energy that has already been absorbed. This is mainly achieved by a thermal dissipation process called Non-Photochemical Quenching (NPQ) in PSII, measured as the quenching of chlorophyll fluorescence.[104,105] The regulatory mechanism of energy dependent NPQ is induced by a decrease of the pH of lumen of the thylakoid that is generated by over-reducing the photosynthetic electron transport chain. The low pH of the lumen protonates certain PSII proteins and activates a xanthophyll cycle (Figure 4.3). A lumen-located violaxanthin de-epoxidase (VDE) enzyme catalyzes the conversion of violaxanthin to zeaxanthin *via* the intermediate antheraxanthin under low pH.[106] Violaxanthin has nine conjugated double bonds, while antheraxanthin and zeaxanthin have ten and eleven, respectively. The increased conjugation will lower the excitation energy level of the carotenoid and facilitate the de-excitation of chlorophyll either directly through triplet formation or indirectly through structural changes. In limiting light, a reversal of the cycle is mediated by the enzyme zeaxanthin epoxidase (ZE). In addition, mutants

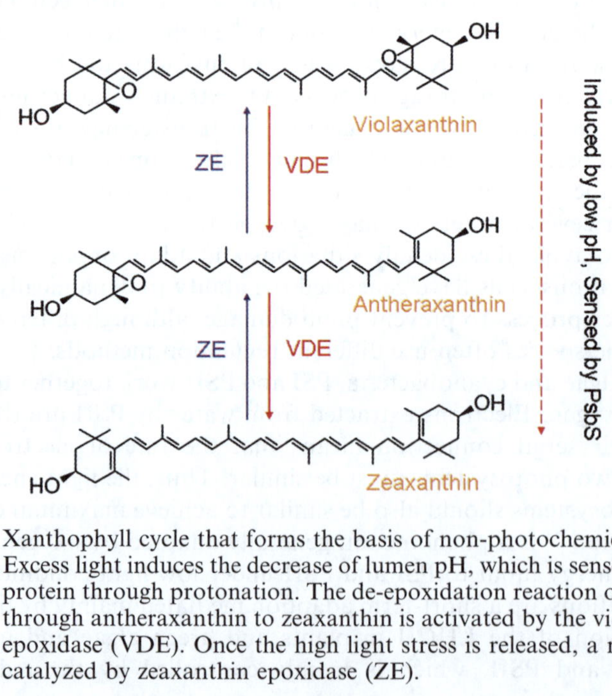

Figure 4.3 Xanthophyll cycle that forms the basis of non-photochemical quenching. Excess light induces the decrease of lumen pH, which is sensed by the PsbS protein through protonation. The de-epoxidation reaction of violaxanthin through antheraxanthin to zeaxanthin is activated by the violaxanthin de-epoxidase (VDE). Once the high light stress is released, a reverse cycle is catalyzed by zeaxanthin epoxidase (ZE).

produced by deleting a component of PSII called the PsbS protein confirmed that this protein is also an essential component for NPQ and is a sensor of the lumen pH.[107] While understanding of the energy dependent NPQ has deepened in the past several years, the most fundamental question that the physical mechanism of chlorophyll deactivation is still not very clear. The identity of the quenching species, in terms of both pigment composition and location, and the means by which the quencher(s) dissipates the excess energy are still under debate.[108–110]

If excess energy is not quenched in a timely manner, reactive oxygen species are produced and photodamage happens. Cells have evolved different mechanisms to repair the damaged proteins. For example, the D1 protein from the PSII core turns over very rapidly in a light dependent manner.[111] Damaged D1 is digested by proteases and new D1 protein is synthesized to repair the PSII. So far intense attention is focused on understanding the recognition of damaged D1 protein by protease and the mechanism of the disassembly and assembly of PSII in the repair cycle.[112] It is hypothesized that the rapid turnover of D1 protein prevents the damage of the entire PSII.

Compared to oxygenic photosynthetic species, the regulation of energy transfer in the anoxygenic prokaryotes, including GSB, FAP, purple bacteria, heliobacteria and the newly discovered photosynthetic acidobacteria, is not well understood.

In GSB, a variety of lines of evidence suggest that a redox-dependent regulation of energy transfer occurs in the chlorosome and FMO antenna protein.[39] When the chlorosome is under oxidizing conditions, the fluorescence emission intensity will dramatically decrease due to induction of quenchers. Such an effect was observed in the isolated chlorosome, native membrane and whole cells, and could be a possible method to protect the cell from transient exposure to oxidizing conditions by efficiently quenching the excitation.[113] The redox active species chlorobiumquinone was proposed to mediate the quenching.[114] The photon energy harvested by the chlorosome is transferred through FMO to the reaction center. A similar quenching phenomenon was also found in the FMO protein, but the molecular mechanism is still unclear.[115] The newly discovered *Cab. thermophilum* also exhibits a similar redox regulation at least in the FMO level, as found in green sulfur bacteria.[116,117] It has also been observed that chlorosomes from GSB and also FAP will modify the composition and morphology along the growth environment to regulate the energy transfer.[118–120] Several species also synthesize different bacteriochlorophylls (*c, d* or *e*) to assemble the chlorosomes, which show different internal structures and optical properties. Formation of triplet excitons were also proposed in chlorosomes owing to triplet–triplet interaction between the closely packed BChls[121] and the energy of these triplet excitons might fall below that of singlet oxygen and triplet carotenoids, thus preventing energy transfer from triplet BChl and serves as an alternative photoprotection mechanism.

Cells of purple photosynthetic bacteria, when exposed to elevated levels of oxygen, will reduce the amount of LH complexes synthesized by mediation through gene regulation processes. Under low light conditions, more antenna complexes are made and a higher LH2 : LH1 ratio is observed. Certain species will synthesize modified antenna complexes in which the spectral properties are changed according to the light intensity and quality.[33] For example, the *Rps. palustris* genome contains at least four different gene pairs that encode putative LH2 αβ-apoproteins and different genes are expressed depending on the growth conditions, which could even result in some complexes having mixed apoprotein types within single complete rings and producing different spectral types.[122,123]

To effectively prevent the production of reactive oxygen species, a common strategy used by almost all the photosynthetic species is synthesizing carotenoids and embedding them into the light-harvesting and reaction center complexes.[3,124] Multiple functions are performed by carotenoids in photosynthesis.[125] They serve as accessory light-harvesting pigments, extending the range of wavelengths over which light can drive photosynthesis, and they are also building blocks of various light-harvesting and RC complexes and help assemble and stabilize proteins.[126] One additional unique property is to protect the chlorophyllous pigments from the harmful photodestructive reaction that occurs in the presence of oxygen. Carotenoids absorb the green and blue light (which is not covered by the chlorophyllous pigment absorption) for photosynthetic energy conversion by a singlet–singlet energy transfer to neighbouring (bacterio)chlorophylls. The protective role against photo-oxidation is achieved by a triplet–triplet energy transfer. Using the triplet state, carotenoids can quench the (B)Chl triplet states which may be formed from the singlet states by inter-system crossing. In this way, carotenoids will prevent the triplet (B)Chl reacting with the ground state triplet oxygen to form highly reactive singlet oxygen species. In addition, carotenoids can directly quench singlet oxygen *via* their triplet states due to their lower energy level of triplet states.[127]

4.6 Conclusions

Photosynthetic organisms have developed a large variety of antenna complexes. These light-harvesting complexes increase the spectral and spatial absorption cross-section of the surrounding reaction center, and let the cells make full use of the solar radiation in a variety of photic environments. However, the increase of the size and/or population of the antennas around the reaction center introduces the challenge of efficient excitation energy transfer by increasing the distance, introducing complex possible transfer pathways and more uncontrollable dynamic fluctuations. Remarkably, nature seems to have been able to circumvent the difficulty through billions of years of evolutionary pressure, and in many cases reach a quantum efficiency of almost unity during the energy transfer process, while at the same time effectively protect the organisms form the potentially destructive effects of excess excitation. A deep understanding of the mechanisms behind this natural process will not only revolutionize our knowledge of the light-harvesting complexes, but the techniques and theories developed by studying model systems may also shine light on the other light–matter interactions, such as nano-materials and semiconductors,[128] on the applications of solar energy conversions,[129] such as artificial photosynthesis, and also on quantum information.[130,131]

References

1. R. E. Blankenship, *Molecular Mechanisms of Photosynthesis*, Blackwell Science, Oxford, UK, 2002.
2. B. Grimm, R. J. Porra, W. Rudiger and H. Scheer, *Chlorophylls and Bacteriochlorophylls*, Springer, Dordrecht, 2006.
3. H. A. Frank, A. J. Young, G. Britton and R. J. Cogdell, *The Photochemistry of Carotenoids*, Kluwer Academic Publishers, Dordrecht, 1999.
4. B. R. Green and W. W. Parson, *Light-Harvesting Antennas in Photosynthesis*, Kluwer Academic Publishers, Dordrecht, 2003.
5. R. Emerson and W. Arnold, *J. Gen. Physiol.*, 1932, **16**, 191.
6. H. Gaffron and K. Wohl, *Naturwissenschaften*, 1936, **24**, 81–103.
7. R. K. Clayton, *Science*, 1965, **149**, 1346.
8. W. Arnold and R. Oppenheimer, *J. Gen. Physiol.*, 1950, **33**, 423.
9. R. S. Knox, *Photosynth. Res.*, 1996, **48**, 35.
10. J. Franck and E. Teller, *J. Chem. Phys.*, 1938, **6**, 861.
11. G. W. Robinson, *Brookhaven Symp. Biol.*, 1967, **19**, 16.
12. T. Förster, *Ann. Phys.*, 1948, **2**, 55.
13. M. Sener, J. Strümpfer, J. Hsin, D. Chandler, S. Scheuring, C. N. Hunter and K. Schulten, *ChemPhysChem*, 2011, **12**, 518.
14. P. E. Jensen, R. Bassi, E. J. Boekema, J. P. Dekker, S. Jansson, D. Leister, C. Robinson and H. V. Scheller, *Biochim. Biophys. Acta*, 2007, **1767**, 335.
15. A. Amunts, O. Drory and N. Nelson, *Nature*, 2007, **447**, 58.
16. E. J. Boekema, P. E. Jensen, E. Schlodder, J. F. L. van Breemen, H. van Roon, H. V. Scheller and J. P. Dekker, *Biochemistry*, 2000, **40**, 1029.
17. F. Klimmek, U. Ganeteg, J. A. Ihalainen, H. van Roon, P. E. Jensen, H. V. Scheller, J. P. Dekker and S. Jansson, *Biochemistry*, 2005, **44**, 3065.
18. E. Wientjes, G. T. Oostergetel, S. Jansson, E. J. Boekema and R. Croce, *J. Biol. Chem.*, 2009, **284**, 7803.
19. K. N. Ferreira, T. M. Iverson, K. Maghlaoui, J. Barber and S. Iwata, *Science*, 2004, **303**, 1831.
20. R. Barbato, H. L. Race, G. Friso and J. Barber, *FEBS Lett.*, 1991, **286**, 86.
21. D. Zheleva, J. Sharma, M. Panico, H. R. Morris and J. Barber, *J. Biol. Chem.*, 1998, **273**, 16122.
22. Z. Liu, H. Yan, K. Wang, T. Kuang, J. Zhang, L. Gui, X. An and W. Chang, *Nature*, 2004, **428**, 287.
23. A. E. Yakushevska, W. Keegstra, E. J. Boekema, J. P. Dekker, J. Andersson, S. Jansson, A. V. Ruban and P. Horton, *Biochemistry*, 2002, **42**, 608.

24. *Photosystem II*, ed. T. J. Wydrzynski and K. Satoh, Springer, Dordrecht, 2005.
25. *Photosystem I*, ed. J. H. Golbeck, Springer, Dordrecht, 2006.
26. L.-N. Liu, X.-L. Chen, Y.-Z. Zhang and B.-C. Zhou, *Biochim. Biophys. Acta*, 2005, **1708**, 133.
27. R. MacColl, *J. Struct. Biol.*, 1998, **124**, 311.
28. A. A. Arteni, G. Ajlani and E. J. Boekema, *Biochim. Biophys. Acta*, 2009, **1787**, 272.
29. A. B. Doust, C. N. J. Marai, S. J. Harrop, K. E. Wilk, P. M. G. Curmi and G. D. Scholes, *J. Mol. Biol.*, 2004, **344**, 135.
30. G. J. Wedemayer, D. G. Kidd and A. N. Glazer, *Photosynth. Res.*, 1996, **48**, 163.
31. E. Collini, C. Y. Wong, K. E. Wilk, P. M. G. Curmi, P. Brumer and G. D. Scholes, *Nature*, 2010, **463**, 644.
32. T. Schulte, S. Johanning and E. Hofmann, *Eur. J. Cell Biol.*, 2010, **89**, 990.
33. R. J. Cogdell, A. Gall and J. Köhler, *Q. Rev. Biophys.*, 2006, **39**, 227.
34. J. N. Sturgis, J. D. Tucker, J. D. Olsen, C. N. Hunter and R. A. Niederman, *Biochemistry*, 2009, **48**, 3679.
35. A. W. Roszak, T. D. Howard, J. Southall, A. T. Gardiner, C. J. Law, N. W. Isaacs and R. J. Cogdell, *Science*, 2003, **302**, 1969.
36. J. Koepke, X. Hu, C. Muenke, K. Schulten and H. Michel, *Structure*, 1996, **4**, 581.
37. G. McDermott, S. M. Prince, A. A. Freer, A. M. Hawthornthwaite-Lawless, M. Z. Papiz, R. J. Cogdell and N. W. Isaacs, *Nature*, 1995, **374**, 517.
38. M. Z. Papiz, S. M. Prince, T. Howard, R. J. Cogdell and N. W. Isaacs, *J. Mol. Biol.*, 2003, **326**, 1523.
39. R. E. Blankenship, K. Matsuura, in *Light-Harvesting Antennas*, ed. B. R. Green and W. W. Parson, Kluwer Academic Publishers, Dordrecht, 2003, p. 195.
40. N.-U. Frigaard and D. Bryant, in *Complex Intracellular Structures in Prokaryotes*, ed. J. M. Shively, Springer, Berlin and Heidelberg, 2006, p. 79.
41. G. Oostergetel, H. van Amerongen and E. Boekema, *Photosynth. Res.*, 2010, **104**, 245.
42. S. Ganapathy, G. T. Oostergetel, P. K. Wawrzyniak, M. Reus, A. Gomez Maqueo Chew, F. Buda, E. J. Boekema, D. A. Bryant, A. R. Holzwarth and H. J. M. de Groot, *Proc. Natl. Acad. Sci., U. S. A.*, 2009, **106**, 8525.
43. M. Pedersen, J. Linnanto, N.-U. Frigaard, N. Nielsen and M. Miller, *Photosynth. Res.*, 2010, **104**, 233.
44. D. Tronrud, J. Wen, L. Gay and R. Blankenship, *Photosynth. Res.*, 2009, **100**, 79.
45. J. Wen, H. Zhang, M. L. Gross and R. E. Blankenship, *Proc. Natl. Acad. Sci., U. S. A.*, 2009, **106**, 6134.
46. G. S. Engel, T. R. Calhoun, E. L. Read, T. K. Ahn, T. Mancal, Y. C. Cheng, R. E. Blankenship and G. R. Fleming, *Nature*, 2007, **446**, 782.

47. D. A. Bryant, A. M. Garcia Costas, J. A. Maresca, A. G. M. Chew, C. G. Klatt, M. M. Bateson, L. J. Tallon, J. Hostetler, W. C. Nelson, J. F. Heidelberg and D. M. Ward, *Science*, 2007, **317**, 523.
48. M. Heinnickel, R. Agalarov, N. Svensen, C. Krebs and J. H. Golbeck, *Biochemistry*, 2006, **45**, 6756.
49. C. Francke, S. C. M. Otte, J. C. van der Heiden and J. Amesz, *Biochim. Biophys. Acta*, 1994, **1186**, 75.
50. H. Brockmann and A. Lipinski, *Arch. Microbiol.*, 1983, **136**, 17.
51. T. Wechsler, R. Brunisholz, F. Suter, R. C. Fuller and H. Zuber, *FEBS Lett.*, 1985, **191**, 34.
52. A. M. Collins, Y. Xin and R. E. Blankenship, *Biochim. Biophys. Acta*, 2009, **1787**, 1050.
53. M. Yamada, H. Zhang, S. Hanada, K. V. P. Nagashima, K. Shimada and K. Matsuura, *J. Bacteriol.*, 2005, **187**, 1702.
54. Y. Xin, S. Lin, G. Montaño and R. Blankenship, *Photosynth. Res.*, 2005, **86**, 155.
55. T. Renger, *Photosyn. Res.*, 2009, **102**, 471.
56. H. van Amerongen, L. Valkunas and R. van Grondelle, *Photosynthetic Excitons*, World Scientific, Singapore, 2000.
57. F. J. Kleima, E. Hofmann, B. Gobets, I. H. M. van Stokkum, R. van Grondelle, K. Diederichs and H. van Amerongen, *Biophys. J.*, 2000, **78**, 344.
58. T. Polivka, R. G. Hiller and H. A. Frank, *Arch. Biochem. Biophys.*, 2007, **458**, 111.
59. G. D. Scholes and G. R. Fleming, *Adv. Chem. Phys.*, 2005, **132**, 57.
60. A. G. Redfield, *IBM J. Res. Dev.*, 1957, **1**, 19.
61. T. Renger, V. May and O. Kühn, *Phys. Rep.*, 2001, **343**, 138.
62. M. Yang and G. R. Fleming, *Chem. Phys.*, 2002, **275**, 355.
63. V. I. Novoderezhkin and R. van Grondelle, *J. Phys. Chem. B*, 2002, **106**, 6025.
64. V. I. Novoderezhkin, J. M. Salverda, H. van Amerongen and R. van Grondelle, *J. Phys. Chem. B*, 2003, **107**, 1893.
65. W. Zhang, T. Meier, V. Chernyak and S. Mukamel, *J. Chem. Phys.*, 1998, **108**, 7763.
66. V. I. Novoderezhkin, M. A. Palacios, H. van Amerongen and R. van Grondelle, *J. Phys. Chem. B*, 2004, **108**, 10363.
67. V. I. Novoderezhkin, D. Rutkauskas and R. van Grondelle, *Biophys. J.*, 2006, **90**, 2890.
68. M. Schröder, U. Kleinekathöfer and M. Schreiber, *J. Chem. Phys.*, 2006, **124**, 084903.
69. Y. C. Cheng and R. J. Silbey, *J. Phys. Chem. B*, 2005, **109**, 21399.
70. M. Mohseni, P. Rebentrost, S. Lloyd and A. Aspuru-Guzik, *J. Chem. Phys.*, 2008, **129**, 174106.
71. A. W. Chin, A. Datta, F. Caruso, S. F. Huelga and M. B. Plenio, *New J. Phys.*, 2010, **12**, 065002.

72. T. Scholak, F. de Melo, T. Wellens, F. Mintert and A. Buchleitner, *arXiv.org, e-Print Arch., Quantum Phys.*, 2009, arXiv:0912.3560v3 [quant-ph].
73. G. D. Scholes, X. Jordanides and G. R. Fleming, *J. Phys. Chem. B*, 2001, **105**, 1640.
74. H. Sumi, *J. Phys. Chem. B*, 1999, **103**, 252.
75. S. Jang, M. D. Newton and R. J. Silbey, *Phys. Rev. Lett.*, 2004, **92**, 218301.
76. S. Jang, *J. Chem. Phys.*, 2007, **127**, 174710.
77. K. Mukai, S. Abe and H. Sumi, *J. Phys. Chem. B*, 1999, **103**, 6096.
78. Z. Fetisova, A. Freiberg, K. Mauring, V. Novoderezhkin, A. Taisova and K. Timpmann, *Biophys. J.*, 1996, **71**, 995.
79. J. Jordanides, G. D. Scholes and G. R. Fleming, *J. Phys. Chem. B*, 2001, **105**, 1652.
80. D. Zigmantas, Y. Z. Ma, E. L. Read and G. R. Fleming, in *Biophysical Techniques in Photosynthesis II*, ed. T. J. Aartsma and J. Matysik, Springer, Dordrecht, 2008, p. 201.
81. A. R. Holzwarth, *Methods Enzymol.*, 1995, **246**, 334.
82. H. van Amerongen and R. van Grondelle, *Methods Enzymol.*, 1995, **246**, 201.
83. R. Berera, R. van Grondelle and J. T. M. Kennis, *Photosynth. Res.*, 2009, **101**, 105.
84. S. Volker, *Annu. Rev. Phys. Chem.*, 1989, **40**, 499.
85. R. Purchase and S. Volker, *Photosyn. Res.*, 2009, **101**, 245.
86. R. J. Cogdell and J. Kohler, *Biochem. J.*, 2009, **422**, 193.
87. Y. Saga and H. Tamiaki, *Cell Biochem. Biophys.*, 2004, **40**, 149.
88. E. L. Read, H. Lee and G. R. Fleming, *Photosynth. Res.*, 2009, **101**, 233.
89. S. Mukamel, *Annu. Rev. Phys. Chem.*, 2000, **51**, 691.
90. D. M. Jonas, *Annu. Rev. Phys. Chem.*, 2003, **54**, 425.
91. M. Cho, *Chem. Rev.*, 2008, **108**, 1331.
92. F. Milota, J. Sperling, A. Nemeth, T. Mančal and H. F. Kauffmann, *Acc. Chem. Res.*, 2009, **42**, 1364.
93. N. S. Ginsberg, Y. C. Cheng and G. R. Fleming, *Acc. Chem. Res.*, 2009, **42**, 1352.
94. T. Brixner, J. Stenger, H. M. Vaswani, M. Cho, R. E. Blankenship and G. R. Fleming, *Nature*, 2005, **434**, 625.
95. D. Zigmantas, E. L. Read, T. Mancal, T. Brixner, A. T. Gardiner, R. J. Cogdell and G. R. Fleming, *Proc. Natl. Acad. Sci. U. S. A.*, 2006, **103**, 12672.
96. E. L. Read, G. Schlau-Cohen, G. S. Enge, J. Wen, R. E. Blankenship and G. R. Fleming, *Biophys. J.*, 2008, **95**, 847.
97. H. Lee, Y. C. Cheng and G. R. Fleming, *Science*, 2007, **316**, 1462.
98. E. Collini, C. Y. Wong, K. E. Wilk, P. M. G. Curmi, P. Brumer and G. D. Scholes, *Nature*, 2010, **463**, 644.
99. A. Ishizaki and G. R. Fleming, *Proc. Natl. Acad. Sci., U. S. A.*, 2009, **106**, 17255.

100. G. Panitchayangkoon, G. Hayes, K. A. Fransted, J. R. Caram, E. Harel, J. Wen, R. E. Blankenship and G. S. Engel, *Proc. Natl. Acad. Sci., U. S. A.*, 2010, **107**, 12766.
101. J. F. Allen, *Science*, 2003, **299**, 1530.
102. E. Rintamaki, M. Salonen, U. M. Suoranta, I. Carlberg, B. Andersson and E. M. Aro, *J. Biol. Chem.*, 1997, **272**, 30476.
103. F. A. Wollman, *EMBO J.*, 2001, **20**, 3623.
104. P. Horton, A. V. Ruban and R. G. Walters, *Plant Physiol.*, 1994, **106**, 415.
105. P. Müller, X. P. Li and K. K. Niyogi, *Plant Physiol.*, 2001, **125**, 1558.
106. B. Demmig-Adams and W. W. Adams III, *Trends Plant Sci.*, 1996, **1**, 21.
107. X. P. Li, A. M. Gilmore, S. Caffarri, R. Bassi, T. Golan, D. Kramer and K. K. Niyogi, *J. Biol. Chem.*, 2004, **279**, 22866.
108. N. E. Holt, D. Zigmantas, L. Valkunas, X. P. Li, K. K. Niyogi and G. R. Fleming, *Science*, 2004, **307**, 433.
109. A. A. Pascal, Z. F. Liu, K. Broess, B. van Oort, H. van Amerongen, C. Wang, P. Horton, B. Robert, W. Chang and A. Ruban, *Nature*, 2005, **436**, 134.
110. A. V. Ruban, R. Berera, C. Ilioaia, I. H. M. van Stokkum, J. T. M. Kennis, A. A. Pascal, H. van Amerongen, B. Robert, P. Horton and R. van Grondelle, *Nature*, 2007, **450**, 575.
111. C. A. Shipton and J. Barber, *Proc. Natl. Acad. Sci., U. S. A.*, 1991, **88**, 6691.
112. Y. Kato and W. Sakamoto, *J. Biochem.*, 2009, **146**, 463.
113. R. E. Blankenship, P. L. Cheng, T. P. Causgrove, D. C. Brune, S. Wang, J. U. Choh and J. Wang, *Photochem. Photobiol.*, 1993, **57**, 103.
114. N. U. Frigaard, S. Takaichi, M. Hirota, K. Shimada and K. Matsuura, *Arch. Microbiol.*, 1997, **167**, 343.
115. J. Wen, J. Harada, K. Buyle, K. Yuan, H. Tamiaki, H. Oh-Oka, R. A. Loomis and R. E. Blankenship, *Biochemistry*, 2010, **49**, 5455.
116. Y. Tsukatani, J. Wen, R. E. Blankenship and D. A. Bryant, *Photosynth. Res.*, 2010, **104**, 201.
117. J. Wen, Y. Tsukatani, W. Cui, H. Zhang, M. L. Gross, D. A. Bryant and R. E. Blankenship, *Biochim. Biophys. Acta*, 2011, **1807**, 157.
118. R. M. Morgan-Kiss, L. K. Chan, S. Modla, T. S. Weber, M. Warner, K. J. Czymmek and T. E. Hanson, *Photosynth. Res.*, 2009, **99**, 11.
119. A. G. Yakovlev, A. S. Taisova and Z. G. Fetisova, *FEBS Lett.*, 2002, **512**, 129.
120. J. Pšenčík, A. M. Collins, L. Liljeroos, M. Torkkeli, P. Laurinmäki, H. M. Ansink, T. P. Ikonen, R. E. Serimaa, R. E. Blankenship, R. Tuma and S. J. Butcher, *J. Bacteriol.*, 2009, **191**, 6701.
121. H. Kim, H. Li, J. A. Maresca, D. A. Bryant and S. Savikhin, *Biophys. J.*, 2007, **93**, 192.
122. M. B. Evans, A. M. Hawthornthwaite and R. J. Cogdell, *Biochim. Biophys. Acta*, 1990, **1016**, 71.

123. V. Moulisová, L. Luer, S. Hoseinkhani, T. H. Brotosudarmo, A. M. Collins, G. Lanzani, R. E. Blankenship and R. J. Cogdell, *Biophys. J.*, 2009, **97**, 3019.
124. K. Yasushi, *J. Photochem. Photobiol., B*, 1991, **9**, 265.
125. H. A. Frank and R. J. Cogdell, *Photochem. Photobiol.*, 1996, **63**, 257.
126. H. P. Lang and C. N. Hunter, *Biochem. J.*, 1994, **298**, 197.
127. A. J. Young and H. A. Frank, *J. Photochem. Photobiol. B*, 1996, **36**, 3.
128. G. D. Scholes and G. Rumbles, *Nature Mater.*, 2006, **5**, 920.
129. N. S. Lewis and D. G. Nocera, *Proc. Natl. Acad. Sci., U. S. A.*, 2006, **103**, 15729.
130. B. Coecke, *Contemp. Phys.*, 2010, **51**, 59.
131. J. P. Dowling and G. J. Milburn, *Philos. Trans. R. Soc., A*, 2003, **361**, 1655.

CHAPTER 5

Structure and Function of Photosynthetic Reaction Centres

JOANNA KARGUL AND JAMES BARBER*

Division of Molecular Biosciences, Faculty of Natural Sciences, Imperial College London, London, SW7 2AZ, UK

5.1 Introduction

Photosynthetic systems are constructed around an exquisitely designed basic blueprint. All contain a reaction centre (RC) protein complex coupled to a light-harvesting (LH) system made up of several hundred pigment molecules. Together the RC and LH system form a "photosynthetic unit" as diagrammatically represented in Figure 5.1. The energy of photons captured by the LH system is rapidly transferred to the RC where it powers the movement of electrical charge, in the form of an electron, across a membrane, generating an electrical gradient as well as a chemical potential gradient in the form of 'redox' energy, as indicated in Figure 5.2. There are two types of RCs which facilitate this light induced charge separation known as Type I and Type II.[1–3] Although different in detail there are sufficient similarities to suggest that they have a common evolutionary origin.[3,4] Structural overlays of both Type I and Type II RCs show a remarkably conserved three-dimensional protein structure and cofactor arrangement, despite only minimal amino acid sequence identity.[5] However, they differ in the chemical nature of their cofactors and thermodynamic properties. Despite this, there is a fundamental arrangement of cofactors such as to favour efficient forward electron transfer with minimal back reactions. The rate of electron tunnelling is dominated by the edge-to-edge distances, free energy and reorganisational energy as discussed by Moser *et al.*[6]

RSC Energy and Environment Series No. 5
Molecular Solar Fuels
Edited by Thomas J. Wydrzynski and Warwick Hillier

Published by the Royal Society of Chemistry, www.rsc.org

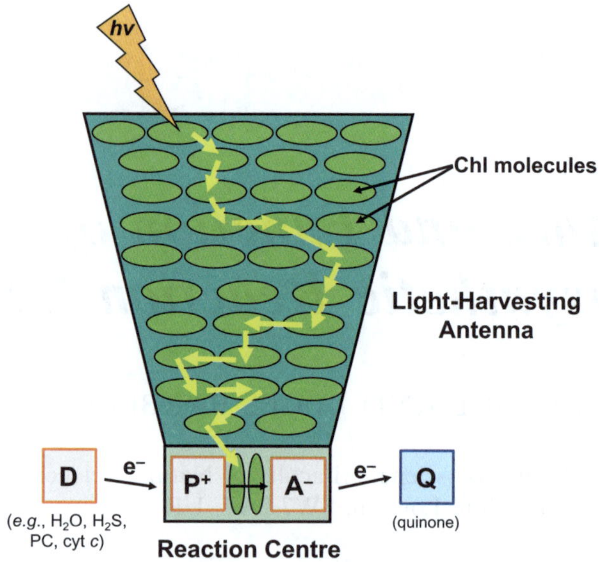

Figure 5.1 Photosynthetic unit. Schematic representation of the photosynthetic unit in which the antenna complex harvests light energy and then channel the excitation energy to the reaction centre. Once the energy reaches the reaction centre, a special pair of (B)Chl molecules (primary electron donor, P) carries out photochemistry and an excited electron is transferred to an adjacent primary electron acceptor "A", which can be either (bacterio)chlorophyll (in Type I RCs) or (bacterio)pheophytin (in Type II RCs). Electrons then are transferred onto the secondary electron acceptor "Q", which is a quinone in both types of RCs. Oxidised "P" is brought to a ground state following its reduction by the electron donor "D", *e.g.*, water, hydrogen sulphide, plastocyanin (PC) or cytochrome (cyt).

Photosynthetic bacteria fall into two classes, distinguished by their RC type: green sulfur bacteria and heliobacteria have Type I RCs, while purple photosynthetic bacteria and green filamentous, non-sulfur photosynthetic bacteria have Type II RCs. Cyanobacteria, algae and higher plants have one RC of each type, known as Photosystem I (PSI) and Photosystem II (PSII), (the former is Type I and the latter Type II), which act in series to drive oxygenic photosynthesis. The primary electron donor, the pigment "P" is always a chlorophyll (Chl) or bacteriochlorophyll (BChl), while the primary electron acceptor "A" can be either (bacterio)chlorophyll (as in Type I RCs) or (bacterio)pheophytin (BPheo) (as in Type II RCs).[7]

Typically the generation of the primary radical pair P^+A^- occurs within a few picoseconds at very high quantum efficiencies.[8,9] The subsequent reduction of P^+ and oxidation of A^- occur on a slower time scale. In both Type I and Type II RCs, A^- is initially re-oxidised by a quinone molecule (Q) (see Figure 5.2). In Type I reaction centres, this quinone is phylloquinone, which typically has a low midpoint redox potential of about $-0.6\,V$ and is strongly

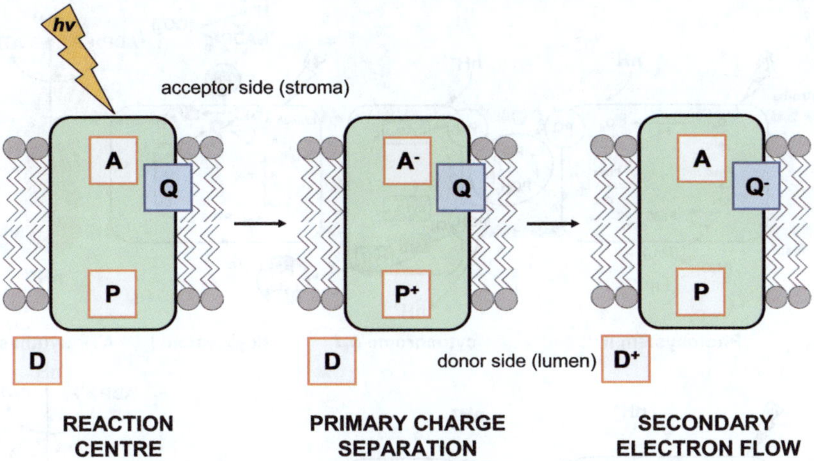

Figure 5.2 Charge separation in photosynthetic reaction centres. The energy of photons captured by the light-harvesting system is rapidly transferred to the reaction centre where it powers the movement of electrical charge in the form of an electron. Primary charge separation between pigment P and A is followed by the secondary electron flow across a membrane, generating an electrical gradient as well as a chemical potential gradient in the form of 'redox' energy.

bound to the RC proteins.[10] In Type II RCs, the quinone electron acceptor (called Q_A) is also tightly bound but has a higher midpoint redox potential of about -0.1 V.[10] The identity of this Q_A quinone depends on the organism: it is plastoquinone in plants and algae, and ubiquinone or menaquinone in purple photosynthetic bacteria and green sulfur bacteria.[7,10] The transfer of electrons from A^- to Q occurs on a timescale of ~200 ps, resulting in the charge transfer state P^+AQ^-.[10] It is the next step in the reductive electron flow that clearly distinguishes Type I and Type II reaction centres. In Type I centres, the electron is passed to an iron–sulfur centre (given the symbol F_X) which is contained within the reaction centre protein, as shown in Figure 5.3. From F_X, the electron proceeds to two further iron–sulfur centres (F_A and F_B) and ultimately to ferredoxin, which as a water-soluble protein, transfers the reducing equivalent away from the membrane. In contrast, Type II reaction centres transfer the electron on Q_A^- to a second loosely bound quinone, Q_B (see Figure 5.3). When Q_B receives a second electron from the next photochemical turnover, it is protonated to form a quinol (QH_2), which diffuses away from the reaction-centre protein into the lipid matrix of the membrane. In plants and algae, Q_B is a plastoquinone, while in purple photosynthetic bacteria it is a ubiquinone or menaquinone, with the latter also present in green sulfur bacteria. These secondary electron transfer events leading to the ejection of reducing equivalents from the reaction centre occur on a timescale stretching from microseconds to milliseconds. Meanwhile, the reduction of P^+ by the electron donor D occurs on the nanosecond to millisecond time scale,

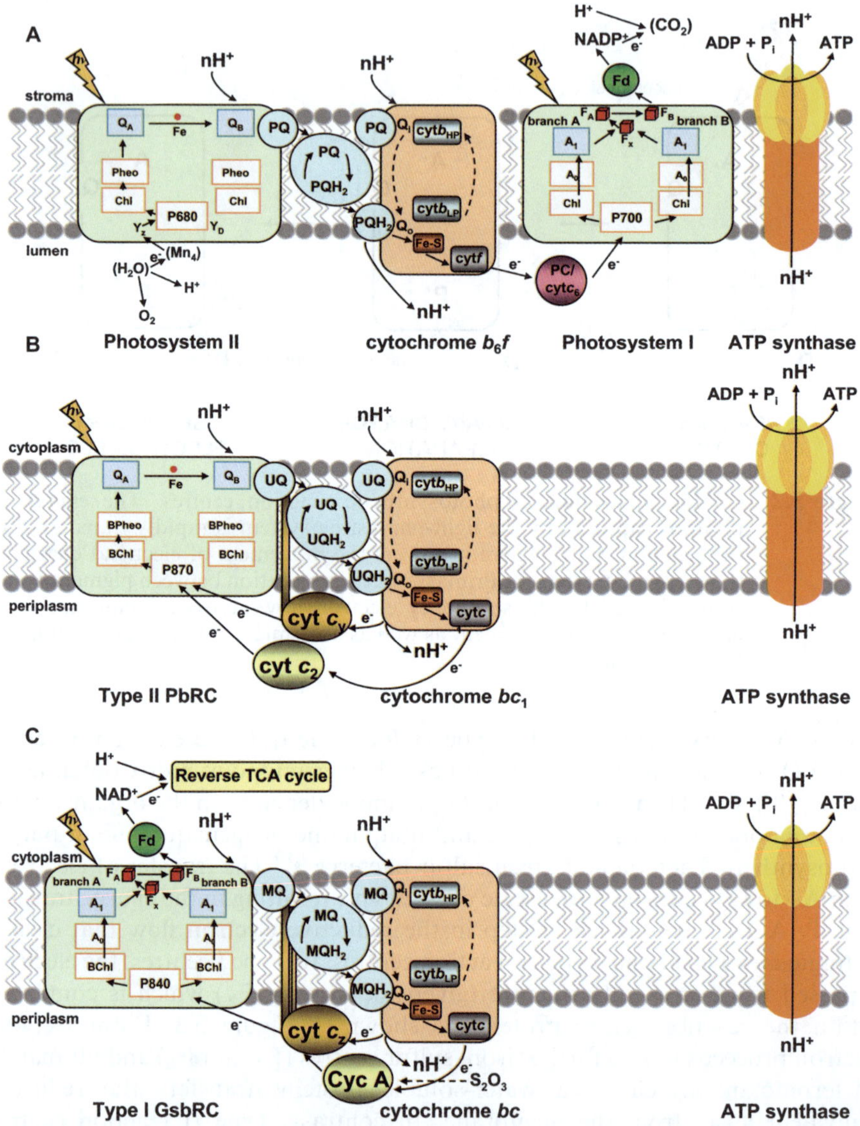

Figure 5.3 Comparison of photosynthetic apparatus and electron transfer pathways between Type II and Type I reaction centres. The electron transport pathways are illustrated for (*A*) cyanobacteria (Type I and Type II RC), (*B*) purple photosynthetic bacteria (PbRC, *e.g.*, *R. sphaeroides*) with Type II RC, and (*C*) green sulfur bacteria (GsbRC) with Type I RC. The N-terminal helices of cytochromes c_y and c_z are anchored into the membrane. Pheo, pheophytin: a primary electron acceptor; Q_A, Q_B: primary and secondary quinone electron acceptors; PQ/PQH_2, plastoquinone/plastoquinol: oxidised and reduced plastoquinone (plastoquinol); UQ/UQH_2, ubiquinone/ubiquinol; MQ/MQH_2, menaquinone/menaquinol; cyt, cytochrome; PC, plastocyanin; A_0, Chl*a*; A_1, phylloquinone; F_x, F_A, F_B: [4Fe-4S] clusters; Fd, ferredoxin; CycA, water-soluble cytochrome *c*-554; TCA, tricarboxylic acid.

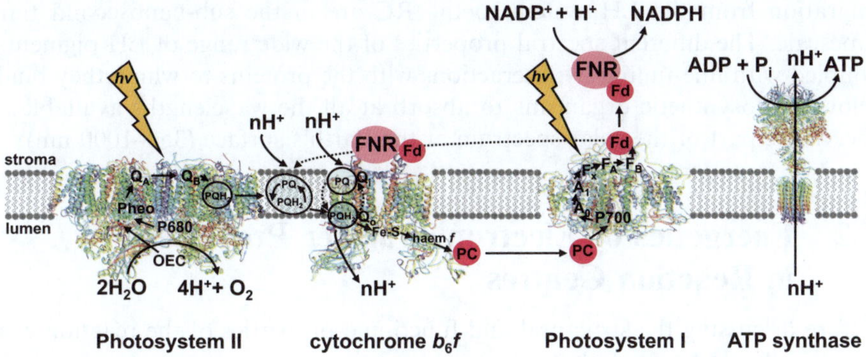

Figure 5.4 Structural representation of the linear photosynthetic electron transfer pathway in oxygenic phototrophs. The diagram shows the X-ray crystal structures of major protein complexes, photosystem II[23], cytochrome b_6f[42] and photosystem I[18], and how they are arranged in the thylakoid membrane to conduct a linear photosynthetic electron transport and generate reducing equivalents and NADPH. Proton gradient concomitantly formed across the thylakoid membrane drives the activity of ATP synthase to produce ATP. Both NADPH and ATP are used for fixation and reduction of CO_2 into carbohydrates. OEC, a Mn_4Ca cluster oxidising substrate water molecules, is indicated with an arrow; Pheo, pheophytin: a primary electron acceptor; Q_A, Q_B: primary and secondary quinone electron acceptors; PQ/PQH_2: plastoquinone/plastoquinol: oxidised and reduced plastoquinone; cyt, cytochrome; PC, plastocyanin; A_0, Chl*a*; A_1, phylloquinone; F_x, F_A, F_B: [4Fe-4S] clusters; Fd, ferredoxin; FNR, ferredoxin-NADP reductase. Figure produced from the pdb files 1S5L[23] (PSII), 1JB0[17] (PSI) and 1VF5[42] (cyt b_6f) using the PyMol molecular graphics system.[146]

depending on conditions.[8] The nature of the electron donor also depends on the particular system. Cytochromes are usually the donors in both Type I and Type II RCs of photosynthetic bacteria, while in plants and algae, water is the electron donor to PSII, and plastocyanin or cytochrome c_6 are electron donors to PSI. In those photosynthetic organisms that evolve O_2 (plants, algae and cyanobacteria), the Type I (PSI) and Type II (PSII) reaction centres are coupled as shown schematically in Figure 5.3 and in more detail in Figure 5.4, so as to use two photons to drive each electron though the system, providing sufficient energy to oxidise water and reduce CO_2. In all cases, the fundamental principle is that energy storage is accomplished by rapidly separating the initial oxidants and reductants of the primary charge transfers so as to avoid wasteful recombination reactions.

The light-harvesting (LH) antenna systems associated with RCs are required to allow them to operate efficiently under relatively low light intensities. The nature of these LH systems varies considerably according to the type of organism, but all function to intercept light and transfer the excitation energy rapidly to the reaction centres. The process is efficient, so the overall transfer rate must be faster than the singlet lifetimes of the pigments, which are typically in the nanosecond time domain. In fact, overall transfer times of energy migration from the LH system to the RC are in the sub-nanosecond time timescale. The different spectral properties of the wide range of LH pigments, coupled with fine-tuning by interactions with the proteins to which they bind, allow photosynthetic organisms to absorb at all the wavelengths available in the visible part of the solar spectrum at the Earth's surface (350–1000 nm).[7]

5.2 Energetics of Electron Transfer Processes in Reaction Centres

Before discussing the structural and functional properties of the reaction centres of different types of photosynthetic organisms, it is necessary to appreciate their specific electron-transfer pathways in terms of redox potentials. Figure 5.5 compares the energetics of Type I and Type II reaction centres of anoxygenic and oxygenic organisms. In purple photosynthetic bacteria (specifically *R. sphaeroides*), the primary electron donor is called P870, because the long wavelength absorption peak of its special pair of bacteriochlorophylls is at 870 nm. Similar notation is used for other primary donors *e.g.*, P800 (heliobacteria), P840 (green sulfur bacteria), P870 (green non-sulfur bacteria), P700 (PSI) and P680 (PSII). The figure emphasizes that the primary electron donors in photosynthetic bacteria and PSI do not generate a redox potential sufficient to oxidise water, while at the same time producing the high energy electrons to reduce CO_2. To achieve the bridging of these two key reactions of oxygenic photosynthesis, PSI is coupled to PSII such that the energy of two photons is used to extract each electron from water and drive it to a redox potential necessary to power CO_2 fixation.

As stated above and shown in Figure 5.3, in green sulfur bacteria and PSI (Type I RCs) the terminal electron acceptor is ferredoxin (Fd) while it is a quinone (Q_B) for purple photosynthetic bacteria and PSII (Type II RCs). In the case of anoxygenic bacteria, some of the reducing potential is used to convert NAD^+ to the NADH needed for CO_2 fixation and some is utilised in cyclic electron flow, whereby the reductant indirectly reduces the oxidised primary donor. This cyclic electron flow involves the cytochrome *bc* complex, which is also embedded in the bacterial chromatophore membrane and which couples the electron flow to the vectorial movement of protons across the membrane (see Figure 5.3B,C). The resulting pH and electrical gradients are then used to drive the conversion of ADP to ATP in accordance with the chemiosmotic mechanism of Peter Mitchell described in 1966,[11] a contribution for which he received the Nobel Prize for Chemistry in 1978.

In the case of oxygenic photosynthetic organisms, PSI and PSII work together to oxidise water and reduce ferredoxin, as depicted in Figures 5.3A and 5.4. PSII functions as the water–plastoquinone oxidoreductase while PSI is a plastocyanin–ferredoxin oxidoreductase. The redox coupling between the two reaction centres is accomplished by a cytochrome *bc* complex rather like that found in anaerobic photosynthetic bacteria but called, for historical reasons, the cytochrome b_6f complex (cyt b_6f). The cytochrome b_6f complex acts as

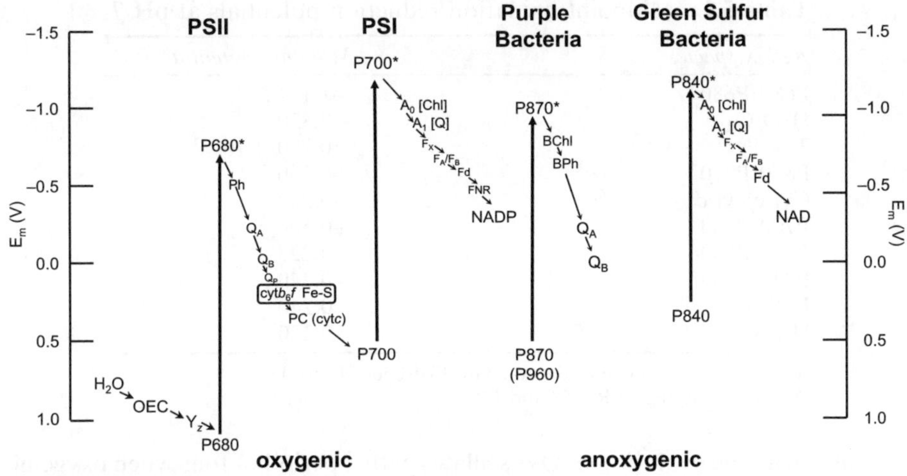

Figure 5.5 Energetics of oxygenic and anoxygenic electron transfer pathways in Type I and Type II reaction centres. Pathways of electron transfer in oxygenic phototrophs (cyanobacteria, algae and plants, all with both types of RCs) and anoxygenic photosynthetic bacteria with only a single type of RC: purple bacteria, with Type II and green sulfur bacteria with Type I RC.

a plastoquinol–plastocyanin oxidoreductase and, like its counterpart in photosynthetic bacteria, facilitates the maintenance of the electrochemical potential gradient of protons across the thylakoid membrane needed to convert ADP to ATP. In oxygenic photosynthesis, the reduced ferredoxin is used to convert nicotinamide adenine dinucleotide phosphate $NADP^+$ to NADPH, which together with ATP is required to convert CO_2 to carbohydrate. Green sulfur bacteria also use reduced ferredoxin in the same way as PSI except that they, like purple photosynthetic bacteria, use non-phosphorylated nicotinamide adenine dinucleotide (NAD^+) rather than its phosphorylated form $NADP^+$.

The similarity in the redox properties and electron transport pathways of the Type I and Type II RCs is evident in Figure 5.5 except for the important fact that $P680^+$ is a much stronger oxidant (with a midpoint potential of ~1.25 V) than $P700^+$, $P840^+$ or $P870^+$ (0.225–0.5 V, see Table 5.1). This is because $P680^+$ must be sufficiently oxidising to remove electrons from water, which is a very stable molecule and difficult to oxidise compared with the substrates oxidised by other reaction centres. This oxidation reaction involves a cluster of 4 Mn atoms and the transfer of electrons and protons from the substrate water molecules is facilitated by a redox-active tyrosine, named Y_Z, positioned between the $(Mn)_4$-cluster and P680, as discussed in detail in Section 5.3.2.3. As the production of dioxygen from water is a four-electron process and a dioxygen molecule is produced at a single PSII reaction centre, the Mn cluster must accumulate four oxidising equivalents.

$$2H_2O \rightarrow O_2 + 4e^- + 4H^+ \quad (1)$$

Table 5.1 Midpoint oxidation/reduction potentials at pH 7.

Redox species	*Midpoint potential*[a,b] *(V)*
$P680/P680^+$	+1.1–1.25
H_2O/O_2	+0.820
Tyr/Tyr^+	+0.700–1.100
$P870/P870^+$	+0.500
Cyt c/cyt c^+	+0.200
$P700/P700^+$	+0.500
$P800/P800^+$	+0.225
$P840/P840^+$	+0.240
H_2S/S^0	−0.240
$H_2/2H^+$	−0.420

[a]Potentials are relative to the standard hydrogen electrode.
[b]Values according to Ref 68 and 145.

This is why the evolution of O_2 oscillates with a period of four when oxygenic organisms are subjected to single turnover flashes of light, as discovered by Pierre Joliot and colleagues in 1969.[12,13] This discovery caused Kok *et al.*[14] to propose the S-state cycle, whereby the absorption of four successive photons drives the series of reactions:

$$S_0 \xrightarrow{h\upsilon} S_1 \xrightarrow{h\upsilon} S_2 \xrightarrow{h\upsilon} S_3 \xrightarrow{h\upsilon} S_4 \qquad (2)$$

When S_4 is formed, dioxygen is released and the cycle resets itself to the S_0-state. Although the precise chemical mechanism of the S-state cycle is unknown, it is generally believed that the two water substrate molecules bind at the S_0-state and that H^+ and electrons are extracted before arriving at the S_4-state. The late Jerry Babcock and colleagues[15] have suggested an attractive 'hydrogen-atom abstraction' hypothesis for the water oxidation mechanism. However, with the emergence of detailed structural information (see Section 5.3.2.3) alternative and compelling reaction schemes are being formulated.

Not surprisingly, the high redox potential of $P680^+$ and the possibility of forming reactive oxygen species during the water-splitting reaction give rise to oxidative damage of the PSII RC. This manifests itself as rapid degradation and regular replacement of the D1 protein of the PSII RC. Plants and other oxygenic organisms have evolved a range of protective strategies that reduce the frequency of photoinduced PSII damage and allow the repair process to cope under normal conditions. The effect of this intrinsic and detrimental property of PSII is, however, observed when organisms are exposed to environmental high light stress, when the rate of repair does not keep up with the rate of damage, resulting in photoinhibition. When this process occurs, the photosynthetic activity and biomass/crop productivity decline.[7]

Whilst $P680^+$ generates the most positive redox potential found in natural systems that is required for oxidisation of water, PSI generates the most negative redox potential found in nature (*e.g.*, reduced F_X); thus, this Type I reaction centre largely determines the global amount of enthalpy achievable in living systems.[16]

5.3 Overview of Structures of Photosynthetic Reaction Centres

5.3.1 General Organisation of Type I and Type II Reaction Centres

Figures 5.3A and 5.4 show how PSI and PSII are functionally coupled with cyt b_6f and ATP synthase in the photosynthetic (thylakoid) membrane. Although a wide range of techniques have been applied to gain structural as well as functional details of photosynthetic RCs, by far the most informative has been X-ray crystallography. There are now medium-to-high resolution crystal structures of PSI[17–20] and PSII[21–26] as well as for the Type II RC of purple photosynthetic bacteria (PbRC).[27–35] In the case of green sulfur bacteria and heliobacteria high resolution structures are lacking, although their close homology with PSI as a Type I RC allows good models to be constructed.[1,36,37]

We now know for certain from X-ray crystallographic studies that all the reaction centres are characterised by a pseudo-2 fold symmetry axis that relates the cofactors and the proteins that bind them. In Type II RCs, this symmetry gives rise to a redox-active branch and an inactive branch, as shown schematically for PSII in Figure 5.3 and, in more detail, in Figure 5.8. Despite intense studies on the purple bacterial RCs it is still not clear how Type II reaction centres are able to differentiate their active and inactive branches. However, this property has distinct advantages when the terminal acceptor (*i.e.* Q_B) requires two electrons to be fully reduced. In Type I reaction centres, where a single, centrally located iron–sulfur centre F_X is a one electron acceptor (as shown for PSI in Figure 5.8), it seems possible that primary charge separation occurs with similar probability in either branch. This must be so in green sulfur bacterial RCs, which are homodimeric[36] while in the case of heterodimeric PSI the situation is less clear.[9,38,39]

The two protein subunits that constitute the RCs of PSII and purple photosynthetic bacteria form a heterodimer in contrast to homodimeric RCs of green sulfur bacteria and heliobacteria. In purple photosynthetic bacteria, the two RC subunits are called L and M, while in PSII the closely related subunits are known as the D1 and D2 proteins, respectively. All four proteins show considerable structural homologies as all have five transmembrane helices (TMs) related to each other in their reaction centres by the same pseudo-2 fold axis that also relates the cofactors.

The two proteins that make up PSI reaction centre, PsaA and PsaB, are also arranged around the pseudo-2 fold axis that relates the cofactors, but in this case they have eleven transmembrane helices, as shown in Figure 5.6. Interestingly, the five transmembrane helices at the C-terminal ends of PsaA and PsaB are arranged in a similar, but not identical, manner as in D1 and D2 (see Figure 5.6).[4,40,41] These structural details will be discussed in the subsequent sections.

The structural arrangements briefly described above have emerged from the Nobel Prize winning X-ray crystallographic studies which began with the elucidation of the structure of a Type II RC isolated from the purple bacterium

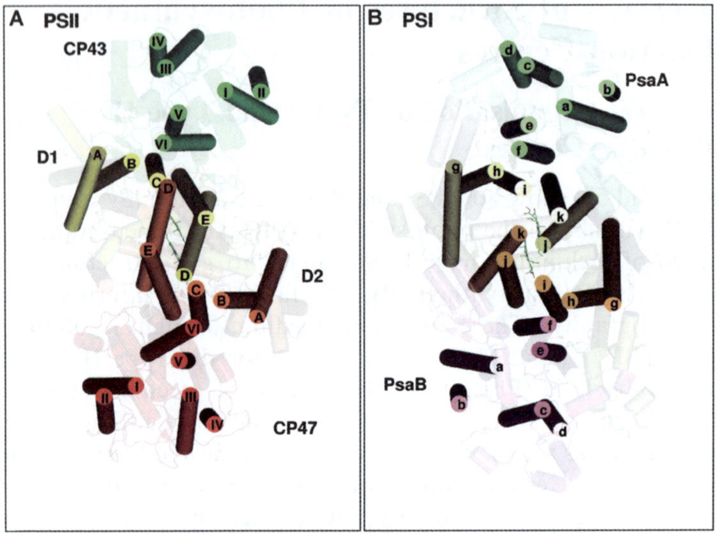

Figure 5.6 Top stromal views derived from X-ray crystallography of (***A***) PSII[23] and (***B***) PSI[17]. The remarkable similarity is clearly visible in the organisation of the transmembrane helices of CP43/CP47 with the 6-TM N-terminal domain of PsaA/PsaB reaction centre of PSI. A striking similarity is also clear in organisation of the transmembrane helices of D1/D2 heterodimer with the C-terminal domains of PsaA/PsaB reaction centre of PSI. The tetrapyrrole head groups of the primary electron donor Chls of PSII (P680) and PSI (P700) (green) are visible in this view from their sides. Figure adapted from Murray *et al.*[143]

Rhodopseudomonas viridis (also known as *Blastochloris viridis*) in the 1980s by Deisenhofer, Huber, Michel and colleagues[28,30] and have advanced to the determination of the structure of PSI at 2.5 Å,[17] and PSII at 3.8-1.9 Å.[21–26] Moreover, the determination of the X-ray structures of the cyt b_6f complex[42,43] and ATP synthase complex of the respiratory membranes,[44] as well as the chloroplast F-ATPase,[45,46] provides a wonderful overall structural model for the whole energy converting machinery of the thylakoid membrane during oxygenic photosynthesis (see Figure 5.4).

5.3.2 Structure and Function of Type II RC of Purple Photosynthetic Bacteria and Photosystem II

5.3.2.1 Protein Subunits and Electron Transfer Cofactors

Bacterial reaction centres convert light energy into a charge-separated state involving an oxidised bacteriochlorophyll (BChl) dimer and reduced secondary quinone Q_B, which serves as a terminal electron and proton acceptor. As mentioned above, the first structures of the Type II reaction centres from two species of purple photosynthetic bacteria, *Rhodopseudomonas viridis* and *Rhodobacter sphaeroides*, were determined by X-ray crystallography in the mid-1980s.[27,29,30] These structures have provided a blueprint that has guided a great deal of research on all of the photosynthetic reaction centres. The purple bacteria RC (PbRC) comprises 3–4 protein subunits, termed L (light), M (medium), H (heavy) and C (cytochrome) (the latter subunit is absent in the *R. sphaeroides* RC),[27–30] as depicted in Figure 5.7A. The L and M subunits and their 14 associated cofactors form the core of PbRC. They both contain five transmembrane helices (TMs) and are related to each other by a twofold pseudo-symmetry axis (see Figure 5.7A). The cofactors include: four BChls (*a* or *b* depending on the species), two BPheo (*a* or *b*), one non-haem iron, two quinones, one carotenoid and four haem groups in the case of *R. viridis* RC. The H subunit is composed of a single TM with its N-terminus near the periplasmic membrane surface and a large C-terminal extramembrane domain oriented towards cytoplasm and binding to the L/M core complex. This subunit has been shown to be essential for stability and maximum efficiency of energy conversion of the PbRC.[47] In *R. viridis*, on the periplasmic side of the membrane the cytochrome subunit with its four covalently bound haem groups is attached to the L/M core complex. The L and M subunits bind the electron transport cofactors that form two branches L and M (also referred to as branch A and B, respectively), each composed of a BChl primary electron donor, an accessory BChl molecule, a bacteriopheophytin (BPheo), and a quinone (ubiquinone in *R. sphaeroides* and menaquinone in *R. viridis*) (see Figure 5.8A). The non-haem iron is located on the symmetry axis that also runs through the primary electron donor BChls. Despite the near symmetry of the two branches of cofactors, electron transfer predominately proceeds unidirectionally along the L-branch[48,49] due to various energetic properties of the different BChls and differences in the amino acid environment of the electron transfer cofactors.[50,51]

The Type II reaction centre of oxygenic photosynthesis, photosystem II (PSII), has an heterodimeric core structure and electron transfer chain arrangement that are similar to those of purple photosynthetic bacteria, but with the addition of the oxygen-evolving complex (OEC) (see Figure 5.7B,C and Figure 5.8A,B). The RC core complex of PS II is a much larger complex than the bacterial reaction centre, containing 16–17 intrinsic and 3 extrinsic subunits[21–25] with approximately 75 cofactors and 20 lipids per monomer, as revealed in the latest atomic X-ray structure of cyanobacterial PSII at 1.9 Å (see Figure 5.7B).[26] Each PSII monomer contains 35 Chl*a* molecules, 11 β-carotenes of all-*trans* type, 2 plastoquinones, 2 haem irons (heam *b* and haem *c*), 1 non-haem iron, 4 Mn atoms (forming a cubane Mn_4Ca-cluster), 3–4 Ca atoms (one of which present within the Mn_4Ca-cluster), 3 Cl^- anions (2 of which are present in the proximity of the Mn_4Ca-cluster), 1 bicarbonate ion (ligand for non-haem iron) and over 20 lipids. Each monomer was found to contain 1300 water molecules organised into 2 layers on the stromal and lumenal surfaces of the complex.[26] Similar to the PbRC, the core of photosystem II has a twofold pseudosymmetry axis which relates the D1, CP47 and PsbI subunits to the D2, CP43 and PsbX polypeptides.[26–25] The core cofactors are divided into two branches, although electron transfer occurs asymmetrically along the D1 branch similar to the L branch in the PbRC, as shown in Figure 5.8. In PSII, five TMs (A–E) from each of the D1 (PsbA) and D2 (PsbD) subunits exhibit structural and functional homology to the 5 TMs of the PbRC L and M subunits (see Figure 5.7C) as well to the 5-TM C-terminal domains of the PSI RC heterodimer (see Figure 5.7D), and provide the scaffold for binding of the electron transfer cofactors. Whilst the organisation of TMs in PbRC and PSII D1/D2 RC is almost identical, the C-terminal domains and the loops between the TMs

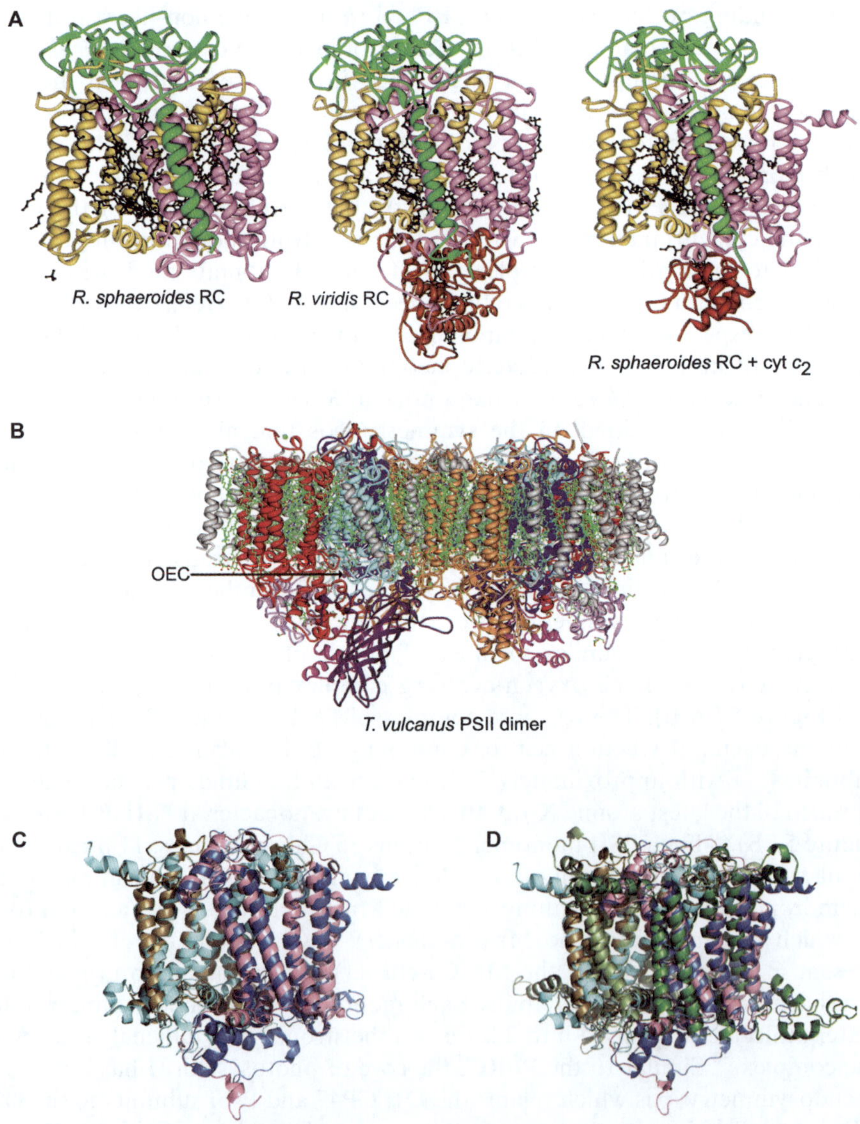

are more extended in the D1 and D2 subunits compared to PbRC L/M subunits, in particular on the lumenal side in the vicinity of the OEC (see Figure 5.7C).

Flanking the opposite sides of the D1/D2 RC heterodimer are the CP43 (PsbC) and CP47 (PsbB) core antenna subunits, each having six transmembrane helices (I to VI) arranged in a picket-like manner similar to the six N-terminal TMs of the PsaA and PsaB core subunits of PSI, as shown in Figure 5.6. CP43 and CP47 bind 13 and 16 light-harvesting Chls, respectively.[24–26] These antenna Chls are distributed in one of the three areas separated by coiled-coil helix dimers of CP43 and CP47 and are coordinated predominantly by the His residues present within the antenna subunits.[24–26] An internal pseudo-threefold symmetry in CP43 and CP47 is disrupted by large lumenal loops between TMs V and VI. This large lumenal domain within CP43 contains two long and three short helices, whereas the corresponding domain of CP47 comprises two long and four short helices as well as three β-sheets, as shown in Figure 5.7B.[23]

The pseudo-symmetric CP43/D1-D2/CP47 core complex is surrounded by 12–13 low molecular weight intrinsic subunits and is capped on the lumenal side by 3 small extrinsic subunits.[23–26] The exact function of the small integral subunits is not clear with the exception of helices of PsbE and PsbF which constitute the α- and β-subunits of cytochrome b_{559} (cyt b_{559})[21–26] involved in photoprotective cyclic electron transport between Q_B, cyt b_{559}, Chl_{ZD2}, β-carotene Car_{D2} and $P680^+$ (see Figure 5.8B).[53] The available X-ray models of

Figure 5.7 Structural comparison of Type II Reaction Centres. ***A***. X-ray crystal structures of the purple bacterial reaction centres (PbRCs) from *R. sphaeroides, R. viridis*, and the reaction centre from *Rhodobacter sphaeroides* with a bound cytochrome c_2. Shown are the backbones of the protein subunits, L (sand), M (pink), H (green), C in *R. viridis* and cytochrome c_2 in *R. sphaeroides* (brown), and the cofactors (black). The views are approximately with the twofold symmetry axis of the protein in the membrane plane. The coordinates are 2J8C[32], 4PRC[28], and 1L9B.[35] ***B***. X-ray crystal structure of the cyanobacterial PSII dimer from *T. vulcanus*. Shown are the helices of several of the protein subunits, D1 (cyan), D2 (blue), CP43 (red), CP47 (orange), and cofactors (green). Helices of the small intrinsic subunits are coloured in grey. The three extrinsic subunits stabilising the OEC (Mn_4Ca-cluster indicated with an arrow) are also shown: PsbO (deep purple), PsbU (magenta) and PsbV (lilac). The view is approximately with the twofold symmetry axis of the protein in the membrane plane. The coordinates are 3ARC.[26] ***C***. Overlay of purple bacteria L/M reaction centre and D1/D2 PSII reaction centre with D1 aligned onto L. The coordinates are 1S5L[23] (PSII) and 2J8C[32] (PbRC). ***D***. Overlay of L/M purple bacteria reaction centre and D1/D2 reaction centre with the C-terminal domain of the PSI RC (composed of 5 TM helices from the C-termini of each PsaA and PsaB subunits). Shown are the backbones of the protein subunits, L (sand), M (pink), D1 (cyan), D2 (blue), C-terminal domain of PsaA (green) and C-terminal domain of PsaB (forest green). The views are approximately with the twofold symmetry axis of the protein in the membrane plane. The coordinates are 2AXT[24] (PSII), 2J8C[32] (PbRC) and 1JB0 (PSI).[17]

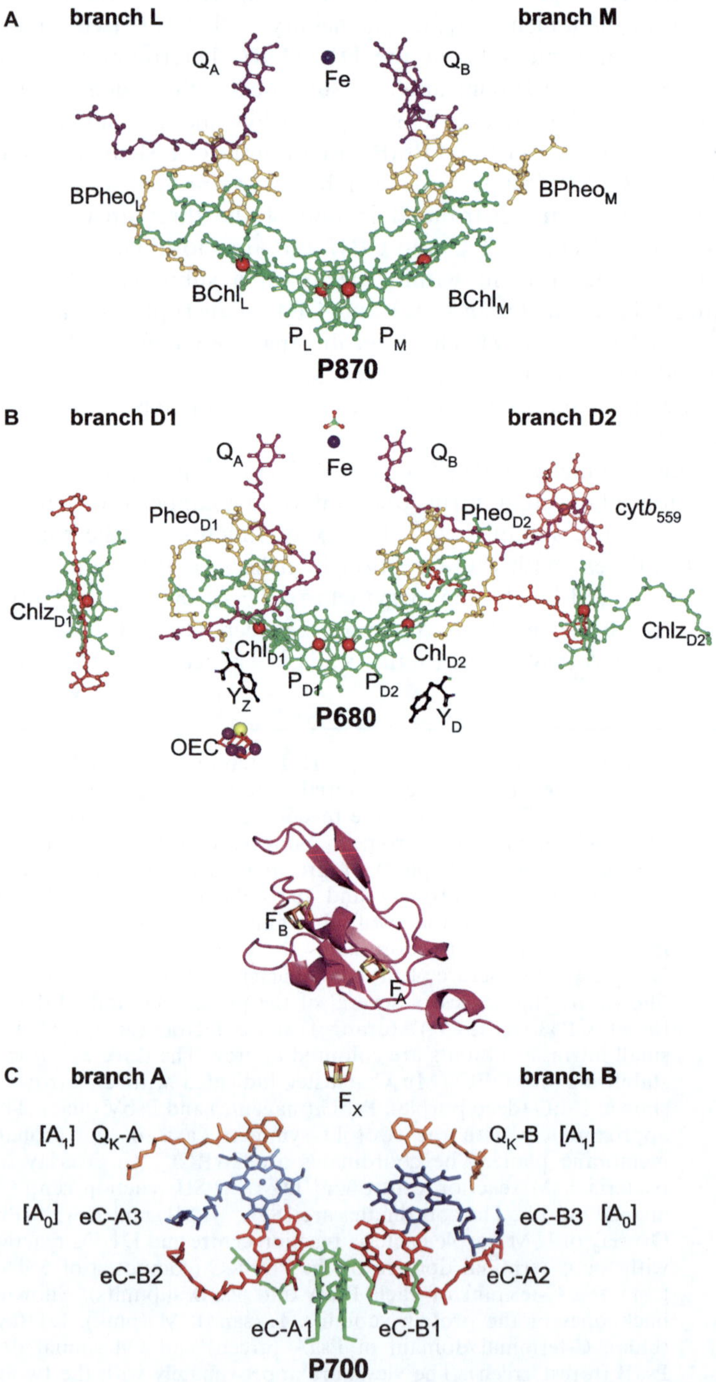
A
branch L
branch M
Q$_A$
Fe
Q$_B$
BPheo$_L$
BPheo$_M$
BChl$_L$
BChl$_M$
P$_L$
P$_M$
P870
B
branch D1
branch D2
Q$_A$
Fe
Q$_B$
Pheo$_{D1}$
Pheo$_{D2}$
cytb_{559}
Chlz$_{D1}$
Chlz$_{D2}$
Chl$_{D1}$
Chl$_{D2}$
P$_{D1}$
P$_{D2}$
Y$_Z$
Y$_D$
P680
OEC
F$_B$
F$_A$
C
branch A
branch B
F$_X$
[A$_1$] Q$_K$-A
Q$_K$-B [A$_1$]
[A$_0$] eC-A3
eC-B3 [A$_0$]
eC-B2
eC-A2
eC-A1
eC-B1
P700

PSII indicate that PsbL, PsbM and PsbT may be involved in dimer formation, whereas PsbI and X are proposed to stabilize binding of the fifth and sixth Chl molecules of the PSII RC, peripheral Chl_{ZD1} and Chl_{ZD2}, that are bound to the D1 and D2 subunits, respectively. Four small subunits (PsbJ, PsbK, ycf12 and PsbZ) clustered near CP43 may facilitate carotenoid binding.[26,23]

Single copies of each of the three extrinsic proteins, PsbO, PsbU and PsbV, are located on the lumenal surface (see Figure 5.7B). In green algae and higher plants, PsbV (cytochrome c_{550} which binds one of the three haem cofactors) and PsbU have been evolutionarily replaced by PsbP and PsbQ. The main role of these extrinsic subunits is to stabilise the oxygen evolving complex (OEC), although none of them provides direct ligands to the OEC. The PsbO protein with the unusual structure of an eight-stranded β-barrel stabilizes the backbone conformation of the AB loop and C-terminus of the D1 protein, which both provide the majority of the direct ligands to the OEC.[23–26] Additionally, a large loop between strands 5 and 6 in PsbO forms a part of a hydrophilic pathway connecting the OEC with the lumenal surface postulated to serve as one of the proton exit channels.[23,54,55]

On the stromal (acceptor) side of the PSII complex, the Q_A binding site is composed of amino acid residues contributed by D2, whereas the Q_B binding pocket is formed by the ligands donated by D1. The overall structure of these sites is quite similar, but not identical, to the quinone binding sites in purple photosynthetic bacteria. The non-haem iron is located midway between the Q_A and Q_B sites and is ligated by 4 His residues donated equally by D1 and D2 and a bicarbonate molecule as the fifth ligand.[23–26] No such bicarbonate ligand is

---

Figure 5.8 Electron transfer cofactors of Type I and Type II Reaction centres. Arrangement of the electron transport cofactors (ETC) in (***A***) PbRC from *R. sphaeroides*, (***B***) PSII from *T. vulcanus*, and (***C***) PSI from *T. elongatus* viewed along the membrane plane. The cofactors of the ETCs are related by the pseudo-symmetry C_2 axis passing through non-haem Fe^{2+} (in ***A*** and ***B***) and F_X (in ***C***), and oriented normal to the paper plane. Cofactors of both Type II ETCs in green (Chl*a*), yellow (Pheo), red (Car), salmon pink (haem), purple (ubiquinone), magenta (plastoquinones), blue sphere (non-haem Fe^{2+} with the bicarbonate ligand), red sphere (Mg ions), yellow sphere (Ca^{2+}), purple sphere (Mn ions) and Tyr_Z/Tyr_D in stick representation. The coordinates are 2J8C[32] (PbRC) and 2ARC[26] (PSII), and 1JB0 (PSI).[17] Nomenclature in (***C***) according to Jordan *et al.*[17] OEC, a Mn_4Ca cluster oxidising substrate water molecules; $P_{L/D1}$, $P_{M/D2}$ (B)Chl and eC-A1/B1: a primary electron donor; (B)$Chl_{L/D1}$, (B)$Chl_{M/D2}$ and eC-A/B-2, accessory Chls; (B)Pheo, (bacterio)pheophytin: a primary electron acceptor in Type II RCs; eC-A/B-3 Chls, A_0 primary electron acceptor in P700; Q_A, Q_B: fixed and mobile quinone secondary electron acceptors in Type II RCs; Q_K-A and Q_K-B, A_1 phylloquinone secondary electron acceptors in P700; F_x, F_A, F_B: [4Fe-4S] clusters, the latter two shown in (***C***) within the backbone of PsaC (pink); Chl_{ZD1} and Chl_{ZD2}, peripheral Chls; cyt, cytochrome. Figure produced from PDB coordinates using the PyMOL molecular graphics system[146] and MBT Protein Workshop viewer.[147]

found in the PbRC which has a glutamate (M-Glu232) instead, although the His ligands are conserved.[28,29]

5.3.2.2 The Primary Electron Donors P870 and P680

The primary electron donor in purple photosynthetic bacteria, P870 in *Rhodobacter sphaeroides* (and P960 in *Rhodopseudomonas viridis*), is a BChl dimer forming a tightly coupled 'special pair' (shown in Figure 5.8A) in which both BChl molecules overlap at the ring A position with a separation of ~3 Å.[28,29] Due to differences in interactions with the immediate protein environment, the energy levels of each BChl are not identical, with 68% of the spin density on the L half of the special pair in wild-type reaction centres.[56] Photosystem II has two Chls of P680 (P_{D1}/P_{D2}) at the equivalent position as a 'special pair' P_L/P_M in PbRC.[21,23,24,26] For both the bacterial and PSII primary donors, the central magnesium atoms are coordinated by histidine residues, L-His173 and M-His202 in PbRC and D1-His198 and D2-His197 in PSII, whose positions are conserved in both structures. While the general orientations of P_{D1} and P_{D2} of P680 are similar to the corresponding BChls of P870, the relative angles between the tetrapyrrole rings are slightly different resulting in a longer separation distance P_{D1}–P_{D2} in PSII (8.2 Å in PSII *versus* 7.5 Å in PbRC).[23,29] These structural differences result in a more monomeric character for $P680^+$ and a weaker excitonic coupling than $P870^+$.[57] It has been demonstrated that hydrogen bonding between the BChl special pair of P870 and nearby L-His ligands affects the electronic structure, thus the redox potential of P870.[58,59] It should be noted that P680 of PSII RC comprises 4 Chl*a* molecules: in addition to P_{D1} and P_{D2}, two accessory Chls, Chl_{D1} and Chl_{D2}, are present with the edge-to-edge distances of 3.3–3.5 Å.[23] The shortest distance, and thus the tightest excitonic coupling, is between P_{D1} and Chl_{D1} (3.3 Å), which may explain the preferential electron transfer on the D1 branch in the PSII RC. In the excited state of P680 the charge is delocalised over the four chlorophylls,[60,61] and Chl_{D1}, which is the chlorophyll molecule closest to the active $Pheo_{D1}$, is thought to be involved in the initial primary charge separation.[62] Unlike accessory BChls of the PbRC, which are ligated by histidines, there are no amino acid ligands for Chl_{D1} and Chl_{D2},[23] with water molecules being identified as ligands in the latest crystal structure.[26]

As discussed in Section 5.2, $P680^+$ must be a very strong oxidant to be able to extract electrons from the chemically stable substrate water molecules. The measure of the ability of a molecule to serve as an oxidant is its redox midpoint potential (E_m). In order to oxidise water, a molecule must have a midpoint potential greater than +0.82 V at pH 7. The midpoint potential of $P680/P680^+$ has been recently estimated as ~1.25 V[63,64] making P680 the strongest most abundant oxidating species known in biology. Conversely, the redox midpoint potential of the PbRC primary electron donors P870 and P960 is ~0.5 V, therefore PbRCs are unable to oxidise water or tyrosine, but only lower potential species such as the haem *c* of cytochrome c_2.[58] Although the overall chemical structures of BChl*a* and Chl*a* are similar, two notable differences exist. Chl*a* has a vinyl group at the 3 position of ring A, whereas BChl*a* has an acetyl group in this place. This structural difference results in a ~0.1–0.2 V gap of midpoint potential values for Chl*a* compared to BChl*a* in organic solvents,[65] and thus may in part contribute to the higher midpoint potential of P680 compared to its bacterial counterpart. However, the overall difference in the midpoint potentials of P870/P960 compared to P680 is a much larger at about 0.75 V. While the specific interactions that give rise to this midpoint potential difference between P680 and P870/P960 have not been identified, this difference most likely arises from the different protein environment of the two primary electron donors.

The other structural difference is that ring B is oxidised in Chl*a* but reduced in BChl*a*, resulting in a single bond rather than the double bond found in Chl*a*. This difference in structure is largely responsible for the absorption of Chl*a* at a shorter wavelength than BChl*a*. By absorbing light at the shorter wavelength of 680 nm compared to 870/960 nm, the energy available for PSII is much greater at 1.82 eV, compared to 1.2 eV for PbRC.[68] Thermodynamic calculations suggest that this amount of light energy is necessary to generate the highly oxidising $P680^+$.[15]

In addition to the chemical structure of the molecules, the properties of electron donors are strongly affected by the protein environment. The hydrogen-bonding interactions between P870 and the nearby amino acid ligands influence the midpoint potential of P870.[58] Removal of the single hydrogen bond found in wild type results in a 0.09 V decrease in potential while the addition of hydrogen bonds increases the potential. This increase is additive, with a 0.26 V increase found when all four possible hydrogen bonds are present. When combined with the replacement of M-Tyr210 with Trp, the midpoint potential of P870 is above 0.8 V, which is a sufficient potential to oxidise tyrosine residues and Mn.[66,67] While these mutants provide the opportunity to investigate the properties of reaction centres that are highly oxidising, they do not represent the evolutionary pathway as Chl*a*, with a vinyl substituent, cannot form the hydrogen bond that is possible to the acetyl group of BChl.[68] It seems more likely that in the case of P680 its midpoint potential is determined by electrostatic interactions between Chls and the surrounding amino acids. The role of such interactions has been investigated by mutagenesis of the charged amino acid residues near P870 in the PbRC. Placing a positively charged residue near P870 should stabilize the ground state relative to the oxidised state and increase the midpoint potential while a negative charge would have the opposite effect. Several such mutants were examined and the midpoint potential of P870 was found to increase or decrease by up to 0.05 V for a single mutation depending on the exact distance, the local value of the dielectric constant and the sign of the charge of the side chain at pH 8.[68,69] These results concerning the influence of the protein environment on the redox properties of P870 support the idea that local electrostatic fields could be key for tuning the redox properties of Chl*a* so that it can drive oxidative photochemistry, namely the oxidation of manganese ions in the catalytic centre for water splitting (see Section 5.3.2.3).

5.3.2.3 Water Oxidation with the Mn_4Ca-cluster and Proton-coupled Electron Transfer

In the 'hydrogen-atom abstraction' model of water oxidation, a critical aspect is the existence of intermediate steps in which electron extraction is coupled to proton transfer.[70] A metal cluster is also required in order to store the four oxidising (electron) equivalents needed for the water splitting reaction during the so-called S-state cycle,[12,14] introduced in Section 5.2. As already stated, in PSII, this cofactor is a Mn cluster consisting of four Mn atoms and one Ca^{2+} cation (Mn_4Ca-cluster) located in the vicinity of the CD lumenal helix of the D1 protein.[23–26] In order to store the oxidising equivalents, the Mn_4Ca-cluster undergoes stepwise increases in its oxidation state, with retention of all intermediate steps, until the final oxidation of water can spontaneously proceed with the fourth oxidation step, the S_4 state according to the S-state cycle model summarised in eqn (2). Electron paramagnetic resonance and time-resolved X-ray absorption analyses established the specific oxidation states of the Mn_4Ca-cluster, with the S_2 state most likely having an electronic state with one Mn(III) and three Mn(IV).[71,72,75,76] By coupling electron and proton transfer, the Mn_4Ca-cluster remains approximately neutral in the S states instead of accumulating positive charges as it becomes increasingly oxidised. Only in the case of the S_1-to-S_2 transition is the electron removal not countered by the loss of a proton.[73,74] The maintenance of almost charge neutrality contributes to the stability of the cluster and the prevention of unwanted back reactions from this highly oxidised cluster due to strong coulombic effects. Thus, by coupling an intrinsically unfavourable process of increasing Mn oxidation state with proton removal, the net yield is an approximately neutral state that is weakly exothermic and hence spontaneous.

The recent 3.8–2.9 Å crystal structures of PSII were invaluable in determining the precise subunit, lipid and cofactor organisation within this complex.[21–25] Unfortunately, due to the intermediate resolution of the diffraction and possibly intricate radiation damage during collection of the X-ray data, the electron densities of the atoms within the Mn_4Ca cluster and its ligating protein/cofactor environment were poorly resolved.[21,23,24] A true breakthrough in revealing the precise geometry of the Mn_4Ca-cluster has recently occurred with the elucidation of the atomic structure of cyanobacterial PSII at 1.9 Å by Shen and colleagues,[26] in which all the individual atoms of the cluster and its ligands were assigned unambiguously. Five oxygen atoms were also identified to serve as oxo bridges linking all five metal atoms of the cluster (see Figure 5.9A). In this atomic model, the Mn_4Ca-cluster is organised as a "distorted chair", with the slightly asymmetric cubane-like structure composed of 3 Mn atoms, 1 Ca^{2+} cation and 4 oxygen atoms in the corners of this structure. The fourth Mn (Mn4, or the 'dangler' Mn) is located outside of the cubane and is linked to Mn1 (with a distance of 5 Å) and Mn3 (distance of 3 Å) atoms of the cubane through two di-μ-oxo bridges involving the O5 and O4 atoms of the cluster (see Figure 5.9A). Interestingly, the general geometry of a cubane has been reported previously by Ferreira *et al.*;[23] however, due to the limited resolution of this

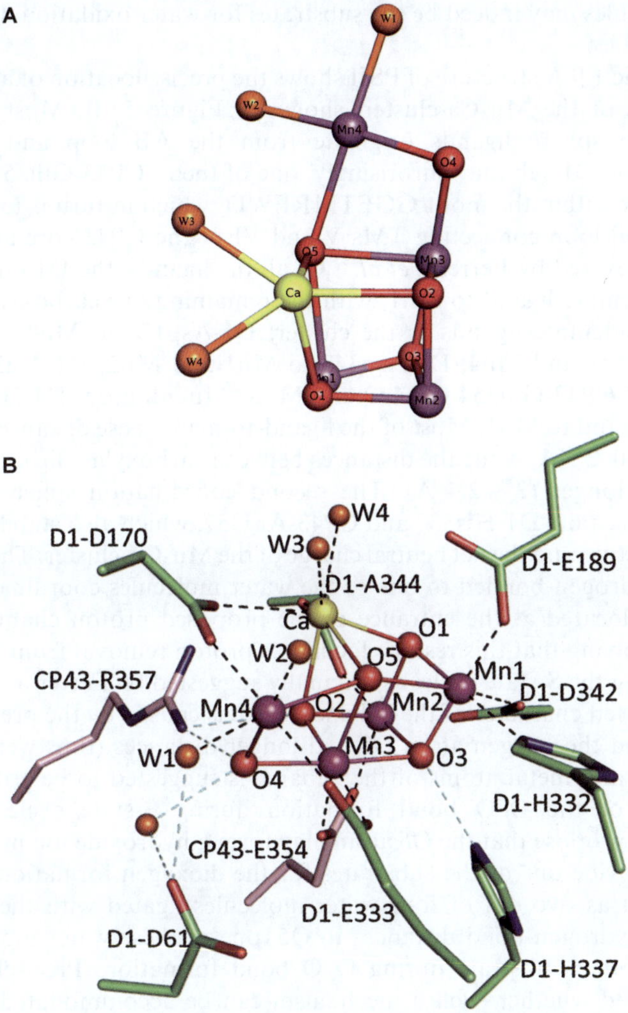

Figure 5.9 Atomic structure of the cyanobacterial Oxygen Evolving Complex at 1.9 Å. ***A***. Geometry of the bonds between metal and oxygen atoms, as well as water (W1-W4, orange) molecules. ***B***. Monophasic view of the Mn_4CaO_5 cluster and its ligand environment. Ligands provided by the D1 and CP43 proteins are coloured in green and lilac, respectively. Manganese, purple; calcium, yellow; oxygen, red; D1, green; CP43, pink. Figure produced from PDB coordinates 3ARC[26] using the PyMol molecular graphics system.[146]

structure the oxo bridges and exact distances between the individual atoms of the cluster could not be determined precisely. Importantly, 4 water molecules were identified in the cluster, 2 of which are ligated to Mn4 (with distances of 2.1 Å and 2.2 Å) and the other two to the calcium ion (with a distance of 2.4 Å) as shown in Figure 5.9B.[26] No other water molecules were found to be associated with the other atoms of the Mn_4Ca-cluster, strongly suggesting that these water molecules may indeed be the substrates for water oxidation during the S-state cycle.

The atomic 1.9 Å structure of PSII shows the precise location of all the amino acid ligands of the Mn_4Ca-cluster, shown in Figure 5.9B. Most of the first coordination sphere ligands originate from the AB loop and C-terminal domain of the D1 subunit. Surprisingly, one of them, CP43-Glu354, is located in a 3_{10} helix within the motif GGETMRFWD, which in turn is located in the large lumenal loop connecting TMs V and VI of the CP43 core antenna subunit, as discovered by Ferreira *et al.*[23] Of all the ligands, the D1-Glu189 serves as a monodentate ligand to Mn1, with the remaining five carboxylate residues serving as bidentate ligands to the cluster: D1-Asp170 to Mn4 and Ca; D1-Glu333 to Mn3 and Mn4; D1-Asp342 to Mn1 and Mn2; D1-Ala344 to Mn2 and Ca; and CP43-Glu354 to Mn2 and Mn3.[26] In addition, D1-His332 serves as a direct ligand to Mn1. Most of the ligand-to-manganese distances are within a range of 2.0–2.3 Å, while the distances between carboxylate ligands and Ca^{2+} are slightly longer (2.3–2.4 Å). The second coordination sphere contains 3 ligands, D1-Asp61, D1-His337 and CP43-Arg357, which also stabilise both the cubane structure and the net neutral charge of the Mn_4Ca-cluster. The D1-Asp61 ligand is hydrogen-bonded to one of the water molecules coordinated to Mn4 and is also located at the entrance of the proposed proton channel,[23,26,54,55] strongly implying that this residue facilitates proton removal from the Mn_4Ca-cluster during the S-state cycle, as originally suggested by Ferreira *et al.*[23]

The distorted character of the cubane, which is caused by the presence of the Ca^{2+} ion and the oxygen atom O5 and longer distances (thus weaker bonds) between O5 and metal atoms of the cubane, is suggested to be critical for the mechanism of the O–O bond formation during S-state cycle. Shen and colleagues[26] propose that the O5 atom may form a hydroxide ion in the S1 state and may provide one of the substrates for the dioxygen formation. They also propose that as two out of four water molecules ligated with the cluster are within the hydrogen-bond distances to O5, one of these water molecules may serve as another substrate during O–O bond formation. Precisely how this will occur and whether such a mechanism can be accommodated with other suggested mechanisms[72,77,78] has yet to be determined. Nevertheless, the presence of a fifth oxo bond to the Mn outside the cubane has been postulated from theoretical considerations.[78]

In photosystem II, a tyrosine residue Y_Z located in the D1 protein (D1-Tyr161 shown in Figure 5.8B) serves as a secondary electron donor to oxidised $P680^+$ during the catalytic water splitting cycle.[79–82] Oxidation of Y_Z proceeds with a concomitant deprotonation of its phenolic group in the process of proton-coupled electron transfer (PCET). As a result, a neutral tyrosyl radical is formed. It is possible that Y_Z is directly involved in the water splitting process not only as an electron donor to $P680^+$ but also as a hydrogen atom abstractor from the manganese complex,[70] although the crystal structures do not give a clear indication of this and no proton is lost during the S1 to S2 transition. The tightly coupled hydrogen-bond network identified in the interfaces between the D1, CP43 and PsbV subunits in the atomic structure of PSII has been proposed as the exit channel for protons generated in PCET *via* Y_Z and the hydrogen-bonded D1-His190.[15,26,70,83] Another redox active tyrosine residue equivalent to Y_Z is Y_D (D2-Tyr160) which is harboured by the D2 protein in the vicinity of P680 (see Figure 5.8B). Similar to Y_Z, Y_D is also oxidised by $P680^+$ to generate a long lived tyrosyl radical;[79,82] however, it is not involved directly in water oxidation but has been proposed to bias the electron transfer reactions to the D1 branch of the PSII RC.[84]

Two Cl^- anions have also been identified in the recent PSII structures, both in the native crystals[26] and crystals substituted with Br^- or I^- for Cl^-.[85,86] As both Cl^- ions are located in the near vicinity of the OEC at the entrance of the hydrogen-bond networks protruding from the OEC to the lumenal side, they are postulated to stabilise the coordination environment of .the Mn_4Ca-cluster as well as facilitate the proton exit from the OEC to the lumenal space.[26,85]

As $P870^+$ in PbRC has a much lower midpoint potential (+0.5 V), it is incapable of utilising a Tyr residue as a secondary electron donor in the wild type bacterial reaction centres, and uses lower potential species, such as the haem *c* of cytochrome c_2, instead.[58]

5.3.3 Structure and Function of Type I Reaction Centres from Anoxygenic Photosynthetic Bacteria and Photosystem I

5.3.3.1 Protein Subunits and Bound Cofactors

Type I reaction centres, which contain iron–sulfur [4Fe-4S] clusters as terminal electron acceptors, are found in anoxygenic green sulfur bacteria and heliobacteria, as well as photosystem I (PSI) of oxygenic phototrophs, such as cyanobacteria, algae and higher plants. The two major breakthroughs in unravelling the molecular organisation of photosystem I (PSI) were the 2.5 Å resolution X-ray structure of trimeric PSI from the cyanobacterium *Thermosynechococcus elongatus*[17] (shown in Figure 5.10A) and the 4.4 Å resolution of the monomeric LHCI-PSI supercomplex from pea (*Pisum sativum*)[18] that was subsequently improved to 3.3 Å (see Figure 5.10B).[19,20] In these structures, a detailed molecular organisation of the protein subunits, inbound cofactors and the antenna systems was dissected.

Cyanobacterial PSI exists predominantly as trimers *in vivo*,[17,87,88] although monomeric form of this complex can also be detected in photosynthetic membranes.[89] In contrast, the eukaryotic PSI exists exclusively as monomers.[90–92] Functionally and structurally PSI is composed of two domains, the core domain, where the bulk of light capturing and charge separation occur, and the external antenna which acts as an additional antenna system that collects solar radiation and transmits the light excitation to the core complex (see Figure 5.10). In higher plants and green algae, the outer antenna is composed of a varying number of the chlorophyll *a*/*b*-binding (C*ab*) Lhca proteins that form the so-called light-harvesting complex I (LHCI).[93,94] In the crystallographic model of a higher plant LHCI-PSI supercomplex, the outer LHCI antenna forms a crescent-like domain composed of 4 Lhca subunits that asymmetrically attach to the core domain on the PsaF side (see Figure 5.10B).[18–20] The higher plant LHCI complex is organised as two heterodimers composed of Lhca1–Lhca4 and Lhca2–Lhca3 subunits[18], although up to 3–4 additional *Cab* subunits may associate with the core domain both in higher plants and green algae.[91,94] In cyanobacteria, the peripheral light-harvesting

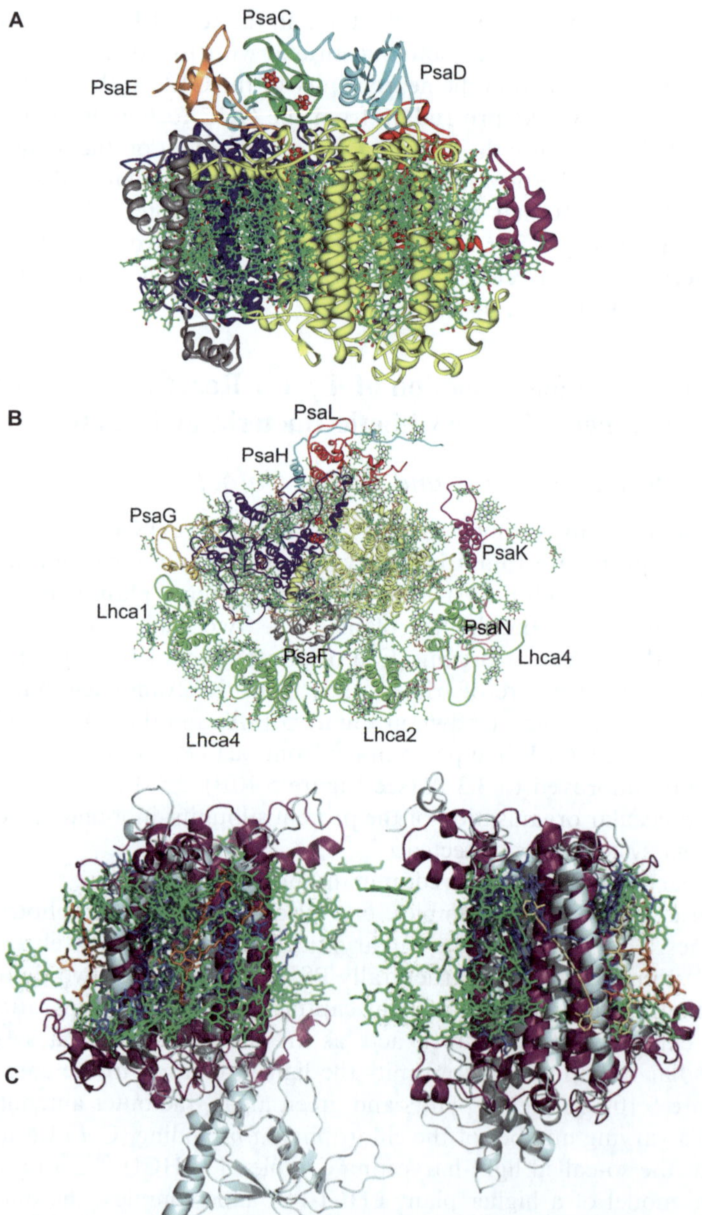

antenna is formed by large water soluble phycobilisome complexes which attach to the PSI core domain on the stromal side of the thylakoid membrane.[95,96] Under some conditions, such as low availability of iron, the peripheral antenna system is dominated by an intrinsic Chl*a*-binding protein complex made up of 18 copies of IsiA subunits.[97–101] This ring of IsiA proteins doubles the size of the absorption cross-section of cyanobacterial PSI.[102,103] A similar antenna system is found in the cyanobacteria known as prochlorophytes which use Chl*a*/Chl*b*-binding Pcb proteins.[104,105] The Pcb and IsiA proteins are structurally related to each other[106–108] and to CP43, CP47 and the N-terminal domains of the PSI RC proteins, PsaA and PsaB. Notably, the Pcb proteins can also act as the light-harvesting antenna for PSII.[105,109]

While the cyanobacterial PSI core complex in the absence of IsiA or Pcb proteins comprises 12 subunits harbouring $\sim$100 chlorophylls,[17] the higher plant PSI is considerably larger containing 18–20 subunits (with the PsaO and PsaP core subunits missing from the X-ray structure) and 173 chlorophyll molecules.[18,20] Most of the additional Chls present in the LHCI-PSI supercomplex over and above those within the cyanobacterial RC core are associated with the four Lhca antenna subunits or belong to the so-called linker and gap Chls facilitating energy transfer within the LHCI antenna and between the LHCI antenna and the RC.[18] The cyanobacterial core domain contains nine intrinsic polytopic protein subunits (PsaA, PsaB, PsaF, PsaI, PsaJ, PsaK, PsaL, PsaM and PsaX) and three extrinsic stromal subunits (PsaC, PsaD and PsaE), as shown in Figure 5.10A. Although the higher plant core domain retains the location and orientation of the electron transfer chain (ETC) cofactors and is

---

Figure 5.10 Structural comparison of Type I Reaction Centres. ***A***. X-ray crystal structure of the cyanobacterial PSI from *T. elongatus*. Shown are the helices of several of the protein subunits, PsaA (yellow), PsaB (blue), PsaL (red), PsaK (magenta), PsaF (grey), PsaC (green), PsaD (cyan), PsaE (orange), and the cofactors (green apart from [4Fe-4S] clusters shown in red). For clarity, some subunits and cofactors are omitted. Only one monomer of the biologically active trimer is shown. The view is approximately with the threefold symmetry axis of the protein in the membrane plane. The coordinates are 1JB0.[17] ***B***. X-ray crystal structure of the higher plant PSI from *P. sativum*. View is from the stromal side. Shown are the helices of several of the protein subunits with colour coding as above, as well as 3 novel core subunits: PsaG (wheat), PsaH (cyan), and PsaN (pink) and the Lhca1-Lhca4 antenna subunits (green). For clarity, the stromal extrinsic subunits and some small core subunits are not shown. The PDB coordinates are 3LW5.[20] ***C***. Overlay of CP43/CP47 inner antenna of PSII with the N-terminal domain of PSI reaction centre (composed of 6 TM helices from the N-termini of each PsaA and PsaB subunits). Shown are the backbones of the protein heterodimers, CP43/CP47 (grey-blue) and PsaA/PsaB N-terminal domains (deep purple). The views are approximately with the twofold symmetry axis of the protein in the membrane plane. Shown are also the pigment cofactors of the N-terminal domain of PSI (Chls, green; carotenoids, orange) and CP43/CP47 inner antenna (Chls, marine; carotenoids, yellow). The coordinates are 3ARC[26] (PSII) and 1JB0 (PSI).[17]

structurally very similar to cyanobacterial PSI, it does not have the X and M subunits. Instead, 4 additional core subunits are present exclusively in higher plants and green algae, namely subunits PsaG, PsaH, PsaN, and PsaO.[90,107]

The largest subunits PsaA and PsaB form a highly conserved heterodimer which binds the majority of the ETC, antenna and lipid cofactors, as depicted in Figure 5.10. Both subunits are conserved with respect to each other in protein sequence and structure. They both contain 11 transmembrane helices that are divided into the N-terminal domain composed of six α-helices (A/B-a to A/B-f) and a C-terminal domain containing five α-helices (A/B-g to A/B-k), as shown in Figure 5.6. The latter form two interlocked semicircles enclosing the ETC cofactors, including 6 Chl*a* molecules, two phylloquinones, and a single [4Fe-4S] iron–sulfur cluster, termed F_X,[17,18] as depicted in Figure 5.8A. Two other [4Fe-4S] clusters (F_A and F_B) are bound to the PsaC subunit located on the stromal side of the complex. This subunit is evolutionarily related to the class of bacterial ferredoxins.[111] Eighteen β-carotenoids were identified in the 3.3 Å structure of the higher plant PSI,[20] whereas 30 β-carotenoids were built into the model of the cyanobacterial PSI.[17] The ETC cofactors are arranged in two symmetric branches along the crystallographic pseudo-C_2 axis. The other intrinsic subunits are peripheral to the PsaA/B heterodimer and coordinate some of the inner antenna cofactors.[17,18]

The majority of the core pigments (87 and 90 Chl*a*, as well as 18 and 22 β-carotenoid molecules in higher plants and cyanobacteria, respectively) act as light-harvesting antenna coordinated mainly by the PsaA/B hererodimer (79 Chls) aided by some small intrinsic core subunits.[17,18,20] The crystal model of plant PSI reveals that the majority of the RC chlorophylls retained the same position and tilting angle as in the cyanobacterial PSI.[17,18] Two types of inner antenna can be distinguished in the cyanobacterial RC: the innermost 'central antenna' of 43 Chls forming a circle around the ETC cofactors at a distance not less than ~18 Å flanked by 2 layers of 'peripheral antenna' of 18 Chls each that are bound predominantly to the N-terminal domains of the PsaA/B here-rodimer.[17] The structures of the N-terminal domains of PsaA and PsaB are equivalent to those of CP43 and CP47, whereas the organisation of the five TM helices of the C-terminal domains of PsaA and PsaB is equivalent to that of the D1 and D2 subunits[4,41] (see Figures 5.6 and 5.10C). The distribution of the peripheral antenna Chls bound to the N-terminal domains of PsaA and PsaB is similar to that of the CP43 and CP47 antenna subunits of PSII. The C-terminal domains of PsaA and PsaB jointly coordinate 25 Chl*a* molecules of the peripheral and central antenna,[17] whereas the D1/D2 heterodimer of PSII coordinates only two peripheral Chls (Chl_{ZD1} and Chl_{ZD2}).[23] The locations of the latter Chls (~24–26 Å from P680)[23,24,26] are not optimized to mediate energy transfer from the CP43/CP47 inner antenna to the RC as is the case for the innermost 'central' antenna system of PSI. This difference probably explains the well-known slow trapping of excitation energy in PSII compared to PSI.[112] However, Chl_{ZD1} and Chl_{ZD2} are photo-oxidisable and seem to be involved in protecting PSII against photoinduced damage.[53]

Interestingly, a single transmembrane helix adjacent to PsaL corresponding to the PsaH subunit has been shown to bind one chlorophyll molecule (see Figure 5.10B).[18] This subunit most likely provides a docking site for the mobile light-harvesting complex II (LHCII) that transiently attaches with the PSI core under conditions favouring excitation of PSII, during the adaptation process of state transitions.[113,114] On the opposite side of the core domain, the 2-TM PsaG subunit provides the contact surface area for the association of the belt-shaped LHCI,[18] as shown in Figure 5.10B.

On the lumenal side of PSI, the docking site for the mobile electron donors, cytochrome c_6 or plastocyanin, is formed by the α-helices of loops A/B-ij containing 2 conserved Trp residues. The plant PsaF subunit contains the N-terminal domain that is longer compared to its cyanobacterial counterpart. This extended domain forms the amphipathic Lys-rich 'helix–loop–helix' motif that enables the strong electrostatic interaction with plastocyanin and, as a consequence, two orders of magnitude faster electron transfer from this mobile electron carrier to $P700^+$ in higher plants compared to cyanobacteria.[18,115,116] On the stromal side, the binding pocket for the mobile electron acceptors ferredoxin or flavodoxin is formed jointly by the PsaC, PsaD and PsaE subunits.[117–121] Notably, the C-terminal domain of the PsaL subunit, which protrudes from the cyanobacterial reaction centre[17] and is responsible for its trimer formation, is absent in plants,[18] in agreement with the exclusively monomeric character of the higher plant PSI.

To date, no high resolution structural information is available for Type I RCs of green sulfur bacteria or heliobacteria with only a low (2 nm) resolution structure of the RC from green sulfur bacterium *Chlorobium tepidum* obtained by electron microscopy coupled with single particle analysis.[122,123] Both types of photosynthetic bacteria contain a homodimeric RC composed of two identical core proteins (PshA in heliobacteria and PscA in green sulfur bacteria).[122–124] Similarly to the RC of PSI, the core proteins PshA and PscA contain 11 TMs, with the 5 C-terminal transmembrane helices binding the bulk of the electron transfer cofactors and 6 N-terminal transmembrane domains serving as the inner light-harvesting antenna.[36,125] The reaction centre of heliobacteria binds 35–40 BChl*g* molecules,[126] whereas 16 BChl*a* and 4 Chl*a* bind to the RC of green sulfur bacteria;[36] a number that is substantially smaller compared to 96 the Chl*a* molecules per PSI core complex of cyanobacteria.[17] The homodimeric character of the heliobacteria and green sulfur bacteria RCs is in contrast to the heterodimeric structure of RCs of purple photosynthetic bacteria and photosystems I and II of oxygenic phototrophs. In the RC of the green sulfur bacteria, reduced compounds such as H_2S or thiosulphate are utilised as electron donors, and ferredoxin and NAD^+ serve are the final electron acceptors. Despite a structural similarity of the green sulfur and heliobacterial RCs, the ETC in the latter organisms varies considerably in that it is based on cyclic electron transfer using only membrane-bound cytochrome *bc* complex containing 2 haem *c* moieties and membrane-anchored cytochrome *c*[127] instead of the linear electron transfer to NAD^+.

5.3.3.2 The Primary Electron Donors and Electron Transfer Chain

In the PSI reaction centre, a cluster of 6 Chl*a* molecules, one of which was suggested to be the 13′-epimer (Chl*a*′),[127,128] functions as the photochemical catalytic centre, and comprises the primary electron donor and primary and secondary electron acceptors. The comparison of the crystallographic structures of the cyanobacterial and higher plant PSI shows similarity in the organisation and specific binding sites of the electron transfer cofactors. This functionally most important part of PSI is formed by six Chl*a* molecules, two phylloquinones and three [4Fe-4S] clusters, as shown in Figure 5.8C. The Chls and phylloquinones are arranged along two branches, A and B, as pairs of pseudo-dimers related by the pseudosymmetry C_2 axis and coordinated to the PsaA and PsaB side chains. The A branch is composed of chlorophylls eC-A1, eC-B2, eC-A3 and phylloquinone Q_K-A, whereas branch B contains chlorophylls eC-B1, eC-A2, eC-B3 and phylloquinone Q_K-B (nomenclature according to Jordan *et al.*[17]). As mentioned earlier, in contrast to unidirectional ETC along branch A in Type II RCs, it is now widely accepted that both the A and B branches in the Type I RC are active in electron transfer, albeit operating with different kinetics, as shown by numerous transient optical spectroscopy measurements coupled to mutagenesis of the ETC coordinating ligands.[9] The rate constants are $35 \times 10^6\ s^{-1}$ and $4.4 \times 10^6\ s^{-1}$ for the electron transfer steps from each phylloquinone to F_X.[129,130] As a point mutation of A-Trp697 ligating the Q_K-A phylloquinone decreases the slower rate, it appears that branch A represents the slower one.[130]

The primary electron donor of PSI RC, the so-called P700, is formed by the 'special' pair of the eC-A1/eC-B1 chlorophylls that are excitonically tightly coupled with a Mg–Mg distance of 6.6 Å.[17] The redox potential of P700 is +0.5 V. The chlorin planes of the P700 chlorophylls are parallel at 3.6 Å interplanar distance and oriented perpendicular to the membrane plane. This organisation varies from the 'special pair' of PbRC P870 BChls where the Mg–Mg distance is larger at 7.6 Å.[28,29] In contrast to the homodimeric bacterial special pair, P700 Chls form a heterodimer, with eC-A1 being the Chl*a*′ 13′-epimer.[17,127,128] The heterodimeric nature of P700 primary electron donor is emphasised by the presence of hydrogen bonds within the binding pocket of eC-A1 and lack of those in the binding site of eC-B1.[17]

The other two Chl*a* pairs are composed of the eC-B2/eC-A2 and eC-A3/eC-B3.[17] The eC-A/B-2 Chls represent the so-called 'accessory' chlorophylls, which, despite being resolved in the X-ray structure, were functionally resolved by spectroscopic methods only recently.[131,132] The eC-A/B-3 Chls are commonly referred to as A_0, which represents the primary electron acceptor reduced in less than ten picoseconds, as observed by the spectroscopic measurements.[9] The A_0 Chls are adjacent to a pair of phylloquinone molecules (often referred to as A_1) termed Q_K-A and Q_K-B according to Jordan *et al.*[17] The A_1 phylloquinones act as secondary electron acceptors that are rapidly reduced to the phyllosemiquinone radicals in ~20–40 ps.[9] All the amino acid ligands coordinating the eC-2 and eC-3 Chl*a* pairs are highly conserved between PsaA and PsaB from cyanobacteria to higher plants, indicating that these interactions are essential for fine-tuning the redox potentials of the ETC cofactors throughout evolution. An interesting exception is the marine cyanobacterium *Acaryochloris marina*, which carries out oxygenic photosynthesis but contains over 95% red-shifted Chl*d* (with an *in vivo* absorbance maximum of ~710 nm) and only trace amounts of Chl*a* in both types of photosystems.[133–135] Chl*d* provides a potential selective advantage because it enables *Acaryochloris* to use infrared light (700–750 nm) that is not absorbed by Chl*a* in the far-red light-enriched habitat of this organism.[136] Consequently, the primary donor of PSI is a dimer of Chl*d* molecules and is known as P740. Similarly, Chl*d* replaces Chl*a* for A_0 and A_1. The situation in PSII is less clear with the four Chls of the primary donor being a mixture of Chl*a* and Chl*d*. However, the long-lived oxidant, corresponding to P_{D1}, seems to be Chl*a*.[137]

The electron acceptor from A_1 is the [4Fe-4S] cluster, the so-called F_X,[138–141] which, similar to P700, is located at the interface of the PsaA/B heterodimer. The F_X cluster (with an unusually low redox potential of −0.7 V) is ligated by 4 strictly conserved Cys residues present in the loop segments A/B-hi.[17] The two terminal [4Fe-4S] iron–sulfur clusters F_A and F_B, which operate in series, are coordinated by Cys residues present within the conserved regions of the stromal PsaC subunit.[17]

In heliobacteria and green sulfur bacteria, the 11-TM homodimeric RC complex coordinates all the cofactors of the ETC arranged in a similar 2-branch way as in PSI. In contrast to the asymmetric character of ETC in PSI, both electron transfer pathways are thought to be fully symmetrical and equally functional in both types of anoxygenic photosynthetic bacteria. In heliobacteria, the primary electron donor, P800, is thought to be a homodimer of BChlg (or *g′* epimer) with a redox potential of +0.225 V, which somewhat lower than that of P700 (+0.5 V) but comparable to that of the BChl*a* P840 homodimer in green sulfur bacteria (+0.240 V).[36,123,142] Membrane-bound cytochrome *c* (PetJ) serves as the electron donor to the oxidised $P800^+$, whereas the soluble form of cytochrome *c* donates electrons to the oxidised $P840^+$ in green sulfur bacteria.[36] As yet, no definite evidence has been obtained for the presence of functional phylloquinone A_1 operating in heliobacteria as the ETC cofactor between A_0 and F_X. The protein harbouring the F_A/F_B clusters is presently unknown in heliobacteria but was identified as PscB in green sulfur bacteria. The latter also provides a binding pocket for ferredoxin.[36,123]

5.4 Evolution of Type I and Type II Reaction Centres

A plethora of evidence, including structural, spectroscopic, thermodynamic, molecular sequence and genetic analyses suggest that all RCs have evolved from a single common ancestor, as they exhibit a similar cofactor and protein structure. As shown in previous sections by structural overlays of Type I and Type II RCs (*e.g.* Figures 5.7C,D and 5.10C), a three-dimensional protein and cofactor structure is remarkably conserved, despite only minimal residual sequence identity.[5] The common design theme in all photosynthetic reaction centres is the charge-separation system that is bound within a protein dimer composed of 5 TMs in each subunit. Each half of the dimer contains a nearly identical set of cofactors. The arrangements of these cofactors exhibit very little variation in all the available crystal structures of RCs, as discussed in previous sections. The Type II RCs in cyanobacteria and eukaryotic phototrophs is flanked by two closely related 6-TM inner antenna complexes (CP43 and CP47) that harbour additional core antenna pigments. These 6-TM complexes are evolutionarily related to the N-terminal part of the Type I RC that forms the core antenna in the Type I RC,[143] as shown by the structural alignment presented in Figure 5.11. In contrast to the cyanobacterial Type II RCs in which 5 TMs and 6 TMs are separate proteins, all known Type I RCs are 11-TM dimers in which the C-terminal 5-TM reaction centre and the N-terminal 6-TM core antenna are present within a single protein. In green sulfur bacteria and heliobacteria, Type I RCs are homodimeric. In contrast, cyanobacterial Type I RCs are heterodimeric, although there appears to be little functional difference between the two electron transport branches that exist within the heterodimer.[132]

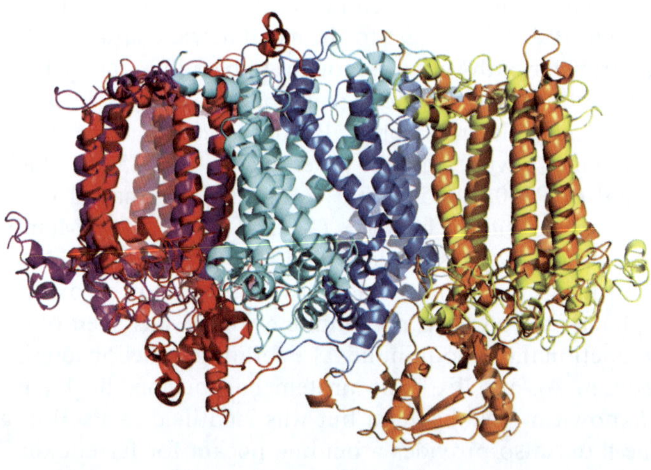

Figure 5.11 Structural alignment of the PSII reaction centre with the inner peripheral antenna domain of PSI. The N-terminal domain of the PSI inner antenna composed of 6 TM helices from the N-termini of each PsaA and PsaB subunits is aligned with the CP43/CP47 antenna complex of PSII. Shown are the backbones of the protein subunits: D1 (cyan), D2 (blue), CP43 (red), CP47 (orange), N-terminal domain of PsaA (deep purple) and N-terminal domain of PsaB (yellow). The views are approximately with the twofold symmetry axis of the protein in the membrane plane. The coordinates are 1S5L[23] (PSII), and 1JB0 (PSI).[17] Figure produced from PDB coordinates using the PyMOL molecular graphics system.[146]

There are good reasons to postulate that the ancestral RC was a homodimer that represented an intermediate between Type I and Type II RCs. It is likely that the heterodimeric forms evolved through gene duplication, modification and divergence events, as depicted in Figure 5.12. However, it is unknown whether that dimer was composed of two 11-TM proteins or two 5-TM proteins, or whether it was a Type I or a Type II RC or possibly an intermediate between both types of RCs.[3] Phylogenetic and structural analyses suggest two independent gene duplication events, one giving rise to the heterodimeric purple RCs and the other to the heterodimeric cyanobacterial RC.[5] This independent origin of two 5-TM RCs suggests that the ancestral RC may indeed have contained 5 TMs, as a split of a hypothetical 11-TM ancestral RC followed by the loss of 6 TMs in purple bacteria and cyanobacteria seems evolutionary unlikely. However, a convergent evolution of a 5-TM core complex with mobile quinones as terminal electron acceptors from an 11-TM ancestral RC is also feasible.[145]

Another important question is how both Type I and Type II RCs evolved in cyanobacteria, while all other anoxygenic photosynthetic prokaryotes contain a single RC of Type I or Type II. The various proposals that have been made to explain this fact can all be divided into either fusion or selective loss scenarios.[144] In the fusion hypothesis, the two types of RCs develop independently in

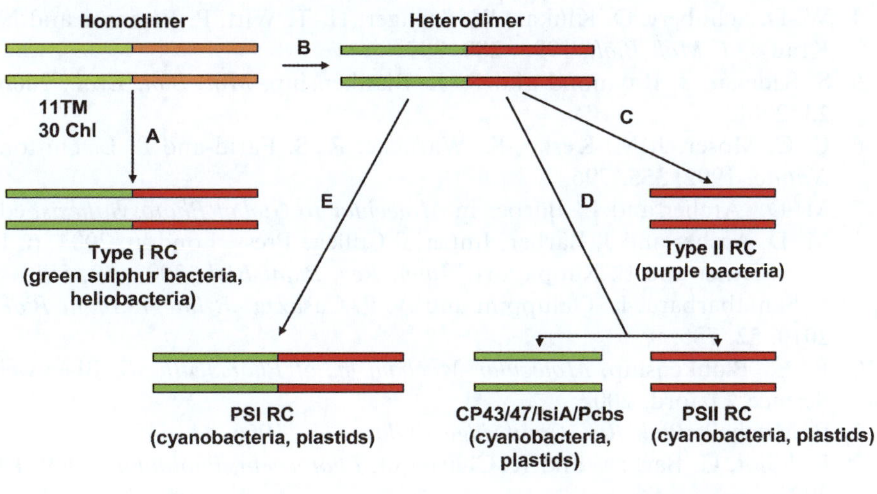

Figure 5.12 Possible evolution of Type I and Type II Reaction Centres. ***A***. A homodimeric 11-TM RC evolves to the current homodimeric Type I RCs found in present green sulfur bacteria and heliobacteria. ***B***. Gene duplication and modification give rise to an ancestral heterodimeric RC. ***C***. The ancestral heterodimeric RC loses 6 N-terminal TMs to form the Type II RC similar to that found in the purple photosynthetic bacteria. ***D***. The ancestral heterodimer splits into a 5-TM RC and a 6-TM light-harvesting subunit to form the cyanobacterial-type PSII. ***E***. The heterodimeric 11-TM RC evolves into the current PSI. Adapted from Nelson and Ben-Shem.[148]

anoxygenic photosynthetic bacteria and are then brought together by a fusion of two organisms, which subsequently developed the ability to oxidise water. In the selective loss hypothesis, the two types of RCs both evolved in an ancestral organism and then loss of one or the other RC gave rise to the organisms with just one RC, while the ability to oxidise water evolved later. The definite scenario remains controversial at present, as discussed by Hohmann-Marriott and Blankenship.[145]

Acknowledgements

We thank Suhail Islam (Imperial College London) for stimulating discussions and invaluable help with structural modelling. We are also grateful to Jian-Ren Shen (Okayama University) for providing a high resolution figure of the manganese cluster. This work is supported by the EPSRC grant EP/F00270X/1.

References

1. P. Heathcote, M. R. Jones and P. K. Fyfe, *Philos. Trans. R. Soc. London*, 2003, **358**, 231.
2. W. Nitschke and A. W. Rutherford, *Trends Biochem. Sci.*, 1991, **16**, 241.
3. R. E. Blankenship, *Plant Physiol.*, 2010, **154**, 434.
4. W.-D. Schubert, O. Klukas, W. Saenger, H.-T. Witt, P. Fromme and N. Krauss, *J. Mol. Biol.*, 1998, **280**, 297.
5. S. Sadekar, J. Raymond and R. E. Blankenship, *Mol. Biol. Evol.*, 2006, **23**, 2001.
6. C. C. Moser, J. M. Keske, K. Warncke, R. S. Farid and P. L. Dutton, *Nature*, 1992, **355**, 796.
7. M. D. Archer and J. Barber in *Molecular to Global Photosynthesis*, ed. M. D. Archer and J. Barber, Imperial College Press, London, 2004, p. 1.
8. B. A. Diner and F. Rappaport, *Annu. Rev. Plant Biol.*, 2002, **53**, 551.
9. S. Santabarbara, L. Galuppini and A. P. Casazza, *J. Integr. Plant Biol.*, 2010, **52**, 735.
10. R. E. Blankenship, *Molecular Mechanisms of Photosynthesis*, Blackwell Science, Oxford, 2002.
11. P. Mitchell, *Biol. Rev. Cambridge Philos. Soc.*, 1966, **41**, 445.
12. P. Joliot, G. Barbieri and R. Charbaud, *Photochem. Photobiol.*, 1969, **10**, 309.
13. P. Joliot, *Photosynth. Res.*, 2002, **76**, 65.
14. B. Kok, B. Forbush and M. McGloin, *Photochem. Photobiol.*, 1970, **11**, 457.
15. C. Tommos and G. T. Babcock, *Biochim. Biophys. Acta*, 2000, **1458**, 199.
16. N. Nelson, *Biochim. Biophys. Acta*, 2010, **1807**, 856–863.
17. P. Jordan, P. Fromme, H. T. Witt, O. Klukas, W. Saenger and N. Krauss, *Nature*, 2001, **411**, 909.
18. A. Ben-Shem, F. Frolow and N. Nelson, *Nature*, 2003, **426**, 630.

19. A. Amunts, O. Drory and N. Nelson, *Nature*, 2007, **447**, 58.
20. A. Amunts, H. Toporik, A. Borovikova and N. Nelson, *J. Biol. Chem.*, 2010, **285**, 3478.
21. A. Zouni, H. T. Witt, J. Kern, P. Fromme, N. Krauss, W. Saenger and P. Orth, *Nature*, 2001, **409**, 739.
22. N. Kamiya and J.-R. Shen, *Proc. Natl. Acad. Sci. U. S. A.*, 2003, **100**, 98.
23. K. N. Ferreira, T. M. Iverson, K. Maghlaoui, J. Barber and S. Iwata, *Science*, 2004, **303**, 1831.
24. B. Loll, J. Kern, W. Saenger, A. Zouni and J. Biesiadka, *Nature*, 2005, **438**, 1040.
25. A. Guskov, J. Kern, A. Gabdulkhakov, M. Broser, A. Zouni and W. Saenger, *Nat. Struct. Mol. Biol.*, 2009, **16**, 334.
26. Y. K. Umena, K. Kawakami, J.-R. Shen and N. Kamiya, *Nature*, 2011, **473**, 55.
27. J. Deisenhofer, O. Epp, K. Miki, R. Huber and H. Michel, *Nature*, 1985, **318**, 618.
28. J. Deisenhofer, O. Epp, I. Sinning and H. Michel, *J. Mol. Biol.*, 1995, **246**, 429.
29. J. P. Allen, G. Fever, T. O. Yeates, H. Komiya and D. C. Rees, *Proc. Natl. Acad. Sci. U. S. A.*, 1987, **84**, 5730.
30. J. P. Allen, G. Feher, T. O. Yeates, H. Komiya and D. C. Rees, *Proc. Natl. Acad. Sci. U. S. A.*, 1987, **84**, 6162.
31. L. Li, D. Mustafi, Q. Fu, V. Tereshko, D. L. Chen, J. D. Tice and R. F. Ismagilov, *Proc. Natl. Acad. Sci. U. S. A.*, 2006, **103**, 19243.
32. J. Koepke, E. M. Krammer, A. R. Klingen, P. Sebban, G. M. Ullmann and G. Fritzsch, *J. Mol. Biol.*, 2007, **371**, 396.
33. C. R. Lancaster, M. V. Bibikova, P. Sabatino, D. Oesterhelt and H. Michel, *J. Biol. Chem.*, 2001, **275**, 39364.
34. A. B. Wöhri, W. Y. Wahlgren, E. Malmerberg, L. C. Johansson, R. Neutze and G. Katona, *Biochemistry*, 2009, **48**, 9831.
35. H. L. Axelrod, E. C. Abresch, M. Y. Okamura, A. P. Yeh, D. C. Rees and G. Feher, *J. Mol. Biol.*, 2002, **319**, 501.
36. G. Hauska, T. Schoedl, H. Remigy and G. Tsiotis, *Biochim. Biophys. Acta*, 2001, **1507**, 260.
37. S. P. Romberger and J. H. Golbeck, *Photosynth. Res.*, 2010, **104**, 333.
38. K. Brettel, *Biochim. Biophys. Acta*, 1997, **1318**, 322.
39. J. H. Golbeck in *Advances in Photosynthesis: The Molecular Biology of Cyanobacteria*, ed. D. A. Bryant, Kluwer Academic, Dordrecht, the Netherlands, 1994, p. 319.
40. K. H. Rhee, E. P. Morris, J. Barber and W. Kühlbrandt, *Nature*, 1998, **396**, 283.
41. J. Barber, J. Nield, E. P. Morris and B. Hankamer, *Trends Biochem. Sci.*, 1999, **24**, 42.
42. G. Kurisu, H. Zhang, J. L. Smith and W. A. Cramer, *Science*, 2003, **302**, 1009.
43. D. Stroebel, Y. Choquet, J. L. Popot and D. Picot, *Nature*, 2003, **426**, 413.

44. J. P. Abrahams, A. G. Leslie, R. Lutter and J. E. Walker, *Nature*, 1994, **370**, 621.
45. G. Groth and E. Pohl, *J. Biol. Chem.*, 2001, **276**, 1345.
46. G. Groth, *Proc. Natl. Acad. Sci. U. S. A.*, 2002, **99**, 3464.
47. R. J. Debus, G. Feher and M. Y. Okamura, *Biochemistry*, 1985, **24**, 2488.
48. C. Kirmaier, D. Holten and W. W. Parson, *Biochim. Biophys. Acta*, 1985, **810**, 33.
49. C. Kirmaier, D. Holten and W. W. Parson, *Biochim. Biophys. Acta*, 1985, **810**, 49.
50. H. Wang, S. Lin, J. P Allen, J. C. Williams, S. Blankert, C. Laser and N. W. Woodbury, *Science*, 2007, **316**, 747.
51. H. Wang, S. Lin, E. Katilius, C. Laser, J. P. Allen, J. C. Williams and N. W. Woodbury, *J. Phys. Chem. B*, 2009, **113**, 818.
52. L. X. Shi and W. P. Schröder, *Biochim. Biophys. Acta*, 2004, **1608**, 75.
53. D. H. Stewart and G. W. Brudvig, *Biochim. Biophys. Acta*, 1998, **1367**, 63.
54. J. W. Murray and J. Barber, *J. Struct. Biol.*, 2007, **159**, 228.
55. J. W. Murray, K. Maghlaoui, J. Kargul, M. Sugiura and J. Barber, *Photosynth. Res.*, 2008, **98**, 523.
56. F. Lendzian, M. Huber, R. A. Isaacson, B. Endeward, M. Plato, B. Bönigk, K. Möbius, W. Lubitz and G. Feher, *Biochim. Biophys. Acta*, 1993, **1183**, 139.
57. L. M. C. Barter, J. R. Durrant and D. R. Klug, *Proc. Natl Acad. Sci. U. S. A.*, 2003, **100**, 946.
58. X. Lin, H. A. Murchison, V. Nagarajan, W. W. Parson, J. P. Allen and J. C. Williams, *Proc. Natl. Acad. Sci. U. S. A.*, 1994, **91**, 10265.
59. F. Müh, F. Lendzian, M. Roy, J. C. Williams, J. P. Allen and W. Lubitz, *J. Phys. Chem. B*, 2002, **106**, 3226.
60. J. R. Durrant, D. R. Klug, S. L. Kwa, R. van Grondelle, G. Porter and J. P. Dekker, *Proc. Natl. Acad. Sci. U. S. A.*, 1995, **92**, 4798.
61. J. P. Dekker and R. van Grondelle, *Photosynth. Res.*, 2000, **63**, 195.
62. B. A. Diner, E. Schlodder, P. J. Nixon, W. J. Coleman, F. Rappaport, J. Lavergne, W. F. Vermaas and D. A. Chisholm, *Biochemistry*, 2001, **40**, 9265.
63. F. Rappaport, M. Guergova-Kuras, P. J. Nixon, B. A. Diner and J. Lavergne, *Biochemistry*, 2002, **41**, 8518.
64. M. Grabolle and H. Dau, *Biochim. Biophys. Acta*, 2005, **1708**, 209.
65. *Chlorophylls and bacteriochlorophylls: biochemistry, biophysics, functions, and applications*, ed. B. Grimm, R. J. Porra, W. Rudiger and H. Scheer, Springer, Dordrecht, the Netherlands, 2006.
66. L. Kálmán, R. LoBrutto, J. P. Allen and J. P. Williams, *Nature*, 1999, **402**, 696.
67. L. Kálmán, J. C. Williams and J. P. Allen in *Photosystem II: The Light-Driven Water Plastoquinone Oxidoreductase*, ed. T. Wydrzynski and K. Satoh, Springer, Dordrecht, The Netherlands, 2005, p. 715.
68. J. P. Allen and J. C. Williams, *Photosynth. Res.*, 2011, **107**, 59.

69. E. T. Johnson, F. Müh, E. Nabedryk, J. C. Williams, J. P. Allen, W. Lubitz, J. Breton and W. W. Parson, *J. Phys. Chem. B*, 2002, **106**, 11859.
70. C. W. Hoganson and G. T. Babcock, *Science*, 1997, **277**, 1953.
71. R. D. Britt, K. A. Campbell, J. M. Peloquin, M. L. Gilchrist, C. P. Aznar, M. M. Dicus, J. Robblee and J. Messinger, *Biochim. Biophys. Acta*, 2004, **1655**, 158.
72. G. W. Brudvig, *Philos. Trans. R. Soc. B*, 2008, **363**, 1211.
73. H. Dau, P. Liebisch and M. Haumann, *Phys. Chem. Chem. Phys.*, 2004, **6**, 4781.
74. J. Clausen, R. J. Debus and W. Junge, *Biochim. Biophys. Acta*, 2004, **1655**, 184.
75. M. Haumann, C. Müller, P. Liebisch, L. Iuzzolino, J. Dittmer, M. Grabolle, T. Neisius, W. Meyer-Klaucke and H. Dau, *Biochemistry*, 2005, **44**, 1894.
76. M. Haumann, P. Liebisch, C. Müller, M. Barra, M. Grabolle and H. Dau, *Science*, 2005, **310**, 1019.
77. E. M. Sproviero, J. A. Gascón, J. P. McEvoy, G. W. Brudvig and V. S. Batista, *Coord. Chem. Rev.*, 2008, **252**, 395.
78. P. E. M. Siegbahn, *J .Photochem. Photobiol. B*, 2011, **104**, 94–99.
79. B. A. Barry and G. T. Babcock, *Proc. Natl. Acad. Sci. U. S. A.*, 1987, **84**, 7099.
80. R. J. Debus, B. A. Barry, I. Sithole, G. T. Babcock and L. McIntosh, *Biochemistry*, 1988, **27**, 9071.
81. J. G. Metz, P. J. Nixon, M. Rögner, G. W. Brudvig and B. A. Diner, *Biochemistry*, 1989, **28**, 6960.
82. B. A. Diner and R. D. Britt in *Photosystem II: The Light-Driven Water: Plastoquinone Oxido-Reductase in Photosynthesis*, ed. T. Wydrzynski and K. Satoh, Springer, Dordrecht, The Netherlands, 2005, p. 206.
83. A. M. Hays, I. R. Vassiliev, J. H. Golbeck and R. J. Debus, *Biochemistry*, 1999, **38**, 11851.
84. P. Faller, R. J. Debus, K. Brettel, M. Sugiura, A. W. Rutherford and A. Boussac, *Proc. Natl Acad. Sci. U. S. A.*, 2001, **98**, 14368.
85. J. W. Murray, K. Maghlaoui, J. Kargul, N. Ishida, T.-L. Lai, A. W. Rutherford, M. Sugiura, A. Boussac and J. Barber, *Energy Environ. Sci.*, 2008, **1**, 161.
86. K. K. Kawakami, Y. Umena, Y. N. Kamiya and J.-R. Shen, *Proc. Natl. Acad. Sci. U. S. A.*, 2009, **106**, 8567.
87. J. Kruip, D. Bald, E. J. Boekema and M. Rögner, *Photosynth. Res.*, 1994, **40**, 279.
88. N. V. Karapetyan, D. Dorra, G. Schweitzer, I. N. Bezsmertnaya and A. R. Holzwarth, *Biochemistry*, 1999, **36**, 13830.
89. E. El-Mohsnawy, M. J. Kopczak, E. Schlodder, M. Nowaczyk, H. E. Meyer, B. Warscheid, N. V. Karapetyan and M. Rögner, *Biochemistry*, 2010, **49**, 4740.
90. H. V. Scheller, P. E. Jensen, A. Haldrup, C. Lunde and J. Knoetzel, *Biochim. Biophys. Acta*, 2001, **1507**, 41.

91. J. Kargul, J. Nield and J. Barber, *J. Biol. Chem.*, 2003, **278**, 16135.
92. M. Germano, A. E. Yakushevska, W. Keegstra, H. J. van Gorkom, J. P. Dekker and E. J. Boekema, *FEBS Lett.*, 2002, **525**, 121.
93. R. Croce, T. Morosinotto, S. Castelletti, J. Breton and R. Bassi, *Biochim. Biophys. Acta*, 2002, **1556**, 29.
94. A. Busch and M. Hippler, *Biochim. Biophys. Acta*, 2010, **1807**, 864–877.
95. A. N. Glazer, *Annu. Rev. Biophys. Biophys. Chem.*, 1985, **14**, 47.
96. R. MacColl, *Biochim. Biophys. Acta*, 2004, **1657**, 73.
97. T. S. Bibby, J. Nield and J. Barber, *Nature*, 2001, **412**, 743.
98. E. J. Boekema, A. Hifney, A. E. Yakushevska, M. Piotrowski, W. Keegstra, S. Berry, K. P. Michel, E. K. Pistorius and J. Kruip, *Nature*, 2001, **412**, 745.
99. T. S. Bibby, J. Nield and J. Barber, *J. Biol. Chem.*, 2001, **22**, 22.
100. J. Nield, E. P. Morris, T. S. Bibby and J. Barber, *Biochemistry*, 2003, **42**, 3180.
101. R. Kouril, A. A. Arteni, J. Lax, N. Yeremenko, S. D′Haene, M. Rögner, H. C. Matthijs, J. P. Dekker and E. J. Boekema, *FEBS Lett.*, 2005, **579**, 3253.
102. A. N. Melkozernov, T. S. Bibby, S. Lin, J. Barber and R. E. Blankenship, *Biochemistry*, 2003, **42**, 3893.
103. E. G. Andrizhiyevskaya, T. M. Schwabe, M. Germano, S. D′Haene, J. Kruip and R. van Grondelle, *Biochim. Biophys. Acta*, 2002, **1556**, 265.
104. T. S. Bibby, J. Nield, F. Partensky and J. Barber, *Nature*, 2001, **413**, 590.
105. T. S. Bibby, J. Nield, M. Chen, A. W. D. Larkum and J. Barber, *Proc. Natl. Acad. Sci. U. S. A.*, 2003, **100**, 9050.
106. J. La Roche, G. W. M. van der Staay, F. Partensky, A. Ducret, R. Aebersold, R. Li, S. S. Golden, R. G. Hiller, P. M. Wrench, A. W. D. Larkum and B. R. Green, *Proc. Natl Acad. Sci. U. S. A.*, 1996, **93**, 15244.
107. B. R. Green and D. G. Durnfold, *Annu. Rev. Plant Physiol. Plant Mol. Biol.*, 1996, **47**, 685.
108. J. Barber, J. Nield, J. Duncan and T. S. Bibby, in *Photosystem I: The Light-Driven Plastocyanin: Ferredoxin Oxidoreductase*, ed. J. H. Golbeck, Springer, Dordrecht, the Netherlands, 2006, p. 99.
109. T. S. Bibby, I. Mary, J. Nield, F. Partensky and J. Barber, *Nature*, 2003, **424**, 1051.
110. J. Knoetzel, A. Mant, A. Haldrup, P. E. Jensen and H. V. Scheller, *FEBS Lett.*, 2002, **510**, 145.
111. M. L. Antonkine, P. Jordan, P. Fromme, N. Krauss, J. H. Golbeck and D. Stehlik, *J. Mol. Biol.*, 2003, **327**, 671.
112. F. L. de Weerd, I. H. van Stokkum, H. van Amerongen, J. P. Dekker and R. van Grondelle, *Biophys. J.*, 2002, **82**, 1586.
113. C. Lunde, P. E. Jensen, A. Haldrup, J. Knoetzel and H. V. Scheller, *Nature*, 2000, **408**, 613.
114. J. Kargul and J. Barber, *FEBS J.*, 2008, **275**, 1056.
115. M. Hippler, J. Reichert, M. Sutter, E. Zak, L. Altschmied, U. Schroer, R. G. Herrmann and W. Haehnel, *EMBO J.*, 1996, **15**, 6374.

116. M. Hippler, F. Drepper, W. Haehnel and J.-D. Rochaix, *Proc. Natl. Acad. Sci. U. S. A.*, 1998, **95**, 7339.
117. N. Fischer, N. M. Hippler, M. P. Sétif, J.-P. Jacquot and J.-D. Rochaix, *EMBO J.*, 1998, **17**, 849.
118. N. Fischer, P. Sétif and J.-D. Rochaix, *J. Biol. Chem.*, 1999, **274**, 23333.
119. H. Bottin, J. Hanley and B. Lagoutte, *Biochem. Biophys. Res. Commun.*, 2001, **287**, 833.
120. S. V. Ruffle, A. O. Mustafa, A. Kitmitto, A. Holzenburg and R. C. Ford, *J. Biol. Chem.*, 2000, **275**, 36250.
121. P. Sétif, N. Fischer, B. Lagoutte, H. Bottin and J.-D. Rochaix, *Biochim. Biophys. Acta*, 2002, **1555**, 204.
122. U. Liebl, M. Mockensturm-Wilson, J. T. Trost, D. C. Brune, R. E. Blankenship and W. Vermass, *Proc. Natl. Acad. Sci. U. S. A.*, 1993, **90**, 7124.
123. H. Remigy, H. Stahlberg, D. Fotiadis, B. Wolpensinger, A. Engel, G. Hauska and G. Tsiotis, *J. Mol. Biol.*, 1999, **290**, 851.
124. G. Tsiotis, C. Hager-Braun, B. Wolpensinger, A. Engel and G. Hauska, *Biochim. Biophys. Acta*, 1997, **1322**, 163.
125. H. Oh-oka, *Photochem. Photobiol.*, 2007, **83**, 177.
126. S. Neerken and J. Amesz, *Biochim. Biophys. Acta*, 2001, **1507**, 278.
127. D. M. Kramer, B. Schoepp, U. Liebl and W. Nitschke, *Biochemistry*, 1997, **36**, 4203.
128. T. Watanabe, M. Kobayashi, A. Hongu, M. Nakazato and T. Hiyama, *FEBS Lett.*, 1985, **235**, 252.
129. P. Joliot and A. Joliot, *Biochemistry*, 1999, **38**, 11130.
130. Guergova-Kuras, B. Boudreaux, A. Joliot, P. Joliot and K. Redding, *Proc. Natl. Acad. Sci. U. S. A.*, 2001, **98**, 4437.
131. C. Slavov, M. Ballottari, T. Morosinotto, R. Bassi and A. R. Holzwarth, *Biophys. J.*, 2008, **94**, 3601.
132. M. G. Müller, C. Slavov, R. Luthra, K. E. Redding and A. R. Holzwarth, *Proc. Natl. Acad. Sci. U. S. A.*, 2010, **107**, 4123.
133. H. Miyashita, H. Ikemoto, N. Kurano, K. Adachi, M. Chihara and S. Miyachi, *Nature*, 1996, **383**, 402.
134. Q. Hu, H. Miyashita, I. Iwasaki, N. Kurano, S. Miyachi, M. Iwaki and S. Itoh, *Proc. Natl. Acad. Sci. U. S. A.*, 1998, **95**, 13319.
135. T. Tomo, T. Okubo, S. Akimoto, M. Yokono, H. Miyashita, T. Tsuchiya, T. Noguchi and M. Mimuro, *Proc. Natl. Acad. Sci. U. S. A.*, 2007, **104**, 7283.
136. M. Chen and R. E. Blankenship, *Trends Plant Sci.*, 2011, doi:10.1016/j.tplants.2011.03.011.
137. E. Schlodder, M. Cetin, H. J. Eckert, F. J. Schmitt, J. Barber and A. Telfer, *Biochim. Biophys. Acta*, 2007, **1767**, 589.
138. M. C. V. Evans and R. Cammack, *Biochem. Biophys. Res. Commun.*, 1975, **63**, 187.
139. M. C. V. Evans, C. K. Sihra and R. Cammack, *Biochem. J.*, 1976, **158**, 71.

140. J. H. Golbeck, A. E. McDermott, W. K. Jones and D. M. Kurtz, *Biochim. Biophys. Acta*, 1987, **891**, 94.
141. A. E. McDermott, V. K. Yachandra, R. D. Guiles, K. Sauer, M. P. Klein, K. G. Parrett and J. H. Golbeck, *Biochemistry*, 1989, **28**, 8056.
142. M. Büttner, D.-L. Xie, H. Nelson, W. Pinther, G. Hauska and N. Nelson, *Biochim. Biophys. Acta*, 1992, **1101**, 154.
143. J. W. Murray, J. Duncan and J. Barber, *Trends Plant Sci.*, 2006, **11**, 152.
144. R. E. Blankenship, S. Sadekar and J. Raymond in *Evolution of Aquatic Photoautotrophs*, ed. P. Falkowski and A. N. Knoll, Academic Press, New York, 2007, p. 21.
145. M. F. Hohmann-Marriott and R. E. Blankenship, *Annu. Rev. Plant Biol.*, 2011, **62**, 515.
146. W. L. DeLano, *The PyMOL Molecular Graphics System*, 2002, DeLano Scientific, San Carlos, CA, USA. http://www.pymol.org.
147. J. L. Moreland, A. Gramada, O. V. Buzko, Q. Zhang and P. E. Bourne, *BMC Bioinf.*, 2005, **6**, 21.
148. N. Nelson and A. Ben-Shem, *BioEssays*, 2005, **27**, 914.

CHAPTER 6

Photophysics of Photosynthetic Reaction Centres

THOMAS RENGER

Institut für Theoretische Physik, Johannes Kepler Universität, Abteilung Theoretische Biophysik, Altenberger Str. 69, 4040 Linz, Austria

6.1 Introduction

In photosynthesis, the energy of sun light absorbed in light-harvesting antenna complexes is transferred with almost 100 percent quantum efficiency to the photosynthetic reaction centre where it is converted into chemical energy by transmembrane electron transfer (ET) reactions. Whereas there is a large diversity in the structure of light-harvesting antennae in different photosynthetic organisms (see Chapter 4), the pigment organization in the reaction centres (RCs) are rather similar (see Figure 6.1 and Chapter 5). Most likely the requirements for performing redox chemistry are more stringent than those for excitation energy transfer. A special feature of all RCs is the arrangement of cofactors in two nearly symmetric branches that start from a central (bacterio) chlorophyll dimer known as the special pair (P_A–P_B and P_{D1}–P_{D2} in Figure 6.1). Two so-called accessory (bacterio) chlorophylls (B_A and B_B, Chl_{D1} and Chl_{D2}, and Acc_A and Acc_B in Figure 6.1) connect the special pair with two (bacterio) pheophytins in the case of type II RCs (H_A and H_B, $Pheo_{D1}$ and $Pheo_{D2}$) and two chlorophyll *a* molecules (A_{0A} and A_{0B}) in the case of type I RCs. The chain of redox-cofactors is continued in type II RCs by two quinones Q_A and Q_B (upper and middle part in Figure 6.1) and in type I by two

RSC Energy and Environment Series No. 5
Molecular Solar Fuels
Edited by Thomas J. Wydrzynski and Warwick Hillier

Published by the Royal Society of Chemistry, www.rsc.org

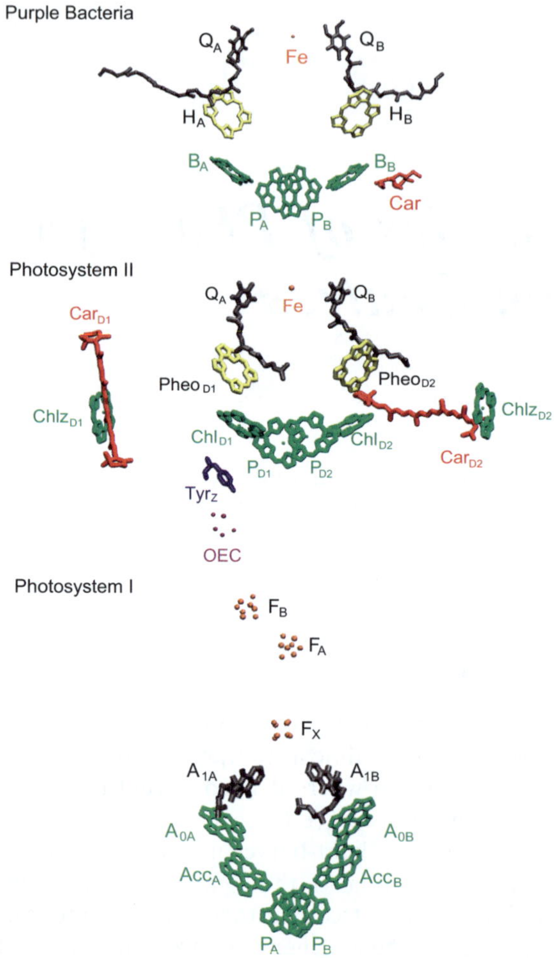

Figure 6.1 Arrangement of cofactors in the reaction centres of purple bacteria[59] (upper), photosystem II[39] (middle) and photosystem I[72] (lower) of cyanobacteria and higher plants. (Bacterio)chlorophylls are shown in green, (bacterio) pheophytins in yellow, carotenoids (Car) in red, quinones in black, non-heme irons and iron–sulfur clusters in orange. The substituents of chlorine pigments have been truncated for clarity.

phylloquinones A_{1A} and A_{1B} and a single chain of three iron–sulfur complexes F_X, F_A, and F_B (lower part in Figure 6.1).

After the excitation energy has arrived in the reaction centre, primary electron transfer creates a charged separated state where the electron is localized at the primary acceptor $Pheo_{D1}(B_A)$ in the case of type II RCs and at A_0 in the case of type I RCs, and the hole stabilizes at the special pair. Whereas in type II RCs only one branch is used for ET (for a review see ref. 1 and 2), in type I RCs most likely both branches are ET active,[3–7] *i.e.* both A_{0A} and A_{0B} act as primary electron acceptors. However, the A-branch is used more frequently.[3–9]

From the two $A_{0'S}$ the electrons are transferred *via* the two phyloquinones A_{1S} and the iron–sulfur clusters F_X, F_A and F_B to the soluble ferredoxin protein, and ultimately NAD^+ or $NADP^+$ is reduced. In this way light energy is stored in a chemical form as a reduction equivalent.

In type two RCs the ET continues from $Pheo_{D1}(B_A)$ *via* the plastoquinone (ubiquinone) Q_A to the secondary plastoquinone (ubiquinone) Q_B. Q_B^- gets protonated and accepts a second electron from the next excitation; and after it gets doubly protonated leaves the binding pocket as plastoquinol (ubiquinol), QH_2. The latter diffuses to the cytb6f (cytbc1) complex, where the reducing power of QH_2 is used to pump protons through the photosynthetic membrane, generating a proton motif force (pmf) that is utilized by the ATPase to generate ATP. In purple bacteria, the electrons extracted from QH_2 are shuttled back to the reaction centre (RC), *via* ET from the cytbc1 protein, to a soluble cytochrome c1 complex, thus completing the ET cycle.

In higher plants, the electrons from QH_2 are shuttled further *via* a soluble plastocyanine to photosystem I (PSI) which works in series with photosystem II (PSII). The electron hole at the special pair in PSII is filled by electron transfer from the oxygen evolving centre (OEC) *via* an intervening Tyr_Z residue. The OEC uses water as an external electron source. The water is split into electrons (shuttled to PSI as described above), molecular oxygen, the basis of oxygenic life on earth, and protons which contribute to the pmf utilized by the ATPase.

Since purple bacteria do not evolve oxygen, their RC does not contain the manganese cluster that serves as the OEC in PSII. Otherwise, the two type II RCs show a striking similarity in arrangement of their redox-active cofactors. We will in the following concentrate on the photophysics of the acceptor site of the type II RCs; for details on the water splitting reaction in PSII see Chapters 5 and 7.

Two important factors for the high photochemical efficiency of a RC are an efficient transfer of excitation energy from the antenna complexes to the RC and the generation of a stable charge separated state in the latter. A key question is: what are the competing side reactions and how are these suppressed? In order to address these questions, we will start with a discussion of some basic photophysical principles of excitation energy and charge transfer reactions in proteins. In a next step, we study the photophysics of the type II RCs of purple bacteria and PSII (cyanobacteria and higher plants) and identify common building principles and important differences. The chapter is finished with a discussion of some unsolved problems.

6.2 Basic Photophysical Principles

The two fundamental interactions that determine the transfer of excitation energy (excitons) and electrons in proteins are the electronic coupling between pigments and the pigment–protein coupling. The electron–vibrational coupling between the electrons of the pigments and the nuclei of the protein allows the protein to dissipate the excess energy of electrons and excitons. The electrostatic pigment–protein coupling is an important factor for the energetics of the reactions.

6.2.1 Trapping of Excitation Energy

In photosynthetic antenna complexes, usually the electronic coupling and the electron–vibrational coupling are of equal magnitude. In this case, the excited states in the antenna (and RC) complexes are delocalized in certain domains of strongly coupled pigments, and the wavefunction $|M_a>$ of the delocalized state (exciton state) M_a in domain *a*

$$|M_a> = \sum_{m_a} c_{m_a}^{(M_a)} |m_a>, \tag{1}$$

is given as a superposition of localized excited states $|m_a>$ in which pigment m_a is excited and all other pigments in domain *a* are in the ground state. The coefficients $c_{m_a}^{(M_a)}$ and the excitation energies E_{Ma} of the delocalized states are obtained by diagonalizing the exciton matrix that contains the site energies of the pigments (defined as the optical transition energies at which the pigments would absorb in their respective binding site in the protein if they were not coupled to other pigments) in the diagonal and the excitonic coupling in the off-diagonal, *i.e.* the electronic coupling between different localized excited states $|m_a>$. In a rough approximation, this coupling can be approximated by the dipole–dipole interaction between the optical transition dipole moments of the pigments (for recent review see ref. 10).

To prevent the oxidation of antenna pigments by pigments in the RC, the interpigment distances between pigments in the antenna and those in the RC are large (Figure 6.2). Hence, the electronic coupling between antenna and RC pigments is much weaker than that between pigments in the same subcomplex and the electron–vibrational coupling. The latter destroys any coherence (delocalization) between excited states in different subcomplexes and restricts the delocalization of excited states to pigments within certain domains. A detailed analysis of the excitonic and the exciton-vibrational coupling in the core complexes of PSI_{82} and PSII_{66} reveals that these core antennae also contain different exciton domains, that is, not all antenna pigments are strongly coupled. In contrast, the LH1-pigments form a single exciton domain.

The excitation energy transfer between the antenna and the RC occurs between excited states that are delocalized in different domains. The transfer of excitation energy between two such delocalised states, say $|M_a>$ and $|N_b>$ is described by generalized Förster theory, giving a rate constant[10,11]

$$k_{M_a \to N_b} = 2\pi \frac{|V_{M_a N_b}|^2}{\hbar^2} \int_{-\infty}^{\infty} d\omega D'_{M_a}(\omega) D_{N_b}(\omega), \tag{2}$$

with the inter-domain excitonic coupling

$$V_{M_a N_b} = \sum_{m_a, n_b} c_{m_a}^{(M_a)} c_{n_b}^{(N_b)} V_{m_a n_b}, \tag{3}$$

that contains the coefficients of the intra-domain exciton states $c_{k_c}^{(K_c)}$ and the inter-domain excitonic couplings $V_{m_a n_b}$ between pigment *ma* in domain *a* and *nb* in domain *b*. If the distance between the centres of the aggregates is large

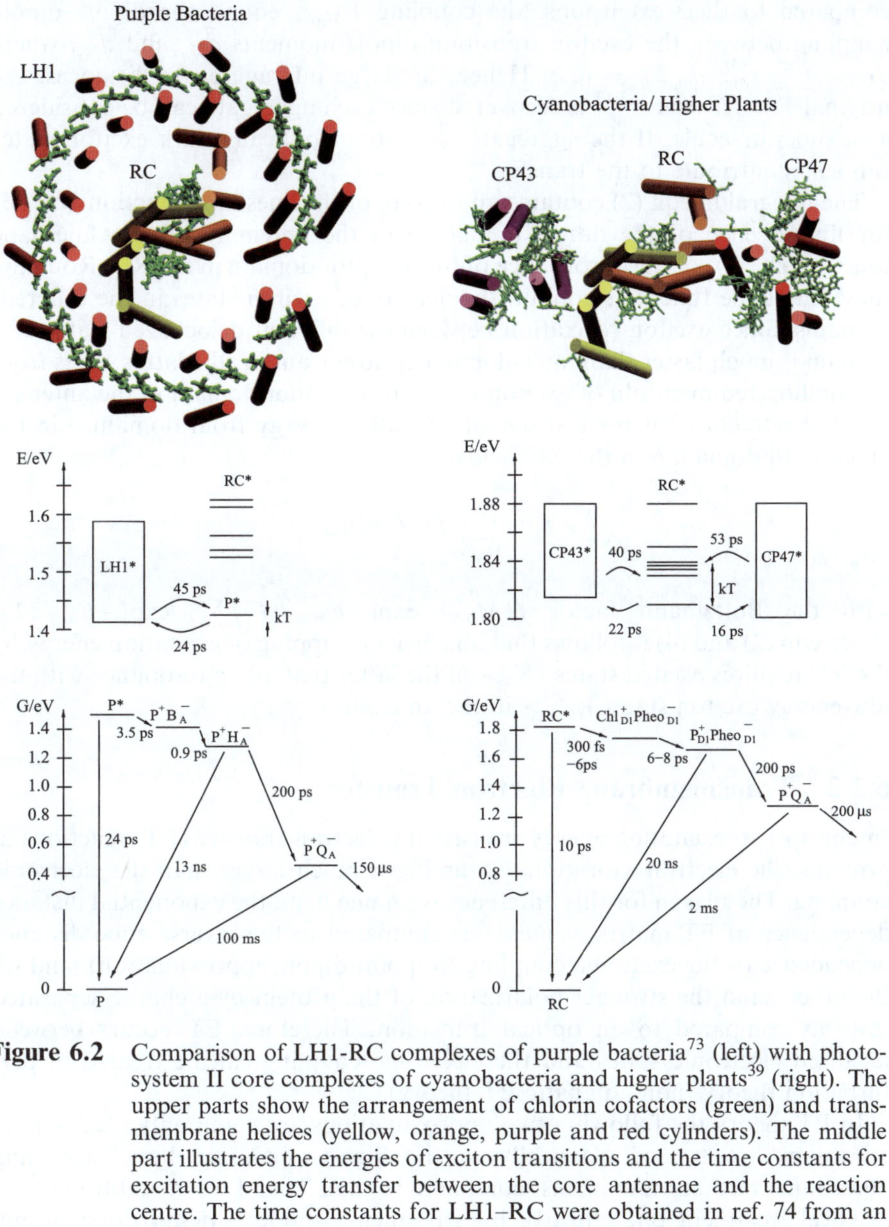

Figure 6.2 Comparison of LH1-RC complexes of purple bacteria[73] (left) with photosystem II core complexes of cyanobacteria and higher plants[39] (right). The upper parts show the arrangement of chlorin cofactors (green) and transmembrane helices (yellow, orange, purple and red cylinders). The middle part illustrates the energies of exciton transitions and the time constants for excitation energy transfer between the core antennae and the reaction centre. The time constants for LH1–RC were obtained in ref. 74 from an analysis of time-resolved experiments and those in PSII were calculated in ref. 66. The lower parts contain free energy differences (relative to the electronic ground state P) of the excited state and the charge separated states and time constants for electron transfer, charge recombination and lifetime of the excited state due to back excitation energy transfer to the antennae (obtained from the time constants in the middle part). The free energies and time constants for the reaction centre of purple bacteria were obtained from ref. 75–77 and ref. 29,32,52,75,78, respectively and those for photosystem II from ref. 77,79,80 and ref. 43,44,66,77, respectively.

compared to their extensions, the coupling $V_{M_a N_b}$ equals the dipole–dipole coupling between the exciton transition dipole moments $\vec{d}_{M_a}$ and $\vec{d}_{N_b}$, where $\vec{d}_{M_c} = \sum_{m_c} c_{m_c}^{(M_c)} \vec{d}_{m_c}$, $c = a,b$. Hence, for large interaggregate distances, the original Förster result[12,13] is recovered since each aggregate can be considered as a supermolecule. If the aggregates are closer, optically dark exciton states can also contribute to the transfer.[14]

The integral in eqn (2) contains an overlap of the lineshape function $D'_{M_a}(\omega)$ for fluorescence of the donor domain a in the antenna and the lineshape function $D_{N_b}(\omega)$ of the absorbance of the acceptor domain in the RC. Roughly, these lineshape functions contain the density of exciton states in the different domains. Since exciton relaxation between the different delocalized states of a domain is much faster than inter-domain exciton transfer, the latter starts from an equilibrated manifold of exciton states of the donor domain in the antenna, and the rate constant for transfer of excitation energy from domain a in the antenna to domain b in the RC reads

$$k_{a \to b} = \sum_{M_a, N_b} f(M_a) k_{M_a \to N_b} \tag{4}$$

with the Boltzmann factor $f(M_a) = \exp(-E_{Ma}/kT) / \sum_{K_c} \exp(-E_{Kc}/kT)$. From eqn (2) and (4) it follows that an efficient trapping of excitation energy by the RC requires excited states $|N_b>$ of the latter that are in resonance with the low-energy exciton states $|M_a>$ in the antenna.

6.2.2 Transmembrane Electron Transfer

In contrast to excitation energy transfer, in electron transfer (ET) reactions in proteins, the electron–vibrational coupling is much larger than the electronic coupling. The reason for this difference is on one hand the exponential distance dependence of ET matrix elements as compared to the inverse cube distance dependence of the excitonic coupling (in point-dipole approximation), and on the other hand the stronger polarization of the protein by a charge separated state as compared to an optical transition. Therefore, ET occurs between localized electronic states and the electronic coupling can be treated in perturbation theory (non-adiabatic ET theory).

In ET theory the following two approximations are commonly used: (i) the vibrations are treated classically in the high-temperature limit, assuming vibrational relaxation to be fast compared to the ET, and (ii) in addition to the classical treatment one effective high frequency mode is described quantum mechanically. The classical result for the rate constant[15]

$$k = \frac{2\pi}{\hbar} |V_{\mathrm{ET}}|^2 \frac{1}{(4\pi\lambda kT)^{1/2}} e^{-\frac{(\Delta G^0 + \lambda)^2}{4\lambda kT}} \tag{5}$$

contains three essential parameters, namely, the charge transfer coupling V_{ET}, the standard free energy difference ΔG^0 and the reorganization energy λ. λ is the

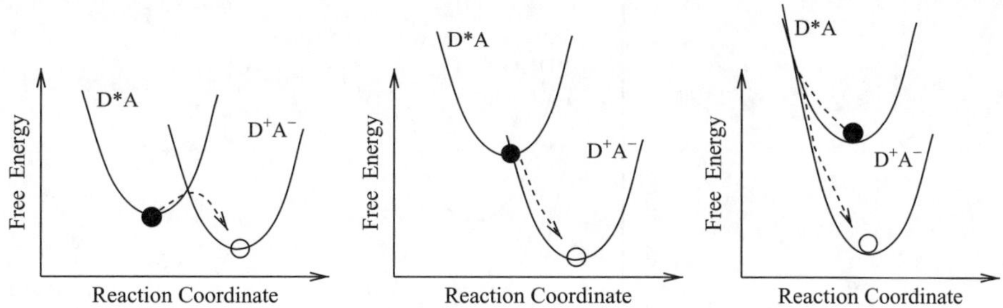

Figure 6.3 Free energy surfaces of the initial and final state of an electron transfer reaction $D^*A \rightarrow D^+A^-$ in the three different regimes of Marcus theory: normal region ($-\Delta G^0 < \lambda$, left), activationless region ($-\Delta G^0 = \lambda$, centre) and inverted region ($-\Delta G^0 > \lambda$, right).

energy that is released when the nuclei relax to the new equilibrium position in response to the charge separated state that is created by the ET. It contains contributions from both intra- and intermolecular rearrangements of the nuclei of the pigments and the protein environment, respectively. Three different regimes of the transfer can be distinguished: the normal, the activationless and the inverted region. The related free energy surfaces for a reaction $D^*A \rightarrow D^+A^-$ are illustrated in Figure 6.3. In the normal region, where $-\Delta G^0 < \lambda$, there is a barrier between the two minima of free energy surfaces and the crossing point has to be reached by thermal fluctuations of nuclei, before the ET can occur. In the activationless region ($-\Delta G^0 = \lambda$), no such thermal activation is necessary and the rate has its maximum value. In the inverted region, a lowering of the free energy of the final state ($-\Delta G^0 > \lambda$) leads to a decrease of the rate constant, because of a barrier created between the two free energy surfaces.

If parts of the vibrations are treated quantum mechanically, the nuclei can tunnel through that barrier and in this way the inverted effect becomes weaker. If one vibrational mode v is treated quantum mechanically, and its energy is sufficiently high, *i.e.*, $\hbar\omega_\nu >> kT$, whereas the remaining modes are treated classically, the rate constant is obtained as[16,17]

$$k_q = 2\pi \frac{|V_{\rm ET}|^2}{\hbar} \frac{1}{(4\pi\lambda kT)^{1/2}} e^{-S_\nu} \sum_{N=0}^{\infty} \frac{(S_\nu)^N}{N!} e^{-\frac{(\Delta G^0 + \lambda + N\hbar\omega_\nu)^2}{4\lambda kT}}. \qquad (6)$$

By comparing this equation with the purely classical result in eqn (5) it is seen that k_q can be expressed as

$$k_q = e^{-S_\nu} \sum_N \frac{(S_\nu)^N}{N!} k_N \qquad (7)$$

where k_N is the classical rate constant obtained by replacing the ΔG^0 in eqn (5) by $\Delta G^0 + N\hbar\omega_\nu$. An illustration of this result in the inverted region of electron

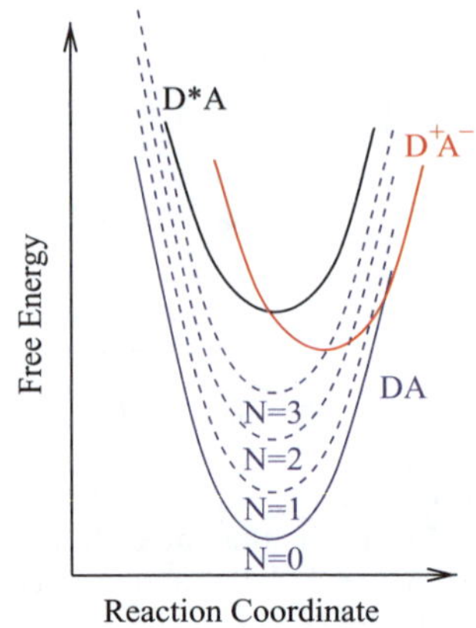

Figure 6.4 Free energy surfaces of ground state DA, excited state D^*A and charge separated state D^+A^-. The electron transfer reaction $D^*A \rightarrow D^+A^-$ occurs in the activationless region, whereas the charge recombination reaction $D^+A^- \rightarrow DA$ occurs in the inverted region of electron transfer. The dashed blue curves illustrate nuclear tunneling, as explained in the text.

transfer is given for a charge recombination reaction $D^+A^- \rightarrow DA$ in Figure 6.4. As can be seen, by excitation of high-frequency vibrational quanta in the ground state DA, the regime of ET changes from inverted for $N = 0$ to activationless for $N = 3$ excited quanta, in the example in Figure 6.4. The relative weight $e^{-S_\nu}(S_\nu)^N/N!$ of the different transitions in the sum in eqn (7) is the Franck–Condon factor that describes the overlap of the vibrational ground state wavefunction of the charge separated state D^+A^- with the different excited vibrational wavefunctions of the electronic ground state DA. The non-zero overlap allows the nuclei to tunnel through the barrier.

How can the parameters of the classical Marcus theory in eqn (5) be determined from experimental data? The driving force ΔG^0 may be estimated from redox potentials, and λ and V_{ET} may be obtained by studying the ET reactions at different driving forces, ΔG^0. From eqn (5) it is seen that the rate constant takes its maximum value if the reorganization energy λ equals ΔG^0. In this case, k becomes $|V_{ET}|^2/\sqrt{4\pi\lambda kT}$. The maximum k and the reorganization energy λ then determine the electronic coupling V_{ET}. Dutton and coworkers[18,19] performed an extensive analysis of ET reactions in different proteins and found that the coupling V_{ET} roughly varies as $-\exp(-\beta R)$ with the donor–acceptor (edge to edge) distance R, and a $\beta \approx 1.4$ Å^{-1}.[18] They converted the Marcus rate constant into a simple empirical expression for the logarithm of the rate constant[18]

$$\log_{10} k = 15 - 0.6R - 3.1(\Delta G^0 + \lambda)^2/\lambda \tag{8}$$

where k is given in units of s^{-1}, R in Å and ΔG^0 and λ in eV.

6.3 Photosynthetic Reaction Centres

6.3.1 The Photosystem of Purple Bacteria

In purple bacteria, ring-shaped pigment–protein complexes serve as peripheral and core light-harvesting complexes (for a review see ref. 20). The peripheral light-harvesting complex LH2 binds two rings of bacteriochlorophyll *a* pigments, containing 18 and 9 pigments, respectively. The ring of 18 pigments that are as closely packed as the 30 BChla of the LH1 ring (Figure 6.2) absorbs light with a wavelength around 850 nm and the 9 pigment ring at 800 nm – that is at a wavelength of monomeric BChl*a*. The large excitonic couplings in the rings with densely packed pigments give rise to a delocalization of the excited state wavefunction in these rings. This delocalization on the one hand shifts the excited state energies and on the other hand redistributes the oscillator strength of the pigments. For the present in-line geometry of local transition dipole moments of the pigments, the low-energy exciton states get most of the oscillator strength. However, the optically dark exciton states, which are in resonance with excited states of the weakly coupled ring, are important for accepting excitation energy from the latter. The LH1 complex absorbance is somewhat red-shifted compared to LH2, at around 875 nm, but also contains dark exciton states for efficient harvesting of excitation energy from the peripheral LH2 complex. After the excitation energy has arrived in the LH1 complex it relaxes within the exciton manifold, and afterwards is transferred to the low energy exciton state of the special pair in the RC absorbing at 860 nm. This low energy excited state of the special pair in the RC is the only one in resonance with the low energy exciton states of LH1 (Figure 6.2). Obviously, the coupling between pigments is utilized in the antenna system of purple bacteria to create an excitation energy funnel.

However, the question may arise why 30 pigments are needed to shift the absorbance from 800 nm for monomeric BChl in LH2 to 875 nm for the low energy exciton states in LH1 and only 2 pigments to obtain a 60 nm red shift in the special pair of the RC? The answer lies in the very small distance between the two special pair pigments. Due to their wavefunction overlap, these two pigments can not only exchange excitation energy but also electrons. This short-range coupling further red-shifts the excited state energies.[21,22] The origin of this additional shift is on one hand a short-range contribution to the excitonic coupling, first proposed by Dexter,[23] and on the other hand a coupling between the exciton states and intramolecular charge transfer states[24] that shifts the local transition energies of the pigments and also contributes to the excitonic coupling. The mixing with charge transfer states, in addition, leads to a large homogeneous broadening of the low-energy exciton state,[25] thereby helping to provide resonance with the low-energy exciton states of the LH1-antenna for efficient energy transfer.[26]

Because of the low excitation energy of the special pair, ET in purple bacteria starts from there. Despite the C2-symmetry of the two branches of cofactors, only the A-branch is ET active, as revealed from absorbance difference spectra involving the oxidized special pair P and a reduced pheophytin H, that was identified as H_A, *i.e.*, the one on the A-branch (for a review see ref. 2). Mutant studies suggest that the differences in free energies of charge separated states are responsible for the exclusive use of the A-branch in ET. In particular, the state $P^+B_A^-$ is lower in free energy than the excited state P*, whereas the free energy of the state $P^+B_B^-$ is higher (ref. 27 and references therein). In addition, the ET matrix elements between cofactors were reported to be larger on the A-branch.[28]

The accessory bacteriochlorophyll B_A serves as an intermediate electron carrier (for a review see ref. 29), *i.e.*, the initial ET scheme reads

$$P^*B_AH_A \overset{3ps}{\rightarrow} P^+B_A^-H_A \overset{0.9ps}{\rightarrow} P^+B_AH_A^- \quad (9)$$

The short lifetime of the intermediate state has provided a challenge for its spectroscopic detection, and therefore alternative models assuming a direct ET from P to H_A, where the intermediate B_A contributes only by mediating the electronic coupling between P and H_A in a superexchange reaction, were discussed initially, but were finally shown to be invalid.[29] Quantum chemical calculations identified the strong electronic coupling between B_A and H_A as an important factor for a fast reaction.[30,31]

From H_A, the electron is transferred in 200 ps to the quinone Q_A,[78] and from there in about 150 μs to the second quinone Q_B[32]

$$P^+H_A^-Q_AQ_B \overset{200ps}{\rightarrow} P^+H_AQ_A^-Q_B \overset{150\mu s}{\rightarrow} P^+H_AQ_AQ_B^-. \quad (10)$$

Whereas the primary ET reactions do not depend on temperature or even accelerate somewhat at lower temperatures, ET between Q_A and Q_B is an activated process that was shown to be independent on the free energy difference between the states $Q_A^-Q_B$ and $Q_AQ_B^-$.[32] A thermally activated gating mechanism, *e.g.*, a proton transfer reaction or a conformational change seems to form the bottleneck of the reaction.

The overall ET in the RC is connected with a substantial decrease in free energy (Figure 6.2, lower). Interestingly, the relaxation of some of the nuclear degrees of freedom was found to be slow compared to the charge transfer (see refs. 33–36 and references therein). We note that the slow relaxation dynamics, which reflects complex protein dynamics, is in conflict with the assumption of a fast vibrational relaxation in non-adiabatic ET theory (see above). In practice this conflict is solved often by introducing time- and temperature-dependent parameters, *i.e.*, free energy differences and reorganization energies in the analysis of experimental data. A more rigorous theoretical treatment is presented by Sumi–Marcus theory,[37,38] that was originally created to include slow solvent diffusion processes in the description of ET reactions.

The competing charge recombination reactions in the RC (Figure 6.2) are a factor of 100–1000 times slower than the competing forward ET reactions.

6.3.2 Photosystem II of Cyanobacteria and Higher Plants

The homodimeric PSII core complexes from the cyanobacterium *Thermosynechococcus elongatus* contains in each monomer the antenna proteins CP43 and CP47, the RC subunits D1 and D2 and 13 additional small subunits.[39] The antenna proteins CP43 and CP47 bind 13 and 16 chlorophyll *a* (Chl *a*) light-harvesting pigments, respectively (Figure 6.2, upper right).

From mutant studies and a theoretical analysis of the optical properties of the RC of PSII it was concluded that the excitation energy sink in the RC of PSII is not formed by the special pair but by the accessory chlorophyll Chl_{D1} of the ET active D1-branch (for a recent review see ref. 40). However, the energy differences between the exciton states are much smaller than in the RC of purple bacteria (Figure 6.2). Therefore, the "special pair" P_{D1}–P_{D2} is also excited at room temperature and in principle could still act as the primary electron donor. At low temperatures, however, it is clear that Chl_{D1} is the primary electron donor in PSII.[41,42] There are different lines of evidence that suggest that Chl_{D1} plays this role also at physiological temperatures: (i) from time-resolved VIS/VIS and VIS/IR pump–probe spectroscopy Holzwarth *et al.*[43] and Groot *et al.*[44] concluded that there is an ultrafast reduction of $Pheo_{D1}$ that seems too fast for ET from the "special pair". However, the time constants deduced in the two studies for the reduction differ by a factor of 7–10, a difference that still needs to be explained. (ii) Ultrafast ET starting at the accessory bacteriochlorophyll B_A in the RC of purple bacteria was found upon direct excitation of B_A,[45,46] with a time constant of less than 400 fs, *i.e.*, an order of magnitude faster than ET starting at the special pair. Such ET does not occur *in vivo*, since the donor states of the LH1 antenna, which supply the RC with excitation energy, are off-resonant to the excited state of B_A (Figure 6.2, middle). Since in PSII the exciton states of the RC are much closer in energy and are all resonant to the exciton states of the core antenna complexes, the excitation energy arriving at the RC will also reside on Chl_{D1}. (iii) From quantum chemical studies of ET matrix elements in purple bacteria, support for a fast ET reaction between B_A and H_A was obtained.[30,31] Fischer and Scherer reported that the ET matrix element for $B_A^*H_A \rightarrow B_A^+H_A^-$ is about three times larger that that of $P_A^*B_A \rightarrow P_A^+B_A^-$.[30] Assuming non-adiabatic ET reactions with similar driving forces and reorganization energies, the ratio of rate constants should scale with the square of the respective matrix elements. Hence, the one order of magnitude larger rate constant of the ET reaction starting at the accessory bacteriochlorophyll in purple bacteria with respect to ET starting at the special pair can be inferred independently from theory[30] and experiment.[45] The almost perfect overlay of pigment macrocycles of the B_A–H_A dimer with its PSII counterpart Chl_{D1}–$Pheo_{D1}$ (Figure 6.5, left) suggests a similarly large ET matrix element in PSII.

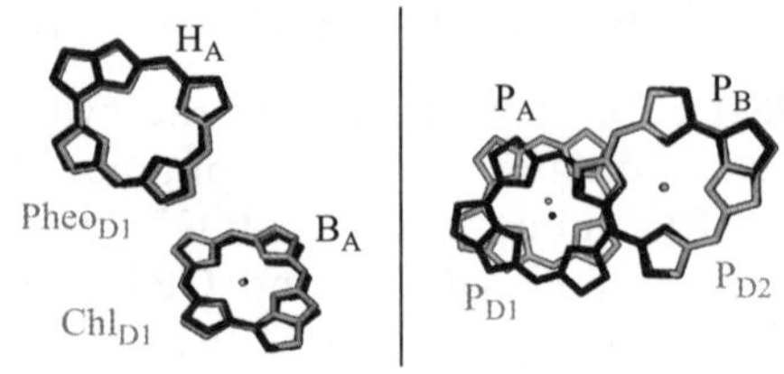

Figure 6.5 Left: Overlay of accessory (bacterio)chlorophylls and pheophytins of the electron transfer active branch of PSII (grey) and the purple bacterial RC (black). Right: Overlay of the special pair (bacterio)chlorophylls of PSII (grey) and the purple bacterial RC (black). The overlays were determined from a minimization of the mean square deviation in positions of equivalent atoms of the two molecules using the program VMD (W. Humphrey, A. Dalke and K. Schulten, *J. Molec. Graphics*, 1996, **14.1**, 33.).

Most likely the primary ET scheme in PSII, therefore, reads

$$P^{*}_{680} \xrightarrow{k_1} P_{D1}Chl^{+}_{D1}Pheo^{-}_{D1}P_{D2}Chl_{D2}Pheo_{D2} \tag{11}$$

$$\xrightarrow{k_2} \quad P^{+}_{D1}Chl_{D1}Pheo^{-}_{D1}P_{D2}Chl_{D2}Pheo_{D2} \tag{12}$$

where $P^{*}{}_{680}$ denotes the excited state of the RC from where ET starts.

A word of caution is necessary concerning this terminology. As discussed above, in the RC of purple bacteria, ET is known to start at the central (bacterio)chlorophyll dimer, the special pair denoted as P, and the cationic state also localizes there (for a review see ref. 29). In the literature about PSII, the spectroscopic term P_{680} is often used to describe the analogous Chl*a* dimer, P_{D1}–P_{D2}, in the structural model. The underlying assumption of equal reaction mechanisms in the two reaction centres is most likely incorrect, as discussed above. The central Chl*a* dimer in PSII should, therefore, be termed as P_{D1}–P_{D2} or as "special pair", solely based on the structural model.

The value deduced for the inverse rate constant $1/k_1$ for primary ET varies between 300 fs,[47] estimated from a fit of time-resolved fluorescence decay data of PSII core particles and 600–800 fs,[44] estimated from VIS/IR pump–probe data, to 6 ps, obtained from VIS/VIS pump-probe data.[43] We note that the rate constant k_1 is related to the intrinsic rate constant k_{intr} for the reaction $Chl^{*}_{D1}Pheo_{D1} \rightarrow Chl^{+}_{D1}Pheo^{-}_{D1}$ by

$$k_1 = P^{(eq)}_{Chl_{D1}} k_{intr} \tag{13}$$

where $P^{(eq)}_{Chl_{D1}}$ is the probability of finding the primary electron donor Chl_{D1} excited,

$$P^{(eq)}_{Chl_{D1}} = \left\langle \sum_{M} f(M) |c^{(M)}_{Chl_{D1}}|^2 \right\rangle_{dis}. \tag{14}$$

where *f(M)* is the Boltzmann factor that expresses the equilibrated population of exciton states and $|c^{(M)}_{Chl_{D1}}|^2$ is the probability of finding the primary electron donor excited in exciton state *M*. For the present system $P^{(eq)}_{Chl_{D1}} \approx 0.3$,[47] *i.e.*, the intrinsic rate constant is about a factor of 3 times larger than the k_1 estimated above.

The value estimated so far for $1/k_2$ varies between 6 ps[44] and 8 ps.[43] The subsequent ET to the quinones Q_A and Q_A resembles closely the kinetics measured in the RC of purple bacteria

$$P^+_{D1}Pheo^-_{D1}Q_AQ_B \overset{200ps}{\rightarrow} P^+_{D1}Pheo_{D1}Q^-_AQ_B \overset{100\mu s}{\rightarrow} P^+_{D1}Pheo_{D1}Q_AQ^-_B \qquad (15)$$

Hence, despite differences in the very early reaction, the overall charge transfer is very similar in the two RCs. Of course, an important difference is the much higher redox potential of P^+ in PSII needed for the water splitting reaction (Figure 2). Three important factors for this increase are (i) the replacement of bacteriochlorophyll *a* by chlorophyll *a*, (ii) the disruption of dimerization in the special pair (see below), and (ii) a change in electrostatic pigment–protein coupling.[48]

6.3.3 Common Design Principles

The quantum efficiency of the primary charge and excitation energy reactions in photosynthesis under low light-conditions is close to unity, *i.e.*, almost every photon absorbed by the photosystem is converted into chemical energy in the RC. At high light intensities, antenna complexes are able to quench excitation energy in order to avoid photochemical damage of the RC. We do not discuss the underlying processes but refer to other reviews on this topic (see ref. 49 and references therein). The key for a high quantum efficiency, of course, is to suppress any competing side reaction. One limit is given by the fluorescence life time of the chlorin pigments which is in the 5 ns range. By limiting the antenna size to a couple of hundred pigments and by arranging them in optimal distances for fast excitation energy transfer it became possible to realize an overall $\approx$ 100 ps decay of excited states in practically all photosystems, *i.e.* there is a factor of 50 difference between the rate constants for excitation energy transfer/trapping and fluorescence.

Once the excitation energy has arrived in the RC it is efficiently trapped by a fast primary charge transfer reaction. To avoid oxidation of antenna pigments the distances between antennae and RC pigments are relatively large, giving rise to a slow transfer to the trap. Due to energetic and entropic (number of pigments) differences, the back transfer from the RC to the antenna is faster by a factor 2–5 than the forward transfer (Figure 6.2). However, the primary ET reactions in the RC of PSII and purple bacteria are still a factor of 30 and 7, respectively, faster than the back excitation energy transfer to the antenna, thus allowing for efficient trapping.

A fast primary charge transfer puts limits on the distance between the primary electron donor and acceptor. According to the simple Dutton ruler (eqn (8)), the maximum inverse rate constant (*i.e.* assuming $-\Delta G^0 = \lambda$) for an edge to edge distance of 5 Å is 1 ps, whereas for twice the distance it is already

100 ns, *i.e.* much slower than even the fluorescence lifetime of the pigments. The exponential distance dependence of the electron transfer matrix element gives rise to this dramatic decrease of the rate constant with increasing distance. Therefore, multiple ET steps are required for transmembrane ET, *i.e.* over a distance of a few tens of Ångstroms. As discussed above, after the second ET step the radical pair P^+Pheo^- is created where P corresponds to P_{D1} in the case of PSI and P_A in purple bacteria, and Pheo is $Pheo_{D1}$ or H_A, respectively. A rough scheme of the energetics and kinetics of the subsequent ET processes that stabilize the radical pair, including also inverse rate constants of the competing recombination reactions, is given in Figure 6.2. As can be seen, the differences in free energy are large compared to the thermal energy, thus suppressing a back ET reaction. In addition, the charge recombination reactions are at least a factor of 100 times slower than the respective forward ET reactions. An interesting question is: what is the mechanism behind the large difference between the forward ET and charge recombination rates? In the absence of any structural reorganization of the cofactors, both rate constants contain the same electronic coupling constant, hence we discuss whether the remaining term in the Marcus expression can be responsible for the large difference. If we use again the simple Dutton ruler, which is based on the Marcus expression, and assume that the forward rate is activationless we obtain a ratio $r = 10^{3.1(\Delta G^0)^2/(4\lambda kT)}$ of rate constants of forward and recombination reactions.

Since charge recombination reactions occur in the inverted region of ET (Figure 6.3), we have $|\Delta G^0| >> \lambda$ and hence a large ratio r. The ratio is particularly large for small γ, which is created by the small polarity of the protein environment inside the membrane.[50] Of course, taking into account nuclear tunneling effects diminishes the inverted effect. However, the coupling of electrons to high-frequency intramolecular vibrations of chlorophylls and bacteriochlorophylls is small,[51] and hence the Huang Rhys factor Sv in eqn (7) is small. The weak temperature dependence of the some recombination reactions[52] suggest that nuclear tunneling is the dominating mechanism (Figure 6.4).

6.3.4 Special Adaptations of Photosystem II for Water Splitting

The water splitting reaction requires a high redox potential of the RC chlorophylls in PSII. Since pigment dimerization, in general, increases the energy of the HOMO, it diminishes the redox potential of the oxidized molecular system for attracting electrons. Therefore, it was necessary to disrupt the wavefunction overlap between the two special pair chlorophylls. Indeed there is a significant tilt of P_{D1} relative to P_{D2} with respect to their purple bacterial counterpart P_A–P_B leading to a more staggered arrangement of the π electron wavefunction than the eclipsed one seen in purple bacteria (Figure 6.5, right).

The decreased wavefunction overlap in the "special pair" diminishes the mixing of excited states with charge transfer states to such an extent that the excitation energy sink is not located at the special pair but at the accessory chlorophyll of the D1-branch Chl_{D1}. The almost perfect overlay of macrocycles of the B_AH_A and the $Chl_{D1}Pheo_{D1}$ dimers suggests that the respective electron transfer matrix elements are similar and therefore, Chl_{D1} is the primary electron donor in PSII.[40,42] Whereas in purple bacteria the different free energies of the two primary radical pair states $P^+B_A^-$ and $P^+B_B^-$ are most likely responsible for unidirectional ET along the A-branch, in PSII the difference in excitation energies between the two accessory chlorophylls Chl_{D1} and Chl_{D2} may be an important factor.[42,47]

The high redox power of the central pigments in PSII RC represents a major challenge for the stability of the system. Under high light intensity the mean *in vivo* lifetime of the D1-protein is only about half an hour.[53,54] During this time the D1-polypeptide is irreversibly damaged and replaced by a new polypeptide. Different molecular mechanisms are discussed for the damage.[54,55] One possibility is an oxidation of pigments or parts of the protein by the highly oxidizing state P_{680}^+ and a subsequent degradation of the unstable cationic states. Another possibility is the formation of Chl triplet states (^{3}Chl) by intersystem crossing/charge recombination reactions and the subsequent reaction with triplet oxygen leading to the formation of highly reactive singlet oxygen which damages the D1-protein. Experimental studies and calculations have revealed that these triplet states accumulate at Chl_{D1} in PSII (for a review see ref. 40).

In the antenna the triplet states ^{3}Chl (^{3}BChl) are quenched by exchange coupling with carotenoids (Cars). There are no Cars found in van der Waals contact of any of the six core pigments of the RC in PSII. The high redox power of these pigments, necessary for the water oxidation, would lead to unwanted oxidation of the Cars. A difference between PSII and the purple bacteria is that in the former the triplet state localizes at the accessory Chl_{D1},[42,56] whereas in the latter at the special pair P_A–P_B.[57] In purple bacteria, the RC triplet energy is quenched *via* energy transfer to B_B which is in van der Waals contact with a carotenoid[58,59] (Figure 6.1). The lifetime of the triplet state in the RC of PSII was found to decrease by two orders of magnitude if Q_A is in the reduced state.[60] This finding led Noguchi to conclude that triplet localization on Chl_{D1} is functionally important for efficient triplet quenching by Q_A^-.[61] The molecular mechanism of this quenching is still unknown.

One possibility to avoid damage by P_{680}^+ is a controlled secondary ET that most likely involves Cars[62] (for a recent review see ref. 63). From calculations of Q_A^- Car^+–Q_ACar optical difference spectra and comparison with experimental data, it was concluded that only Car_{D2} and not Car_{D1} is involved in secondary ET in PSII.[64] Secondary ET involving Car_{D2} leads to a controlled reduction of P_{680}^+ by cyt*b*559, that may be completed by a subsequent charge recombination with the reduced Q_B.[65] A second photoprotective function of Car_{D2} discussed[55,63] is the quenching of singlet oxygen, which is generated by reaction of triplet oxygen with triplet states of Chl.

Another protection mechanism concerns the slowing down of primary charge transfer by the reduced Q_A, and thereby the escape of excitation energy from the RC to the antenna, utilizing the entropic factor in the free energy difference, discussed above.[66] This effect is enhanced in PSII because of the change in primary electron donor from the special pair in purple bacteria to Chl_{D1} in

PSII, due to the close distance between the latter and Q_A (as discussed above for the triplet quenching).

Finally we note that the antenna system of PSII had to be reorganized strongly compared to purple bacteria in order to avoid the low-energy antenna states caused by the strong exciton delocalization in LH1 and LH2 of purple bacteria, thereby allowing excitation energy transfer from the antenna to the special chlorophyll monomer Chl_{D1}, serving as the primary electron donor. The reorganization of the core antenna is a nice example for the more flexible building principles of antenna systems as compared to those of the RCs. The slight tilt of the special pair in the RC (Figure 6.5), which was necessary to increase the oxidation potential for the water splitting, increased the excitation energy of the special pair so much that it could not have acted as excitation energy donor from a LH1-type antenna.

6.4 Challenging Questions

Whereas in the case of the RC of purple bacteria most of the spectroscopic properties have been explained in detail on the basis of the structural data, there are many open questions concerning the structure–function relationships in PSII. A major difference of the optical properties of the two RCs is the clear separation of optical bands in purple bacteria and the strong overlap of bands in PSII. Mutant studies in combination with exciton calculations are the most promising tools to solve this "spectral congestion" problem (for a recent review see ref. 40).

An important parameter type of the exciton Hamiltonian is given by the site energies of the pigments. So far these energies were inferred indirectly from a fit of optical spectra of the subunits.[66] However, a direct structure-based calculation is desired. First successful attempts were reported for the Fenna–Matthews–Olson protein, a light-harvesting complex of green sulfur bacteria,[67,68,81] photosystem I,[82] and the light-harvesting complex LHC-II of higher plants.[83] Recently, a crystal structure of photosystem II with an improved resolution of 1.9 Å was reported by Shen and coworkers.[84] This study will allow structure-based calculations of site energies in PSII to be performed, in order to unravel structure–function relationships. For example, it will be exciting to find out which protein–cofactor-interactions are responsible for the low transition energy of Chl_{D1} in the RC. Also, a calculation of short-range effects on the excitonic coupling of the "special pair" and of the coupling of excited states to intradimer charge transfer states is possible now. A method exists to extract these effects from monomer and dimer quantum chemical calculations in a vacuum.[22] The finding of Krausz and coworkers,[69,70] that charge separation can be initiated at low temperatures with long-wavelength excitation of a strongly homogeneously broadened excited state raises the question about a low-lying charge transfer state in the RC of PSII. For an accurate calculation of its energy, most likely solvent and protein environment effects have to be included in the calculations. The findings of Krausz and coworkers[69,70] also raise fundamental questions about the mechanism and the pathway of primary charge transfer in PSII. Since the wavefunction overlap between pigments is by far the largest in the "special pair", it seems primary charge transfer can also start from an excited state of the "special pair" that may be mixed with a charge transfer state. Hence, primary charge transfer from Chl_{D1} seems to compete with exciton relaxation into a low-lying charge transfer (and/or excited) state of the "special pair". Finally, we note that the PSII of *A. marina* represents an interesting model system to investigate the primary photophysical processes in PSII, since its RC binds both Chl*a* and Chl*d* and therefore the "spectral congestion" is much weaker in *A. marina*, making the interpretation of optical experiments much easier.[71]

Acknowledgement

It is a pleasure to acknowledge support from the German Research Foundation (DFG) through Collaborative Research Centers (SFB) 498 (TP A7) and SFB 429 (TPA9) as well as through the Emmy Noether program.

References

1. F. Rappaport and B. A. Diner, *Coord. Chem. Rev.*, 2008, **252**, 259–272.
2. W. W. Parson, *Photosynth. Res.*, 2003, **76**, 81–92.
3. M. Guergova-Kuras, B. Boudreaux, A. Joliot, P. Joliot and K. Redding, *Proc. Natl. Acad. Sci. U. S. A.*, 2001, **98**, 4473–4442.
4. J. A. Bautista, F. Rappaport, M. Guergova-Kuras, R. O. Cohen, J. H. Golbeck, J. Y. Wang, D. Beal and B. A. Diner, *J. Biol. Chem.*, 2005, **280**, 20030–20041.
5. Y. Li, A. van der Est, M. G. Lucas, V. M. Ramesh, F. Gu, A. Petrenko, S. Lin, A. N. Webber, F. Rappaport and K. Redding, *Proc. Natl. Acad. Sci. U. S. A.*, 2006, **103**, 2144–2149.
6. F. Rappaport, B. A. Diner and K. Redding, in *The Light-Driven Water–Plastocyanin Ferredoxin Oxidoreductase*, ed. J. H. Golbeck, Springer, Dordrecht, 2006, p. 223.
7. K. Redding and A. van der Est, in *The Light-Driven Water–Plastocyanin Ferredoxin Oxidoreductase*, ed. J. H. Golbeck, Springer, Dordrecht, 2006, p. 413.
8. R. O. Cohen, G. Shen, J. H. Golbeck, W. Xu, P. R. Chitnis, A. I. Valieva, A. van der Est, Y. Pushkar and D. Stehlik, *Biochemistry*, 2004, **43**, 4741–4754.
9. N. Dashdorj, W. Xu, R. O. Cohen, J. H. Golbeck and S. Savikhin, *Biophys. J.*, 2005, **88**, 1238–1249.
10. T. Renger, *Photosynth. Res.*, 2009, **102**, 471–485.
11. K. Mukai, S. Abe and H. Sumi, *J. Phys. Chem. B*, 1999, **103**, 6096–6102.
12. T. Förster, *Ann. Phys. Leipzig*, 1948, **2**, 55–75.
13. T. Förster, *Modern Quantum Chemistry, Part III, Action of Light and Organic Crystals*, Academic Press, New York, 1965, pp. 93–137.

14. H. Sumi, *J. Phys. Chem. B*, 1999, **103**, 252–260.
15. R. A. Marcus and N. Sutin, *Biochim. Biophys. Acta*, 1985, **811**, 265–322.
16. V. May and O. Kühn, *Charge and Energy Transfer Dynamics in Molecular Systems: A Theoretical Introduction*, Wiley–VCH, Berlin, 2000.
17. J. Jortner, *J. Chem. Phys.*, 1976, **64**, 4860–4867.
18. C. C. Moser, J. M. Keske, K. Warncke, R. S. Farid and P. L. Dutton, *Nature*, 1992, **355**, 796–802.
19. C. P. Page, C. C. Moser, X. Chen and P. L. Dutton, *Nature*, 1999, **402**, 47–52.
20. R. J. Cogdell, A. T. Gardiner, A. W. Roszak, C. J. Law, J. Southall and N. W. Isaacs, *Photosynth. Res.*, 2004, **81**, 207–214.
21. A. Warshel and W. W. Parson, *J. Am. Chem. Soc.*, 1987, **109**, 6143–6151.
22. M. E. Madjet, F. Müh and T. Renger, *J. Phys. Chem. B*, 2009, **113**, 12603–12614.
23. D. L. Dexter, *J. Chem. Phys.*, 1953, **21**, 836–850.
24. R. D. Harcourt, K. P. Ghiggino, G. D. Scholes and S. Speiser, *J. Chem. Phys.*, 1996, **105**, 1897–1901.
25. T. Renger, *Phys. Rev. Lett.*, 2004, **93**, 188101.
26. H. Sumi, *J. Phys. Chem. B*, 2004, **108**, 11792–11801.
27. C. Kirmaier, C. He and D. Holton, *Biochemistry*, 2001, **40**, 12132–12139.
28. L. Y. Zhang and R. A. Friesner, *Proc. Natl. Acad. Sci. U.S.A.*, 1998, **95**, 13603–13605.
29. W. Zinth and J. Wachtveitl, *ChemPhysChem*, 2005, **6**, 871–880.
30. S. F. Fischer and P. O. J. Scherer, *Chem. Phys.*, 1987, **115**, 151–158.
31. A. Warshel, S. Creighton and W. W. Parson, *J. Phys. Chem.*, 1988, **92**, 2696–2701.
32. M. S. Graige, G. Feher and M. Y. Okamura, *Proc. Natl. Acad. Sci. U. S. A.*, 1998, **95**, 11679–11684.
33. J. M. Peloquin, J. C. Williams, X. Lin, R. G. Alden, A. K. W. Taguchi, J. P. Allen and N. W. Woodbury, *Biochemistry*, 1994, **33**, 8089–8100.
34. W. W. Parson, Z. T. Chu and A. Warshel, *Biophys. J.*, 1998, **74**, 182–191.
35. J. M. Kriegl and G. U. Nienhaus, *Proc. Natl. Acad. Sci. U. S. A.*, 2004, **101**, 123–128.
36. H. Wang, S. Lin, J. P. Allen, J. C. Williams, S. Blankert, C. Laser and N. W. Woodbury, *Science*, 2007, **316**, 747–750.
37. H. Sumi and R. A. Marcus, *J. Chem. Phys.*, 1986, **84**, 4894–4914.
38. H. Wang, S. Lin, E. Katilius, C. Laser, J. P. Allen, J. C. Williams and N. W. Woodbury, *J. Phys. Chem. B*, 2009, **113**, 818–824.
39. A. Guskov, J. Kern, A. Gabdulkhakov, M. Broser, A. Zouni and W. Saenger, *Nat. Struct. Mol. Biol.*, 2009, **16**, 334–342.
40. T. Renger and E. Schlodder, *ChemPhysChem*, 2010, **11**, 1141–1153.
41. B. A. Diner, E. Schlodder, P. J. Nixon, W. J. Coleman, F. Rappaport, J. Laverge, W. F. J. Vermaas and D. A. Chisholm, *Biochemistry*, 2001, **40**, 9265–9281.
42. G. Raszewski, W. Saenger and T. Renger, *Biophys. J.*, 2005, **88**, 986–998.

43. A. R. Holzwarth, M. G. Müller, M. Reus, M. Nowaczyk, J. Sander and M. Rögner, *Proc. Natl. Acad. Sci. U. S. A.*, 2006, **103**, 6895–6900.
44. M. L. Groot, N. P. Pawlowicz, L. J. G. W. Van Wilderen, J. Breton and I. H. M. Van Stokkum, *Proc. Natl. Acad. Sci. U. S. A.*, 2005, **102**, 13087–13092.
45. M. E. Van Brederode, M. R. Jones, F. Van Mourik, I. H. M. Van Stokkum and R. Van Grondelle, *Biochemistry*, 1997, **36**, 6855–6861.
46. M. E. Van Brederode, F. Van Mourik, I. H. M. Van Stokkum, M. R. Jones and R. Van Grondelle, *Proc. Natl. Acad. Sci. U. S. A.*, 1999, **96**, 2054–2059.
47. G. Raszewski, B. A. Diner, E. Schlodder and T. Renger, *Biophys. J.*, 2008, **95**, 105–119.
48. H. Ishikita, W. Saenger, J. Biesiadka, B. Loll and E.-W. Knapp, *Proc. Natl. Acad. Sci. U. S. A.*, 2006, **103**, 9855–9860.
49. P. Horton, M. P. Johnson, M. L. Perez-Bueno, A. Z. Kiss and A. V. Ruban, *FEBS J.*, 2008, **275**, 1069–1079.
50. A. Warshel and D. W. Schlosser, *Proc. Natl. Sci., U. S. A.*, 1981, **78**, 5564–5568.
51. M. Rätsep, J. Pieper, K. D. Irrgang and A. Freiberg, *J. Phys. Chem.*, 2008, **112**, 110–118.
52. J. M. Ortega, P. Mathis, J. C. Williams and J. P. Allen, *Biochemistry*, 1996, **35**, 3354–3361.
53. J. H. Kim, J. A. Nemson and A. Melis, *Plant Physiol.*, 1993, **103**, 181–189.
54. J. Barber, *Aust. J. Plant Physiol.*, 1994, **22**, 201–208.
55. A. Telfer, *Photochem. Photobiol. Sci.*, 2005, **4**, 950–956.
56. E. Schlodder, T. Renger, G. Raszewski, W. J. Coleman, P. J. Nixon, R. O. Cohen and B. A. Diner, *Biochemistry*, 2008, **47**, 3143–3154.
57. H. Frank, J. D. Bolt, S. M. B. Costa and K. Sauer, *J. Am. Chem. Soc.*, 1980, **102**, 4893–4898.
58. C. C. Schenck, P. Mathis and M. Lutz, *Photochem. Photobiol.*, 1984, **39**, 407–417.
59. T. O. Yeates, H. Komiya, A. Chirino, D. C. Rees, J. P. Allen and G. Feher, *Proc. Natl. Acad. Sci. U. S. A.*, 1988, **85**, 7993–7997.
60. F. Van Mieghem, K. Brettel, B. Hillmann, A. Kamlowski, A. W. Rutherford and E. Schlodder, *Biochemistry*, 1995, **34**, 4789–4813.
61. T. Noguchi, *Plant Cell Physiol.*, 2002, **43**, 1112–1116.
62. C. C. Schenck, B. Diner, P. Mathis and K. Satoh, *Biochim. Biophys. Acta*, 1982, **680**, 216–227.
63. P. Faller, C. Fufezan and A. W. Rutherford, Side-path electron donors: cytochrome b559, chlorophyll Z and β-carotene in photosystem II, in: *The Light-Driven Water–Plastoquinone Oxidoreductase*, ed. K. Wydrzynski and K. Satoh, Springer, Dordrecht, 2005, pp. 347–365.
64. E. Schlodder, T. Renger, G. Raszewski, W. J. Coleman, P. J. Nixon, R. O. Cohen and B. A. Diner, *Biochemistry*, 2008, **47**, 3143–3154.
65. C. A. Buser, B. A. Diner and G. W. Brudvig, *Biochemistry*, 1992, **31**, 11449–11459.
66. G. Raszewski and T. Renger, *J. Am. Chem. Soc.*, 2008, **130**, 4431–4446.

67. F. Müh, M. E. Madjet, J. Adolphs, A. Abdurahman, B. Rabenstein, H. Ishikita, E.-W. Knapp and T. Renger, *Proc. Natl. Acad. Sci. U. S. A.*, 2007, **104**, 16862–16867.
68. J. Adolphs, F. Müh, M. E. Madjet and T. Renger, *Photosynth. Res.*, 2008, **95**, 197–209.
69. E. Krausz, J. L. Hughes, P. Smith, R. Pace and S. P. Årsköld, *Photochem. Photobiol. Sci.*, 2005, **4**, 744–753.
70. J. L. Hughes, P. Smith, R. Pace and E. Krausz, *Biochim. Biophys. Acta*, 2006, **1757**, 841–851.
71. T. Renger and E. Schlodder, *J. Phys. Chem. B*, 2008, **112**, 7351–7354.
72. P. Jordan, P. Fromme, O. Klukas, H. T. Witt, W. Saenger and N. Krauß, *Nature*, 2001, **411**, 909–917.
73. W. Roszak, T. D. Howard, J. Southall, A. T. Gardiner, C. J. Law, N. W. Isaacs and R. J. Cogdell, *Science*, 2003, **302**, 1969–1972.
74. Z. Katiliene, E. Katilius and N. W. Woodbury, *Biophys. J.*, 2003, **84**, 3240–3251.
75. M. Volk, G. Aumeier, T. Langenbacher, R. Feick, A. Ogrodnik and M. E. Michel-Beyerle, *J. Phys. Chem. B*, 1998, **102**, 735–751.
76. N. W. Woodbury and W. W. Parson, *Biochim. Biophys. Acta*, 1986, **850**, 197.
77. F. Rappaport, M. Guergova-Kuras, P. J. Nixon, B. A. Diner and J. Lavergne, *Biochemistry*, 2002, **41**, 8518–8527.
78. D. Holten, M. W. Windsor, W. W. Parson and J. P. Thornber, *Biochim. Biophys. Acta*, 1978, **501**, 112.
79. P. J. Booth, B. Crystall, L. B. Giorgi, J. Barber, D. R. Klug and G. Porter, *Biochim. Biophys. Acta*, 1990, **1016**, 141.
80. B. G. De Grooth and H. J. van Gorkom, *Biochim. Biophys. Acta*, 1981, **635**, 445–456.
81. M. Schmidt am Busch, F. Müh, M. E. Madjet and T. Renger, *J. Phys. Chem. Lett.*, 2011, **2**, 93–98.
82. J. Adolphs, F. Müh, M. E. Madjet, M. Schmidt am Busch and T. Renger, *J. Am. Chem. Soc. U. S. A.*, 2010, **132**, 3331–3343.
83. F. Müh, M. E. Madjet and T. Renger, *J. Phys. Chem. B*, 2010, **114**, 13517–13535.
84. Y. Umea, K. Kawakami, J.-R. Shen and N. Kamiya, *Nature*, 2011, **473**, 55–60.

CHAPTER 7

Photosynthetic O_2 Evolution

JOHANNES MESSINGER,*[a] TAKUMI NOGUCHI[b] AND JUNKO YANO[c]

[a] Department of Chemistry, Kemiskt Biologiskt Centrum (KBC), Umeå University, S-90 187 Umeå, Sweden; [b] Division of Material Science, Graduate School of Science, Nagoya University, Furo-cho, Chikusa-ku, Nagoya, 464-8602, Japan; [c] Lawrence Berkeley National Laboratory, Physical Biosciences Division, Berkeley 94720, USA

7.1 Introduction

In a process known as photosynthesis plants, algae and certain bacteria use the energy of sunlight to split water into molecular oxygen and bound hydrogen, which finally reduces carbon dioxide to carbohydrates. This process evolved about 2–3 billion years ago, and is the energetic basis for life as we know it. Basically, all the oxygen in the Earth's atmosphere was generated in this way as well as most of the biomass, part of which was converted by geological processes into coal, oil and natural gas. Researchers are on their way to unravelling the details of this unique biological energy conversion process. This will facilitate building manmade devices that, by employing the same principles, are able to store solar energy within fuels such as molecular hydrogen or ethanol.

In biological systems the conversion of light into chemical (free) energy occurs in two types of large protein complexes, photosystem I (PSI) and photosystem II (PSII). These are part of the thylakoid membrane inside the chloroplasts or cells. They are connected to light-harvesting antenna complexes that bind most of the pigments that collect the light and transfer the created excitation energy to the reaction centers in the photosystems. In the reaction

RSC Energy and Environment Series No. 5
Molecular Solar Fuels
Edited by Thomas J. Wydrzynski and Warwick Hillier

Published by the Royal Society of Chemistry, www.rsc.org

center, the energized electron can be transferred from the excited chlorophyll molecule to a suitable electron acceptor, so that a positive and a negative charge are created. This initial charge separated state needs to be stabilized by a sequence of electron transfer reactions that reduce the free energy difference and increase the distance between the two opposite charges. This stabilization allows the reaction centers to couple the fast one-electron charge separations with chemical reactions that are several orders of magnitude slower.

In photosystem II (PSII), the created holes are used to oxidize a metal-oxide cluster stepwise; this cluster consists of four Mn ions, one Ca ion and five oxygen's (Mn_4CaO_5 cluster).[1–4] After four-fold oxidation the cluster splits two water molecules at a time into one molecule of dioxygen, and distributed over the reaction cycle a total of four protons are released (Reaction (1)):

$$2H_2O \xrightarrow{4\ \text{photons}} O_2 + 4H^+ + 4e^- \qquad (1)$$

In this way, the energetically demanding one-electron oxidation step of water to the hydroxylradical and the accumulation of more than one positive charge at the cluster are circumvented. The four electrons liberated in Reaction (1) reduce the inorganic cluster to its original state so that the cycle can start anew.

On the other side of PSII, the sequentially created electrons with moderate reducing power are transferred *via* an intermediate step to a bound plastoquinone molecule (Q_B) that, after double reduction and protonation, leaves PSII and is replaced by another plastoquinone molecule. From here, the water-derived electrons travel further along the photosynthetic electron transfer chain, where they gain further reducing power by the photoreaction within PSI so that finally carbon dioxide is reduced to carbohydrates in the Calvin–Benson cycle. By coupling these two photoreactions in series it is possible for these biological systems to generate the large redox potential difference required for driving water oxidation and CO_2 reduction with visible photons.

In this chapter we will summarize the latest results regarding the structure of the water-oxidizing complex in PSII and the mechanism of photosynthetic water-splitting, and highlight the concepts that may be important for developing manmade systems for converting solar energy into fuels. To put these latest findings into perspective, we will start with a brief historical overview, followed by the description of the structure and overall function of photosystem II. To achieve clarity we have to be selective and concentrate mostly on recent findings. For other views and citations of earlier data we refer the reader to excellent earlier reviews on this topic.[5–19]

7.2 Brief Historical Overview

Research into oxygenic photosynthesis can be dated back to experiments performed in 1754 by Bonnet,[20] who showed that submerged, illuminated leaves produce bubbles (later shown to be O_2), and to the work of Priestley (1772),[21] in which he demonstrated that an illuminated mint plant can regenerate the air trapped in a closed glass container in which a burning candle had changed the gas composition. As a detection device for what we now know is molecular oxygen he employed mice. These first fundamental experiments were followed by many more discoveries in photosynthesis,[22] but it was only in 1937 that the requirement of manganese for oxygenic photosynthesis was realized by Pirson.[23–25] The first real breakthrough in studying the mechanism of photosynthetic water-splitting by PSII was made by the groups around Pierre Joliot and Bessel Kok in the late 1960's and early 1970's.[26–29] Their flash-induced oxygen evolution experiments revealed that a periodicity of four governs this reaction. This discovery led to the postulate of four stable and of one highly reactive reaction intermediates during water-splitting, which were termed S_i states ($i = 0,\ldots,4$). These findings will be described in detail in the next section.

Since then the task has been identifying the components that constitute the S_i states, and revealing how the molecular structure and the oxidation and protonation states of the water-oxidizing complex (WOC) change during the water-splitting cycle. Furthermore, the kinetics and the energetics of the S_i state transitions, the roles of Ca^{2+}, Cl^- and HCO_3^-, and the modes of substrate water binding were investigated in great detail with an ever growing arsenal of techniques and degree of sophistication. Researchers in the field employed basically every newly available biophysical technique in the hope of revealing another facet of this intriguing reaction.[5–19] Examples are (in approximate chronological order) polarographic/amperometric oxygen evolution measurements, proton release measurements, various types of (time-resolved) optical spectroscopy, electron and nuclear magnetic resonance spectroscopy, X-ray spectroscopy, Fourier transform infrared spectroscopy, time-resolved mass spectrometry, X-ray crystallography, QM/MM and DFT calculations. These techniques only give useful information when combined with sophisticated biochemical sample treatments that include highly active PSII preparations from various organisms, the use of substrate/cofactor analogs, stable isotope labeling, and point mutations. Details on the biophysical techniques employed in photosynthesis research were recently described in two books and in a special issue in the journal Photosynthesis Research.[30–33]

7.3 Structure and Overall Reaction Sequence of Photosystem II

Good diffracting single crystals have so far been obtained only from the two thermophilic cyanobacteria *Thermosynechococcus elongatus* [34–41] and *Thermosynechococcus vulcanus* [1,42,43] that have almost identical PSII complexes. Until very recently the best published resolution was 2.9 Å, but a new structure with 1.9 Å resolution has been presented.[1] Isolated PSII core complexes from these organisms crystallize as dimers (see top and side view in Figure 7.1 a and b). Each complex comprises 20 protein subunits, 14 lipids, and the following cofactors: 35 chlorophylls (Chl), 12 carotenoids (Car), 2 pheophytins, 2 cytochromes, 3 plastoquinone molecules (Q_A,Q_B and Q_C), 1 non-heme Fe, 4 Mn,

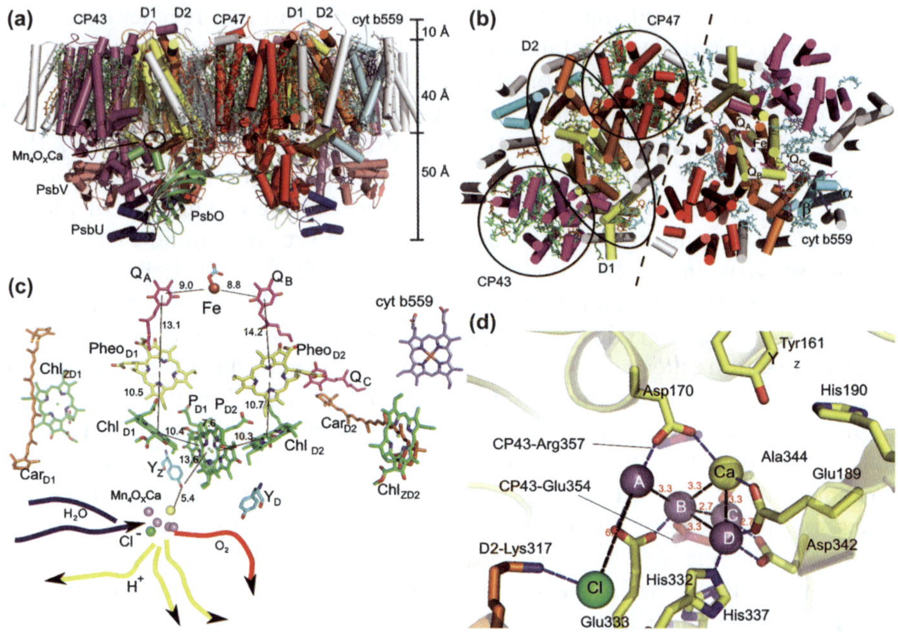

Figure 7.1 Crystal structure of photosystem II at 2.9 Å resolution.[34] **a**: Side view of the PSII dimer. The major protein subunits and the Mn_4CaO_5 cluster are labeled. Prior to isolation for crystallography the complex is part of the thylakoid membrane, and the position and thickness of that membrane corresponds to the 40 Å label on the right side. **b**: Top view on the PSII dimer. The separation into the central D1/D2 proteins and the attached inner antenna proteins CP43 and CP47 can be seen. The dashed line indicates the border between the two monomers. **c**: Side view of the cofactor arrangement in the D1/D2 and cytb559 proteins. Primary charge separation occurs after receiving the excitation energy from the antenna between Chl_{D1} and $Pheo_{D1}$; for further details see text. Chl, chlorophyll; Car, carotenoid; Pheo, pheophytin; Y, tyrosine side chain; Q, plastoquinone; $P_{D1}P_{D2}$, dimer of Chl a molecules. Channels through the surrounding protein (see Figure 7.1 a) for water access to, and protons and oxygen release from, the Mn_4CaO_5 cluster are indicated with blue, yellow and red arrows, respectively. **d**: Mn_4CaO_5 cluster and its ligands that together form the water-oxidizing/oxygen evolving cluster. The five oxygen bridges between the Mn and Ca ions are not shown because they were not resolved in this study (see however the new crystal structure at 1.9 Å resolution[1]). The cluster is therefore referred to as Mn_4O_xCa cluster in this Figure. For more details see Section 7.5.

1 Ca and 1–2 Cl^-.[34,38,42] While the Chl and Car cofactors responsible for capturing light energy and excitation energy transfer are bound by the inner antenna proteins CP43 and CP47, and by additional light harvesting complexes not shown in Figure 7.1, the cofactors involved in electron transfer reactions are bound by the central D1 and D2 proteins. These electron transfer cofactors are displayed in Figure 7.1c.

The electron transfer cofactors are arranged almost symmetrically (Figure 7.1c). Despite that, charge separation in photosystem II starts, after receiving excitation energy from the antenna (Reaction (2)), on the left side branch between Chl_{D1} and $Pheo_{D1}$ (Reaction (3)).

$$Chl_{antenna}^{*}\ P680 \rightarrow Chl_{antenna}\ {}^{1}P680^{*} \tag{2}$$

$$^{1}P680^{*}\ Pheo_{D1} \rightarrow P680^{\bullet+}\ Pheo_{D1}^{\bullet-} \tag{3}$$

This side is therefore referred to as the active branch.[44] Because the pigments $Chl_{D1}P_{D1}P_{D2}Chl_{D2}$ are tightly coupled, we collectively describe them in Reaction (3) and in the text below by the classical term P680 that refers to the wavelength of the red absorption maximum of the primary donor. For a more detailed discussion of these reactions please see ref. 5 and 45.

The additional Chl_Z and carotene (Car) molecules, as well as the heme of cytb559, are thought to have protective functions (protective branch) when the electron transfer from the Mn_4CaO_5 cluster is interrupted, or chlorophyll triplet states are formed due to charge recombination. The quenching of the latter minimizes the formation of highly reactive singlet oxygen.[46–49]

Charge separation is then stabilized by spatially separating the charges and reducing their free energy difference according to the sequence of reactions outlined in Reactions (4)–(9). The electron transfer on the acceptor side involves the following reaction sequences:

$$Pheo_{D1}^{\bullet-}\ Q_A \rightarrow Pheo\ Q_A^{\bullet-} \tag{4}$$

$$Q_A^{\bullet-}\ Fe\ Q_B \rightarrow Q_A\ Fe\ Q_B^{\bullet-} \tag{5}$$

$$Q_A^{\bullet-}\ Fe\ Q_B^{\bullet-} + 2H^{+}_{stroma} \rightarrow Q_A\ Fe\ Q_BH_2 \tag{6}$$

$$Q_A\ Fe\ Q_BH_2 + PQ \rightarrow Q_A\ Fe\ Q_B + PQH_2 \tag{7}$$

Here, PQ symbolizes a plastoquinone molecule from the PQ pool in the thylakoid membrane that is already pre-bound in the Q_C position (see Figure 7.1c) and replaces the loosely bound Q_BH_2. Reactions (6) and (7) occur after a second charge separation in the reaction center. So while Q_A acts due to its special protein environment as a tightly bound one-electron acceptor, Q_B is only transiently connected with PSII and is exchanged after double reduction and protonation.

The Mn_4CaO_5 cluster at the donor side of photosystem II is stepwise oxidized by $P680^{\bullet+}$. However, this oxidation does not occur directly, but *via* a redox-active tyrosine side chain of the D1 protein that is known as Y_Z. This is important, since Y_Z can donate its electron much faster to $P680^{\bullet+}$ than the

Mn_4CaO_5 cluster, thereby minimizing losses due to charge recombination reactions. The donor side reactions can be denoted as follows:

$$S_i\ Y_Z\ P680^{\bullet+} \rightarrow S_i\ Y_Z^{\bullet}\ P680 \rightarrow S_{i+1}\ Y_Z\ P680 \quad (8)$$

$$2H_2O + S_3\ Y_Z^{\bullet} \rightarrow 2H_2O + S_4\ Y_Z \rightarrow O_2 + 4H^+\ S_0\ Y_Z \quad (9)$$

Here S_i represents the oxidation states of the Mn_4CaO_5 cluster, which are described in detail in the next section. The reactions 8 and 9 are not fully charge balanced, since Y_Z transfers its proton to a nearby histidine (D1-His 189) that for simplicity is not included in the above notation. Similarly, most oxidation reactions of the Mn_4CaO_5 cluster are coupled to proton release into the lumen (see Figure 7.2 and Section 7.8) so that in reality the waters in Equation 9 are partially or fully deprotonated, and less protons are released in the last step. This equation also ignores for simplicity the required rebinding of substrate water.

Figure 7.1d shows the Mn_4CaO_5 cluster and its ligand environment that includes six carboxylates (D1-Asp-170, D1-Glu189, D1-Glu333, D1-Asp342, D1-C-term Ala344, CP43-Glu354) and one histidine (D1-His332).[35,41] One or two Cl^- ions are found about 6.5 Å away from the nearest Mn, from which the more tightly bound one is shown in Figure 7.1d.[34,38,42] Tyrosine Y_Z is closest to Ca with a distance of 5.4 Å. It also forms a hydrogen bridge to D1-His189. The oxygens that bridge between the Mn and Ca ions are not shown in Figure 7.1 since they were not resolved at 2.9 Å resolution (the cluster is therefore referred to as Mn_4O_xCa in this Figure). The Mn–Mn and Mn–Ca distances are likely somewhat lengthened due to radiation damage.[50] For more details see Section 7.5.

Due to the large extrinsic loops of CP43 and CP47, and the extrinsic proteins (33 kDa, 23 kDa and C550) the Mn_4CaO_5 cluster is located close to the center of the PSII core complex. Channels have therefore been proposed for H_2O entry and O_2 and H^+ release even before the first crystal structures were

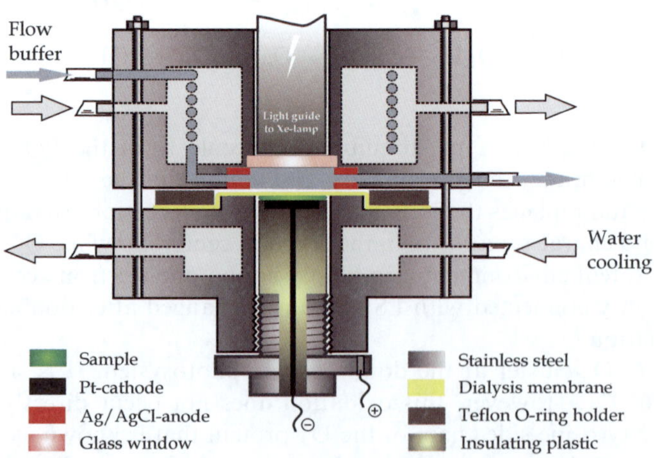

Figure 7.2 Joliot type electrode. For details see text.

published.[10,51–54] Such channels were subsequently identified in the crystals structures based on Xe experiments, and by various computational methods.[34,39,55] Cl^- has been suggested to be an important part in one of the proton channels. It is also noteworthy that one water molecule (not shown) near Mn_A maybe hydrogen bonded to the more firmly bound Cl^- ion.[34]

7.4 Basic Mechanism of Photosynthetic O_2 Evolution

In the 1960's, *i.e.* long before the above discussed structural details were known, Pierre Joliot developed an oxygen electrode that consisted of a bare platinum surface, which was polarized against a Ag/AgCl counter electrode.[56,57] A thin layer of dark-adapted PSII sample was applied directly onto the Pt surface and was held in place by a transparent dialysis membrane. The electrical contact between the Pt and Ag/AgCl electrodes was made *via* a Cl^- containing buffer of defined pH. The PSII sample could be illuminated with µs Xe-lamp flashes *via* a slit through the Ag block, usually at flash frequencies of 2 Hz. In modern devices the temperature can be controlled *via* a thermostat (Figure 7.2).

With such a device, he made the astonishing finding that the first maximal O_2 release for dark-adapted PSII samples occurs after three saturating light flashes, while thereafter always four successive flashes were required until the next maximum was observed (Figure 7.3A).[28,58] Using a similar setup, Bessel Kok showed that the period-four oscillation in O_2 production remained essentially unchanged when a large fraction of the PSII centers was inhibited by adding a herbicide. Since this latter observation excludes co-operativity during water-splitting between separate PSII complexes, he proposed the presence of a unit, the water-oxidizing complex (WOC), that is present in each PSII complex and that stores the oxidizing equivalents ('charges') that are generated during the

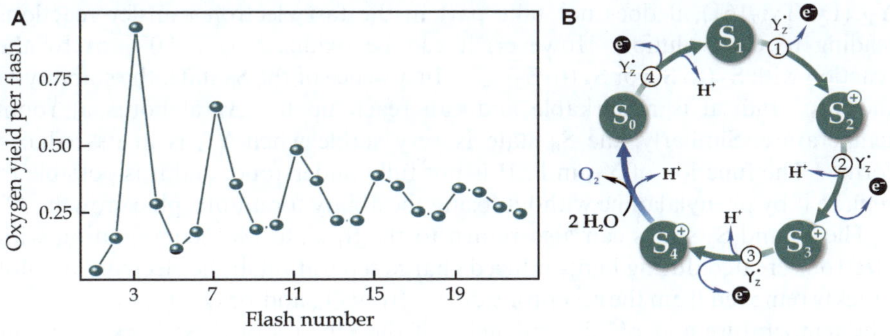

Figure 7.3 **A**: Flash-induced oxygen-release pattern (FIOP) of dark-adapted spinach thylakoids at pH 6.3 and 10 °C. A similar pattern was originally obtained by Pierre Joliot using an electrode like that displayed in Figure 7.2. **B:** Kinetic scheme (Kok cycle) describing the redox and protonation state changes of the water-oxidzing complex during light-induced catalysis. The numbers in circles on the arrows indicate the flash number. Note that depending on the reaction step either the electron or the proton is assumed to be released first. For further details see this Section and Section 7.8.

light-induced charge separations in the reaction center.[27,59,60] By this process, the very fast one-electron photochemistry can be linked to the several orders of magnitude slower four-electron four proton chemistry of water-splitting. Once the WOC accumulates four oxidizing equivalents, two 'water molecules' are oxidized according to Reaction (9). Kok and coworkers summarized this idea in the S_i state model, which is also known as the Kok model (Figure 7.3B). In this model, the S stands for 'state' and the index gives the number of stored oxidizing equivalents relative to the lowest oxidation state that the WOC attains during the water-splitting cycle (S_0 state).

If PSII samples are incubated with low concentrations of reductants like NH_2OH or NH_2NH_2 the occurrence of the first maximum of oxygen evolution can be shifted back to the 7th flash.[61] This shows that the WOC can be reduced without decomposition by 3 electrons below the S_0 state. This state is denoted as the S_{-3} state. Further reduction gives indications of S_{-4} and S_{-5} states, but, if present, they are very unstable.[62] It is thought that this reduction process might be the reverse to the light-driven photo activation process that is required for the assembly of the active Mn_4CaO_5 cluster.[10,63–66] Together with spectroscopic data (see below) these observations demonstrate that the Mn oxidation states in the S_1 state of the WOC are at least Mn_4(III,III,III,III), but more likely Mn_4(III,III,IV,IV).

It is a surprising finding of Joliot that the first maximum of O_2 evolution is induced in dark-adapted PSII samples by the 3rd flash, and not the fourth flash. This shows that the singly oxidized S_1 state, and not the S_0 state, is the most stable state in the dark. It is possible to synchronize basically 100% of the centers in the S_1 state. The main dark process that converts the S_0 into the S_1 state with halftimes of about 30 min is electron transfer to the oxidized form of tyrosine D, Y_D^{ox}.[67–72] Y_D is an amino acid side chain of the D2 protein (D2-Tyr 160)), which is at a homologous position to Y_Z in the D1 protein (Figure 7.1c). In contrast to Y_Z (D1-Tyr-161), it does not take part in the fast electron transfer reactions leading to O_2 evolution. However, it can be oxidized by $P680^{\bullet+}$ or by the reaction with S_2 (to S_1) or S_3 (to S_2).[73–78] In absence of the S_0 state, the stability of the Y_D^{ox} radical is remarkable and can reach up to several hours at room temperature. Similarly, the S_0 state is very stable when Y_D is in its reduced form.[79] The function of Y_D in PSII is not fully understood, and it is possible to replace it by phenylalanine without losing the ability for autotrophic growth.[75,80]

The S_2 and S_3 states can also return to the S_1 state *via* recombination with electrons created during light-induced charge separation, if the electrons are not quickly removed from the acceptor side, *i.e.* from Q_A and/or Q_B (Figure 7.1).[81–84] The temperature and pH dependencies of the kinetics of these back-reactions were studied in great detail. At standard conditions (neutral pH and 20 °C) the S_3 and S_2 state reductions occur with characteristic times of about a few seconds (Y_D) and about one minute (acceptor side).[70–72,78,85]

The S_4 state is a postulated intermediate that the WOC attains during the reaction $S_3Y_Z^{ox} \rightarrow S_0Y_Z$. In principle one should expect several intermediates during the O_2 formation, but due to their very high reactivity none of these have been detected yet (see Section 7.8).

7.5 Structure of the Water-Oxidizing/Oxygen-Evolving Complex

The WOC/OEC is a unit within PSII that is defined by its function; it is not a specific, isolatable protein subunit, or just the inorganic Mn_4CaO_5 cluster. No size can be defined at present, since it is not clear how many amino acids are required for water-splitting. On the basis of X-ray crystallography (see above), all but one of the direct ligands to the inorganic cluster of the WOC are suggested to be provided by the D1 subunit of PSII. The seventh protein ligand, Glu354, is from the CP43 subunit.[34–37,41,43] It is likely that at least parts of the second and third co-ordination spheres, which include at least one Cl^- ion,[34,38,42] are also important for tuning the redox potentials of the Mn_4CaO_5 cluster, ensuring its stability, and/or regulation of proton removal, water access and O_2 release.[86,87] The WOC is protected from the aqueous lumen by extrinsic loops of the CP43 (and CP47) protein and by three extrinsic proteins, which are slightly different for cyanobacteria and plants. Manipulations of this shield often lead to instability of the Mn_4CaO_5 cluster and/or to reduced water-splitting activity.[88] Channels likely exist that regulate water-entry, and the release of O_2 and protons (Figure 7.1c).[10,34,39,51,52,89,90]

The WOC is electronically coupled to the photo-active, chlorophyll-containing reaction center (P680/Pheo) *via* the redox-active Y_Z of the D1 protein (Figure 7.1c,d and Section 7.3). It is *via* this relay that the fast (ps) charge separating reactions are linked to the much slower (μs to ms) kinetics of the S_i state redox transitions in the WOC (this will be discussed in more detail in Section 7.8). Since Y_Z is intimately connected to the reactions of the WOC, Y_Z is considered to be part of the WOC. However, because it is spectroscopically easily distinguishable from the Mn_4CaO_5 cluster, we will treat it as a separate redox component and do not include its oxidation state in the S_i state notation. The distance of Y_Z to the Ca is about 5 Å and to the nearest Mn about 7 Å (Figure 7.1c,d).[34,36,41]

The inorganic core of the WOC consists of four manganese ions, one calcium ion and several connecting μ-oxo bridges. Until very recently, the crystal structures were limited to a resolution of 2.9 Å,[34] and suffered from specific radiation damage to the WOC (reduction of manganese accompanied with loss of structural integrity) during data collection,[50,91] and therefore did not allow the determination of the precise structure of this cluster. As described in detail below, spectroscopic measurements and DFT calculations suggested the presence of 4–6 oxo bridges, and the inorganic core was therefore denoted as a Mn_4O_xCa cluster. Similarly, the number of water-derived terminal ligands (not included in $x = 4$–6) was uncertain and may vary with S_i state. The newest crystal structure of photosystem II has a resolution of 1.9 Å and the amount of radiation damage appears to be significantly lower than in the previous studies.[1] This study identifies five oxo bridges, in good overall agreement with structures derived by polarized EXAFS spectroscopy and a DFT-based model of the WOC by Siegbahn.[3,92,93] Theoretical calculations of EPR and ENDOR parameter of the S_2 state and polarized EXAFS data on PSII single crystals in the S_1 state show that the Siegbahn model is consistent with spectroscopic data.[94–96] Based on our current knowledge, the inorganic core is therefore very likely a Mn_4CaO_5 cluster. Nevertheless, further testing of this model is required. Below, we discuss in detail various recent proposals for the structure of the WOC. Since these models have different numbers of oxygen bridges, we use the older Mn_4O_xCa cluster notation for this section.

Apart from X-ray crystallography, several spectroscopic methods provide invaluable information on the geometric and electronic structures, not only of the dark-stable S_1 state, but also of the other S_i states, except for S_4. Extended X-ray absorption fine structure (EXAFS) spectroscopy can provide highly precise geometric constraints for the Mn_4O_xCa cluster, especially when performed on single crystals.[3,97] It is noteworthy that polarized EXAFS on single crystals does not directly reveal the structure of the Mn_4O_xCa cluster but allows for collecting of data that are powerful tests for proposed models. The precision of this filtering is in part dependent on the EXAFS simulation procedures. EXAFS simulations require some experience and calibration based on model studies. Advanced electron paramagnetic resonance (EPR) techniques, such as pulsed ^{55}Mn ENDOR (electron nuclear double resonance) deliver very valuable constraints for suggested models by determining the hyperfine and exchange coupling constants, but are unable to provide directly the overall structure of the Mn_4O_xCa cluster.[98,99] Similarly, Fourier transform infrared (FTIR) difference spectra[100–103] and X-ray absorption near edge spectroscopy (XANES) spectra[104,105] for the S_i state transitions provide constraints for discriminating model structures for the Mn_4O_xCa cluster. However, these latter data have not yet been quantitatively explored in theoretical calculations. The most direct information on the S_i state dependence of the binding of the two substrate 'water' molecules (for simplicity we refer to them as water independent of their protonation state) comes from time resolved membrane inlet mass spectrometry (TR-MIMS).[106–108] These water-exchange data are discussed in detail in Section 7.7, since they provide strong constraints for DFT models and mechanistic proposals.

Computational methods such as QM/MM (quantum mechanics/molecular mechanics) and DFT (density functional theory) are currently employed by a number of groups in efforts to bridge the gaps between the various techniques.[94,95,109,110] The aim is to unify the information within testable models. Because of the size and complexity of the WOC, these calculations are very challenging, and often require significant method development and testing on well characterized models.

Figure 7.4 presents 8 recently proposed QM/MM and/or DFT based models for the WOC in the S_2 state of PSII. They all have in common that the inorganic core consists of a μ_3-oxo bridged Mn_3 unit, one 'outer' Mn (in Figure 7.4 shown on the left side of the Mn_3 trimers), and one Ca^{2+} ion. Differences exist in the number of oxo-bridges ($x = 4$, 5 or 6) and the bridging pattern (Mn–Mn distances), the position and binding mode of Ca^{2+} relative to the four Mn ions, orientation of the core within the protein, position of Cl^-, and the arrangements of the amino acid ligands around the metal ions. The required structural

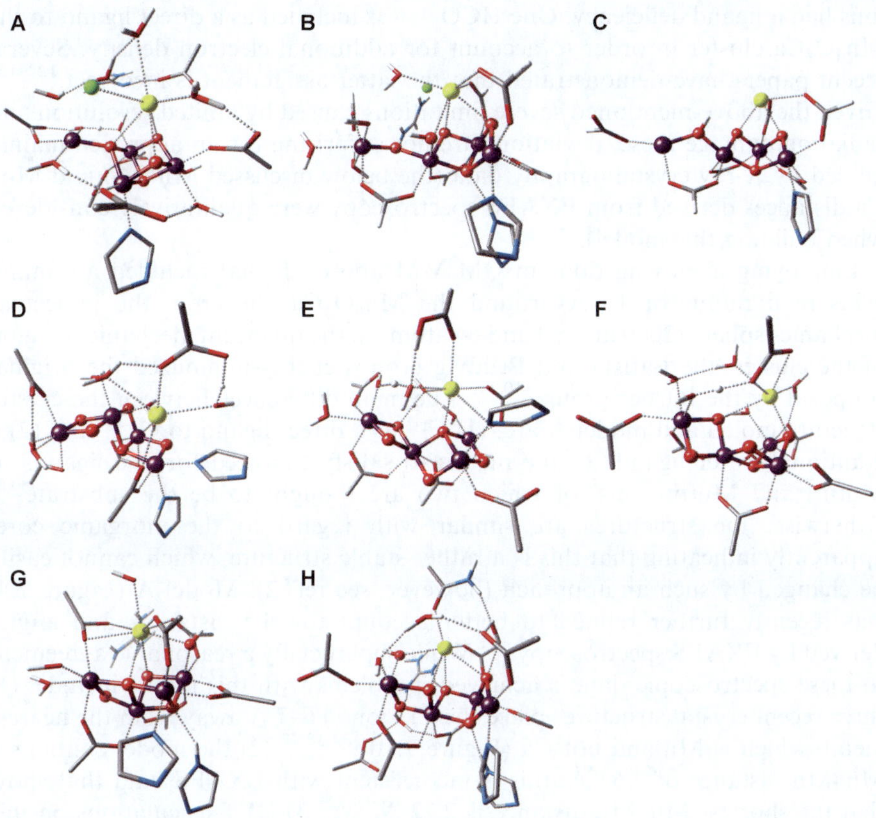

Figure 7.4 Molecular mechanics (MM) and density functional theory (DFT) models of the WOC in the S_2 state. Model A, Batista/Brudvig;[111–113] Model B, Kusunoki;[114] Model C, Petrie/Stranger/Pace;[115–117] Models D-F, Yano/Yachandra/Neese/Lubitz/Messinger;[3,118,119] Model G, Dau/Haumann;[14,120] Model H, Siegbahn.[93] For some models the protein environment was truncated for clarity of presentation. All models are oriented in the same way in a side view onto PSII and the thylakoid membrane, so that the stromal side is on top, and the lumenal side towards the bottom.

starting information for the calculations came in some cases from the PSII crystal structural models, in other cases more importance was put on the spectroscopic parameters (EXAFS, EPR/ENDOR) of the Mn_4O_xCa cluster, or to the minimum O–O bond formation energy.

Model A. The starting point for model A proposed by Batista *et al.*[111,121] was the 3.5 Å resolution crystal structure of PSII from *T. elongatus* that was published by Barber and colleagues in 2004.[41] Barber *et al.* proposed a Mn_4O_4Ca cluster (not shown) that consisted of an almost regular Mn_3O_4Ca cube to which the forth Mn is attached through one of the bridging μ_3-oxygens (making it μ_4). In their structure, most ligands to Mn and Ca^{2+} were thought to be monodentate to a single metal, which resulted in a situation where most Mn ions had a ligand deficiency. One HCO_3^- was included as a direct ligand to the Mn_4O_4Ca cluster in order to account for additional electron density. Several recent papers have demonstrated that the latter assignment is incorrect.[122–124] Given the above-mentioned severe limitations caused by limited resolution and radiation damage these, deviations are not surprising for an approach mainly guided by X-ray crystallography data (the below discussed Mn–Mn and Mn–Ca distances derived from EXAFS spectroscopy were qualitatively considered when building this model).

Employing a moving domain QM/MM approach that included all amino acids in a radius of 15 Å around the Mn_4O_xCa cluster in the molecular mechanics sphere (2000 atoms) and 64 atoms in the quantum mechanical region of the cluster, the Batista and Brudvig groups energy-minimized the original proposal by the Barber group.[110,113] The main differences between the crystal structure model and model A are: (1) Cl^- as a direct ligand to Ca^{2+} and (2) 6 additional water ligands to the metals to satisfy the need for 5–6 ligands to Mn(III) and Mn(IV) ions, of which two are thought to be the substrate.[121] Otherwise, the structures are similar with regard to the inorganic core, apparently indicating that this is a rather stable structure, which cannot easily be changed by such an approach (however, see ref. 2). Model A (Figure 7.4) was recently further refined to better account for the distances and angles derived by EXAFS spectroscopy.[111] While superficially a reasonable agreement to these spectroscopic data is achieved, problems with this model include: (1) three recent crystal structures place the Cl^- ion(s) 6–7 Å away from the nearest metal, which is Mn and not Ca (Figure 7.1d),[34,38,42] (2) the model contains a Mn–Mn distance of 2.63 Å; this is inconsistent with EXAFS data that show that the shortest Mn–Mn distance is 2.72 Å,[18,125] 3) DFT calculations on this model yield a ground spin state of $S = 7/2$ for the S_2 state as opposed to $S = 1/2$ found experimentally.[94] In this model, the slowly exchanging substrate water[11,106] is suggested to bind to Ca^{2+}, while the fast one is thought to bind to the outer Mn.[121] While the proposed structures for the S_1 and S_2 states (Figure 7.1A) are hard to reconcile with the 1.9 Å crystal structure,[1] the proposed S_0 state configuration shows some similarities with regard to the oxygen bridges.[121,126]

Model B. This model was developed by Kusunoki on the basis of the crystal structure by Loll *et al.*[35] and further constrained by time resolved ^{18}O substrate water exchange experiments of Messinger, Hillier and Wydrzynski.[106,127] The starting structure was optimized and evaluated with regard to water-exchange kinetics by employing a hybrid DFT method. The derived $Mn_4CaO_4(OH)$ cluster is reminiscent of an earlier model by the author that was based on EPR simulations of the S_2 state.[128] Cl^- is proposed in model B to be bound to CP43-Arg357, *i.e.* 3.8 Å away from Ca^{2+}, which is in conflict with subsequent crystallographic data.[34,38,42] One interesting aspect of this work is that the author suggests that for each S_i state at least two tautomers of similar total energy exist, which differ in the position of one or more protons. As a consequence, the Mn oxidation state distribution and/or electron spin distribution over the cluster varies in some cases between the tautomers. In total, the models include five terminal water/hydroxo ligands, of which two are ligated to Ca and three to the outer Mn. In addition, there is a hydroxo bridge between the central Mn and Ca (in contrast to the author we consider the 2.6 Å distance of this OH to the Mn on the right side to be too long for a bond). It is concluded that two of the water-derived ligands at the outer Mn represent the fast and slow exchanging substrate water molecules revealed in the time resolved $H_2{}^{18}O$/$H_2{}^{16}O$ exchange experiments, because those are found to provide the best match for the S_i state dependence of the exchange rates and activation energies (Section 7.7). The model includes two long Mn–Mn distances of 3.3 Å and 3.2 Å, and two short Mn–Mn distances of 2.76 Å and 2.82 Å. In contrast to model A, but in agreement with the Loll *et al.* crystal structure and the polarized EXAFS data discussed below (models D–F), one of the long distances is within the trimeric part, breaking its symmetry. Detailed comparison with EXAFS data was published recently.[295]

Model C: Petrie, Stranger and Pace constructed three models with the common formula $[CaMn_4C_9N_2H_{10}]^+ \cdot nH_2O$ (n = 0–7) that reflect the different geometries suggested in three earlier crystal structures.[35,41,43,117] These starting models were then subjected to energy minimization by DFT calculations. Depending on the hydration level, similar total energies were found for all three structures. This was interpreted by Pace *et al.* to suggest a possible structural flexibility of the Mn_4O_xCa cluster in the S_1 state, as an alternative explanation to radiation damage.[117] However, the relatively small size of the models that contain only few selected ligands may have prevented meaningful conclusions with regard to the total energies of the different configurations within PSII. The model most favored by Pace *et al.* is depicted in Figure 7.4C. It is a Mn_4CaO_4 cluster that is mostly based on the 3.0 Å structure by the Zouni group[35] and closely resembles the more complete model B.[114] In a subsequent paper, the Mn–Mn distances and the electronic structure were investigated.[116] In their most recent study, they further constrained the model by calculating the S_i state dependence of the activation energies of substrate water exchange. On that basis, they suggest that one water is bound along the long Mn–Mn distance within the distorted cuboidal part.[115] While this binding site is similar to the one proposed by Kusunoki (model B), they suggest that this water is the fast exchanging substrate, not the slow one. Like Brudvig and Sproviero, Pace and coworkers identify W_s as a ligand to Ca^{2+}.

Models D, E and F: This family of models has two starting points: the information on the protein environment comes from X-ray crystallography and the structural information about the Mn_4O_xCa cluster from extended X-ray absorption fine structure (EXAFS) spectroscopy. They are extensions of previous simpler models for the core by DeRose[129] and Robblee.[4]

As a model for the protein environment the 3.4 Å crystal structure of Zouni *et al.* was chosen.[35] Because of the above discussed uncertainties stemming from limited resolution and Mn-specific radiation damage, Zouni and coworkers refrained from proposing a detailed structural model for the

Mn_4O_xCa cluster; only the most likely positions for Mn and Ca ions and the ligands were indicated. An important difference to the above discussed crystal structure model of Barber *et al.*[41] is that in the crystal structure of Zouni *et al.* most carboxylate amino acid ligands were assumed to bridge between two metal ions.

The starting point for the structure of the Mn_4O_xCa cluster were four models that were derived by Yano *et al.*[3] These models were built on the basis of information on Mn–Mn and Mn–Ca distances and on Mn coordination numbers obtained from solution and oriented EXAFS studies on Mn and Ca/Sr, and on the trends of the dichroism observed in polarized EXAFS experiments on PSII single crystals.[3] These four models are structurally highly related and they have defined orientations within the PSII coordinates. They all have a Mn_4CaO_5 core structure, *i.e.* contain five oxo bridges. This core structure of models D–F deviates substantially from that of model A, because the 3.3 Å Mn–Mn distance is not between the outer Mn and the Mn_3 unit, but within the μ_3-oxo bridged Mn-trimer, and the outer Mn is connected to the Mn_3 unit *via* two μ-oxo bridges. The Ca position with respect to the Mn_4O_5 core was left somewhat ambiguous due to the uncertainty of the Mn–Ca dichroism in the single crystal XAS that arises from the small contribution of Ca to Mn EXAFS and its overlap with the Mn–Mn contribution at ~3.3 Å. However, accurate Mn–Ca distances were obtained by Ca(Sr) EXAFS[130,131] and therefore incorporated into models D–F. Placing these models in the 3.4 Å crystal structure such that the geometric centers of their Mn_4 units coincide with that of the proposed Mn_4 unit in the crystal structure resulted in the models of the WOC reported by Yano *et al.*[3] It is noted that the contributions of the terminal ligands were omitted in the above EXAFS simulation, mostly due to a small contribution of the numerous terminal ligands to the trend of EXAFS dichroism, and partly due to the uncertainty of the ligand positions in the reported crystal structures that arises from the aforementioned moderate resolution and the effects of radiation damage.[3,50] Therefore, the relative positions of the WOC and the ligand positions were unsatisfactorily resolved in the EXAFS models. In order to achieve chemically satisfying ligand positions and ligation modes, a preliminary set of 10 models was constructed in a separate study, in which the truncated amino acids were initially moved to feasible positions without taking possible backbone restrictions into account. During DFT-based energy minimization of the ligand positions, all the models converged.[118] They displayed only minor additional adjustments when the metal ions and the bridging oxygens were also included in the optimization procedure. This suggests that these models are reasonably stable structures (see however, ref. 2). The models shown in Figures 4D–F are three of these 10 structures. These models were then further tested against spectroscopic data by calculating the ground spin state for the paramagnetic S_2 state (Mn_4(III,IV,IV,IV)) and of the Mn hyperfine constants.[132] All models displayed either a $S = 1/2$ or a $S = 5/2$ ground spin state, which is consistent with the EPR S_2 multiline signal and the $g = 4.1$ signal of the S_2 state, respectively.[133–138] Several models also had hyperfine coupling constants reasonably close to the experimental ones determined by ^{55}Mn ENDOR spectroscopy.[119,139,140]

Total energy calculations performed by Siegbahn for some of the models displayed in Figure 7.4 show that model A and models D–F have significantly higher total energies than model H discussed below. Note that models B, C and G were not included in this comparison.[2]

Model G: Model G was subsequently proposed by Dau *et al.* on the basis of the crystal structure of Loll *et al.*[35] and a model for the Mn_4O_xCa cluster that was constructed to match the EXAFS results obtained in this group with isotropic and unidirectional oriented PSII samples (the quality of this agreement was however not presented).[14,120] Molecular mechanics calculations were employed to bring the surrounding amino acids into suitable ligand positions. This process was used, as outlined above, to compensate for radiation-induced modifications in the ligand environment. The obtained $Mn_4CaO_4(OH)$ model is the most compact of all, containing a μ_4-oxo that connects the outer Mn with Ca and with two of the Mn ions in the Mn_3 unit. An interesting feature is a μ-OH bridge between one Mn of the Mn_3 unit and Ca^{2+}. This OH bridge is assumed to be the site of proton release during the structural change in the $S_2 \rightarrow S_3$ transition. The only Mn(III) ion in the S_2 state (the one with the His ligand, hereafter named Mn_{His}) is assumed to be nearly 5 coordinated (the Mn–OH distance is >2.4 Å) in the S_1 and S_2 states and six coordinated in the S_3 state. Similar to models D–F the cuboidal part is incomplete, but in variation of these models, the long site is between the two Mn ions depicted on the right side in Figure 7.4G, not towards to the central Mn. In addition to this 3.3 Å Mn–Mn distance, a second Mn–Mn distance of about 3.3 Å is proposed to exist within the trimer in the S_0–S_2 states, so that the $Mn_4CaO_4(OH)$ cluster contains only two short Mn–Mn distances. Similar to other models, two water/hydroxo ligands are placed on the outer Mn. Presently, no simulations of the polarized EXAFS or calculations of the electronic structure of this model are published.

Model H: Siegbahn was the first to apply DFT calculations to the WOC.[141,142] As described above, DFT calculations are being applied to understand the geometric and electronic structure of the WOC and to elucidate its mechanism of water-splitting. While the above described studies started mainly from structural information provided by crystallography and EXAFS on the well characterized S_1 and S_2 states, Siegbahn opted to start with the search for the geometry and electronic structure that results in the lowest energy barrier for O–O bond formation, and only later incorporated crystallographic and spectroscopic information. Therefore, his starting point was the spectroscopically uncharacterized S_4 state to which he then added electrons and protons to derive structures for the lower S_i states.[92,93,143,144]

The bridging pattern for the metal ions in model H is basically a mirror image to model F (open edge of the Mn_3 unit is in the front and not in the back) and is also closely related to model D (here the bridging motif is the same, but the geometry around the central Mn differs). It is therefore not surprising that model H, like models D–F, has a ground spin state of $S = 1/2$ and reasonable

Mn hyperfine couplings in DFT calculations.[132] The latest study by Cox *et al.* demonstrates that the Siegbahn model is currently the only model that is in agreement with all EPR and ^{55}Mn ENDOR spectroscopy data.[94] The good agreement of this model with EPR/ENDOR data has been also demonstrated in a recent study by the Kaupp group. They showed that model H has the best agreement with experimental ^{14}N and ^{13}C couplings to the Mn_4CaO_x cluster.[95] Compared to models D–F, model H has one extra hydroxo group that forms a bridge between Ca and the outer Mn ion. As a result, the cluster in model H has a $Mn_4CaO_5(OH)$ stoichiometry. The ligand arrangement around this $Mn_4CaO_5(OH)$ cluster is very similar in models G and H, because they are both based on the arrangement suggested in the crystal structure of Loll *et al.*[35] It is also noteworthy that the cuboidal parts of structures B and H are very similar: model B (Kusunoki) can be transformed into model H (Siegbahn) by simply connecting the outer Mn to the OH bridge between the central Mn and Ca of model B. Unpublished polarized EXAFS simulations on model H show a good agreement with the data of Yano *et al.*[3,96]

At the International Photosynthesis Congress in Beijing (August 2010) the 1.9 Å crystal structure of PSII was presented.[1] Since the coordinates were not published at the time of writing this review, a detailed comparison could not be included. In general, this new model appears to be most similar to model H. However, two differences are obvious. Firstly, Asp170 bridges the outer Mn and Ca instead of the hydroxo bridge. This agrees with the arrangement shown Figure 7.1d.[34] Secondly, the central oxygen bridge in the front is thought to be in a more central position so that it also connects to Mn_{His}. Therefore, in the 1.9 Å crystal structure all Mn are six coordinated. In terms of the number and length of Mn–Mn interactions, the 1.9 Å model is in agreement with the Berkeley EXAFS parameters,[3,4] showing one long ($\sim$3.2 Å) Mn–Mn and three short ($\sim$2.7 Å) distances, where the long Mn–Mn distance is within the cuboidal part, breaking its symmetry.

While there are some remaining discrepancies between X-ray structure and spectroscopic and DFT-derived information, the structure of the dark-stable S_1 state appears now to be quite well characterized. Significant further efforts are required to determine the structures for the other S_i states. According to the solution EXAFS data, the Mn_4CaO_5 cluster appears to have almost identical structures in the S_1 and S_2 states, while clear structural changes can be detected for the $S_0 \rightarrow S_1$ and $S_2 \rightarrow S_3$ state transitions that are reversed during the $S_3 \rightarrow S_4 \rightarrow S_0$ transition.[4,105,145] In the S_0 state one of the three 2.7 Å distances seen in the S_1/S_2 state models is elongated to 2.85 Å. This is consistent with a deprotonation of one μ-hydroxo-bridge bridge in a Mn(–OH–)(–O–)Mn unit during the $S_0 \rightarrow S_1$ transition. Based on ^{55}Mn ENDOR simulations, it was suggested that one of the two bridges to the outer Mn is the site of deprotonation (see models D–F and H in Figure 7.4).[4,16,146,147] The $S_2 \rightarrow S_3$ transition occurs with a high reorganization energy (1.6 eV)[148,149] and one EXAFS study suggested that an elongation of some Mn–Mn distances accompanies this transition.[130,145] This was confirmed in recent extended EXAFS results (unpublished data by Yano *et al.*). In contrast, in another EXAFS report, the formation of an additional 2.7 Å Mn–Mn distance is suggested, probably by losing the 3.3 Å Mn–Mn distance within the 'cube'.[105] It will be important to solve this latter discrepancy, since the structure for the S_3 state is crucial for deriving the mechanism of O–O bond formation. A detailed mechanism for water oxidation (Section 7.9) based on model H that includes structures for all the S_i states has been proposed by Siegbahn.[92,93]

7.6 Storage of Oxidizing Equivalents by the Mn_4CaO_5 Cluster

For a detailed understanding of how the oxidizing equivalents are first stored by the Mn_4CaO_5 cluster in the WOC and how the electrons are subsequently removed from the substrate molecules and the O–O bond is formed, the electronic structures of the Mn_4CaO_5 cluster need to be known for all the intermediates of the Kok cycle. Initially, UV spectroscopy was employed, but the spectra are quite featureless and difficult to deconvolute from other contributions, such as the absorption changes by the acceptor side quinones. No consistent interpretation on the Mn oxidation state changes during the catalytic cycle was obtained.[150–154] On the other hand, transient UV experiments provide useful and reliable information on the kinetics of the S_i state turn over by following absorption changes due to the oxidation and reduction of Y_Z.[154–156] The discovery of the S_2 EPR multiline signal by Dismukes and Siderer in 1981 and Hansson and Andreasson in 1982 opened the door towards the detailed elucidation of the electronic structure of the S_2 state.[133,157] These data showed that Mn was oxidized during the S state cycle, and that the S_2 state is a mixed valent state containing a coupled $Mn^{III}Mn^{IV}$ ion pair. This result was soon after confirmed by the use of Mn X-ray near edge spectroscopy (XANES; see below).[158] The S_2 multiline EPR signal (total spin $S = 1/2$) allowed detailed work on the interaction of small molecules (including water, see below) with the Mn_4CaO_5 cluster in the S_2 state and allowed for assessment of the effects of various biochemical treatments.[9,15,134,159] Subsequently, EPR signals for higher spin states of the S_2 state,[160] and for all other S_i states were found that allowed the determination of their total spin states (for review see ref. 134).[138,161–163] The discovery of the S_0 state EPR multiline signal [164–167] and its subsequent exploitation for ^{55}Mn ENDOR experiments[140,146] led to the conclusion that the S_0 state does not contain a Mn^{2+} ion, and that accordingly all S state transitions up to S_2 represent Mn^{III} to Mn^{IV} oxidations. This finding excluded some models popular at that time that assumed that only one of the four Mn ions is redox active.[13,168]

While the EPR data contain in principle all the information on the geometric and electronic structures of the paramagnetic S_i states, traditional simulations of the data rarely give unique answers. This dilemma is caused by the large number of parameters that govern the shape of the EPR spectra, especially in cases of multi-nuclear transition metal complexes of unknown geometry. Nevertheless, information about the spin states, *g*-factors, and hyperfine interactions[15,119,139,169–171] are highly important, specific, and sensitive constraints for QM/MM- and DFT-based models of the WOC.[94,132,172]

Alternative approaches for determining the electronic structure of the Mn_4CaO_5 cluster are X-ray absorption and X-ray emission spectroscopy.[97,173] The X-ray absorption near-edge structure ('XANES' region) is sensitive to the Mn oxidation states and the ligand environment. In the case of Mn, it probes the forbidden and therefore weak transitions of 1s electrons into unoccupied 3d orbitals (pre-edge) and the allowed 1s → 4p transitions (main edge). During the S state transitions, a clear shift in the energy of the XANES is observed in the $S_0 \rightarrow S_1$ and $S_1 \rightarrow S_2$ transitions, indicating Mn-centered oxidation state changes. However, a much smaller energy shift and a clear change in edge shape was observed for the $S_2 \rightarrow S_3$ state transition, raising the possibility of a ligand-centered oxidation state change during this step. This was reported in the initial work by Klein and co-workers,[158,174,175] and was subsequently confirmed by further studies of the group at Berkeley.[176,177] Other groups have reported XANES data that showed comparable shifts for the $S_1 \rightarrow S_2$ and $S_2 \rightarrow S_3$ transitions.[178,179] The recent revised data from Dau *et al.*[105] is, however, practically identical to that by Messinger *et al.* (Figure 7.5).[177] Dau and coworkers suggested an alternative interpretation for this finding. They interpret the small edge shift in the $S_2 \rightarrow S_3$ transition to reflect a Mn-centered oxidation state change that is accompanied by structural changes. Specifically, a change from a five coordinated Mn^{III} to a six coordinated Mn^{IV} ion is suggested.[104,105]

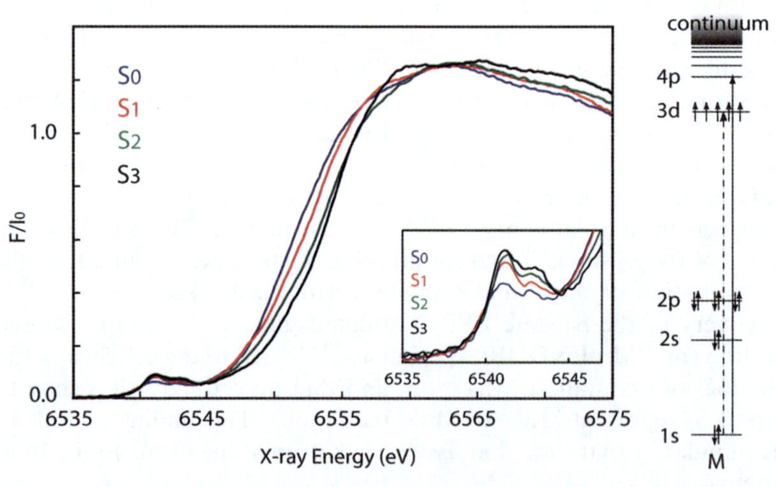

Figure 7.5 Mn K-edge X-ray absorption near edge spectroscopy (XANES) spectra of the S_0, S_1, S_2 and S_3 states of the water-oxidizing complex in photosystem II.[177] The inset shows an enlargement of the pre-edge region. The scheme on the right side indicates the energy levels in Mn^{2+} and their occupancy with electrons (short arrows). The long dashed vertical arrow indicates the forbidden (low intensity) pre-edge transition, while the long full arrow signifies a transition contributing to the main edge.

More detailed studies and theoretical interpretations, especially of the dichroism of the pre-edge spectra observed in PSII single crystals, need to be performed on the basis of the best available structural models of the WOC to fully understand the nature of the $S_2 \rightarrow S_3$ transition.[3]

Complementary data about the oxidation state changes of the WOC can be obtained by X-ray emission experiments. These techniques monitor the energy of photons that are emitted when core holes (*i.e.* missing electrons in the 1s shell, created by X-ray irradiation) are filled with electrons from higher occupied orbitals.[173] Of special interest are the transitions of 3p electrons into the 1s level ($K\beta_{1,3}$ and $K\beta'$ transitions). As 3p electrons are close to the valence (3d) electrons, the transition energy reflects primarily the charge density of 3d orbitals rather than the ligand symmetry.[104] These data confirmed the earlier conclusion by Roelofs *et al.*[176] that the $S_2 \rightarrow S_3$ transition is clearly different from the other transitions; compared to the $S_0 \rightarrow S_1$ or $S_1 \rightarrow S_2$ transitions, the charge density on Mn changes significantly less in the $S_2 \rightarrow S_3$ transition.

A further extension of this approach is to obtain two dimensional resonant inelastic X-ray scattering spectra (RIXS). This technique measures L-edge like spectra and K-edge pre-edge spectra simultaneously. While K-edge pre-edge spectra (metal 1s to 3d transitions) are sensitive to the charge density of metals and to the ligand symmetry, L-edge spectra (metal 2p to 3d transitions) are also sensitive to the spin density of 3d orbitals by spin-orbit couplings. In general, shorter attenuation length in the soft X-ray energy compared to the hard X-ray energy makes transition metal L-edge spectroscopy much harder due to two orders of magnitude faster radiation damage. The RIXS spectroscopy overcomes this problem by using K-edge energy as an excitation energy, and L-edge-like spectra are collected through two electron processes.[173] In PSII, the RIXS data for the S_1 and S_2 states show that a strong charge delocalization occurs within the Mn_4CaO_5 cluster that includes oxygen bridges.[180,181]

In summary, the present data suggest that there are Mn^{III} to Mn^{IV} type oxidations on the $S_0 \rightarrow S_1$ and $S_1 \rightarrow S_2$ transitions, *i.e.* the 'formal' Mn oxidation states are: S_0 = Mn_4(III,III,III,IV), S_1 = Mn_4(III,III,IV,IV), and S_2 = Mn_4(III,IV,IV,IV). The nature of the subsequent $S_2 \rightarrow S_3$ transition is at present not yet fully understood. For sure the Mn K-edge shift is different from the earlier transitions, and it remains to be shown if this difference can be quantitatively explained by the known concomitant structural change, or whether a partial or full ligand (oxo-bridge) centered oxidation needs to be invoked.

To fully describe the electronic structure, not only do the overall Mn oxidation states of the Mn_4CaO_5 cluster need to be known, but also the oxidation states of the individual Mn ions have to be assigned. Here the controversy exists, whether at room temperature the oxidation states are fully localized, or instead delocalized over the cluster. While based on liquid helium temperature EPR/ENDOR spectroscopy the spin states in S_0 and S_2 appear to be trapped on the Mn ions, the charge appears to be delocalized over the metals and ligands according to the RIXS measurements.

Nevertheless, based on general considerations and DFT calculations, specific predictions about which Mn is oxidized in which S_i state transition have been

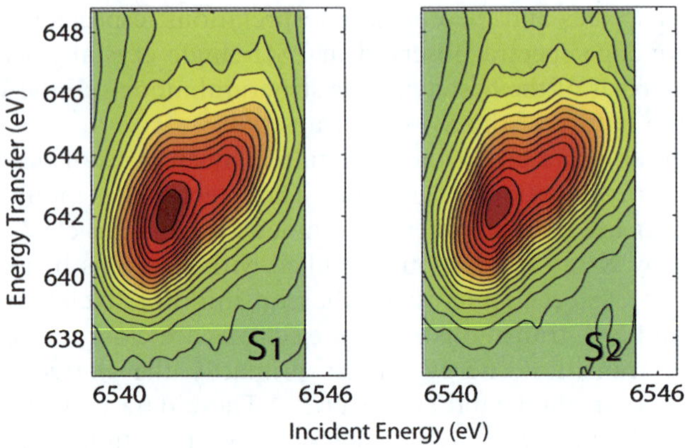

Figure 7.6 Resonant inelastic X-ray scattering spectra (RIXS) of the S_1 and S_2 states of the WOC in photosystem II.

made.[92,94,118,119,147,182] However, it is beyond the scope of the present review to discuss these proposals in detail. Important test cases for these types of predictions are FTIR S_{i+1}–S_i difference spectroscopy studies on PSII samples, especially those in which ligands to the Mn_4CaO_5 cluster have been exchanged by point mutations.[103] For example, significant frequency shifts of the vibrations of the carboxylate ligands from D1-Ala344 (α-COO^-)[183–187] and CP43-E354[188,189] in the $S_1 \rightarrow S_2$ transition have suggested the oxidation of the Mn ion(s) ligated by these carboxylate groups. While the interpretation of the FTIR difference spectra of various mutants that replace the putative ligands are not straightforward,[111,190] the correct model for the WOC should form the basis for their quantitative analysis. Further important inputs for deriving the electronic structures of the Mn_4CaO_5 cluster in the different S_i states are provided by polarized XANES Mn pre-edge and K-edge data,[3] and proton, nitrogen and ^{13}C hyperfine couplings determined by advanced EPR techniques for the S_0 and S_2 states.[170,191–195]

7.7 Substrate Water Binding to the Mn_4CaO_5 Cluster

Studying substrate binding to the WOC is not an easy task, since aqueous buffers are required to keep PSII in a functional state.[196] Therefore, isotope labeling, often in combination with time-resolved methods, is required to understand this important aspect of the mechanism of photosynthetic water-splitting. Our current information on water-binding to the OEC was derived by the following methods: time-resolved isotope ratio membrane-inlet mass spectrometry, advanced EPR techniques, such as ESEEM, ENDOR and HYSCORE sensitive to the hyperfine coupling of the nuclear spins of H, D and ^{17}O with the electron spin on the Mn_4CaO_5 cluster in the S_2 and S_0 states,[98] ^1H-NMR relaxation enhancement measurements monitoring the increased T1 relaxation rates of bulk water protons due to the binding and exchange at the Mn_4CaO_5 cluster,[197] and FTIR difference spectroscopy.[100–102] Alternative approaches involving the study of the effects of water analogs such as NH_3, H_2S, CH_3OH, NH_2OH, and NH_2NH_2 on water-splitting and binding have been reviewed previously, for example in ref. 9 and 198. Those latter studies will not be covered here.

TR-MIMS. The most direct information about *substrate* water binding to the WOC is obtained by time-resolved membrane inlet mass spectrometry studies (TR-MIMS), since the isotopic composition of the *product* is studied with this technique. In brief, PSII samples suspended in normal ($H_2{}^{16}O$) buffer are flashed into a certain S_i state within the mass spectrometer sample cell. Then the binding of water to this S_i state is probed by rapid (few ms) injecting and mixing of $H_2{}^{18}O$ (free of dissolved O_2) into the TR-MIMS cell at various times before the O_2-evolving flash sequence. Simultaneous detection of mass peaks 32, 34 and 36 allows determining the percentage of exchanged substrate water for each mixing time.[106,127] Such measurements demonstrate that the two substrate molecules bind at chemically different sites so that one can distinguish between a fast (W_f) and a slow (W_s) exchanging substrate 'water' molecule (for convenience we refer to the substrate molecules as water independent of their actual protonation state).[106,199] The slow exchanging substrate water, W_s, is bound in all S_i states and W_f is bound in at least the S_2 and S_3 states.[200,201] It was also shown that the binding affinities of W_f and W_s vary with S_i state in a non-intuitive way. Importantly, the exchange rate of W_s practically does not change during the $S_2 \rightarrow S_3$ transition, despite the structural changes that are known to occur in this transition (Section 7.3).[10,11,108,200] In addition to answering the question of how many substrate molecules are bound in a certain S_i state, it is also possible with this technique to reveal qualitative information about the binding sites. For example, it was shown that the exchange of Ca *vs.* Sr in the WOC weakens the binding of W_s, which suggests the involvement of Ca in binding this substrate molecule (Ca/Sr exchange leads to, at best, very minor structural changes in the inorganic core and retains water-splitting activity at slightly reduced rates).[94,130,202] Similarly, several point mutations of ligands around the Mn_4CaO_5 cluster that retain some O_2-activity were studied.[103] It was found that in some of such mutants, the exchange rates of one or both substrate molecules were affected.[203,204] No final picture about the sites of substrate water-binding arises from these data yet, but they provide valuable constraints for models of the WOC.[10,108]

It is important to note that the TR-MIMS data do not give direct information about whether the substrate molecules are bound as water, hydroxo or oxo, and if binding occurs in the terminal or bridging position. Comparison of exchange rates measured in PSII to those determined for Mn(III) or Mn(IV) model complexes is important, but very difficult.[205–208] Based on such comparisons, it was argued for example that the substrate cannot be bound in a bridging position.[206] However, exchange rates are highly sensitive to the precise geometry, environment and geometric flexibility of the complex. Detailed theoretical calculations of water exchange rates based on realistic models of the

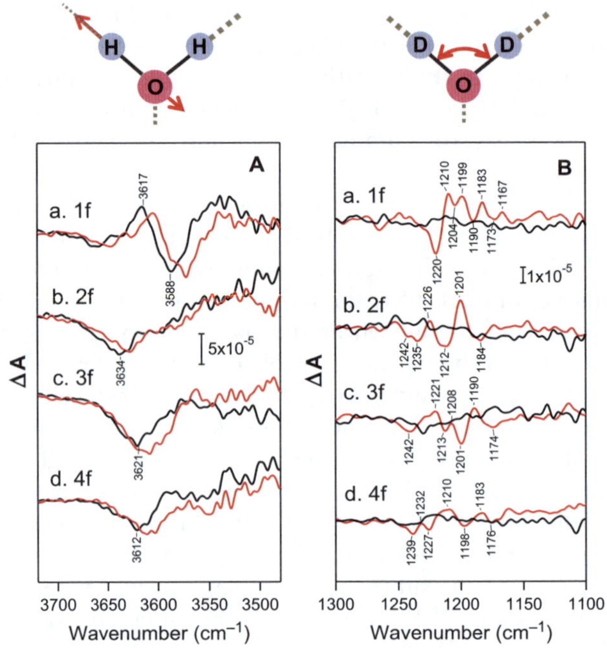

Figure 7.7 (A) Flash-induced FTIR difference spectra in the OH stretching region during the S-state cycle of WOC in PSII core complexes moderately hydrated using $H_2{}^{16}O$ (black lines) and $H_2{}^{18}O$ (red lines). Difference spectra were recorded upon the 1st (a), 2nd (b), 3rd (c), and 4th (d) flash. (B) DOD bending bands of D_2O coupled with the S-state transitions revealed as $D_2{}^{16}O$-minus-$D_2{}^{18}O$ double difference spectra (red lines) during the S-state cycle at the 1st (a), 2nd (b), 3rd (c), and 4th (d) flash, in comparison with the corresponding $H_2{}^{16}O$-minus-$H_2{}^{18}O$ double difference spectra (black lines).

WOC (Section 7.5) will be required to determine the binding modes and sites based on TR-MIMS and other data.[209]

FTIR. An independent powerful approach is FTIR difference spectroscopy, which can directly detect the vibrations of water molecules in the WOC, if combined with isotope labeling and mutagenesis.[102,210,211] It was demonstrated that the high-frequency region (3700–3500 cm^{-1}) of FTIR difference spectra during the S-state cycle (Figure 7.7A) includes the OH stretching vibrations of water having relatively weak hydrogen bonds.[196,212] A differential signal around 3600 cm^{-1} in the $S_1 \rightarrow S_2$ transition and negative bands at 3640–3610 cm^{-1} in the $S_2 \rightarrow S_3$, $S_3 \rightarrow S_0$, and $S_0 \rightarrow S_1$ transitions all downshifted by several wavenumbers upon $H_2{}^{18}O$ substitution (Figure 7.7A, red lines) and by $\sim$1000 cm^{-1} to the 2700–2650 cm^{-1} region upon H/D exchange, providing definitive assignments of water vibrations. The upshift of the peak in the $S_1 \rightarrow S_2$ transition indicates the change in the hydrogen bond strength of water with a highly asymmetric hydrogen bond structure, while the negative OH bands observed in the other transitions imply either proton release from water or a significant change in the hydrogen bond strength from a weak to strong hydrogen bond.[196,211,212] CP43-Glu354Gln mutation significantly altered the water OH bands in the $S_1 \rightarrow S_2$ transition, and it was proposed that this water molecule is attached to the Mn ion ligated by this carboxylate residue.[189]

Water molecules in WOC were also detected as the DOD bending vibrations of D_2O,[213] which was used instead of H_2O to avoid a severe overlap the HOH bending vibration with the protein amide I bands at $\sim 1650\,cm^{-1}$. DOD bending bands coupled to the S_i state transitions were detected at 1250–1150 cm^{-1} as ^{16}O-minus-^{18}O double difference spectra (Figure 7.7B). The spectra at individual S_i state transitions exhibited 6–8 peaks, indicating that at least two water molecules, not in any deprotonated form, participate in the reaction at each transition throughout the cycle. This may indicate that some water molecules other than substrate water exist in the vicinity of the Mn_4CaO_5 cluster as constituents of the WOC, consistent with structural quantum chemical models and the new 1.9 Å crystal structure of PSII (Section 7.5).[1,14,114,118,126,132] Detailed analysis of the DOD data suggested that substrate water is inserted to the WOC in the $S_2 \rightarrow S_3$ and $S_3 \rightarrow S_0$ transitions from a water cluster within the proteins. These water insertion steps are consistent with the observation that inhibition of the S_i state transition by dehydration is more significant in the $S_2 \rightarrow S_3$ and $S_3 \rightarrow S_0$ transitions than in the $S_1 \rightarrow S_2$ and $S_0 \rightarrow S_1$ transitions.[196] This conclusion on water insertions is in agreement with the theoretical water-splitting mechanism of Siegbahn,[93,190,214,215] but seems to be in conflict with TR-MIMS that suggests that both waters are bound in the S_2 state already (see above).[200] However, the fast rate of exchange determined in the S_2 state is close to the detection limit, so a careful reinvestigation of the fast exchange in S_2 will be required.

The low-frequency region below 1000 cm^{-1} in the FTIR difference spectra of the WOC contains vibrations of the Mn_4CaO_5 cluster core.[216] The bands at 606/625 cm^{-1} in the S_2/S_1 difference spectra, which downshifted by about 10 cm^{-1} in $H_2{}^{18}O$ but were not affected by $^{44}Ca^{2+}$ substitution, were assigned to the Mn–O–Mn vibrations of the Mn_4CaO_5 cluster.[217] The low-frequency spectra during the S-state cycle were further analyzed using water isotopes ($H_2{}^{18}O$, $D_2{}^{16}O$, and $D_2{}^{18}O$).[218] Several bands sensitive to both $^{16}O/^{18}O$ and H/D exchanges were found at 670–540 cm^{-1} and attributed to the vibrations including water-derived intermediates associated with the Mn_4CaO_5 cluster. A quantitative interpretation of these data, which may reflect the binding of W_s,[219] is difficult at present. Such data will become highly important once advanced theoretical calculations on detailed structural models of the WOC allow testing of different hypotheses for the binding modes and sites of substrate in the various S_i states.

Magnetic resonance. It was realized early on that paramagnetic relaxation can be used in NMR relaxation experiments to study water-binding.[220–223] While the approach is very good, the complexity of the WOC has thus far not allowed drawing strong conclusions. However, the observation of a S_i state dependent relaxation effect is suggestive of at least one water molecule or hydroxo ion being close to Mn in the S_2 state (direct binding is not strictly required for inducing the effect). Other magnetic resonance techniques are EPR and ENDOR.[98,99,224] Both can be used in cw and pulse modes to detect the coupling between paramagnetic states of the Mn_4CaO_5 cluster with the nuclear spins of nuclei such as ^{17}O or H/D. The most studied state in this regard is the S_2 state.

Attempts were made to detect the line broadenings of the S_2 multiline EPR signal caused by $H_2{}^{17}O$ binding by cw X-band EPR.[225] However, the effects are rather small and remain controversial in the field. Similarly, ^{17}O-ESEEM studies were performed with varying results.[226,227] Recently, HYSCORE spectroscopy was employed and the hyperfine coupling of one ^{17}O oxygen with Mn in the S_2 state was suggested on that basis. While the reported coupling strength of 10–11 MHz is consistent with an assignment to a Mn–O–Mn bridge,[228] further experiments have shown that the assignment was incorrect and that the signals instead arise from ^{14}N couplings within oxidized cytochrome b559.[229]

H/D ESEEM, HYSCORE and ENDOR spectroscopy have been used in several studies to probe the protonation state of water bound to the Mn_4CaO_5 cluster.[15,170,171,219,230–233] Since the above described water-exchange experiments show that both substrate water molecules exchange on second time scales in all S_i states, a minimum requirement for the assignment of signals to substrate is their rapid response to H/D exchange.[170,219] As a further criterion, it can be expected that substrate protons have some anisotropic coupling *via* oxygen to the Mn_4CaO_5 cluster, if the substrate (H_2O, HO^-) molecule is directly bound to Mn. For the S_2 state, ESEEM measurements revealed two D-couplings with a moderately strong isotopic coupling of 0.45 MHz and a dipolar coupling strength of 0.64 MHz.[170,219,234] Quite similar couplings were suggested in preliminary fits for the S_0 state.[15] For both states, several more H/D exchange sensitive signals with only dipolar coupling were observed from nuclei further away from the cluster. Similar proton couplings were also observed for the S_0 and S_2 states in proton matrix ENDOR measurements.[233]

X-ray spectroscopy. For all techniques, calibration to data obtained with suitable model systems is essential for the interpretation.[235] One good example with regard to substrate water binding are X-ray absorption measurements; it was demonstrated with model complexes that terminal oxo groups at Mn^{IV} or Mn^{V} ions lead to large pre-edge peaks in Mn K-edge XANES edges. Such features are not seen in the edges obtained for states S_0 to S_3 in PSII samples. Therefore, the binding of substrate water in the form of a terminal oxo group, $Mn=O$, is highly unlikely up to the S_3 state.[236] Time resolved EXAFS spectra also did not reveal any evidence for such a species during the $S_3 \rightarrow S_4 \rightarrow S_0$ transition.[237]

DFT. Only few attempts exist so far to theoretically model the exchange kinetics, and no studies have been published with a theoretical analysis of the above discussed FTIR and magnetic resonance data concerning water binding in the WOC. A study by Siegbahn and coworkers supports the binding of W_s in a bridging position,[209] while calculations by Batista and coworkers favor the binding of W_s as H_2O to Ca, and of W_f as a terminal OH ligand to the outer Mn^{IV} in the S_3 state.[121]

In summary, most data presently indicate that W_s is bound in a bridging position between two Mn ions. In this position, W_s is also either directly or indirectly coupled to Ca^{2+}. A possible position of W_s is the bis μ-oxo bridge to the outer Mn (Figure 7.4H). It may be identified as Mn–OH–Mn in S_0, and as Mn–O–Mn in S_1, S_2 and S_3. Much less information is available on the binding of W_f. In the S_3 state the exchange rate of W_f is clearly less affected by Ca^{2+}/Sr^{2+} substitution. W_f may therefore be bound as a terminal water/hydroxo ligand to Mn. If it is already bound to the WOC in the S_2 state, then possibly to the last Mn^{III} or to Ca. However, the majority of the presently available data and calculations favors the idea that the second substrate molecule (W_f) binds to the Mn_4CaO_5 cluster during the $S_2 \rightarrow S_3$ transition. As discussed above, water-binding during the $S_2 \rightarrow S_3$ state transition can be suggested to occur to Mn_{His} (Figure 7.4H) during its oxidation from a 5-coordinated Mn^{III} to a 6-coordinated Mn^{IV}.[93,190]

7.8 Product Release and the Energetics of Water-splitting

From Reaction (1), it is clear that water-splitting is a complex chemical reaction: it involves the orchestrated extraction of four-electrons and four-protons. This reaction is also energetically very demanding, because it requires strong oxidants to extract electrons out of stable water molecules. As we will see below, this reaction can only be driven by visible light if a suitable catalyst avoids the formation of high energy intermediates, and if a buildup of charge at the catalytic center is limited by suitable couplings of electron and proton transfers.

To drive the water-splitting reaction, PSII has tuned the potential of its chlorophyll-based primary electron donor P680/P680$^{\bullet+}$ to the highest oxidizing power known in Nature: +1.25 V.[238,239] This is probably the limit for biological systems, since even in PSII amino acids and cofactors surrounding the P680 chlorophyll molecules are damaged over time; under full sunlight, the D1 protein has to be replaced every 20–30 min. The midpoint potential, E_m, for Reaction 1 is a temperature, pH and O_2 pressure dependent quantity. For atmospheric O_2 pressure (0.21 bar) and room temperature (298 K), E_m = 1.22 V − 0.06 V × pH, *i.e.* at pH 7.0 $E_{m,7}$ = +0.82 V and at pH 5.0 $E_{m,5}$ = +0.92 V.[6,14] The above considerations show that charge separations leading to P680$^{\bullet+}$ formation provide on average enough energy for water-splitting down to at least the pH 5 reached in the lumen under continues illumination. However, this type of analysis does not allow conclusions about the individual steps leading to O_2 formation.

Uncatalyzed water-oxidation in aqueous solution is characterized by very different potentials for each electron abstraction, the largest being greater than 2 eV for the formation of the hydroxyl radical from water.[240] Therefore, it is clear that P680$^{\bullet+}$ is not able to drive water-oxidation by extracting one electron at a time directly from water. It rather requires a storage unit, the Mn_4CaO_5 cluster, that first accumulates the oxidizing equivalents and then catalyzes a concerted oxidation of two water molecules to molecular oxygen. If, however, the Mn_4CaO_5 cluster accumulates charges during this process, this would lead to increasing potentials for each subsequent electron removal due to the columbic attraction of the remaining electrons. Since each charge separation between P680 and Pheo generates the same driving force, this would imply that in the first S_i state transitions a lot of energy would be wasted and that on the other hand there would be not enough oxidizing potential provided by $P680^{\bullet+}$ to drive the last step(s). The most straightforward way to release charges from the Mn_4CaO_5 cluster is to couple the electron removals with deprotonation reactions of either bound substrate molecules or by deprotonating other groups in the direct vicinity.[241] In all cases, the oxidation of the Mn_4CaO_5 cluster occurs *via* the intermediate tyrosine Z, $Y_Z/Y_Z^{\bullet}$, which has an estimated operating midpoint potential of about 1.2 V.[239]

The proton release pattern of the WOC has been examined by measuring flash-induced pH changes using pH-indicating dyes and a sensitive pH electrode.[242–245] Although the 1: 0: 1: 2 proton pattern for the $S_0 \rightarrow S_1 \rightarrow S_2 \rightarrow S_3 \rightarrow S_0$ transitions was first accepted by experiments using thylakoids,[246–248] it was later found that the proton pattern was generally non-integer and strongly dependent on pH and type of PSII preparation.[249–255] In these proton release experiments, pH shifts were monitored in media with very low concentration buffer. In such low concentration buffer, however, there is a concern that proton detection could be interfered with by the buffering effect of PSII proteins themselves. Recently, isotope-edited FTIR spectroscopy was applied to proton release measurement during water oxidation.[256] In this method, PSII core complexes from a thermophilic cyanobacterium were suspended in a high concentration buffer, which trapped virtually all protons from the WOC. The buffer signals showed a clear period-four oscillation, providing a proton release pattern consistent with 1: 0: 1: 2 for the $S_0 \rightarrow S_1 \rightarrow S_2 \rightarrow S_3 \rightarrow S_0$ transitions. A perturbation from this pattern was ascribed to the partial protonation/deprotonation of protonatable side groups located near the WOC.[249,256]

These results show that one positive charge is added to the overall net charge of the WOC only during the $S_1 \rightarrow S_2$ transition. The potential differences involved in the individual S_i state transitions can be estimated to be approximately 0.85, 1.10, 1.15 and 1.0 eV for the $S_0 \rightarrow S_1$, $S_1 \rightarrow S_2$, $S_2 \rightarrow S_3$ and $S_3 \rightarrow S_4 \rightarrow S_0$ transitions.[6,239] It has been suggested that up to the S_1 state the proton leaves after the electron has been removed from the cluster, while after the positive charge has been accumulated during the $S_1 \rightarrow S_2$ transition the proton has to leave the WOC prior to the oxidation of the Mn_4CaO_5 cluster to the next S_i state.[7,14] Such a notion is, for example, supported by the finding that the $S_2 \rightarrow S_3$ transition is especially sensitive to biochemical treatments such as low temperatures or Ca^{2+} or Cl^- removal that are believed to limit or prevent proton release from the WOC (but might also affect water binding).[87,257] Further support comes from the discovery of a lag phase of about 200 μs between the oxidation of Y_Z and the onset of the oxidation of the Mn_4CaO_5 cluster to the S_4 state. This lag phase is accompanied by electrochromic changes and has been ascribed to the release of one proton.[154,237,258]

In addition to the energetic, the kinetics of the S_i state transitions are also important: if the potentials are correct, but the high energy species are too short lived, then charge recombination will win and no chemistry can occur. Light-induced charge separation between P680 and Pheo takes place with time-dependent rate constants in the order of 3 ps^{-1}.[5] In contrast, the S_i state transitions occur with characteristic times of about 50, 100, 220 and 1300 μs (S_0...S_3);[6] a difference of almost nine orders of magnitude in time. Stabilization of the initial charge separation for such long time periods is only possible if the initially formed charges are separated in space, and if part of the initially stored free energy is dissipated as heat. On the acceptor side, where most of this free energy loss occurs, the electron is transferred first from $Pheo_{D1}$ to the one-electron acceptor Q_A, and then further to Q_B. The transfer to Q_A results in a distance increase of 13 Å (Figure 7.1c), which results in a total distance of about 28 Å between the negative charge and $P680^{\bullet+}$.[259] It also leads to an energy drop of about − 500 mV as compared to the state of $^1P680^*$ (excited singlet state of P680, which lies 1.83 eV above the ground state and about 150 meV above $P680^{\bullet+}/Pheo^{\bullet-}$). On the donor side of PSII, tyrosine Z, Y_Z, is the electron transfer cofactor located between $P680^{\bullet+}$ and the Mn_4CaO_5 cluster. Oxidation of Y_Z leads to a further potential drop of about − 100 meV, and the 'through cofactor' distance between the charges is increased by about 14 Å. This now allows $Y_Z^{\bullet}$ to drive the slow S_i state transitions, including O_2 evolution. Interestingly, the rate of O_2 evolution is usually limited by electron transfer to Q_B and further to the PQ pool, which occurs with characteristic times of 0.5–10 ms, and not by the 1–2 ms half-time for O_2 formation and release.[5,14,147,239,260,261]

Considering that PSII converts light energy directly into chemical energy and does not 'only' generate electricity, the estimated energy yield of 16% of the full solar spectrum is remarkable. Even more impressive is the maximum quantum yield for water-splitting that PSII can achieve under optimal conditions. This value is closely related to the miss factor of water-splitting and can be as high as about 90%.[147,260]

The kinetics and energetics involved in the reactions of PSII are summarized in Figure 7.8. It is important to note that the O_2 release kinetic is practically identical to the reduction kinetic of $Y_Z^{\bullet}$ in the $S_3 \rightarrow S_4 \rightarrow S_0$ transition. This shows that all possible intermediates of the water-splitting reaction are extremely short lived and decay faster than they are formed. Consequently, various attempts to detect the transient formation of S_4 have failed so far. The only resolvable kinetic intermediate is the above discussed lag phase during the $S_3 \rightarrow S_4 \rightarrow S_0$ transition that is attributed to the $S_3Y_Z^{\bullet}$ formation with concomitant proton release from the WOC.[154,237,258]

Based on Reaction (1) it can be expected that O_2 pressure can block the water-splitting *via* product inhibition. The only reaction step of the Kok cycle that can be sensitive to oxygen pressure is the $S_4 \rightarrow S_0$ transition. UV absorption experiments under O_2 or N_2 pressure were therefore performed in the hope to trap the S_4 state. Clausen and Junge found that a half-inhibition of the $S_4 \rightarrow S_0$ transition occurs at the surprisingly low partial pressures of 2–3 bar O_2, but not up to 20 bar N_2.[263,264] This was confirmed in one subsequent delayed fluorescence

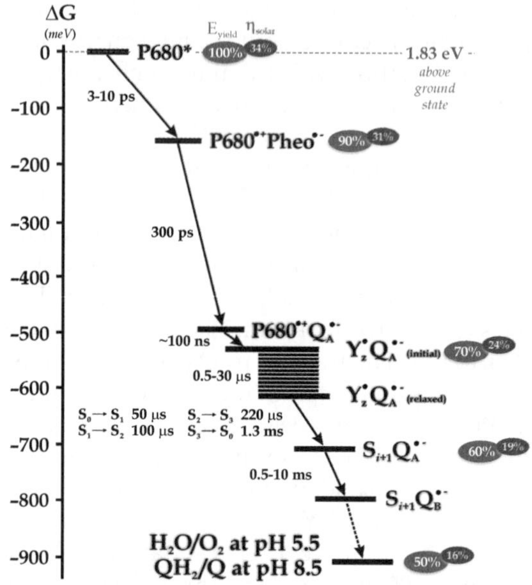

Figure 7.8 Kinetics and energetic of PSII. The E_{yield} represents the free energy taken up by P680 when forming ^{1}P680*: 100% corresponds to 1.83 eV. The overall efficiency of PSII for converting solar light into chemical (free) energy, η_{solar}, is given in the green circles. This number refers to the total spectrum of the sun and low light intensities. At high light intensity, PSII actively reduces the light harvesting efficiency to minimize damaging side reactions. The figure is modified after ref. 239, 260 and 262.

study,[265] but no inhibition was found in a time-resolved X-ray spectroscopy experiment[266] or by visible chlorophyll fluorescence experiments.[267] Very recent MIMS measurements that allow detection of light-induced $^{36}O_2$ production from $H_2{}^{18}O$ water under $^{32}O_2$ pressure up to 20 bars conclusively demonstrate that the WOC is not inhibited by 50-times elevated O_2 concentrations in solution. This shows that PSII has a driving force for O_2 production of at least 160–220 meV at pH 6.5,[268] in agreement with an earlier proposal.[269]

Since water oxidation involves the breaking of O–H bonds and because deprotonation reactions are so important for the energetic of water-oxidation (see above), one might expect to find strong H/D isotope effects on the kinetics of S_i state turnovers. However, the observed kinetic isotope effects are rather small, with about 2.2 for the $Y_Z{}^{\bullet}$ reduction during the $S_2 \rightarrow S_3$ transition, and about 1.3–1.4 for the $Y_Z{}^{\bullet}$ reduction on all other S_i state transitions. Similarly, an H/D isotope effect of 1.4 was reported for O_2 release.[270–273]

7.9 Recent Mechanistic Proposals

Until a few years ago, mechanistic models were built on very selected and sparse information (phase 1). Only more recently was it attempted to incorporate as many experimental information as possible into mechanistic suggestions in order to derive at reasonable working models. Such mechanistic models were constructed at first by employing a qualitative approach (phase 2). With the advent of QM/MM and DFT calculations to PSII research, a new era has begun that holds the promise to eventually resolve – in a tight interplay with experimental results – the remaining issues concerning water oxidation by photosystem II (phase 3).

Mechanistic models of phase 1 and 2 have been summarized recently.[10,13] In general, three types of mechanisms were proposed: (a) Mn is oxidized in all S_i state transitions, and the O–O bond formation occurs *via* nucleophilic attack of a free or Ca-bound water or hydroxo onto a terminal Mn^V-oxo group. (b) During the $S_2 \rightarrow S_3$ transition a ligand centered oxidation occurs, and the O–O bond formation occurs *via* a radical mechanism, possibly involving a bridging oxygen. (c) The O–O bond is already formed within a minority population of a dynamic equilibrium in the S_3 or $S_3Y_Z^{\bullet}$ state, and it is only this fraction of centers that can react with $Y_Z^{\bullet}$ to S_4Y_Z and further to S_0 under the release of O_2. While the phase 1 and 2 models only allow qualitative comparisons to experimental observations, they are important since they illustrated very clearly the different mechanistic possibilities and enabled the design of new experiments for testing them. To date, in phase 3, the border between nucleophilic attack mechanisms and radical mechanisms is not so strict anymore, since it is often assumed that there is no Mn(V) in the S_4 state, but a terminal or bridging Mn(IV)–$O^{\bullet}$ radical. These mechanisms, therefore, mostly differ in the point in time when the radical is formed, *i.e.* in the S_3 state or the S_4 state. Further differences exist over the question of whether the O–O bond is formed between terminal or bridging oxygens.

The first computational mechanistic models for water oxidation in photosystem II were derived by Siegbahn.[141,142,274,275] His recent mechanism[93] is described in detail in Chapter 13 of this book, and the S_4 state is displayed in Figure 7.9, right side. The crucial point for a low activation barrier for O–O bond formation is the correct sequence of spins, βαβα, between Mn–$O^{\bullet}$O–Mn.[92] Siegbahn's calculations show that this type of O–O bond formation is best achieved within model H (Figure 7.4) between a terminal $^{\bullet}$O-radical formed upon S_4 state formation from the hydroxo that binds to the Mn_4CaO_5 cluster during the $S_2 \rightarrow S_3$ transition (Section 7.7) and the μ_3-oxo bridge that connects the outer two Mn on the left and Ca. As discussed above, this proposal is based on the current best model of the WOC, and many mechanistic details are consistent with experimental data.

The only other DFT-based mechanistic proposal currently available is that of the groups of Batista and Brudvig.[182] It is based on model A in Figure 7.4. In contrast to the mechanism suggested by Siegbahn, the O–O bond is proposed to be formed between a Ca-bound water molecule and a terminal Mn(IV)–$O^{\bullet}$ radical. It is assumed that the slow exchanging water (W_s) binds to Ca during the $S_4 \rightarrow S_0$ transition, while the fast exchanging W_f binds to the outer Mn during the $S_0 \rightarrow S_1$ transition. The oxidation states in S_0 are suggested to be Mn_4(III,III,III,IV), and all S state transitions up to the S_3 state are proposed to be

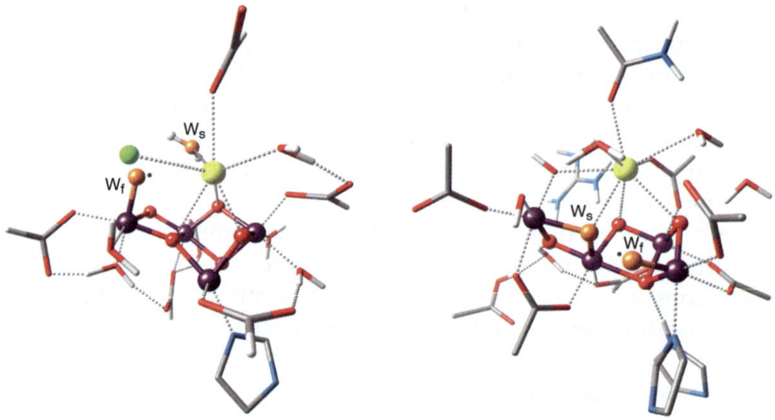

Figure 7.9 Theoretical models for the S_4 state just prior to the O–O bond formation. The model on the left is from Batista and co-workers,[182] that on the right from Siegbahn.[92,190] Oxygens thought to originate from the substrate water molecules are shown in orange and labeled W_s and W_f for slow and fast exchanging (Section 7.7). The grey dot at W_f in both models indicates the radical character of this oxygen in the S_4 state. The model of Siegbahn was slightly simplified for better presentation. The corresponding S_2 state structures are displayed in Figure 7.4 A and H.

Mn(III) to Mn(IV) oxidation state changes. Similar to the Siegbahn mechanism, the $S_2 \rightarrow S_3$ transition involves a structural change: here, however, the formation of an extra μ-oxo bridge to the outer Mn is suggested (compare Figure 7.4A with Figure 7.9, left side) instead of the binding of the second substrate water molecule. While the Ca-bound W_s remains fully protonated up to S_4, W_f is proposed to be stepwise deprotonated during the $S_2 \rightarrow S_3$ and $S_3 \rightarrow S_4$ transitions. During S_4 state formation, W_f is suggested to be oxidized to the radical state Mn(IV)–O$^\bullet$.

7.10 Lessons for Artificial Systems

The light reactions of photosystem II start with the harvesting of sunlight by suitable pigments in a light collecting antenna system. This promotes an electron within a pigment into a singlet excited state. The excitation energy is then transferred *via* many pigment molecules to a reaction center, where the proximity of a chlorophyll molecule to a suitable acceptor allows the excited electron to jump from Chl to pheophytin. The initially formed charge pair is subsequently stabilized by electron transport steps that involve additional cofactors. This stabilization is important to gain time for driving comparatively slow chemical reactions. Stabilization is achieved by reducing the driving force ('wasting' free energy) for charge recombination and by increasing the physical separation between the charge pair. The most interesting chemical reaction driven by PSII is the splitting of water. This reaction requires a high oxidation-potential, the highest known in biology (approx. +1.25 V).[239] It is remarkable that biological systems build the rather complex looking photosystems in a modular way, *i.e.* highly similar proteins and cofactors are also used in non-oxygenic photosynthesis. The high oxidizing potential of PSII was simply achieved by specific changes in the protein environment, not by synthesizing a specific new chromophore.[276–278]

One notable exception is the water-oxidizing complex in PSII. The WOC contains the Mn_4CaO_5 cluster that is unique to all oxygenic photosynthetic organisms. Indeed, it is the only biological catalyst that splits water to molecular oxygen, and it appears that it has remained basically unchanged during 2 billion years of evolution. The success of oxygenic photosynthesis shows that the metals like Mn and Ca are, if placed in the right environment and geometry, highly effective in water splitting. Importantly, they are abundant in the earth crust and easily extractable.

The WOC has two important functions: it stores up to four oxidizing equivalents, and it binds the substrate water molecules in a suitable geometry for O–O bond formation. The storage of oxidizing equivalents allows for the oxidation of water in a concerted reaction, circumventing the formation of energetically very unfavourable hydroxylradicals.[6] Importantly, all but one ($S_1 \rightarrow S_2$) of the oxidation steps of the WOC are electroneutral, which is achieved by coupling electron transfer with proton release. The released protons likely originate from the deprotonation of substrate water,[92,190] but the deprotonation of other groups within the WOC is also a possibility.[14,241] In this way, the oxidation potential of the WOC remains almost constant during the reaction cycle, and tuned to the constant driving power of the light-driven charge separations in the reaction center. Besides providing substrate binding sites with favourable geometry for O–O bond formation, the spin states at the substrate oxygens also need to be adjusted *via* the Mn_4CaO_5 cluster to achieve efficient oxygen production from water.[215] It is remarkable that the Mn_4CaO_5 cluster is not a static template, but that it changes its structure during the reaction cycle and thereby likely controls the binding of substrate water. This may be one factor that helps minimize side reactions that could lead to the release of peroxide or other reactive oxygen species (ROS).

In addition, the following principles of photosynthetic water splitting might be useful for developing artificial devices for the conversion of solar light to fuels:

1. *Specific reaction environment.* The WOC in PSII is located in a sequestered environment; channels exist for controlled substrate entry and product release. This is thought to stabilize the Mn_4CaO_5 cluster, to minimize side reactions, and to facilitate proton coupled electron transfer.
2. *Regulation of excitation energy and repair.* The components of the primary photo-reactions are surrounded by protective components. Once a pigment is destroyed or the protein is damaged by side reactions, a

specific repair mechanism exists to replace the damaged parts.[279] Most photosynthetic organisms compete for light and therefore have large antenna systems to harvest light efficiently. To reduce damage to the photosystems that can be caused by strong light, they have many short and long term strategies for controlling the amount of excitation that reaches the photosystems.[280–282]

3. *Coupling of two reaction centers with different redox potentials.* To bridge the large free energy gap between water oxidation and proton reduction, oxygenic photosynthetic organisms employ two photosystems (PSII and PSI) that work in series and have quite different redox-potentials. This allows driving these reactions with visible photons down to 680–700 nm.[277]
4. *Coupling of electron transfer with proton movements across a membrane.* In oxygenic photosynthesis all complexes required for light-induced water splitting and $NADP^+$ reduction are part of the thylakoid membrane that encloses an inner volume. This arrangement not only reduces the diffusion path for mobile electron carriers connecting the photosystems, but also allows coupling of the electron transport between complexes with proton translocations across the membrane. The resulting proton motif force is utilized by ATPsynthase to store free energy by converting ADP and inorganic phosphate to ATP.[283,284]

While many features of PSII are very attractive, the limited stability poses the largest drawback. The turnover number (oxygen molecules produced until inactivation) for PSII can be estimated to be in the order of 10^5–10^6.[147] This is much better than for most current artificial system, but insufficient for an industrial catalyst. Since repair mechanisms are not easy to implement in artificial systems, chemists will have to find very stable ligands or have to employ simple inorganic systems. One further drawback is the low turn-over frequency (produced O_2 molecules per second). While O_2 formation and release during the $S_3 \rightarrow S_4 \rightarrow S_0$ transition occurs with characteristic times of 1–2 ms, the rate of O_2 evolution in PSII is limited by the removal of electrons from the acceptor side to about 20 ms per cycle.[147,285]

Even more than PSII, most present coordination compounds employing organic ligands for complexing the catalytic metal ions have proven to be very unstable, many surviving only a few reaction cycles.[286] While the stability can be likely improved by proper ligand design, promising alternatives can be found in fully inorganic ligands[287] or simple Mn oxides or MnCa oxides. In this regard, it is interesting to note that in analogy to PSII, the MnCa oxides show a significantly higher oxygen-evolution activity than Mn oxides.[288] Besides Mn (plus Ca), also Co, Ru and Ir complexes/oxides were found to be promising artificial catalysts for water-splitting.[289–294]

We hope that this book chapter will help to inspire many young scientists to take up the challenge to solve the remaining questions regarding photosynthetic water-splitting and to develop artificial systems that allow the production of 'solar' fuels like H_2 or ethanol from water.

Acknowledgements

The authors gratefully acknowledge the ongoing discussions with Alain Boussac, Warwick Hillier, Wolfgang Junge, Jan Kern, Wolfgang Lubitz, Frank Neese, Dimitrios Pantazis, Fabrice Rappaport, Gernot Renger, Per Siegbahn, Tom Wydrzynski and Vittal Yachandra. We thank Per Siegbahn, Victor Batista, Eduardo Sproviero and Simon Petrie for providing us with coordinates of their structures, and Dmitriy Shevela, Dimitrios Pantazis and Jan Kern for preparing Figures. JM is thankful for financial support by the Knut och Alice Wallenbergs stiftelse, Kempestiftelserna and by Vetenskapsrådet. JY is supported by the Department of Energy, Director, Office of Science, Office of Basic Energy Sciences, Division of Chemical Sciences, Geosciences, and Biosciences under Contract DE-AC02-05CH11231.

References

1. Y. Umena, K. Kawakami, J. R. Shen and N. Kamiya, *Nature*, 2011, **473**, 55–60.
2. P. E. M. Siegbahn, *J. Am. Chem. Soc.*, 2009, **131**, 18238–18239.
3. J. Yano, J. Kern, K. Sauer, M. J. Latimer, Y. Pushkar, J. Biesiadka, B. Loll, W. Saenger, J. Messinger, A. Zouni and V. K. Yachandra, *Science*, 2006, **314**, 821–825.
4. J. H. Robblee, J. Messinger, R. M. Cinco, K. L. McFarlane, C. Fernandez, S. A. Pizarro, K. Sauer and V. K. Yachandra, *J. Am. Chem. Soc.*, 2002, **124**, 7459–7471.
5. G. Renger and T. Renger, *Photosynth. Res.*, 2008, **98**, 53–80.
6. J. Messinger and G. Renger, in *Primary Processes of Photosynthesis – Part 2: Basic Principles and Apparatus* ed. G. Renger, The Royal Society of Chemistry, Cambridge, UK, 2008, pp. 291–349.
7. G. Renger, *Photosynth. Res.*, 2007, **92**, 407–425.
8. G. Renger, *Biochim. Biophys. Acta, Bioenerg.*, 2004, **1655**, 195–204.
9. R. J. Debus, *Biochim. Biophys. Acta, Bioenerg.*, 1992, **1102**, 269–352.
10. W. Hillier and J. Messinger, in *Photosystem II. The Light-Driven Water:Plastoquinone Oxidoredutase*, eds. T. Wydrzynski and K. Satoh, Springer, Dordrecht, 2005, **vol. 22**, pp. 567–608.
11. W. Hillier and T. Wydrzynski, *Biochim. Biophys. Acta, Bioenerg.*, 2001, **1503**, 197–209.
12. Ö. Hansson and T. Wydrzynski, *Photosynth. Res.*, 1990, **23**, 131–162.
13. J. P. McEvoy and G. W. Brudvig, *Chem. Rev.*, 2006, **106**, 4455–4483.
14. H. Dau and M. Haumann, *Coord. Chem. Rev.*, 2008, **252**, 273–295.
15. R. D. Britt, K. A. Campbell, J. M. Peloquin, M. L. Gilchrist, C. P. Aznar, M. M. Dicus, J. Robblee and J. Messinger, *Biochim. Biophys. Acta, Bioenerg.*, 2004, **1655**, 158–171.
16. S. Zein, L. V. Kulik, J. Yano, J. Kern, Y. Pushkar, A. Zouni, V. K. Yachandra, W. Lubitz, F. Neese and J. Messinger, *Philos. Trans. R. Soc., B*, 2008, **363**, 1167–1177.

17. J. Yano and V. K. Yachandra, *Inorg. Chem.*, 2008, **47**, 1711–1726.
18. V. K. Yachandra, K. Sauer and M. P. Klein, *Chem. Rev.*, 1996, **96**, 2927–2950.
19. C. Goussias, A. Boussac and A. W. Rutherford, *Philos. Trans. R. Soc., B*, 2002, **357**, 1369–1381.
20. C. Bonnet, *Recherches sur L'usage des feuilles dans les plantes*, Elie Luzac et fils, Göttingen/Leiden, 1754.
21. J. Priestley, *Philos. Trans. R. Soc. London*, 1772, **62**, 147–264.
22. Govindjee and D. Krogmann, *Photosynth. Res.*, 2004, **80**, 15–57.
23. A. Pirson, *Z. Bot.*, 1937, **31**, 193–267.
24. A. Pirson, *Photosynth. Res.*, 1994, **40**, 209–221.
25. T. J. Wydrzynski, *Photosynth. Res.*, 2004, **80**, 125–135.
26. G. M. Cheniae, *Photosynth. Res.*, 1993, **38**, 225–227.
27. B. Kok, B. Forbush and M. McGloin, *Photochem. Photobiol.*, 1970, **11**, 457–476.
28. P. Joliot, G. Barbieri and R. Chabaud, *Photochem. Photobiol.*, 1969, **10**, 309–329.
29. P. Joliot, *Photosynth. Res.*, 2003, **76**, 65–72.
30. *Biophysical Techniques in Photosynthesis*, ed. J. Amesz and A. J. Hoff, Kluwer Academic Publishers, Dordrecht, 1996.
31. *Biophysical Techniques in Photosynthesis Volume II*, ed. T. J. Aartsma and J. Matysik, Springer, Dordrecht, 2008.
32. J. Messinger, A. Alia and Govindjee, *Photosynth. Res.*, 2009, **102**, 103–106.
33. J. Messinger, A. Alia and Govindjee, *Photosynth. Res.*, 2009, **101**, 89–92.
34. A. Guskov, J. Kern, A. Gabdulkhakov, M. Broser, A. Zouni and W. Saenger, *Nat. Struct. Biol. Mol. Biol.*, 2009, **16**, 334–342.
35. B. Loll, J. Kern, W. Saenger, A. Zouni and J. Biesiadka, *Nature*, 2005, **438**, 1040–1044.
36. J. Biesiadka, B. Loll, J. Kern, K.-D. Irrgang and A. Zouni, *Phys. Chem. Chem. Phys.*, 2004, **6**, 4733–4736.
37. A. Zouni, H. T. Witt, J. Kern, P. Fromme, N. Krauß, W. Saenger and P. Orth, *Nature*, 2001, **409**, 739–743.
38. J. W. Murray, K. Maghlaouri, J. Kargul, N. Ishida, Thanh-Lan-Lai, A. W. Rutherford, M. Sugiura, A. Boussac and J. Barber, *Energy Environ. Sci.*, 2008, **1**, 161–166.
39. J. W. Murray, K. Maghlaoui, J. Kargul, M. Sugiura and J. Barber, *Photosynth. Res.*, 2008, **98**, 523–527.
40. J. Kargul, K. Maghlaoui, J. W. Murray, Z. Deak, A. Boussac, A. William Rutherford, I. Vass and J. Barber, *Biochim. Biophys. Acta, Bioenerg.*, 2007, **1767**, 404–413.
41. K. N. Ferreira, T. M. Iverson, K. Maghlaoui, J. Barber and S. Iwata, *Science*, 2004, **303**, 1831–1838.
42. K. Kawakami, Y. Umena, N. Kamiya and J.-R. Shen, *Proc. Natl. Acad. Sci. U. S. A.*, 2009, **106**, 8567–8572.
43. N. Kamiya and J.-R. Shen, *Proc. Natl. Acad. Sci. U. S. A.*, 2003, **100**, 98–103.

44. G. Renger, *Curr. Sci.*, 2010, **98**, 1305–1319.
45. G. Renger, *J. Photochem. Photobiol., B*, 2011, **104**, 35–43.
46. I. Vass, S. Styring, T. Hundal, A. Koivuniemi, E. M. Aro and B. Andersson, *Proc. Natl. Acad. Sci. U. S. A.*, 1992, **89**, 1408–1412.
47. F. Van Mieghem, K. Brettel, B. Hillmann, A. Kamlowski, A. W. Rutherford and E. Schlodder, *Biochemistry*, 1995, **34**, 4798–4813.
48. T. G. Truscott, *J. Photochem. Photobiol., B*, 1990, **6**, 359–371.
49. V. Martinez-Junza, M. Szczepaniak, S. E. Braslavsky, J. Sander, M. Nowaczyk, M. Rogner and A. R. Holzwarth, *Photochem. Photobiol. Sci.*, 2008, **7**, 1337–1343.
50. J. Yano, J. Kern, K. D. Irrgang, M. J. Latimer, U. Bergmann, P. Glatzel, Y. Pushkar, J. Biesiadka, B. Loll, K. Sauer, J. Messinger, A. Zouni and V. K. Yachandra, *Proc. Natl. Acad. Sci. U. S. A.*, 2005, **102**, 12047–12052.
51. J. M. Anderson, *FEBS Lett.*, 2001, **488**, 1–4.
52. T. Wydrzynski, W. Hillier and J. Messinger, *Physiol. Plant.*, 1996, **96**, 342–350.
53. G. Renger, in *Concepts in Photobiology: Photosynthesis and Photomorphogenesis*, ed. G. S. Singhal, G. Renger, S. K. Sopory, K.-D. Irrgang and Govindjee, Narosa Publishing House, New Delhi, 1999, pp. 292–329.
54. F. Ho, *Photosynth. Res.*, 2008, **98**, 503–522.
55. F. M. Ho and S. Styring, *Biochim. Biophys. Acta, Bioenerg.*, 2008, **1777**, 140–153.
56. P. Joliot and A. Joliot, *Biochim. Biophys. Acta, Bioenerg.*, 1968, **153**, 625–634.
57. G. Renger and B. Hanssum, *Photosynth. Res.*, 2009, **102**, 487–498.
58. P. Joliot and B. Kok, in *Bioenergetics of Photosynthesis*, ed. Govindjee, Academic Press, New York, NY, 1975, pp. 387–412.
59. B. Forbush, B. Kok and M. P. McGloin, *Photochem. Photobiol.*, 1971, **14**, 307–321.
60. P. Joliot, A. Joliot, B. Bouges and G. Barbieri, *Photochem. Photobiol.*, 1971, **14**, 287–289.
61. J. Messinger, G. Seaton, T. Wydrzynski, U. Wacker and G. Renger, *Biochemistry*, 1997, **36**, 6862–6873.
62. J. Messinger, J. H. Robblee, C. Fernandez, R. M. Cinco, H. Visser, U. Bergmann, P. Glatzel, S. P. Cramer, K. A. Campbell, J. M. Peloquin, R. D. Britt, K. Sauer, M. P. Klein and V. K. Yachandra, in *Photosynthesis: Mechanisms and Effects*, ed. G. Garab, Kluwer Academic Publishers, Dordrecht, Editon edn., 1998, pp. 1279–1282.
63. G. M. Ananyev, L. Zaltsman, C. Vasko and G. C. Dismukes, *Biochim. Biophys. Acta, Bioenerg.*, 2001, **1503**, 52–68.
64. N. Tamura, Y. Inoue and G. M. Cheniae, *Biochim. Biophys. Acta, Bioenerg.*, 1989, **976**, 173–181.
65. N. Tamura and G. Cheniae, *Biochim. Biophys. Acta, Bioenerg.*, 1987, **890**, 179–194.
66. T.-A. Ono, *Biochim. Biophys. Acta, Bioenerg.*, 2001, **1503**, 40–51.

67. G. T. Babcock and K. Sauer, *Biochim. Biophys. Acta, Bioenerg.*, 1973, **325**, 483–503.
68. S. Styring and A. W. Rutherford, *Biochemistry*, 1987, **26**, 2401–2405.
69. I. Vass and S. Styring, *Biochemistry*, 1991, **30**, 830–839.
70. J. Messinger, W. P. Schröder and G. Renger, *Biochemistry*, 1993, **32**, 7658–7668.
71. S. Isgandarova, G. Renger and J. Messinger, *Biochemistry*, 2003, **42**, 8929–8938.
72. J. Messinger and G. Renger, *Biochemistry*, 1994, **33**, 10896–10905.
73. P. Faller, R. J. Debus, K. Brettel, M. Sugiura, A. W. Rutherford and A. Boussac, *Proc. Natl. Acad. Sci. U. S. A.*, 2001, **98**, 14368–14373.
74. W. F. J. Vermaas, G. Renger and G. Dohnt, *Biochim. Biophys. Acta, Bioenerg.*, 1984, **764**, 194–202.
75. W. F. J. Vermaas, A. W. Rutherford and Ö. Hansson, *Proc. Natl. Acad. Sci. U. S. A.*, 1988, **85**, 8477–8481.
76. J. H. A. Nugent, C. Demetriou and C. J. Lockett, *Biochim. Biophys. Acta, Bioenerg.*, 1987, **894**, 534–542.
77. A. W. Rutherford and S. Styring, *Cytochrome Syst.: Mol. Biol. Bioenerg., [Proc. UNESCO Int. Symp.]*, 1987, 541–547.
78. I. Vass, Z. Deak and E. Hideg, *Biochim. Biophys. Acta, Bioenerg.*, 1990, **1017**, 63–69.
79. J. Messinger, J. H. Robblee, W. O. Yu, K. Sauer, V. K. Yachandra and M. P. Klein, *J. Am. Chem. Soc.*, 1997, **119**, 11349–11350.
80. R. J. Debus, B. A. Barry, G. T. Babcock and L. Mcintosh, *Proc. Natl. Acad. Sci. U. S. A.*, 1988, **85**, 427–430.
81. B. A. Diner, *Biochim. Biophys. Acta, Bioenerg.*, 1977, **460**, 247–258.
82. A. W. Rutherford, A. R. Crofts and Y. Inoue, *Biochim. Biophys. Acta, Bioenerg.*, 1982, **682**, 457–465.
83. H. H. Robinson and A. R. Crofts, *FEBS Lett.*, 1983, **153**, 221–226.
84. A. W. Rutherford and Y. Inoue, *FEBS Lett.*, 1984, **165**, 163–170.
85. D. Shevela, B. Nöring, H. J. Eckert, J. Messinger and G. Renger, *Phys. Chem. Chem. Phys.*, 2006, **8**, 3460–3466.
86. P. H. Homann, *Photosynth. Res.*, 2002, **73**, 169–175.
87. C. F. Yocum, *Coord. Chem. Rev.*, 2008, **252**, 296–305.
88. T. Kuwabara and N. Murata, *Plant Cell Physiol.*, 1982, **23**, 533–539.
89. J. W. Murray and J. Barber, *J. Struct. Biol.*, 2007, **159**, 228–237.
90. F. M. Ho and S. Styring, *Biochim. Biophys. Acta, Bioenerg.*, 2008, **1777**, 140–153.
91. M. Grabolle, M. Haumann, C. Müller, P. Liebisch and H. Dau, *J. Biol. Chem.*, 2006, **281**, 4580–4588.
92. P. E. M. Siegbahn, *Acc. Chem. Res.*, 2009, **42**, 1871–1880.
93. P. E. M. Siegbahn, *Chem.–Eur. J.*, 2008, **14**, 8290–8302.
94. N. Cox, L. Rapatskiy, J.-H. Su, D. A. Pantazis, M. Sugiura, L. Kulik, P. Dorlet, A. W. Rutherford, F. Neese, A. Boussac, W. Lubitz and J. Messinger, *J. Am. Chem. Soc.*, 2011, **133**, 3635–3648.

95. S. Schinzel, J. Schraut, A. V. Arbuznikov, P. E. M. Siegbahn and M. Kaupp, *Chem.–Eur. J.*, 2010, **16**, 10424–10438.
96. X. Li, E. M. Sproviero, U. Ryde, P. E. M. Siegbahn, V. S. Batista and G. Chen, *private communication*, 2011.
97. J. Yano and V. K. Yachandra, *Photosynth. Res.*, 2009, **102**, 241–254.
98. M. van Gastel, *Photosynth. Res.*, 2009, **102**, 367–373.
99. L. Kulik and W. Lubitz, *Photosynth. Res.*, 2009, **102**, 391–401.
100. C. Berthomieu and R. Hienerwadel, *Photosynth. Res.*, 2009, **101**, 157–170.
101. T. Noguchi and C. Berthomieu, in *Photosystem II: The Light-Driven Water:Plastoquinone Oxidoreductase*, ed. T. Wydrzynski and K. Satoh, Springer, Dordrecht, 2005, pp. 367–387.
102. T. Noguchi, *Coord. Chem. Rev.*, 2008, **252**, 336–346.
103. R. J. Debus, *Coord. Chem. Rev.*, 2008, **252**, 244–258.
104. J. Messinger, J. H. Robblee, U. Bergmann, C. Fernandez, P. Glatzel, H. Visser, R. M. Cinco, K. L. McFarlane, E. Bellacchio, S. A. Pizarro, S. P. Cramer, K. Sauer, M. P. Klein and V. K. Yachandra, *J. Am. Chem. Soc.*, 2001, **123**, 7804–7820.
105. M. Haumann, C. Müller, P. Liebisch, L. Iuzzolino, J. Dittmer, M. Grabolle, T. Neisius, W. Meyer–Klaucke and H. Dau, *Biochemistry*, 2005, **44**, 1894–1908.
106. J. Messinger, M. Badger and T. Wydrzynski, *Proc. Natl. Acad. Sci. U. S. A.*, 1995, **92**, 3209–3213.
107. K. Beckmann, J. Messinger, M. R. Badger, T. Wydrzynski and W. Hillier, *Photosynth. Res.*, 2009, **102**, 511–522.
108. W. Hillier and T. Wydrzynski, *Coord. Chem. Rev.*, 2008, **252**, 306–317.
109. M. Orio, D. A. Pantazis and F. Neese, *Photosynth. Res.*, 2009, **102**, 443–453.
110. E. M. Sproviero, M. B. Newcomer, J. A. Gascon, E. R. Batista, G. W. Brudvig and V. S. Batista, *Photosynth. Res.*, 2009, **102**, 455–470.
111. E. M. Sproviero, J. A. Gasco, J. P. McEvoy, G. W. Brudvig and V. S. Batista, *J. Am. Chem. Soc.*, 2008, **130**, 6728–6730.
112. E. M. Sproviero, J. A. Gascon, J. P. McEvoy, G. W. Brudvig and V. S. Batista, *Curr. Opin. Struct. Biol.*, 2007, **17**, 173–180.
113. E. M. Sproviero, J. A. Gascon, J. P. McEvoy, G. W. Brudvig and V. S. Batista, *J. Chem. Theory Comput.*, 2006, **2**, 1119–1134.
114. M. Kusunoki, *Biochim. Biophys. Acta, Bioenerg.*, 2007, **1767**, 484–492.
115. S. Petrie, R. Stranger and R. J. Pace, *Angew. Chem., Int. Ed.*, 2010, **49**, 4233–4236.
116. S. Petrie, R. Stranger and R. L. Pace, *Chem.–Eur. J.*, 2008, **14**, 5482–5494.
117. S. Petrie, R. Stranger, P. Gatt and R. J. Pace, *Chem.–Eur. J.*, 2007, **13**, 5082–5089.
118. S. Zein, L. V. Kulik, J. Yano, J. Kern, Y. Pushkar, A. Zouni, V. K. Yachandra, W. Lubitz, F. Neese and J. Messinger, *Philos. Trans. R. Soc., B*, 2008, **363**, 1167–1177.

119. L. V. Kulik, B. Epel, W. Lubitz and J. Messinger, *J. Am. Chem. Soc.*, 2007, **129**, 13421–13435.
120. H. Dau, A. Grundmeier, P. Loja and M. Haumann, *Philos. Trans. R. Soc., B*, 2008, **363**, 1237–1243.
121. E. M. Sproviero, K. Shinopoulos, J. A. Gascon, J. P. McEvoy, G. W. Brudvig and V. S. Batista, *Philos. Trans. R. Soc., B*, 2008, **363**, 1149–1156.
122. G. Ulas, G. Olack and G. W. Brudvig, *Biochemistry*, 2008, **47**, 3073–3075.
123. C. Aoyama, H. Suzuki, M. Sugiura and T. Noguchi, *Biochemistry*, 2008, **47**, 2760–2765.
124. D. Shevela, J. H. Su, V. Klimov and J. Messinger, *Biochim. Biophys. Acta, Bioenerg.*, 2008, **1777**, 532–539.
125. V. K. Yachandra, V. J. DeRose, M. J. Latimer, I. Mukerji, K. Sauer and M. P. Klein, *Science*, 1993, **260**, 675–679.
126. E. M. Sproviero, J. A. Gascon, J. P. McEvoy, G. W. Brudvig and V. S. Batista, *Coord. Chem. Rev.*, 2008, **252**, 395–415.
127. W. Hillier and T. Wydrzynski, *Phys. Chem. Chem. Phys.*, 2004, **6**, 4882–4889.
128. K. Hasegawa, T.-A. Ono, Y. Inoue and M. Kusunoki, *Bull. Chem. Soc. Jpn.*, 1999, **72**, 1013–1023.
129. V. J. DeRose, I. Mukerji, M. J. Latimer, V. K. Yachandra, K. Sauer and M. P. Klein, *J. Am. Chem. Soc.*, 1994, **116**, 5239–5249.
130. Y. Pushkar, J. Yano, K. Sauer, A. Boussac and V. K. Yachandra, *Proc. Natl. Acad. Sci. U. S. A.*, 2008, **105**, 1879–1884.
131. R. M. Cinco, K. L. McFarlane Holman, J. H. Robblee, J. Yano, S. A. Pizarro, E. Bellacchio, K. Sauer and V. K. Yachandra, *Biochemistry*, 2002, **41**, 12928–12933.
132. D. A. Pantazis, M. Orio, T. Petrenko, S. Zein, W. Lubitz, J. Messinger and F. Neese, *Phys. Chem. Chem. Phys.*, 2009, **11**, 6788–6798.
133. G. C. Dismukes and Y. Siderer, *Proc. Natl. Acad. Sci. U. S. A.*, 1981, **78**, 274–278.
134. A. Haddy, *Photosynth. Res.*, 2007, **92**, 357–368.
135. A. Haddy, W. R. Dunham, R. H. Sands and R. Aasa, *Biochim. Biophys. Acta, Bioenerg.*, 1992, **1099**, 25–34.
136. A. Haddy, K. V. Lakshmi, G. W. Brudvig and H. A. Frank, *Biophys. J.*, 2004, **87**, 2885–2896.
137. A. V. Astashkin, Y. Kodera and A. Kawamori, *J. Magn. Reson., Ser. B*, 1994, **105**, 113–119.
138. O. Horner, E. Rivière, G. Blondin, S. Un, A. W. Rutherford, J.-J. Girerd and A. Boussac, *J. Am. Chem. Soc.*, 1998, **120**, 7924–7928.
139. J. M. Peloquin, K. A. Campbell, D. W. Randall, M. A. Evanchik, V. L. Pecoraro, W. H. Armstrong and R. D. Britt, *J. Am. Chem. Soc.*, 2000, **122**, 10926–10942.
140. L. V. Kulik, B. Epel, W. Lubitz and J. Messinger, *J. Am. Chem. Soc.*, 2005, **127**, 2392–2393.

141. M. R. A. Blomberg, P. E. M. Siegbahn, S. Styring, G. T. Babcock, B. Åkermark and P. Korall, *J. Am. Chem. Soc.*, 1997, **119**, 8285–8292.
142. P. E. M. Siegbahn and R. H. Crabtree, *J. Am. Chem. Soc.*, 1999, **121**, 117–127.
143. P. E. M. Siegbahn, *Dalton Trans.*, 2009, 10063–10068.
144. P. E. M. Siegbahn and M. R. A. Blomberg, *Dalton Trans.*, 2009, 5832–5840.
145. W. Liang, T. A. Roelofs, R. M. Cinco, A. Rompel, M. J. Latimer, W. O. Yu, K. Sauer, M. P. Klein and V. K. Yachandra, *J. Am. Chem. Soc.*, 2000, **122**, 3399–3412.
146. L. V. Kulik, B. Epel, W. Lubitz and J. Messinger, *J. Am. Chem. Soc.*, 2007, **129**, 13421–13435.
147. W. Lubitz, E. J. Reijerse and J. Messinger, *Energy Environ. Sci.*, 2008, **1**, 15–31.
148. G. Renger and B. Hanssum, *FEBS Lett.*, 1992, **299**, 28–32.
149. G. Renger, *Physiol. Plant.*, 1997, **100**, 828–841.
150. P. J. van Leeuwen, H. J. van Gorkom and J. P. Dekker, *J. Photochem. Photobiol., B*, 1992, **15**, 33–43.
151. J. P. Dekker, in *Manganese Redox Enzymes*, ed. P. V. L., VHC Publishers, New York, 1992, pp. 85–101.
152. J. Lavergne, *Biochim. Biophys. Acta, Bioenerg.*, 1991, **1060**, 175–188.
153. H. Kretschmann, J. P. Dekker, Ö. Saygin and H. T. Witt, *Biochim. Biophys. Acta*, 1988, **932**, 358–361.
154. F. Rappaport, M. Blanchard-Desce and J. Lavergne, *Biochim. Biophys. Acta, Bioenerg.*, 1994, **1184**, 178–192.
155. J. P. Dekker, J. J. Plijter, L. Ouwehand and H. J. van Gorkom, *Biochim. Biophys. Acta, Bioenerg.*, 1984, **767**, 176–179.
156. G. Renger and W. Weiss, *Biochem. Soc. Trans.*, 1986, **14**, 17–20.
157. Ö. Hansson and L.-E. Andréasson, *Biochim. Biophys. Acta*, 1982, **679**, 261–268.
158. D. B. Goodin, V. K. Yachandra, R. D. Britt, K. Sauer and M. P. Klein, *Biochim. Biophys. Acta*, 1984, **767**, 209–216.
159. A.-F. Miller and G. W. Brudvig, *Biochim. Biophys. Acta, Bioenerg.*, 1991, **1056**, 1–18.
160. A. Boussac, S. Un, O. Horner and A. W. Rutherford, *Biochemistry*, 1998, **37**, 4001–4007.
161. A. Boussac, M. Sugiura, A. W. Rutherford and P. Dorlet, *J. Am. Chem. Soc.*, 2009, **131**, 5050–5051.
162. H. Matsuoka, K. Furukawa, T. Kato, H. Mino, J. R. Shen and A. Kawamori, *J. Phys. Chem. B*, 2006, **110**, 13242–13247.
163. C. Teutloff, S. Kessen, J. Kern, A. Zouni and R. Bittl, *FEBS Lett.*, 2006, **580**, 3605–3609.
164. J. Messinger, J. H. A. Nugent and M. C. W. Evans, *Biochemistry*, 1997, **36**, 11055–11060.
165. J. Messinger, J. H. Robblee, W. O. Yu, K. Sauer, V. K. Yachandra and M. P. Klein, *J. Am. Chem. Soc.*, 1997, **119**, 11349–11350.

166. K. A. Åhrling, S. Peterson and S. Styring, *Biochemistry*, 1997, **36**, 13148–13152.
167. A. Boussac, H. Kuhl, E. Ghibaudi, M. Rögner and A. W. Rutherford, *Biochemistry*, 1999, **38**, 11942–11948.
168. T. J. Meyer, M. H. V. Huynh and H. H. Thorp, *Angew. Chem., Int. Ed.*, 2007, **46**, 5284–5304.
169. H. Yamada, H. Mino and S. Itoh, *Biochim. Biophys. Acta, Bioenerg.*, 2008, **1767**, 197–203.
170. C. P. Aznar and R. D. Britt, *Philos. Trans. R. Soc., B*, 2002, **357**, 1359–1365.
171. R. Fiege, W. Zweygart, R. Bittl, N. Adir, G. Renger and W. Lubitz, *Photosynth. Res.*, 1996, **42**, 227–244.
172. D. A. Pantazis, M. Orio, T. Petrenko, S. Zein, E. Bill, W. Lubitz, J. Messinger and F. Neese, *Chem.–Eur. J.*, 2009, **15**, 5108–5123.
173. U. Bergmann and P. Glatzel, *Photosynth. Res.*, 2009, **102**, 255–266.
174. R. D. Guiles, V. K. Yachandra, A. E. McDermott, J. L. Cole, S. L. Dexheimer, R. D. Britt, K. Sauer and M. P. Klein, *Biochemistry*, 1990, **29**, 486–496.
175. R. D. Guiles, J.-L. Zimmermann, A. E. McDermott, V. K. Yachandra, J. L. Cole, S. L. Dexheimer, R. D. Britt, K. Wieghardt, U. Bossek, K. Sauer and M. P. Klein, *Biochemistry*, 1990, **29**, 471–485.
176. T. A. Roelofs, W. Liang, M. J. Latimer, R. M. Cinco, A. Rompel, J. C. Andrews, K. Sauer, V. K. Yachandra and M. P. Klein, *Proc. Natl. Acad. Sci. U. S. A.*, 1996, **93**, 3335–3340.
177. J. Messinger, J. H. Robblee, U. Bergmann, C. Fernandez, P. Glatzel, H. Visser, R. M. Cinco, K. L. McFarlane, E. Bellacchio, S. A. Pizarro, S. P. Cramer, K. Sauer, M. P. Klein and V. K. Yachandra, *J. Am. Chem. Soc.*, 2001, **123**, 7804–7820.
178. T.-A. Ono, T. Noguchi, Y. Inoue, M. Kusunoki, T. Matsushita and H. Oyanagi, *Science*, 1992, **258**, 1335–1337.
179. L. Iuzzolino, J. Dittmer, W. Dörner, W. Meyer-Klaucke and H. Dau, *Biochemistry*, 1998, **37**, 17112–17119.
180. P. Glatzel, U. Bergmann, J. Yano, H. Visser, J. H. Robblee, W. W. Gu, F. M. F. de Groot, G. Christou, V. L. Pecoraro, S. P. Cramer and V. K. Yachandra, *J. Am. Chem. Soc.*, 2004, **126**, 9946–9959.
181. P. Glatzel, J. Yano, U. Bergmann, H. Visser, J. H. Robblee, W. W. Gu, F. M. F. de Groot, S. P. Cramer and V. K. Yachandra, *J. Phys. Chem. Solids*, 2005, **66**, 2163–2167.
182. E. M. Sproviero, J. A. Gascon, J. P. McEvoy, G. W. Brudvig and V. S. Batista, *J. Am. Chem. Soc.*, 2008, **130**, 3428–3442.
183. Y. Kimura, N. Mizusawa, A. Ishii, S. Nakazawa and T. Ono, *J. Biol. Chem.*, 2005, **280**, 37895–37900.
184. H. A. Chu, W. Hillier and R. J. Debus, *Biochemistry*, 2004, **43**, 3152–3166.
185. M. A. Strickler, L. M. Walker, W. Hillier and R. J. Debus, *Biochemistry*, 2005, **44**, 8571–8577.

186. N. Mizusawa, T. Yamanari, Y. Kimura, A. Ishii, S. Nakazawa and T. A. Ono, *Biochemistry*, 2004, **43**, 14644–14652.
187. N. Mizusawa, Y. Kimura, A. Ishii, T. Yamanari, S. Nakazawa, H. Teramoto and T. Ono, *J. Biol. Chem.*, 2004, **279**, 29622–29627.
188. M. A. Strickler, H. J. Hwang, R. L. Burnap, J. Yano, L. M. Walker, R. J. Service, R. D. Britt, W. Hillier and R. J. Debus, *Philos. Trans. R. Soc., B*, 2008, **363**, 1179–1187.
189. Y. Shimada, H. Suzuki, T. Tsuchiya, T. Tomo, T. Noguchi and M. Mimuro, *Biochemistry*, 2009, **48**, 6095–6103.
190. P. E. M. Siegbahn, *Philos. Trans. R. Soc., B*, 2008, **363**, 1221–1228.
191. G. J. Yeagle, M. L. Gilchrist, L. M. Walker, R. J. Debus and R. D. Britt, *Philos. Trans. R. Soc., B*, 2008, **363**, 1157–1166.
192. G. J. Yeagle, M. L. Gilchrist, R. M. McCarrick and R. D. Britt, *Inorg. Chem.*, 2008, **47**, 1803–1814.
193. J. A. Stull, T. A. Stich, R. J. Service, R. J. Debus, S. K. Mandal, W. H. Armstrong and R. D. Britt, *J. Am. Chem. Soc.*, 2010, **132**, 446–447.
194. R. J. Debus, C. Aznar, K. A. Campbell, W. Gregor, B. A. Diner and R. D. Britt, *Biochemistry*, 2003, **42**, 10600–10608.
195. R. J. Debus, K. A. Campbell, W. Gregor, Z. L. Li, R. L. Burnap and R. D. Britt, *Biochemistry*, 2001, **40**, 3690–3699.
196. T. Noguchi and M. Sugiura, *Biochemistry*, 2002, **41**, 2322–2330.
197. L. Helm, G. M. Nicolle and A. E. Merbach, *Adv. Inorg. Chem.*, 2005, **57**, 327–379.
198. B. Nöring, D. Shevela, G. Renger and J. Messinger, *Photosynth. Res.*, 2008, **98**, 251–260.
199. W. Hillier, J. Messinger and T. Wydrzynski, *Biochemistry*, 1998, **37**, 16908–16914.
200. W. Hillier and T. Wydrzynski, *Biochemistry*, 2000, **39**, 4399–4405.
201. G. Hendry and T. Wydrzynski, *Biochemistry*, 2002, **41**, 13328–13334.
202. G. Hendry and T. Wydrzynski, *Biochemistry*, 2003, **42**, 6209–6217.
203. M. Sugiura, F. Rappaport, W. Hillier, P. Dorlet, Y. Ohno, H. Hayashi and A. Boussac, *Biochemistry*, 2009, **48**, 7856–7866.
204. S. Singh, R. J. Debus, T. Wydrzynski and W. Hillier, *Philos. Trans. R. Soc., B*, 2008, **363**, 1229–1234.
205. R. Tagore, R. H. Crabtree and G. W. Brudvig, *Inorg. Chem.*, 2007, **46**, 2193–2203.
206. R. Tagore, H. Y. Chen, R. H. Crabtree and G. W. Brudvig, *J. Am. Chem. Soc.*, 2006, **128**, 9457–9465.
207. C. A. Ohlin, R. Brimblecombe, L. Spiccia and W. H. Casey, *Dalton Trans.*, 2009, 5278–5280.
208. D. Lieb, A. Zahl, T. E. Shubina and I. Ivanovic-Burmazovic, *J. Am. Chem. Soc.*, 2010, **132**, 7282–7284.
209. M. Lundberg, M. R. A. Blomberg and P. E. M. Siegbahn, *Theor. Chem. Acc.*, 2003, **110**, 130–143.
210. T. Noguchi, *Photosynth. Res.*, 2007, **91**, 59–69.
211. T. Noguchi, *Philos. Trans.R. Soc., B*, 2008, **363**, 1189–1194.

212. T. Noguchi and M. Sugiura, *Biochemistry*, 2000, **39**, 10943–10949.
213. H. Suzuki, M. Sugiura and T. Noguchi, *Biochemistry*, 2008, **47**, 11024–11030.
214. P. E. M. Siegbahn and M. Lundberg, *Photochem. Photobiol. Sci.*, 2005, **4**, 1035–1043.
215. P. E. M. Siegbahn, *Chem.–Eur. J.*, 2006, **12**, 9217–9227.
216. H. A. Chu, W. Hillier, N. A. Law and G. T. Babcock, *Biochim. Biophys. Acta, Bioenerg.*, 2001, **1503**, 69–82.
217. H.-A. Chu, H. Sackett and G. T. Babcock, *Biochemistry*, 2000, **39**, 14371–14376.
218. Y. Kimura, A. Ishii, T. Yamanari and T. A. Ono, *Biochemistry*, 2005, **44**, 7613–7622.
219. J. H. Su and J. Messinger, *Appl. Magn. Reson.*, 2010, **37**, 123–136.
220. T. Wydrzynski, N. Zumbulyadis, P. G. Schmidt and Govindjee, *Biochim. Biophys. Acta, Bioenerg.*, 1975, **408**, 349–354.
221. T. Wydrzynski, N. Zumbulyadis, P. G. Schmidt, H. S. Gutowsky and Govindjee, *Proc. Natl. Acad. Sci. U. S. A.*, 1976, **73**, 1196–1198.
222. H. H. Robinson, R. R. Sharp and C. F. Yocum, *Arch. Biochem. Biophys.*, 1981, **207**, 1–8.
223. R. R. Sharp, in *Manganese Redox Enzymes*, ed. V. L. Pecoraro, VCH Publishers, New York, 1992, pp. 177–196.
224. Y. Deligiannakis, M. Louloudi and N. Hadjiliadis, *Coord. Chem. Rev.*, 2000, **204**, 1–112.
225. Ö. Hansson, L.-E. Andréasson and T. Vänngård, *FEBS Lett.*, 1986, **195**, 151–154.
226. M. C. W. Evans, J. H. A. Nugent, R. J. Ball, I. Muhiuddin and R. J. Pace, *Biochemistry*, 2004, **43**, 989–994.
227. S. Turconi, D. J. MacLachlan, P. J. Bratt, J. H. A. Nugent and M. C. W. Evans, *Biochemistry*, 1997, **36**, 879–885.
228. J. H. Su, W. Lubitz and J. Messinger, *J. Am. Chem. Soc.*, 2008, **130**, 786–787.
229. J. H. Su, W. Lubitz and J. Messinger, *J. Am. Chem. Soc.*, 2011, **133**, 12317–12317.
230. A. Kawamori, T. Inai, T. Ono and Y. Inoue, *FEBS Lett.*, 1989, **254**, 219–223.
231. X.-S. Tang, M. Sivaraja and G. C. Dismukes, *J. Am. Chem. Soc.*, 1993, **115**, 2382–2389.
232. M. C. W. Evans, A. M. Rich and J. H. A. Nugent, *FEBS Lett.*, 2000, **477**, 113–117.
233. H. Yamada, H. Mino and S. Itoh, *Biochim. Biochim. Biophys. Acta, Bioenerg.*, 2007, **1767**, 197–203.
234. K. A. Åhrling, M. C. W. Evans, J. H. A. Nugent, R. J. Ball and R. J. Pace, *Biochemistry*, 2006, **45**, 7069–7082.
235. S. Mukhopadhyay, S. K. Mandal, S. Bhaduri and W. H. Armstrong, *Chem. Rev.*, 2004, **104**, 3981–4026.

236. T. C. Weng, W. Y. Hsieh, E. S. Uffelman, S. W. Gordon-Wylie, T. J. Collins, V. L. Pecoraro and J. E. Penner-Hahn, *J. Am. Chem. Soc.*, 2004, **126**, 8070–8071.
237. M. Haumann, P. Liebisch, C. Müller, M. Barra, M. Grabolle and H. Dau, *Science*, 2005, **310**, 1019–1021.
238. F. Rappaport, M. Guergova-Kuras, P. J. Nixon, B. A. Diner and J. Lavergne, *Biochemistry*, 2002, **41**, 8518–8527.
239. F. Rappaport and B. A. Diner, *Coord. Chem. Rev.*, 2008, **252**, 259–272.
240. A. B. Anderson and T. V. Albu, *J. Am. Chem. Soc.*, 1999, **121**, 11855–11863.
241. L. I. Krishtalik, *Biochim. Biophys. Acta, Bioenerg.*, 1986, **849**, 162–171.
242. J. Lavergne and W. Junge, *Photosynth. Res.*, 1993, **38**, 279–296.
243. F. Rappaport and J. Lavergne, *Biochim. Biophys. Acta, Bioenerg.*, 2001, **1503**, 246–259.
244. M. Haumann and W. Junge, in *Oxygenic Photosynthesis: The Light Reactions*, ed. D. R. Ort and C. F. Yocum, Kluwer, Dordrecht, 1996, **vol. 4**, pp. 165–192.
245. W. Junge, M. Haumann, R. Ahlbrink, A. Mulkidjanian and J. Clausen, *Philos. Trans. R. Soc., B*, 2002, **357**, 1407–1417.
246. C. F. Fowler, *Biochim. Biophys. Acta, Bioenerg.*, 1977, **462**, 414–421.
247. S. Saphon and A. R. Crofts, *Z. Naturforsch., C: J. Biosci.*, 1977, **32**, 617–626.
248. V. Förster and W. Junge, *Photochem. Photobiol.*, 1985, **41**, 183–190.
249. E. Schlodder and H. T. Witt, *J. Biol. Chem.*, 1999, **274**, 30387–30392.
250. F. Rappaport and J. Lavergne, *Biochemistry*, 1991, **30**, 10004–10012.
251. P. Jahns, J. Lavergne, F. Rappaport and W. Junge, *Biochim. Biophys. Acta, Bioenerg.*, 1991, **1057**, 313–319.
252. P. Jahns and W. Junge, *Biochemistry*, 1992, **31**, 7398–7403.
253. K. Lübbers, M. Haumann and W. Junge, *Biochim. Biophys. Acta, Bioenerg.*, 1993, **1183**, 210–214.
254. M. Haumann and W. Junge, *Biochemistry*, 1994, **33**, 864–872.
255. G. Renger, *Photosynthetica*, 1987, **21**, 203–224.
256. H. Suzuki, M. Sugiura and T. Noguchi, *J. Am. Chem. Soc.*, 2009, **131**, 7849–7857.
257. V. Petrouleas, D. Koulougliotis and N. Ioannidis, *Biochemistry*, 2005, **44**, 6723–6728.
258. M. R. Razeghifard and R. J. Pace, *Biochemistry*, 1999, **38**, 1252–1257.
259. S. G. Zech, J. Kurreck, G. Renger, W. Lubitz and R. Bittl, *FEBS Lett.*, 1999, **442**, 79–82.
260. H. Dau and I. Zaharieva, *Acc. Chem. Res.*, 2009, **42**, 1861–1870.
261. G. Renger, in *Primary Processes of Photosynthesis – Part 2: Basic Principles and Apparatus* ed. G. Renger, The Royal Society of Chemistry, Cambridge, UK, 2008, pp. 237–290.
262. G. Renger, *Curr. Sci.*, **98**, 1305–1319.
263. J. Clausen and W. Junge, *Photosynth. Res.*, 2008, **98**, 229–233.
264. J. Clausen and W. Junge, *Nature*, 2004, **430**, 480–483.

265. J. Clausen, W. Junge, H. Dau and M. Haumann, *Biochemistry*, 2005, **44**, 12775–12779.
266. M. Haumann, A. Grundmeier, I. Zaharieva and H. Dau, *Proc. Natl. Acad. Sci. U. S. A.*, 2008, **105**, 17384–17389.
267. D. R. J. Kolling, T. S. Brown, G. Ananyev and G. C. Dismukes, *Biochemistry*, 2009, **48**, 1381–1389.
268. D. Shevela, K. Beckmann, J. Clausen, W. Junge and J. Messinger, *Proc. Natl. Acad. Sci. U. S. A.*, 2011, **108**, 3602–3607.
269. G. Renger, *FEBS Lett.*, 1977, **81**, 223–228.
270. M. Haumann, O. Bögershausen, D. Cherepanov, R. Ahlbrink and W. Junge, *Photosynth. Res.*, 1997, **51**, 193–208.
271. M. Karge, K.-D. Irrgang and G. Renger, *Biochemistry*, 1997, **36**, 8904–8913.
272. M. Karge, K.-D. Irrgang, S. Sellin, R. Feinaugle, B. Liu, H. J. Eckert, H. J. Eichler and G. Renger, *FEBS Lett.*, 1996, **378**, 140–144.
273. J. Sinclair and T. Arnason, *Biochim. Biophys. Acta, Bioenerg.*, 1974, **368**, 393–400.
274. P. E. M. Siegbahn, *Inorg. Chem.*, 2000, **39**, 2923–2935.
275. P. E. M. Siegbahn and M. R. A. Blomberg, *Chem. Rev.*, 2000, **100**, 421–437.
276. *Primary Processes of Photosynthesis. Principles and Apparatus.*, ed. G. Renger, The Royal Society of Chemistry, Cambridge, 2008.
277. R. E. Blankenship, *Molecular mechanisms of photosynthesis*, Blackwell Publishing, Oxford, 2002.
278. *Photosystem II. The Light-Driven Water:Plastoquinone Oxidoreductase*, ed. T. Wydrzynski and K. Satoh, Springer, Dordrecht, 2005.
279. J. Barber and B. Andersson, *Trends Biochem. Sci.*, 1992, **17**, 61–66.
280. P. Horton, A. V. Ruban and R. G. Walters, *Annu. Rev. Plant Physiol. Plant Mol. Biol.*, 1996, **47**, 655–684.
281. P. Müller, X. P. Li and K. K. Niyogi, *Plant Physiol.*, 2001, **125**, 1558–1566.
282. A. M. Gilmore and X.-P. Li, in *Primary Processes of Photosynthesis. Principles and Apparatus – Part 1*, ed. G. Renger, The Royal Society of Chemistry, Cambridge, 2008, pp. 369–390.
283. P. Mitchell, *Biochem. J.*, 1961, **79**, 23–25.
284. W. Junge, in *Primary Processes of Photosynthesis – Principle and Apparatus – Part 2*, ed. G. Renger, The Royal Society of Chemistry, Cambridge, 2008, pp. 447–487.
285. H. H. Stiehl and H. T. Witt, *Z. Naturforsch.*, 1968, **23B**, 220–224.
286. K. Beckmann, H. Uchtenhagen, G. Berggren, M. F. Anderlund, A. Thapper, J. Messinger, S. Styring and P. Kurz, *Energy Environ. Sci.*, 2008, **1**.
287. Z. Huang, Z. Luo, Y. V. Geletii, J. W. Vickers, Q. Yin, D. Wu, Y. Hou, Y. Ding, J. Song, D. G. Musaev, C. L. Hill and T. Lian, *J. Am. Chem. Soc.*, **133**, 2068–2071.
288. M. M. Najafpour, T. Ehrenberg, M. Wiechen and P. Kurz, *Angew. Chem., Int. Ed.*, 2010, **49**, 2233–2237.

289. M. W. Kanan and D. G. Nocera, *Science*, 2008, **321**, 1072–1075.
290. F. Jiao and H. Frei, *Energy Environ. Sci.*, 2010, **3**, 1018–1027.
291. N. D. McDaniel, F. J. Coughlin, L. L. Tinker and S. Bernhard, *J. Am. Chem. Soc.*, 2008, **130**, 210–217.
292. S. W. Gersten, G. J. Samuels and T. J. Meyer, *J. Am. Chem. Soc.*, 1982, **104**, 4029–4030.
293. J. J. Concepcion, J. W. Jurss, M. K. Brennaman, P. G. Hoertz, A. O. T. Patrocinio, N. Y. M. Iha, J. L. Templeton and T. J. Meyer, *Acc. Chem. Res.*, 2009, **42**, 1954–1965.
294. I. Romero, M. Rodriguez, C. Sens, J. Mola, M. R. Kollipara, L. Francas, E. Mas-Marza, L. Escriche and A. Llobet, *Inorg. Chem.*, 2008, **47**, 1824–1834.
295. M. Kusunoki, *J. Photochem. Photobiol. B*, 2011, **104**, 100–110.

CHAPTER 8

Substrate and Product Channels in Photosystem II

FELIX M. HO

Department of Photochemistry & Molecular Science, The Ångström Laboratory, Box 523, Uppsala University, 752 10 Uppsala, Sweden

8.1 Photosystem II and Channels

Water oxidation is the first step in a chain of reactions that allows oxygenic photosynthetic organisms to utilise sunlight as their energy source. As such, understanding how it takes place and is optimised is of fundamental importance. Beyond the pure academic interest in discovering how Nature is able to utilise the apparently simple ingredients of sunlight, water and CO_2 to sustain life, this knowledge can provide us with a template for the design and development of artificial means of harnessing solar energy for fuel production.

Photosystem II (PSII) is a large, transmembrane multi-subunit enzyme found in the thylakoid membranes in cyanobacteria, algae and higher plants. It is responsible for catalysing water oxidation, the first step in oxygenic photosynthesis. The absorption of a photon by antenna chlorophyll molecules leads to the transfer of excitation energy to the primary donor, a special assembly of four chlorophyll molecules known as P_{680}. At this point, charge separation takes place, whereby an electron is sent *via* a pheophytin molecule and two plastoquinones Q_A and Q_B down the rest of the electron transfer chain. During this process, a proton gradient is established across the thylakoid membrane, providing the driving force for ATP synthesis by ATP-synthase. The ultimate source of the electrons that are transferred in this chain of reactions is water.

RSC Energy and Environment Series No. 5
Molecular Solar Fuels
Edited by Thomas J. Wydrzynski and Warwick Hillier

Published by the Royal Society of Chemistry, www.rsc.org

Water oxidation takes place at the oxygen evolving centre (OEC), which consists of a μ-oxo-bridged cluster of four Mn ions and a Ca ion ($CaMn_4$) and a nearby redox-active tyrosine residue D1-Tyr161, known as Y_Z. During water oxidation, the $CaMn_4$ cluster accumulates four oxidation equivalents during the so-called S-cycle (also known as the Kok cycle), which consists of five intermediate states, labelled Sn, $n = 0–4$. It is believed that two water molecules bind to the $CaMn_4$ cluster, and their oxidation is completed during the $S_4 \rightarrow S_0$ transition. O_2 is thus released, and the $CaMn_4$ cluster is reset. For general reviews of PSII and oxygenic photosynthesis, see for example Barber,[1] McEvoy and Brudvig,[2] Nelson and Ben-Shem,[3] Nelson and Yocum,[4] and Wydrzynski and Satoh.[5]

The publication of the fully-refined, medium-resolution crystal structures of PSII at 3.5, 3.0 and 2.9 Å resolutions (PDB 1S5L,[6] 2AXT,[7] and 3BZ1/3BZ2,[8] respectively) represented major advances in the field of PSII research. Much of the discussions and debates have since then focused on the structure of the $CaMn_4$ cluster, the arrangement of its associated ligands and the detailed mechanism of the S-cycle (especially in comparison to data obtained using X-ray spectroscopic techniques).[5,9–11] The recent announcement of a 1.9 Å resolution crystal structure by Shen and co-workers[188] represents an enormous step forward, and will not doubt prove extremely valuable for structural and mechanistic investigations of PSII.

One aspect of the PSII structure that has received increasing attention is the possible existence of channels in the PSII protein complex.[12–15] As the $CaMn_4$ cluster at which water oxidation takes place is buried within the protein D1 protein subunit, it was proposed even prior to the availability of the crystal structures that specific channels exist in the protein matrix to allow substrate water to access the catalytic site, and for the protons and O_2 that are produced to leave.

Why do certain proteins contain channels? Two main reasons can be advanced in response to this question, one or both of which may be applicable to any given channel-containing protein. Firstly, channels can be used to facilitate the movement of molecules, whether they be substrates, co-factors or products, through a protein. Such movement may be necessary to allow a reaction to take place. The most obvious example of this would be the delivery of the substrate molecule(s) to a catalytic site, with the channel helping to optimise the delivery of the molecules both spatially and temporally, *i.e.* in the right orientation, at the right place, and at the right time/rate. Well-known examples in the literature are acetylcholinesterase,[16,17] cytochrome *c* oxidase,[18–20] and bacteriorhodopsin.[21,22] Alternatively, the movement could itself *be* the main function of the protein, as is the case for the multitude of membrane channels in living organisms. Aquaporin[23–25] is a well-known and well-studied example of this. In such cases, the channel defines both the path and potentially the rate of the movement. In either case, the channel's properties and its interactions with the molecules in question can greatly influence the transport characteristics.

A second possible reason for the existence of channels is to keep certain molecules that must traverse the protein matrix away from certain parts of the protein, or to only selectively allow molecules with certain characteristics (*e.g.* size, shape) to pass. In doing so, undesirable reactions that impede the correct functioning of the protein are prevented from taking place. This is a corollary of having a channel defining the pathway along which the transported molecules move. By providing a favourable and optimised pathway for the molecule, undesired access to other parts of the protein is disfavoured, and properties of the channels themselves (*e.g.* diameter, hydrophobicity/hydrophilicity) can have a filtering effect. Channels can therefore serve a protective function for a protein. Good examples of this are the gas channels in a number of hydrogenases from *Ralstonia eutropha* that exhibit O_2-tolerance, which is an unusual property among [NiFe]-hydrogenases.[26] Site-directed mutations of residues involved in channels leading to the active site from O_2 access have been found to alter the sensitivity of the hydrogenases to oxygen, at least partly as a result of changes to the diameter of channels.[27,28]

With this background in mind, this chapter will present an overview of the substrate water and product proton and O_2 channels that have been proposed for PSII.* For each of these molecules, the background and arguments for the existence of channels for their transport are summarised, followed by a review of the current literature proposals and experimental data relevant for the identification of these putative channels. Far from being passive "scaffolding" to hold the electron transport co-factors in place, the protein matrix itself can serve an important function in optimising and regulating the water oxidation reaction.

8.2 Water Channels

8.2.1 Reasoning for the Existence of Water Channels

The motivation for proposing the existence of water channels falls into the first category of the reasons above, or more specifically, to control the rate of water access to the OEC. It may at first glance seem to be paradoxical to regulate substrate water access in an enzyme that oxidises ~100 water molecules per second. However, relative to the rate of water (self-)diffusion, this frequency of water oxidation is not so high. For example, compared to the rate of water permeation rate through aquaporin, the rate of water oxidation rate in PSII is approximately 10^5–10^6 times slower (for further discussions, see Ho[29]).

It has been argued that the amount of water delivered to the catalytic site must be optimised to avoid "short-circuiting" of the S-cycle. Based on studies of the effects of Ca^{2+}- and Cl^--depletion and the removal of the extrinsic

*The existence of a diffusion pathway on the acceptor side of PSII for the entry and exit of Q_B^- has also been proposed by Loll *et al.*[7] and Guskov *et al.*[8] have suggested the existence of a third quinone, Q_C, which is also connected to the Q_B site by a channel. However, these will not be covered in the present review. A briefer version of the present treatment of channels in PSII, presented together with other computational mechanistic and structural studies of PSII, can be found in Ho. (See notes added in proof.)

protein subunits (17, 23, 33 kDa) on the water oxidation capabilities of PSII (for a general review of the Ca^{2+} and Cl^- cofactors in PSII, see van Gorkom and Yocum[30]), Rutherford[31] suggested that the extrinsic polypeptides may protect the Mn cluster from external reducing reagents, and also help to improve binding of Ca^{2+} and Cl^- ions to the cluster. In particular, it was suggested that Cl^- binding could be involved in preventing water oxidation taking place too early during the S-cycle, which would lead to destabilisation and loss of the Mn cluster through the formation of Mn^{2+} ions. Ca^{2+}-depletion has been shown to increase access by NH_2OH and other larger extrinsic electron donor molecules to the Mn ions in the $CaMn_4$ cluster.[32,33] An unusually stable, modified S_2-state that gives rise to an altered S_2 multiline EPR signal is also found in Ca^{2+}-depleted samples,[34–37] which has been attributed as resulting from the hydrolysis of the $CaMn_4$ cluster due to the presence of excessive amounts of water, leading to disproportionation between two Mn(III) ions.[33] A number of workers[31,33–35,37] have suggested that Ca^{2+} takes part in regulating access to the Mn cluster, acting as a "gatekeeper". The absence of Ca^{2+} has been suggested to cause structural perturbations to the cluster that lead to the various effects observed.

Building upon these observations, Wydrzynski *et al.*[38] suggested that specific channels in the PSII protein matrix may provide the means to control and limit substrate water access to the $CaMn_4$ cluster. These channels would play a role in optimising water oxidation in PSII, and avoid the formation of harmful intermediates, such as H_2O_2, through undesired side-reactions. Previous experimental evidence had shown that such H_2O_2 formation was possible from the donor side of PSII after certain pre-treatments, such as depletion of extrinsic subunits,[39] osmotic stress,[40] treatment with the lipid analogue lauroylcholine chloride,[41] and Cl^--depletion.[42–44] Wydrzynski and co-workers[38,40] have argued that the common factor in these treatments is that the structural integrity of the protein around the manganese cluster is likely to have been disturbed, and that a "tight" hydrophobic pocket is needed around the catalytic site. Excessive water access due to a "leaky" environment would increase the likelihood of incomplete water oxidation and thus reduced O_2 evolution. From this crystallised the idea of controlled water accessibility to the catalytic site through the presence of (a) specific water channel(s). It should be noted, however, that more recent studies suggest that not all extrinsic subunits are involved in limiting water access to the $CaMn_4$ cluster. ESEEM and EXAFS experiments have found that water binding to the cluster was unchanged upon removal of the 33 kDa (PsbO) subunit, at least in the S_1 state,[45] and ^{18}O substrate water exchange rates for the S_3 state were also not found to be significantly affected[46] (for a review of PSII extrinsic protein subunits, see Bricker and Burnap[47]).

A long line of ^{18}O experiments have demonstrated the existence of two distinct and defined substrate water binding sites at the $CaMn_4$ cluster, each with its own S-state dependent affinity for water. At least one substrate water is bound from the S_0 state, and the other is bound by the S_2 state.[48–54] Such specific binding behaviour are consistent with the suggestion of some mechanism for the optimisation of both the delivery and the conformations of substrate waters.

Finally, the very nature of the channels inherently places a size limitation on the transported molecules. For chemically analogous molecules, this should be reflected in an inverse correlation between the size of the molecule and the access to the $CaMn_4$ cluster due to steric hindrance. Such a correlation was observed in an ESEEM study of binding to the cluster by a series of four alcohols of increasing bulk (methanol, ethanol, *n*-propanol and 2-propanol),[55] which led the authors to propose the existence of an access channel of 3–4 Å in diameter.

8.2.2 Studies in Search of Water Channels

Three studies so far have looked in detail for the presence of channels in the PSII crystal structure and examined their characteristics.[12–14] In each of these studies, continuous/contiguous spaces which are able to accommodate a water-/O_2-sized probe were sought, although slightly different methodologies were employed. The locations and orientations of different channels identified with respect to the PSII complex, and a comparison of the channels reported in the different studies, are summarised in Figure 8.1 and Table 8.1.

In Ho and Styring,[13] all possible solvent accessible regions around the $CaMn_4$ were calculated from the crystal structure (2AXT), and the full solvent contact surface was then examined in detail to locate the channels. Three sets of channels that were found to connect the $CaMn_4$ cluster to the lumen were

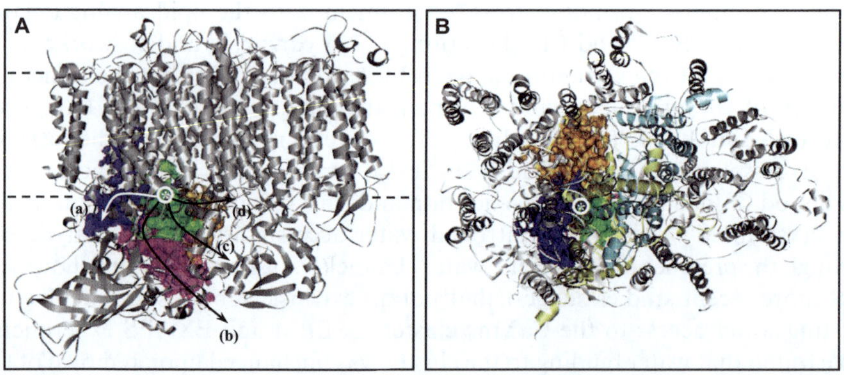

Figure 8.1 The locations and orientations of the various channels in PSII. The position of the $CaMn_4$ is shown with an asterisk. (**A**) Viewed along the thylakoid membrane plane (dashed lines). (a) The back channel/channel (i)/Channels A1-A2 (blue surface); (b) the narrow channel/Channels D, E, F (pink surface); (c) the broad channel/channel (iii)/Channel C (green surface); (d) the large channel system/channel (ii)/Channels B1-B2 (orange surface). (**B**) Viewed from the stromal side, along the membrane normal down towards the lumen. The channels are shown according to the colour scheme in (**A**). The narrow channel (pink) is obscured by the other channels from this viewing angle. The D1 and D2 subunits are shown with yellow and light blue ribbons, respectively.

Table 8.1 Residues near the $CaMn_4$ cluster involved with the channels identified in the PSII crystal structure.

	Name	*Transport of*		*Name*	*Transport of*		*Name*	*Transport of*		*Name*	*Transport of*	
Ho & Styring[13]	*back*	*Water*		*Narrow*	H^+		*broad*	H^+		*large*	O_2	
Murray & Barber[12]	*(i)*	O_2		---	---		*(iii)*	*Water*/H^+		*(ii)*	*Water*/H^+	
Gabdulkhakov et al.[14]	*A1, A2*	*Water*		*D, E, F*	H^+		*C, (G)*	H^+		*B1, B2*	O_2	
Residues within 15Å of the	D1	-Asn	87	D1	-Asp	61	D1	-Ile	60	D1	-Glu	189
$CaMn_4$ cluster, based on		-Ala	88		-Gly	62		-Asp	61		-Glu	329
channels in Ho & Styring[13]		-Ile	89		-Ile	63		-Ile	63		-His	332
		-Gly	90		-Asn	87		-Glu	65		-Pro	340
		-Leu	91		-Glu	333		-Val	67		-Leu	341
See text for discrepancies between		-Tyr	161		-Ser	169		-Pro	84		-Asp	342
the studies, and see the original		-Ile	163		-Asn	335		-Tyr	161		-Leu	343
publications for full residue		-Gly	164		-Ala	336		-Ser	169		-Ala	344
listings.		-Gln	165		-Asn	338		-Asp	170	CP47	-Arg	384
		-Gly	166	CP43	-Pro	334		-Gly	171	CP43	-Glu	354
		-Glu	189		-Thr	335		-Met	172		-Met	396
		-His	190		-Leu	337		-Pro	173		-Ala	399
		-Asn	296		-Met	342		-Asn	181		-Leu	401
		-Asn	298		-Gly	353		-Phe	182		-Gly	409
		-Asp	342		-Glu	354		-Val	185		-Val	410
		-Leu	343		-Met	356		-Phe	186		-Thr	412
		-Ala	344		-Arg	357		-Met	331		-Glu	413
	CP43	-Trp	291					-His	332	D2	-Arg	348
		-Phe	292					-Glu	333		-Ala	351
		-Gly	306					-Arg	334		-Leu	352
		-Phe	307				D2	-Glu	312	PsbU	-Tyr	133

Table 8.1 (*Continued*)

	Name	*Transport of*	*Name*	*Transport of*	*Name*	*Transport of*	*Name*	*Transport of*
Ho & Styring[13]	*back*	*Water*	*Narrow*	H^+	*broad*	H^+	*large*	O_2
Murray & Barber[12]	*(i)*	O_2	---	---	*(iii)*	*Water*/H^+	*(ii)*	*Water*/H^+
Gabdulkhakov et al.[14]	*A1, A2*	*Water*	*D, E, F*	H^+	*C, (G)*	H^+	*B1, B2*	O_2
		-Ala 309				-Phe 314	PsbV	-Lys 160
		-Met 356				-Lys 317		
		-Arg 357				-Leu 320		
		-Phe 358				-Leu 321		
		-Ala 399						
		-Pro 400						
		-Leu 401						

identified in this study, and they were named the "back channel", "narrow and broad channels" and the "large channel system". Of these, the back channel (Figure 8.2A) was proposed to function as a water channel, though perhaps not exclusively, given that the large channel system (Figure 8.2C) is also likely to be filled with water. An interesting feature of the back channel was that, while it is

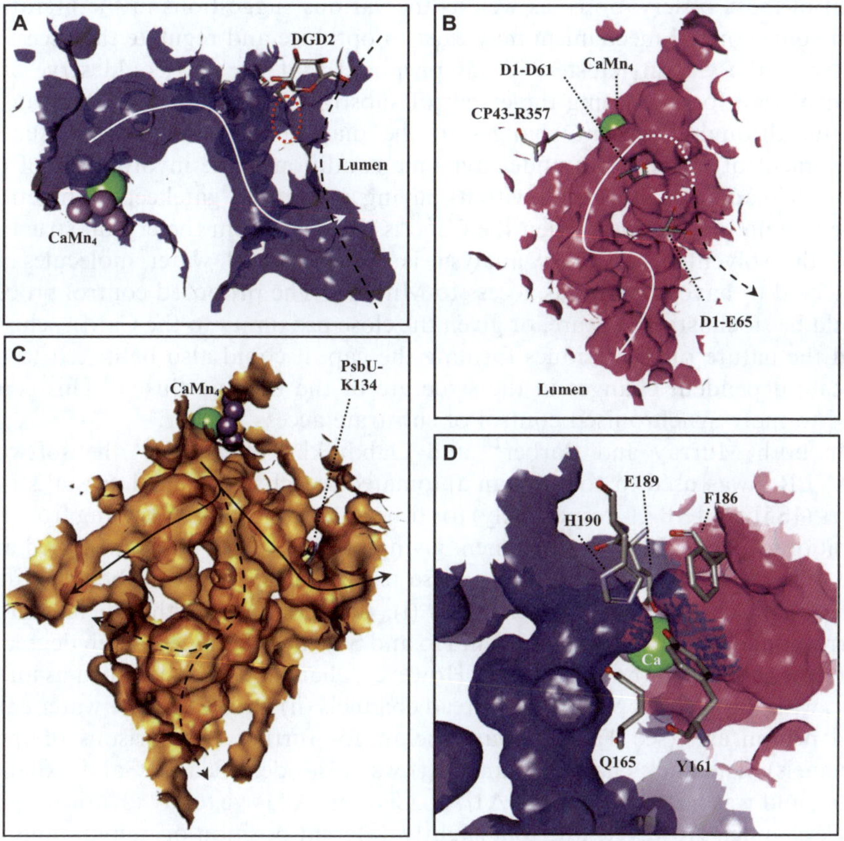

Figure 8.2 The individual channel systems in PSII as found in Ho and Styring,[13] with the pathways marked with arrows. See the text for comparison to channel designations in other studies. **A**: The back channel. The DGD2 molecule is shown in stick representation. The bottleneck that opens to give an extra exit in Gabdulkhakov *et al.*[14] is marked with a dashed oval. **B**: The narrow/broad channels, with the residues D1-D61, -E65 and CP43-R357 shown in stick representation. The path of the narrow channel out to the lumen is shown with the solid white arrow, the (closed) broad channel is shown with a dotted white arrow, and the continuation of the broad channel past the bottleneck block (see text) is shown with the dashed black arrow. **C**: The large channel system, with the different branches and exits shown with arrows. The residue PsbU-K134 is shown in stick representation. **D**: The gap between the back (blue) and broad (pink) channels at the $CaMn_4$ suggested as a possible control gate for water access. All residues are on the D1 subunit.

in contact with the Ca^{2+}, it would be necessary for water molecules to cross a gap formed of the residues D1-Tyr161, -His190, -Glu189, -Phe186, -Gln165 and the Ca^{2+} in order to reach the Mn ions *via* the distal section of the broad channel (Figure 8.2D). These same residues have been proposed to be of special importance in water oxidation (acting variously as oxidisers, proton acceptors and being part of a hydrogen bonding network near the $CaMn_4$ cluster).[56–58] In light of these observations, as well as the various suggestions in the literature that some control mechanism may exist to optimise and regulate the access of water to the $CaMn_4$ cluster, it was proposed that the gap could serve as a control gate for regulating the extent of substrate water access to the $CaMn_4$ cluster through transient changes in the diameter of the opening due to movement of the residues under dynamic conditions. The involvement of the Ca^{2+} is in addition consistent with its having a putative "gatekeeper" function. The gap no longer exists when the Ca^{2+} is removed from the crystal structure and the solvent accessibility analysis is repeated, and water molecules are observed to have continuous access to Mn ions. The proposed control process could be stochastic in nature, or given the close proximity to the $CaMn_4$ cluster and the nature of the residues forming the gap, it could also be linked to the S-state dependent changes in the structure of the $CaMn_4$ cluster. This could lead to more synchronised control of substrate access.

In both Murray and Barber[12] and Gabdulkhakov *et al.*,[14] the software CAVER[59] was used to conduct an automated search of the PSII crystal structures (1S5L and 3BZ1, respectively) for trajectories of channels leading from the position of the $CaMn_4$ to the lumen, giving profiles of the paths identified and the diameter of the channel along these paths. Murray and Barber identified three main channels, named channels (i), (ii) and (iii). Of these, channel (i) corresponded to the back channel in Ho and Styring, with a very high degree of agreement between the two studies. However, channel (i) was assigned as an O_2 exit, with the authors preferring instead channels (ii) and/or (iii) for water entry and proton exit (see Table 8.1 and below for further comparisons of these channels). The back channel/channel (i) was also identified in Gabdulkhakov *et al.*, and was named Channels A1/A2. Channel A2 is virtually identical to the back channel. However, due to a slightly different position of a lipid molecule (DGD2, Figure 8.2A) in the 2.9Å-resolution crystal structure (3BZ1, which was obtained from reprocessing of the X-ray diffraction data set used to produce the 2AXT structure), a new exit was found, with the bifurcation of the main channel occurring close to the luminal surface (Figure 8.2A). Channel A1 is essentially a subset of A2, with 18 out of its 24 residues also found in A2 and the remaining residues describing the new exit. In contrast to Murray and Barber, and in agreement with Ho and Styring, Channels A1/A2 were assigned as water channels. Additionally, channels B1/B2, C, D, E, F and G were also reported in Gabdulkhakov *et al.* and assigned to O_2 and proton removal functions (Table 8.1). These are considered further in later sections.

A significant limitation of these studies based on the crystal structures of PSII is that only a particular static conformation of the protein complex is considered. As discussed in Ho,[29] there is significant movement or "breathing" in the protein structure under dynamic physiological conditions. Important information about dynamic processes (movement of water molecules and amino acid residues, transient interactions and binding *etc.*) is not taken into account. Molecular dynamics (MD) simulations, where classical Newtonian laws of motion are used to simulate the thermal motions of a system over time, would be particularly useful for overcoming this limitation.

MD simulations have played a key part in elucidating pathways and conformational changes in many biological molecules.[60] A particularly relevant example to substrate channels and gating mechanisms is studies of acetylcholinesterase (AChE). The active site of AChE, one of the fastest enzymes known, is buried deep within the protein and is connected to the enzyme surface by a ~20 Å long channel. When the crystal structure was first solved,[17] it was found from this static view of the enzyme that this channel was at one point too narrow to allow the passage of the substrate ACh to the catalytic site. MD simulations revealed, however, that thermal fluctuations of the protein caused the channel to transiently open and close, thus permitting substrate access.[16,61,62] Such "dynamic selectivity"[16] ensures that substrate specificity can be achieved without significant loss of efficiency, and is relevant to the potential existence of control gates in PSII. More discussions of studies of channels in other enzyme systems and computational simulation strategies relevant to different kind of channels can be found in Ho.[29]

Vassiliev *et al.*[15] have recently reported a 10-ns MD simulation of a PSII core complex in the presence of explicit solvent water molecules in order to identify water channels from a dynamic view point. A diffusion tensor tracking method was used to generate fibre tracks or "streamlines" to analyse the movement of water molecules, thus revealing the most heavily trafficked paths along which water molecules moved during the simulation. This was a novel implementation of an analysis and visualisation method that until now has mainly been restricted to medical imaging.[63] A large number of water paths were found from this MD simulation. There was a high level of agreement between these paths and the various channels described in the other studies, but the MD simulations also revealed a number of new tracks and exit points not found in earlier analyses. For example, in addition to confirmation of the presence of streamlines corresponding to the narrow channel, additional streamlines not corresponding to any previous channel proposals were found to lead from this channel and connect to the large channel system. On the other hand, at certain points of some water-filled channels (*e.g.* back channel), there were discontinuities in the water streams, suggesting that some other factors such as hydrophobicity affected or disrupted potential water flow. Vassiliev *et al.* found that water could flow from the lumen to the $CaMn_4$ *via* numerous paths, with the large channel system supplying the most. This is reasonable given its size and numerous exits, and as pointed out in earlier studies,[12–14] the suggested channel functions may not be exclusive, and water transport could occur through a number of channels. Furthermore, the MD simulation identified an additional group of streamlines leading from the large channel system to new exits at the luminal surface. Overall, the distribution, exact trajectories and the dynamic behaviour of the water molecules and protein were much more complex than could be predicted from the static structure alone.

Given the multitude of paths available for water movement between the lumen and the $CaMn_4$ cluster, Vassiliev *et al.*[15] suggested that the availability of substrate water is unlikely to be a problem. However, some sort of regulatory mechanism may nevertheless exist to control water flow or access. Particularly relevant to this is the observations of transient opening and closing of pathways and exits under dynamic conditions in the MD simulation. Especially interesting is the confirmation of possible control of water access to the $CaMn_4$ cluster *via* the back channel[13]/channel (i)[12]/Channel A1-A2.[14] Synchronous movement of a number of residues with some correspondence to those in the water control gate proposal in Ho and Styring[13] led to an extension of the back channel streamlines all the way to the $CaMn_4$ cluster around 20% of the time. Though not additionally leading to a connection between the back and broad channels as previously suggested, the MD results confirm the possible existence of such a dynamic control mechanism for water access. Additionally, the MD simulation also revealed transient and rapid opening and closing of a number of exits of the large channel system (as identified by the streamlines) that could intermittently prevent water exit/entry.

These results clearly demonstrate the point that the presence of open channels in the static structure does not give the whole picture of how and where water would flow under dynamic conditions, and illustrate the advantage of taking into account the motion of the protein and water molecules, as well as the interactions (hydrogen bonding, electrostatic interactions *etc.*) between them.

8.3 Proton Pathway in PSII

8.3.1 The Need for a Proton Exit Pathway

Proton release during water oxidation is a vitally important aspect of the S-cycle mechanism. Far from just being a structural question of getting the protons that are produced from substrate water oxidation to the lumen, it is intrinsically connected to the energetics of the stepwise oxidation of the $CaMn_4$ cluster. The defined pattern of proton release during the different steps of the S-cycle[64] directly influences the ability of the cluster to accumulate the four oxidation equivalents required to oxidise water, all the while having the same set of photooxidant and electron acceptors that conduct the necessary charge separation reactions. In particular, for each S-state transition, the $CaMn_4$ is oxidised by the same $Y_Z^{\bullet}$ radical with the same reduction potential (around $+1.0$ V). Yet considerations of the energetics of pure oxidation steps of Mn complexes would suggest that an extra 0.5–1.0 V of potential would be required for each successive electron removal, thus leading to a "redox-potential problem".[65] A central concept for resolving this issue is the need for redox potential levelling,[10] which has its basis in the concept of "configurational potential" introduced by Krishtalik.[66] Amongst other things, it was recognised that the release of protons also needs to be accounted for when considering the energetics of water oxidation. Only by including the release of protons from the catalytic site (considered then to be Mn-bound substrate water molecules, but could also be the $CaMn_4$ itself given current models), to be immediately bound by a nearby strong base, could the restricted potentials available for water oxidation be satisfied to give an overall energetically favourable reaction.[10,67] More recently, Dau and co-workers considered the abundant experimental data on the kinetics and energetics of the S-state transitions under different conditions, and put forward a unifying model involving a strictly alternating sequence of proton and electron release from the $CaMn_4$ cluster during the individual steps of the S-cycle.[10,68–70] While each electron removal would raise the reduction potential of the cluster beyond the reach of the $Y_Z^{\bullet}/Y_Z$ couple, the subsequent release of a proton returns the potential of the $CaMn_4$ cluster to a level below that of the $Y_Z^{\bullet}$, thus allowing the next electron removal step to take place. Even prior to this, such an alternating model had been included in the DFT-based mechanistic proposals by Siegbahn and co-workers (*e.g.* Siegbahn and Lundberg[71]) and this continues to be a feature in subsequent theoretical studies of the S-cycle (see below). Consideration of the intermediate proton and electron release steps thus takes into account the issue of redox-levelling.

Any proton exit pathway for PSII would therefore be expected to be not only efficient, but also fulfil the need for a sufficiently strong base to accept the protons released. "Naked" protons do not exist in aqueous environments, and their movement through the protein must be conducted *via* water and/or amino acid side chains. The generally accepted model for this is the Grotthuss mechanism, which has seen much theoretical development since it was first proposed early in the 19th century[72] (for reviews, see Agmon,[73] Brzezinski and Ädelroth,[18] Cukierman,[74] Marx,[75] Swanson *et al.*[76] and Wraight[77]). Through the interconversion of covalent bonds and hydrogen bonds of water molecules around the hydrated excess proton, the proton is effectively passed through a network of hydrogen-bonded water molecules (often referred to as "proton hopping"). More precisely, there is an interconversion of the hydrated proton across a continuum of two conformational extremes, namely an Eigen cation $[(H_2O)_3H_3O]^+$ and a Zundel cation $[H_2O–H^+–OH_2]$ (for more detailed discussions of the complex nature of these cation complexes and their interconversion, see Marx *et al.*,[78] Swanson and Simons,[79] Wraight,[77] and references therein). For a one-dimensional (1D) case of proton transport, the process can be thought of as a hydrated proton starting at one end of a chain of hydrogen-bonded water molecules, and a net transfer of the proton taking place through the bond interconversions (Figure 8.3). The proton that exits is not the same as the one that entered in the first place, and in principle, (de)protonatable amino acid side chains can also participate in the proton transfer.

Importantly for the context of biological proton transfer, 1D proton transport such as along a single, well-defined chain of water molecules is much more efficient than the 3D case (*e.g.* hydrated proton in bulk water phase). This has

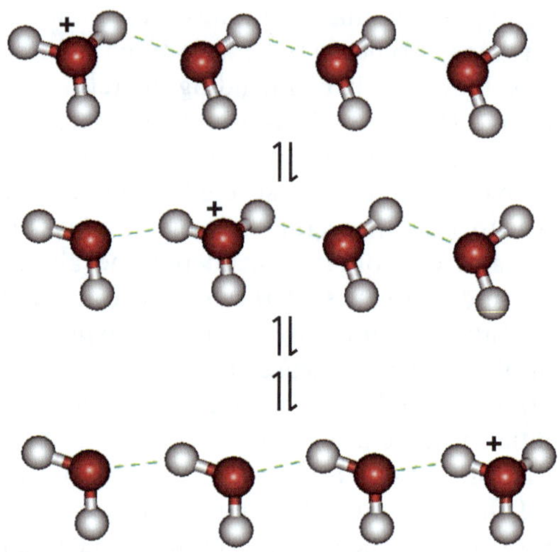

Figure 8.3 Schematic diagram of the Grotthuss mechanism. Hydrogen bonds are shown as dashed lines. Note that the dipole moment of the water chain is rotated as a result.

been demonstrated by simulations of a model system consisting of a chain of water molecules confined within carbon nanotubes.[80,81] In the 1D case, the Grotthuss-type hopping only requires a local rearrangement of water structure, and a certain path is defined. By contrast, the movement of a positive charge through a 3D system requires more substantial disruption and hence rearrangement of the water structure around what is in effect a structural defect, and directionality is less well-defined. As such, a well-ordered water chain promotes efficient proton transfer, with the rate being affected by the presence of orientational defects along that chain.[80] It is therefore reasonable that there be a well-ordered and defined (1D) pathway of water molecules/amino acids in PSII to ensure efficient removal of protons from the $CaMn_4$ cluster. Indeed, it was found in the simulation of a model system that increased spatial constriction of water molecules within a channel can lead to an increased proton diffusion rate with a decrease in the diffusion of the water molecules themselves along the channel.[82] In other words, stabilising the positions of the water molecules enhances proton transport rate.

It is not likely that just any random path in the protein would suffice. Specific and well-tuned proton transfer pathways are known in a number of other biological systems, including cytochrome *c* oxidase and bacteriorhodopsin and carbonic anhydrase (see Swanson *et al.*[76] and Ho[29] for reviews). Studies of how aquaporin is able to prevent proton transfer despite the existence of a water chain[24] have also provided valuable information about the mechanisms of proton transfer. Apart from efficiency, there is also the question of directionality. The synthesis of ATP by ATP synthase at the end of the photosynthetic electron transport chain is driven by the proton gradient that is established across the thylakoid membrane during the light-driven reactions. Therefore, there must be directed transport of the protons released from water oxidation in PSII to the luminal side in order to maintain this gradient. Proton leakage to the stromal side must be avoided, as abolition of this gradient (*e.g.* by uncouplers) abolishes ATP synthesis.[83] The released protons must be moved against the proton gradient in order to establish and maintain it. As a result, some local tuning of electrostatic potentials is likely to be required to ensure directionality of proton transport, and to avoid proton back-flow. A large body of literature exists concerning the influence of pH, changes in pK_a of amino acid residues with respect to individual S-state transitions, the proton release pattern during the S-cycle, proton-coupled electron transfer during water oxidation, as well as numerous other related energetics considerations.[10,66,67,84–92]

8.3.2 Identification of a Proton Pathway

In the following section, the identification of possible proton exit pathways based on the computational studies of the crystal structure is first outlined. This is followed by a review of biochemical and spectroscopic studies that have implicated certain amino acid residues in such a pathway.

8.3.2.1 Computational Studies

From a direct examination of the amino acids residues around the $CaMn_4$ cluster as revealed in their fully-refined 3.5 Å resolution crystal structure, Barber and co-workers[1,6] proposed that the proton exit channel started at the D1-D61 residue, and proceeded through the D2 subunit and then across the neck region of the extrinsic PsbO protein, involving the residues D1-E65, D2-K317, -E312, PsbO-D158, -D222, -D223, -D224, -H228 and -E229 (residue numbering from the 2AXT structure is used throughout). The possibility that this channel could additionally act as a water channel was also raised. Although PsbO residues proximal to the D2 subunit are involved in this proposed channel, the alternative suggestion that the β-barrel PsbO itself could function as a channel was dismissed, as it is essentially a solid protein at its core.[93]

A proton exit pathway and a water entry channel were also proposed by Ishikita *et al.*[94] on the basis of calculated S-state dependent changes in the pK_a of residues near the $CaMn_4$ cluster. A monotonic increase in pK_a of these residues going away from the $CaMn_4$ cluster was found. A number of residues (D1-D61, -D59, D2-K317 and CP43-R357) underwent particularly large changes during the $S_4 \rightarrow S_0$ transition. The absolute magnitudes of change in pK_a values may not necessarily be realistic, however, as the Mn ions accumulated very high charges (+4 between the S_0 and S_4 states) without explicit consideration of charge compensation by proton release. Nevertheless, observed trends in S-state dependent changes may still be valuable, giving clues as to which residues may be involved in the proton exit pathway.

Beyond an examination of the identity of the residues around the $CaMn_4$ cluster, the three studies[12–14] that have looked in detail for channels in the PSII structure based calculated accessibility for solvent water have yielded similar, but not identical proposals for the proton channels (Table 8.1, Figure 8.2B).

The structural search using CAVER by Murray and Barber[12] assigned a channel designated channel (iii) as the proton channel on the basis of its narrowness, hydrophilicity and the similarity of the residues in the proposed proton channel in Ferreira *et al.*[6] Starting from D1-D61, D2-K317, D1-D59, D2-E312 and D1-E65, it passes across the top of the PsbO barrel (R152, D158, D222, D224, H228, E229) on its way to the lumen. This is essentially in agreement with the earlier proposal by Barber and co-workers (above). In the analysis by Ho and Styring,[13] this channel corresponded at the $CaMn_4$ to the "broad channel", which was connected to the "narrow channel". These channels made contact at their proximal ends with the Mn ions (especially ^{4}Mn) of the $CaMn_4$ cluster, in the vicinity of the residues D1-D59, -D61, -E65, D2-K317 and CP43-R357. The narrow channel was then found to extend all the way to the lumen, passing through mainly the CP43, PsbO and PsbU subunits, with the exit formed by CP43 and PsbU residues. This narrow/broad channel system was also suggested as serving as a proton exit pathway.

There are two key differences between the narrow/broad channels and channel (iii). Firstly, the narrow channel was not reported in Murray and Barber, but the presence of this channel and the movement of water within it have been confirmed in the recent MD simulations.[15] Secondly, whereas channel (iii) and the broad channel coincided near the $CaMn_4$ cluster, the broad channel was found to have a bottleneck at D1-E65, D1-P66, D1-V67 and D2-E312 that was too narrow for passage. Beyond this point, however, an open channel towards the lumen was observed, as in Murray and Barber. It is likely that these differences in results are due to the different simulation methodologies used (see Ho and Styring for a more detailed comparison of these two studies). Note however, that whether channel (iii)/broad channel is open is not critical here, given that proton movement does not necessarily require a fully open channel for Grotthuss-type proton transfer, if intervening protonatable amino acids that are able to participate in the transfer are present. In fact, in the MD simulations by Vassiliev *et al.*,[15] water streamlines were found to flow around the blockage in the broad channel under dynamic conditions, and reaching the luminal surface. However, no streamlines were found in the distal part of channel (iii).

In both Murray and Barber[12] and Ho and Styring,[13] reliance was placed on literature experimental data suggesting that D1-D61 and D1-E65 could be involved in a proton exit pathway, as well as on calculated S-state dependent pK_a changes.[94] Murray and Barber pointed out further that a Ca^{2+} binding site in PsbO was located near the exit of channel (iii), and therefore could play an electrostatic role in assisting proton transfer into the lumen. Ho and Styring additionally relied on the fact that CP43-R357 is part of the narrow channel, and that this residue has been proposed to be important in deprotonation steps during the S-cycle.

These assignments find agreement in the higher level calculations by Batista and co-workers using hybrid density function theory quantum mechanics/molecular mechanics (DFT-QM/MM).[57,95,96] While these studies mainly dealt with the elucidation of the catalytic mechanism of the S-cycle and water oxidation, with focus on the $CaMn_4$ cluster (*e.g.* structural changes, oxidation states, water binding sites *etc.*), the question of the proton exit pathway proposals was also addressed. In particular, from the results of the hydrated DFT-QM/MM structural models, it was proposed that protons are translocated from the $CaMn_4$ cluster to the lumen *via* a network of hydrogen bonded residues and water molecules, starting from water molecules between the cluster and D1-61, then further *via* D1-E65 and out towards the lumen (Figure 8.4). CP43-R357 was also found to be important for the hydrogen bonding to, and proton removal from, substrate water molecules. A number of other recent DFT studies by Siegbahn and co-workers[97–100] have also focused on the S-cycle mechanism cycle using a cluster model of the OEC. As a recent addition to their detailed investigations of the energetics of the individual S-states and of O–O bond formation, the proton release step to reach the dark-stable S_1 state after the initial oxidation of the S_0 state (denoted the $S_1^{-1} \rightarrow S_1^{-2}$ transition) was considered from both structural and energetic points of view (Figure 8.5).[100] Again, D1-D61 and CP43-R357 were found to play crucial roles in proton removal. The released proton is first passed to D1-D170 *via* a bound water

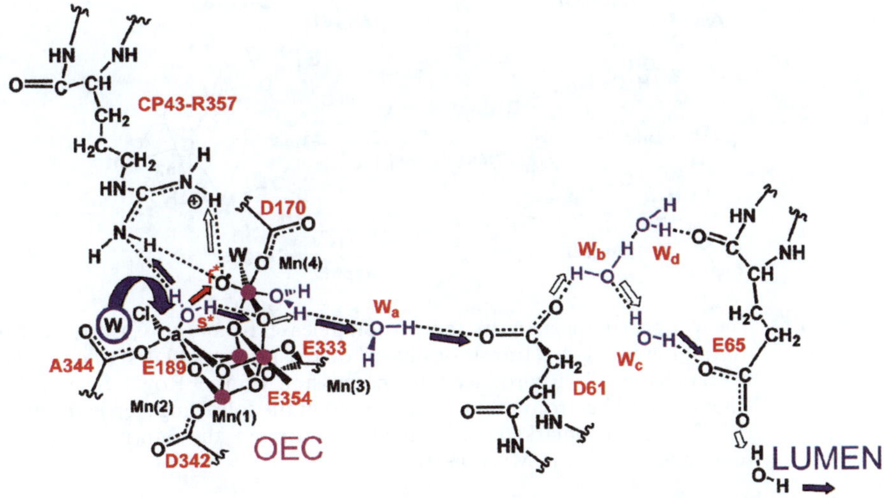

Figure 8.4 Proton exit pathways as proposed by Sproviero *et al.*[95] based on DFT-QM/MM calculations. The OEC in the S_4 state prior to O–O bond formation is shown. Proton transfer events are marked by blue and white arrows, and the O–O bond formation step by a red arrow. The fast and slow exchanging substrate water molecules are marked with "s*" and "f*", respectively. (Reprinted with permission from Sproviero *et al.*[95] Copyright 2008 American Chemical Society.)

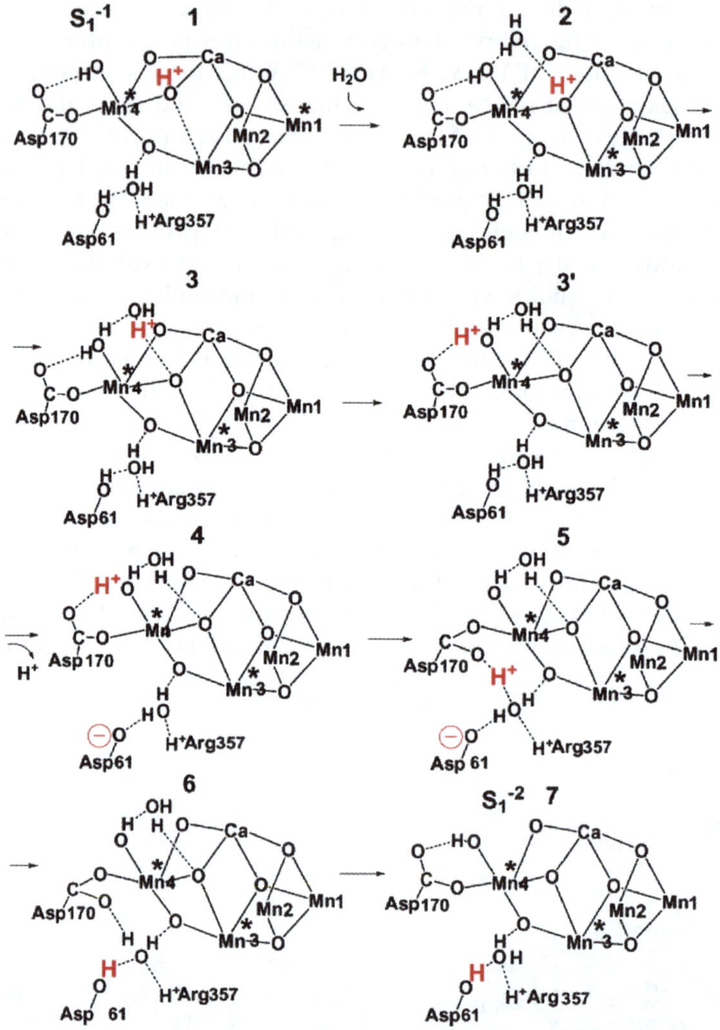

Figure 8.5 The proposed mechanism for proton release from the $CaMn_4$ cluster during the $S_1^1 \rightarrow S_1^2$ transitions based on cluster model DFT calculations by Siegbahn.[100] (Reproduced by permission of the Royal Society for Chemistry.) The proton that is transferred in the following step is coloured red. Mn atoms marked with an asterisk are in the Mn(IV) state, the remainder are in the Mn(III) state.

molecule, then *via* another water that is bound between CP43-R357 and D1-D61 to reach D1-D61. D1-D61 is directly involved in the reception and release of the proton, and there is a close association between CP43-R357 and D1-D61 throughout the proposed mechanism.

In the more recent analysis of the of the 2.9 Å crystal structure[8] by Gabdulkhakov *et al.*,[14] again using CAVER, a smaller probe radius criterion

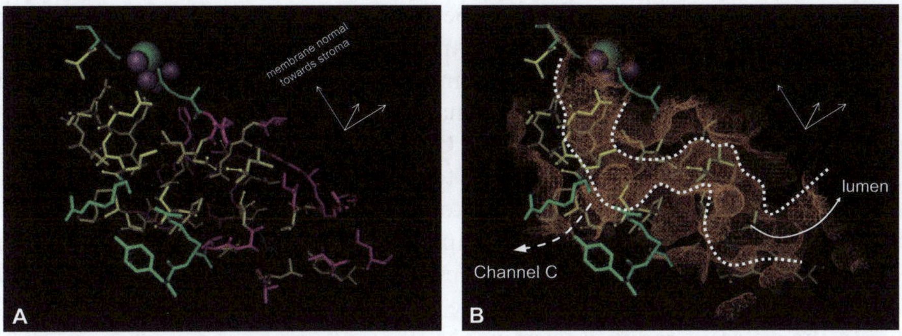

Figure 8.6 Comparison of the residues involved in the narrow channel[13] with those involved in the Channels D–F.[14] **A**: Yellow residues – part of narrow channel as well as Channels D, E, and/or F; Purple residues – narrow channel only; Green residues – Channels D, E and/or F only. **B**: Residues forming Channels D, E, and/or F overlaid on the surface of the narrow channel (outlined with dotted lines for clarity). The connection to Channel C is marked.

than that in Murray and Barber[12] and Ho and Styring[13] was used to identify possible proton channels. All channels which were not wide enough for transporting O_2 or water molecules, but which could accommodate the presence of water molecules, were regarded as possible candidates. Breaks in continuity not longer than about 3.5 Å (maximum distance between heavy atoms in a hydrogen bond) were permitted. A set of channels named Channels C–G were assigned as proton exit pathways. A closer examination of these channels reveals that they are essentially in agreement with the putative proton channels identified in Murray and Barber and Ho and Styring, with the exception of the distal section of Channel G.

Channel C is essentially the same as channel (iii) and the blocked broad channel,[†] with minor differences with the definitions of the proximal and distal limits of the channel. The proximal section of Channel G is the same as that of Channel C/channel (iii)/broad channel, but then diverts to a new path at D1-R334, D2-E312, -T316, -L320, leading to an exit at the lumen (CP47-R422, PsbO-N212, -E205) not described in the other two studies.

Channels D, E and F overlap with each other very considerably, and arguably form a channel system rather than strictly separated channels. Together, they correspond almost fully to the narrow channel (Figure 8.6). All except three (D1-D170, PsbU-T60, -N61) of the 29 residues listed for Channel E are part of the narrow channel, and in their turn, Channels D and F are essentially parts of Channel E, with small deviations in path. Twelve out of the 17 residues (70%) listed for Channel D are part of Channel E (made up of 29 residues), and 18 out of the 26 residues (70%) listed for Channel F are also part of Channel E.

[†] There would not have been as a blockage in the broad channel had the smaller radius probe in Gabdulkhakov *et al.* been used.

There is also a large degree of overlap between Channels D and F (12 out of the 17 Channel D resides are found in Channel F). Channel D corresponds to the proximal section of the narrow channel reported in Ho and Styring, and the three residues belonging only to Channel D (D1-D64, PsbO-Y177, -R178) correspond to a connection between Channel D and Channel C, which in Gabdulkhakov *et al.* combine and lead to a common exit. The remainder of the Channel E and F residues corresponds to a large extent to the rest of the narrow channel (Figure 8.6B). Therefore, Channels C–F are in good agreement with the previous studies, and the differences in designation of channel paths and thus channel count lie essentially in the definition of the channels' trajectories due to the different radius criteria used.

Two Cl^- binding sites were identified by X-ray crystallography on Br^- and I^- substituted PSII crystals.[101,102] One site is located between the residues D1-N181, D1-E333 and D2-K317, and the other between CP43-E354 and D1-N338. Electron density consistent with the presence of Cl^- ions at these sites have also been found in the most recent PSII crystal structures at 2.9[8] and 1.9 Å resolutions.[189] Significantly, the first site is located near the start of the proton exit pathway suggested by Murray and Barber (channel (iii)),[12] and in the "broad channel" in the analysis of Ho and Styring.[13] This Cl^- was found in Gabdulkhakov *et al.*[14] to block Channels C and G, and this was regarded as further support for the assignment of these channels as proton exit pathways. Given the similarity between the proposals across the three studies, it is not surprising that Cl^- is also found to be associated with the putative proton exit pathways identified in all three studies.

Despite differences in details, the different proposals for the proton exit pathway are in good general agreement, and it would seem that they do describe the most likely region where the pathway lies. Indeed, our recent 10 ns MD study of the region corresponding to the broad channel/channel (iii) has shown that there is a hydrogen-bonded chain of residues and water molecules leading from the $CaMn_4$ cluster to the lumen.[103] By analysing the residues and water molecules involved over the course of the MD trajectory, it was found that there is a "hotspot" of a network of highly interconnected residues and water molecules around D1-D61 and D1-E65. Not only were these two residues found to be connected by water bridges almost 100% of the time, in agreement with their proposed role in H^+ transport, it was also found that the stability of this chain was greatly influenced by the balancing of hydrogen bonding interactions with other residues in the channel that would not otherwise be expected to participate in H^+ transport (Figure 8.7). In particular, the D1-R334 residue was found to be highly water bridged to the D1-E65 residue, and parallel MD simulation showed that mutations to R334V and R334E *in silico* led to a very significant disruption of this putative H^+ pathway. Substantial changes to the conformations of the residues in the pathway were observed, as well as a lengthening and destabilisation of the connection between D1-D61 and D1-E65. This is in good agreement with experimental mutagenesis data that has also shown reduced O_2 evolution activity, slowed down flash O_2 release kinetics, and disrupted S-cycle behaviour due to these mutations.[104]

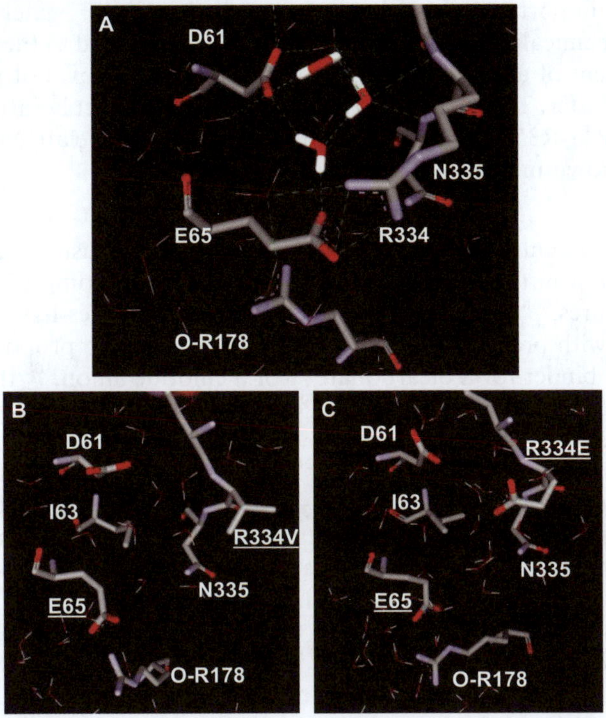

Figure 8.7 The "hot-spot" of highly interconnected residues and water molecules around D1-D61 and D1-E65, as identified by MD simulations.[103] **A**: Native PSII. The residues are in the D1 subunit unless otherwise indicated. **B**: D1-R334V and **C**: D1-R334E mutants, generated *in silico*.

In particular, the slowed O_2 release could be an indicator of reduced H^+ removal efficiency, especially since the $S_3 \rightarrow [S_4] \rightarrow S_0$ transitions, during which O_2 is evolved, involve the two H^+ release steps. Impaired H^+ removal may lead to the reported slower O_2 release kinetics. It can also be further noted that the end of the putative H^+ pathway studied by MD connected nicely to a cluster of carboxylic acid residues in PsbO that Shutova *et al.*[105] have proposed to act as a buffer for receiving released protons as they are released into the bulk. Given that slowed flash O_2 release has also been observed in ΔPsbO,[106] there is further circumstantial evidence that the putative pathway is involved in proton removal.

8.3.2.2 Biochemical and Spectroscopic Studies

Even before the crystal structure of PSII was available, a large number of site-directed mutants of PSII were constructed to determine the ligands of the $CaMn_4$ cluster.[9,56] A number of residues, while not apparently direct ligands to the $CaMn_4$ cluster (now confirmed by crystallography), were found to be nonetheless important for the proper function of water oxidation.‡ Further biochemical and spectroscopic studies have pointed to the possibility of the involvement of a number of these residues in the removal of protons from the $CaMn_4$ cluster. Three residues that have received particular attention in this regard are CP43-R357, D1-D61, D1-E65. Their positions relative to the $CaMn_4$ cluster are shown in Figure 8.2B.

CP43-R357

While not apparently a direct ligand of the $CaMn_4$ cluster, this conserved residue is found in the second coordination sphere according to the available crystal structures,[6–8] and recent QM/MM and DFT studies have suggested its involvement with proton removal (see above). Apart from proposals of CP43-R357 being a binder for a bicarbonate[107] or a chloride anion,[108] there has been much interest in the possibility that CP43-R357 could be the base that accepts the proton released during the aforementioned coupled proton and electron transfers during the S-cycle.[10,57,68,70,95,109–111] Haumann *et al.*[68] have proposed that the 200 μs lag phase during the $S_3 \rightarrow S_0$ transition in their time-resolved EXAFS measurements may be attributable to just such a proton release event. These workers regard the CP43-R357 as a plausible, if tentative assignment of the base facilitating this deprotonation.

Mutation studies have demonstrated the importance of this residue for water oxidation. Neither the CP43-R357S[112] nor the more conservative CP43-R357K[113] was able to grow photoautotrophically, and O_2 evolution activity was severely reduced in both mutants (Table 8.2). No O_2 evolution was observed at all in the CP43-R357S mutant unless DCMU was present in the medium (possibly providing a photoprotective effect). PSII assembly was not affected to the same extent as O_2 evolution, but not all assembled centres contained photooxidisable Mn ions (Table 8.2). For CP43-R357S, it was found that with water as the electron donor, the rise time of variable fluorescence was about 8 times slower in the mutant compared to the wild type (WT), whereas there was only a slight decrease in the rise time when hydroxylamine was used as the electron donor.[112] This indicated a severe decrease in the transfer of electrons from the $CaMn_4$ cluster, while electron transfer on the acceptor side was not seriously impaired.

Similarly, the decay kinetics of variable fluorescence in the CP43-R357K mutant in the presence and absence of DCMU showed that acceptor side electron transfer behaved in a similar way to WT PSII, while electron transfer on the donor side was significantly impaired, with apparently much reduced advancement to the S_2 state.[113] Consistent with this, the flash O_2 yield measured in thylakoid membranes of the CP43-R357K mutant revealed a total lack of the classic period-of-four oscillation pattern in O_2 evolution. This could be attributed to a very large miss factor for S-state transitions (estimated to be

‡ While ligands of the $CaMn_4$ cluster could be involved in proton removal (e.g. D1-D170), the effects observed upon site-directed mutagenesis at these positions are confounded by possible effects on the $CaMn_4$ cluster itself.

Table 8.2 Effects of site-directed mutagenesis of three proposed H^+ exit pathway residues.

	O_2 Evolving activity (% of WT activity)	*Assembled PSII (% of WT)*	*Fraction of PSII containing photo-oxidisable Mn*
CP43-R357[112,113]			
Arg → Ser	0% (10% in presence of DCMU in medium)	60%	71%
Arg → Lys	18% (cells);15% (membranes)	82%	58%
D1-D61[118,119]			
Asp → Glu	61%	80%	~100%
Asp → Asn	17%	92%	~100%
Asp → Ala	19%	97%	~100%
Asp → Val	~5%	~75%	N/A
D1-E65[118,119]			
Glu → Asp	93%	86%	~100%
Glu → Gln	17%	94%	~100%
Glu → Ala	21%	97%	~100%
Glu → Leu	~10%	100%	N/A

46%). Measurements of the kinetics of the rise time of flash-induced O_2 evolution on the millisecond time scale showed an increase from 0.9 ms in WT thylakoid membranes to 3.2 ms.

Given these observations, it would seem to be a reasonable suggestion that CP43-R357 could act as strong base near the $CaMn_4$ cluster to accept protons released during the S-cycle, thereby facilitating redox-levelling, as discussed above. The drastic reduction in O_2 evolution in the mutants cannot simply be accounted for by a reduction in the assembly of functional PSII and $CaMn_4$, and there was a clear slowing down in the flash-induced O_2 evolution kinetics of the CP43-R357K mutant. The kinetics data from variable fluorescence, as well as the increased miss factor for S-state transitions are also consistent with a reduction in the efficiency of electron transfer from $CaMn_4$ to $P_{680}{}^+$, resulting in an increase in the recombination of the $Q_A{}^- \cdots P_{680}{}^+$ following charge separation.

In order to act as a base, the Arg residue would need to be in its deprotonated state at the time of proton release. But given its high pK_a (~12.5), it would normally be expected to be protonated under physiological conditions. It has been suggested[109,110] that there is a temporary deprotonation of the CP43-R357 as a result of a shift in its pK_a, due to the fact that a positive charge is localised on a base (most likely D1-H190) during the proton-coupled electron transfer from Y_Z to $P_{680}{}^+$.[85,114–117] This was regarded as particularly relevant in the S_2 and S_3 states, when the $CaMn_4$ cluster carries an extra positive charge relative to the S_0 and S_1 states (though more recent schemes from DFT-QM/MM calculations[95,111] have also shown Arg as a protonated sidechain throughout the S-cycle). It can be noted that shifts in pK_a near the $CaMn_4$ cluster were also found in other studies. For example, D1-D61 was protonated in the cluster model DFT calculations of proton release by Siegbahn.[99] Also the pK_a for the oxidation of Y_Z (D1-Y161), which involves a proton transfer to the nearby base D1-H190, shifts from ~4.5 in the presence of the $CaMn_4$ cluster to ~7.5 in Mn-depleted PSII.[114,117]

In any case CP43-R357 appears to be important in hydrogen bonding to the bound substrate water molecules to orient and stabilise them for proton removal. This seems especially true for the S_4 state in the model by Sproviero *et al.*[95] (Figure 8.4), where CP43-R357 is hydrogen bonded to both substrate water molecules in the S_4 state during the final proton removal and O–O bond formation. Interestingly, according to this model, *two* deprotonation pathways are envisaged during the $S_4 \rightarrow S_0$ transition, one *via* the D1-D61 and D1-E65, and another *via* CP43-R357[95,111] (Figure 8.4). Significantly, Siegbahn[100] has also commented that in the ongoing calculations of proton release during the $S_2 \rightarrow S_3$ transition, the same proton pathway as for reaching the S_1 state (Figure 8.5) could not be used. The energies barriers would be too high. Therefore, it would appear that there is still much more to be discovered concerning proton removal, and the choice of pathways may be S-state dependent.

Another interesting point to note is that, despite the replacement Lys residue in the CP43-R357K mutant also having a basic ($pK_a = 10.5$) and (de)protonatable side chain, it is nevertheless severely impaired in O_2 evolution and electron transfer from the $CaMn_4$ cluster. O_2 evolution activity was reduced by almost the same extent as the more radical CP43-R357S mutation, which has a polar but neutral side chain (compare also with the effects of the D1-D61E/N/A and D1-E65D/Q/A mutations,[118] below, and Table 8.2). This would suggest that the basic character of the CP43-R357 side chain is not the only property that is important. Perhaps *two* hydrogen-bond donor sites is also an essential requirement for the sidechain at this position, so that the replacement of Arg with Lys falls foul of this. This would, for example, be consistent with the DFT-QM/MM-derived model by Batista and co-workers (see above), where the residue at this position must have *two* distinct hydrogen-bond donor sites available in order to bind to *both* substrate waters simultaneously in the S_4 state to create the favourable conformation needed for O–O bond formation (Figure 8.4). While this proposal is speculative and dependent on just one mechanistic model, it presents an alternative and perhaps additional consideration for the role of CP43-R357.

D1-D61

The mutants D1-D61E, -D61N and -D61A[118] as well as -D61V[119] have been constructed. All four are capable of photoautotrophic growth (though slow in mutants except for D61E), and all showed a reduction in O_2 evolution (Table 8.2). The activities of the D61N and D61A mutants were dramatically reduced to approximately the same extent, and that of D61V was even more reduced. However, in all mutants, the apparent PSII content was very high, and essentially all PSII centres in the mutants contained photooxidisable Mn ions.

A range of kinetic data have suggested that the D61X mutations cause an increase in the midpoint potential of the various S-state transition couples S_2/S_1, S_3/S_2 and S_3/S_0. In support of this the following points can be made. (1) The decay of variable fluorescence in the presence of DCMU was found to be faster in the mutants than in the WT. This represented faster $S_2 \cdots Q_A^-$ recombination kinetics, and was consistent with a slight increase in the midpoint potential of the S_2/S_1 couple.[118] (2) Transient absorption measurements of the D61N mutant showed that the $S_1 \rightarrow S_2$ and $S_2 \rightarrow S_3$ transitions slowed down by around 2.4-fold compared to WT,[120] whereas the $S_3 \rightarrow S_0$ transition slowed down by around 10-fold.[121] The former implied an increase in the S_2/S_1 and S_3/S_2 midpoint potentials by around 30 mV.[120] (3) Studies of the rise time of O_2 release during the $S_3 \rightarrow S_0$ transition measured in whole cells showed that while the D61E mutant gave a very similar half-time to the WT organism, oxygen release in the D61N and D61A mutants was slowed by 10- and 8-fold, respectively.[120] Note, however, that a moderate decrease in O_2 release rate was observed even in the D61E mutant in Qian *et al.*,[106] where thylakoid membranes were used instead. (4) The lifetimes of the S_2 and S_3 states were shorter in the D61E mutant than in WT, again consistent with an increase in the S_2/S_1 and S_3/S_2 midpoint potentials.[106]

Flash-induced O_2 evolution studies of the whole cells of the D61N mutant showed that there was a marked dampening of the period-of-four oscillation pattern compared to WT, consistent with a larger miss factor for S-state transitions in the mutant.[120] This can be contrasted with the thylakoid membranes from the D61E mutant, where a period-of-four oscillation and a calculated miss factor very similar to those for WT thylakoid membranes were obtained.[106] Finally, ^{18}O water exchange studies[122] showed that the D61N mutation led to slower water exchange rates in both the fast and slow phases of exchange compared to WT PSII (6.5- and 3-fold reductions, respectively). This was consistent with stronger substrate binding in the mutant and the mutation affecting both water binding sites. The possibility that slowing of the fast exchange phase could be related to a slowing down of substrate entry by the D61N mutation was also noted.

If D1-D61 is indeed involved in proton exit pathway, then the moderate reduction in O_2 evolution activity in the D61E mutation is reasonable, bearing in mind the Grotthuss-type mechanism for proton transfer. The (de)protonatable and charged carboxylate side chain is preserved and available for proton transfer. The additional $-CH_2-$ link in the Glu probably leads to a conformationally less optimised hydrogen bonding network, leading to the reduction in activity. This can be contrasted with the non-protonatable D61N and D61A mutations, where the reduction in O_2 evolution activity is more severe. Particularly striking also is the similar extent to which O_2 evolution activity, O_2 rise-time kinetics and recombination kinetics are affected in the D61N and D61A mutants. Although Asn is a neutral side chain under normal conditions, it is isosteric with the Asp reside that it replaces, and it is also able to hydrogen bond at two sites. This is in contrast to the Ala residue, which is much smaller than Asp and cannot form hydrogen bonds. It is clear then that it is not enough for the residue at the D1-61 position to be able to form a hydrogen bonding network with surrounding water molecules and amino acids. It must also be protonatable. This is consistent with the participation of D1-D61 in a proton exit pathway functioning in a Grotthuss-type manner.

The effect of the presence or absence of a negative charge at the D1-61 position is by contrast not as straightforward. On the one hand, it is clear that O_2 evolution activity in the mutant with the negative charge-containing D61E mutation is much higher in than any other mutant. Also, unlike the neutral D61N mutation, there is no significant change in the miss factor for transitions due to the D61E mutation when compared to WT. On the other hand, the increase in the rate of $S_2 \cdots Q_A^-$ recombination for each of the D61E, D61N and D61A mutants suggests that while the midpoint potential of the S_2/S_1 couple is sensitive to mutation at this position, it is insensitive to whether or not the negative charged is retained at this position. This insensitivity may be true even in the S_2 and S_3 states, where both the D61E and D61N mutations led to kinetic effects consistent with an increase in midpoint potentials (see points (2) and (4), above). Nevertheless, the rise-time for O_2 release during the $S_3 \rightarrow S_0$ transition is affected much less, if at all, in the D61E mutant compared to the other mutants.

Looked at in this way, if there is an electrostatic effect originating from the residue at the D1-61 position, then it is perhaps mechanistically most important for the $S_3 \rightarrow S_0$ transition. By contrast, the effects for the other transitions are less related to the presence or absence of a charge at this position. The process of going from the S_3 to the S_0 state is complex. The final oxidation of the $CaMn_4$ complex takes place, the O–O bond is formed, and two protons as well as O_2 are released. It is difficult to speculate with any precision what such an effect of charge at the D1-D61 position would be. Possibilities could include fine tuning of S_4/S_3 midpoint potential, or a polarising effect that ensures the correct orientation of the bound substrate, ligands around the $CaMn_4$ cluster, and/or those water molecules/amino acids that will accept the released protons for transfer to the lumen during the $S_3 \rightarrow S_4 \rightarrow S_0$. Such a $S_3 \rightarrow S_0$ transition-specific electrostatic effect may be behind the reduction in ^{18}O water exchange rate in the S_3 state for the D61N mutant.[122] However, in the absence of exchange rate data for the other S-states, as well as for the D61E mutant for comparison purposes, it is unclear whether this specificity holds. Further investigations are clearly warranted.

D1-E65

This residue has been proposed to be second in the chain of the proton exit pathway, again based on its position to in the crystal structure and its closeness to D1-D61. The mutants D1-E65D, -E65Q, -E65A[118] and -E65L[119] have been constructed. The effects of these mutations were very similar to those observed for the D61X mutants (Table 8.2). All the mutants were photoautotrophic, with the D1-E65Q and -E65A showing slow growth. O_2 evolution activity in the D1-E65D mutant was virtually unaffected, whereas the -E65Q and -E65A mutants showed a severe reduction in activity, to approximately the same extent as the D61N and D61A mutants. Despite this, the E65 mutants showed similar PSII contents as WT PSII, and all Mn ions were photooxidisable. This was also observed for the D61 mutants. In contrast to the D61 mutants, the decay kinetics of variable fluorescence were essentially the same in the WT and E65D mutant, and the E65Q and E65A showed only slightly faster kinetics. It was concluded that the midpoint potential for the S_2/S_1 couple is not significantly affected by these E65X mutations.[118] However, faster recombination kinetics were reported for the E65L mutant.[119]

Although literature mutation data is somewhat more sparce for the D1-E65 residue, the similarities to the results for D1-D61 mutants are clear. The pattern of effects for replacement of the negatively charged Glu residue with the smaller analogue Asp, the isosteric, hydrogen bonding but uncharged Gln, and the hydrophobic residues Ala and Leu mirrors those for the D61E, D61N and D61A/V mutants, respectively. The one exception is the lack of effect on the S_2/S_1 midpoint potential, at least for the D61E/Q/A mutants. Overall, the available data are consistent with the proposal that D1-E65 is a part of a proton exit pathway. In particular, the need for the residue to be both hydrogen bonding and deprotonatable can again be seen in the equally detrimental effects of both the E65Q and E65A mutation, as opposed to the E65D mutation. This is consistent with the characteristics required for a Grotthuss-type proton transfer.

The Cl^- ion

The presence of Cl^- ion is necessary for the proper functioning of PSII. However, what the effects of Cl^--depletion are and the origins of such effects have long been a matter of debate.[30,31,123–128] Briefly, while it is generally agreed that Cl^--depletion in PSII centres stripped of their extrinsic subunits inhibits the S-cycle at the $S_2 \rightarrow S_3$ transition, there are numerous discussions in the literature about whether Cl^- is necessary for oxygen evolution even in nominally intact PSII samples, or whether observed inhibitions originate from overly-harsh sample treatment. There has also been a debate between advocates of a one-site, two-state model for the number and affinity of Cl^- binding, and proponents of a two-site model. The actual role of the Cl^- ion itself is also not fully established, and it has been variously suggested that Cl^- could act to structurally stabilise the $CaMn_4$ cluster, regulate the potential of a Mn ion bound to Cl^-, or function as a bridge between Mn ions. The recent crystallographic data confirm the large number of spectroscopic studies suggesting Cl^- being located near the $CaMn_4$ cluster, and show that the two Cl^- ions found are not bound to any Mn ion.

Another function that has been suggested for Cl^- is to facilitate proton transport from the $CaMn_4$ cluster as part of a proton-relay network leading towards the lumen. The Cl^- would be bound to a protonatable amino acid sidechains forming the proton pathway, stabilising the network, and its depletion would disrupt proton removal to the bulk and thus impair the water oxidation reaction.[123,125] The presence of this negative charge could of course also change the pK_a of ligating and nearby residues. Olesen and Andréasson suggested that Cl^- may be bound to lysines with an unusually low pK_a at around 7.5. As already mentioned above, two Cl^- binding sites have been identified by X-ray crystallography and one of these sites is found to be associated with the current proposals for proton exit pathways. Note, in particular, that this Cl^- was found to be bound to D2-K317, which is in good agreement of the predictions by Olesen and Andréasson.[125] Neverthless, the recent DFT calculations by Siegbahn[100] suggested that the favourable energetics effects of the presence of this Cl^- was small, and around the same magnitude as uncertainties for such calculations. However, it was noted that the effect was more noticeable in the higher S-states, and would also be expected to be more obvious if membrane potential gradient was included in the simulations.

In summary, the combination of biochemical and computational studies has given important indications of where the H^+ removal pathway(s) in PSII may lie. The exact pathway(s) for proton exit may not yet be established, but further studies that take protein dynamics and energetics into account[29,79] will no doubt shed further light on this issue.

8.4 O_2 Channels

8.4.1 Rationale for the Existence of an O_2 Channel

The existence of an O_2 channel was first suggested as a means to prevent oxidative damage to PSII resulting from the formation of reactive singlet oxygen $^1O_2$.[129,130] $^1O_2$ can be produced as a result of a reaction between ordinary triplet O_2 and a triplet Chl (^{3}Chl) species, such as $^3P_{680}{}^*$. One mechanism for $^3P_{680}$ formation is charge recombination of the radical pair $P_{680}{}^+Pheo^-$ after the initial photoexcitation of P_{680} and charge separation. The $^3P_{680}$ thus formed can then react with O_2 to give rise to singlet oxygen production: $^3P_{680}{}^* + {}^3O_2 \rightarrow {}^1P_{680} + {}^1O_2$. ^{3}Chl can also originate from antenna Chl molecules, but evidence suggests that carotenoids in the chlorophyll-rich light harvesting complexes of PSII play a protective role by quenching any excited triplet chlorophylls or $^1O_2$ that form,[131–135] and the crystal structures of PSII confirm that carotenoids are found near such Chl molecules. By contrast, there are no carotenoid molecules within quenching distance (*i.e.* van der Waals distance) of P_{680}. This is perfectly reasonable, since the presence of a carotenoid close to the highly oxidising $P_{680}{}^+$ formed during charge separation would compete with electron transfer from Y_Z, thereby hindering water oxidation.[135] However, this means that any $^1O_2$ that is generated is only quenched if it happens to diffuse and make contact with a carotenoid. Until then, it has the potential to cause oxidative damage to the protein and other co-factors. Channels preventing access of O_2 to P_{680} would be, therefore, one method to avoid $^1O_2$ formation in the first place. The question of the existence of such channels, or even the need for such channels, is still a matter of debate.

Whether or not permanent channels are required for the movement of O_2 through proteins has been discussed in the literature for other systems. Some workers have argued that O_2 pathways can consist of a series of hydrophobic cavities that fluctuate in and out of existence under dynamic conditions, with the O_2 molecules moving between these cavities over time.[136–138] On the other hand, other workers have argued that permanent channels are required to direct movement of O_2 in some proteins, and have experimentally demonstrated changes in enzyme activity as a result of point mutations that have altered the dimensions of such channels.[27,28,139,140] An active gating mechanism linked to the enzymatic function has even been proposed for cholesterol oxidase, where the sub-atomic resolution crystal structure revealed two conformers that respectively closed and opened a putative O_2 channel.[141] Subsequent mutagenesis studies have demonstrated the functional importance of these gating residues.[142] It has been suggested[29] that whether O_2 channels are transient or permanent may be linked to the function of the enzyme. Permanent channels may be needed where control of the rate and/or location of O_2 access are vital for the enzyme's proper function, whereas transient channels may be sufficient if it is functionally less critical. Arguably PSII falls in the former category, given the need to keep O_2 away from ^{3}Chl species. Nevertheless, the possibility of transient O_2 channels not obvious from the crystal structure cannot be excluded.

8.4.2 Proposals Based on the Crystal Structure of Native PSII

In each of the studies locating channels on the basis of the PSII crystal structures, possible O_2 channels were proposed (Table 8.1, Figure 8.2C). In Murray and Barber,[12] channel (i) was assigned such a function. This channel starts at the $CaMn_4$ from D1-E189, -D342 and -A344 and ends around D1-L91, CP43-G220 and -F292. This channel corresponds to the back channel in Ho and Styring,[13] where this channel was instead assigned the function of water transport. By contrast, the large channel system was proposed in Ho and Styring as being preferable for O_2 exit. Of the channels identified in that study, this channel system was the largest in volume, consisting of three main branches leading to four exits (Figure 8.2C). Since O_2 should be removed efficiently to avoid $^1O_2$ formation, this channel system was geometrically speaking a better candidate for O_2 exit. The presence of the non-polar O_2 molecules in a larger, water-rich cavity would also lead to a less significant disruption of hydrogen bonding between water molecules compared to O_2 molecules in narrower channels with fewer water molecules present. The branches of this channel system were found to radiate from the $CaMn_4$ cluster and along the surface of the PsbV subunit, akin to ski slopes down a hill. Although suggested as being favourable for O_2 transport, the possibility that the large channel system could provide substrate water access to the $CaMn_4$ cluster was not excluded, since it is also most likely to be filled with water. Indeed, in the MD simulation study by Vassiliev *et al.*,[15] all the branches of the large channel system were found, containing an abundance of water molecules (see Section 8.2.2, above). Some water molecules were observed to enter from the lumen *via* different entrances to the system (some described for the first time), and reach the $CaMn_4$ cluster.

In Murray and Barber, the channel designated channel (ii) partially overlapped with the large channel system, corresponding to the left and right branch as shown in Figure 8.2C. The central "trunk" of the system (dashed arrows) was, however, not reported. The same result was reported in Gabdulkhakov *et al.*,[14] which was also based on a CAVER analysis. Two related channels designated Channels B1 and B2 were reported, which correspond to the left and right branches in Figure 8.2C, respectively. Again, the central trunk of the system was not reported. By contrast, as mentioned above (Section 8.2.2), this central trunk was found in MD simulations[15] to be replete with water and to contain a number of water streamlines.

While Murray and Barber assigned channel (ii) to water and proton transport, Gabdulkhakov *et al.* were in agreement with Ho and Styring (2008) in assigning an O_2 transport function to these channels. It was also found in PSII crystals co-crystallised with DMSO that this larger molecule was able to penetrate into these channels. In addition, the authors suggested that the residue PsbU-K134 could open or close Channel B2 depending on its conformational orientation (Figure 8.2C). The manual rotation of this residue performed there (like the manual placement of O_2 and water molecules elsewhere in that study) is nevertheless based solely on available space, and does not yet take into account the energetics or interactions between water molecules and amino acid residues. Molecular dynamic studies with direct inclusion of O_2 molecules would provide better information, though special attention would then need to be paid to the issue of sufficient sampling.[29] Finally, Gabdulkhakov *et al.*[14] performed noble gas perfusion experiments to give further support to their assignment, and these are discussed in more detail below.

8.4.3 Crystallography of Xe Perfused-PSII

Three recent crystallographic studies[8,14,143] have used Xe perfusion in PSII crystals in an attempt to identify how O_2 may be transported away from the site of water oxidation. While the number of Xe sites identified differed between these studies (25 sites per PSII monomer in Murray *et al.*,[143] 10 in Guskov *et al.*,[8] and 26 in Gabdulkhakov *et al.*[14]), in all three cases virtually all of the Xe densities were found in the transmembrane region of the PSII dimer complex, with one isolated site being found in the centre of the β-barrel of the PsbO subunit (Figure 8.8). The majority of the Xe sites were found to be relatively well separated from the co-factors involved in electron transfer, but there was no obvious chain of sites leading from the $CaMn_4$ cluster to either the luminal or stromal side of the thylakoid membrane, in a direction normal to the membrane plane. In particular, the Xe sites were not located in the O_2 channels proposed in any of the studies discussed above, with closest being a Xe site located CP43-C288, which is in the vicinity of channel (i)[12]/back channel[13]/Channels A1-A2.[14]

Gabdulkhakov *et al.* performed additional Kr-perfusion experiments to see if the smaller size and atomic weight of Kr would reveal binding sites not seen with Xe. Two Kr binding sites were located, one in each of Channels B1 and B2. The authors argued that this supported the assignment of these channels as O_2 exit channels. To reconcile the proposed channels with the Xe binding sites, the authors proposed a (rather complex) O_2 release mechanism whereby O_2 is first guided away from the $CaMn_4$ side to the luminal side through the proposed O_2 channels, where it may accumulate. O_2 would then diffuse

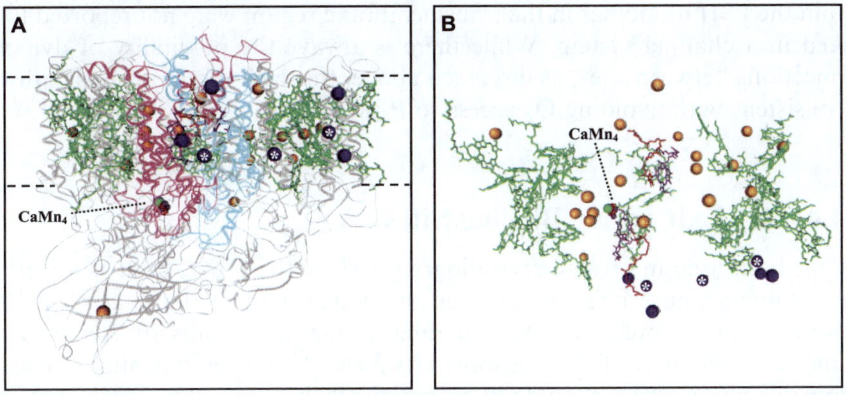

Figure 8.8 Xe binding sites in PSII as found in Murray *et al.*,[143] showing sites in one monomer. Xe atoms sites are shown as orange and blue atoms in space filling representation within the 1S5L crystal structure.[6] Blue atoms represent Xe atoms in the inter-monomer interface, and the three Xe atoms in common with the inter-monomer interface sites in Gabdulkhakov *et al.*[14] are marked with asterisks. Chlorophylls are shown in green. Pheophytin and the plastoquinone cofactors are shown in purple and red, respectively. **A**: PSII monomer viewed along the thylakoid membrane plane (dashed lines). The D1 and D2 subunits are shown as pink and light blue ribbons, respectively. **B**: PSII monomer viewed from the stromal side, along the membrane normal down towards the lumen. (With thanks to Dr James Murray for providing the Xe binding site coordinates.)

back towards the transmembrane region of PSII *via* the same channels, and subsequently diffuse into the lipid region of the thylakoid membrane *via* a number of selected Xe binding sites. From there, O_2 would eventually diffuse out of the thylakoid membrane.

It should be noted that the Xe binding sites proposed by Gabdulkhakov *et al.* to be involved in the exit pathway in the transmembrane region PSII lie in the interface *between* two PSII monomers, not *within* the monomers proper. As such, the physiological significance of these sites for directing O_2 away from the $CaMn_4$ cluster after water oxidation is not entirely clear. They may simply be sites between the PSII monomers that happen to be hydrophobic, and wide enough to accommodate O_2 due to the nature of the packing. In Gabdulkhakov *et al.*, 7 of the 8 Xe sites listed as being in the inter-monomer region were proposed to be involved in this proposed O_2 exit path. However, while 7 Xe-binding sites were also found in Murray *et al.*,[143] only 3 Xe binding sites were in agreement across these two studies (Sites 5, 12, 16 in Murray *et al.* corresponding to Xe14, Xe5 and Xe24 in Gabdulkhakov *et al.*, respectively; Figure 8.8). Given the fluid nature of PSII monomers/dimers in thylakoid membranes, and that PSII monomers are also functional,[144] it is not clear how essential the proposed inter-monomer path is physiologically. In agreement with the various examinations of the crystal structure that found no channels leading from the $CaMn_4$ cluster towards the stromal side of PSII, the Xe sites within the PSII monomer in the transmembrane region were not reported to be linked in a channel system. While there is always the possibility of dynamic connections between sites, as discussed above, this lack of an obvious pathway is consistent with avoiding O_2 access to P_{680}.

8.4.4 O_2 Exit *vs.* Xe Binding Sites

Xe has been used in protein crystallography since the 1960s, both as a method for obtaining phase information, and as indicator of O_2 binding sites.[145,146] However, care should be taken in interpreting the results of Xe-perfusion studies as indicators of O_2 transport channels. A lack of crystallographically resolvable binding sites should not automatically be equated to a lack of O_2 exit pathways. Considering the underlying basis of the perfusion experiments more carefully, the absence of Xe binding sites can be quite consistent with (though certainly not proof of) efficient removal of O_2 through certain pathways or channels. The very ability of Xe to indicate O_2 binding is not necessarily compatible with the aim of identifying channels for O_2 exit in PSII.

Xe-perfused crystals, obtained from the application of pressurised Xe, leads to preferential binding of Xe to hydrophobic cavities of a protein. This property has been used to identify O_2 binding sites in a large number of protein systems, including sperm whale myoglobin,[147–149] horse haemoglobin,[150] cytochrome *c* oxidase,[151] cholesterol oxidase,[142] copper amine oxidase,[152–154] methane monooxylase hydroxylase[155], [NiFe] hydrogenase,[156] hydroxybenzoate hydroxylase,[157] and urate oxidase.[145,158] As Xe binds to such perfused proteins with sufficient order for the atoms to be resolved crystallographically, these are sites where O_2 may bind with high affinity. In the case of channels for O_2 exit in PSII, by contrast, rapid O_2 exit rather than high affinity binding is desired. A channel system through which Xe (O_2) is able to move efficiently will not necessarily give rise to ordered binding of the gas molecules – this is most likely not even desirable. The resulting disorder in Xe positions would not give resolved electron densities in the diffraction data, and therefore would not been visible.

It is interesting to note that the majority of instances where Xe perfusion has been used to study O_2 translocation have involved enzyme systems with O_2 as a *substrate* (see the examples above). This means that the enzyme needs to scavenge dissolved O_2 from the bulk solvent environment and translocate it to the active site to sufficiently provide for the enzyme reaction in question (or in the case of myoglobins/haemoglobins, to allow binding and transport of O_2 throughout the organism). Therefore, it is reasonable that high affinity O_2 binding sites would be present to concentrate/sequester O_2 inside the protein matrix, as a ready supply of substrate. Such a concentration mechanism has indeed been proposed for cytochrome *c* oxidase,[151] methane monoxygenase hydroxylase[155] and copper amine oxidase.[153]

An additionally interesting example is carbon monoxide dehydrogenase/acetyl-CoA synthase,[159] where Xe was used as an analogue for substrate CO.

The Xe-binding data showed very well-ordered and long chains of Xe binding sites in the deeper regions of the enzyme, leading towards the active site.[159] However, no Xe sites were found towards the surface of the enzyme, where the channel entrances were located. The authors noted that this was mostly likely due to the dynamic nature of the Xe around these entry/exit points in the enzyme, which would therefore not be resolved in the X-ray diffraction data. Such consideration of disorder/mobility is directly applicable to interpretation of noble gas perfusion data for identifying efficient O_2 exit channels in PSII. Furthermore, some studies have shown that the presumed one-to-one correspondence between Xe binding and O_2 binding may not be valid,[152,153] and O_2 could also bind to hydrophilic sites not amenable to detection using Xe.[158,160] There are also gas-specific differences in the energetics of migration and binding of gas molecules such as Xe, O_2, CO and NO,[136] and the movement of the diatomic gases studied (including O_2) are not restricted to those cavities accessible to Xe. The apparent absence of Xe binding cavities does not necessarily imply that favourable paths for the transport of other gases do not exist.

In summary, although Xe binding studies have been very useful as an O_2 analogue for the identification of O_2 binding sites in a variety of proteins, the absence of such binding sites in the proposed O_2 channels in PSII does not exclude the possibility of their having such a function, especially if O_2 mobility is high, and binding to the protein undesirable. The proposed large channel system[13]/Channels B1-B2[14] appear to be well suited as favourable routes for O_2 exit, and once in the lumen, O_2 can escape through the thylakoid membrane. EPR spin-label studies have shown that lipid bilayers are not significant barriers to O_2 diffusion,[161–163] and it does not appear to be necessary to invoke the involvement of the inter-monomer Xe-binding sites. As for the Xe-binding sites found within the transmembrane regions of the PSII monomers, it can be speculated that they may play an alternative role, namely as sequestration sites. If O_2 molecules do happen to enter this region, these higher affinity binding sites could help to keep them away from the electron transport co-factors, including P_{680}. Examination of the locations of the Xe-binding sites in both Murray *et al.*[143] and Gabdulkhakov *et al.*[14] shows that these sites are generally located away from the electron transport co-factors, and could therefore be suitable for trapping "wayward" O_2 molecules.

8.4.5 Are O_2 Channels "Gas-Tight"?

As much as it would be intuitively desirable to prevent O_2 from ever accessing P_{680}, $^1O_2$ production in PSII is experimentally observed, under certain conditions. The majority of these studies have concerned the mechanisms of photoinhibition, where high light conditions can inhibit the O_2 evolving activity of PSII. Degradation of the protein complex, mostly the D1 subunit, can be observed under aerobic conditions. The mechanistic details behind photoinhibition are highly controversial and under intense study and debate, and there is as yet no consensus.[164–183]

Whatever the mechanism, one outcome is the generation of $^3P_{680}$, and the subsequent production of $^1O_2$ suggests that O_2 is in fact able to reach the P_{680} to some extent, at least under photoinhibiting conditions. This puts into question the effectiveness of O_2 channel, or perhaps even their existence. It has been proposed that when PSII becomes photoinhibited and non-functional, structural changes occur such that the protein complex opens up to a certain extent, allowing O_2 to diffuse to P_{680}.[130] As support of this proposition, reference was made to the observation that $^1O_2$ production *in vivo* was proportional to the amount of structurally intact, but inactive PSII centres.[184] It would therefore be centres that are already inhibited by the highlight conditions that produces $^1O_2$, triggering the observed D1 degradation.[§] Indeed, experiments showing $^1O_2$ production performed in so-called reaction centre preparations (D1/D2/Cyt b_{559}, non-water oxidising preparations lacking the acceptors Q_A and Q_B)[169,173,176,185] could also conceivably be affected by structural changes, as could samples were the $CaMn_4$ has been lost due to photoinhibition. Significantly, very substantial changes in the protein structure of PSII as a result of $CaMn_4$ depletion has been observed in the MD simulations by Vassiliev *et al.*[15]

As yet there is no direct experimental evidence for such an opening up of the protein structure leading to increased O_2 access to P_{680}. Regardless of whether this occurs, the observation of $^1O_2$ production is not fatal to the idea that O_2 channels exist. Taking into account the dynamic nature of the protein,[29] it is possible that O_2 molecules do occasionally make it into the denser protein matrix towards P_{680}. The ability of Xe to enter the transmembrane region of PSII demonstrates this possibility for gas penetration. The motion of O_2 through the protein is essentially stochastic, and even if not "gas-tight", an open, permanent channel provides a more favourable pathway along which O_2 could more easily escape into the lumen. As important as the presence of channels leading from $CaMn_4$ towards the lumen is the absence of channels leading from the cluster towards P_{680}. Furthermore, the amount of O_2 reaching $^3P_{680}$ may not have to be very high before damage to D1 can be observed experimentally. An apparent cascading effect has been reported for $^1O_2$-related D1 degradation,[186] where initial damage to a D1 subunit releases Chl, decoupling them from quenching carotenoids. Such free Chl could then act as photosensitisers, giving rise to a cascade of ^{3}Chl and $^1O_2$ formation, then further D1 degradation. This releases more free Chl, quite independent of $^3P_{680}$ formation. This is consistent with observations that oxygen uptake has been observed with thylakoid membranes isolated from early greening leaves, where pigment organisation is not complete, and this is thought to reflect the production of $^1O_2$ through photosensitisation by

[§]It is important to note that photoinhibition, *i.e.* reduction in PSII activity due to high light conditions, should not be equated with D1 degradation. Under anaerobic conditions, PSII can still become photoinhibited without degradation of D1. This is one of the arguments for $^1O_2$ being the cause of D1 damage.

^{3}Chl.[187] Therefore, while the observation of $^1O_2$ formation does suggest that O_2 is able to reach P_{680} under certain conditions, it does not exclude the possibility that channels exist to enhance the efficiency of O_2 exit, and reduce the likelihood of $^1O_2$ formation.

8.5 Concluding Remarks

Far from being a mere frame for keeping the electron transfer co-factors in place, it is becoming clear that the protein matrix of PSII can have a direct role in optimising the catalytic activity of this remarkable enzyme. The identification of channel systems within the protein complex has raised the possibility that the flow of substrate and products in and out of PSII is regulated. This regulation may be passive, in terms of "merely" steering the directionality of the movement of the different species, but there may also be active control elements built into the protein framework. This has been shown to be especially likely in the case of proton transport, but control mechanisms may also exist for water and O_2. There is still much room for investigations, both experimentally and computationally, to verify the channel proposals that have so far been presented. Better understanding of the role of the protein environment may provide key insights into how water oxidation could be achieved using artificial but biomimetic systems.

Acknowledgements

The author would like to thank Dr Clyde Cady and Prof. Stenbjörn Styring for valuable discussions. The financial support of the Carl Tryggers Foundation for Scientific Research, the Knut & Alice Wallenberg Foundation, the Solar-H_2 program of the European Community, the Swedish Energy Agency, and the Swedish Research Council is gratefully acknowledged.

References

1. J. Barber, K. Ferreira, K. Maghlaoui and S. Iwata, *Phys. Chem. Chem. Phys.*, 2004, **6**, 4737–4742.
2. J. P. McEvoy and G. W. Brudvig, *Chem. Rev.*, 2006, **106**, 4455–4483.
3. N. Nelson and A. Ben-Shem, *Nat. Rev. Mol. Cell Biol.*, 2004, **5**, 971–982.
4. N. Nelson and C. F. Yocum, *Annu. Rev. Plant Biol.*, 2006, **57**, 521–565.
5. *Photosystem II: The light-driven water:plastoquinone oxidoreductase*, ed. T. J. Wydrzynski and K. Satoh, Springer, The Netherlands, 2005.
6. K. N. Ferreira, T. M. Iverson, K. Maghlaoui, J. Barber and S. Iwata, *Science*, 2004, **303**, 1831–1838.
7. B. Loll, J. Kern, W. Saenger, A. Zouni and J. Biesiadka, *Nature*, 2005, **438**, 1040–1044.

8. A. Guskov, J. Kern, A. Gabdulkhakov, M. Broser, A. Zouni and W. Saenger, *Nat. Struct. Mol. Biol.*, 2009, **16**, 334–342.
9. R. J. Debus, *Coord. Chem. Rev.*, 2008, **252**, 244–258.
10. H. Dau and M. Haumann, *Coord. Chem. Rev.*, 2008, **252**, 273–295.
11. K. Sauer, J. Yano and V. K. Yachandra, *Coord. Chem. Rev.*, 2008, **252**, 318–335.
12. J. W. Murray and J. Barber, *J. Struct. Biol.*, 2007, **159**, 228–237.
13. F. M. Ho and S. Styring, *Biochim. Biophys. Acta*, 2008, **1777**, 140–153.
14. A. Gabdulkhakov, A. Guskov, M. Broser, J. Kern, F. Muh, W. Saenger and A. Zouni, *Structure*, 2009, **17**, 1223–1234.
15. S. Vassiliev, P. Comte, A. Mahboob and D. Bruce, *Biochemistry*, 2010, **49**, 1873–1881.
16. T. Y. Shen, K. H. Tai, R. H. Henchman and J. A. McCammon, *Acc. Chem. Res.*, 2002, **35**, 332–340.
17. J. L. Sussman, M. Harel, F. Frolow, C. Oefner, A. Goldman, L. Toker and I. Silman, *Science*, 1991, **253**, 872–879.
18. P. Brzezinski and P. Adelroth, *Curr. Opin. Struct. Biol.*, 2006, **16**, 465–472.
19. J. P. Hosler, S. Ferguson-Miller and D. A. Mills, *Annu. Rev. Biochem.*, 2006, **75**, 165–187.
20. M. Wikström and M. I. Verkhovsky, *Biochim. Biophys. Acta*, 2007, **1767**, 1200–1214.
21. C. Kandt, K. Gerwert and J. Schlitter, *Proteins: Struct. Funct. Bioinf.*, 2005, **58**, 528–537.
22. J. K. Lanyi, *Biochim. Biophys. Acta*, 2006, **1757**, 1012–1018.
23. B. L. de Groot and H. Grubmuller, *Science*, 2001, **294**, 2353–2357.
24. B. L. de Groot and H. Grubmüller, *Curr. Opin. Struct. Biol.*, 2005, **15**, 176–183.
25. N. Chakrabarti, B. Roux and R. Pomès, *J. Mol. Biol.*, 2004, **343**, 493–510.
26. H. Ogata, W. Lubitz and Y. Higuchi, *Dalton Trans.*, 2009, 7577–7587.
27. T. Buhrke, O. Lenz, N. Krauss and B. Friedrich, *J. Biol. Chem.*, 2005, **280**, 23791–23796.
28. M. Ludwig, J. A. Cracknell, K. A. Vincent, F. A. Armstrong and O. Lenz, *J. Biol. Chem.*, 2009, **284**, 465–477.
29. F. M. Ho, *Photosynth. Res.*, 2008, **98**, 503–522.
30. H. J. van Gorkom and C. F. Yocum, in *Photosystem II: The light-driven water:plastoquinone oxidoreductase*, ed. T. Wydrzynski and K. Satoh, Springer, The Netherlands, 2005, pp. 307–327.
31. A. W. Rutherford, *Trends Biochem. Sci.*, 1989, **14**, 227–232.
32. R. Mei and C. F. Yocum, in *Current Research in Photosynthesis*, ed. M. Baltscheffsky, Kluwer Academic Publishers, The Netherlands, 1990, vol. 1, pp. 729–732.
33. J. Tso, M. Sivaraja and G. C. Dismukes, *Biochemistry*, 1991, **30**, 4734–4739.
34. A. Boussac, J. L. Zimmermann and A. W. Rutherford, *Biochemistry*, 1989, **28**, 8984–8989.

35. A. Boussac, J. L. Zimmermann and A. W. Rutherford, *FEBS Lett.*, 1990, **277**, 69–74.
36. T.-A. Ono and Y. Inoue, *Biochim. Biophys. Acta*, 1989, **973**, 443–449.
37. M. Sivaraja, J. Tso and G. C. Dismukes, *Biochemistry*, 1989, **28**, 9459–9464.
38. T. Wydrzynski, W. Hillier and J. Messinger, *Physiol. Plant.*, 1996, **96**, 342–350.
39. W. P. Schröder and H.-E. Åkerlund, *Biochim. Biophys. Acta*, 1986, **848**, 359–363.
40. T. Wydrzynski, J. Ångström and T. Vänngård, *Biochim. Biophys. Acta*, 1989, **973**, 23–28.
41. G. Ananyev, T. Wydrzynski, G. Renger and V. Klimov, *Biochim. Biophys. Acta*, 1992, **1100**, 303–311.
42. R. L. Bradley, K. M. Long and W. D. Frasch, *FEBS Lett.*, 1991, **286**, 209–213.
43. P. L. Fine and W. D. Frasch, *Biochemistry*, 1992, **31**, 12204–12210.
44. A. Arató, N. Bondarava and A. Krieger-Liszkay, *Biochim. Biophys. Acta*, 2004, **1608**, 171–180.
45. W. G. Gregor, R. M. Cinco, H. Yu, V. K. Yachandra and R. D. Britt, *Biochemistry*, 2005, **44**, 8817–8825.
46. W. Hillier, G. Hendry, R. L. Burnap and T. Wydrzynski, *J. Biol. Chem.*, 2001, **276**, 46917–46924.
47. T. M. Bricker and R. L. Burnap, in *Photosystem II: The light-driven water:plastoquinone oxidoreductase*, ed. T. Wydrzynski and K. Satoh, Springer, The Netherlands, 2005, pp. 95–120.
48. G. Hendry and T. Wydrzynski, *Biochemistry*, 2003, **42**, 6209–6217.
49. G. Hendry and T. Wydrzynski, *Biochemistry*, 2002, **41**, 13328–13334.
50. W. Hillier, J. Messinger and T. Wydrzynski, *Biochemistry*, 1998, **37**, 16908–16914.
51. W. Hillier and T. Wydrzynski, *Biochemistry*, 2000, **39**, 4399–4405.
52. W. Hillier and T. Wydrzynski, *Phys. Chem. Chem. Phys.*, 2004, **6**, 4882–4889.
53. W. Hillier and T. Wydrzynski, *Coord. Chem. Rev.*, 2008, **252**, 306–317.
54. J. Messinger, M. Badger and T. Wydrzynski, *Proc. Natl. Acad. Sci. U. S. A.*, 1995, **92**, 3209–3213.
55. D. A. Force, D. W. Randall, G. A. Lorigan, K. L. Clemens and R. D. Britt, *J. Am. Chem. Soc.*, 1998, **120**, 13321–13333.
56. R. J. Debus, in *Photosystem II: The light-driven water:plastoquinone oxidoreductase*, ed. T. Wydrzynski and K. Satoh, Springer, The Netherlands, 2005, pp. 261–284.
57. E. M. Sproviero, J. A. Gascon, J. P. McEvoy, G. W. Brudvig and V. S. Batista, *Coord. Chem. Rev.*, 2008, **252**, 395–415.
58. C. Tommos, *Philos. Trans. R. Soc. B*, 2002, **357**, 1383–1394.
59. M. Petřek, M. Otyepka, P. Banáš, P. Košinová, J. Koča and J. Damborský, *BMC Bioinf.*, 2006, **7**, 316.
60. M. Karplus and J. A. McCammon, *Nat. Struct. Biol.*, 2002, **9**, 646–652.

61. S. T. Wlodek, T. W. Clark, L. R. Scott and J. A. McCammon, *J. Am. Chem. Soc.*, 1997, **119**, 9513–9522.
62. H. X. Zhou, S. T. Wlodek and J. A. McCammon, *Proc. Natl. Acad. Sci. U. S. A.*, 1998, **95**, 9280–9283.
63. H. Johansen-Berg and M. F. S. Rushworth, *Annu. Rev. Neurosci.*, 2009, **32**, 75–94.
64. W. Junge, M. Haumann, R. Ahlbrink, A. Mulkidjanian and J. Clausen, *Philos. Trans. R. Soc. B*, 2002, **357**, 1407–1418.
65. H. Dau and M. Haumann, *Photosynth. Res.*, 2005, **84**, 325–331.
66. L. I. Krishtalik, *Biochim. Biophys. Acta*, 1986, **849**, 162–171.
67. F. Rappaport and B. A. Diner, *Coord. Chem. Rev.*, 2008, **252**, 259–272.
68. M. Haumann, P. Liebisch, C. Muller, M. Barra, M. Grabolle and H. Dau, *Science*, 2005, **310**, 1019–1021.
69. H. Dau and M. Haumann, *Science*, 2006, **312**, 1471–1472.
70. H. Dau and M. Haumann, *Biochim. Biophys. Acta*, 2007, **1767**, 472–483.
71. P. E. M. Siegbahn and M. Lundberg, *Photochem. Photobiol. Sci.*, 2005, **4**, 1035–1043.
72. C. J. T. V. Grotthuss, *Ann. Chim.*, 1806, **58**, 54–74.
73. N. Agmon, *Chem. Phys. Lett.*, 1995, **244**, 456–462.
74. S. Cukierman, *Biochim. Biophys. Acta*, 2006, **1757**, 876–885.
75. D. Marx, *ChemPhysChem*, 2006, **7**, 1848–1870.
76. J. M. J. Swanson, C. M. Maupin, H. N. Chen, M. K. Petersen, J. C. Xu, Y. J. Wu and G. A. Voth, *J. Phys. Chem. B*, 2007, **111**, 4300–4314.
77. C. A. Wraight, *Biochim. Biophys. Acta*, 2006, **1757**, 886–912.
78. D. Marx, M. E. Tuckerman, J. Hutter and M. Parrinello, *Nature*, 1999, **397**, 601–604.
79. J. M. J. Swanson and J. Simons, *J. Phys. Chem. B*, 2009, **113**, 5149–5161.
80. C. Dellago, M. M. Naor and G. Hummer, *Phys. Rev. Lett.*, 2003, **90**, 105902.
81. C. Dellago and G. Hummer, *Phys. Rev. Lett.*, 2006, **97**, 245901.
82. M. L. Brewer, U. W. Schmitt and G. A. Voth, *Biophys. J.*, 2001, **80**, 1691–1702.
83. J. D. Mills, in *Photosynthesis Energy Transduction: a Practical Approach*, ed. M. F. Hipkins and N. R. Baker, IRL Press, Oxford, 1986.
84. G. Bernát, F. Morvaridi, Y. Feyziyev and S. Styring, *Biochemistry*, 2002, **41**, 5830–5843.
85. K. G. V. Havelius and S. Styring, *Biochemistry*, 2007, **46**, 7865–7874.
86. R. Hienerwadel, B. A. Diner and C. Berthomieu, *Biochim. Biophys. Acta*, 2008, **1777**, 525–531.
87. M. H. V. Huynh and T. J. Meyer, *Chem. Rev.*, 2007, **107**, 5004–5064.
88. L. I. Krishtalik, *Bioelectrochem. Bioenerg.*, 1990, **23**, 249–263.
89. J. Lavergne and W. Junge, *Photosynth. Res.*, 1993, **38**, 279–296.
90. A. Y. Mulkidjanian, *FEBS Lett.*, 1999, **463**, 199–204.
91. F. Rappaport and J. Lavergne, *Biochim. Biophys. Acta*, 2001, **1503**, 246–259.

92. H. Suzuki, M. Sugiura and T. Noguchi, *Biochemistry*, 2005, **44**, 1708–1718.
93. J. De Las Rivas and J. Barber, *Photosynth. Res.*, 2004, **81**, 329–343.
94. H. Ishikita, W. Saenger, B. Loll, J. Biesiadka and E. W. Knapp, *Biochemistry*, 2006, **45**, 2063–2071.
95. E. M. Sproviero, J. A. Gascón, J. P. McEvoy, G. W. Brudvig and V. S. Batista, *J. Am. Chem. Soc.*, 2008, **130**, 3428–3442.
96. E. M. Sproviero, K. Shinopoulos, J. A. Gascón, J. P. McEvoy, G. W. Brudvig and V. S. Batista, *Philos. Trans. R. Soc. B*, 2008, **363**, 1149–1156.
97. E. M. Siegbahn, *Chem.–Eur. J.*, 2008, **14**, 8290–8302.
98. P. E. M. Siegbahn and M. R. A. Blomberg, *Dalton Trans.*, 2009, 5832–5840.
99. P. E. M. Siegbahn, *Acc. Chem. Res.*, 2009, **42**, 1871–1880.
100. P. E. M. Siegbahn, *Dalton Trans.*, 2009, 10063–10068.
101. K. Kawakami, Y. Umena, N. Kamiya and J. R. Shen, *Proc. Natl. Acad. Sci. U. S. A.*, 2009, **106**, 8567–8572.
102. J. W. Murray, K. Maghlaoui, J. Kargul, N. Ishida, T. L. Lai, A. W. Rutherford, M. Sugiura, A. Boussac and J. Barber, *Energy Environ. Sci.*, 2008, **1**, 161–166.
103. F. M. Ho, *Proceedings of the 15th International Congress of Photosynthesis*, Beijing, China, 2011, (submitted).
104. Z. L. Li and R. L. Burnap, *Photosynth. Res.*, 2002, **72**, 191–201.
105. T. Shutova, V. V. Klimov, B. Andersson and G. Samuelsson, *Biochim. Biophys. Acta*, 2007, **1767**, 434–440.
106. M. Qian, L. A. Dao, R. J. Debus and R. L. Burnap, *Biochemistry*, 1999, **38**, 6070–6081.
107. G. Ananyev, T. Nguyen, C. Putnam-Evans and G. C. Dismukes, *Photochem. Photobiol. Sci.*, 2005, **4**, 991–998.
108. M. Haumann, M. Barra, P. Loja, S. Loscher, R. Krivanek, A. Grundmeier, L. E. Andreasson and H. Dau, *Biochemistry*, 2006, **45**, 13101–13107.
109. J. P. McEvoy and G. W. Brudvig, *Phys. Chem. Chem. Phys.*, 2004, **6**, 4754–4763.
110. J. P. McEvoy, J. A. Gascon, V. S. Batista and G. W. Brudvig, *Photochem. Photobiol. Sci.*, 2005, **4**, 940–949.
111. E. M. Sproviero, J. P. McEvoy, J. A. Gascón, G. W. Brudvig and V. S. Batista, *Photosynth. Res.*, 2008, **97**, 91–114.
112. N. Knoepfle, T. M. Bricker and C. Putnam-Evans, *Biochemistry*, 1999, **38**, 1582–1588.
113. H. J. Hwang, P. Dilbeck, R. J. Debus and R. L. Burnap, *Biochemistry*, 2007, **46**, 11987–11997.
114. R. Ahlbrink, M. Haumann, D. Cherepanov, O. Bögershausen, A. Mulkidjanian and W. Junge, *Biochemistry*, 1998, **37**, 1131–1142.
115. M. Haumann, A. Mulkidjanian and W. Junge, *Biochemistry*, 1999, **38**, 1258–1267.

116. F. Rappaport and J. Lavergne, *Biochemistry*, 1997, **36**, 15294–15302.
117. J. Sjöholm, K. G. V. Havelius, F. Mamedov and S. Styring, *Biochemistry*, 2009, **48**, 9393–9404.
118. H. A. Chu, A. P. Nguyen and R. J. Debus, *Biochemistry*, 1995, **34**, 5839–5858.
119. P. J. Nixon, D. A. Chisholm and B. A. Diner, in *Plant Protein Engineering*, ed. P. Shewry and S. Gutteridge, Cambridge University Press, Cambridge, 1992, pp. 93–141.
120. M. Hundelt, A. M. A. Hays, R. J. Debus and W. Junge, *Biochemistry*, 1998, **37**, 14450–14456.
121. M. Hundelt, A. M. A. Hays, R. J. Debus and W. Junge, in *Photosynthesis: Mechanisms and Effects*, ed. G. Garab, Kluwer Academic Publishers, Dordrecht, 1998, vol. 2, pp. 1387–1390.
122. S. Singh, R. J. Debus, T. Wydrzynski and W. Hillier, *Philos. Trans. R. Soc. B*, 2008, **363**, 1229–1234.
123. W. J. Coleman, *Photosynth. Res.*, 1990, **23**, 1–27.
124. K. Lindberg and L. E. Andréasson, *Biochemistry*, 1996, **35**, 14259–14267.
125. K. Olesen and L. E. Andréasson, *Biochemistry*, 2003, **42**, 2025–2035.
126. S. M. Theg and P. H. Homann, *Biochim. Biophys. Acta*, 1982, **679**, 221–234.
127. P. van Vliet and A. W. Rutherford, *Biochemistry*, 1996, **35**, 1829–1839.
128. C. F. Yocum, *Coord. Chem. Rev.*, 2008, **252**, 296–305.
129. J. M. Anderson and W. S. Chow, *Philos. Trans. R. Soc. B*, 2002, **357**, 1421–1430.
130. J. M. Anderson, *FEBS Lett.*, 2001, **488**, 1–4.
131. P. Faller, C. Fufezan and A. W. Rutherford, in *Photosystem II: The Light-driven Water:plastoquinone Oxidoreductase*, ed. T. Wydrzynski and K. Satoh, Springer, The Netherlands, 2005, pp. 347–365.
132. H. A. Frank and G. W. Brudvig, *Biochemistry*, 2004, **43**, 8607–8615.
133. J. Hanley, Y. Deligiannakis, A. Pascal, P. Faller and A. W. Rutherford, *Biochemistry*, 1999, **38**, 8189–8195.
134. H. Ishikita, B. Loll, J. Biesiadka, J. Kern, K. D. Irrgang, A. Zouni, W. Saenger and E. W. Knapp, *Biochim. Biophys. Acta*, 2007, **1767**, 79–87.
135. A. Telfer, *Photochem. Photobiol. Sci.*, 2005, **4**, 950–956.
136. J. Cohen, A. Arkhipov, R. Braun and K. Schulten, *Biophys. J.*, 2006, **91**, 1844–1857.
137. J. Cohen, K. Kim, P. King, M. Seibert and K. Schulten, *Structure*, 2005, **13**, 1321–1329.
138. J. Cohen and K. Schulten, *Biophys. J.*, 2007, **93**, 3591–3600.
139. J. Cohen, K. Kim, M. Posewitz, M. L. Ghirardi, K. Schulten, M. Seibert and P. King, *Biochem. Soc. Trans.*, 2005, **33**, 80–82.
140. L. Salomonsson, A. Lee, R. B. Gennis and P. Brzezinski, *Proc. Natl. Acad. Sci. U. S. A.*, 2004, **101**, 11617–11621.
141. P. I. Lario, N. Sampson and A. Vrielink, *J. Mol. Biol.*, 2003, **326**, 1635–1650.

142. L. Chen, A. Y. Lyubimov, L. Brammer, A. Vrielink and N. S. Sampson, *Biochemistry*, 2008, **47**, 5368–5377.
143. J. W. Murray, K. Maghlaoui, J. Kargul, M. Sugiura and J. Barber, *Photosynth. Res.*, 2008, **98**, 523–527.
144. F. Mamedov, M. M. Nowaczyk, A. Thapper, M. Rogner and S. Styring, *Biochemistry*, 2007, **46**, 5542–5551.
145. T. Prangé, M. Schiltz, L. Pernot, N. Colloc'h, S. Longhi, W. Bourguet and R. Fourme, *Proteins*, 1998, **30**, 61–73.
146. M. Schiltz, R. Fourme and T. Prange, *Method. Enzymol.*, 2003, **374**, 83–119.
147. B. P. Schoenborn, *J. Mol. Biol.*, 1969, **45**, 297–298.
148. B. P. Schoenborn and C. L. Nobbs, *Mol. Pharmacol.*, 1966, **2**, 495–498.
149. B. P. Schoenborn, H. C. Watson and J. C. Kendrew, *Nature*, 1965, **207**, 28–30.
150. B. P. Schoenborn, *Nature*, 1965, **208**, 760–762.
151. M. Svensson-Ek, J. Abramson, G. Larsson, S. Tornroth, P. Brzezinski and S. Iwata, *J. Mol. Biol.*, 2002, **321**, 329–339.
152. A. P. Duff, D. M. Trambaiolo, A. E. Cohen, P. J. Ellis, G. A. Juda, E. M. Shepard, D. B. Langley, D. M. Dooley, H. C. Freeman and J. M. Guss, *J. Mol. Biol.*, 2004, **344**, 599–607.
153. B. J. Johnson, J. Cohen, R. W. Welford, A. R. Pearson, K. Schulten, J. P. Klinman and C. M. Wilmot, *J. Biol. Chem.*, 2007, **282**, 17767–17776.
154. P. Pirrat, M. A. Smith, A. R. Pearson, M. J. McPherson and S. E. V. Phillips, *Acta Crystallogr. Sect. F: Struct. Biol. Cryst. Commun.*, 2008, **64**, 1105–1109.
155. D. A. Whittington, A. C. Rosenzweig, C. A. Frederick and S. J. Lippard, *Biochemistry*, 2001, **40**, 3476–3482.
156. Y. Montet, P. Amara, A. Volbeda, X. Vernede, E. C. Hatchikian, M. J. Field, M. Frey and J. C. FontecillaCamps, *Nat. Struct. Biol.*, 1997, **4**, 523–526.
157. T. Hiromoto, S. Fujiwara, K. Hosokawa and H. Yamaguchi, *J. Mol. Biol.*, 2006, **364**, 878–896.
158. N. Colloc'h, L. Gabison, G. Monard, M. Altarsha, M. Chiadmi, G. Marassio, J. Santos, M. El Hajji, B. Castro, J. H. Abraini and T. Prangé, *Biophys. J.*, 2008, **95**, 2415–2422.
159. T. I. Doukov, L. C. Blasiak, J. Seravalli, S. W. Ragsdale and C. L. Drennan, *Biochemistry*, 2008, **47**, 3474–3483.
160. D. Roeser, A. Preusser-Kunze, B. Schmidt, K. Gasow, J. G. Wittmann, T. Dierks, K. von Figura and M. G. Rudolph, *Proc. Natl. Acad. Sci. U. S. A.*, 2006, **103**, 81–86.
161. B. G. Dzikovski, V. A. Livshits and D. Marsh, *Biophys. J.*, 2003, **85**, 1005–1012.
162. A. Missner and P. Pohl, *ChemPhysChem*, 2009, **10**, 1405–1414.
163. W. K. Subczynski, J. S. Hyde and A. Kusumi, *Proc. Natl. Acad. Sci. U. S. A.*, 1989, **86**, 4474–4478.
164. K. Asada, *Plant Physiol.*, 2006, **141**, 391–396.

165. A. Krieger-Liszkay, C. Fufezan and A. Trebst, *Photosynth. Res.*, 2008, **98**, 551–564.
166. E. Tyystjarvi, *Coord. Chem. Rev.*, 2008, **252**, 361–376.
167. I. Vass and E. M. Aro, in *Primary Processes of Photosynthesis, Part 2 Principles and Apparatus*, ed. G. Renger, Royal Society of Chemistry, Cambridge, 2008, pp. 393–427.
168. I. Vass and K. Cser, *Trends Plant Sci.*, 2009, **14**, 200–205.
169. J. R. Durrant, L. B. Giorgi, J. Barber, D. R. Klug and G. Porter, *Biochim. Biophys. Acta*, 1990, **1017**, 167–175.
170. I. Vass, S. Styring, T. Hundal, A. Koivuniemi, E. M. Aro and B. Andersson, *Proc. Natl. Acad. Sci. U. S. A.*, 1992, **89**, 1408–1412.
171. I. Vass and S. Styring, *Biochemistry*, 1992, **31**, 5957–5963.
172. I. Vass and S. Styring, *Biochemistry*, 1993, **32**, 3334–3341.
173. F. J. E. van Mieghem, W. Nitschke, P. Mathis and A. W. Rutherford, *Biochim. Biophys. Acta*, 1989, **977**, 207–214.
174. E. Hideg, C. Spetea and I. Vass, *Photosynth. Res.*, 1994, **39**, 191–199.
175. A. Krieger-Liszkay and A. W. Rutherford, *Biochemistry*, 1998, **37**, 17339–17344.
176. A. Telfer, S. M. Bishop, D. Phillips and J. Barber, *J. Biol. Chem.*, 1994, **269**, 13244–13253.
177. M. Edelman and A. K. Mattoo, *Photosynth. Res.*, 2008, **98**, 609–620.
178. M. Hakala, I. Tuominen, M. Keranen, T. Tyystjarvi and E. Tyystjarvi, *Biochim. Biophys. Acta*, 2005, **1706**, 68–80.
179. N. Ohnishi, S. I. Allakhverdiev, S. Takahashi, S. Higashi, M. Watanabe, Y. Nishiyama and N. Murata, *Biochemistry*, 2005, **44**, 8494–8499.
180. E. Tyystjarvi, M. Hakala and P. Sarvikas, *Photosynth. Res.*, 2005, **84**, 21–27.
181. N. Keren, I. Ohad, A. W. Rutherford, F. Drepper and A. Krieger-Liszkay, *Photosynth. Res.*, 2000, **63**, 209–216.
182. N. Keren, A. Berg, P. J. M. VanKan, H. Levanon and I. Ohad, *Proc. Natl. Acad. Sci. U. S. A.*, 1997, **94**, 1579–1584.
183. A. Szilárd, L. Sass, E. Hideg and I. Vass, *Photosynth. Res.*, 2005, **84**, 15–20.
184. E. Hideg, T. Kalai, K. Hideg and I. Vass, *Biochemistry*, 1998, **37**, 11405–11411.
185. A. N. Macpherson, A. Telfer, J. Barber and T. G. Truscott, *Biochim. Biophys. Acta*, 1993, **1143**, 301–309.
186. C. Fufezan, A. W. Rutherford and A. Krieger-Liszkay, *FEBS Lett.*, 2002, **532**, 407–410.
187. V. Caspi, S. Malkin and J. B. Marder, *Photochem. Photobiol.*, 2000, **71**, 441–446.
188. Y. Umena, K. Kawakami, J.-R. Shen and N. Kamiya, *Nature*, 2011, **473**, 55–60.
189. F. M. Ho, *Biochim. Biophys. Acta*, 2011 (in press, doi: 10:1016/j.bbabio.2011.04.009).

CHAPTER 9

Energy from Photosystem II: Manganese Water Oxidation Catalysts

ROBIN BRIMBLECOMBE,[a] G. CHARLES DISMUKES,[c] GERHARD F. SWIEGERS[b] AND LEONE SPICCIA[a]

[a] ARC Centre of Excellence for Electromaterials Science and School of Chemistry, Monash University, 3800, Victoria, Australia; [b] ARC Centre of Excellence for Electromaterials Science and Intelligent Polymer Research Institute, University of Wollongong, Wollongong, NSW 2522, Australia; [c] Rutgers University, Department of Chemistry and Chemical Biology, Piscataway, NJ 08854, USA

9.1 Need for Fuels

Photosystem II is the only known catalyst in nature capable of harvesting sunlight to split water and evolve oxygen. Photosystem II performs the first light step in photosynthesis, which is the overall process whereby sunlight is captured and through a series of biochemical processes, water and carbon dioxide are converted into simple carbohydrates and O_2. Oxygen is a by-product and is released into the atmosphere, while carbohydrates are used to build and sustain the organism, finally ending up as what is collectively referred to as biomass. Because of the abundance of water on the surface of our planet, it is the preferred reductant. However, given the unfavorable energetics of water oxidation, the ability of Photosystem II to efficiently utilize it as a source of electrons and protons is a remarkable evolutionary achievement. As described in detail in previous chapters, the structure and operation of photosystem II is still under intense investigation. However, there is general consensus that the water oxidizing complex (WOC),

RSC Energy and Environment Series No. 5
Molecular Solar Fuels
Edited by Thomas J. Wydrzynski and Warwick Hillier

Published by the Royal Society of Chemistry, www.rsc.org

also known as the oxygen evolving complex (OEC), is a Mn_4CaO_x cluster, supported by a series of amino acid and oxygen based ligands.

The majority of fuel that the human race has consumed to date has been generated from ancient solar energy. Life is thought to have evolved on Earth around 3.8 billion years ago, and some time after this event, the ability to convert solar energy to chemical energy evolved. As the name describes, fossil fuels are the fossilised remains of biological matter that lived millions of years ago.[1] The energy accumulated in this fossilized biomass was originally derived from the solar radiation collected by photosynthetic organisms and converted into chemical energy. Over several billion years of photosynthesis, a vast amount of organic matter has accumulated on the earth.[2] Through the utilization of fossil fuels we are effectively consuming ancient solar energy, in the form of highly concentrated biomass. It has been estimated that about 25 tonnes of ancient plant life are required to form 1 litre of petrol.[2] The efficiency of photosynthesis in converting solar energy into stored chemical energy is typically below 1% and over the course of history only a small portion of this energy has been converted to fossil fuels. Due to this slow build up, it is estimated that current annual global fossil fuel consumption is equivalent to 400 years of ancient photosynthesis.[2]

Despite the large amount of fossil fuels that has already been consumed, large reserves still remain.[3] However, these reserves, and in particular oil, are typically difficult to access, making them increasingly expensive to extract and are concentrated in a relatively small number of countries.[3,4] Further, the combustion of these fuels is effectively reversing millions of years of carbon assimilation, which has well documented implications for the atmosphere.[5]

Two thirds of the global energy demand is in the form of liquid fuels, primarily used for transport.[3] In the future, some of this demand can be met sustainably by using electricity generated from a range of renewable technologies, such as wind, geothermal, photovoltaics and solar thermal. These technologies could be coupled with extensions to electric rail networks which would allow more freight transport by rail and the replacement of some air travel with high-speed electric trains. Within cities (where most people now live), grid connected electric trains and light rail are excellent forms of mass transport. Further, with battery technologies rapidly improving, electric vehicles including buses, cars and scooters are becoming efficient forms of personal transport, that have the potential to meet the daily travel requirements of the majority of the population.[4,6] However, the demand for mobile fuels is enormous and for long-range transport (such as international shipping) electricity and battery technology are unlikely to be able to meet the large energy output demands. Thus, there is a clear and urgent need for renewable low carbon transport fuels.

In addition to having provided the majority of our past energy reserves, photosynthesis has great potential to contribute to our future fuel requirements. One option is to utilize photosynthetic organisms as biomass feedstock to produce biofuels, or to utilize photosynthetic organisms to produce hydrogen gas. Another option is to use photosystem II as a blueprint for the design of efficient water oxidation catalysts that can be incorporated into photoelectrochemical cells for the renewable and sustainable production of hydrogen gas.

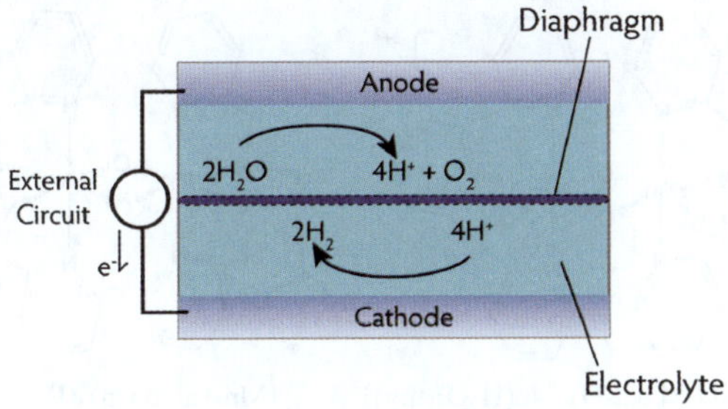

Figure 9.1 Basic representation of a water electrolysis cell.

9.2 Electrolysis of Water

The use of electricity to split water into oxygen and hydrogen is an extremely convenient means of producing hydrogen. Ideally, renewable technologies can be used to produce hydrogen from water at the point of use, minimising the need for hydrogen transport. Water splitting can be achieved in an electrolyser, made up of an inert anode, where water is oxidised to oxygen, protons and electrons, and an inert cathode, where protons and electrons are combined to form hydrogen (Figure 9.2). Both electrodes are in contact with the aqueous electrolyte and are typically separated by a gas tight diaphragm (commonly a polymer membrane or a porous ceramic material), that allows the flow of protons from the anode to the cathode while restricting evolved oxygen to the anode half cell.[7] The overall process is referred to as the splitting of water into its elements (eqn (1c)).

The electrical potential required to oxidise water (eqn (1a)) is 1.23 V under standard conditions (potential *vs.* standard hydrogen electrode (SHE), pH 0).[8]

$$2H_2O \rightleftarrows O_2 + 4H^+ + 4e^- \quad 1.23\ \text{V}\ vs.\ \text{SHE} \tag{1a}$$

$$4H^+ + 4e^- \rightleftarrows 2H_2 \quad 0.0\ \text{V}\ vs.\ \text{SHE} \tag{1b}$$

$$2H_2O \rightleftarrows O_2 + 2H_2 \tag{1c}$$

To drive the water oxidation reaction at rates useful for solar energy conversion, additional energy is required to overcome the activation energy for chemical bond rearrangements. Thus, the minimum energy required to oxidise water corresponds to 1.48 V (pH 0),[7] and is typically larger for electrodes in use today. It is important to note that as the oxidation of water produces protons, the potential required to drive this process increases by 59 mV per pH unit decrease as the reaction proceeds.[8]

In practice, however, an additional overpotential is needed which varies depending on the ability of the electrode to form the highest energy intermediates during water oxidation; hence, the choice of material at the electrode-water interface is critical. In commercial electrolysers, other factors must also be overcome including: (i) cell resistance (proton and electron current

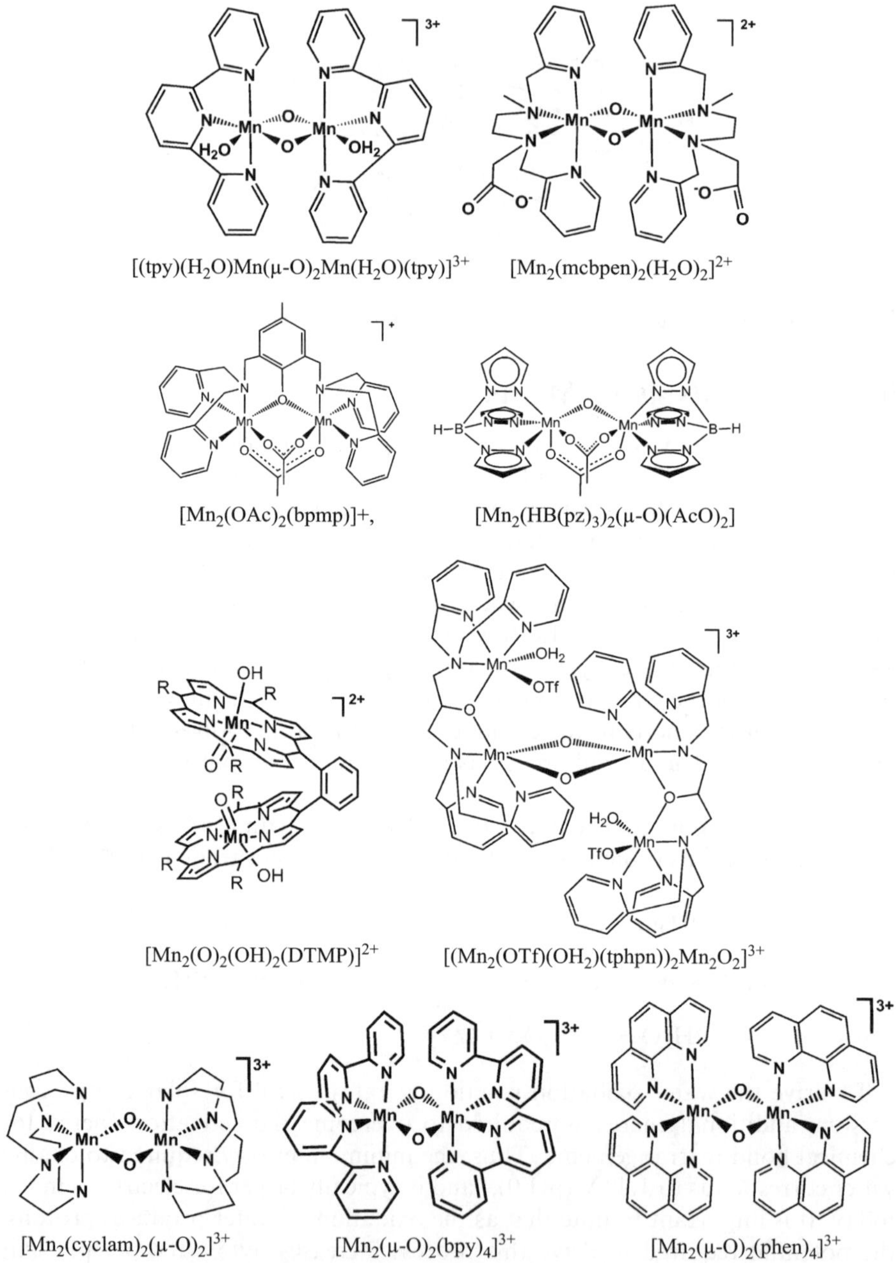

Figure 9.2 Schematic representations of a range of manganese complexes shown to catalytically oxidize water and/or evolve oxygen.

resistances, electrolyte conductivity, gas permeability); and (ii) irreversible processes (generation of inactive intermediates). Due to these factors, the actual voltage required to achieve water electrolysis is typically 1.85–2.05 V,[7,9] which

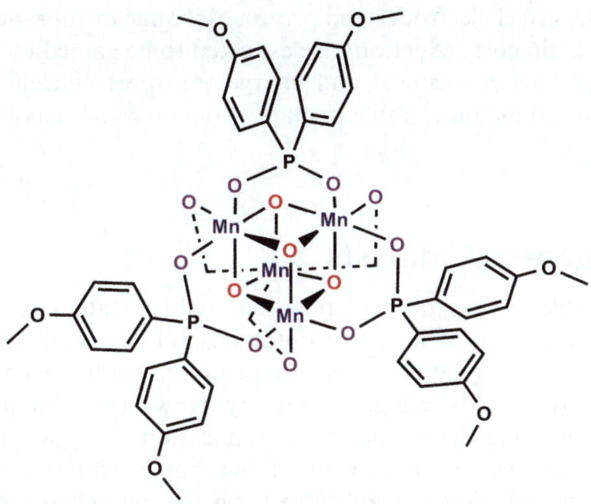

$[Mn_4O_4((MeOPh)_2PO_2)_6]$

Figure 9.2 Continued.

corresponds to an overpotential of 0.5–0.7 V, *cf.*, thermodynamic potential. A range of approaches have been utilized to minimise the potential required for water oxidation. These include: (1) increasing the temperature of electrolysis to lower the activation potential; (2) minimising the resistance of the cell by increasing the conductivity of the electrolyte, typically achieved using acidic or basic conditions;[7] (3) using catalysts to lower the activation energy and improve the specificity of water oxidation at the electrode surfaces. Indeed, efficient and robust catalysts for the oxidation of water are seen as vital to the successful development of water splitting devices for hydrogen production.[10]

Modern electrolyzers convert electrical energy from conventional grid power sources to hydrogen using 55 kWh kg^{-1} H_2 (2003 AY)[11] or 49–53 kWh kg^{-1} H_2 (2004 AY).[12,13] Total consumption of hydrogen in the USA was 7.74 billion kg in 2003 (mainly industrial uses). If all of this were generated by electrolysis using 2004 electrolyzer technology it would amount to 410 000 GWh yr^{-1}. This significant amount of electricity represents 12% of total electricity usage in the US. Addition of LHV transportation vehicles would add further to this total.

According to the most recent study reported by the Electric Power Research Institute (EPRI) in 2006,[11] the cost to produce hydrogen by commercial electrolyzers (with capacity 1000 kg H_2 per day) is \$3–7 per kg H_2 at a mean power cost of 5.5¢ per kWh. This cost excludes the cost to upgrade the electric power grid to that would be needed to supply the increased capacity. Electricity is the major cost for electrolysis, followed by capital cost for the electrolyzer, and the electrolyzer operating costs (electrode corrosion and deterioration in efficiency). These use nickel or platinum supported on carbon-fiber, titanium mesh, or stainless steel electrodes and proton exchange membranes between the half-cells. The main cost reductions are expected to be gained by: (1) reduction in power usage (electrochemical and charge transport efficiencies); (2) elimination of low abundance noble metals; (3) corrosion passivation of the electrodes.

9.3 Manganese Catalysts

Historically, noble metal catalysts have been used for water electrolysis as they are chemically inert and can have overpotentials as low as 0.32 V.[7,14] Solid films of these catalysts are relatively easy to produce, readily incorporated into devices and many demonstrate good stability. However, noble metals are rare and expensive, limiting their widespread application.[15–17] Furthermore, water oxidation requires the rearrangement of five bonds with a requirement for atomic precision and, thus, flexibility to form the high energy reaction intermediates, which may be difficult to achieve on a rigid solid surface. The development of molecular catalysts arguably offers advantages in terms of the ability to exquisitely control atomic structure and composition, while utilizing abundant and cheap materials.

To date the most successful molecular water oxidation catalysts utilize Mn, Ru and Ir metal centres.[18,19] In homogeneous solutions, many ruthenium and iridium catalysts have achieved higher turnovers numbers and turnover frequencies than the reported manganese catalysts.[17–19] However, in the majority of previous studies chemical oxidants have been used rather than electrochemical or renewable energy sources. Additionally, Ru and Ir are relatively rare which means that their availability and high cost will be a major limiting factor in large scale applications. In contrast, manganese is 10^6–10^7 fold more abundant and cheap. For the development of molecular catalysts, manganese has the favorable features of being able to access a range of oxidation states, readily forms mixed-oxidation state oxo clusters and displays ligand lability in catalytic oxidation which can facilitate rapid catalytic turnover.[20] Most importantly, nature has provided us with a highly efficient manganese water oxidation catalyst that can be used as a blueprint for the development of abiotic manganese catalysts. As such, a broad range of manganese complexes have been developed and many tested for catalytic potential. However, while many Mn complexes are described in the literature as "catalysts" for water oxidation, few have exhibited unequivocal evidence of oxygen evolution from water and even fewer have achieved multiple turnovers.[18,19,21] Table 1 provides a summary of some of the manganese complexes that have demonstrated oxygen evolution from water.

A diverse range of dinuclear Mn complexes have been investigated, but only two have been reported to achieve multiple catalytic turnovers (Table 1).[22,29,30] These catalysts, $[Mn_2^{II}(mcbpen)_2(H_2O)_2]^{2+}$ (ref. 22) and $[(tpy)(H_2O)Mn(O)_2$-$Mn(H_2O)(tpy)]^{3+}$ (ref. 23,24) (Figure 9.2), have been studied in homogenous solution and catalysis was dependent on the use of oxygen containing chemical oxidants. In addition, $[(tpy)(H_2O)Mn(\mu\text{-}O)_2Mn(H_2O)(tpy)]^{3+}$, has been shown

Table 9.1 Water oxidation properties of molecular manganese clusters reported in the literature (Structures of catalysts can be found in Figures 2–4).[28]

Complex	*Oxidant*	*Support*	*Mn Centres*	*TON*	*$TOF_{(O2)}$*
PSII WOC	Sunlight/ Chlorophyll	Protein	4	600000	$144000\ h^{-1}$
$[Mn_2(mcbpen)_2(H_2O)_2]^{2+}$ (ref. 22)	$BuOOH^{b}$	Homogeneous	2	13	$1.5\ h^{-1}$
$[((tpy)(H_2O)Mn)_2(\mu\text{-}O)_2]^{3+}$ (ref. 23, 24)	HSO_5^{-b} or ClO^{-b}	Homogeneous	2	4	$12\ h^{-1}$
	Ce^{4+a}	Kaolin	$2+2^{c}$	17	N/A
$[Mn_4O_4((MeOPh)_2PO_2)_6]^{+}$ (ref. 25)	1.2V *vs.* NHE	Nafion	4	> 1000	$20\text{–}270 h^{-1}$
$[Mn_2\ (O)_2(OH)_2\ (DTMP)]^{2+}$ (ref. 26)	*m*-Chloroper benzoic acid	Homogeneous	2	≤1	N/A
$[Mn_2(OAc)_2(bpmp)]^{+}$ (ref. 27)	HSO_5^{-} or Pb^{4+}	Homogeneous	2	<1	N/A
$[Mn_2(cyclam)_2(\mu\text{-}O)_2]^{3+}$ (ref. 28)	HSO_5^{-b}	Kaolin	2	<1	N/A
$[Mn_2(HB(pz)_3)_2(\mu\text{-}O)(AcO)_2]$ (ref. 28)	Ce^{4+a}	Kaolin	$2+2^{c}$	<1	N/A
$[Mn_2(\mu\text{-}O)_2(phen)_4]^{3+}$ (ref. 28)	HSO_5^{-b}	Kaolin	2	<1	N/A
$[Mn_2(\mu\text{-}O)_2(bpy)_4]^{3+}$	Ce^{4+a}	Kaolin	$2+2^{c}$	<1	N/A
$[Mn_2(pmpa)_2(\mu\text{-}O)_2]^{n+}$ (ref. 28)	HSO_5^{-b}	Kaolin	2	<1	N/A

[a]oxidative potential 1.7 V *vs.* NHE;
[b]2 electron oxygen donors; TON = number of catalytic turnovers; $TOF_{(O2)}$ = the turnover frequency;
[c]dimer–dimer interaction.

to generate O_2 using Ce^{IV} as a chemical oxidant when immobilised in a clay support.[19,23] Cerium(IV) may be problematic as oxidant in water oxidation studies because it readily hydrolyzes to CeO_2, which is itself is a well known catalyst for water oxidation. A range of other dimanganese complexes (Figure 9.2) have been shown to evolve oxygen, in stoichiometric or lesser amounts, in the presence of chemical oxidants.[26–28,31] The final example is the tetra-manganese complex, $[Mn_4O_4((MeOPh)_2PO_2)_6]$ (Figure 9.2), which when immobilised in a Nafion membrane is able carry out the electro-oxidation of water to O_2 at the surface of an electrode (Pt, glassy carbon, ITO) poised at potentials above 1.2 V (*vs.* NHE) under illumination with visible light.[25]

Despite the success of this diverse array of complexes, their turnover numbers and frequency are dwarfed by those of the WOC of PSII (Table 9.1). As highlighted in preceding chapters, there is still much debate over the precise structure and mechanism of water oxidation by PSII; however, a range of key features have been identified which contribute to its unrivalled success. These features include the concentration of high energy oxidation states of multiple manganese centres, the ability to facilitate di-oxygen bridge formation, a dynamic supporting environment that prevents dissociation of the complex and assists in electron and proton removal and coupling to a photoactive charge separation centre. In the remainder of this chapter, we explore these key features with reference to the complexes described above and summarise the key lessons learned from these examples.

9.4 Concentration of High Valent States

A feature essential for the success of PSII is the range of oxidation states accessible to the four manganese centres. The multiple valence states allow the storage and delivery of large units of energy, in the form of metal centred charge and metal–ligand bond energy. The energy gaps between these states can be tuned by altering the ligand environment about the metal centres, allowing the matching of substrate orbitals and intermediate geometries.[32] Electron delocalisation between Mn^{III} and Mn^{IV} ions, under the control of the ligand field, is a further means for lowering the energy required to reach these high oxidation states in closely spaced metal centres. Changes in ligand coordination to Mn have been observed to occur during light-induced cycling of the PSII-WOC and both Mn^{III} and Mn^{IV}. Some proposed mechanisms invoke the generation of a Mn^{V} intermediate, although no such species has been detected to date.[33] The exact electrochemical potential of the PSII-WOC cluster varies through the various intermediates of the S-state (Kok) cycle. However, it is well established that all $4e^-$ are removed using the oxidative potential provided by the chlorophyll pigment P680 (1.25 V *vs.* SHE).[34]

The redox potential of Mn complexes is strongly influenced by the supporting ligands, and in particular the electron donating/withdrawing properties of their coordinating atoms.[35] For example, in the $[Mn_2O_2(bipyridine)_4]^{3+}$ complex, electron donation from two oxo ligands and four N lone pairs from bipyridine facilitates oxidation of the $Mn^{III} \cdot Mn^{IV}$ core to $Mn^{IV} \cdot Mn^{IV}$ at 1.5 V *vs.* SHE.[36] This is equivalent to the potential required to oxidise aqueous Mn^{II} to Mn^{III}. In the PSII-WOC crystal structure, the Mn centres are supported by a range of ligands that include bridging oxo groups, water, Cl^- and carboxylates from amino acids.[35,37] The adjustable ligand field surrounding the Mn_4 cluster permits variable electron donation and allows the PSII-WOC to occupy the Mn valence states required to oxidise water at or below 1.25 V *vs.* SHE.

Just as the higher oxidation states of manganese are supported by electron donating ligands, these high oxidation states are essential for supporting the oxo intermediates involved in the oxidation of water. The pK_a of a coordinated H_2O molecule is approximately 10.5 for Mn(II), approximately 0 for Mn(III) and the remaining proton is lost upon oxidation to Mn(IV) generating and stabilizing O^{2-}.[20] The abstraction of protons and stabilization of these oxo intermediates are crucial to the formation of the O–O bond which can subsequently be oxidized to O_2 by the oxidizing capacity stored within the manganese ions of the cluster. Thus, the close juxtaposition of high valent manganese centres to stabilize reaction intermediates is crucial to successful catalysis.

As the oxidation state of manganese influences the strength of ligand binding, the delivery of four electrons to form O_2 can rapidly alter the stability of the complex upon oxygen release. This is an important consideration in the choice of supporting ligands and number of manganese centres. For example, if the release of oxygen from a complex results in the reduction of the Mn centres to Mn^{II}, nitrogen donor ligands may be more desirable than oxygen donor ligands, which tend to form more labile bonds to Mn^{II}. Thus, the use of multidentate nitrogen donors can limit the loss of the metal ion into solution, if Mn^{II} forms during the catalytic cycle. An alternative approach is to utilize a larger number (4) of high valent manganese centres, so that the oxidation state of the complex is sufficiently high to maintain stable binding of the supporting ligands throughout the catalytic process.

Amongst the model manganese complexes, the very strong ligand donation afforded by porphyrin ligands, makes the porphyrin complex the only one capable of stabilising Mn^V (Figure 9.6), and then only under basic conditions.[26] A single Mn^V-porphyrin is only able to coordinate water/oxo groups in *trans* axial sites, and upon reduction to Mn^{II} it can abstract three electrons from water indicating that at least two Mn^V centres are required to achieve the four electron oxidation of water.

Of the two dinuclear Mn complexes reported to catalyse multiple turnovers, the $[Mn_2{}^{II}(mcbpen)_2(H_2O)_2]^{2+}$ complex is proposed to pass through Mn^{II}, Mn^{III} and Mn^{IV} in the water oxidation process.[22] The $[(tpy)(H_2O)Mn(O)_2Mn(H_2O)(tpy)]^{3+}$ catalyst is also proposed to pass through a range of oxidation states from Mn(II,III) to Mn(IV,V). However, as in PSII, to date no Mn^V states have been observed for either of these complexes.[38] In their high valent Mn states, these complexes have the potential to extract the four electrons required to achieve water oxidation. The challenge is how to engineer di-manganese complexes to achieve this, releasing oxygen, whilst maintaining their stability during catalytic cycling. For these two complexes, in solution O_2

formation involves the extraction of two electrons from water and an oxygen atom from a chemical oxidant. The requirement of an oxygen donating oxidant suggests that for these $\{Mn_2O_2\}$ complexes, two high valent centres are not sufficient or are too unstable to deliver the required charge to oxidize two water molecules simultaneously, as required for O_2 evolution.

For the $[Mn_2(OAc)_2(bpmp)]^+$ dinuclear complex, water oxidation has been reported using a non-oxygen donor chemical oxidant. In this system, the complex is proposed to be oxidised to a $2Mn^{IV}$ state, *via* two sequential 2 electron oxidations, by Pb^{4+} (Figure 9.3). The release of oxygen reduces the complex to $2Mn^{II}$. In the absence of the bridging acetato ligands (which are displaced in the generation of O_2), the two labile Mn^{II} centres are exposed to solvation, resulting in the rapid break down of the complex and preventing multiple turnovers.[27]

In the Mn-oxo cubane model complexes, $[Mn_4O_4L_6]$, each metal centre is bound to three μ_3-oxo ligands and three phosphinate oxygen donors providing a strong electron donating field around the metals, which facilitate transition from a $2Mn^{III}2Mn^{IV}$ to a $Mn^{III}3Mn^{IV}$ state at potentials as low as 0.97 V *vs.* NHE. This process yields a highly reactive $Mn_4O_4^{\ 7+}$ core, which can be utilized in the oxidation of water.[39] The reduction potentials of the core can be tuned by several tenths of volts by varying the X functional groups of bis(X-phenyl)-phosphinate ligands. For example, addition of para-methoxy groups lowers the potential of the $2Mn^{III}2Mn^{IV}$ to $Mn^{III}3Mn^{IV}$ oxidation by 0.15 V relative to the phenyl derivative.[40] By coupling together four manganese centres, the $[Mn_4O_4]^{n+}$ central core achieves a formal charge of 6+ and 7+, respectively,

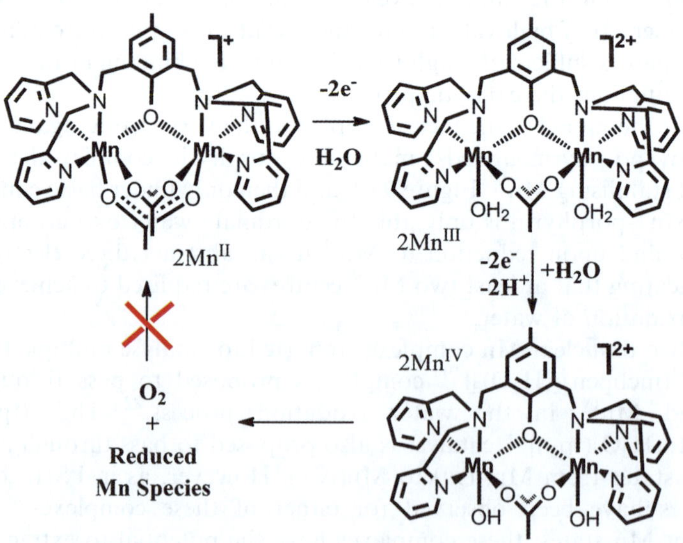

Figure 9.3 Proposed reaction of water with $[Mn_2(OAc)_2(bpmp)]^+$ and Pb^{4+}, to produce O_2, Pb^{2+} and Mn^{II}. Reaction occurs in homogeneous solution, containing 5% water.[27]

and thus has a higher energy than, for example, the $[Mn_2O_2]^{3+}$ core. This feature allows uptake of four electrons together with four protons (effective H atom transfers) at a near constant high electrochemical potential. This higher electrical charge causes expansion of the core yielding longer (weaker) Mn–O bonds and reduced electronic–nuclear coupling (smaller Jahn–Teller distortions). Consequently, on transfer of the electrons/protons the manganese centres do not undergo the dramatic change in structure, as discussed above.

9.5 Formation of the O–O Bond

The oxygen evolving centre of PSII is unique in nature for its ability to form an O–O bond from two water molecules. Not surprisingly, replicating this feat in the laboratory has proved extremely difficult. This topic is explored in detail in Chapter 13, thus here we only consider the characteristics of the few manganese complexes that have been shown to successfully form O_2.

One example of non-catalytic dioxygen formation by a manganese complex, was reported by Natutra *et al.*[26] Under basic conditions in the presence of 5% water, bis-porphyrin–$[HO\text{-}Mn^{V}{=}O]_2$ was generated by chemically (chloroperbenzoic acid) oxidizing the bis-porphyrin-Mn_2^{III} complex (Figure 9.4). The double bond found in manganyl oxo complexes (*i.e.*, $Mn^{V}{=}O$) appears to be capable of forming only in high oxidation states (V,VI,VII) and with strong ligand fields (porphyrins or strongly alkaline conditions). It has not been detected in any studies of PSII-WOCs to-date. The few known examples are exclusively synthetic molecular complexes. Although the $Mn^{V}{=}O$ unit in these complexes has been shown to be a precursor to stoichiometric O_2 evolution, multiple turnovers have not been demonstrated thus far. Moreover, O_2 evolution does not appear to occur in mono-manganese complexes of this type, but it does in dimeric complexes. This requires acidification to drive the presumed protonation reaction: $[HOMn^{V}{=}O]_2 \rightarrow [H_2OMn^{V}{=}O]_2$ (*i.e.*, weakening of the ligand field). Because dinuclear complexes are required, it has been suggested to occur *via* homolytic coupling of two electron deficient oxyl groups to form the O–O bond.[26]

Non-catalytic (stoichiometric) oxygen evolution using non-oxygen donating chemical oxidants was also observed for the $[Mn_2(OAc)_2(bpmp)]^+$ complex,[27] and, as described above, two equivalents of the two electron oxidant Pb^{4+} were required. In this example, the $[Mn_2(OAc)_2(bpmp)]^+$ complex was oxidized in two sequential steps in the presence of water to a Mn_2^{IV} species reported to generate oxygen. Oxygen evolution results in the degradation of the complex.[27] It should be noted that 2-electron chemical oxidants, such as Pb^{4+}, may contribute to the formation of oxygen as they are able to generate metal oxo and/or hydroxo species in water.[41] As a result, they maybe non-innocent integral components required for catalyst turnover.

These two examples of oxygen evolution from dinuclear complexes species highlight the challenges of achieving water oxidation with only two manganese centres. For the bis-porphyrin system, the manganese centres are tightly bound

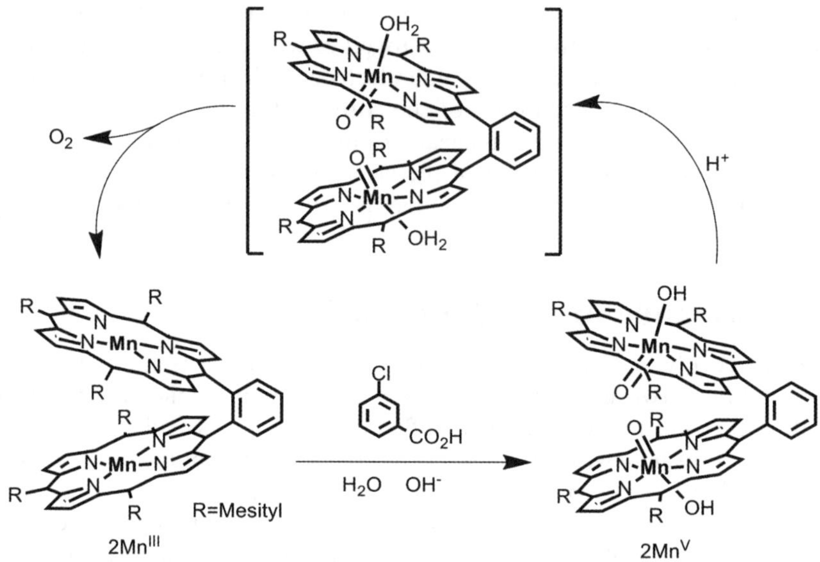

Figure 9.4 Proposed reaction of water with $[Mn_2(DTMP)]^{2+}$, *m*-chloroperbenzoic acid and tetrabutylammonium hydroxide followed by CF_3SO_3H to produce O_2 and reform $[Mn_2(DTMP)]^{2+}$. Reaction carried out in homogenous solution containing 10% water.[26]

in a strong ligand field potential, which facilitates the generation of highly oxidized states and prevents their dissociation into solution. The series of chemical treatments and specific solvent conditions required to achieve oxygen evolution by this system limit its practicality for catalytic turnover. In contrast, the evolution of oxygen from $[Mn_2(OAc)_2(bpmp)]^+$ exposes the resulting Mn^{II} centres to the bulk solution, facilitating rapid dissociation of the complex and limiting the oxygen yield.[27]

For complexes containing a $[Mn_2O_2]$ core, the di-(μ_2-oxo) linkage between two Mn^{III} or Mn^{IV} ions appears to be too stable to serve as the high energy precursor to intramolecular O–O bond formation for O_2 production. This could be due to the relatively strong Mn–O bonds, arising from the multiple bond character in the planar Mn_2O_2-rhombohedral core. Furthermore, because valence bonding involves highly directional 3d orbitals on Mn atoms, the oxygen atoms are forced to remain at distances beyond their van der Waals radii, preventing appreciable bonding from occurring. For complexes such as $[(tpy)(H_2O)Mn(\mu\text{-}O)_2Mn(H_2O)(tpy)]^{3+}$ the water molecules are bound at either end of the $[Mn_2O_2]$ core. This means that the two potential reactive sites are structurally remote, limiting the catalytic potential of this structure. In solution, this complex is proposed to undergo a series of chemical oxidations resulting in the generation of a $\{H_2O/HO\text{-}Mn^{IV}(\mu\text{-}O)_2Mn^{V}{=}O\}$ intermediate. As the terminal oxo species is structurally isolated there is no opportunity to form an intramolecular O–O bond. However, oxygen evolution can be achieved by the

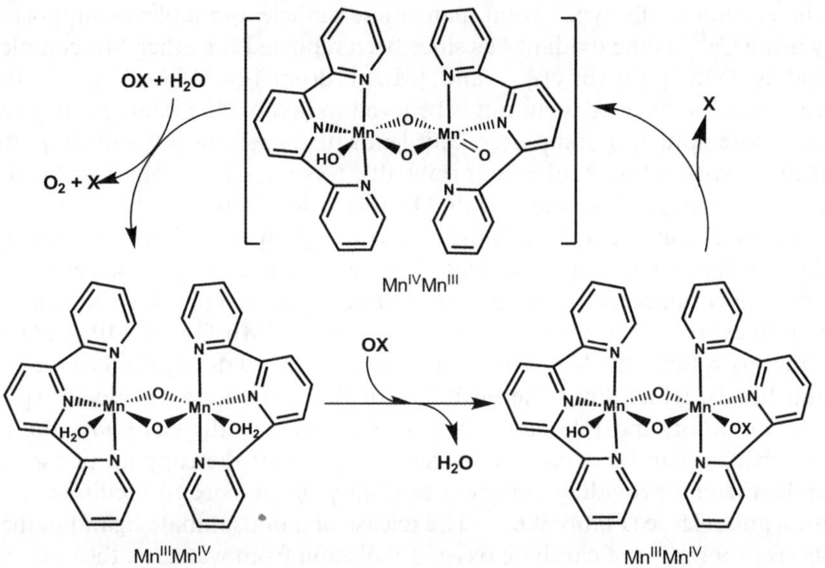

Figure 9.5 Proposed catalytic reaction of water with $[(tpy)(H_2O)Mn(\mu\text{-}O)_2Mn(H_2O)(tpy)]^{3+}$, HSO_5^- or ClO^- (OX) to produce O_2 in a homogenous aqueous solution.[24]

formation of an O–O bond between the $Mn^V{=}O$ unit and a chemical oxidant, such as OCl^- or HSO_5^-, which donates two electrons and an oxo group (Figure 9.5).[29] Thus, the chemical oxidant effectively provides half of the resulting oxygen molecule.[42]

When supported within a ceramic matrix $[(tpy)(H_2O)Mn(\mu\text{-}O)_2Mn(H_2O)(tpy)]^{3+}$ has been shown to evolve oxygen using Ce^{IV} as the chemical oxidant.[30] The activity of the complex in the matrix was found to increase with the square of the concentration of the complex, which indicated that oxygen evolution was achieved through interaction between two $[(tpy)(H_2O)Mn(\mu\text{-}O)_2Mn(H_2O)(tpy)]^{3+}$.[30] Thus, rather than the chemical oxygen donor completing the reaction, two dimers each ionize water to form a reactive oxo species (possibly $Mn^V{=}O$), which come together to facilitate the formation of an O_2 molecule. This mechanism is analogous to that proposed for the bis-porphyrin complex. However, unlike the porphyrin model, this system does not require acid treatment to facilitate O_2 formation and release, presumably due to the greater flexibility afforded by the electrical coupling between the di-(μ-oxo) Mn centres. As this dimer–dimer interaction is only observed within the pores of the clay support, it suggests that the probability of two reactive intermediates interacting increases as the concentration of dimers within the pores increases. In solution, where the probability of dimer–dimer interactions is lower, the $\{H_2O/HO\text{–}Mn^{IV}(\mu\text{-}O)_2Mn^V{=}O\}$ intermediate is unstable and readily forms MnO_4^- and Mn^{III}.[24] This process appears to be suppressed within the clay support.

The evolution of oxygen from manganese dinuclear complexes supported in clay using Ce^{IV} as the oxidant has since been reported for other Mn complexes, including $[Mn_2(\mu\text{-}O)_2(bpy)_4]^{3+}$ and $[Mn_2(HB(pz)_3)_2(\mu\text{-}O)(AcO)_2]$.[19,28] In all three examples, oxygen evolution is believed to involve the interaction between two or more dinuclear complexes. This lends further support for the hypothesis that each [Mn–(μ-O)$_2$–Mn] core is only able to successfully ionise and stabilise one reactive oxygen species, with O_2 formation achieved when two such intermediates come together. Thus, catalytic oxygen evolution from two water molecules appears to require at least four high valent manganese centres.

The only example of a molecular tetra-manganese complex that has been shown to evolve oxygen is the manganese cubane, $[Mn_4O_4((MeOPh)_2PO_2)_6]$.[43] Within this system, the Mn centres are restricted by relatively rigid (directional) ligand bonds, which limit the flexibility of the oxo core and prevent spontaneous formation and release of oxygen. However, in the gas phase the large kinetic barrier can be overcome by removing one of the capping phosphinate ligands, thereby providing sufficient flexibility in the core to facilitate O_2 formation and release (Figure 9.6).[44] The release of a phosphinate ligand in the gas phase is prohibitive of catalytic oxygen evolution from water, as the complex is unable to reform. As described below, this challenge was overcome by suspending the cubane within a cation exchange membrane, Nafion, and multiple turnovers of oxygen from water were achieved.[25]

DFT calculations of thermally induced reaction pathways that may lead to O_2 release from the Mn-cubane complex, $[Mn_4O_4(H_2PO_2)_6]$, in the gas phase have recently been described.[45] The facially capping bidentate phosphinates prevent intramolecular O_2 formation apparently by eliminating the flexibility required for this reaction. However, upon removal of one phosphinate ligand to produce an open face cation, $[Mn_4O_4(H_2PO_2)_5]^+$, a low energy intramolecular reaction pathway emerges that reveals the nature of the barrier to O_2 formation. The first step has the highest energy and involves formation of a side-on peroxo ($\mu^2{:}\eta^2\text{-}O_2^{2-}$) intermediate, $[Mn_4O_2(O_2)(H_2PO_2)_5]^+$, that passes through an asymmetric transition state at 28.3 kcal mol^{-1} above the precursor. This transition state involves two-electron transfer from the distal pair of Mn^{4+} ions, resulting in reduced Mn–O bonding and forces out-of-plane displacement of the oxo pair concurrent with formation of the peroxide bond.[45]

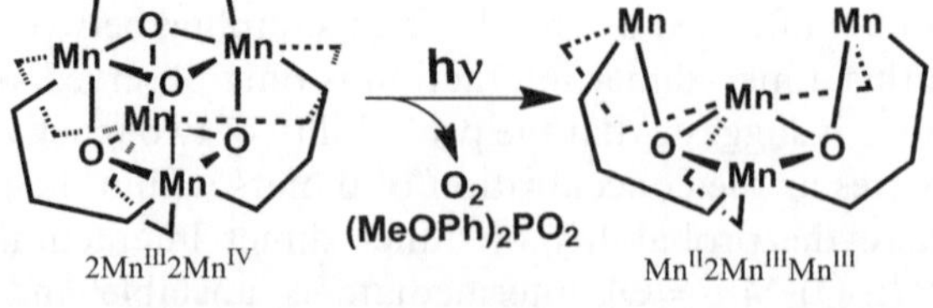

Figure 9.6 Oxygen evolution from $[Mn_4O_4((MeOPh)PO_2)_6]$ on excitation with light in the gas phase.[44]

Although the systems are very different, both the bis-porphyrin complex and the cubane require activation to facilitate the formation of the oxygen–oxygen bond. In both cases, high valent Mn centres stabilize two oxo species in close proximity, and O_2 is only formed when the ligand field is disrupted. In the porphyrin case, this is achieved by the protonation of the *trans*-hydroxide groups. For the cubane, the complex is activated by light, which is thought to cause a charge transfer to one Mn centre, changing the binding strength of the surrounding ligands, and thus facilitating the conformational changes required to achieve O_2 formation.

9.6 Supporting Environment

The PSII-WOC is encapsulated within a protein structure that holds the four Mn ions in place and provides for the needed conformational flexibility and chemical inertness to unwanted oxidation. The support structure also facilitates efficient removal of protons and provides the electrical contact between the cluster and the P680 photo-conversion complex (see preceding chapters). A crucial feature of enzyme active sites is their ability to self-assemble. Particularly for complex clusters, such as the PSII WOC, after a dynamic reaction process the cluster must be able to reassemble directed only by the supporting ligands. Within the active site, a series of cofactors and peptide bound ligands facilitate this process.[32]

The homogeneous Mn catalyst, $[Mn_2^{II}(mcbpen)_2(H_2O)_2]^{2+}$, employs two tetradentate mcbpen ligands (Figure 9.7).[22] These ligands retard the loss of the Mn centres to solution during the reaction, allowing the complex to facilitate water oxidation for 13 turnovers before deactivation. Cycling is achieved using *tert*-butylhydrogen peroxide.[22] The proposed reaction mechanism suggests that the dimer effectively dissociates into monomers, enabling the binding and ionization of a water molecule. The two monomers appear to come together and accept an oxo group from *tert*-butylhydrogen peroxide to form an O–O bridge between them. The subsequent release of oxygen causes the complex to dissociate into two monomers. In homogeneous solution, the requirement for the monomers to diffuse together limits catalytic stability and results in rapid loss of activity as individual monomers migrate away from each other once formed.[22] Ligands that are not able to hold the metal subunits together for the duration of the catalytic cycle are, therefore, insufficient to ensure sustained catalysis. The dynamic nature of these reactions places importance on the supporting environment that facilitates self-assembly and reassembly for sustained water oxidation.

Kaolin and mica are materials consisting of layers of negatively charged aluminosilicate sheets whose charge is balanced by cations, such as H^+, Li^+, Na^+, K^+, located in the interlayers.[23,46] Positively charged metal complexes can be introduced by ion-exchange of the cations present within the interlayer space, making these materials ideal hosts for cationic manganese clusters. For $[(tpy)(H_2O)Mn(\mu\text{-}O)_2Mn(H_2O)(tpy)]^{3+}$, the pores of the ceramic support

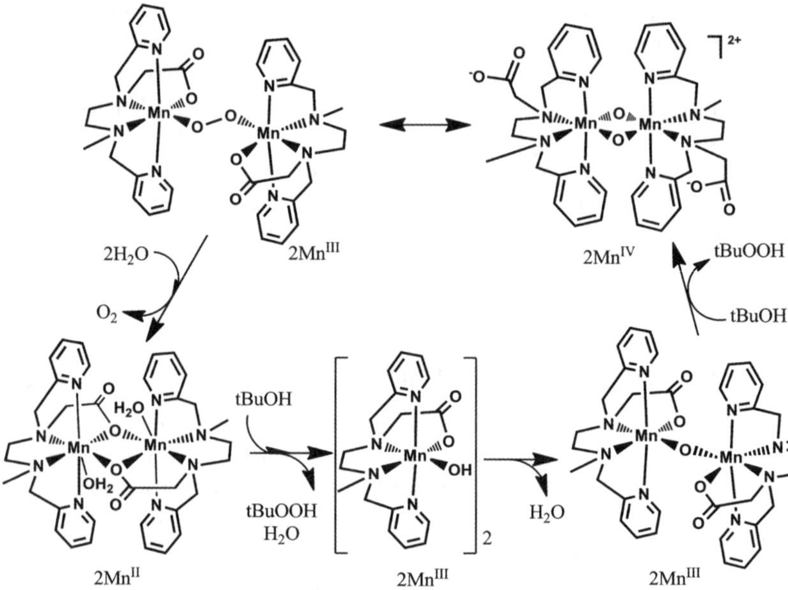

Figure 9.7 Catalytic reaction of water with $[Mn_2(mcbpen)_2(H_2O)_2]^{2+}$ and tBuOOH to produce O_2 in homogenous aqueous solution proposed by McKenzie and co-workers.[22]

appear to facilitate dimer–dimer interactions, with higher turnover rates observed in mica due to the higher density/closer proximity of the cation exchange sites.[23] From these findings, it was concluded that oxygen evolution from water was facilitated through interaction between the dimers, with catalysis proposed to be achieved by clusters of four or more manganese.[23,28] In homogenous solution, reaction of $[(tpy)(H_2O)Mn(\mu\text{-}O)_2Mn(H_2O)(tpy)]^{3+}$ with excess Ce^{IV} results in decomposition of the complex to MnO_4^-.[19] This process is suppressed within the clay matrix, facilitating a greater number of turnovers before catalyst degradation. The protective nature of the clay matrix and its ability to facilitate interactions between reactive intermediates is analogous to the active site of PSII. In addition to the complexes described herein, O_2 evolution has been observed for the following complexes when supported in ceramic matrix; $[Mn_2(cyclam)_2(\mu\text{-}O)_2]^{3+}$, $[Mn_2(pmpa)_2(\mu\text{-}O)_2]^{n+}$ and $[Mn_2(\mu\text{-}O)_2(phen)_4]^{3+}$, using HSO_5^- as chemical oxidant.[23,28] Although only sub-stoichiometric amounts of O_2 were observed, these systems demonstrate the importance of immobilizing complexes for facilitating water oxidation.

An alternative approach to improve catalyst stability and proton conduction is to use a supporting matrix, such as Nafion. Nafion membranes can be formed in which the sulfonic acid groups generate cation permeable hydrophilic channels that are ~20 nm wide.[47,48] Applications of Nafion include proton exchange membranes in various fuel cells, electrochromic devices based on viologens and electrolysers for H_2 generation.[49,50–52] In studies utilizing Nafion to support Ru water oxidation catalysts, it was found that in some cases the Nafion support resulted in improved catalytic stability and activity compared to catalysis in homogeneous solution.[53–57]

As noted above, light activation of the cubanes in the gas phase causes the release of one of the bridging ligands, which, in turn, allows for the formation and release of oxygen.[47] Catalytic cycling could not be achieved in condensed liquid media because: (i) the cubane is insoluble in open solution when the reactant, water, is present; and (ii) a suitable support to facilitate the release of O_2 and the interaction with water was absent. In offering a mixture of hydrophilic and hydrophobic domains, whilst simultaneously allowing interaction across an interface with water, Nafion provided the environment to facilitate catalytic water oxidation. In a similar fashion to cation exchange into clays, the oxidized form of the cubane, $Mn_4O_4((MeOPh)PO_2)_6]^+$, can be ion-exchanged into Nafion membranes deposited on conductive surfaces. With an applied electrical potential of 1.2 V (*vs.* NHE) and under illumination by white light, over 1000 catalytic turnovers were achieved.[25]

Analogous to the active site of PSII, the Nafion provides an inert support structure for the catalyst where it is able to interact with water while being protected within the hydrophobic lattice of the polymer. The sulfonate groups may also act in a similar supporting fashion to the carboxylate ligands within the enzyme active site; that is, they may stabilize reaction intermediates. Furthermore, the sulfonate groups may play a critical role in sequestering protons from catalyst-bound water molecules, and providing a pathway for removing them from the reaction site.

In a similar fashion to the pores of the clay, the substructure of the Nafion appears to be crucial for holding the reaction intermediates in close proximity to facilitate reassembly of the catalytic species following oxygen evolution. This system also has the advantage of allowing the clusters to be localized in close proximity to the electrode surface, where the manganese clusters can be electrochemically oxidized.

9.7 Coupling Catalysts to Light Sensitisers

For all of the molecular catalysts described above, the energy required to drive the oxidation of water is provided either by a sacrificial oxidant or a polarised electrode surface. The use of sacrificial oxidants is not practical for the application of these catalysts for sustained water oxidation. In the case of systems driven by an electrical potential, photo-oxidation of water can be achieved by coupling them to photovoltaic cells, *via* an electrical circuit. However, the application of photovoltaic devices is typically prohibitively expensive, as multiple solar cells connected in series are needed to achieve the required potential.[16] In PSII, the required oxidative potential is generated by the light absorbing P680 complex. P680 is made up of a network of chlorophyll molecules that on excitation by light pass electrons to charge carriers, such as phenophytin, which transfer the electrons to Photosystem I. The resulting electron hole in $P680^+$ has an oxidative potential of ~1.25 V *vs.* SHE.[34] This potential is utilized to oxidise the manganese cluster *via* the Y_z tyrosine linker. The complete oxidation of two water molecules to O_2 and four protons is driven by the absorption of four photons of light corresponding to the extraction of $4e^-$ from the WOC. Coupling the light oxidising complexes directly to the water oxidation catalyst minimises the energy loses and facilitates the oxidation process at remarkably low over potentials.

The process of light sensitizing manganese complexes by covalently coupling them to light absorbing dyes has been reported.[58] In one example, a $[Ru^{II}(bipy)_3]$ unit has been coupled *via* a tyrosine link to a manganese dinuclear complex (Figure 9.8). Under irradiation, the photo-excited ruthenium complex was able to drive the three electron oxidation of the manganese complex from a $2Mn^{II}$ state to Mn^{III} Mn^{IV}, using methyl viologen as the terminal electron acceptor.[58] Although the photo-driven water oxidation is yet to be demonstrated with this system, it represents an important step forward in the quest for light sensitized manganese catalysts. The other major challenge for this system is harvesting the electrons, which for application in a photoelectrochemical cell, must be collected to facilitate reduction of protons to hydrogen at the counter electrode.

Successful light-driven charge transfer between a Mn complex and TiO_2 nanoparticles has been achieved using an acetylacetonate group to attach the complex to the TiO_2 surface (Figure 9.9(a)). By utilizing this anchoring group, the researchers report rapid charge transfer upon light excitation.[59] Furthermore, the link was found to be stable in aqueous environments, even under oxidative conditions. Unfortunately, recombination of the excited electron with the Mn complex is very fast, resulting in a high probability of back electron transfer after excitation and a low efficiency of electron injection into the semiconductor. In this case, a mononuclear Mn complex was utilized, which as discussed above is unlikely to be a successful water oxidation catalyst. These researchers have since coupled the known catalysts $[(tpy)(H_2O)Mn(O)_2Mn(H_2O)(tpy)]^{3+}$ to TiO_2 (Figure 9.9(b)).[60] As discussed above, this complex is able to oxidize water without an oxygen donor but only when a matrix is used which allows the interaction between two or more complexes. The pores of the

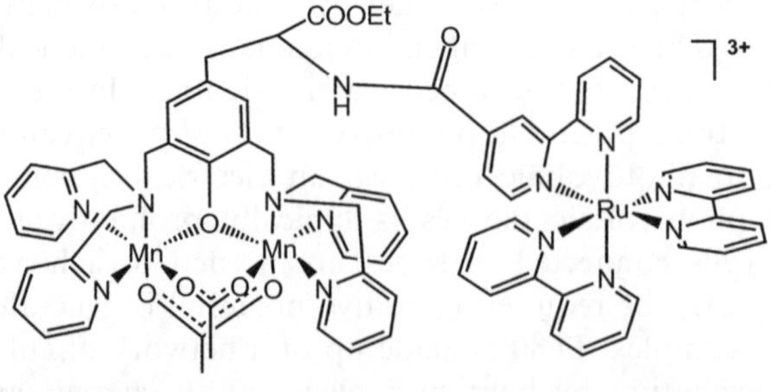

Figure 9.8 Coupling of manganese dinuclear complex to a Ru(II) photosensitiser.

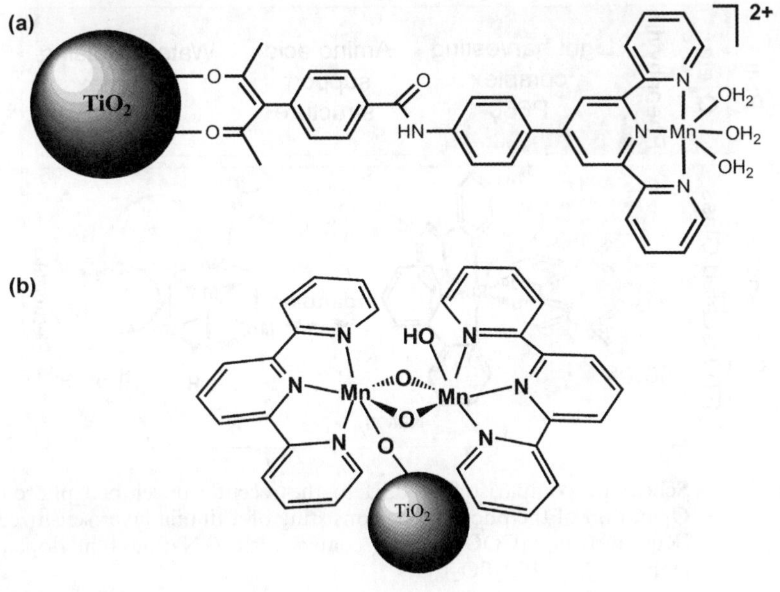

Figure 9.9 Manganese clusters anchored to titania.

nano-structured TiO_2 film could be conducive to such interactions. Interestingly, in this study the formation of tetramers at the TiO_2 surface resulting from dimerisation of two $[(tpy)(H_2O)Mn(O)_2Mn(H_2O)(tpy)]^{3+}$ complexes was observed. The reactivity of this complex on TiO_2 is yet to be reported. This approach offers the possibility of achieving high surface concentrations and efficient charge transfer for molecular catalysts.

Very recently, the $[Mn_4O_4((MeOPh)PO_2)_6]^+$ doped Nafion system was successfully coupled to a dye-sensitized titania film and water oxidation was observed using light as the sole energy source.[61,65] The construction of this device, which was based on the dye sensitized solar cell (DSSC) developed by Graetzel and coworkers,[15] involved firstly the attachment of a Ru(II) dye, $[Ru^{II}(bipy)_2(bipy(COO)_2)]$, onto a titania-coated FTO conductive glass anode. This half-cell was over-coated with a Nafion membrane into which $[Mn_4O_4((MeOPh)PO_2)_6]^+$ was introduced by ion-exchange. As described above, when doped within a Nafion film this catalyst was shown to oxidize water when provided with light and an oxidizing potential of > 1.2 V (*vs.* SHE). For the dye coupled device, light excitation of the $[Ru^{II}(bipy)_2(bipy(COO)_2)]$ dye injects an electron into the titania film, generating $[Ru^{III}(bipy)_2(bipy(COO)_2)]^+$, which has an oxidative potential of ~ 1.4 V (*vs.* SHE). The oxidized dye was found to be able to oxidize the manganese cluster, which in turn facilitates water oxidation. The release of O_2 gas and protons was observed simultaneously with photocurrent generation.[62] The complete photoanode represents, in principle, a functional analog of the PSII-WOC that contains the

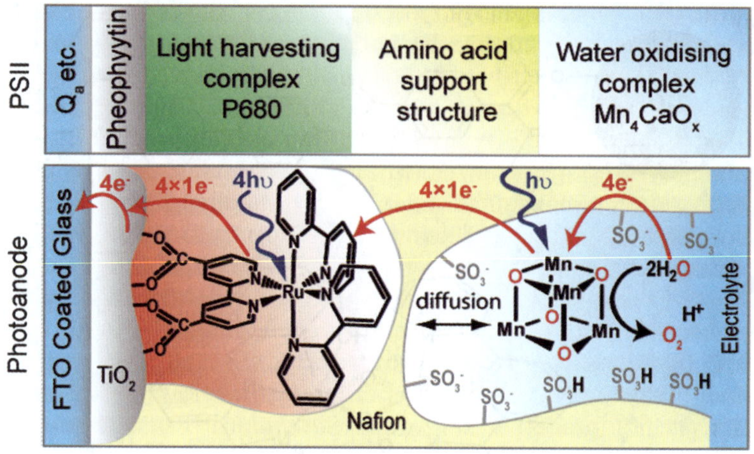

Figure 9.10 Schematic comparison of PSII to the recently developed photoanode. Operation of the photoanode consisting of a titania layer sensitized with $[Ru(bipy)_2(bipy(COO)_2)]$, and coated with a Nafion film doped with $[Mn_4O_4((MeOPh)_2PO_2)_6]^+$.

same types of elements, albeit in this case configured in a different and less efficient way.

9.8 Conclusions

In our opinion, a number of key lessons have been learned from these studies.

(1) At least four manganese ions are required to successfully catalyze the rapid oxidation of water molecules to form O_2 and cycling between the Mn^{III} and Mn^{IV} oxidation states appears to be sufficient to achieve rapid catalytic water oxidation.
(2) The coordinating ligands need to allow attainment of these oxidation states at low electrochemical potentials (0.82–1.5 V) and at a pH where the catalysts are stable, and to provide sufficient flexibility for molecular distortions (without release of Mn) to occur that facilitate formation of the O–O bond (peroxo intermediate) along the reaction coordinate to O_2 formation. Rapid removal of the substrate-derived protons from the catalytic centre to the bulk solution is critical for catalyst stability and high TOF (*i.e.*, sulfonate groups in Nafion).
(3) The Mn_2O_2-rhombohedral core formed by the linking of two Mn^{III} or Mn^{IV} ions with two (μ_2-oxo) bridges appears to be too stable to serve as the high energy precursor to intramolecular O–O bond formation for O_2 production. Furthermore, because valence bonding involves highly directional 3d orbitals on the Mn atoms, the oxygen atoms are forced to remain at distances beyond their van der Waals radii restricting O–O bond formation. In contrast, much longer (weaker) Mn–O bonds occur

in the oxo core of the $[Mn_4(\mu_3\text{-}O)_4]^{6+/7+}$ complexes where π-bonding is essentially eliminated by the cubical core topology (σ-bonds, d^2sp^3 hybrids). This weaker bonding allows greater flexing of the cubical core, which can occur when at least one of the facially capping bidendate phosphinate ligands is removed, facilitating O_2 evolution.

(4) Manganyl oxo species (*i.e.*, $Mn^V{=}O$) are capable of forming only for high oxidation manganese states (V, VI, VII) and with strong ligand fields (porphyrins or strongly alkaline conditions) but the Mn^V state has not been detected in any studies of PSII-WOCs to-date. Although dinuclear $Mn^V{=}O$ porphyrin complexes evolve O_2 on acidification, presumably *via* homolytic coupling of two electron deficient oxyl groups to form the O–O bond, catalysis is yet to be achieved.

(5) The supporting environment is crucial to facilitate reassembly of the catalyst. It appears that four or more Mn centres are required to achieve successful O_2 formation from two water molecules. The pores of cation exchange matrices appear to offer the ideal supporting environment for the dynamic interactions between multiple Mn centres.

(6) Promising advances have been made towards achieving water oxidation using solar energy, which range from the direct coupling of Mn complexes to Ru dyes or TiO_2 to demonstrate successful oxidation of Mn centres using light, to the successful coupling of a Nafion supported catalyst to a Ru-dye sensitized TiO_2 electrode to achieve the first example of light driven water oxidation by a Mn cluster.

The systems described herein embody one or more of the key are features observed in the PSII WOC; the concentration of high energy oxidation states of multiple manganese centres, the ability to facilitate di-oxygen bridge formation, a dynamic supporting environment that prevents dissociation of the complex and assists in electron and proton removal and coupling to a photoactive charge separation centre, with the most successful examples incorporating most or all of these key features.

Acknowledgments

This work was supported by the Australian Research Council Discovery Program (LS/GCD/GFS), the ARC Centre of Excellence for Electromaterials Science (LS/GFS), the US National Institutes of Health and Air Force Office of Scientific Research (GCD), a Lemberg Fellowship (GCD), an Australian Academy of Sciences Travel Fellowship (GFS), an Australian Postgraduate Award (RB), a Fullbright Postgraduate Award (RB) and a Monash University Postgraduate Publication Award (RB).

References

1. T. Flannery, *The Weather Makers*, The Text Publishing Company, Melbourne, 2004.
2. J. S. Dukes, *Climate Change*, 2003, **61**, 31.

3. G. Boyle, B. Everett and J. Ramage, *Energy systems and sustainability*, Oxford, Oxford University Press, Oxford, 2003.
4. A. B. Lovins, E. K. Datta, O.-E. Bustnes, J. G. Koomey and N. J. Glasgow, *Winning the Oil End Game*, Rocky Mountain Institute, Boulder, Colarado, 2004.
5. IPCC, *IPCC Fourth Assessment Report*, 2007.
6. J. Vincent and S. Teske, *Energy Revolution*, Greenpeace Australia, Canberra, Australia, 2008.
7. R. H. Jones, *Materials for the Hydrogen Economy*, Taylor & Francis Ltd, Hoboken, 2007.
8. A. J. Bard and L. R. Faulkner, *Electrochemical Methods: Fundamentals and Applications*, John Wiley & Sons, Inc., New York, 1990.
9. J. R. Wilson and G. Burgh, *Energizing Our Future: Ratinal Choices for the 21st Centruy*, John Wily & Sons, Inc., Hoboken, USA, 2008.
10. M. Dresselhaus, ed. DOE, Office of Basic Energy Sciences (OBES), 2003.
11. D. Rastler, *The roles and opportunities for power companies in the hydrogen-electric economy*, EPRI, 2006.
12. D. H2A, DOE Hydrogen Analysis Team (H2A), 2005, http://www.hydrogen.energy.gov/h2a_analysis.html.
13. C. Yang, *Int. J. Hydrogen Energy*, 2008, **33**, 1977.
14. F. Jiao and H. Frei, *Angew. Chem., Int. Ed.*, 2009, **121**, 1873.
15. M. Gratzel, *Nature*, 2001, **414**, 338.
16. M. Gratzel, *Chemistry Lett.*, 2005, **34**, 8.
17. R. Brimblecombe, G. C. Dismukes, G. F. Swiegers and L. Spiccia, *Dalton Trans.*, 2009, **43**, 9374.
18. C. W. Cady, R. H. Crabtree and G. W. Brudvig, *Coord. Chem. Rev.*, 2008, **252**, 444.
19. M. Yagi, A. Syouji, S. Yamada, M. Komi, H. Yamazaki and T. S., *Photochem. Photobiol. Sci.*, 2009, **8**, 139.
20. F. A. Armstrong, *Phil. Trans. R. Soc. B*, 2008, **363**, 1263.
21. C. S. Mullins and V. L. Pecoraro, *Coord. Chem. Rev.*, 2008, **252**, 416.
22. A. K. Poulsen, A. Rompel and C. J. McKenzie, *Angew. Chem., Int. Ed.*, 2005, **44**, 6916.
23. K. Narita, T. Kuwabara, K. Sone, K. i. Shimizu and M. Yagi, *J. Phys. Chem. B*, 2006, **110**, 23107.
24. J. Limburg, J. S. Vrettos, L. M. Liable-Sands, A. L. Rheingold, R. H. Crabtree and G. W. Brudvig, *Science*, 1999, **283**, 1524.
25. R. Brimblecombe, G. F. Swiegers, G. C. Dismukes and S. Spiccia, *Angew. Chem. Int. Ed.*, 2008, **120**, 7445.
26. Y. Shimazaki, T. Nagano, H. Takesue, B.-H. Ye, F. Tani and Y. Naruta, *Angew. Chem. Int. Ed.*, 2004, **43**, 98.
27. K. Beckmann, H. Uchtenhagen, G. Berggren, M. F. Anderlund, A. Thapper, J. Messinger, S. Styring and P. Kurz, *Energy Environ. Sci.*, 2008, **1**, 668.
28. P. Kurz, *Dalton Trans.*, 2009, 6103.
29. J. Limburg, J. S. Vrettos, H. Y. Chen, J. C. de Paula, R. H. Crabtree and G. W. Brudvig, *J. Am. Chem. Soc.*, 2001, **123**, 423.

30. M. Yagi and K. Narita, *J. Am. Chem. Soc.*, 2004, **126**, 8084.
31. P. Kurz, G. Berggren, M. F. Anderlund and S. Styring, *Dalton Trans.*, 2007, **38**, 4258.
32. J. Dasgupta, G. M. Ananyev and G. C. Dismukes, *Coord. Chem. Rev.*, 2008, **252**, 347.
33. M. Haumann, P. Liebisch, C. Muller, M. Barra, M. Grabolle and H. Dau, *Science*, 2005, **310**, 1019.
34. F. Rappaport, M. Guergova-Kuras, P. J. Nixon, B. A. Diner and J. Lavergne, *Biochemistry*, 2002, **41**, 8518.
35. R. Manchanda, G. W. Brudvig and R. H. Crabtree, *Coord. Chem. Rev.*, 1995, **144**, 1.
36. S. R. Cooper and M. Calvin, *J. Am. Chem. Soc.*, 1977, **99**, 6623.
37. K. N. Ferreira, T. M. Iverson, K. Maghlaoui, J. Barber and S. Iwata, *Science*, 2004, **303**, 1831.
38. H. Chen, R. Tagore, G. Olack, J. S. Vrettos, T.-C. Weng, J. Penner-Hahn, R. H. Crabtree and G. W. Brudvig, *Inorg. Chem.*, 2007, **46**, 34.
39. R. Brimblecombe, A. M. Bond, G. C. Dismukes, G. F. Swiegers and L. Spiccia, *Phys. Chem. Chem. Phys.*, 2009, **11**, 6441.
40. J.-Z. Wu, F. D. Angelis, T. G. Carrell, G. P. A. Yap, J. Sheats, R. Car and G. C. Dismukes, *Inorg. Chem.*, 2006, **45**, 189.
41. C. F. Base and R. E. Mesmer, *The Hydrolysis of Cations*, Wiley, New York, 1976.
42. R. Tagore, R. H. Crabtree and G. W. Brudvig, *Inorg. Chem.*, 2008, **47**, 1815.
43. G. C. Dismukes, R. Brimblecombe, G. A. N. Felton, R. S. Pryadun, J. E. Sheats, L. Spiccia and G. F. Swiegers, *Acc. Chem. Res.*, 2009, **42**, 1935.
44. W. Ruettinger, M. Yagi, K. Wolf, S. Bernasek and G. C. Dismukes, *J. Am. Chem. Soc.*, 2000, **122**, 10353.
45. A. E. Kuznetsov, Y. V. Geletii, C. L. Hill and D. G. Musaev, *J. Phys. Chem. A*, 2010, **43**, 11417.
46. M. Ogawa and K. Kuroda, *Chem. Rev.*, 2002, **95**, 399.
47. A. Rollet, O. Diat and G. Gebel, *J. Phys. Chem. B*, 2002, **106**, 3033.
48. N. P. Blake, M. K. Petersen, G. A. Voth and H. Metiu, *J. Phys. Chem. B*, 2005, **109**, 24244.
49. R. J. Mortimer and J. L. Dillingham, *J. Electrochem. Soc.*, 1997, **144**, 1549.
50. J. Zhu, R. Sattler, A. Garsuch, O. Yepez and P. Pickup, *Electrochim. Acta*, 2006, **51**, 4052.
51. Y. Park and Y. Yamazaki, *Solid State Ionics*, 2005, **176**, 1079.
52. H. Park and W. Choi, *Langmuir*, 2006, **22**, 2906.
53. H. Hagiwara, N. Ono, T. Inoue, H. Matsumoto and T. Ishihara, *Angew. Chem. Int. Ed.*, 2006, **45**, 1420.
54. M. A. Gondal, A. Hameed and Z. H. Yamani, *Energy Sources*, 2005, **27**, 1151.
55. R. Bandyopadhyay, S. K. Maiti and R. Bhattacharyya, *Inorg. Chem. Commun.*, 2002, **5**, 452.

56. K. Sayama, K. Yase, H. Arakawa, K. Asakura, A. Tanaka, K. Domen and T. Onishi, *J. Photochem. Photobiol. A*, 1998, **114**, 125.
57. K. Hashimoto, H. Irie and A. Fujishima, *Jpn. J. Appl. Phys.*, 2005, **44**, 8269.
58. P. Huang, A. Magnuson, R. Lomoth, M. Abrahamsson, M. Tamm, L. Sun, B. van Rotterdam, J. Park, L. Hammarström, B. Åkermark and S. Styring, *J. Inorg. Biochem.*, 2002, **91**, 159.
59. S. G. Abuabara, C. W. Cady, J. B. Baxter, C. A. Schmuttenmaer, R. H. Crabtree, G. W. Brudvig and V. S. Batista, *J. Phys. Chem.*, 2007, **111**, 11982.
60. G. Li, E. M. Sproviero, R. C. Snoeberger, N. Iguchi, J. D. Blakemore, R. H. Crabtree, G. W. Brudvig and V. S. Batista, *Energy Environ. Sci.*, 2009, **2**, 230.
61. R. A. Service, *Science*, 2009, **325**, 1200.
62. R. Brimblecombe, A. Koo, G. C. Dismukes, G. F. Swiegers and L. Spiccia, *J. Am. Chem. Soc.*, 2010, **132**, 2892.

CHAPTER 10

Molecular Ru and Ir Complexes Capable of Acting as Water Oxidation Catalysts

X. SALA,[a] L. ESCRICHE[a] AND A. LLOBET*[a,b]

[a] Universitat Autònoma de Barcelona, Departament de Química, Cerdanyola del Vallès, E-08193 Barcelona, Spain; [b] Institute of Chemical Research of Catalonia (ICIQ), Av. Països Catalans 16, E-43007 Tarragona, Spain

10.1 Introduction

The evolution of humanity on planet earth is directly linked to the access of energy resources. In recent decades, the importance of oil access and processing has been strongly projected in our modern society, basically dictating the world economics and politics and the welfare of countries.[1] The extraordinarily high rate of fossil fuel consumption together with the realization that these fossil fuel resources are not limitless and the consequences of substantially increasing the CO_2 concentration in the atmosphere, clearly advocate for a change of energy source. Sunlight is a highly desirable one since this is the only truly sustainable option in the long term.

Nature has been harvesting sunlight as a source of energy for a few billion years through photosynthetic processes such as the ones carried out by green plants. Nature's strategy consists of the oxidation of H_2O as the source of oxidative equivalents and the reduction of CO_2 to generate carbohydrates as exemplified in the following equation,[2]

RSC Energy and Environment Series No. 5
Molecular Solar Fuels
Edited by Thomas J. Wydrzynski and Warwick Hillier

Published by the Royal Society of Chemistry, www.rsc.org

Table 10.1 Selected redox potentials for the oxidation of water and the reduction of CO_2.

Water Oxidation	*E°, V (vs. SSCE), pH = 7.0*		*CO_2 Reduction*
$^{\cdot}OH + 1H^+ + 1e^- \rightarrow H_2O$	2.15	–2.14	$CO_2 + 1e^- \rightarrow (CO_2)\cdot$
$HO\text{-}OH + 2H^+ + 2e^- \rightarrow 2H_2O$	1.13	–0.85	$CO_2 + 2H^+ + 2e^- \rightarrow HCO_2H$
$HO\text{-}O\cdot + 3H^+ + 3e^- \rightarrow 2H_2O$	1.02	–0.72	$CO_2 + 4H^+ + 4e^- \rightarrow HCHO + H_2O$
$O{=}O + 4H^+ + 4e^- \rightarrow 2H_2O$	0.58	–0.62	$CO_2 + 6H^+ + 6e^- \rightarrow CH_3OH + H_2O$
		–0.48	$CO_2 + 8H^+ + 8e^- \rightarrow CH_4 + 2H_2O$

$$3H_2O + 3CO_2 \xrightarrow{hv} 3O_2 + C_3H_6O_3 \tag{1}$$

From a chemical perspective this overall reaction can be described as two redox half-reactions involving the reduction of CO_2 to carbohydrates and the oxidation of water to molecular oxygen. Both reactions involve multiple electron/proton transfer as indicated in Table 10.1. For the particular case of the oxidation of water to O_2, at the oxygen evolving complex of photosystem II (OEC-PSII), occurs *via* a $4H^+/4e^-$ process,

$$2H_2O \rightarrow O_2{+}4H^+{+}4e^- \tag{2}$$

Understanding this reaction has been the focus of major research efforts for the last two decades. A strong impulse in this field comes from the recently improved mono-crystal X-ray structures and its further refinement by spectroscopic and theoretic methodologies.[3] This has created significant advances in the understanding of the operating mechanisms, although there is a long way to go before is fully understood. At this stage, low weight molecular analogues of the OEC-PSII become a crucial subject not only to test potential proposals but also to understand the different parameters involved in such a complex reaction, and how to manipulate them.

Water splitting by sunlight is an attractive alternative to photosynthesis because it is a very clean reaction and because of its greater simplicity from a chemical viewpoint,

$$2H_2O \xrightarrow{hv} H_2{+}O_2 \tag{3}$$

This reaction again can be viewed as two redox half reactions. The oxidation of water to O_2 in a similar manner as the OEC-PSII (see eqn (2)) and the reduction of protons to hydrogen,

$$2H^+{+}2e^- \rightarrow H_2 \tag{4}$$

The discovery of both good catalysts for water oxidation (WOC) and proton reduction (PRC) to speed up the reactions shown in eqn (2) and (4) is fundamental in order to be able to build devices for energy conversion schemes. A key issue in this topic is the difficulty for a catalyst to deal with multiple proton and electron management. The path to technologically useful catalyst will be unraveled only through deep understanding at a molecular level of the mechanisms involved.

At the moment there is a significant number of molecular transition metal complexes that have shown activity with regard to the oxidation of water to oxygen, mainly based on Ru,[4] but also with other metals such as Ir,[5] Co,[6] Fe,[7] or Mn.[8] The aim of the present paper is not to carry out a thorough description of all reported catalysts but to underline the main reports where fundamental understanding of the catalyzed water oxidation process is described concerning active species, intermediates and transition states in potential catalytic cycles. The main theme of the present report will be based on Ru complexes and, where relevant, the parallel chemistry related to Ir will also be mentioned.

For simplicity purposes and for easy electron counting, only formal oxidation states for the metal centers will be used. Further, in the following equations, auxiliary ligands that do not undergo change during the reaction will not be written, nor will the formal charges of their respective complexes. All redox potentials reported here are referred to SSCE.

10.2 Mononuclear-Monoaqua Ru/Ir complexes

Ru–OH_2 type complexes have been described over the last three decades and their characterization and reactivity with regard to organic substrates has been thoroughly described.[9] An example of typical Ru–OH_2 complex with octahedral type of geometry around the metal center is $[Ru^{II}(tpm)(bpy)(H_2O)]^{2+}$, **1**$^{2+}$,[10] (tpm is the tridentate facial ligand tris(1-pyrazolyl)methane and bpy is the 2,2′-bpy ligand; see Scheme 10.1 for a diagram of these ligands and all the ligands mentioned in the present chapter) and is depicted in Figure 10.1.

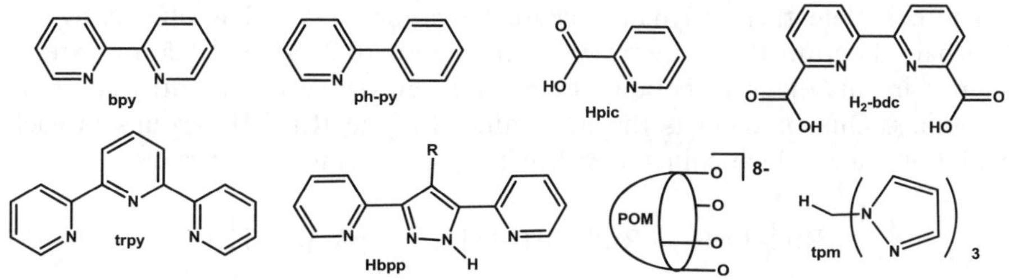

Scheme 10.1 Ligands discussed in this chapter.

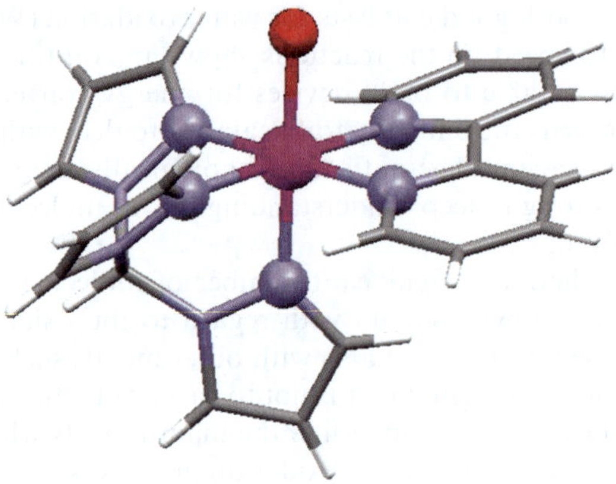

Figure 10.1 Drawing of the molecular structure of complex $\mathbf{1}^{2+}$, $[Ru^{II}(tpm)(bpy)(H_2O)]^{2+}$. Color codes: Ru, cyan; O, red; N, blue; C, gray; H, white.

10.2.1 Thermodynamic Properties

Ru–OH_2 complexes are of interest because their corresponding higher oxidation states can be reached within a relatively narrow potential range by sequential electron and/or proton loss, as shown in eqn (5)–(7) for $\mathbf{1}^{2+}$,[11]

$$\begin{aligned} &Ru^{III}\text{–}OH + H^+ + 1e^- \rightarrow Ru^{II}\text{–}OH_2; \\ &E_{1/2}\left(Ru^{III/II}\ \text{at pH} = 7\right) = 0.40\ \text{V} \end{aligned} \tag{5}$$

$$\begin{aligned} &Ru^{IV}{=}O + 1H^+ + 1e^- \rightarrow Ru^{III}\text{–}OH; \\ &E_{1/2}\left(Ru^{IV/III}\ \text{at pH} = 7\right) = 0.71\ \text{V} \end{aligned} \tag{6}$$

$$Ru^{V}{=}O + 1e^- \rightarrow Ru^{IV}{=}O; \qquad E_{1/2}\left(Ru^{V/IV}\ \text{at pH} = 7\right) = 1.50\ \text{V} \tag{7}$$

The complete thermodynamic characterization over pH is offered by the Pourbaix diagram that is represented in Figure 10.2. Several features are of interest in this diagram besides the precise knowledge of the intrinsic redox potentials. One of them is the pK_a values for the Ru–OH_2 groups in each oxidation state, which coincide with the corresponding slope brakes,

$$Ru^{II}\text{–}OH_2 \rightarrow Ru^{II}\text{–}OH + H^+; \qquad pK_{a,II} = 11.1 \tag{8}$$

$$Ru^{III}\text{–}OH_2 \rightarrow Ru^{III}\text{–}OH + H^+; \qquad pK_{a,III} = 1.2 \tag{9}$$

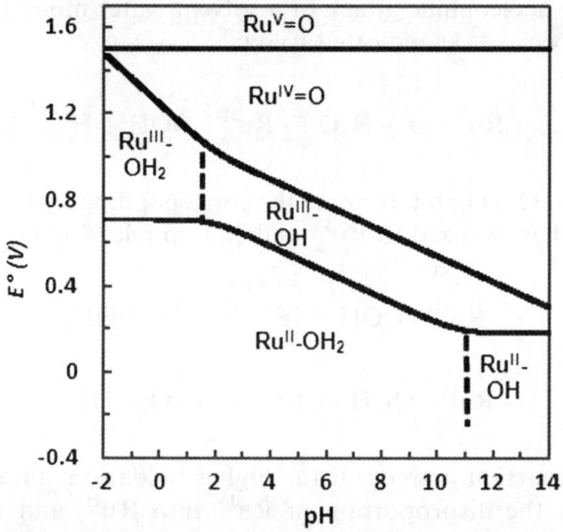

Figure 10.2 Pourbaix diagram for $\mathbf{1}^{2+}$, $[Ru^{II}(tpm)(bpy)(H_2O)]^{2+}$. The zones of stability of the different species generated by proton and electron loss or gain are indicated.

Further, it is also of interest to note that at a sufficiently high pH, the redox potentials for the IV/III and III/II couples will end up merging, and thus a two-electron process would take place,

$$Ru^{IV}{=}O + 2e^- + 1H^+ \rightarrow Ru^{II}–OH \qquad (10)$$

Similarly, at sufficiently low pH, the potentials for the V/IV and the IV/III couples will also merge generating a two-electron wave as indicated in the following equation,

$$Ru^{V}{=}O + 2e^- + 2H^+ \rightarrow Ru^{III}–OH_2 \qquad (11)$$

The easy access to higher oxidation states is rationalized in terms of the strong σ- and π-donating character of the oxido ligands, together with the benefit of low energy PCET pathways.[12]

10.2.2 Reactivity

The access to higher oxidation states for the Ru is important since it can trigger a series of reactions that end up making oxygen. One of the most invoked pathways is the nucleophilic attack of a solvent water molecule (WNA) to an active Ru=O species,[13] as indicated below,

$$Ru^{V}{=}O + H_2O \rightarrow Ru^{III}{-}OOH + H^+ \quad (12)$$

that generates an O–O bond, forming the corresponding hydroperoxide species that can be further oxidized to Ru^{IV} and end up releasing O_2,

$$Ru^{III}{-}OOH - 1e^- \rightarrow Ru^{IV}{-}OOH \quad (13)$$

$$Ru^{IV}{-}OOH \rightarrow Ru^{II} + O{-}O + 1H^+ \quad (14)$$

Given the important activity of the higher oxidation states of Ru, and in particular Ru^{V}, the disproportion of Ru^{IV} into Ru^{III} and Ru^{V} might also become a viable path for water oxidation, as has been recently put forward in a recent work for the mononuclear *out*-$[Ru^{II}(Hbpp)(trpy)(H_2O)]^{2+}$, $\mathbf{2}^{2+}$, complex.[14]

Given the complexity of the water oxidation catalytic cycle, other related minor reaction pathways have also been described.[15] Other potential side reactions have also been proposed when CAN (cerium ammonium nitrate) is used as a chemical oxidant, namely the formation of CeO· which can interact with RuO and form an O–O bond,[11b] and the formation of NO_2 from the reaction of Ce(IV) and the catalyst.[15a] Furthermore, a number of complexes have also been reported where the non-aqua auxiliary ligand also plays a redox role and thus is fundamental for the water oxidation catalytic cycle.[16] Mono-aqua Ir analogues are proposed to make oxygen *via* a WNA pathway in a similar manner to the case of the Ru complexes.[17]

10.2.3 Coordination Expansion

Most of the Ru complexes described so far possess an octahedral coordination, however in some cases the reactive species have been proposed to undergo expansion to seven coordination. This phenomenon has been proposed based on: (a) the analysis of initial and final products with complexes that do water oxidation with apparently no labile bonds;[18] (b) based on DFT analysis of reactive intermediates where a $Ru(\eta^2\text{-}OO)$[15b] is proposed, and (c) based on an X-ray crystal structure of a seven coordinated Ru^{IV}–OH complex, $\{[Ru^{IV}(bdc)(pic)_2]_2(\mu\text{-}HOHOH)\}^{3+}$, $\mathbf{3}^{3+}$, (bdc^{2-} is the [2,2′-bipyridine]-6,6′-dicarboxylato tetradentate dianionic ligand and pic is 4-methyl picolinate; See Scheme 10.1).[19] The latter has shown that when the $Ru^{V}{=}O$ species is reached, it dimerizes following an I2M (interaction of two M–O units) mechanism to generate a peroxo bridged intermediate, that finally evolves oxygen as shown in the following equations,

$$2Ru^{V}-O \rightarrow Ru^{IV}-OO-Ru^{IV} \quad (15)$$

$$Ru^{IV}-OO-Ru^{IV} \rightarrow 2Ru^{III} + O_2 \quad (16)$$

The seven coordination in this case is favored by the geometrical constrains imposed by the bdc^{2-} ligand that forces an angle of 123° for ORuO, and thus leaving sufficient room for seven coordination with a pentagonal bypyramide type of geometry. In a very crude but intuitive manner, the electronic nature of the $Ru^{V}{=}O$ species can also be described as $Ru^{IV}{-}O\cdot$ as a limiting structure. The former should favor a WNA type of mechanism, whereas the latter will favor I2M. The precise nature of this high oxidation state species will be basically determined by the auxiliary ligands that complete the coordination sphere, hence their paramount importance.

10.3 Mononuclear-Diaqua Ru/Ir Complexes

Mononculear Ru complexes containing two aqua groups are of interest because potentially they can lose $4H^+/4e^-$. However, this is a very stringent condition for a metal center because it implies that the metal center has to be accessible in five different oxidation states. One of the complexes that meet this requirement is *trans*-$[Ru^{II}(bpy)_2(H_2O)_2]^{2+}$, $\mathbf{4}^{2+}$, that can undergo four consecutive PCET process to generate $Ru^{VI}(O)_2$, as can be shown in the Pourbaix diagram depicted in Figure 10.3.

$$Ru^{II}(H_2O)_2 - 4H^+ - 4e^- \rightarrow Ru^{VI}(O)_2 \quad (17)$$

The Pourbaix diagram of $\mathbf{4}^{2+}$ is much more complex than its mono-aqua complexes, reflecting the increased number of oxidation states with different degrees of protonation. Although complex $\mathbf{4}^{2+}$ is of interest from a mechanistic perspective it is only capable of performing a few catalytic cycles before it decomposes. The active species that triggers the formation of oxygen in this case is $Ru^{VI}(O)_2$. ^{18}O labeling experiments nicely demonstrate that once the $Ru^{VI}(O)_2$ species is generated, it undergoes a WNA mechanism that produces a hydroperoxide that finally evolves dioxygen,[20] as indicated in the following equations,

$$\left[Ru^{IV}(O)_2\right] + H_2O \rightarrow \left[Ru^{IV}(OH)(OOH)\right] \quad (18)$$

$$\left[Ru^{IV}(OH)(OOH)\right] + H_2O \rightarrow \left[Ru^{II}(H_2O)\right] + O{-}O \quad (19)$$

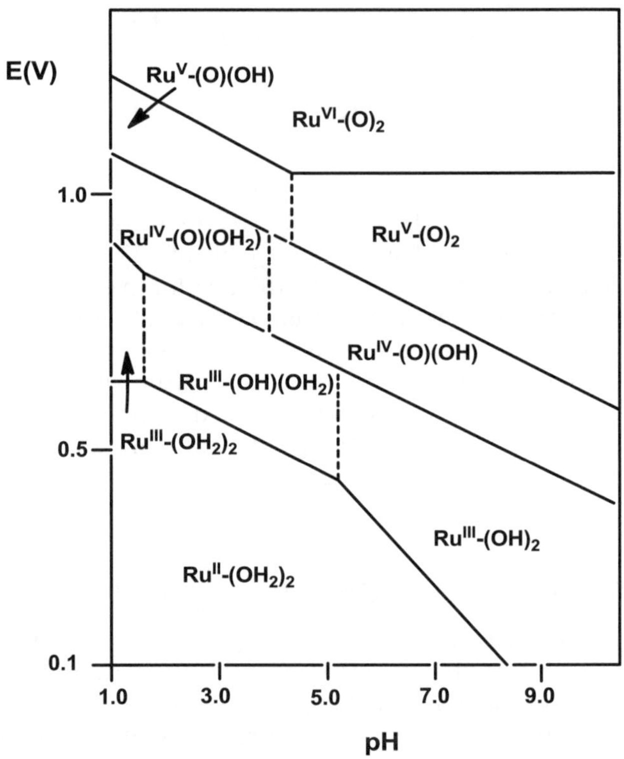

Figure 10.3 Pourbaix diagram for *cis*-$[Ru^{II}(bpy)_2(H_2O)_2]^{2+}$, $\mathbf{4}^{2+}$.

There are Ir analogues that are capable of acting as effective WOCs, one example is the *cis*-$[Ir^{III}(phpy)_2(H_2O)_2]^+$, $\mathbf{5}^+$, (phpy is 2-phenylpyridine) although the mechanism in this case has not been investigated.[21]

10.4 Polynuclear $[Ru(OH_2)]_n$ Complexes

A few polynuclear complexes containing multiple Ru–OH_2 groups have been described that are capable of oxidizing water to O_2.[22] Those complexes contain two or more Ru–OH_2 groups bonded through a bridging ligand (BL). The BL has two crucial functions: (a) it dictates the degree of electronic coupling between the metal centers and (b) it controls the relative space disposition of the Ru–H_2O groups with regard to one another.

Compared with the mononuclear *cis*-$Ru^{II}(H_2O)_2$ complex described in the previous section, dinuclear complexes containing two Ru–H_2O groups with a BL that enables a sufficient degree of electronic coupling, allow sharing of the burden of multiple PCET process between the two metal centers. For instance, for the $\{[Ru^{II}(trpy)(H_2O)]_2(\mu\text{-bpp})\}^{3+}$, $\mathbf{6}^{3+}$, case 4 PCET process are found as depicted in the Pourbaix diagram (see Figure 10.4) but the metal centers only

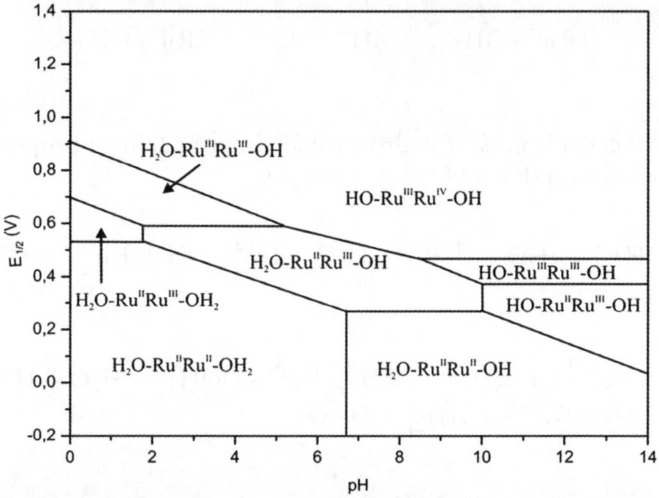

Figure 10.4 Pourbaix diagram for $\{[Ru^{II}(trpy)(H_2O)]_2(\mu\text{-bpp})\}^{3+}$, $\mathbf{6}^{3+}$.

have to cycle between three redox sates, as shown globally in the following equation,

$$[H_2O - Ru^{II}(\mu\text{-BL})Ru^{II} - OH_2] - 4H^+ - 4e^- \\ \rightarrow [O{=}Ru^{IV}(\mu\text{-BL})Ru^{IV}{=}O] \quad (20)$$

Due to the geometry of the bridging ligand together with the nature of the $Ru^{IV}{=}O$ unit in this complex, the mechanism found is an intramolecular I2M, *via* the formation of a peroxo-bridged intermediate that finally evolves dioxygen, as indicated in the following equation,

$$[O{=}Ru^{IV}(\mu\text{-BL})Ru^{IV}{=}O] \rightarrow [Ru^{III}(\mu\text{-BL})(\mu\text{-OO})Ru^{III}] \quad (21)$$

The replacement of the meridional trpy ligand by the facial bpea[22f] produces a geometrical distribution that forces the mechanism to become either a WNA or an intermolecular I2M. The WNA mechanism is also the main mechanism proposed with the so called blue dimer[22a] that contains an oxido ligand as BL, $\{[Ru^{III}(bpy)_2]_2(\mu\text{-O})\}^{4+}$, $\mathbf{7}^{4+}$. Another interesting and efficient WOC is the complex $[Ru^{IV}_4(\mu\text{-O})_4(\mu\text{-OH})_2(H_2O)_4(\gamma\text{-SiW}_{10}O_{36})_2]^{10-}$, $\mathbf{8}^{10-}$, (POM is the polyoxometalate ligand $\gamma\text{-SiW}_{10}O_{36}{}^{8-}$), that basically contains four Ru–OH_2 groups interconnected by multiple oxido and hydroxido ligands. In this case, only two different oxidation states are needed to generate the active species that finally evolve dioxygen,

$$[(Ru^{IV}\text{–}OH_2)_4] - 4H^+ - 4e^- \rightarrow [(Ru^{V}\text{–}OH)_4] \quad (22)$$

A WNA mechanism *via* a hydroperoxide intermediate is proposed for this complex, based on DFT and kinetic analysis,

$$[(Ru^{V}-OH)_4] + H_2O \rightarrow [(Ru^{V}-OH)_2(Ru^{IV}-OH_2)(Ru^{IV}-OOH)] + H^+ \quad (23)$$

$$[(Ru^{V}-OH)_2(Ru^{IV}-OH_2)(Ru^{IV}-OOH)] + H_2O + H^+ \rightarrow [(Ru^{IV}-OH_2)_4] + O-O \quad (24)$$

Other dinculear complexes containing dinucleating ligands have been described and presumably operate by multiple Ru–OH_2 groups, but their nature has not been established yet.[23] Given the large number of bridging ligands available, the potential for complex design is basically limitless.

10.5 Light induced WOC and Water Splitting Devices

In the long run, water oxidation catalysts will have to be incorporated into devices where the main energy input comes from solar energy. In this regard, exploration of light induced water oxidation is a very important process that needs to be understood to build up complete water splitting devices in a modular manner. Examples of light induced water oxidation catalysis have been reported using $[Ru^{II}(bpy)_3]^{2+}$ or related molecules as photosensitizers (Ru_P) in the presence of sacrificial electron acceptors, such as Co(III) or persulfate, together with a WOC. Scheme 10.2 illustrates the combination of reactions that have to occur, where Ru_C stands for the WOC in the lower oxidation sate and Ru_{AC} for the active species,

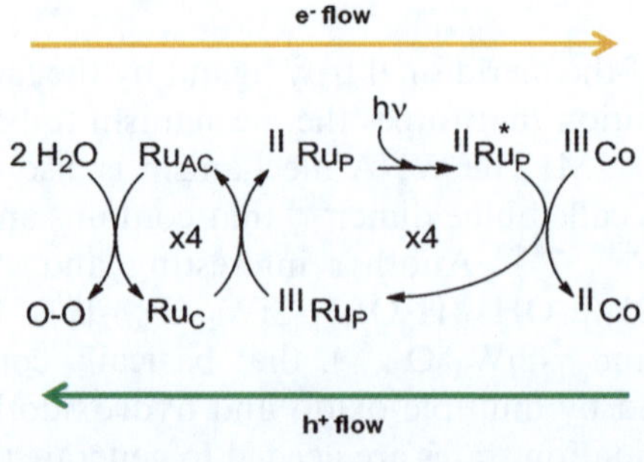

Scheme 10.2 Light driven water oxidation.

A few examples with mononuclear,[14,24] dinulcear[25] and the tetranuclear[26] Ru WOC have been shown to successfully carry out this series of reactions in a harmonious manner, and thus generate O_2 upon shining light. However, the performance of the whole system under photochemically induced conditions needs to be largely improved. Further work needs to focus on the optimization of parameters such as pH, the nature of the photosensitizer, nature of sacrificial electron acceptor, *etc.* and on the elucidation of deactivation pathways.

So far, successful systems have been constructed with independent molecules performing the role of WOC and light harvesting. However, both roles can be combined in a single molecule, a *diad*, where the active units are linked by a bridging ligand. This strategy can potentially achieve the same or even better results provided the electronic coupling between the metal centers is understood and mastered. So far diads, containing both Ru_P and Ru_C have been prepared but have not been reported to be able to carry out WOC photochemically, although they have been shown to act as electron relays.[27] Other interesting combinations useful from the point of view of building solid devices is the attachment of WOCs into solid surfaces and/or conducting electrodes, since it gives valuable information about the electron dynamics in the overall process.[28]

The current understanding of the water oxidation and the high efficiencies obtained for some of the WOCS allow for the building up of complete water splitting devices with a certain degree of success. This in turn, allows for the envisaging of the problems associated with the cell design and come up with potential solutions. Models have been proposed inspired by the so-called Grätzel cell, where the dye consists of a phosophonate derivatized $Ru(bpy)_3^{2+}$ that is attached at the surface of FTO transparent electrode. The cathode consists of Pt wire and is separated from the anode with a Nafion membrane, as shown schematically in Figure 10.5. A successful example of a related device containing a Ru-based WOC but with the pthotosensitizer directly attached at

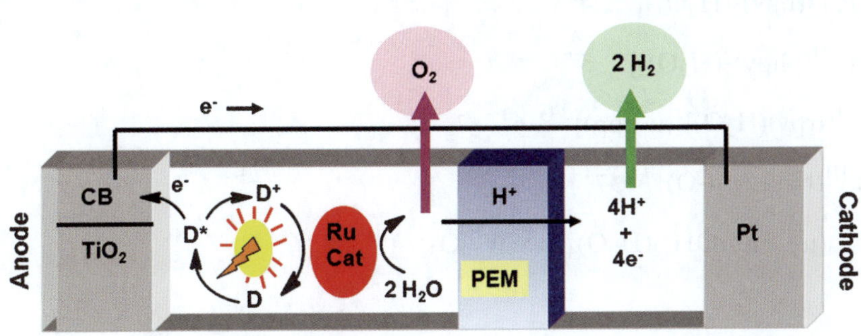

Figure 10.5 Water-splitting photochemical cell with three components: (1) a light-harvesting molecule, D, attached to a semiconductor photoanode, (2) a water-oxidation catalyst labeled Ru-Cat and (3) a Pt cathode where hydrogen is evolved. The anodic and cathodic compartments are physically separated by a proton exchange membrane, PEM.

the surface of an ITO electrode has been recently reported.[24b] Other devices have been reported using solid oxides of Ir[29] and Mn[30] as WOCs.

10.6 Conclusions

Over the last 5 years immense progress has been made at a molecular level in the field of water oxidation catalysis. Notably, there has been the discovery of a significant number of WOCs that have been prepared and tested; some of them with impressive efficiencies. The different pathways under which the catalytic cycles proceed is now emerging and even though there is still a long way to go, the basic bricks have already been fixed. The basic chemistry related to light induced WOC is now also starting to appear, and a huge amount of work lies ahead to understand and control the different reactions involved so that it can be carried out in an efficient manner. Finally, the building of integrated devices for water splitting is basically at an embryonic stage, but a successful one has already been reported.

The pace of progress in this field during the last five years have been impressive, with a degree of development that is really fascinating at all levels. Exciting new discoveries also lie ahead with a high degree of potential payback to our society.

10.7 Complex Numbering

$[Ru^{II}(tpm)(bpy)(H_2O)]^{2+}$, $\mathbf{1}^{2+}$

out-$[Ru^{II}(Hbpp)(trpy)(H_2O)]^{2+}$, $\mathbf{2}^{2+}$

$\{[Ru^{IV}(bdc)(pic)_2]_2(\mu\text{-HOHOH})\}^{3+}$, $\mathbf{3}^{3+}$

cis-$[Ru^{II}(bpy)_2(H_2O)_2]^{2+}$, $\mathbf{4}^{2+}$

cis-$[Ir^{III}(phpy)_2(H_2O)_2]^{+}$, $\mathbf{5}^{+}$

$\{[Ru^{II}(trpy)(H_2O)]_2(\mu\text{-bpp})\}^{3+}$, $\mathbf{6}^{3+}$

$\{[Ru^{III}(bpy)_2]_2(\mu\text{-O})\}^{4+}$, $\mathbf{7}^{4+}$

$[Ru^{IV}{}_4(\mu\text{-O})_4(\mu\text{-OH})_2(H_2O)_4(\gamma\text{-SiW}_{10}O_{36})_2]^{10-}$, $\mathbf{8}^{10-}$

Acknowledgments

Funding support for this research by SOLAR-H2 (EU 212508) and the MICINN (Consolider Ingenio 2010 (CSD2006-0003), CTQ2010-67918 and CTQ2010-21532-C02-02) is gratefully acknowledged.

References

1. *Oil Wars*, ed. M. Kaldor, T. L. Karl and Y. Said, Pluto Press, London, 2007.
2. *Photosystem II: The Light-DrivenWater: Plastoquinone Oxidoreductase*, ed. T. Wydrzynski and K. Satoh, Springer, Dordrecht, 2005.
3. (a) K. N. Ferreira, T. M. Iverson, K. Maghlaoui, J. Barber and S. Iwata, *Science*, 2004, **303**, 1831; (b) J. Yano, J. Kern, K. Sauer, M. J. Latimer, Y. Pushkar, J. Biesiadka, B. Loll, W. Saenger, J. Messinger, A. Zouni and V. K. Yachandra, *Science*, 2006, **314**, 821; (c) M. Haumann, P. Liebisch, C. Müller, M. Barra and M. Grabolle, *Science*, 2005, **310**, 1019; (d) E. M. Sproviero, J. A. Gascon, J. P. McEvoy, G. W. Brudvig and V. S. Batista, *J. Am. Chem. Soc.*, 2008, **130**, 3428.
4. (a) X. Sala, M. Rodriguez, I. Romero, L. Escriche and A. Llobet, *Angew. Chem. Int. Ed.*, 2009, **48**, 2842; (b) S. Romain, L. Vigara and A. Llobet, *Acc. Chem. Res.*, 2009, **42**, 1944.
5. (a) J. F. Hull, D. Balcells, J. D. Blakemore, C. D. Incarvito, O. Eisenstein, G. W. Brudvig and R. H. Crabtree, *J. Am. Chem. Soc.*, 2009, **131**, 8730; (b) N. D. McDaniel, F. J. Coughlin, L. L. Tinker and S. Bernhard, *J. Am. Chem. Soc.*, 2008, **130**, 210; (c) R. Lalrempuya, M. D. McDaniel, H. Müller-Bunz, S. Bernhard and M. Albrecht, *Angew. Chem., Int. Ed.*, 2010, **49**, 9765.
6. Q. Yin, J. M. Tan, C. Besson, Y. V. Geletii, D. G. Musaev, A. E. Kuznetsov, Z. Luo, K. I. Hardcastle and C. L. Hill, *Science*, 2010, **328**, 342.
7. W. C. Ellis, N. D. McDaniel, S. Bernhard and T. J. Collins, *J. Am. Chem. Soc.*, 2010, **132**, 10990.
8. J. Chen, J. W. Faller, R. H. Crabtree and G. W. Brudvig, *J. Am. Chem. Soc.*, 2004, **126**, 7345.
9. T. J. Meyer and M. H.-V. Huynh, *Inorg. Chem.*, 2003, **42**, 8140.
10. A. Llobet, P. Doppelt and T. J. Meyer, *Inorg. Chem.*, 1988, **27**, 514.
11. (a) A. Llobet, *Inorg. Chim. Acta*, 1994, **221**, 125; (b) M. Yoshida, S. Masaoka, J. Abe and K. Sakai, *Chem.–Asian. J.*, 2010, **5**, 2369.
12. M. H.-V. Huynh and T. J. Meyer, *Chem. Rev.*, 2007, **107**, 5004.
13. (a) S. Romain, L. Vigara and A. Llobet, *Acc. Chem. Res.*, 2009, **42**, 1944; (b) J. J. Concepcion, J. W. Jurss, J. L. Templeton and T. J. Meyer, *J. Am. Chem. Soc.*, 2008, **49**, 16462.
14. S. Roeser, P. Farràs, F. Bozoglian, M. Martínez-Belmonte, J. Benet-Buchholz and A. Llobet, *ChemSusChem.*, 2011, **4**, 197.
15. (a) D. J. Wasylenko, C. Ganesamoorthy, M. A. Henderson, B. D. Koivisto, H. D. Osthoff and C. P. Berlinguette, *J. Am. Chem. Soc.*, 2010, **132**, 16094; (b) J. J. Concepcion, M.-K. Tsai, J. T. Muckerman and T. J. Meyer, *J. Am. Chem. Soc.*, 2010, **132**, 1545.
16. J. T. Muckerman, D. E. Polyansky, T. Wada, K. Tanaka and E. Fujita, *Inorg. Chem.*, 2008, **47**, 1787.
17. J. F. Hull, D. Balcells, J. D. Blakemore, C. D. Incarvito, O. Eisenstein, G. W. Brudvig and R. H. Crabtree, *J. Am. Chem. Soc.*, 2010, **132**, 16017.

18. H.-W. Tseng, R. Zong, J. T. Muckerman and R. Thummel, *Inorg. Chem.*, 2008, **47**, 11763.
19. L. Duan, A. Fischer, Y. Xu and L. Sun, *J. Am. Chem. Soc.*, 2009, **131**, 10397.
20. X. Sala, M. Z. Ertem, L. Vigara, T. K. Todorova, W. Chen, R. C. Rocha, F. Aquilante, C. J. Cramer, L. Gagliardi and A. Llobet, *Angew. Chem., Int. Ed.*, 2010, **49**, 7745.
21. N. D. McDaniel, F. J. Coughlin, L. L. Tinker and S. Bernhard, *J. Am. Chem. Soc.*, 2008, **130**, 210.
22. (a) J. A. Gilbert and D. A. Eggleston, W. A. Murphy Jr. D. A. Geselowitz, S. W. Gersten, D. J. Hodgson and T. J. Meyer, *J. Am. Chem. Soc.*, 1985, **107**, 3855; (b) J. J. Concepcion, J. W. Jurss, J. L. Templeton and T. J. Meyer, *Proc. Natl. Acad. Sci. U. S. A.*, 2008, **105**, 17632; (c) C. Sens, I. Romero, M. Rodriguez, A. Llobet, T. Parella and J. Benet-Buchholz, *J. Am. Chem. Soc.*, 2004, **126**, 7798; (d) S. Romain, F. Bozoglian, X. Sala and A. Llobet, *J. Am. Chem. Soc.*, 2009, **131**, 2768; (e) F. Bozoglian, S. Romain, M. Z. Ertem, T. K. Todorova, C. Sens, J. Mola, M. Rodriguez, I. Romero, J. Benet-Buchholz, X. Fontrodona, C. J. Cramer, L. Gagliardi and A. Llobet, *J. Am. Chem. Soc.*, 2009, **131**, 15176; (f) J. Mola, C. Dinoi, X. Sala, M. Rodríguez, I. Romero, T. Parella, X. Fontrodona and A. Llobet, *Dalton Trans.*, 2011, **40**, 3640; (g) A. Sartorel, M. Carraro, G. Scorrano, R. De Zorzi, S. Geremia, N. D. McDaniel, S. Bernhard and M. Bonchio, *J. Am. Chem. Soc.*, 2008, **130**, 5006; (h) Y. V. Geletii, B. Botar, P. Kogerler, D. A. Hillesheim, D. G. Musaev and C. L. Hill, *Angew. Chem., Int. Ed.*, 2008, **47**, 3896.
23. R. Zong and R. P. Thummel, *J. Am. Chem. Soc.*, 2005, **127**, 12802.
24. (a) L. Duan, Y. Xu, P. Zhang, M. Wang and L. Sun, *Inorg. Chem.*, 2010, **49**, 209; (b) L. Li, L. Duan, Y. Xu, M. Gorlov, A. Hagfeld and L. Sun, *Chem. Commun.*, 2010, **46**, 7307.
25. L. Duan, Y. Xu, M. Gorlov, L. Tong, S. Andersson and L. Sun, *Chem.–Eur. J.*, 2010, **16**, 4659.
26. (a) M. Orlandi, R. Argazzi, A. Sartorel, M. Carraro, G. Scorraro, M. Bonchio and F. Scandola, *Chem. Commun.*, 2010, **46**, 3152; (b) F. Puntoriero, G. La Ganga, A. Sartorel, M. Carraro, G. Scorrano, M. Bonchio and S. Campagna, *Chem. Commun.*, 2010, **46**, 4725; (c) Y. V. Geletii, Z. Huang, Y. Hou, D. G. Musaev, T. Lian and C. L. Hill, *J. Am. Chem. Soc.*, 2009, **131**, 7522.
27. J. J. Concepcion, J. W. Jurss, P. G. Hoertz and T. J. Meyer, *Angew. Chem., Int. Ed.*, 2009, **48**, 9473.
28. (a) Z. Chen, J. J. Concepcion, J. W. Jurss and T. J. Meyer, *J. Am. Chem. Soc.*, 2009, **131**, 15580; (b) J. Mola, E. Mas-Marza, X. Sala, I. Romero, M. Rodríguez, C. Vinas and A. Llobet, Angew. Chem., *Int. Ed.*, 2008, **47**, 5830; (c) L. Francas, X. Sala, J. Benet-Buchholz, L. Escriche and A. Llobet, *ChemSusChem.*, 2009, **2**, 321; (d) F. M. Toma, A. Sartorel, M. Iurlo, M. Carraro, P. Parisse, C. Maccato, S. Rapino, B. R. Gonzalez, H. Amenitsch, T. D. Ros, L. Casalis, A. Goldoni, M. Marcaccio, G. Scorrano,

G. Scoles, F. Paolucci, M. Prato and M. Bonchio, *Nature Chem.*, 2010, **2**, 826; (e) A. Llobet, *Nature Chem.*, 2010, **2**, 804.

29. (a) W. J. Youngblood, S.-H. A. Lee, Y. Kobayashi, E. A. Hernandez-Pagan, P. G. Hoertz, T. A. Moore, A. L. Moore, D. Gust and T. E. Mallouk, *J. Am. Chem. Soc.*, 2009, **191**, 926; (b) W. J. Youngblood, S.-H. A. Lee, K. Maeda and T. E. Mallouk, *Acc. Chem. Res.*, 2009, **42**, 1966.
30. R. Brimblecombe, A. Koo, G. C. Dismukes, G. F. Swiegers and L. Spiccia, *J. Am. Chem. Soc.*, 2010, **132**, 2892.

CHAPTER 11

Structure and Function of Hydrogenase Enzymes

WOLFGANG LUBITZ,[a] HIDEAKI OGATA,[a] EDUARD REIJERSE[a] AND YOSHIKI HIGUCHI[b]

[a] Max-Planck-Institut für Bioanorganische Chemie, Stiftstr. 34–36, 45470 Mülheim an der Ruhr, Germany; [b] Department of Life Science, Graduate School of Life Science, University of Hyogo, Himeji Institute of Technology, 3-2-1 Koto Kamigori-cho, Ako-gun Hyogo, Japan

11.1 Introduction

Our current energy infrastructure is dominated by the use of fossil fuels, which are limited and lead to the emission of greenhouse gases. Hydrogen is considered the ideal non polluting energy carrier of the future[1–3] if produced in a sustainable way, *e.g.* from sunlight driven photolysis of water.[4,5] Existing large-scale processes to produce hydrogen are almost exclusively based on fossil resources, are relatively energy inefficient and release carbon dioxide. Nature has developed its own set of catalysts to produce hydrogen – or utilize it as an energy source – the hydrogenases.[1,6–8] The genesis of hydrogenase enzymes is dated back to the early stages of life on earth, when the atmosphere was hydrogen-rich and the existing primitive organisms relied on this molecule as energy source.

Hydrogenases are found in many microorganisms, in particular in archaea and bacteria, but also in some eukaryotes.[8] Three phylogenetically distinct classes are known today; these are the [NiFe], the [FeFe], and the [Fe] hydrogenases.[8–12] The first two classes contain binuclear metal cores[8–11] with an unusual ligand sphere as catalytic centers, whereas the third class harbors a

RSC Energy and Environment Series No. 5
Molecular Solar Fuels
Edited by Thomas J. Wydrzynski and Warwick Hillier

Published by the Royal Society of Chemistry, www.rsc.org

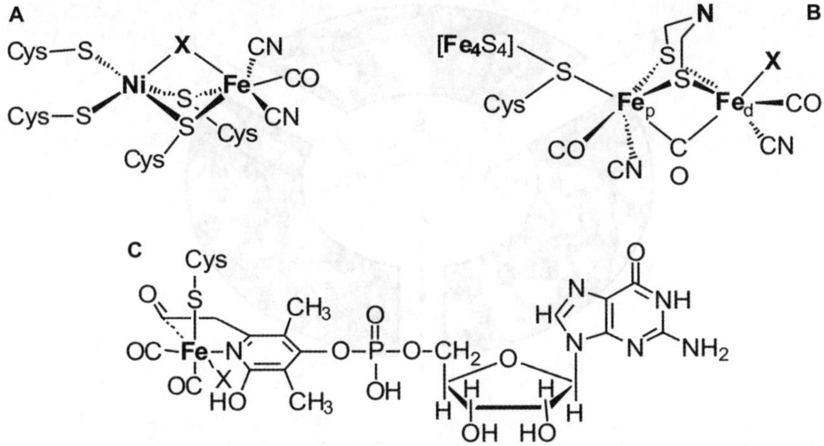

Figure 11.1 The active sites of hydrogenases. (A) [NiFe] hydrogenase; (B) [FeFe] hydrogenase; (C) [Fe] hydrogenase. The possible hydrogen binding sites are indicated by the label 'X'.

mononuclear iron next to a special organic cofactor.[12,13] The structures of the active metal sites of these enzymes are shown in Figure 11.1.

Nature's concept of hydrogen conversion – or the reverse process of hydrogen generation – at these highly optimized catalytic centers of hydrogenases is based on a heterolytic mechanism

$$H_2 \rightleftharpoons H^- + H^+ \rightleftharpoons 2H^+ + 2e^- \quad (1)$$

Thereby the acidity of H_2, which is extremely low, is dramatically increased by binding to a metal. Many of the currently used catalysts for anthropogenic utilization of hydrogen involve precious metals such as platinum, while Nature's catalysts are based on cheap and abundant first row transition metals. An important feature of the native systems is the very high turnover frequency of the hydrogenases, which often reach numbers close to $10^4\ s^{-1}$.[14] Moreover, hydrogenases from (hyper)thermophilic organisms[15–17] show a good stability at high temperatures (up to ~100 °C), while the pertinent problem of the high oxygen-sensitivity of many hydrogenases has been solved by species with specifically modified catalytic sites that show a greatly enhanced tolerance *versus* the deleterious effects of molecular oxygen.[17–20] It is therefore important to study and understand Nature's concept of hydrogenase structure and function when designing biomimetic (or bioinspired) systems for efficient hydrogen generation or utilization. This chapter will give insight into this class of enzymes.

11.2 Structure of Hydrogenases

Since 1995, several crystal structures of hydrogenases have been determined, five periplasmic [NiFe] hydrogenases from sulfate-reducing bacteria,[9,10]

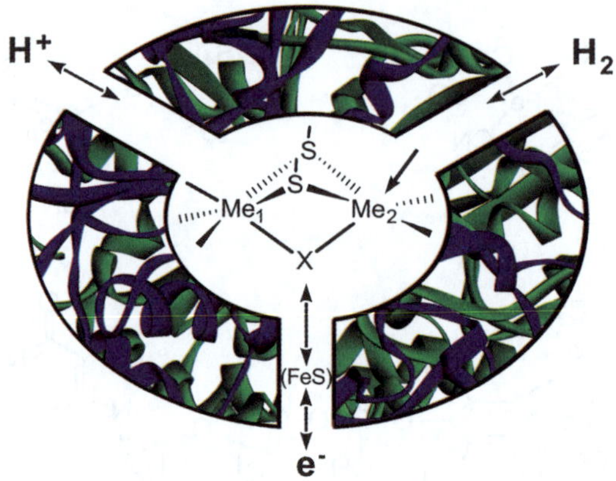

Figure 11.2 Schematic representation of a [NiFe]- or [FeFe]-hydrogenase showing the S-bridged bimetallic catalytic center, the H^+ and H_2 channels and the electron transport chain. The open coordination site at one metal is indicated by an arrow.

one from a photosynthetic sulfur bacterium,[21] two [FeFe] hydrogenases from the sulfate-reducing bacteria *Desulfovibrio desulfuricans* ATCC7757[9,22] and *Clostridium pasteurianum*[9,11] and two [Fe] hydrogenases from methanogenic archaea.[12,13] In this section the structure of each class of hydrogenases will be discussed.

In the case of the [NiFe]/[FeFe] hydrogenases the active site is deeply buried in the molecules. The hydrogenases possess three types of channels (gas-access, proton transfer and electron transfer) that can transfer the substrate and the educts between the active site and the molecular surface.[9,10,23] Xenon gas has been used to probe the hydrophobic cavities in the molecule.[10,23] The protons produced by the heterolytic cleavage of the molecular hydrogen at the active site are transferred to the surface of the protein *via* several hydrogen-bonding networks, and the electrons are transported from and to the active site *via* a chain of iron–sulfur clusters (*cf.* Figure 11.2). In the case of [Fe] hydrogenase the reaction mechanism is different and the molecular hydrogen reacts indirectly with the iron metal center in the presence of the substrate. A common feature of all hydrogenases is that they contain at least one Fe atom in the active site that carries non-protein ligands, *e.g.* carbon monoxide and cyanide.

11.2.1 [NiFe] Hydrogenase

So far, crystal structures of [NiFe] hydrogenase have been determined from sulfate-reducing bacteria (*Desulfovibrio*)[9,10] and the photosynthetic purple sulfur bacterium *Allochromatium vinosum*.[21] In the subclass [NiFeSe] hydrogenase the crystal structure is known from *Desulfomicrobium baculatum* and *D. vulgaris* Hildenborough.[24,25]

11.2.1.1 Overall Structure

The [NiFe] hydrogenase from *Desulfovibrio* sp. contains two subunits, a small (~38 kDa) and a large one (~65 kDa). The overall structure is shown in Figure 11.3A. The total size of the heterodimeric molecule is about 66 × 67 × 68 Å. The small subunit comprises two domains. The N-terminal domain of the small subunit has a flavodoxin-like fold, which coordinates one of the iron–sulfur clusters (proximal [4Fe4S] cluster). The medial and distal iron–sulfur clusters are located at the C-terminal domain in the small subunit.[9]

The large subunit comprises two αβ-domains, a helical domain and two less structured domains. In the large subunit, two metal centers are found in the crystal structure. One of them is the [NiFe] active site, which is deeply buried in the center of the molecule. The other metal center has been assigned to a magnesium ion located at the C-terminus of the large subunit.[10,27,28] The contact between the small and large subunit is made by hydrophobic interactions. The amino acid sequences are not conserved in these regions.[9]

11.2.1.2 The [NiFe] Active Site

The [NiFe] cluster is considered the catalytic site of the enzyme. Four thiolates from cysteine residues are coordinated to the nickel and two of them are bridging to the iron atom. The other cysteines are bound to the nickel in a terminal fashion. The iron has an octahedral coordination; three non-protein diatomic ligands are coordinated to the iron and two of the cysteines bridging to the nickel. Furthermore, a third bridging ligand (X) is coordinated between the iron and the nickel (Figure 11.3B), whose type depends on the redox state of the enzyme.

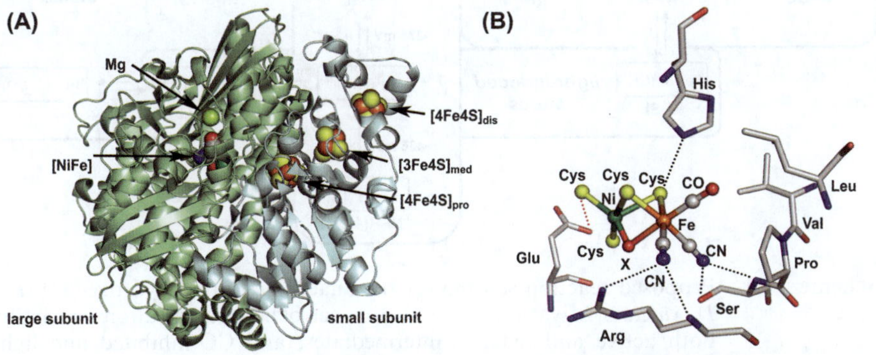

Figure 11.3 (A) Three-dimensional structure of the [NiFe] hydrogenase from *Desulfovibrio vulgaris* Miyazaki F. The small and large subunits are shown in light blue and green, respectively. (B) The [NiFe] active site and its surrounding amino acid residues. The possible hydrogen bonds are depicted with black dotted lines. The red dotted line represents the start of the putative proton transfer pathway. The third bridging ligand between Ni and Fe is denoted by 'X' [PDB 1WUI].[26]

The diatomic ligands coordinated to the iron could not be determined by X-ray crystallography. However, FTIR spectroscopy revealed that one CO and two CN^- ligands are present.[29] In the case of [NiFe] hydrogenase from *D. vulgaris* Miyazaki F, the possibility of an SO species for one of the diatomic ligands has been proposed based on the results from the crystallographic analysis[27] and pyrolysis mass spectroscopy.[30] The cyanides are well known as strong σ-donor ligands and can form hydrogen bonds to the surrounding amino acids, *i.e.* to arginine and serine, in the vicinity of the NiFe active site (Figure 11.3B). On the other hand, the CO ligand is not H-bonded and the CO stretching bands in the FTIR spectra are thus much more sensitive than those of the structurally stabilized CN^- ligands.[31,32]

In the catalytic cycle of the enzyme and during activation/inactivation and inhibition the [NiFe] hydrogenase passes through several (redox) states [33–35] (see Scheme 11.1). Crystal structures have been reported for several intermediates, *e.g.* oxidized (Ni-A, and Ni-B), reduced (Ni-C/Ni-R) and CO inhibited (Ni-SCO) states.[10]

In the oxidized states (Ni-A, Ni-B), an oxygen-based third bridging ligand is present between the nickel and iron. In the most oxidized state, Ni-A, the X-ray crystal structural data were interpreted to show a peroxide ligand (OOH^-)[26,36] in this site. In the Ni-B state, a monatomic oxygen species is bound in the

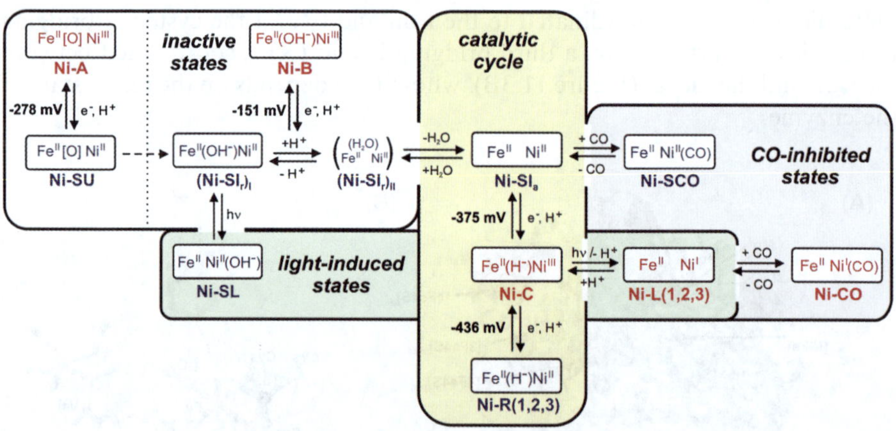

Scheme 11.1 Proposed reaction scheme for the anaerobic [NiFe] hydrogenase from *D. vulgaris* Miyazaki F based on electrochemical titrations, including both active and inactive intermediates, and CO-inhibited and light-induced states. The formal oxidation states of the Ni and the Fe are given, furthermore the type of bridging ligand that changes identity and its position (in some cases) in the cycle is indicated (see text for details). In Ni-A and Ni-SU the bridging oxygenic ligand presented by [O] has not been finally determined. The midpoint potential for the redox couples are given for pH 7.4, except for the Ni-A to Ni-SU transition, which is given for pH 8.2. The paramagnetic (EPR-active) states are in red, the EPR-silent states in blue.

bridge. The hydroperoxy or sulfur species in the Ni-A state at the bridging ligand site is still a matter of discussion.[33,35] In the structure of the [NiFe] hydrogenase of *A. vinosum*, a monoatomic bridging ligand has recently been reported.[21]

One electron reduction of the Ni-B state leads to the Ni-SI_r state. Since carbon monoxide does not bind to the active site it is suggested that the bridging ligand still remains in this state and prevents binding of the inhibitor to the nickel. Upon further reduction, the enzyme converts to the active form. The most oxidized active state is called Ni-SI_a. This state can be inhibited by carbon monoxide, which shows that the bridging ligand is absent, and the carbon monoxide can bind to the open coordination site of the nickel.[37] Direct evidence of the binding of carbon monoxide has been presented by the crystal structure of [NiFe] hydrogenase from *D. vulgaris* Miyazaki F.[38] The results clearly showed that the carbon monoxide is bound to the nickel in a bent form.

The Ni-C state is reached by one electron reduction of the Ni-SI_a state or upon reaction with molecular hydrogen, (the third bridging ligand X is removed). EPR and ENDOR studies[39–41] have shown that a hydride is bridging between the nickel and the iron in the Ni-C state, see below. The NiFe distance becomes slightly shorter than that in the oxidized Ni-A/Ni-B states.[42] When the enzyme in the Ni-C state is illuminated at cryogenic temperatures (below 110 K), Ni-C converts to the light induced state, Ni-L.[43–45] In this process, the hydride is removed from the bridging position and moved as a proton to a nearby base. Furthermore, carbon monoxide can bind to the Ni-L state, leading to the paramagnetic inhibited state termed Ni-CO.[37,46]

The most reduced state is called Ni-R. It is suggested that at least one or more hydride species are bound between the nickel and the iron in this state. Structural data are not available due to the difficulties in observing the hydrogen that has very little electron density.

11.2.1.3 The Iron–Sulfur Clusters and the Electron Transfer Pathway

The electrons produced as a result of the cleavage of the hydrogen at the active site are transferred from the NiFe center to the exogenous electron acceptor *via* iron–sulfur clusters.[9,10] The physiological partners for the hydrogenases are classified depending on the location of the enzyme. For example, the soluble periplasmic hydrogenases are hydrogen uptake enzymes and the electrons are transferred to cytochrome c_3 or a high-molecular weight cytochrome (HMC).[47] Three iron-sulfur clusters are located in an almost linear fashion in the small subunit within a distance of $\sim$13 Å. *Desulfovibrio* hydrogenases possess one [3Fe4S] and two [4Fe4S] clusters.[9,10] The [3Fe4S] cluster located in the middle of the iron–sulfur cluster chain is termed the medial cluster, while the two [4Fe4S] clusters are called the distal and proximal clusters, according to their distance from the [NiFe] active site (Figure 11.3A). In the case of the distal cluster, three cysteine residues and one histidine residue are coordinated to the iron atoms, and this histidine residue is exposed to the solvent accessible region.

Between the [NiFe] active site and the proximal cluster, two cysteine residues are located within distances of 10.7 Å; these are coordinated to the [NiFe] active site and the proximal cluster. The medial [3Fe4S] cluster has a higher potential than the [4Fe4S] clusters, which has been discussed by Dutton *et al.*[48]

11.2.1.4 C-Terminus Metal Site and Proton Transfer

A Mg ion was located in several crystal structures,[9,10] coordinated by His, Glu, an oxygen from the peptide backbone and three water molecules. It has been discussed that the Mg metal site is related to the proton transfer pathway between the [NiFe] active site and the molecular surface. The glutamate in the vicinity of the [NiFe] active site and the Mg site, several amino acid residues and some water molecules could form a hydrogen-bonding network. Recent theoretical calculations suggest that another possible proton transfer pathway could exist from the glutamate to the molecular surface *via* amino acids closely located to the proximal iron–sulfur cluster.[49] Site directed mutagenesis of the glutamate residue located near the terminal cysteine, which is bound to the nickel, showed that the H/D exchange rate and the activity were both decreased in the case of the mutant, indicating the importance of the glutamate residue (Figure 11.3B).[50]

11.2.1.5 Gas-access Channels

Since the active site is deeply buried in the center of the molecule, gas-access channels are necessary for catalytic function so that small molecules, like hydrogen – but also oxygen and carbon monoxide – can reach the active site. In the crystal structures of these enzymes this has been probed by xenon-binding.[10,23] The crystallographic results revealed that several channels to the molecular surface exist and their cavities are combined to only one channel in the vicinity of the active site. It has also been shown that the gas-access channel ended near the nickel in the active site. Site directed mutants reducing the size of the gas channel of the [NiFe] hydrogenase from *D. fructosovorans* have been characterized and showed that the gas diffusion rate was decreased.[51]

11.2.1.6 [NiFeSe] Hydrogenase

The subclass of [NiFeSe] hydrogenases contains a selenocysteine replacing one of the terminal cysteines coordinating the Ni in the active site. The ratio of H_2 production to H_2 oxidation activity of this enzyme (in the presence of 100% H_2) is higher than that for the other [NiFe] hydrogenases, which makes this enzyme particularly interesting for biohydrogen technologies. Furthermore, the [NiFeSe] hydrogenases exhibit a higher oxygen-tolerance and can be more easily activated than the [NiFe] hydrogenases. The oxidized state of [NiFeSe] hydrogenase is EPR-silent. The crystal structure of [NiFeSe] hydrogenase from *Desulfomicrobium baculatum* has been determined in the reduced state.[24] The overall structure of this [NiFeSe] hydrogenase is very similar to that of the other *Desulfovibrio* [NiFe] hydrogenases (Figure 11.3), except for the presence of three [4Fe4S] clusters in the electron transfer chain. Furthermore, an additional metal site has been found. Crystallographic analysis using the Fe-MAD (multiwavelength anomalous dispersion) method showed that an iron instead of a magnesium is coordinated to the C-terminal histidine residue. The iron has an octahedral coordination as found for the Mg in the *Desulfovibrio* [NiFe] hydrogenases.

In the active site, one selenocysteine was clearly assigned to the terminal coordinated position. The selenol is more acidic than the thiol (the pK_a of SeCys is 5.2, whereas that of the cysteine thiol is 8.0). The distance between the nickel and the selenium was 2.46 Å, which is in agreement with XAS studies. The temperature factor (indication of the oscillation or the vibration of the atoms around a specified position) exhibited a slightly higher value of 35.8 $Å^2$ as compared with the other atoms (14.7 $Å^2$). One glutamic acid (Glu18) could form a hydrogen-bond to the selenium atom. The distance between the nickel and the iron was determined to be 2.5 Å in the reduced state, which is comparable to the one found in *D. vulgaris* Miyazaki F [NiFe] hydrogenase in the reduced state (2.57 Å).

Interestingly, the recent analysis of the as-isolated [NiFeSe] hydrogenase form *D. vulgaris* Hildenborough showed no extra bridging ligand (**X**) in its oxidized state.[25] Furthermore, the selenocysteine ligand was modified and showed three different conformations including an S–SeCys conformer. The terminal bound cysteine was present in the doubly oxidized state (SO_2), with the sulfur atom still bound to the nickel atom.

11.2.2 [FeFe] Hydrogenase

Up to now, crystal structures of the periplasmic [FeFe] hydrogenase from *D. desulfuricans* (DdH)[22] and the cytoplasmic [FeFe] hydrogenase from the Gram-positive bacterium, *Clostridium* (*C.*) *pasteurianum*,[11] have been determined (Figure 11.4A and 11.4B).

11.2.2.1 Overall Structure

The [FeFe] hydrogenases comprise the di-iron active site (H-cluster) and additional iron–sulfur clusters (F-clusters). The H-cluster is considered to be the catalytic center of these hydrogenases. The [FeFe] hydrogenases differ in the number of iron–sulfur centers. The [FeFe] hydrogenase from algae contains only the H-cluster and no other iron–sulfur cluster.[52,53] The first crystal structure of a [FeFe] hydrogenase was determined from *C. pasteurianum* in 1998[11] in the oxidized state. The 'core' domain holds the active site (H-cluster), and the other domains contain the iron–sulfur clusters, namely three [4Fe4S] and one [2Fe2S] (Figure 11.4A). The *D. desulfuricans* hydrogenase[22] showed a similar structure but has only 2 additional [4Fe4S] clusters in a ferredoxin-like domain and comprises a second protein subunit with five helices and a loop region (Figure 11.4B).

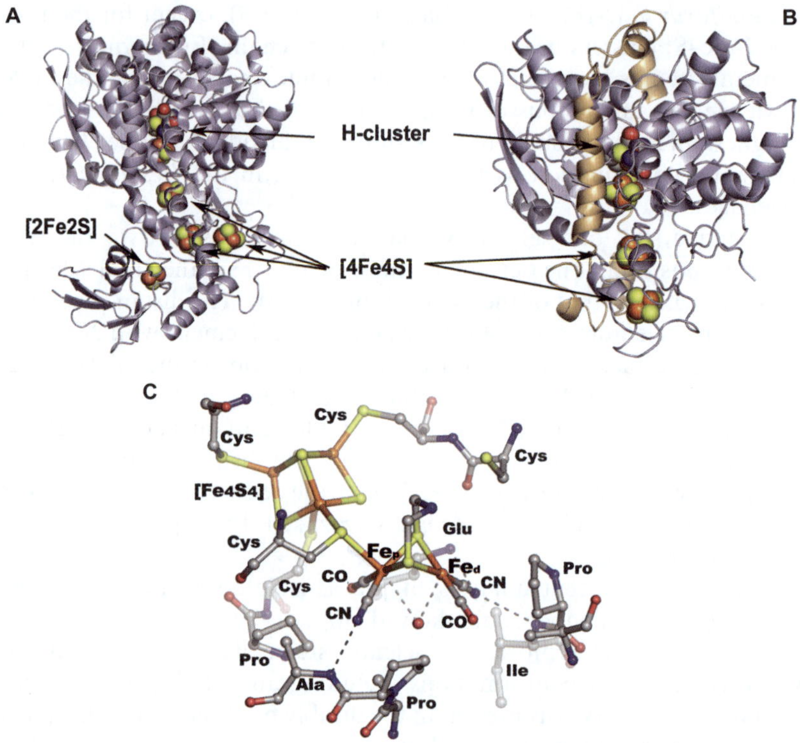

Figure 11.4 Three-dimensional representation of (A) the [FeFe] hydrogenase from *C. pasteurianum*[11] and (B) [FeFe] hydrogenase from *D. desulfuricans*.[22] (C) Ball and stick representation of the H-cluster and the surrounding amino acid residues based on the *D. desulfuricans* hydrogenase in the oxidized form. [PDB 1HFE].

11.2.2.2 H-Cluster

By comparison of the active sites of two [FeFe] hydrogenases (*C. pasteurianum* and *D. desulfuricans*) a consensus model of the active site could be built. The H-cluster consists of a dinuclear Fe–Fe center, $[2Fe]_H$, which is connected to one $[4Fe4S]_H$ cluster by a thiolate of a cysteine residue (Figure 11.4C). The $[2Fe]_H$ subcluster consists of two octahedrally ligated Fe atoms and five diatomic non-protein ligands, *i.e.* CO, CN^-. An assignment to CO and CN^- was achieved on the basis of the surrounding amino acid residues (Figure 11.4C). The CO ligands are located in a hydrophobic pocket, whereas the CN^- ligands form hydrogen bonds to amino acid residues in their surroundings. In the case of *D. desulfuricans* hydrogenase, amino acid residues provide the H-bonds to the CN^- ligands (Figure 11.4C). (It is noteworthy that the $[2Fe]_H$ subcluster is not bound to the protein – except for H-bonds to the CN^- ligands.) Furthermore, a dithiolate bridge exists between the two iron atoms. The dithiolate ligand was assigned in *D. desulfuricans* to a propanedithiolate, which was later revised to a dithiomethylamine. In the dithiolate bridge of *D. desulfuricans*, a nitrogen was confirmed as the apical atom that could be involved in the catalytic cycle of the enzyme.[54] The distance between the distal Fe and the [4Fe4S] cluster is *ca.* 4 Å, which is much shorter than the respective distance in the [NiFe] hydrogenase. This is reasonable for a fast electron transfer process and for the high turnover rate of H_2 production of the [FeFe] hydrogenases.

11.2.2.3 Catalytic Cycle of the [FeFe] Hydrogenase

As in the case of [NiFe] hydrogenase, the crystal structure of the CO-bound state of the [FeFe] hydrogenase from C. *pasteurianum* revealed that the extrinsic CO binds to the active site,[55] it is ligated to the coordination site in which a water molecule or OH^- is bound at the distal Fe atom in the H-cluster, where most probably the substrate hydrogen binding takes place.

The crystal structure of [FeFe] hydrogenase from *D. desulfuricans* has been reported in two different oxidation states (the oxidized and the reduced forms).[22,56] In the oxidized state, no H_2O/OH^- binding to the open coordination site of the distal Fe was observed. In the reduced form, the bridging CO ligand moved toward the distal Fe site.

11.2.2.4 Gas-access Channel

The calculation of the hydrophobic gas-access channel indicated the existence of a short tunnel from the molecular surface to the distal Fe of the active site. This is reasonable since the extrinsic CO binds to the distal Fe at the end of the tunnel. The gas-access channel is much shorter than that of the [NiFe] hydrogenase. This fact could contribute to the higher activity of the [FeFe] hydrogenases.

11.2.2.5 The Electron Transfer Chains

All iron–sulfur clusters are located within a distance of approximately 11 Å. Both *C. pasteurianum* and *D. desulfuricans* hydrogenase have two [4Fe4S] clusters at the ferredoxin-like domain. In *C. pasteurianum* additional iron–sulfur clusters exist, one [4Fe4S] and one [2Fe2S]. This distal [4Fe4S] cluster is coordinated by three cysteine residues and one histidine, as found in the distal [4Fe4S] cluster of the [NiFe] hydrogenases. The electrons provided by the heterolytic cleavage of H_2 in the H-cluster are transferred *via* the [FeS] clusters to the molecular surface and to the electron acceptor, *e.g.* cytochrome c_3, which further transfers the electrons to be used in subsequent reactions of the metabolic pathway.

11.2.3 [Fe] Hydrogenases

11.2.3.1 Overall Structure

[Fe] hydrogenases contain an iron-cofactor (Figure 11.1C) and no iron–sulfur clusters. The enzyme catalyses the reversible heterolytic cleavage of dihydrogen, H^+/H_2-exchange, and *para*/*ortho*-H_2 conversion, as found in the [NiFe]/[FeFe] hydrogenases, but the catalytic reaction mechanism is very different. The activity of this enzyme is dependent on the substrate, methenyl-H_4MPT^+. [Fe] hydrogenases catalyze the reaction of methenyltetrahydromethanopterin (methenyl-H_4MPT^+) with H_2 to methylene-H_4MPT and a proton by transferring a hydride ion to the *proR* position of the C14a carbon of methylene-H_4MPT.

Recently the crystal structures of the [Fe] hydrogenase from *Methanocaldococcus jannaschii* and *Methanopyrus kandleri* have been determined.[12,57,58] The [Fe] hydrogenase is a homodimeric protein with dimensions of about $90 \times 50 \times 40$ Å (Figure 11.5).[59] The crystal structure revealed that the [Fe] hydrogenase comprises two peripheral units bound to the central globular unit in a linear manner. The N-terminal domain (each peripheral unit) is composed of an α/β structure which belongs to the Rossmann[60] fold protein family. Each C-terminal region, which composes the central globular unit, contains four α-helices. These subunits form an intertwined intersubunit helix bundle. The crystal structures showed that the *M. jannaschii* [Fe] hydrogenase is in a closed form, while the *M. kandleri* enzyme is in an open state. The rotation angle between the peripheral units of these two forms is 35°. The cleft between the peripheral unit and the central globular unit has a depth of 13 Å.

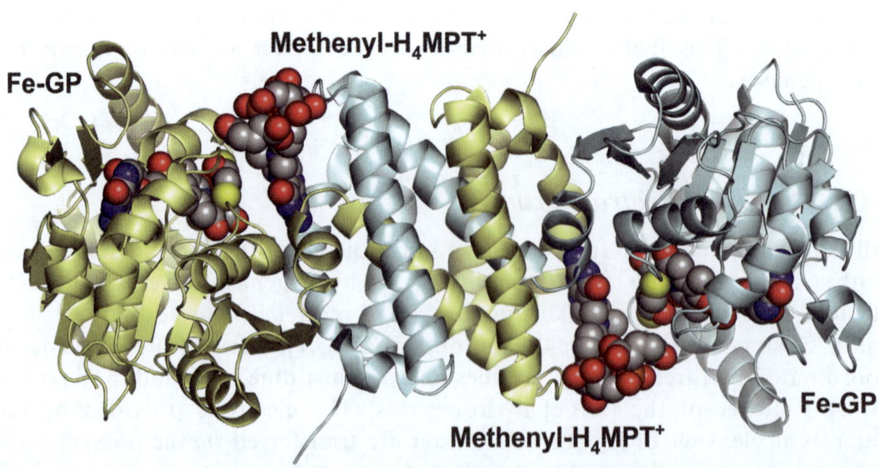

Figure 11.5 The crystal structure of [Fe] hydrogenase from *M. jannaschii* with the substrate bound, [PDB 3F46].[57]

11.2.3.2 The Mono-iron Active Site

The crystallographic analysis and earlier spectroscopic studies revealed a mono-iron active site of [Fe] hydrogenases (Figure 11.1C). The enzyme harbors a unique iron-cofactor (iron-guanylylpyrodinol, Fe-GP). The iron is five-coordinated; ligands are a thiolate from a cysteine, two CO molecules, a nitrogen (sp^2 hybridized) and an acyl-carbon from the pyridinol of the GP. The open coordination position is supposed to be the site of dihydrogen activation. The putative substrate-binding site is assigned to the *trans* position of the pyridinol nitrogen. So far no direct evidence exists from the crystallographic analysis for hydrogen or inhibitor (carbon monoxide) binding states. However, a complex with extrinsic cyanide has been determined at high resolution, in which the cyanide is bound to the sixth ligation site.[12]

The modeled closed form indicates that the [Fe] hydrogenase also possesses a hydrophobic gas-access channel. It is located between the peripheral and the central units. The channel is connected to the iron site and its diameter is approximately 4 Å.

11.3 Functional Spectroscopy of Hydrogenases

To understand catalytic processes of metalloenzymes, knowledge of the X-ray crystallographic structure alone is not sufficient to unravel the complex reaction mechanism, but additional information from spectroscopic methods is required. The redox chemistry of hydrogenases shows several intermediate states that are involved in the enzyme's activation/inactivation, inhibition, light-sensitivity and in the actual catalytic cycle. These can often be trapped or stabilized for sufficiently long times to allow their spectroscopic characterization. If these states are paramagnetic, EPR techniques can be employed yielding detailed information about the electronic and also the geometrical structure.[33] Both EPR-active and -inactive states are amenable to infrared spectroscopy, in which the CO and CN^- ligands at the Fe serve as structural probes.[29,34,61,62] The combination with electrochemical techniques allows the pH-dependent determination of the redox transitions.[34] Together with quantum chemical calculations, these and other techniques, like Mössbauer[63] or XAS[64,65] spectroscopies, are helpful in elucidating the details of the reaction mechanism of the hydrogenases.[66–69]

11.3.1 [NiFe] Hydrogenase

11.3.1.1 Characterization by Spectroscopic Methods

The active site of the enzyme is shown in Figure 11.1A. Whereas the Fe is 6-coordinated, the Ni carries only 5 ligands (or less). The nickel is considered to be the business-end of the enzyme, where the substrate H_2 is attached, the redox chemistry takes place and the catalytic process is initiated. This is corroborated by the fact that the hydrophobic gas channel ends at the nickel (Figure 11.2) and also the CO inhibitor binds terminally to this metal.[38] EPR spectra obtained from [NiFe] hydrogenases are typical for Ni-centered species (Figure 11.6A), with the single unpaired electron ($S = 1/2$) in the d_{z^2} orbital.[70] Based on EPR experiments, the different intermediate states of the [NiFe] hydrogenase have been called Ni-A, Ni-B (oxidized states), Ni-S (EPR-silent

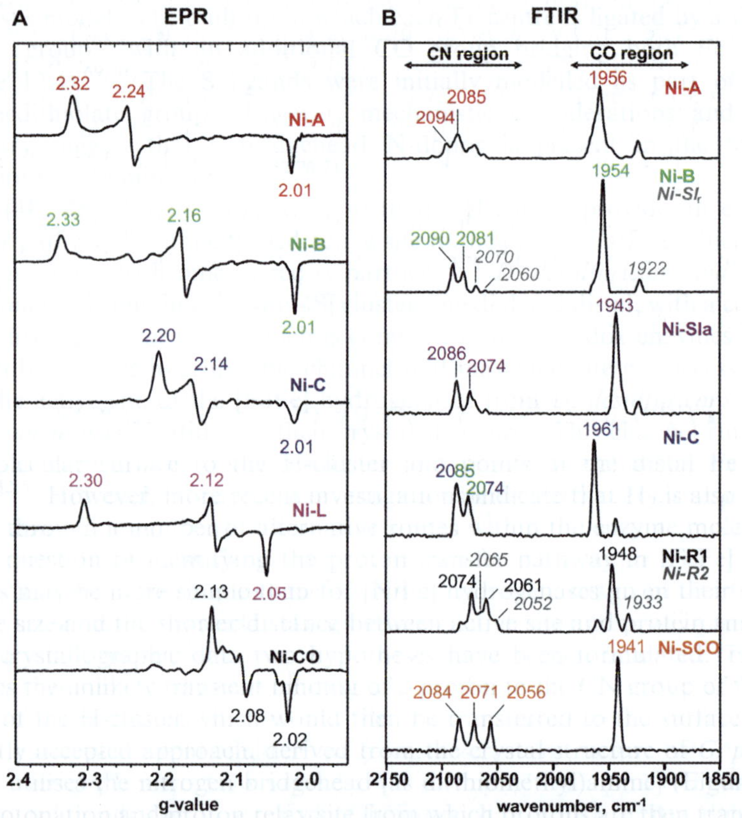

Figure 11.6 (A) CW EPR spectra (X-band) of the paramagnetic states of the anaerobic [NiFe] hydrogenase from *D. vulgaris* Miyazaki F. The 'unready' inactive Ni-A, the 'ready' active Ni-B state, the hydride carrying Ni-C state, the light-induced Ni-L state (Ni-L2 according to accepted nomenclature) and the CO-inhibited paramagnetic Ni-CO complex are shown (refer to text for details). In the spectrum of Ni-CO, lines from an additional redox state are present, usually observed upon different treatments of the *D. vulgaris* Miyazaki F enzyme. The principal *g* tensor components g_x, g_y, g_z (from left) are given at the respective positions. See refs. 33 and 35 for details. (B) FTIR spectra of the different intermediate states of the [NiFe] hydrogenase from *D. vulgaris* Miyazaki F. Each state is characterized by 1 CO and 2 coupled CN^- vibrational bands for which the vibrational frequencies are given. Note that in most cases, mixtures of states are obtained with one dominant species (redox titrations, pH 7.4). For further details see refs. 35 and 74.

states), Ni-C and Ni-R (reduced catalytically active states) and Ni-L (light induced states).

EPR on ^{61}Ni (nuclear spin $I = 3/2$) labeled samples[71,72] proved the existence of paramagnetic Ni states in the enzyme and showed that there is large spin density at the nickel ($\geq 50\%$).[73]

EPR and related techniques, such as electron nuclear double resonance (ENDOR) can thus be used together with isotope labeling to probe the Ni as well as its surrounding ligand sphere.[33]

The application of magnetic resonance techniques is of particular importance in hydrogenase research, since these methods are very sensitive to hydrogen nuclei that cannot be detected easily by X-ray crystallography; the latter technique is practically blind for the substrate and product of this enzyme.

The existence of Fe in the active site was only discovered with the advent of the first X-ray crystallographic structure. Due to the cyanide and CO ligands attached to the Fe, this metal is kept in a low spin and low oxidation state ($S = 0$, Fe^{2+}). It is thus not paramagnetic and carries very little spin density in all states. EPR and Mössbauer studies on ^{57}Fe labeled [NiFe] hydrogenase corroborate these findings.[35,75,76] The iron appears to play essentially only a structural role in this enzyme, whereas much of the important chemistry is happening at the nickel and in the 'bridge' between Ni and Fe (see below).

Changes in the charge density of the catalytic center, (*i.e.* mainly at the iron) during the activation, inhibition and the enzymatic cycle, can be detected in a sensitive way *via* changes in the vibrational stretching frequencies of the CN^- and CO ligands by FTIR spectroscopy (see Figure 11.6B). Fourier transform infrared spectroscopy monitors the IR frequencies corresponding to the intrinsic CO, CN^- ligands at the Fe that are very sensitive probes for electronic changes in the active site of hydrogenases. Each redox state is characterized by a set of three bands, one of high intensity in the range 1900–1970 cm^{-1} corresponding to the π-back bonded CO and two less intense bands in the range 2040–2100 cm^{-1} corresponding to the vibrationally coupled CN^- ligands. FTIR spectra of several states of a standard catalytic [NiFe] hydrogenase are collected in Figure 11.6B.

Whereas EPR can only be employed for paramagnetic states, FTIR can detect all intermediates. In addition, the use of an electrochemical cell offers the possibility to shuffle between the various redox intermediate in a well defined way under a variety of conditions. Scheme 11.1 shows an overview of the intermediate states and their interconversions, as observed for the standard [NiFe] hydrogenase of *D. vulgaris* Miyazaki F.

11.3.1.2 The Intermediate States of [NiFe] Hydrogenase

The most oxidized states of the [NiFe] hydrogenase are called Ni-A and Ni-B. Ni-B, also called "ready", can be activated within seconds but Ni-A ("unready") requires prolonged activation times. In standard (anaerobic) hydrogenases a mixture of these states is often obtained in the as-isolated enzyme, whereas in oxygen-tolerant (aerobic) hydrogenases, the Ni-A state is not observed in the intact enzyme, *i.e.* only Ni-B is present. Ni-A has thus been considered the 'oxygen-inhibited' state. Whereas the IR stretching vibrations are rather similar (Figure 11.6B), the two states are clearly distinguished by their different EPR spectra (Figure 11.6A).

It has been proposed that the difference between Ni-A and Ni-B is based on a different bridging ligand X, which is oxygen-based according to ^{17}O labeling experiments.[46,77] EPR/ENDOR experiments showed that the oxygenic ligand in Ni-B is a hydroxide (μ-OH^-) bridging the Ni and the Fe.[78] In the case of the Ni-A state,[26] the two metal centers Ni and Fe are found to have a slightly longer distance (2.8 Å) compared to Ni-B (2.7 Å). To date, the chemical identity of the oxo-based ligand in Ni-A has not been determined with confidence. X-ray structural studies[26,36] have postulated a di-oxo or an SO species, whereas the recent crystallographic structure of the hydrogenase from *A. vinosum* showed only a monomeric (oxygen) species in the bridge.[21] In addition modifications of the cysteines have been observed near the NiFe center (S=O or SOH)[26,36] both for Ni-A and Ni-B.

In the as-isolated state of anaerobic [NiFe] hydrogenases, the iron–sulfur clusters are in their most oxidized forms; $[3Fe4S]^{1+}$ ($S=1/2$) and $[4Fe4S]^{2+}$ ($S=0$),[79,80] and therefore at low temperatures the EPR spectrum of an $[3Fe4S]^{1+}$ cluster can be observed. However, for aerobic hydrogenases the spectra of the FeS clusters in the as-isolated enzyme are more complex; this has been ascribed to the fact, that the proximal cluster is paramagnetic and couples both to the Ni^{3+} and the $[3Fe4S]^{1+}$ cluster.[18,81]

One-electron reduction of Ni-A and Ni-B leads to the "EPR-silent" states Ni–SU (silent unready) and Ni–SI_r (silent ready). These could be either diamagnetic ($S=0$, low spin) or paramagnetic ($S=1$, high spin). They are catalytically inactive,[82–84] but have distinct FTIR spectroscopic signatures allowing their identification. The Ni-SI_r state exists in two isoelectronic acid–base forms, $(Ni\text{-}SI_r)_I$ and $(Ni\text{-}SI_r)_{II}$, which most likely differ in protonation of the OH^- bridge.[31,83]

The terminal Cys546[38] that is located close to the suggested proton transfer pathway is proposed to act as proton acceptor.[50,85] This cysteine is replaced by a selenocysteine in the subgroup of [NiFeSe] hydrogenases.[24,86]

Further activation of the enzyme leads to the "EPR-silent" Ni-SI_a state, where the oxygenic species is no longer bound.[35] Ni-SI_a has an open bridge; *i.e.* nickel is only 4- and iron 5-coordinated. The Ni-Fe distance is elongated to 2.8 Å according to theoretical studies.[87] Ni-SI_a is considered a central intermediate in the catalytic cycle (Scheme 11.1). A high-spin Ni^{2+} species with a distorted tetrahedral geometry is in agreement with the one-electron reduced Ni-$SI_{r/a}$ states according to recent calculations[67,88] and Ni L-edge X-ray circular magnetic dichroism (XCMD) measurements.[89]

Further reduction of the hydrogenase molecules in the Ni-SI_a state yields the paramagnetic (EPR-active) catalytic Ni-C intermediate. The crystal structure of the reduced enzyme from *D. vulgaris* Miyazaki F (probably a mixture of Ni-C/Ni-R) did not show any additional electron density at the position of the third bridging ligand, but the two metals are now at a closer distance of 2.6 Å.[42] The Ni-C state has been characterized by FTIR and EPR spectroscopy (Figure 11.6A,B). The EPR spectrum of Ni-C is highly characteristic and almost identical in all [NiFe] hydrogenases, indicating a structurally well-defined species. The theoretical analysis of the single crystal EPR spectra showed a Ni^{3+} ($S=\frac{1}{2}$) with a $d^1_{z^2}$ ground state and suggested that the bridging ligand is a hydride,[40] which is derived from the heterolytic splitting of the substrate dihydrogen. Pulse EPR (^1H and ^2H HYSCORE) and ENDOR experiments, first on the regulatory hydrogenase from *Ralstonia (R.) eutropha*[39] and later on *D. vulgaris* Miyazaki F[41] using H/D exchange, uniquely identified the hydride ligand (H^-, D^-) in the bridging position between Ni and Fe (see Figure 11.7).

Its location was estimated from the analysis of the hyperfine tensor and a point-dipole approximation, and is in good agreement with DFT calculations on geometry optimized models of Ni-C.[90] The Q-band ^2H ENDOR study on *D. gigas*[91] showed the presence of two types of exchangeable protons in the Ni-C state; the second one having a smaller hyperfine coupling. This second coupling was proposed to correspond to a water molecule near the active site, however it has not yet

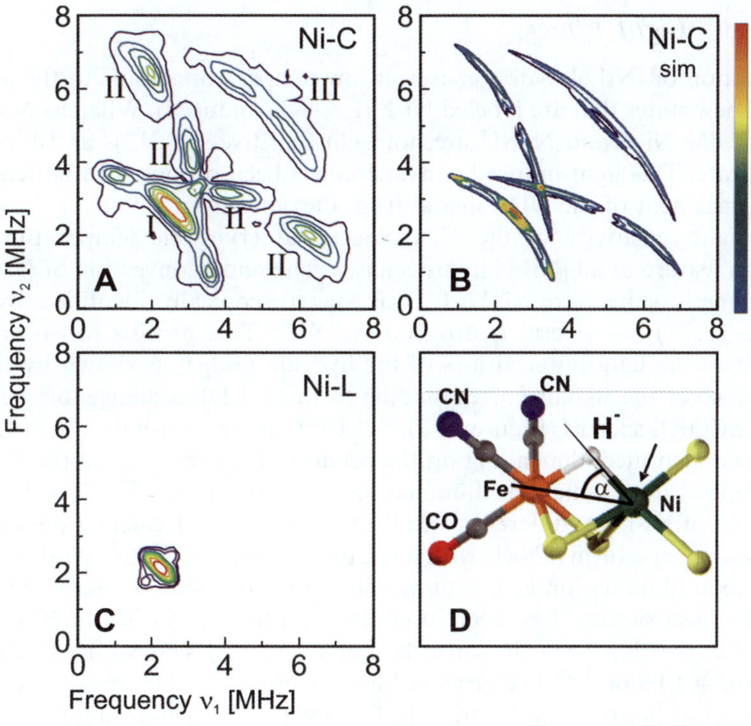

Figure 11.7 ^2H HYSCORE (X-band at position g_y) of the active Ni-C and Ni-L states of regulatory [NiFe] hydrogenase from *Ralstonia eutropha*: (A) Ni-C, experiment; (B) Ni-C, simulation; (C) Ni-L after illumination of Ni-C, experiment; (D) structural model of the Ni-C state with a bridging hydride, based on the EPR, ENDOR and HYSCORE data.[39]

been uniquely assigned. This is interesting in view of proposed structural schemes of Ni-C involving more than one hydrogenic species in the active site.[92,93]

Since the proximal [4Fe4S] cluster also becomes reduced in this process, the paramagnetic Ni-C is magnetically coupled to the $[4Fe4S]^{1+}$ state, which leads to a splitting of the EPR signals at low temperatures. Information about distances and relative orientations of the spin centers have been estimated from this interaction.[94]

The most reduced state of the enzyme, Ni-R, is reached by one-electron reduction of Ni-C. The FTIR spectrum of the Ni-R state in many standard hydrogenases, is a mixture of two or more isoelectronic forms that differ in their degree of protonation.[34] During catalytic turnover, Ni-C is the only state in equilibrium with Ni-R under H_2,[82,84,95] suggesting that the hydride also remains bound in this state.[66,69,92,96] Ni-R is EPR-silent, with a divalent nickel proposed by theoretical calculations to be in a high spin state (Ni(II), $S = 1$). All geometry optimized models are in line with a hydride bridge and a NiFe distance of 2.6 Å, fully compatible with X-ray data.[24,42]

11.3.1.3 Light Effects

Illumination of [NiFe] hydrogenases at low temperatures leads to the appearance of new states that are labeled Ni-L (L = light-induced). Whereas Ni-A and in particular Ni-B and Ni-SU are not light-sensitive, Ni-SI_r is and forms the Ni-SL state. This light-induced conversion has been proposed to be related to the displacement of the OH^- ligand from the active site.[31]

The light sensitivity of the Ni-C species at cryogenic temperatures is a common feature of all [NiFe] hydrogenases. The photoconversion of Ni-C to a paramagnetic state, termed Ni-L, was spectroscopically identified by EPR (Figure 11.6A) for several hydrogenases.[41,43–45] This process has been associated with the light-induced loss of the hydride bridge, as shown by the disappearance of the signals corresponding to this (H/D) exchangeable ligand in the HYSCORE spectra (Figure 11.7).[39,41] Up to three light-induced Ni-L states have been identified, depending on the origin of the enzyme, and the duration and temperature at which the illumination was performed.[39,45,97] The electronic properties of Ni-L are different from those of Ni-C; DFT calculations on Ni-L models support a formal Ni(I) with a vacant bridge, proposing that the hydride is lost upon illumination as a proton that is transferred to a nearby base.[90]

FTIR spectroscopy has been used to monitor the Ni-L to Ni-C back-conversion that has been shown to be a first-order process with an activation barrier of 46 kJ mol^{-1}.[98] The primary kinetic isotope effect on the re-association rate constants lies between 5 and 7 (KIE = 5.9 at 155 K) and demonstrates that transfer of a proton is the rate-limiting step.

The Ni-R state in *D. vulgaris* Miyazaki F has also been shown to be light-sensitive.[99] The photo-induced transition is suggested to correspond to the displacement of a ligand bound to the active site, which on the basis of proposed structural schemes is most likely a hydride ligand.

11.3.1.4 Enzyme Inhibition

The [NiFe] hydrogenase can be inhibited by various molecules that block access or turnover of the enzyme or change the electronic or geometrical structure of the cofactors. An important effect in these highly active enzymes is product inhibition, which is related to the efficiency of the transport of H_2 (and/or H^+/e^-) away from the active site.[100] In sulfate-reducing bacteria, inhibition can also occur by the product H_2S (or HS^-, S^{2-}) that binds at the [NiFe] site.[101] Interaction with NO seems to irreversibly damage the [NiFe] hydrogenase.[102]

Carbon monoxide inhibits enzyme activity.[103,104] FTIR and EPR experiments with ^{13}CO labelled gas, as well as X-ray crystallographic studies, have demonstrated that the exogenous CO ligand binds to the nickel ion at the active site.[38,46] This indicates that the nickel is indeed the site of hydrogen attachment and conversion. Two electronically distinct CO-inhibited states have been observed: an EPR-silent (Ni-SCO) and a paramagnetic (Ni-CO) species (see Scheme 11.1).

Ni-SCO is formed only from Ni–SI_a, *i.e.* from the four-coordinated Ni(II) electronic state, lacking the bridging ligand X. The paramagnetic CO inhibited state can be formed by treating a solution containing Ni-C with carbon monoxide followed by illumination to generate Ni-L, in which the nickel ion is in the formal Ni(I) state (monovalent).[105,106] Dark adaption leads predominatly to the generation of Ni-CO as CO rebinds faster than the proton at the active site.

It is interesting to note that inhibition by CO is only possible for specific states of the [NiFe] hydrogenase reaction cycle. CO does not react with the Fe(II) and only with nickel in a low oxidation state (Ni(II), Ni(I)) yielding a diamagnetic and a paramagnetic product, respectively. Both species Ni-CO and Ni-SCO are light sensitive and reversibly lose the extrinsic CO ligand at low temperatures. No CO bridge is formed in [NiFe] hydrogenase, as has been found in the [FeFe] enzyme.

The inhibition of the enzyme by molecular oxygen has been mentioned above. In contrast to many of the strictly anaerobic [FeFe] and [Fe] hydrogenases, which are irreversibly damaged or destroyed by contact with O_2, the [NiFe] hydrogenases have developed a mechanism to cope with the deleterious effects of O_2. The effect of oxygen on the catalytic process of hydrogen conversion performed by these enzymes can best be followed by protein film electrochemistry (PFE).[100,107] It has been proposed by Vincent and Armstrong that [NiFe] hydrogenase can act as an 'oxygenase', *i.e.* reduce the oxygen to two water molecules, one retained as OH^- forming the bridging ligand in Ni-B. The process depends on the potential, *i.e.* availability of reducing equivalents (electrons) and can, when electrons are not abundant, lead to peroxide, thus creating a powerful and potentially damaging oxidant to attack the active site. This produces the Ni-A state (with a proposed OOH^- bridging ligand) and other products, *e.g.* oxidized cysteines (S=O, S–OH groups).[107]

Ni-A (and perhaps also Ni-B) can be considered the oxygen-inhibited states of the enzyme. It is interesting that there are a few [NiFe] hydrogenases, *e.g.* from Knallgas[108] or hyperthermophilic[16,17] bacteria which are oxygen-tolerant.

These do not show the Ni-A state. Unfortunately, X-ray crystallographic structures of these hydrogenases have not yet been reported.

Various reasons have been discussed for the enhanced oxygen-tolerance of these enzymes, including a narrower gas channel[109] and the existence of additional (diatomic) ligands at the catalytic site.[110] More recently, it has been found that *R. eutropha* and *A. aeolicus* hydrogenases all operate at significantly higher (more positive) redox potentials.[17,19] Redox titrations of the FeS clusters[81] reveal an additional high potential iron–sulfur species. In EPR, the Ni-B state shows a magnetic coupling to the proximal FeS cluster, which is ligated by two additional cysteines.[18] Replacement of the respective cysteines in *R. eutropha* MBH leads to a decreased oxygen-tolerance. Although a final interpretation leading to a detailed mechanism is still under discussion[18,81] it is clear that a modification of the clusters in the electron transport chain are likely to be a major factor for the oxygen-tolerance of these enzymes.

11.3.1.5 Catalytic Reaction Mechanism

In Scheme 11.1, the different states found for the anaerobic hydrogenase are presented together with the midpoint potentials for the interconversion of these states. This scheme also includes the enzyme activation, the CO-inhibition and the light-induced states, as discussed above. A detailed description of this scheme is found in refs. 10 and 35. The actual catalytic cycle encompasses only 3 states: Ni-SI_a, Ni-C and Ni-R, which are interconverted by one electron/one proton equilibria (Scheme 11.2).[82,84,105] In the catalytic process, the approaching H_2 is attached to the Ni and the bond is polarized, followed by base assisted heterolytic cleavage of the H_2 molecule, leading to a bridging hydride species. One of the candidates for acting as a base is the terminal cysteine (Cys530 in *D. gigas*). Alternatively, a water molecule bound to the iron has been proposed to act as base.[96] Concomitant electron transfer to the proximal FeS cluster then leads to the Ni-C state, which has been shown to carry the hydride bridge between Ni(III) and Fe(II).[39–41] Further reduction of Ni-C by another H_2 molecule leads to the Ni-R state which still carries the hydride.[92] A proposal for the mechanistic details of the reaction with H_2 in the actual catalytic cycle has recently been made, which is based on earlier experiments.[82] The final step would then be the release of another proton and electron to give rise to the initial Ni-SI_a state with an open bridge, ready for the next turnover. This last step could be preceded by a relocation of the proton from the bridge to the terminal cysteine, creating a state that resembles Ni-L.[111]

Two proposed mechanisms are shown in Scheme 11.2, which are based on spectroscopic results and DFT calculations. In both cycles, H_2 is initially attached to Ni. In A, the H_2 is polarized at a Ni(III), and the proton is accepted by water bound to Fe. Release of H_3O^+ leads to the hydride carrying Ni-C state, which is converted by a second proton and electron to Ni-R. Here the proton is bound either by Ni itself or by the terminal cysteine. Release of H_2 and introduction of H_2O completes the cycle. In an alternative mechanism, cysteine acts as base and the hydride is formed in the Ni-R state before electron

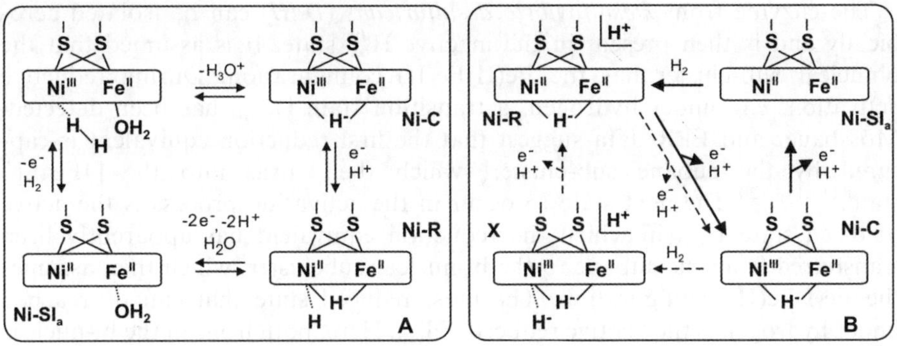

Scheme 11.2 Two models showing the catalytic cycle of [NiFe] hydrogenase based on DFT calculations. (A) Model by Stein and Lubitz.[96] (B) Model proposed by Fontecilla-Camps *et al.*[66] for the catalytic cycle of [NiFe] hydrogenase (see text).

release leads to Ni-C, which is then converted back to the initial Ni-SI_a, state.[69,112,113]

In another model (B), proposed by Fontecilla-Camps *et al.*,[66] the Ni-R state is formed in the first step from Ni-SI_a. In the catalytic cycle, the hydride remains in the bridge between Ni and Fe and acts as a base for the incoming H_2. In this mechanism the hydrogenase cycles between the Ni-R, Ni-C and a transient Ni-X state, the latter has a second hydride bound to the Ni (see Scheme 11.2 B). This model assumes that the two protons are released in two subsequent oxidation steps. Here, H_2 splitting is described, but it is assumed that the reversed reaction (H_2 production) will occur through the same reverse pathway.

11.3.2 [FeFe] Hydrogenase

The active site of the [FeFe] hydrogenase (the so-called "hydrogen-converting cluster" or H-cluster) is shown in Figure 11.1B. The 5-coordinate distal iron of the 2Fe subcluster, Fe_d, is believed to be the site where H_2 initially binds. The hydrophobic (gas) channel (see Section 11.2.2) ends at this position and the inhibiting CO is attached at this site. Since only the H_{ox} state (and H_{ox}-CO) is paramagnetic, less information on the electronic structure is available from EPR on the [FeFe] hydrogenase as compared to [NiFe] hydrogenase. In particular, the hydride-carrying species (H_{red}) cannot be characterized by EPR techniques. However, FTIR and Mössbauer spectroscopy have given important information on all intermediate states of the H-cluster.[114–118]

11.3.2.1 Functional Redox States of the H-cluster

Several redox and catalytic intermediate states have been identified or postulated for the H-cluster in [FeFe] hydrogenases.[33] In Scheme 11.3, these intermediates and the relationships between them are depicted.

The enzyme from *Desulfovibrio desulfuricans* (*DdH*) can be isolated aerobically and is then present in the inactive H_{ox}^{air} state. It is assumed that the binuclear sub-cluster has the Fe(II)Fe(II) configuration. During reductive activation, *e.g.* under hydrogen, a transition state H_{trans} has been detected. Mössbauer and EPR data suggest that the first reduction equivalent is captured by the cubane sub-cluster, which then turns into the $[4Fe4S]^{+}$ state.[117,118,120] The next state to occur in the activation process is the active oxidized state H_{ox}, in which the reduction equivalent has apparently been transferred from the cubane to the bi-nuclear subcluster, which then assumes the Fe(I)Fe(II) configuration. The most reduced state that can be reached under hydrogen is the "active reduced" H_{red}. Here, both irons in the bi-nuclear subcluster are reduced to Fe(I). As is indicated in Scheme 11.3, the H_{ox} and

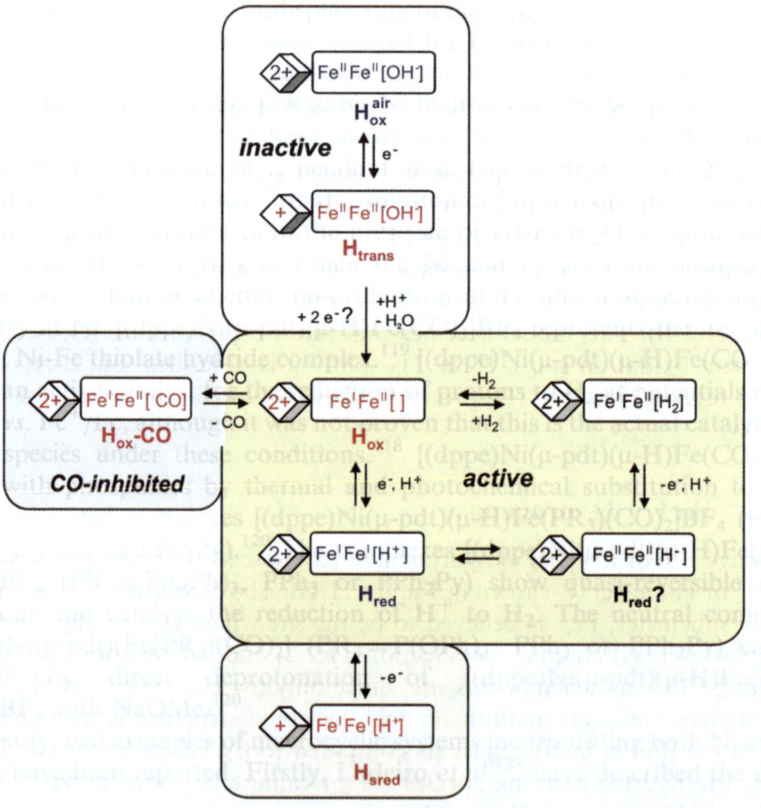

Scheme 11.3 Intermediate states identified in the H-cluster. Proposed ligands to the exchangeable site at Fe_d are indicated between brackets. The formal oxidation states of the irons in the [2Fe] and the charge of the [4Fe4S] subcluster are given. The paramagnetic (EPR-active) states are in red, the EPR-silent states in blue, and transient states in black. Note that H_{trans} has not yet been observed by EPR, H_{sred} is only observed as stable intermediate in hydrogenase from algae.[119] See text for further details.

H_{red} states are considered to take part in the catalytic cycle. The short lived intermediates which close the reaction cycle are not identified yet, but clearly these have to involve species with a bound dihydrogen and/or hydride. These considerations are further discussed below in Section 11.3.2.6. *In vitro* manipulation of the H-cluster allows inducing additional stable intermediates that may or may not play a physiological role in the organism. For *Chlamydomonas (C.) reinhardtii* hydrogenase HydA1, an additional reduced state H_{sred} ("super reduced"), has been identified using FTIR monitored spectroelectrochemistry.[119] It is suggested that here, similar to the H_{trans} state, the cubane sub-cluster is reduced to $[4Fe4S]^+$. These experiments are discussed in more detail in Section 11.3.2.3.

It is known that CO is a strong inhibitor for virtually all hydrogenases. The *D. desulfuricans* and *C. pasteurianum* hydrogenases can both be transformed to the H_{ox}-CO state under CO atmosphere. This state is very stable because the external CO ligand is strongly bound at the exchangeable site, thus completely inhibiting the enzyme activity. For most enzymes the inhibition seems to be virtually irreversible.[120] On the other hand, the stability of the H_{ox}-CO state and the fact that it is paramagnetic makes it an attractive object for spectroscopic investigations. Since most CO ligands in the H-cluster are photolabile, several low temperature light induced species could be induced and investigated (see Section 11.3.2.4). In addition, it was shown that the CO ligands could be partly exchanged by ^{13}C labeled CO ligands by simply adding external ^{13}CO and illuminating the sample at room temperature (Section 11.3.2.2).

11.3.2.2 Electronic Structure of the Paramagnetic States: Identification of the Bridging Ligand

The oxidized states H_{ox} and H_{ox}-CO are the only physiological intermediates that are accessible by EPR. By studying the hyperfine interactions of the ^{57}Fe nuclei of the cluster-core as well as the ^{14}N and ^{13}C nuclei of the ligands, important information can be obtained on the unpaired spin distribution and electronic structure of the H-cluster. It turns out that the CO inhibited state of the enzyme, H_{ox}-CO, yields additional information on the redistribution of spin density in the H-cluster upon binding an electron donating substrate (like CO) in the exchangeable site at Fe_d.

The H_{ox} state, which for the [FeFe] hydrogenase of *D. desulfuricans* is stable in the potential range from –100 to –320 mV (NHE), shows a rhombic EPR signal ($S=\frac{1}{2}$) characterized by the *g* tensor principal values 2.10, 2.04, and 2.00 (Figure 11.10). This EPR signal, which is found in all [FeFe] hydrogenases studied to date, differs fundamentally from spectra of ferredoxin type clusters [4Fe4S]. The intensity of this signal is maximized at a potential of about –300 mV. Below –320 mV, the signal rapidly vanishes. The iron atoms in the H-cluster can be exchanged to ^{57}Fe by growing the organism on ^{57}Fe enriched medium. Since ^{57}Fe has a nuclear spin of $I=1/2$ the magnetic interactions between the unpaired electron spin and ^{57}Fe can be used as a reporter on the spin distribution over the H-cluster in the H_{ox} and H_{ox}-CO state.

Advanced multi-frequency EPR experiments (X- and Q-band Davies ENDOR) on the ^{57}Fe labeled H_{ox} state of *D. desulfuricans*[121] detected several ^{57}Fe HF couplings around 12 MHz. It was concluded that all six iron atoms are interacting with the unpaired electron. This means that the unpaired spin density in H_{ox} apparently is equally distributed over both subclusters of the H-cluster.

The H_{ox}-CO state shows a much more axial EPR spectrum (Figure 11.8B). Also, this species is observed in all currently studied [FeFe] hydrogenases. Multifrequency pulse ENDOR and ESEEM experiments on the ^{57}Fe labeled H_{ox}-CO state[121] are depicted in Figure 11.9.

For this state, Q-band pulse ENDOR experiments revealed the ^{57}Fe HF couplings of all four iron atoms in the cubane subcluster, whereas ^{57}Fe Q-band HYSCORE experiments turned out to be sensitive to the much smaller hyperfine interactions attributed to the irons in the bi-nuclear subcluster. Here,

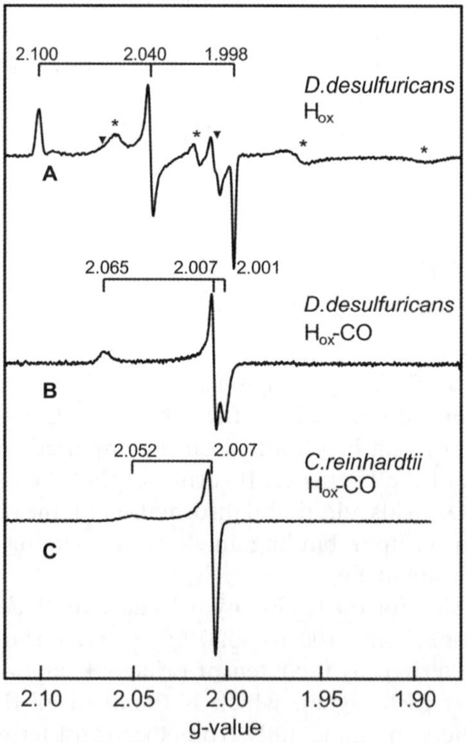

Figure 11.8 Q-band EPR spectra of the (A) H_{ox} and (B) H_{ox}-CO state of *D. desulfuricans* [FeFe] hydrogenase as well as (C) the H_{ox}-CO state of *C. reinhardtii*. The signals marked with * in (A) are due to residual reduced F-clusters present in the enzyme. For details see references 121, 142.

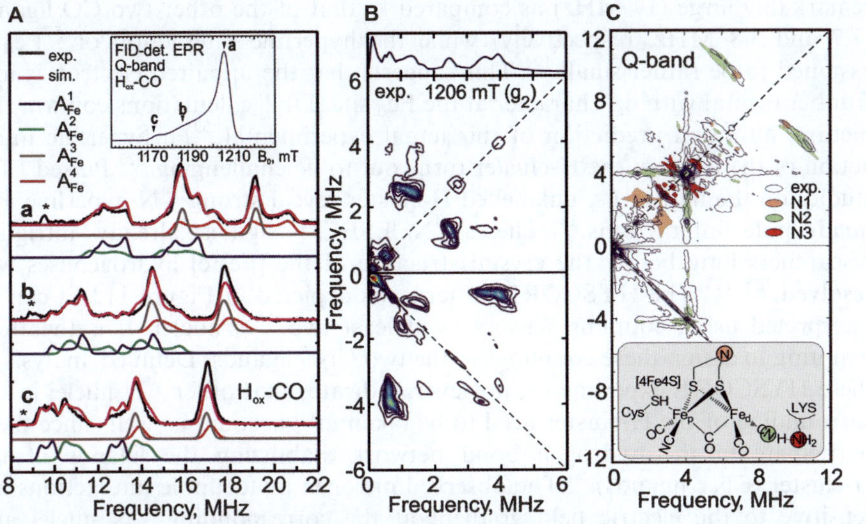

Figure 11.9 Q-band ENDOR and ESEEM spectra of [FeFe] hydrogenase from *D. desulfuricans*. (A) Davies ENDOR of H_{ox}-CO showing four ^{57}Fe hyperfine interactions from the [4Fe4S] subcluster; (B) HYSCORE of H_{ox}-CO showing the two ^{57}Fe hyperfine couplings from the binuclear [2Fe] subcluster; (C) HYSCORE of H_{ox} revealing three ^{14}N couplings; the assignment is indicated in the inset. For all spectra the simulations are shown. For details see references 54, 121.

the apparent spin density on the iron atoms has been shifted to the cubane sub-cluster. This behavior points to large changes in the electronic structure of the H-cluster upon binding of the external CO ligand. The remarkable large spin density found on the (formally diamagnetic) cubane sub-cluster has been shown to originate from the extensive exchange interaction between the iron sub-clusters.[115,121,122] This exchange interaction is strongly dependent on the distance between Fe_p and the cubane sub-cluster. It has been shown that this distance shortens slightly when going from the H_{ox} state to H_{ox}-CO.[55,123] The binding of the external CO ligand into the exchangeable site thus strongly affects the electronic and geometrical structure of the H-cluster, leading to a much stronger apparent spin density on the cubane. This effect illustrates the strong electronic contact between the iron atoms in the H-cluster. It is very likely that hydrogen binding into the exchangeable site will have similar effects on the electronic structure, a property that must be essential for the catalytic cycle.

As was already mentioned, the light sensitivity of the H_{ox}-CO state can be used to exchange three of the four CO ligands with ^{13}CO.[120] X-Band HYSCORE experiments were able to map the hyperfine interactions of the extrinsic CO bound to the exchangeable site as well as the bridging CO and the CO bound to Fe_d.[124] The ^{13}C hyperfine coupling of the extrinsic CO is remarkably large (17 MHz) as compared to that of the other two CO ligands (7.7 and 3.8 MHz, respectively) while the hyperfine interaction of $^{57}Fe_d$ is assumed to be rather small.[121] This suggests that the unpaired electron is in a frontier orbital with d_{z^2} character at the Fe_d site. DFT calculations confirm this picture, although prediction of the actual experimental ^{57}Fe hyperfine interaction of the irons in the H-cluster turns out to be challenging.[123] Pulsed EPR studies on the native, *i.e.* unlabelled H-cluster reveal strong ^{14}N hyperfine and quadrupole interactions.[33] These ^{14}N ESEEM signals already intrigued researchers long before the crystal structure of the [FeFe] hydrogenases was resolved.[125–127] The HYSCORE patterns, as depicted in Figure 11.9C, can be interpreted using coupling tensors of at least two ^{14}N nuclei. It is therefore tempting to assign these couplings to the two CN^- ligands. Detailed analysis of these HYSCORE experiments, however, indicates that other ^{14}N nuclei in the surrounding of the H-cluster need to be taken into account as well, since these participate in the hydrogen bond network stabilizing the ligands of the H-cluster (CN^- ligands).[54] The observed nitrogen quadrupole interactions are sensitive to the electric field gradient at the corresponding ^{14}N nuclei and strongly depend on their chemical bonding characteristics. The quadrupole interactions can therefore be used to identify the chemical nature of the magnetically coupled nitrogen species.[128] The nuclear quadrupole assignments to the specific nitrogens around the H-cluster were verified using DFT calculations, including the hydrogen bonded protein residues. Figure 11.9C shows the assignment of all three observed ^{14}N interactions. The most significant finding is that coupling N1 must belong to the bridging dithiol ligand DTN, assuming it has a central amino group. This interpretation seems to resolve a long standing debate on the nature of the central atom in the bridging dithiol ligand.[129] As will be discussed in a later section, the presence of an amino group in the bridging dithiol moiety has implications for the possible catalytic mechanism, since the amino group plays an essential role in acting as proton donor/acceptor in the reaction cycle of the [FeFe] hydrogenase.

11.3.2.3 Light Sensitivity and Inactivation of the H-cluster

From EPR and IR investigations it is known that upon light excitation at temperatures between 20 K and 70 K the external CO ligand can be reversibly removed.[62,120,130] During illumination an additional state of the H-cluster can be obtained. This state (H_{L2}) is characterized by a broad rhombic EPR spectrum with *g*-tensor principal values of 2.22, 2.13, and 2.05 for *D. desulfuricans*[120] and 2.26, 2.12, and 1.89 for hydrogenase I from *C. pasteurianum*.[131] FTIR studies on both hydrogenase I from *C. pasteurianum* and *D. desulfuricans* hydrogenase show that for the H_{L2}-state the signature of the bridging CO is lost. This indicates a complete dissociation of this CO ligand or a conversion to terminal coordination.[118,132] This is interesting in view of the reaction mechanism (see Section 11.3.2.6). Albracht *et al.*[120] have recently shown that the H-cluster of *D. desulfuricans* is very light sensitive in general at room temperature in the active H_{ox} and H_{red} states. Even under ambient light conditions the enzyme can be severely damaged during reductive activation and incubation under Ar. One H-cluster may lose its CO ligands under irradiation and be irreversibly inactivated. Other H-clusters will pick up the dissociated CO ligands and become inhibited. This explains the occurrence of the H_{ox}-CO EPR signals in preparations of hydrogenases which were never exposed to CO. The enzyme is, however, light stable when in the H_{ox}^{air} or H_{ox}-CO state. A detailed FTIR study on H_{ox}-CO, in which the extrinsic CO ligand was labeled with ^{13}C, showed that all three CO ligands associated with Fe_d can be exchanged under illumination at room temperature.[118]

Inhibition by O_2 is believed to be irreversible. Thin film electrochemical studies by Vincent *et al.*[133] show that the activity of the enzyme is strongly reduced after exposure to oxygen. The catalytic current is irreversibly lost when the enzyme is exposed to oxygen while in an active state. Binding of oxygen to the free coordination site (in the H_{ox} and H_{red} state) may oxidize the binuclear iron cluster to Fe(III) causing the loss of the CO and CN^- ligands, thereby destroying the H-cluster.[134,135]

11.3.2.4 *Comparison of Bacterial and Algal [FeFe] Hydrogenases*

In the last decades, the [FeFe] hydrogenases of several algae have been identified, although at present still no X-ray structure has been obtained. The HydA1 protein from *Chlamydomonas reinhardtii* has been the subject of many biochemical and spectroscopic studies.[136–139] It turns out that this hydrogenase lacks all accessory FeS clusters (F-clusters) which, in the case of prokaryotic [FeFe] hydrogenases, take care of the electron transport. This function is taken over by the ferredoxin PetF, which directly docks at the hydrogenase near the H-cluster.[140,141] EPR investigations of the H_{ox} and H_{ox}-CO state of HydA1 indicate that the overall electronic structure of this hydrogenase is very similar to that of the one from *D. desulfuricans*.[142] The most significant differences were observed for the H_{ox}-CO state, in which the observed *g*-values vary somewhat between the different species. Spectroelectrochemical investigations, however, indicate that there maybe more fundamental differences between the two types of hydrogenases.[119] Figure 11.10 shows the FTIR bands observed in an *in situ* electrochemical titration of HydA1 as compared to those of *D. desulfuricans*.

In both hydrogenases, the H_{ox} and H_{red} state can be readily identified. The midpoint potential of this transition is similar for the two proteins, (–400 mV for HydA1 and –395 for *D. desulfuricans* at pH = 8). However, the FTIR of HydA1 shows in the H_{red} state an intact bridging CO ligand, whereas for the *D. desulfuricans* hydrogenase the bridging CO is shifted toward the terminal position at Fe_d in this state. Moreover, further reduction of the HydA1 protein (E_m = –455 mV) produces a second reduced species (H_{sred}) which, in contrast to that of *D. desulfuricans* (E_m = –540 mV) is stable and fully reversible. For the H_{sred} state the FTIR signature of the bridging CO (1793 cm^{-1}) is lost

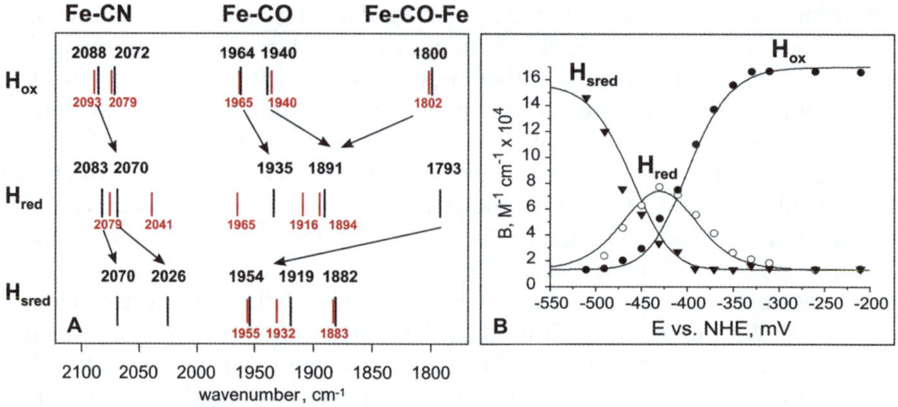

Figure 11.10 Spectroelectrochemistry of the H-cluster in *Chlamydomonas reinhardtii*. (A) FTIR frequencies of the H_{ox}, H_{red} and H_{sred} states. The FTIR band of the corresponding states from *D. desulfuricans* are indicated in red. (B) Redox titration of the respective states shown in (A) for HydA1 of *Chlamydomonas reinhardtii*.[119]

and instead a new band appears consistent with a terminal CO ligand (Figure 11.10A). It is suggested that for the H_{sred} state the cubane sub-cluster is reduced, while the bi-nuclear subcluster stays in the Fe(I)Fe(I) configuration. The observation that the H_{red} state of HydA1 can exist with an intact bridging CO, suggests that the hydrogen species is not bound in the bridge but instead at the terminal position in the exchangeable site. This finding is relevant for the possible catalytic mechanism of hydrogen oxidation and evolution at the H-cluster.

11.3.2.5 Catalytic Reaction Mechanism

As was already discussed in the previous sections, the ^{57}Fe hyperfine pattern for the H_{ox} state suggests an almost equal spin distribution over the two iron atoms of the bi-nuclear subcluster. Upon binding of the external CO (H_{ox}-CO state) the observed spin density is shifted towards the proximal iron and the cubane subcluster, but the IR data show that the formal oxidation state of the Fe_p is not changed, *i.e.* it remains Fe(I). DFT calculations[123] on the H_{ox} state favor a valence localized Fe(I)Fe(II) pair in which the unpaired electron is located on the distal Fe, while the H_{ox}-CO state is characterized by a valence delocalized FeFe pair. These calculations, however, do not reproduce the observed ^{57}Fe hyperfine interaction of the H-cluster very accurately.[121] Nevertheless, various DFT calculations of the H-cluster and its possible transition states (*e.g.* involving the binding of molecular hydrogen) offer very useful insights in the possible scenarios of the catalytic cycle.

The overall consensus oxidation/reduction reaction is depicted in Scheme 11.3, where it is assumed that electron and proton transport will take place in (almost) simultaneous steps. Several studies involve the stability of the hydride intermediate, in which a hydride is attached to the Fe(II)Fe(II) subcluster, either bound terminally to the distal iron[143–145] or in a bridging position between the two irons.[146–149] It turns out that both configurations are feasible but the bridging configuration is thermodynamically more stable. This would suggest that the hydrogen splitting and proton reduction will occur in the bridging position. At the same time, however, it has been demonstrated by Hall *et al.*[143] that the amine group in the DTN bridging ligand would provide an excellent proton acceptor in the catalytic cycle. In addition, the crystal structure provides a proton pathway through the protein leading to the DTN bridging ligand.[56] If the hydride would bind in the bridging position, the dithiolate sulfur ligands could act as proton acceptors but it would also require the bridging CO ligand to flip over to the terminal position on the distal iron. Thermodynamically this rearrangement would be feasible but it is not clear if the active site provides sufficient flexibility to allow such a ligand exchange process at high rates (*i.e.* compatible with the high turnover rates of the enzyme). The two possible scenarios for the catalytic cycle are depicted in Scheme 11.4. In the "bridging hydride scenario" (A), it is assumed that H_{ox} with its closed CO bridge is in equilibrium with an "open" conformation where ligand rotation of the bridging CO has occurred.

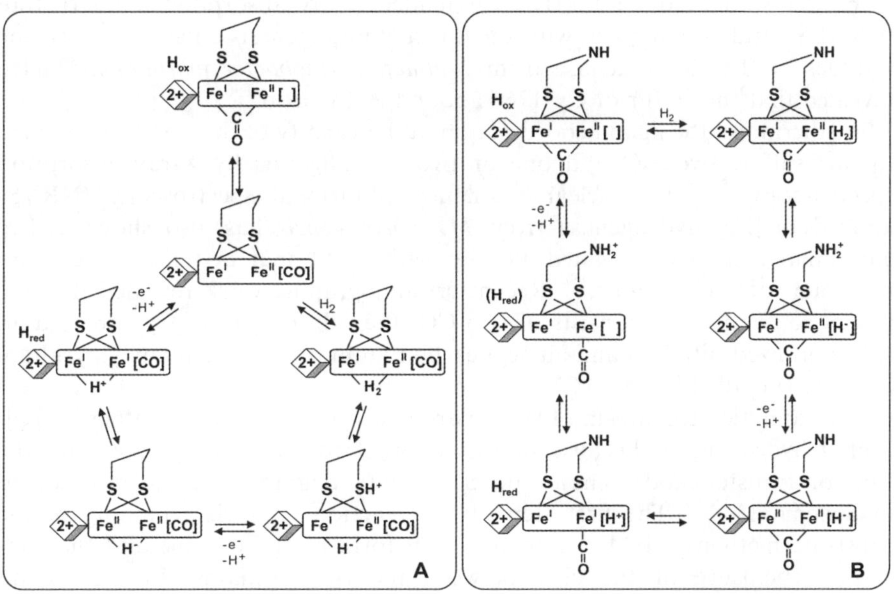

Scheme 11.4 Proposed mechanisms for the reversible heterolytic splitting of H_2 by [FeFe] hydrogenases. Based on DFT calculations assuming hydride binding in (A) the bridging position[146–149] or (B) terminal.[59,143–145,150]

11.3.3 [Fe] Hydrogenase

[Fe] hydrogenase, found in only methanogenic archaea, contains a mono iron-cofactor (iron-guanylylpyrodinol, FeGP) and no iron–sulfur cluster. Its catalytic reaction mechanism differs fundamentally from that of the other types of hydrogenases. [Fe] hydrogenase catalyzes the reversible reduction of methenyltetrahydromethanopterin (methenyl-H_4MPT^+) with H_2 to methylene-H_4MPT and a proton by transferring a hydride ion to the *proR* position of the C14a carbon of methylene-H_4MPT. [Fe] hydrogenase can catalyze H^+/H_2-exchange and *para/ortho*-H_2 conversion. Its activity is dependent on the substrate, methenyl-H_4MPT^+. These observations suggested that the H_2 activation occurs only in the presence of the substrate.[151]

The purified enzyme can catalyze the hydride transfer reaction in an atmospheric environment (20% O_2). It is sensitive to the superoxide anion radical and copper ions. The crystal structures of [Fe] hydrogenase from *Methanocaldococcus jannaschii* and *Methanopyrus kandleri* have been reported.[12,58,59] The crystallographic analysis revealed that [Fe] hydrogenase comprises two peripheral units bound to the central globular unit in a linear manner (see Section 11.2.3). Their two domains are connected *via* a linker region, and large conformational changes ('open' and 'closed' forms) occur during the catalytic cycle.

Spectroscopic studies revealed the ligands of the mono-iron active site. FTIR studies showed that two CO molecules are bound to the iron.[152] From the results of Mössbauer spectroscopy, the oxidation state of the iron was identified to be low spin Fe(0) or Fe(II).[153] Extended X-ray absorption fine structure (EXAFS) studies combined with site directed mutagenesis on all three cysteine residues of [Fe] hydrogenase from *Methanothermobacter marburgensis* have revealed that the sulfur of Cys176 is bound to the iron.[150]

Furthermore, the ligand coordination in the vicinity of the iron was assigned to one sulfur, two CO and one or two O/N ligands by X-ray absorption spectroscopy.[153] ^{57}Fe nuclear resonance vibrational spectroscopy (NRVS) applied to [Fe] hydrogenase from *M. marburgensis* has also shown a five coordinated *cis*-$(CO)_2$ ligated Fe site with additional cysteine, water, and pyridone cofactor ligands.[154] Recent crystallographic work revealed that the mono-iron active site consists of two CO, the sulfur atom of Cys176, and an sp^2-hybridized nitrogen and an acyl carbon atom of an iron-guanylylpyridinol cofactor (FeGP, Figure 1.1C).

The catalytic mechanism of [Fe] hydrogenase differs from the other [NiFe]/[FeFe] hydrogenases because of the different redox-active partner and the electron transfer mode. The proposed catalytic reaction mechanism is shown in Scheme 11.5.[57] The first step of the reaction comprises binding of the substrate methenyl-H_4MPT^+ to the open form of the enzyme and the concomitant closure of the cleft between the two domains. In the second step, the FeGP-cofactor is activated and the hydrogen binds to the open coordination site of the iron, probably in a side-on manner. The hydrogen is heterolytically cleaved and the adjacent C14a carbocation of the

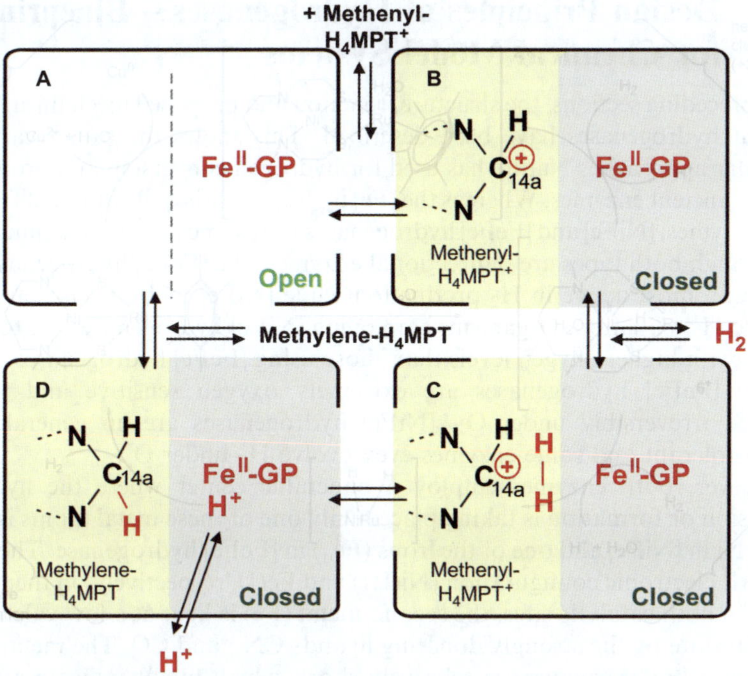

Scheme 11.5 Reaction cycle of [Fe] hydrogenase. (A) Open form. (B) Upon binding of methenyl-H_4MPT^+, the active-site cleft is closed. (C) H_2 binds side-on to the iron. (D) The carbocationic C14a of methenyl-H_4MPT^+ accepts the hydride on its Re-face and thereby generates methylene-H_4MPT. The peripheral and central units are shown in grey and yellow in the closed form, respectively.[57]

methenyl-H_4MPT^+, acting as a Lewis acid, accepts the hydride. The proton is probably transferred *via* the Cys176 thiol or the pyridinol hydroxy group. A similar situation is proposed in the case of [NiFe] hydrogenase that might use cysteine residues as a base for accepting the proton. The pyridinol hydroxy group is another possible candidate since it is located near the hydrogen binding site and possibly interacts with the His14 residue. The H14A mutant showed a decrease of activity compared to the wild type (<1%). The proton derived from the heterolytic cleavage of H_2 is then exchanged quickly with protons of bulk solvent.

At first sight, this mechanism of [Fe] hydrogenase looks quite different from that of the other two bimetallic hydrogenases. However, it has recently been pointed out by Shima and coworkers that the heterolytic hydrogen splitting step is quite similar. In all cases, it occurs at a dinuclear catalytic center Fe-X, in which the iron is a low-spin Fe(II). In the case of [FeFe] hydrogenase the second active metal is also an iron, in case of the [NiFe] enzyme it is a different metal (Ni), and in the Fe hydrogenase X is a carbocation. This similarity suggests a convergent evolution of the H_2-activation machinery.

11.4 Design Principles of Hydrogenases – Blueprints for Chemical Model Systems

In the preceding sections, the structure, function and proposed mechanisms of the different hydrogenases have been described. This shows the broad variety of molecular approaches Nature has used for hydrogen conversion and production by these ancient enzymes. Whereas the [Fe] hydrogenase is quite different from the other enzymes, [NiFe] and [FeFe] hydrogenases bear some remarkable similarities.

Although both types are bidirectional enzymes, the [FeFe] hydrogenases are, in general, most active in H_2 production while [NiFe] hydrogenases are more tuned to H_2 oxidation. Organisms employing [NiFe] hydrogenases are found in regions with higher oxygen levels than those using [FeFe] hydrogenases. This is because [FeFe] hydrogenases are extremely oxygen sensitive and will be inhibited irreversibly under O_2. [NiFe] hydrogenases are, in general, more oxygen tolerant and some enzymes even evolve H_2 under O_2.

However, both enzymes employ a bimetallic center where the hydrogen conversion or formation is taking place. Only one of these metal atoms is redox active: Ni in [NiFe] and one of the irons (Fe_d) in [FeFe] hydrogenase. They both have a d^7 electronic configuration (Ni(III) and Fe(I), respectively) in their active states. In both catalytic sites, the second metal (Fe) is kept at a low valence and low spin state by the strongly donating ligands CN^- and CO. The metal–metal distance in both structures is relatively short, which might indicate a metal–metal bond. Both active sites have a "butterfly-shaped" core in which the two metals are bridged by soft sulfur ligands (thiolates). One metal with an open coordination site can be identified in both active states. This is the site where H_2 is believed to bind – or is being released. The observed H/D-isotope effect shows that in both cases the H_2 splitting is heterolytic. A sulfur or nitrogen/oxygen ligand probably acts as base to accept or donate the H^+. In both enzymes the catalytic activity is inhibited by O_2 and CO, although exceptions to this rule are known for the [NiFe] hydrogenases.

The features described above can serve as guidelines for the construction of biomimetic hydrogenase model systems.

Acknowledgement

M.-E. Pandelia is gratefully acknowledged for critically reading the manuscript and R. Groever and B. Deckers for their help with the text and artwork. The authors own work cited in the references was financially supported by the EU/Energy Network project SOLAR-H2 (FP7 contract 212508) and BMBF (03SF0355C), and Max Planck Society.

References

1. R. Cammack, M. Frey and R. Robson, *Hydrogen as a fuel*, Taylor & Francis, London, 2001.
2. W. Lubitz and W. Tumas (eds), *Chem. Rev.*, 2007, **107**, 3899–4435.

3. D. A. J. Rand and R. M. Dell, *Hydrogen Energy*, RSC Publishing, Cambridge, 2007.
4. W. Lubitz, E. J. Reijerse and J. Messinger, *Energy Environ. Sci.*, 2008, **1**, 15–31.
5. N. S. Lewis and D. G. Nocera, *Proc. Natl. Acad. Sci. U. S. A.*, 2006, **103**, 15729–15735.
6. P. M. Vignais, B. Billoud and J. Meyer, *FEMS Microbiol. Rev.*, 2001, **25**, 455–501.
7. P. M. Vignais and A. Colbeau, *Curr. Issues Mol. Biol.*, 2004, **6**, 159–188.
8. P. M. Vignais and B. Billoud, *Chem. Rev.*, 2007, **107**, 4206–4272.
9. J. C. Fontecilla-Camps, A. Volbeda, C. Cavazza and Y. Nicolet, *Chem. Rev.*, 2007, **107**, 4273–4303.
10. H. Ogata, W. Lubitz and Y. Higuchi, *Dalton Trans.*, 2009, 7577–7587.
11. J. W. Peters, W. N. Lanzilotta, B. J. Lemon and L. C. Seefeldt, *Science*, 1998, **282**, 1853–1858.
12. S. Shima, O. Pilak, S. Vogt, M. Schick, M. S. Stagni, W. Meyer-Klaucke, E. Warkentin, R. K. Thauer and U. Ermler, *Science*, 2008, **321**, 572–575.
13. S. Shima, E. J. Lyon, M. S. Sordel-Klippert, M. Kauss, J. Kahnt, R. K. Thauer, K. Steinbach, X. L. Xie, L. Verdier and C. Griesinger, *Angew. Chem., Int. Ed.*, 2004, **43**, 2547–2551.
14. R. Cammack, *Nature*, 1999, **397**, 214–215.
15. J. O. Eberly and R. L. Ely, *Crit. Rev. Microbiol.*, 2008, **34**, 117–130.
16. M. Guiral, P. Tron, V. Belle, C. Aubert, C. Léger, B. Guigliarelli and M. T. Giudici-Orticoni, *Int. J. Hydrogen Energy*, 2006, **31**, 1424–1431.
17. M. E. Pandelia, V. Fourmond, P. Tron-Infossi, E. Lojou, P. Bertrand, C. Léger, M. T. Giudici-Orticoni and W. Lubitz, *J. Am. Chem. Soc.*, 2010, **132**, 6991–7004.
18. O. Lenz, M. Ludwig, T. Schubert, I. Bürstel, S. Ganskow, T. Goris, A. Schwarze and B. Friedrich, *ChemPhysChem*, 2010, **11**, 1107–1119.
19. K. A. Vincent, J. A. Cracknell, O. Lenz, I. Zebger, B. Friedrich and F. A. Armstrong, *Proc. Natl. Acad. Sci. U. S. A.*, 2005, **102**, 16951–16954.
20. M. J. Lukey, A. Parkin, M. M. Roessler, B. J. Murphy, J. Harmer, T. Palmer, F. Sargent and F. A. Armstrong, *J. Biol. Chem.*, 2010, **285**, 3928–3938.
21. H. Ogata, P. Kellers and W. Lubitz, *J. Mol. Biol.*, 2010, **402**, 428–444.
22. Y. Nicolet, C. Piras, P. Legrand, C. E. Hatchikian and J. C. Fontecilla-Camps, *Struct. Folding Design*, 1999, **7**, 13–23.
23. Y. Montet, P. Amara, A. Volbeda, X. Vernede, E. C. Hatchikian, M. J. Field, M. Frey and J. C. Fontecilla-Camps, *Nat. Struct. Biol.*, 1997, **4**, 523–526.
24. E. Garcin, X. Vernede, E. C. Hatchikian, A. Volbeda, M. Frey and J. C. Fontecilla-Camps, *Structure*, 1999, **7**, 557–566.
25. M. C. Marques, R. Coelho, A. L. De Lacey, I. A. Pereira and P. M. Matias, *J. Mol. Biol.*, 2010, **396**, 893–907.

26. H. Ogata, S. Hirota, A. Nakahara, H. Komori, N. Shibata, T. Kato, K. Kano and Y. Higuchi, *Structure*, 2005, **13**, 1635–1642.
27. Y. Higuchi, T. Yagi and N. Yasuoka, *Structure*, 1997, **5**, 1671–1680.
28. A. Volbeda, E. Garcin, C. Piras, A. L. De Lacey, V. M. Fernandez, E. C. Hatchikian, M. Frey and J. C. Fontecilla-Camps, *J. Am. Chem. Soc.*, 1996, **118**, 12989–12996.
29. R. P. Happe, W. Roseboom, A. J. Pierik, S. P. Albracht and K. A. Bagley, *Nature*, 1997, **385**, 126.
30. Y. Higuchi, F. Toujou, K. Tsukamoto and T. Yagi, *J. Inorg. Biochem.*, 2000, **80**, 205–211.
31. M. E. Pandelia, H. Ogata, L. J. Currell, M. Flores and W. Lubitz, *J. Biol. Inorg. Chem*, 2009, **14**, 1227–1241.
32. A. L. De Lacey, V. M. Fernandez, M. Rousset, C. Cavazza and E. C. Hatchikian, *J. Biol. Inorg. Chem.*, 2003, **8**, 129–134.
33. W. Lubitz, E. Reijerse and M. van Gastel, *Chem. Rev.*, 2007, **107**, 4331–4365.
34. A. L. De Lacey, V. M. Fernandez, M. Rousset and R. Cammack, *Chem. Rev.*, 2007, **107**, 4304–4330.
35. M. E. Pandelia, H. Ogata and W. Lubitz, *ChemPhysChem*, 2010, **11**, 1127–1140.
36. A. Volbeda, L. Martin, C. Cavazza, M. Matho, B. W. Faber, W. Roseboom, S. P. J. Albracht, E. Garcin, M. Rousset and J. C. Fontecilla-Camps, *J. Biol. Inorg. Chem*, 2005, **10**, 239–249.
37. M. E. Pandelia, H. Ogata, L. J. Currell, M. Flores and W. Lubitz, *Biochim. Biophys. Acta*, 2010, **1797**, 304–313.
38. H. Ogata, Y. Mizoguchi, N. Mizuno, K. Miki, S. Adachi, N. Yasuoka, T. Yagi, O. Yamauchi, S. Hirota and Y. Higuchi, *J. Am. Chem. Soc.*, 2002, **124**, 11628–11635.
39. M. Brecht, M. van Gastel, T. Buhrke, B. Friedrich and W. Lubitz, *J. Am. Chem. Soc.*, 2003, **125**, 13075–13083.
40. S. Foerster, M. Stein, M. Brecht, H. Ogata, Y. Higuchi and W. Lubitz, *J. Am. Chem. Soc.*, 2003, **125**, 83–93.
41. S. Foerster, M. van Gastel, M. Brecht and W. Lubitz, *J. Biol. Inorg. Chem.*, 2005, **10**, 51–62.
42. Y. Higuchi, H. Ogata, K. Miki, N. Yasuoka and T. Yagi, *Structure*, 1999, **7**, 549–556.
43. J. W. van der Zwaan, S. P. J. Albracht, R. D. Fontijn and E. C. Slater, *FEBS Lett.*, 1985, **179**, 271–277.
44. M. Medina, R. Williams, R. Cammack and E. Hatchikian, *J. Chem. Soc., Faraday Trans.*, 1994, **90**, 2921–2924.
45. M. Medina, E. C. Hatchikian and R. Cammack, *Biochim. Biophys. Acta*, 1996, **1275**, 227–236.
46. J. W. van der Zwaan, J. M. C. C. Coremans, E. C. M. Bouwens and S. P. J. Albracht, *Biochim. Biophys. Acta*, 1990, **1041**, 101–110.
47. T. Yagi, *J. Biochem.(Tokyo)*, 1970, **68**, 649–657.

48. C. C. Page, C. C. Moser, X. X. Chen and P. L. Dutton, *Nature*, 1999, **402**, 47–52.
49. V. H. Teixeira, C. M. Soares and A. M. Baptista, *Proteins: Struct., Funct., Bioinf.*, 2008, **70**, 1010–1022.
50. S. Dementin, B. Burlat, A. L. De Lacey, A. Pardo, G. Adryanczyk-Perrier, B. Guigliarelli, V. M. Fernandez and M. Rousset, *J. Biol. Inorg. Chem.*, 2004, **279**, 10508–10513.
51. F. Leroux, S. Dementin, B. Burlatt, L. Cournac, A. Volbeda, S. Champ, L. Martin, B. Guigliarelli, P. Bertrand, J. C. Fontecilla-Camps, M. Rousset and C. Léger, *Proc. Natl. Acad. Sci. U. S. A.*, 2008, **105**, 11188–11193.
52. L. Florin, A. Tsokoglou and T. Happe, *J. Biol. Inorg. Chem.*, 2001, **276**, 6125–6132.
53. M. Winkler, B. Heil, B. Heil and T. Happe, *Biochim. Biophys. Acta, Gene Struct. Expression*, 2002, **1576**, 330–334.
54. A. Silakov, B. Wenk, E. Reijerse and W. Lubitz, *Phys. Chem. Chem. Phys.*, 2009, **11**, 6592–6599.
55. B. J. Lemon and J. W. Peters, *Biochemistry*, 1999, **38**, 12969–12973.
56. Y. Nicolet, A. L. De Lacey, X. Vernede, V. M. Fernandez, E. C. Hatchikian and J. C. Fontecilla-Camps, *J. Am. Chem. Soc.*, 2001, **123**, 1596–1601.
57. T. Hiromoto, E. Warkentin, J. Moll, U. Ermler and S. Shima, *Angew. Chem., Int. Ed.*, 2009, **48**, 6457–6460.
58. T. Hiromoto, K. Ataka, O. Pilak, S. Vogt, M. S. Stagni, W. Meyer-Klaucke, E. Warkentin, R. K. Thauer, S. Shima and U. Ermler, *FEBS Lett.*, 2009, **583**, 585–590.
59. O. Pilak, B. Mamat, S. Vogt, C. H. Hagemeier, R. K. Thauer, S. Shima, C. Vonrhein, E. Warkentin and U. Ermler, *J. Mol. Biol.*, 2006, **358**, 798–809.
60. M. G. Rossmann and P. Argos, *Annu. Rev. Biochem.*, 1981, **50**, 497–532.
61. K. A. Bagley, E. C. Duin, W. Roseboom, S. P. J. Albracht and W. H. Woodruff, *Biochemistry*, 1995, **34**, 5527–5535.
62. A. J. Pierik, M. Hulstein, W. R. Hagen and S. P. J. Albracht, *Eur. J. Biochem.*, 1998, **258**, 572–578.
63. B. Huynh, D. S. Patil, I. Moura, M. Teixeira, J. J. G. Moura, D. V. Dervartanian, M. H. Czechowski, B. C. Prickril, H. D. Peck and J. Legall, *J. Biol. Inorg. Chem*, 1987, **262**, 795–800.
64. C. Bagyinka, J. P. Whitehead and M. J. Maroney, *J. Am. Chem. Soc.*, 1993, **115**, 3576–3585.
65. Z. J. Gu, J. Dong, C. B. Allan, S. B. Choudhury, R. Franco, J. J. G. Moura, J. LeGall, A. E. Przybyla, W. Roseboom, S. P. J. Albracht, M. J. Axley, R. A. Scott and M. J. Maroney, *J. Am. Chem. Soc.*, 1996, **118**, 11155–11165.
66. J. C. Fontecilla-Camps, P. Amara, C. Cavazza, Y. Nicolet and A. Volbeda, *Nature*, 2009, **460**, 814–822.
67. H. Wu and M. B. Hall, *C. R. Chim.*, 2008, **11**, 790–804.

68. M. Stein and W. Lubitz, *Curr. Opin. Chem. Biol.*, 2002, **6**, 243–249.
69. M. Bruschi, G. Zampella, P. Fantucci and L. De Gioia, *Coord. Chem. Rev.*, 2005, **249**, 1620–1640.
70. J. C. Salerno, *The Bioinorganic Chemistry of Nickel*, VCH Publishers, Inc., Weinheim, 1988, pp. 53–71.
71. S. P. J. Albracht, E. G. Graf and R. K. Thauer, *FEBS Lett.*, 1982, **140**, 311–313.
72. M. Teixeira, I. Moura, A. V. Xavier, B. H. Huynh, D. V. Dervartanian, H. D. Peck, J. Legall and J. J. G. Moura, *J. Biol. Inorg. Chem*, 1985, **260**, 8942–8950.
73. M. Flores, A. G. Agrawal, M. van Gastel, W. Gärtner and W. Lubitz, *J. Am. Chem. Soc.*, 2008, **130**, 2402–2403.
74. C. Fichtner, C. Laurich, E. Bothe and W. Lubitz, *Biochemistry*, 2006, **45**, 9706–9716.
75. J. E. Huyett, M. Carepo, A. Pamplona, R. Franco, I. Moura, J. J. G. Moura and B. M. Hoffman, *J. Am. Chem. Soc.*, 1997, **119**, 9291–9292.
76. K. K. Surerus, M. Chen, J. W. van der Zwaan, F. M. Rusnak, M. Kolk, E. C. Duin, S. P. J. Albracht and E. Münck, *Biochemistry*, 1994, **33**, 4980–4993.
77. M. Carepo, D. L. Tierney, C. D. Brondino, T. C. Yang, A. Pamplona, J. Telser, I. Moura, J. J. G. Moura and B. M. Hoffman, *J. Am. Chem. Soc.*, 2002, **124**, 281–286.
78. M. van Gastel, M. Stein, M. Brecht, O. Schroeder, F. Lendzian, R. Bittl, H. Ogata, Y. Higuchi and W. Lubitz, *J. Biol. Inorg. Chem.*, 2006, **11**, 41–51.
79. R. Cammack, D. S. Patil, E. C. Hatchikian and V. M. Fernandez, *Biochim. Biophys. Acta*, 1987, **912**, 98–109.
80. M. Teixeira, I. Moura, A. V. Xavier, J. J. G. Moura, J. LeGall, D. V. Dervartanian, H. D. Peck and B. H. Huynh, *J. Biol. Inorg. Chem*, 1989, **264**, 16435–16450.
81. M. E. Pandelia, W. Nitschke, P. Infossi, M. T. Giudici-Orticoni, E. Bill and W. Lubitz, *Proc. Natl. Acad. Sci. U. S. A.*, 2011, **108**, 6097–6102.
82. S. Kurkin, S. J. George, R. N. F. Thorneley and S. P. J. Albracht, *Biochemistry*, 2004, **43**, 6820–6831.
83. B. Bleijlevens, F. A. van Broekhuizen, A. L. De Lacey, W. Roseboom, V. M. Fernandez and S. P. J. Albracht, *J. Biol. Inorg. Chem*, 2004, **9**, 743–752.
84. S. J. George, S. Kurkin, R. N. F. Thorneley and S. P. J. Albracht, *Biochemistry*, 2004, **43**, 6808–6819.
85. A. L. De Lacey, A. Pardo, V. M. Fernandez, S. Dementin, G. Adryanczyk-Perrier, E. C. Hatchikian and M. Rousset, *J. Biol. Inorg. Chem*, 2004, **9**, 636–642.
86. N. K. Menon, H. D. Peck, J. LeGall and A. E. Przybyla, *J. Bacteriol.*, 1987, **169**, 5401–5407.
87. P. Jayapal, D. Robinson, M. Sundararajan, I. H. Hillier and J. J. W. McDouall, *Phys. Chem. Chem. Phys.*, 2008, **10**, 1734–1738.

88. A. Pardo, A. L. De Lacey, V. M. Fernandez, H. J Fan, Y. B. Fan and M. B. Hall, *J. Biol. Inorg. Chem.*, 2006, **11**, 286–306.
89. H. Wang, D. S. Patil, C. Y. Ralston, C. Bryant and S. P. Cramer, *J. Electron Spectrosc. Relat. Phenom.*, 2001, **114**, 865–871.
90. M. Stein and W. Lubitz, *Phys. Chem. Chem. Phys.*, 2001, **3**, 5115–5120.
91. C. L. Fan, M. Teixeira, J. Moura, I. Moura, B. H. Huynh, J. Legall, H. D. Peck and B. M. Hoffman, *J. Am. Chem. Soc.*, 1991, **113**, 20–24.
92. A. Volbeda and J. C. Fontecilla-Camps, *Top. Organomet. Chem.*, 2006, **17**, 57–82.
93. M. B. Hall, S. Q. Niu and L. Thomson, *J. Inorg. Biochem.*, 1999, **74**, 152.
94. B. Guigliarelli, C. More, A. Fournel, M. Asso, E. C. Hatchikian, R. Williams, R. Cammack and P. Bertrand, *Biochemistry*, 1995, **34**, 4781–4790.
95. L. M. Roberts and P. A. Lindahl, *J. Am. Chem. Soc.*, 1995, **117**, 2565–2572.
96. M. Stein and W. Lubitz, *J. Inorg. Biochem.*, 2004, **98**, 862–877.
97. F. Dole, M. Medina, C. More, R. Cammack, P. Bertrand and B. Guigliarelli, *Biochemistry*, 1996, **35**, 16399–16406.
98. P. Kellers, M. E. Pandelia, L. J. Currell, H. Görner and W. Lubitz, *Phys. Chem. Chem. Phys.*, 2009, **11**, 8680–8683.
99. K. A. Bagley, Personal Communication, 2009.
100. F. A. Armstrong, N. A. Belsey, J. A. Cracknell, G. Goldet, A. Parkin, E. Reisner, K. A. Vincent and A. F. Wait, *Chem. Soc. Rev.*, 2009, **38**, 36–51.
101. K. A. Vincent, N. A. Belsey, W. Lubitz and F. A. Armstrong, *J. Am. Chem. Soc.*, 2006, **128**, 7448–7449.
102. M. E. Pandelia, H. Ogata, M. Flores and W. Lubitz, EPR and FTIR study on the inhibition of the [NiFe] hydrogenase from *D. vulgaris* Miyazaki F by nitric oxide (NO), unpublished work, 2011.
103. L. Purec, A. I. Krasna and D. Rittenberg, *Biochemistry*, 1962, **1**, 270–275.
104. G. Fauque, Y. Berlier, E. S. Choi, H. D. Peck, J. LeGall and P. A. Lespinat, *Biochem. Soc. Trans.*, 1987, **15**, 1050–1051.
105. A. L. De Lacey, E. C. Hatchikian, A. Volbeda, M. Frey, J. C. Fontecilla-Camps and V. M. Fernandez, *J. Am. Chem. Soc.*, 1997, **119**, 7181–7189.
106. R. P. Happe, W. Roseboom and S. P. J. Albracht, *Eur. J. Biochem.*, 1999, **259**, 602–608.
107. K. A. Vincent, A. Parkin and F. A. Armstrong, *Chem. Rev.*, 2007, **107**, 4366–4413.
108. M. Saggu, I. Zebger, M. Ludwig, O. Lenz, B. Friedrich, P. Hildebrandt and F. Lendzian, *J. Biol. Inorg. Chem.*, 2009, **284**, 16264–16276.
109. T. Buhrke, O. Lenz, N. Krauss and B. Friedrich, *J. Biol. Inorg. Chem.*, 2005, **280**, 23791–23796.
110. E. van der Linden, T. Burgdorf, A. L. De Lacey, T. Buhrke, M. Scholte, V. M. Fernandez, B. Friedrich and S. P. J. Albracht, *J. Biol. Inorg. Chem.*, 2006, **11**, 247–260.
111. A. L. De Lacey, V. M. Fernandez and M. Rousset, *Coord. Chem. Rev.*, 2005, **249**, 1596–1608.

112. L. De Gioia, P. Fantucci, B. Guigliarelli and P. Bertrand, *Inorg. Chem.*, 1999, **38**, 2658–2662.
113. L. De Gioia, P. Fantucci, B. Guigliarelli and P. Bertrand, *Int. J. Quantum Chem.*, 1999, **73**, 187–195.
114. G. Wang, M. J. Benecky, B. H. Huynh, J. F. Cline, M. W. W. Adams, L. E. Mortenson, B. M. Hoffman and E. Münck, *J. Biol. Inorg. Chem.*, 1984, **259**, 4328–4331.
115. F. M. Rusnak, M. W. W. Adams, L. E. Mortenson and E. Münck, *J. Biol. Inorg. Chem.*, 1987, **262**, 38–41.
116. E. Münck and C. V. Popescu, *Hyperfine Interact.*, 2000, **126**, 59–67.
117. A. S. Pereira, P. Tavares, I. Moura, J. J. G. Moura and B. H. Huynh, *J. Am. Chem. Soc.*, 2001, **123**, 2771–2782.
118. W. Roseboom, A. L. De Lacey, V. M. Fernandez, E. C. Hatchikian and S. P. J. Albracht, *J. Biol. Inorg. Chem.*, 2006, **11**, 102–118.
119. A. Silakov, C. Kamp, E. Reijerse, T. Happe and W. Lubitz, *Biochemistry*, 2009, **48**, 7780–7786.
120. S. P. J. Albracht, W. Roseboom and E. C. Hatchikian, *J. Biol. Inorg. Chem.*, 2006, **11**, 88–101.
121. A. Silakov, E. J. Reijerse, S. P. J. Albracht, E. C. Hatchikian and W. Lubitz, *J. Am. Chem. Soc.*, 2007, **129**, 11447–11458.
122. C. V. Popescu and E. Münck, *J. Am. Chem. Soc.*, 1999, **121**, 7877–7884.
123. A. T. Fiedler and T. C. Brunold, *Inorg. Chem.*, 2005, **44**, 9322–9334.
124. A. Silakov, B. Wenk, E. Reijerse, S. P. J. Albracht and W. Lubitz, *J. Biol. Inorg. Chem.*, 2009, **14**, 301–313.
125. H. Thomann, M. Bernardo and M. W. W. Adams, *J. Am. Chem. Soc.*, 1991, **113**, 7044–7046.
126. R. Williams, R. Cammack and E. C. Hatchikian, *J. Chem. Soc, Faraday Trans.*, 1993, **89**, 2869–2872.
127. M. Fedurco, C. J. Sartoretti and J. Augustynski, *J. Electrochem. Soc.*, 2001, **148**, D19–D23.
128. E. A. C. Lucken, *Nuclear Quadrupole Coupling Constants*, Academic Press, London and New York, 1969.
129. A. S. Pandey, T. V. Harris, L. J. Giles, J. W. Peters and R. K. Szilagyi, *J. Am. Chem. Soc.*, 2008, **130**, 4533–4540.
130. D. S. Patil, B. H. Huynh, S. H. He, H. D. Peck, D. V. Dervartanian and J. LeGall, *J. Am. Chem. Soc.*, 1988, **110**, 8533–8534.
131. A. T. Kowal, M. W. W. Adams and M. K. Johnson, *J. Biol. Inorg. Chem.*, 1989, **264**, 4342–4348.
132. Z. J. Chen, B. J. Lemon, S. Huang, D. J. Swartz, J. W. Peters and K. A. Bagley, *Biochemistry*, 2002, **41**, 2036–2043.
133. K. A. Vincent, A. Parkin, O. Lenz, S. P. J. Albracht, J. C. Fontecilla-Camps, R. Cammack, B. Friedrich and F. A. Armstrong, *J. Am. Chem. Soc.*, 2005, **127**, 18179–18189.
134. S. T. Stripp, G. Goldet, C. Brandmayr, O. Sanganas, K. A. Vincent, M. Haumann, F. A. Armstrong and T. Happe, *Proc. Natl. Acad. Sci. U. S. A.*, 2009, **106**, 17331–17336.

135. M. T. Stiebritz and M. Reiher, *Inorg.Chem.*, 2009, **48**, 7127–7140.
136. A. Hemschemeier, A. Melis and T. Happe, *Photosynth. Res.*, 2009, **102**, 523–540.
137. S. T. Stripp and T. Happe, *Dalton Trans.*, 2009, 9960–9969.
138. M. L. Ghirardi, M. C. Posewitz, P. C. Maness, A. Dubini, J. P. Yu and M. Seibert, *Annu. Rev. Plant Biol.*, 2007, **58**, 71–91.
139. S. E. McGlynn, S. S. Ruebush, A. Naumov, L. E. Nagy, A. Dubini, P. W. King, J. B. Broderick, M. C. Posewitz and J. W. Peters, *J. Biol. Inorg. Chem.*, 2007, **12**, 443–447.
140. M. Winkler, S. Kuhlgert, M. Hippler and T. Happe, *J. Biol. Inorg. Chem.*, 2009, **284**, 36620–36627.
141. J. Jacobs, S. Pudollek, A. Hemschemeier and T. Happe, *FEBS Lett.*, 2009, **583**, 325–329.
142. C. Kamp, A. Silakov, M. Winkler, E. J. Reijerse, W. Lubitz and T. Happe, *Biochim. Biophys. Acta*, 2008, **1777**, 410–416.
143. H. J. Fan and M. B. Hall, *J. Am. Chem. Soc.*, 2001, **123**, 3828–3829.
144. Z. X. Cao and M. B. Hall, *J. Am. Chem. Soc.*, 2001, **123**, 3734–3742.
145. Z. P. Liu and P. Hu, *J. Am. Chem. Soc.*, 2002, **124**, 5175–5182.
146. M. Bruschi, P. Fantucci and L. De Gioia, *Inorg. Chem.*, 2002, **41**, 1421–1429.
147. M. Bruschi, P. Fantucci and L. De Gioia, *Inorg. Chem.*, 2003, **42**, 4773–4781.
148. T. J. Zhou, Y. R. Mo, A. M. Liu, Z. H. Zhou and K. R. Tsai, *Inorg. Chem.*, 2004, **43**, 923–930.
149. T. J. Zhou, Y. R. Mo, Z. H. Zhou and K. Tsal, *Inorg. Chem.*, 2005, **44**, 4941–4946.
150. M. Korbas, S. Vogt, W. Meyer-Klaucke, E. Bill, E. J. Lyon, R. K. Thauer and S. Shima, *J. Biol. Inorg. Chem.*, 2006, **281**, 30804–30813.
151. S. Vogt, E. J. Lyon and S. Shima R. K. Thauer, *J. Biol. Inorg. Chem.*, 2008, **13**, 97–106.
152. E. J. Lyon, S. Shima, R. Boecher, R. K. Thauer, F. W. Grevels, E. Bill, W. Roseboom and S. P. J. Albracht, *J. Am. Chem. Soc.*, 2004, **126**, 14239–14248.
153. S. Shima, E. J. Lyon, R. K. Thauer, B. Mienert and E. Bill, *J. Am. Chem. Soc.*, 2005, **127**, 10430–10435.
154. Y. S. Guo, H. X. Wang, Y. M. Xiao, S. Vogt, R. K. Thauer, S. Shima, P. I. Volkers, T. B. Rauchfuss, V. Pelmenschikov, D. A. Case, E. E. Alp, W. Sturhahn, T. Yoda and S. P. Cramer, *Inorg. Chem.*, 2008, **47**, 3969–3977.

CHAPTER 12

Metal Complex of Hydrogenase Active Sites

JOE DAWSON, FABIO GHIOTTO, JONATHAN MCMASTER AND MARTIN SCHRÖDER

School of Chemistry, University of Nottingham, Nottingham NG7 2RD, UK

12.1 Introduction

Reactions that are catalysed by metal–ligand complexes span the chemical and biological disciplines, and impact upon medicinal, organic, materials and polymer chemistry.[1–7] The current industrial uses for metal complexes include the catalysis of nitrogen fixation,[8] hydrogenation,[9] and CO insertion[10,11] reactions, and these underpin multibillion dollar industries.[12,13] Metal complex catalysts often operate at elevated temperatures and/or pressure regimes and in organic solvents.[13] Thus, an industrial catalyst must often be stable under harsh environmental conditions, but should preferably be relatively cheap to synthesise *via* a readily accessible route. These factors often prevent the replacement of these chemical systems with biological catalysts. However, biological enzymes provide key advantages as they usually function at ambient temperature and pressure, and operate in water as a solvent.[12,14] Moreover, biological catalysts can function at near catalytic perfection to provide extremely high turnovers.[14] High regio-, stereo- and enantioselectivity[4] are displayed by active biosites[14] and their directed specificity generally removes the need for protecting groups.[12]

RSC Energy and Environment Series No. 5
Molecular Solar Fuels
Edited by Thomas J. Wydrzynski and Warwick Hillier

Published by the Royal Society of Chemistry, www.rsc.org

The advantages of both *in vivo* and *in vitro* catalysts may be incorporated into a unified system *via* the design, synthesis and study of low molecular weight metal complex analogues. This approach can provide an optimum balance of properties and also bring insight into structural, mechanistic and spectroscopic features and changes that take place at the active biosite.[12,15] This approach affords routes to the design and synthesis of new metal complexes and clusters *via* new ligands and variations of metal centre(s) towards new catalysts unprecedented either *in vivo* or *in vitro*.[12]

12.2 Introduction to Hydrogenases

The primordial H_2-rich atmosphere of the early Earth is presumed as the point of 'invention' of the hydrogenase enzymes with some of the earliest bacterial life utilising H_2 as an energy source.[16,17] Hydrogenases were first identified by Stephenson and Stickland in 1931,[18] and were first isolated from prokaryotes of the domain bacteria.[19] There are thirteen known families of hydrogenases, and most play a role directly or indirectly in the metabolism of various bacteria and algae.[20,21] This function can be categorised into two main physiological functions: i. catalysis of H_2 consumption linked to energy conservation (*e.g.* respiration, NAD(P)H formation and methanogenesis) and ii. catalysis of H_2 evolution coupled to the disposal of excess reducing potential through the re-oxidation of reduced pyridine nucleotides and electron carriers.[20] Throughout the aerobic and anaerobic microbial world, this role is mediated in enzymes as the catalysis of the cleavage of H_2 into protons or the evolution of H_2 from reduction of protons (Scheme 12.1).[17,22,23]

Although most, if not all, known hydrogenases can catalyse (*in vitro*) the oxidation of H_2 in either direction (Scheme 12.1), they are usually committed to catalyse either H_2 uptake or evolution *in vivo*, depending on the demands of the host organism and on the redox potential of the components interacting with the enzyme.[17,21] In the presence of H_2 and an electron acceptor, the enzyme will oxidise H_2, while in the presence of an electron donor of low potential, reduction of protons from water to H_2 will occur.[17]

The two main groups of hydrogenase enzymes are the [NiFe] hydrogenases, which incorporate Ni and one Fe atom at the active site,[23–31] and [FeFe] hydrogenases, which possess two Fe atoms in their active sites.[32,33] All hydrogenases have at least one carbon monoxide (CO) group and most also contain one or more cyanide (CN^-) ligand(s) in the Fe co-ordination sphere.[34] A third type of hydrogenase discovered by Zirngibl *et al.* in methanogens (*e.g. Methanothermobacter marburgensis*) are [Fe] hydrogenases.[35–37] [Fe] hydrogenases were initially classified as metal-free as they contain no Fe-S cluster and they were believed to function as purely organic catalysts.[36] However, it was

$$H_2 \rightleftharpoons 2H^+ + 2e^-$$

Scheme 12.1 The reversible two electron oxidation of H_2.

later established that their function was dependent on an Fe-containing cofactor.[38–40] Shima *et al.* later re-named these enzymes as Fe-S cluster-free hydrogenases or [Fe] hydrogenases.[41] Further discussion of the active site of [Fe] hydrogenases is provided in Section 12.9 and the following sections will focus on [NiFe] and [FeFe] hydrogenases.

12.2.1 [NiFe] Hydrogenases

In 1995 Volbeda *et al.* reported the first X-ray crystal structure of a [NiFe] hydrogenase which was obtained from the oxidised form of *Desulfovibrio gigas*.[23] Further X-ray crystal structures of *Desulfovibrio* hydrogenases have confirmed the common nature of the binuclear [NiFe] active site.[24–29,31,42] The protein environment is found to be virtually identical in all known [NiFe] hydrogenase structures and the available amino acid sequence data are highly conserved.[43]

The enzyme comprises of an αβ heterodimer with the α (large) subunit of *ca.* 60 kDa incorporating the [NiFe] bimetallic active site, with the Fe-S clusters residing in the β (small) subunit of *ca.* 30 kDa. These two subunits interact extensively through a large contact surface to form a globular heterodimer (Figure 12.1).[17,24–31]

In [NiFe] hydrogenase the Ni centre is bound by four S donors from cysteine, two of which bridge the Ni and Fe centres. Infrared (IR) spectroscopy has

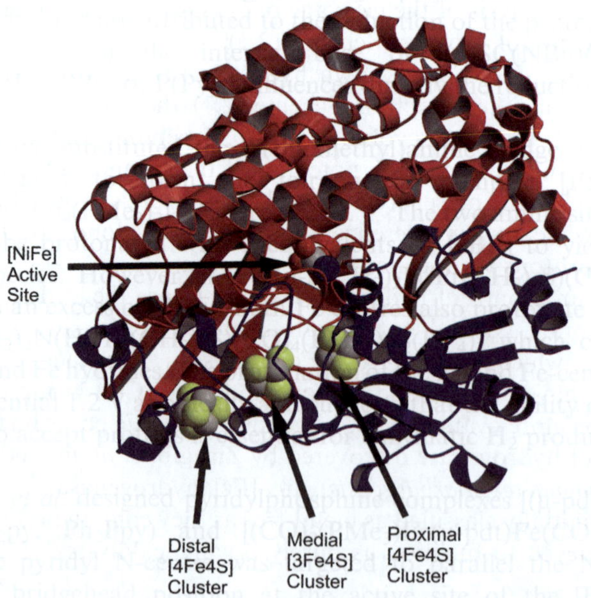

Figure 12.1 The X-ray crystal structure of the [NiFe] hydrogenase of *Desulfovibrio gigas*. The α (large) subunits are depicted in red and the β (small) subunits in blue.[23]

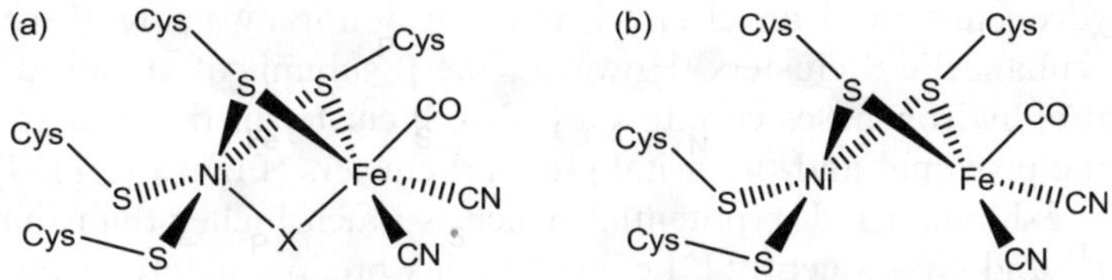

Figure 12.2 Schematic view of the active sites of [NiFe] hydrogenase *D. vulgaris* Myazaki F in the (a) oxidised and (b) reduced forms.[24,27,28] The oxidised form of the active site shows the X bridging species which is currently assigned as peroxide (OOH^-) or hydroxide (OH^-).[48,49]

shown the Fe centre is further ligated by three varying non-protein ligands,[17] either two cyanide groups and one carbonyl group,[25,44,45] three carbonyl groups,[28] or a S monoxide, a carbonyl and a cyanide group.[24,46] In each example the Fe centre is found in a distorted square pyramidal geometry.[43] The FTIR and EPR spectroscopy of the [NiFe] hydrogenase centre in *Ralstonia eutropha* suggest the presence of two additional CN^- ligands, with one CN^- bound to Ni within a proposed $[(NC)NiFe(CN)_3(CO)]$ moiety.[47]

[NiFeSe] hydrogenases, a subclass of [NiFe] hydrogenases found in the sulfate reducing *Desulfomicrobium baculatum*, show the same active site structure as most [NiFe] hydrogenases but are defined by the replacement of one terminal cysteine with a selenocysteine bound to the Ni centre.[26]

The oxidised forms of the [NiFe] hydrogenase enzymes (*D. gigas*,[23,25] *D. vulgaris*,[24,27,28] *D. fructosovorans*,[30] *D. desulfuricans*[29]) show an exogenous species that bridges the Ni and Fe centres (denoted as X in Figure 12.2). This bridging species results in an increase in the Ni–Fe distance and rearrangement of the Ni and Fe centres to distorted square pyramidal and distorted octahedral geometries, respectively. Originally, it was thought that the nature of this bridging species varied according to the organism and/or the redox state in question and was likely an oxo or S ligand.[23,25,43] More recently, crystallographic evidence supports a peroxide, or hydroxide bridging species.[48,49] The reduced form of *D. vulgaris* has also been isolated[26,27] and the conformation of the cysteine ligands in the structure suggests that the bridging site is occupied by a hydride H^- donor.[28]

In the metalloenzyme H_2, protons and electrons shuttle to and from the active site. Thus, access of the substrate to the active site, proton-transfer to solution, and external electron-transfer are all factors affecting the rate of the reversible oxidation of H_2.[50] Hydrophobic channels within the enzyme are believed to link the active site, which is *ca.* 30 Å within the protein, to the surface of the molecule to facilitate gas access.[23–31,51,52] X-Ray diffraction data of crystals of the [NiFe] hydrogenase of *D. fructosovorans* exposed to a Xe atmosphere (at 9 bar) showed ten Xe atoms located within hydrophobic channels. This observation supports the argument that these channels are used for gas (H_2) transfer.[20]

In the [NiFeSe] hydrogenases from *D. baculatum*,[26] and *D. vulgaris*[53] and *Methanococcus voltae*,[54] electrons are conducted between the active site of

[NiFe] hydrogenase and an electron acceptor/donor by up to three linearly arranged cubane Fe-S clusters. However, the β subunit of standard *Desulfovibrio* [NiFe] hydrogenases contains a [3Fe-4S] cluster in the median position between the proximal and the distal [4Fe-4S] clusters (Figure 12.1). The [3Fe-4S] cluster exhibits a redox potential which is much higher than that for H_2 oxidation[55] and consequently [3Fe-4S] remains predominately in the reduced state during enzyme turnover.[50] The centre-to-centre distance of these Fe-S cluster redox centres is *ca.* 12 Å, which is appropriate for electron-transfer. However, direct electron-transfer from the proximal to the distal [4Fe-4S] clusters is more difficult due to the 20 Å intercluster distance.[56,57] It has been proposed that the proximal [4Fe-4S] cluster is essential to H_2 activation as it is within *ca.* 14 Å of the active site.[23,51] Furthermore, electron-transfer from the active site to the distal [4Fe-4S] must be efficient as only the latter is exposed at the molecular surface[50] where it mediates electron-transfer to a redox partner electron donor/acceptor, such as cytochrome c_3.[49,52]

Several proton-transfer pathways have been proposed for [NiFe] hydrogenases based on their respective three-dimensional structures. However, the problem in defining proton-transfer is that as one moves away from the active site a multitude of pathways become plausible.[50] However, *His536* has been postulated to play a role in proton-transfer (proton-relay) due to its proximity to the active site.[24]

A Mg(II) ion ligated by three waters is also present in all [NiFe] hydrogenase crystal structures, but this only became apparent in a 2.8 Å resolution structure of an oxidised form of the protein from *D. gigas*.[50] Metal-bound waters may be well suited to proton-transfer due to their lower pK_a compared to free water. The Mg(II) ion, found near the protein surface, may be instrumental in allowing proton transfer to a redox partner, *e.g.* cytochrome c_3.[29] Internal water molecules may also be involved in proton transfer.[50] One possible mechanism for this process is the Grotthuss mechanism in which water is protonated and subsequently undergoes rearrangement and releases a proton at a different site.[58] Furthermore, nearby residues with suitable pK_a values for transient protonation may also be required, such as histidine and carboxylate groups.[50]

12.2.2 [FeFe] Hydrogenases

Unlike the [NiFe] hydrogenases, which are composed of at least two subunits, many [FeFe] hydrogenases are monomeric enzymes containing up to four Fe-S centres in addition to the catalytic Fe-Fe unit and the proximal cluster (collectively named the H-cluster) (Figure 12.3). However, dimeric, trimeric, and tetrameric enzymes are also known.[21,32,52,59–63]

The catalytic subunits of [FeFe] hydrogenases vary considerably in size, with the smallest [FeFe] hydrogenases being *ca.* 45–48 kDa.[17] The first reported X-ray crystal structure of the [FeFe] hydrogenases of *D. desulfuricans*[33] and *Clostridium pasteurianum*[32] showed the central H-cluster as a binuclear Fe-Fe structure where each Fe centre is ligated by two terminal CO ligands, and three

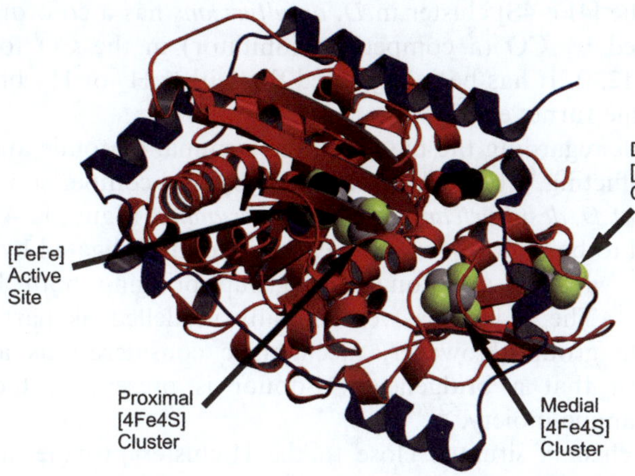

Figure 12.3 The X-ray crystal structure of the [FeFe] hydrogenase of *Desulfovibrio desulfuricans*.[52]

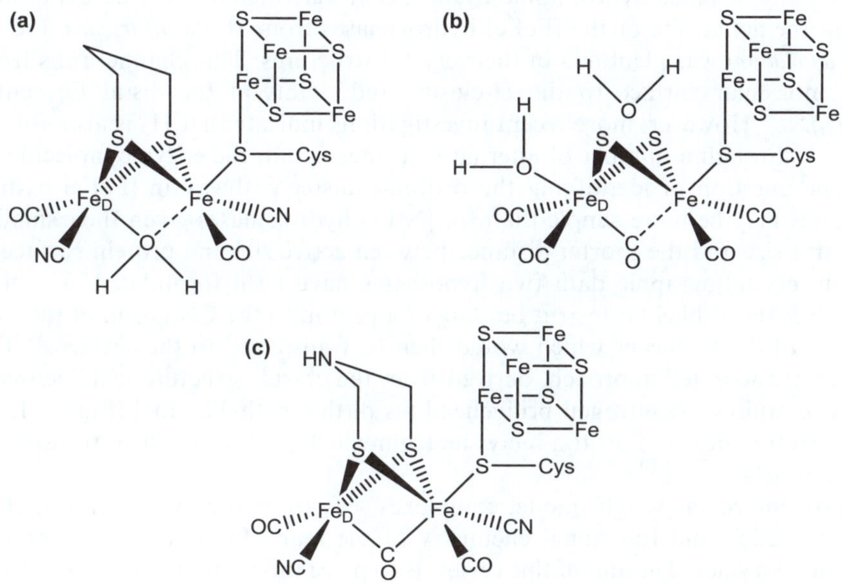

Figure 12.4 The proposed structures of the active sites of [FeFe] hydrogenases from (a) *D. desulfuricans* and (b) *C. pasteurianum* hydrogenases.[64–67] View of (c) a consensus proposal for the active sites of the [FeFe] hydrogenases from *D. desulfuricans* and *C. pastuerianum* containing a di(thiomethyl)-amine bridge.[68–70]

bridging ligands (Figure 12.4). The bridging ligands comprise of two S groups and a small molecule.[64–67] A [4Fe-4S] cluster is also connected to the Fe-Fe centre *via* the thiolate of a cysteine belonging to the protein.[32,33] The Fe atom

(Fe_D) distal to the [4Fe-4S] cluster in *D. desulfuricans* has a co-ordination site which is occupied by CO (a competitive inhibitor) in the CO form of the enzyme (Figure 12.4). It has been suggested that either H_2 or H^- binds at this site during enzyme turnover.[17]

Early proposals regarding the nature of the terminal diatomic and bridging ligands were conflicting,[69] but subsequent analysis and comparison of the two enzymatic cores of *D. desulfuricans* and *C. pastuerianum* (Figure 12.4) allowed a consensus model to be built up in which each Fe centre is ligated by a CO and a CN^- group[68] with an additional CO group bridging both Fe centres (Figure 12.4).[69,70] The S ligands were initially modelled as part of a 1,3-propanedithiolate group. However, mechanistic considerations and model chemistry suggest that a bridgehead N-donor is present in the form of a di(thiomethyl)amine moiety.[52,70,71]

The [4Fe-4S] clusters situated close to the H-cluster provide an electron-transfer pathway from the buried active site to the protein surface which relays electrons to a cytochrome c_3 redox partner.[33,72] *D. desulfuricans and C. pastuerianum* both contain two [4Fe-4S] clusters, medial and distal, with a centre to centre distance of *ca.* 11 Å, which is common to many redox enzymes.[32,33]

Initially, a single hydrophobic channel that may facilitate gas access to and from the active site of the [FeFe] hydrogenases from *D. desulfuricans* and *C. pasteurianum* was identified in their crystal structures. This channel runs from the molecular surface to the H-cluster and points at the distal Fe centre (Fe_D).[32,52] However, more recent investigations indicate that H_2 is also able to diffuse through a number of alternative routes within the enzyme molecule.[73]

The question of identifying the proton-transfer pathway in [FeFe] hydrogenases may be more simple than for [NiFe] hydrogenases given their smaller relative size and the shorter distance between active site and protein surface.[50] From crystallographic data two hypotheses have been formulated. The first involves the unlikely transient binding of a proton to the CN group of the Fe_D centre of the H-cluster which would then be transferred to the surface.[33] The currently accepted approach, derived from the crystal structure of *C. pasteurianum*, utilises the nitrogen bridgehead [as di(thiomethyl)amine] (Figure 12.4) as a protonation and proton relay site from which protons are then transferred to the surface.[33,59,70]

Low molecular weight model complexes seek to mimic the structural, electronic, redox and functional chemistry of the many facets of these complex metalloenzymes. The aim of this review is to place this work into context and to highlight recent advances in developing new *in vitro* catalysts for reversible H_2 oxidation and production.

12.3 Nickel Thiolate Complexes

Early work on the modelling of [NiFe] hydrogenase pre-dates the solution of the first crystal structure of the enzyme.[74] Initially it was believed that the active site of the enzyme consisted of a mono- or binuclear Ni-thiolate system, with

EXAFS studies suggesting that the ligand environment about the Ni centre was a tetragonal $[Ni(SR)_4]^{2-}$ unit and Fe was believed to be present in [4Fe-4S] electron-transfer relays.[75–79] Therefore, much research focussed at that time on the preparation of complexes of type $[Ni(SR)_4]^{2-}$,[20,80–82] which were compared to the known features of the hydrogenases.[75]

Rosenfield *et al.* reported a series of mononuclear Ni(II) complexes that incorporated four monodentate thiolate ligands to form a distorted tetrahedral species of the type $(R_4N)_2[Ni(SAr)_4]$ (Ar = C_6H_5, *p*-C_6H_4Cl, *p*-$C_6H_4CH_3$ and *m*-C_6H_4Cl) (Figure 12.5).[83] The distorted complex has two S–Ni–S angles that are close to 90° and an average Ni–S distance of 2.281(1) Å which closely resemble the observed co-ordination at the Ni site in the [NiFe] hydrogenase enzyme.[84] This work was extended *via* the complexation of chelating bidentate thiolate ligands such as ethane-1,2-dithiol (H_2L^1),[77,85–87] butane-2,3-dithiol (H_2L^2),[86] and norbornane dithiol (H_2L^3)[79] (Figure 12.6) to Ni(II).

The complexes $(Ph_4P)_2[Ni(L^1)_2]\cdot 4H_2O$ and $K_2[Ni(L^2)_2]\cdot C_2H_5OH$ can be synthesised by reaction of $NiCl_2\cdot 6H_2O$ with an excess of the respective thiolate ligands, and the crystal structure of $[Ni(L^1)_2]^{2-}$ confirms this to be a square planar complex (Figure 12.7).[86] This contrasts with the tetrahedral Ni(II)

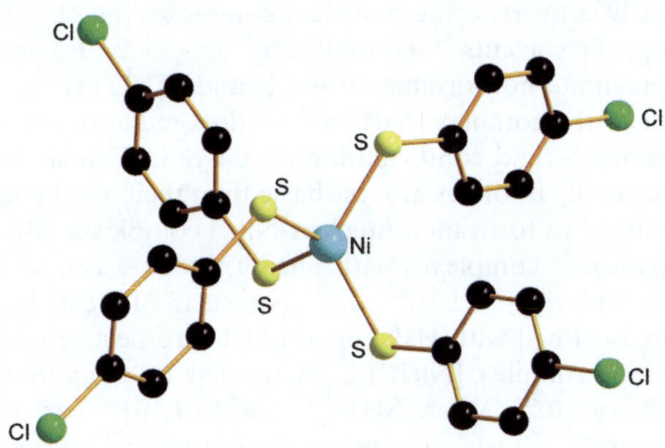

Figure 12.5 View of the single crystal X-ray structure of $[Ni(S\text{-}p\text{-}C_6H_4Cl)_4]^{2-}$ from the series of complexes $(R_4N)_2[Ni(SAr)_4]$ (Ar = C_6H_5, *p*-C_6H_4Cl, *p*-$C_6H_4CH_3$ and *m*-C_6H_4Cl).[83]

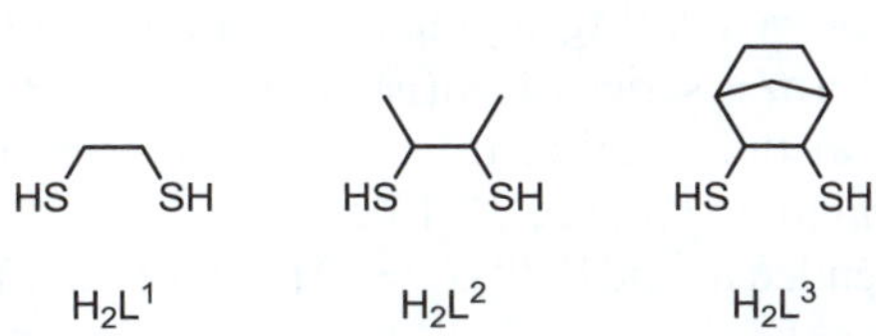

Figure 12.6 The chelating bidentate thiol ligands ethane-1,2-dithiol (H_2L^1),[77,85–87] butane-2,3-dithiol (H_2L^2),[86] and norbornane dithiol (H_2L^3).[79]

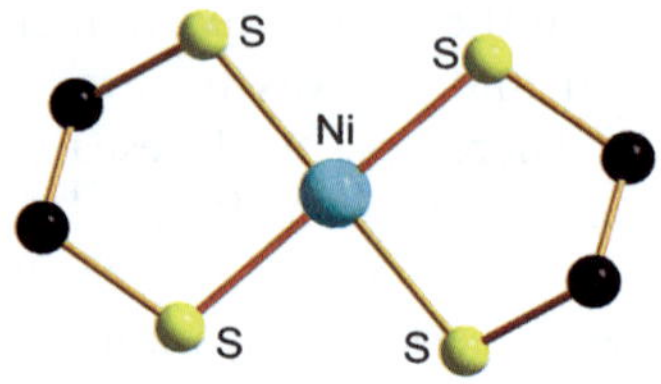

Figure 12.7 View of the single crystal X-ray structure of $[Ni(L^1)_2]^{2-}$.[86]

complexes of thiophenolates reported above, which may reflect the lower basicity and weaker ligand field of thiophenolates compared to alkyl thiolates.[83] The redox properties of the complexes $[Ni(L^1)_2]^{2-}$ and $[Ni(L^2)_2]^{2-}$ are significantly different.[86] $[Ni(L^1)_2]^{2-}$ degrades at the electrode surface upon oxidation in the range –0.60 to –0.46 V *vs.* SHE to the formal Ni(III) state, while $[Ni(L^2)_2]^{2-}$ undergoes reversible oxidation from Ni(II) to Ni(III) at –0.51 V *vs.* SHE.

Thiolate groups have a strong tendency to form bridges between two or three metal centres, and therefore simple alkyl thiolates readily form oligonuclear or polynuclear complexes with Ni(II) salts.[77,85] Consistent with this, the complexes $[Ni(L^n)_2]^{2-}$ rapidly convert to the binuclear complexes $[Ni_2(L^n)_3]^{2-}$ ($n = 1, 2$) in protic or wet aprotic solvents.[86] To inhibit this reactivity, Fox *et al.* developed the chelating bidentate norbornane dithiol ligand, $[H_2L^3]$ (Figure 12.6),[79] and oligomerisation of the complex $[Ni(L^3)_2]^{2-}$ is thus reduced due to the norbornane unit enforcing a rigid conformation on the S–C–C–S chelate.

In general, aromatic thiolates are less basic than their alkyl counterparts and are thus more likely to form mononuclear Ni(II) complexes, although the formation of polynuclear complexes with phenyl thiolates can still occur.[88] The mononuclear complexes $[Ni(L^4)_2]^{2-}$ and $[Ni(^tBu\text{-}L^4)_2]^{2-}$ can be generated by reaction of $[NiCl_2(PPh)_3]$ with H_2L^4 or $^tBu\text{-}H_2L^4$, respectively (Figure 12.8).[89] The formal Ni(III) complex $[Ni(L^4)_2]^-$ is formed by oxidation of the Ni(II) precursor at $E_{1/2} = -0.23$ V *vs.* SHE.[90] $[Ni(^tBu\text{-}L^4)_2]^{2-}$ can be oxidised to $[Ni(^tBu\text{-}L^4)_2]^-$ and $[Ni(^tBu\text{-}L^4)_2]^0$, and crystallographic data suggest that these oxidations are '$[NiS_4]$' based.[91] However, the aromatic ligand H_2L^4 reacts with Ni(II) salts to give the neutral trinuclear cluster $\{Ni_3[\mu_2\text{-}L^4][\mu,\mu'\text{-}L^4]_2(PPh_3)_2\}$ (Figure 12.9), which in co-ordinating solvents readily fragments to charged species and other polynuclear Ni-thiolate complexes.[89]

To avoid undesired oligomerisation, Sellmann *et al.* adopted an alternative approach involving polychelating ligands containing both thiophenolate and thioether donors to form a series of Ni(II) complexes.[78,91–95] The observed co-ordination geometry at the metal centres in these species are controlled by the flexible polydentate ligands (Figure 12.10).

Reaction of the extended ligand $[L^n]^{2-}$ ($n = 5$) (Figure 12.10) with Ni(II) forms the binuclear complex $[Ni_2(L^5)_2]$ (Figure 12.11) with the ligands bridging M(II) ions rather than chelating a single metal centre. A similar binuclear complex $[Ni_2(L^6)_2]$ was isolated from the reaction of $[L^6]^{2-}$ with Ni(II) salts (Figure 12.11).[78]

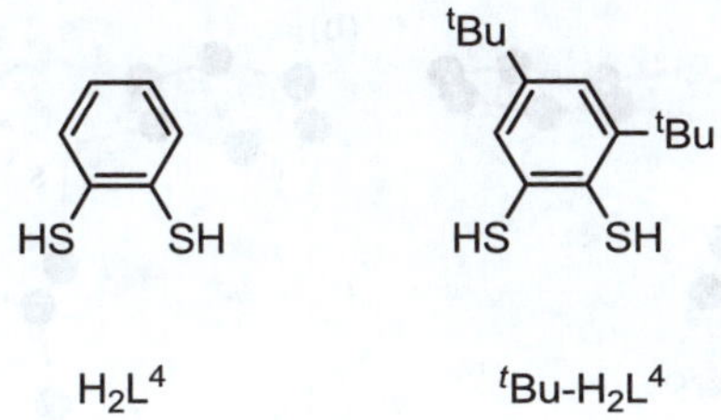

Figure 12.8 H_2L^4 and ^{t}Bu-H_2L^4 reported by Cha *et al.*[89,90]

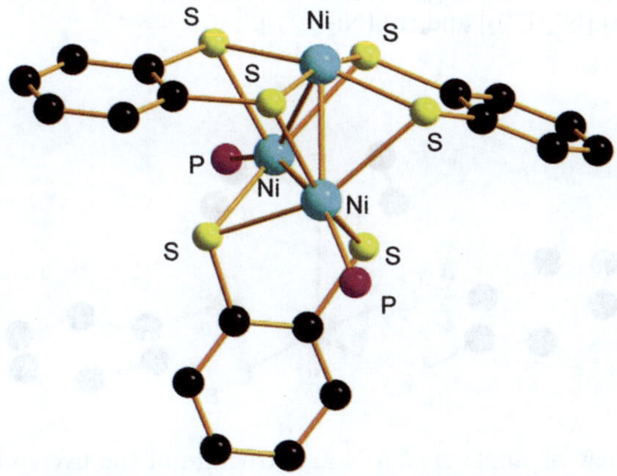

Figure 12.9 View of the single crystal X-ray structure of $\{Ni_3[(\mu)_2\text{-}L^4][(\mu,\mu')\text{-}L^4]_2(PPh_3)_2\}$; phenyl rings of PPh_3 have been omitted for clarity.[89]

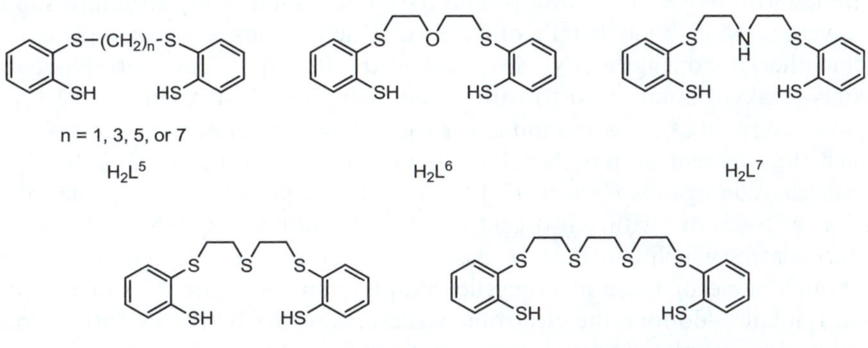

Figure 12.10 Selection of polychelate ligands used by Sellmann *et al.*[78,91–95]

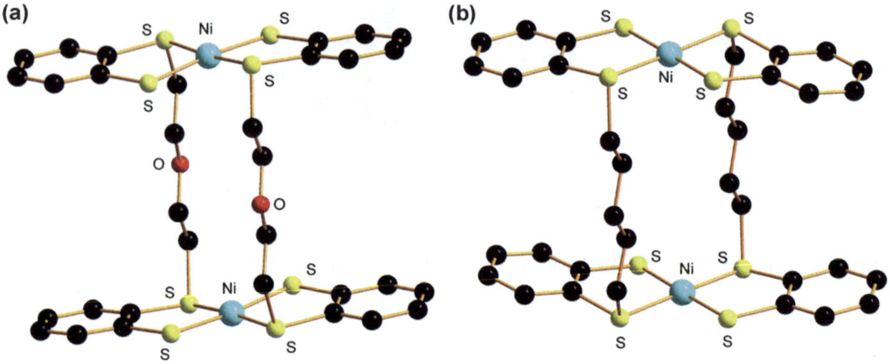

Figure 12.11 Views of the single crystal X-ray structures of the binuclear complexes (a) $[Ni_2(L^6)_2]$ and (b) $[Ni_2(L^5)_2]$.[78]

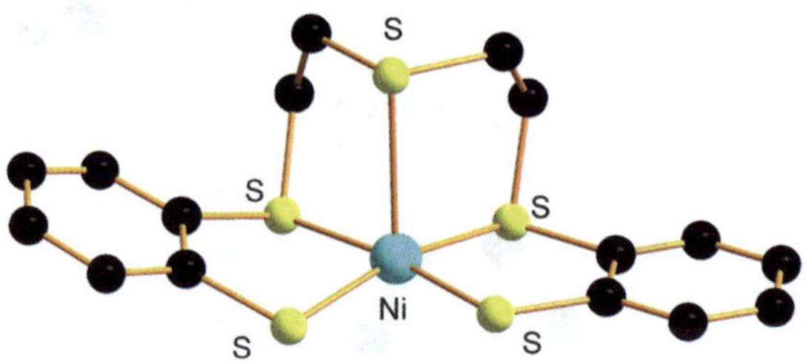

Figure 12.12 View of single crystal X-ray structure of the five co-ordinate complex $[Ni(L^8)]$.[78]

In contrast, the Ni(II) complexes $[Ni(L^7)]$ and $[Ni(L^8)]$ are both mononuclear five co-ordinate species, with the central NH and S centres of the polychelate involved in binding to the metal ion.[78] $[Ni(L^8)]$ was the first example of a Ni(II) complex with five S-donor atoms, and its single crystal X-ray structure shows an average Ni–S bond length of 2.21 Å, which is shorter than typical Ni–S(thioether) bond lengths ($|Ni–S| = 2.36$ Å; $\sigma = 0.11$ Å).[96] The central S-centre is only weakly bound to Ni(II) with a bond length of 2.74 Å (Figure 12.12).[78] Conversely, $[Ni(L^9)]$ is a mononuclear square planar diamagnetic complex in which the two central thioether S atoms do not bind to the metal centre.[78]

Macrocyclic ligands (Figure 12.13) have also been used to study the interaction of S-donors with Ni(II) centres.[97–101] These ligands stabilise Ni(I) and Ni(III) centres which can then be assessed structurally and spectroscopically. Although some of these macrocyclic complexes incorporate thioether rather than thiolate S-donors, the electronic structures of the Ni(I) and Ni(III) redox products can be related to the paramagnetic Ni(I) and Ni(III) intermediate states of the active forms of [NiFe] hydrogenase.[102] Thus, the complex $[Ni([9]aneS_3)_2]^{3+}$ (Figure 12.14) was the first homoleptic octahedral Ni(III)

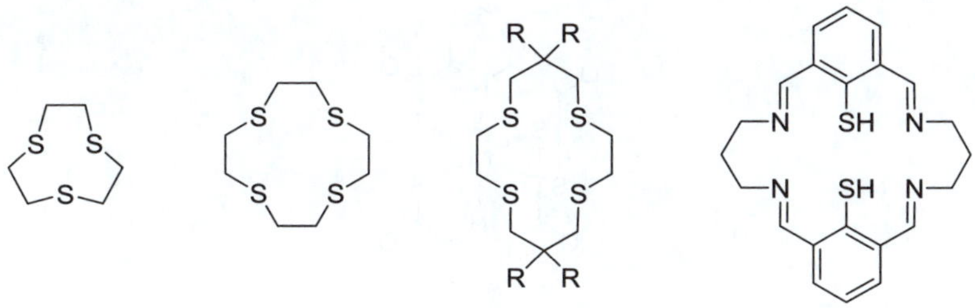

[9]aneS_3 [12]aneS_4 R_4[16]aneS_4 H_2L^{10}

Figure 12.13 A selection of sulfur containing macrocyclic ligands.[97–101]

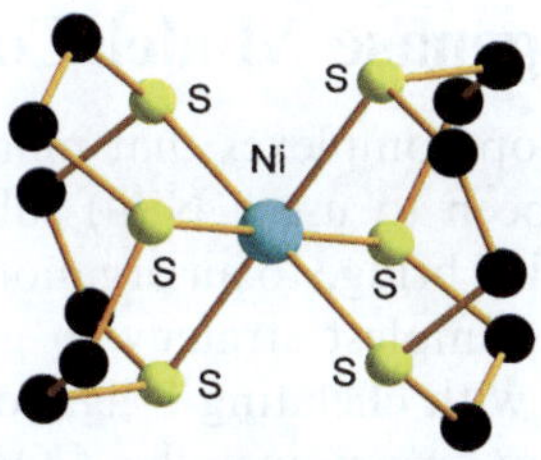

Figure 12.14 View of the single crystal X-ray structure of $[Ni([9]aneS_3)_2]^{3+}$.[99]

species to be reported.[99] It shows an average Ni–S bond length of 2.3129(25) Å with the EPR spectrum showing $g_1 = 2.093$, $g_2 = 2.075$, $g_3 = 2.027$.

The mixed valence Ni(II)/Ni(III) complex $[Ni_2(L^{10})(MeCN)_4]^{3+}$ has been prepared by electrochemical and chemical oxidation of the Ni(II)Ni(II) precursor[103,104] and analysed by multi-frequency and pulsed EPR spectroscopy. This confirms that 91% spin density is found at the Ni(III) centre, of which 72% is in the $3dz^2$ orbital and 16% in the $3dxy$ orbital with the Ni(II) containing 3.5% spin density, and the axial MeCN ligands 7.5% spin density.[101]

To probe the unusual Ni(III) state found within [NiFe] hydrogenases, Krüger and Holm investgated[105] the reaction of pyridine-2,6-bis(monothiocarboxylic acid) (H_2pdtc) with $[Ni(OAc)_2 \cdot 4H_2O]$ in a mixture of EtOH and Et_4NOH to give $(Et_4N)_2[Ni(pdtc)_2]$ (Figure 12.15). $(Et_4N)_2[Ni(pdtc)_2]$ contains two tridentate pdtc^{2-} ligands bound meridionally to give a planar NiS_4 centre, with two mutually *trans* axial pyridyl N-donors to give an overall tetragonally distorted octahedral stereochemistry at Ni(II). $(Et_4N)_2[Ni(pdtc)_2]$ is readily oxidised by iodine to form $(Et_4N)[Ni(pdtc)_2]$, and EPR studies confirm this complex to incorporate Ni(III) in a d_{z^2} ground-state electronic configuration similar to the oxidised [NiFe] hydrogenase enzyme.[105] The metal-centred redox processes of $(Et_4N)[Ni(pdtc)_2]$ were found to be strongly influenced by the ligand properties which results in the relatively anodic potential of –0.085 V *vs.* Fc^+/Fc for the Ni(II)/Ni(III) couple in DMF. Moreover, $(Et_4N)_2[Ni(pdtc)_2]$ and

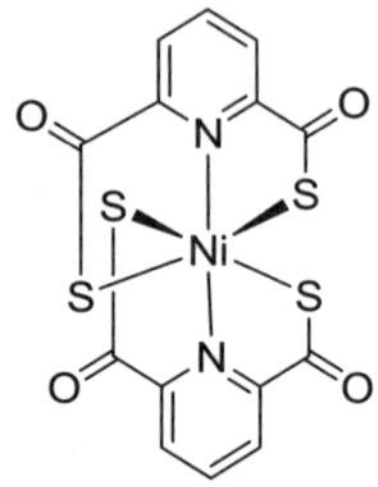

Figure 12.15 View of the complex $(Et_4N)_2[Ni(pdtc)_2]$.[105]

$(Et_4N)[Ni(pdtc)_2]$ were the first pair of structurally defined Ni(II)/Ni(III) complexes incorporating anionic S-donors.

12.4 [NiFe] Hydrogenase Model Complexes

The main strategy to develop complexes that mimic the active site in [NiFe] hydrogenase enzymes has been to use a Ni(II) subunit ligated by thiolate or related S-donors, two of which bridge to an organometallic CO/CN^- ligated Fe subunit.[23–29,31,42] Thus, the simplest strategy to generate these systems is to synthesise a Ni(II) complex with chelating S ligands and use this subunit as a metallothiolate ligand to an organometallic CO/CN^- ligated Fe centre.[106] However, this approach is troublesome as metal-thiolate complexes tend to form homometallic oligomers and clusters and, therefore, mixed donor environments were used in early synthetic models.[107–110] Nevertheless, relatively few examples of structures possessing [NiFe] motifs with pertinent ligation around the metal centres are known, which is in stark contrast to the extensive number of structural models for [FeFe] hydrogenase.[111] Pickett *et al.* have recently published[111] a comprehensive review of [NiFe] model complexes that have been described in the literature, and there have been a number of related reviews.[80,84,106,107,112,113] In the following sections we discuss a selection of the various mixed donor [NiFe] systems that have been reported.

12.4.1 Amine $[N_2Ni(\mu\text{-}S_2)Fe]$ Complexes

Darensbourg *et al.* synthesised the first thiolate bridged Ni-Fe compound $[Ni(L)Fe(CO)_4]$ [L = N,N′-bis(2-mercaptoethyl)-1,5-diazacyclooctane] in 1996 (Figure 12.16a).[109] This complex confirms that a heterobimetallic complex can be generated using simple N_2S_2-co-ordination at Ni(II) with a terminal thiolate binding in a monodentate fashion to a $Fe(CO)_4$ fragment. The single thiolate bridge results in a relatively long Ni-Fe separation of 3.76 Å, too long for any direct interaction between the metal centres, and longer than that observed for the biosites which shows a Ni-Fe separation of 2.5–2.6 Å in the reduced form[26] and 2.9 Å in the oxidised form.[23,25] Pohl *et al.* reported the first example of a Ni-Fe complex with two thiolate bridges in 1997 (Figure 12.16b).[114] This unique double thiolate bridge results in a Ni-Fe bond length of 2.8 Å which is similar to that observed in the active site of [NiFe] hydrogenase from *D. gigas* (*ca.* 2.9 Å).[23,25]

Schröder *et al.* have reported two structural models based upon mixed amine-thiolate polychelate scaffolds. The complex [Ni(tsalen)Fe(CO)$_3$][115] (Figure 12.16c) uses a Schiff-base ligand to introduce imine π-bonds and bridging thiolates, while an alternative approach protects the Fe centre within a Fe(cp)$^+$ moiety which can bind to a Ni(II) thiolate complex to give a distorted square-pyramidal geometry at Ni(II) (Figure 12.16d).[110] Both of these complexes show relatively short Ni-Fe bond lengths of 2.539(4) and 2.8924(6) Å, respectively, suggesting potential metal–metal interactions in these systems.

12.4.2 Phosphine [P$_2$Ni(μ-S$_2$)Fe] Complexes

The replacement of the N$_2$S$_2$ ligation about the Ni(II) centre with mixed soft P and S-donors to better replicate the cysteinate environment of the enzymatic centre has been reported. Evans *et al.*[116,117] have reported the synthesis and characterisation of complex [{Fe(*NS$_3$*)(CO)$_2$-*S,S*'}NiCl(dppe)] (*NS$_3$* = [N(CH$_2$CH$_2$S)$_3$]$^{3-}$; dppe = 1,2-diphenylphosphinoethane; Figure 12.17a), which has structural features similar to the active site of [NiFe] hydrogenase with a CO stretching frequency (1944 cm^{-1}) at a similar frequency (1947 cm^{-1}) to that in the inactive form of *D. Gigas* [NiFe] hydrogenase.[25] Reaction of [Ni(dppe)(pdt)] (H$_2$pdt = 1,3-propanedithiol) (Figure 12.17b) with [Fe(CO)$_3$(bda)]

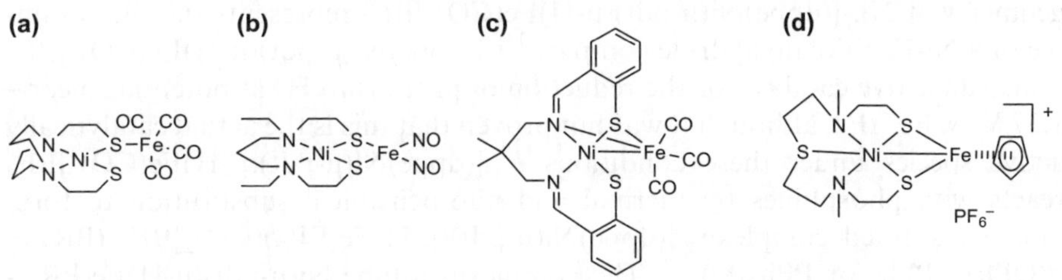

Figure 12.16 View of structures incorporating the [N$_2$Ni(μ-S$_2$)Fe] ligand motifs reported by (a) Darensbourg *et al.*,[109] (b) Pohl *et al.*[114] and (c), (d) Schröder *et al.*[110,115]

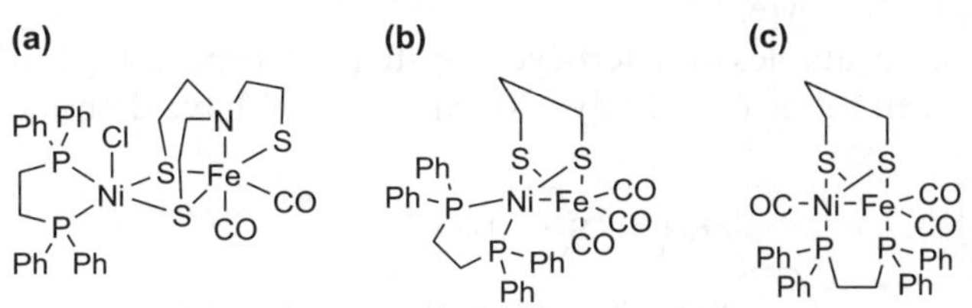

Figure 12.17 View of (a) [{Fe(*NS$_3$*)(CO)$_2$-*S,S*'}NiCl(dppe)] (*NS$_3$* = [N(CH$_2$CH$_2$S)$_3$]$^{3-}$; dppe = 1,2-diphenylphosphinoethane) reported by Evans *et al.*,[116,117] (b) [(dppe)Ni(μ-pdt)Fe(CO)$_3$] and (c) [(OC)Ni(μ-dppe)(μ-pdt)Fe(CO)$_2$] reported by Schröder *et al.*[110]

(bda = benzylidene acetone) or $[Fe_3(CO)_{12}]$ forms $[(dppe)Ni(\mu\text{-}pdt)Fe(CO)_3]$.[110] In this complex Ni(II) adopts a distorted tetrahedral P_2S_2 geometry bridging to an Fe(0) centre in a square-based pyramidal geometry (Figure 12.17b). The complex $[(dppe)Ni(\mu\text{-}pdt)Fe(CO)_3]$ converts to $[(OC)Ni(\mu\text{-}dppe)(\mu\text{-}pdt)Fe(CO)_2]$ (Figure 12.17c) in solution. Both $[(dppe)Ni(\mu\text{-}pdt)Fe(CO)_3]$ and $[(OC)Ni(\mu\text{-}dppe)(\mu\text{-}pdt)Fe(CO)_2]$ show butterfly arrangements at the binuclear core[110] and the Ni-Fe separation of 2.5 Å in these complexes, supported by DFT calculations, confirm the presence of a formal Ni-Fe bond suggesting that a similar direct metal-metal bonds may be important at the reduced active form of the enzyme. Furthermore, the acute dihedral angle $<Ni\text{–}(\mu\text{-}S)_2\text{–}Fe$ of 80.3° found in $[(dppe)Ni(\mu\text{-}pdt)Fe(CO)_3]$ is comparable to that in the natural enzyme in its reduced active form (88.95°).[26]

Although more than fifty [NiFe] hydrogenase models have been reported,[111] very few have been shown to display functional activity. Often it is not clear whether these products have been assayed for H_2 oxidation or proton reduction. One reason for this may be that it is generally believed that H_2 activation occurs at the Ni site and very few stable Ni hydrides are known with any ligand set.[74] However, Dubois *et al.* have shown that Ni can be an excellent hydride acceptor in the presence of a pendant base (see Section 12.7). Moreover, Rauchfuss *et al.* have shown[118] that a solution of $[(dppe)Ni(\mu\text{-}pdt)Fe(CO)_3]$[110] in CH_2Cl_2 readily protonates in the presence of $HBF_4.Et_2O$ to form $[(dppe)Ni(\mu\text{-}pdt)(\mu\text{-}H)Fe(CO)_3]BF_4$ in which the Ni and Fe sites are bridged by a hydride, with quasi-octahedral co-ordination at Fe and a square-pyramidal geometry at Ni. $[(dppe)Ni(\mu\text{-}pdt)(\mu\text{-}H)Fe(CO)_3]BF_4$ represents the first example of a Ni-Fe thiolate hydride complex.[119] $[(dppe)Ni(\mu\text{-}pdt)(\mu\text{-}H)Fe(CO)_3]BF_4$ is also an active catalyst for the reduction of protons to H_2 at potentials near −1.37 V *vs.* Fc^+/Fc, although it was not proven that this is the actual catalytically active species under these conditions.[118] $[(dppe)Ni(\mu\text{-}pdt)(\mu\text{-}H)Fe(CO)_3]BF_4$ reacts with phosphines by thermal and photochemical substitution to form monosubstituted complexes $[(dppe)Ni(\mu\text{-}pdt)(\mu\text{-}H)Fe(PR_3)(CO)_2]BF_4$ ($PR_3 = P(OPh)_3$, PPh_3 or PPh_2Py).[120] The complexes $[(dppe)Ni(\mu\text{-}pdt)(\mu\text{-}H)Fe(PR_3)(CO)_2]BF_4$ ($PR_3 = P(OPh)_3$, PPh_3 or PPh_2Py) show quasi-reversible redox behaviour and catalyse the reduction of H^+ to H_2. The neutral complexes $[(dppe)Ni(\mu\text{-}pdt)(Fe(PR_3)(CO)_2]$ ($PR_3 = P(OPh)_3$, PPh_3 or PPh_2Py) can be isolated by direct deprotonation of $[(dppe)Ni(\mu\text{-}pdt)(\mu\text{-}H)Fe(PR_3)(CO)_2]BF_4$ with NaOMe.[120]

Recently, two examples of macrocyclic systems incorporating both Ni and Fe centres have been reported. Firstly, Lodeiro *et al.*[121] have described the use of

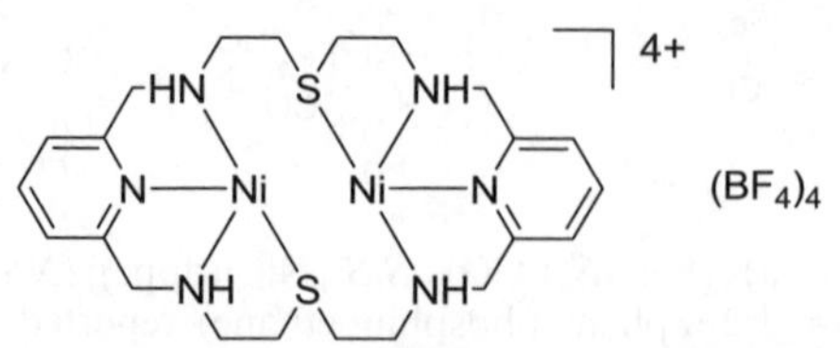

Figure 12.18 View of $[Ni_2L](BF_4)_4$ ($L = Py_2N_4S_2$) reported by Lodeiro *et al.*[121]

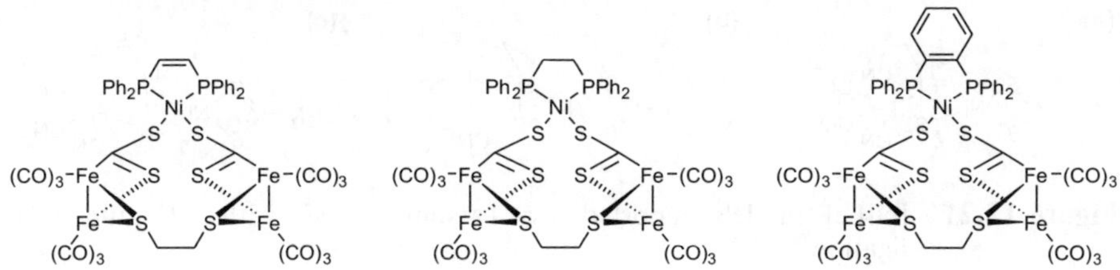

Figure 12.19 The series of [$NiFe_4$] macrocyclic compounds reported by Song *et al.*[123]

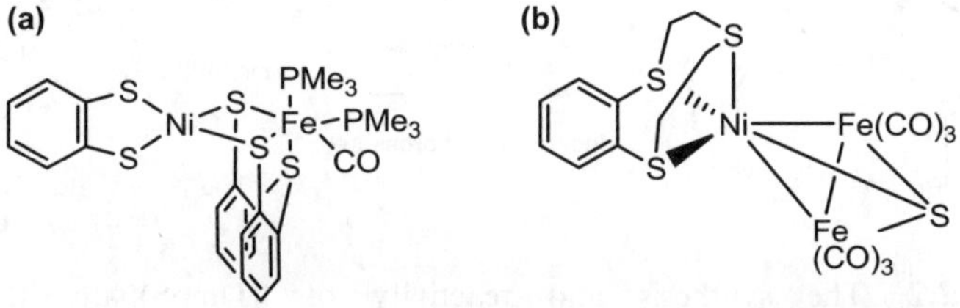

Figure 12.20 [NiFe] models reported by (a) Sellmann *et al.*[124] and (b) Perra *et al.*[125]

the known $Py_2N_4S_2$ macrocycle[122] to successfully bind two metallic centres, either as [Ni_2L] (Figure 12.18), [Fe_2L] or [NiFeL] complex moieties. The ligand is bound to the metal ions *via* mixed N/S donation and each metal centre is co-ordinated further by solvent and anions in a distorted octahedral geometry.

Song *et al.* have reported[123] the synthesis of several novel macrocyclic [$NiFe_4$] complexes (Figure 12.19) that show thiolate bridged quadruple-butterfly structures that are similar to the [$NiFe_2S$] butterfly motif found in the active site of [NiFe] hydrogenase. However, these complexes have not been assayed for catalytic activity.

12.4.3 Thiolate [S_xNi(μ-S_y)Fe] Complexes

Sellmann *et al.* reported[124] the first low spin Ni(II) centre bound in a S-only environment in which the [$Ni(L^8)L]^{2-}$ (L^{2-} = benzene-1,2-dithiolate) subunit is bound by two thiolate bridges to a [$Fe^{II}CO(PMe_3)_2]^{2+}$ fragment to give [(L)Ni(L^8)FeCO($PMe_3)_2$] (Figure 12.20a). It was suggested that the two PMe_3 ligands donate electron density in an analogous way to two CN^- ligands. Perra *et al.* have reported[125] the cluster [(L)NiSFe$_2$(CO)$_6$] (Figure 12.20b), which incorporates a bridging sulfide ligand and has parameters similar to those observed in the active site of the [NiFe] hydrogenase from *D. vulgaris*[24,27,28] and *D. desulfuricans*.[29]

Two structural models of the [NiFe] hydrogenase biosites have been reported by Tatsumi *et al.* in which square planar dithiocarbamate Ni(II) derivatives are bridged by pdt^{2-} to a $Fe(CN)_2(CO)_2$ fragment (Figure 12.21a and b).[126] Importantly, this is a rare system in which the Fe site is bound by both CO and CN^- ligation, with Ni-Fe and M-S bond lengths closely matching those of the oxidised form of the [NiFe] hydrogenase enzymes of *D. gigas* and

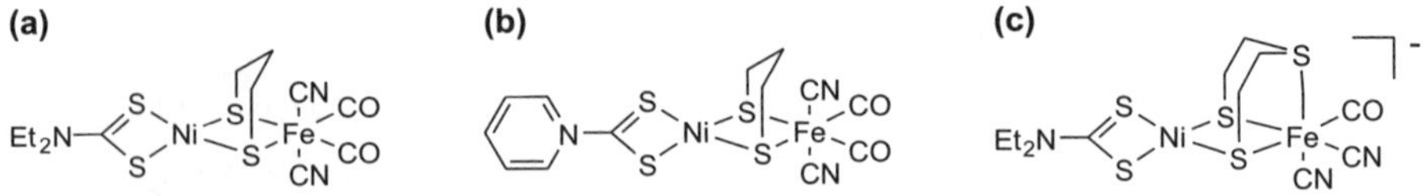

Figure 12.21 [NiFe] models reported by Tatsumi *et al.* with CO and CN^- ligation.[126,129]

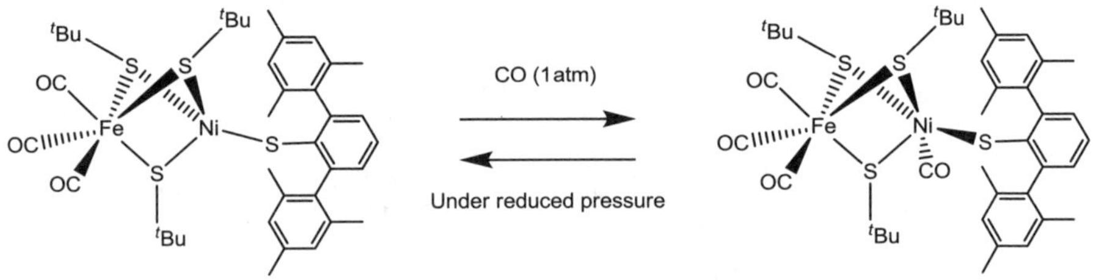

Scheme 12.2 The synthesis and reactivity of [(DmpS)Ni(μ-S^{t}Bu)$_3$Fe(CO)$_3$], [Dmp = C_6H_3-2,6-(mesityl)$_2$] reported by Tatsumi *et al.*[132]

D. fructosovorans.[25,127,128] More recently, the same group reported the synthesis of a dithiolate-bridged binuclear complex, [(Et_2NCS_2)Ni(μ-tpdt) Fe(CO)$(CN)_2$]$^-$ (tpdt^{2-} = 3-thiapentanedithiolate) (Figure 12.21c),[129] which shows the desired *fac*-Fe$(CN)_2$(CO) geometry at the Fe centre. Furthermore, this complex reproduces the CN/CO stretching vibrations in the IR spectrum associated with the Ni-A, Ni-B, and Ni-SU states of the enzyme. However, the square planar Ni(II) centre in [(Et_2NCS_2)Ni(μ-tpdt)Fe(CO)$(CN)_2$]$^-$ is not an accurate structural match for the Ni-A, Ni-B and Ni-SU states of [NiFe] hydrogenase, which are thought to contain penta-co-ordinated Ni(III) and Ni(II) centres.[127,130,131]

Tatsumi *et al.* have also reported[132] the thiolate-bridged binuclear NiFe complex [(DmpS)Ni(μ-S^{t}Bu)$_3$Fe(CO)$_3$] [Dmp = C_6H_3-2,6-(mesityl)$_2$] (Scheme 12.2) comprising a Fe(CO)$_3$ unit coupled to a trigonal-pyramidal Ni centre. The reversible binding of CO occurs at the Ni site to form [(DmpS)(CO)Ni(μ-S^{t}Bu)$_3$-Fe(CO)$_3$] (Scheme 12.2) which is structurally related to the CO-inhibited form of [NiFe] hydrogenase from *C. vinosum* and *D. fructosovorans*.[132] However, this system incorporates 3 bridging metal-thiolates between the Ni and Fe centres, has no CN^- binding at Fe, and involves an elongated Ni-Fe separation of 3.17 Å, substantially longer than that in the protein.[28] These disparities mean it is likely that the formal oxidation states and electronic structures of the Fe and/or the Ni centres are different from those found at the active sites.[132]

12.4.4 Polymetallic [Ni(μ-S)$_z$Fe$_y$] Complexes and Functional Models

Two polymetallic [Ni(μ-S)$_z$Fe$_y$] systems have been reported to show electrocatalytic activity and in both instances this activity involves the reduction of

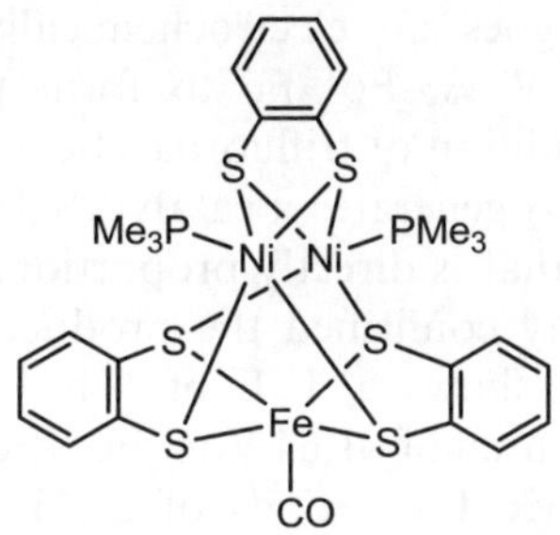

Figure 12.22 View of the complex {('S_2')[Ni(PMe_3)]$_2$Fe-(CO)('S_2')$_2$} reported by Sellmann *et al.*[133,134]

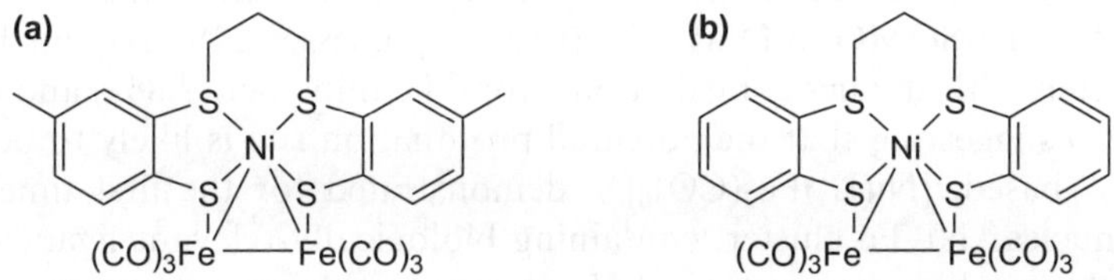

Figure 12.23 View of (a) [Ni(L^1)$Fe_2(CO)_6$] and (b) [Ni($L^{1'}$)$Fe_2(CO)_6$] reported by Schröder *et al.*[135]

protons to H_2 rather than H_2 oxidation.[111] Sellmann *et al.* reported {('S_2')[Ni(PMe_3)]$_2$Fe-(CO)('S_2')$_2$} (Figure 12.22), a trinuclear [Ni_2Fe] cluster that shows structural similarities to the [NiFe] hydrogenase enzyme with Fe and Ni centres in a co-ordination sphere dominated by thiolate S-donors and with CO bound to Fe.[133] {('S_2')[Ni(PMe_3)]$_2$Fe-(CO)('S_2')$_2$} shows a reversible electrochemical process at 0.095 V *vs.* SHE in CH_2Cl_2 and a low activity of 0.1 h^{-1} turnover frequency (ToF) in bulk electrolysis experiments.[134] It was proposed that this relatively anodic reduction potential was associated with the protonation of the system at the S-centre bound to Ni, an argument supported by DFT calculations. In addition, the synthesis and characterisation of the complex [Ni(dppe)(μ_3-S)$Fe_2(CO)_6$] has been reported by Song *et al.*[134b] and Evans *et al.*[134c] but the reactivity of this species with acid was not discussed.

Schröder *et al.* have reported two novel [NiFe] hydrogenase analogues [Ni(L^1)$Fe_2(CO)_6$] and [Ni($L^{1'}$)$Fe_2(CO)_6$] ($L^1 = (CH_3C_6H_3S_2)_2(CH_2)_3$), $L^{1'} = (C_6H_4S_2)_2(CH_2)_3$) (Figure 12.23).[135] These heterotrinuclear clusters are synthesised from [Ni(L^1)] and [Ni($L^{1'}$)], respectively, by reaction with [$Fe_3(CO)_{12}$]. In both complexes the Ni(II) ion is co-ordinated to two thioether and two thiolate donors, which bridge the Ni and Fe atoms. Each Fe centre has a six co-ordinate distorted octahedral co-ordination geometry involving direct interactions with a Ni and Fe centre, one thiolate donor and three terminal CO ligands. The Ni-Fe bond lengths are in the range 2.4789(9) Å to 2.4936(8) Å[135] and are similar to those found in the active site of the [NiFe] hydrogenase from *D. baculatum* (2.5 Å in the reduced form) and *D. vulgaris* Myazaki F (2.55 Å and 2.59 Å for the oxidised and reduced forms, respectively).[23,24,27,135]

$[Ni(L^1)Fe_2(CO)_6]$ undergoes an electrochemically reversible one-electron reduction at $E_{1/2} = -1.31$ V *vs.* Fc^+/Fc to form the paramagnetic species $[Ni(L^1)Fe_2(CO)_6]^-$. The addition of trifluoroacetic acid (TFA) to a solution of $[Ni(L^1)Fe_2(CO)_6]$ in CH_2Cl_2 generates a catalytic cathodic current response in the cyclic voltammogram that is directly proportional to the concentration of TFA. Gas chromatography confirmed the production of H_2 in this experiment.[136] $[Ni(L^1)Fe_2(CO)_6]$ shows a ToF of 6 h^{-1} in the presence of TFA (50 mM) at an applied potential of −1.64 V *vs.* Fc^+/Fc,[136] and a steady catalytic current could be maintained for periods of *ca.* 1 h, after which continued electrolysis resulted in a slow decolourisation of the reaction mixture due to decomposition of $[Ni(L^1)Fe_2(CO)_6]$. Detailed EPR, IR spectroscopic and DFT studies of $[Ni(L^1)Fe_2(CO)_6]^-$ confirm an in-plane delocalised SOMO that is anti-bonding with respect to the metal–metal bonds within the $[NiFe_2]$ framework.[135] The SOMO in $[Ni(L^1)Fe_2(CO)_6]^-$ possesses 24% Ni and 16–17% Fe character, with smaller contributions from the thioether (2.8%) and thiolate (1.4–1.7%) suggesting that the potential protonation site is likely to be metal- and not S-based. $[Ni(L^1)Fe_2(CO)_6]^{0/-}$ demonstrated for the first time that a heterobinuclear Ni–Fe cluster, containing biologically relevant ligands, could support the catalytic production of H_2 at comparable rates of turnover to the synthetic analogues of the [FeFe] hydrogenases, but at potentials that were considerably more anodic (*ca.* 900 mV).[136]

It has been suggested that the presence of several metal–metal interactions in $\{(\text{'}S_2\text{'})[Ni(PMe_3)]_2Fe(CO)(\text{'}S_2\text{'})_2\}$ and $[Ni(L^1)Fe_2(CO)_6]$ may stabilise their respective redox products, $\{(\text{'}S_2\text{'})[Ni(PMe_3)]_2Fe\text{-}(CO)(\text{'}S_2\text{'})_2\}^+$ and $[Ni(L^1)Fe_2(CO)_6]^-$.[106] This inherent stability may account for the capability of these complexes to reduce protons to H_2. Conversely, the lack of reported catalytic activity of heterobinuclear [NiFe] analogues may be related to their general instability and the lability of these complexes upon redox switching during catalysis due to the lack of multiple metal–metal interactions.[106]

12.5 [NiRu] Complexes

12.5.1 Heterobinuclear Complexes

A series of NiRu complexes have been prepared and tested for activity in order to probe the binding and reactivity of H_2 at a diamagnetic d^6 metal ion that would be more stable and less labile than complexes of Fe(II).[106,111] Ru(II) is able to form water-stable compounds and bind H_2O ligands, properties that are relevant to the commercial use of catalysts in fuel cells.[137] Moreover, Ru(II) complexes are well known to bind H_2 and are active catalysts for hydrogenation and hydrogen transfer reactions.[138] The metal centre can readily accommodate both hard and soft ligands such as H_2 and hydrides, and may therefore stabilise hydridic and H_2 complexes within its co-ordination sphere. Rauchfuss *et al.* have shown that binuclear RuRu complexes can activate H_2 and catalyse proton reduction to H_2, even if they are less effective than di-iron analogues (see

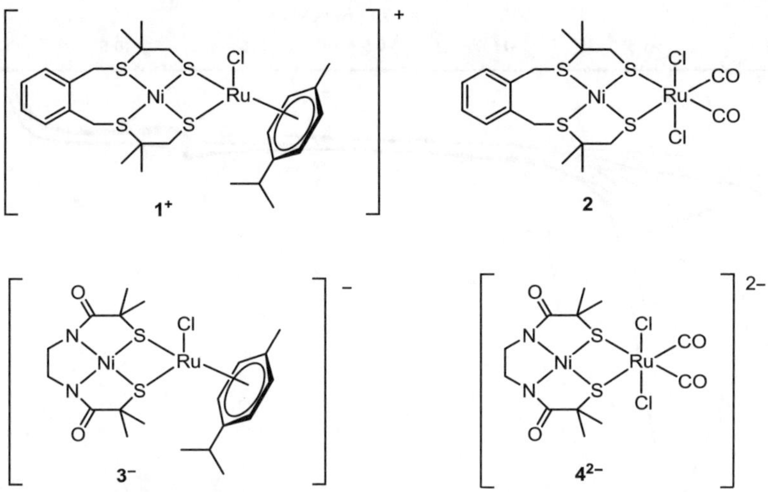

Figure 12.24 View of [NiRu] compexes prepared by Fontecave *et al.*[147]

Section 12.6.4).[139] Although commercially less feasible than Fe analogues, Ru complexes can give an insight into potential mechanisms and intermediates within catalytic cycles, since reactions at Ru(II) are slower and intermediates more stable than for Fe(II). Thus, the isolation of reactive intermediates is of particular interest in this work.[106]

The first NiRu complex catalytically-active towards H_2 production incorporating the unsaturated centres necessary to co-ordinate H^+ or H^- and to initiate catalysis was reported in 2006 by Fontecave *et al.*[140,141] $[Ru(CO)_2Cl_2]$ was employed as the Ru-containing fragment since CO ligands can stabilise hydridic species, for example in $[Ru(d^tbpe)(CO)_3H](OTf)]$ ($d^tbpe = {}^tBu_2PCH_2CH_2P^tBu_2$), through charge delocalisation,[142,143] while incorporation of $[Ru(p\text{-cymene})Cl]^+$ groups introduces the possibility of ring-slippage of the arene ring to generate vacant sites at the Ru centre.[144–146]

Compounds **1–4** (Figure 12.24) are catalytically active towards proton reduction to give H_2 (2–8 h^{-1} ToF).[147] **1–3** show a chemically irreversible reduction process between –0.8 and –1.30 V *vs.* Ag/AgCl, assigned to the irreversible loss of a Cl^- ligand to yield a co-ordinatively unsaturated Ru(I) species. On scanning by cyclic voltammetry, a catalytic wave develops at potentials *ca.* –1.5/–1.6 V *vs.* Ag/AgCl, which is more cathodic than that required for the initial reduction of these compounds. The current increases and shifts to more cathodic potentials upon addition of acid (Figure 12.25). An ECEC mechanism (E: electron-transfer, C: chemical reaction) was suggested for this system with the complex losing a Cl^- ligand on reduction to give an unsaturated species that reacts with protons to give a hydride complex, which then undergoes protonation to release H_2.[147]

This overall mechanism was verified for complex 2 by the simulation of the cyclic voltammetric response (Scheme 12.3).[147]

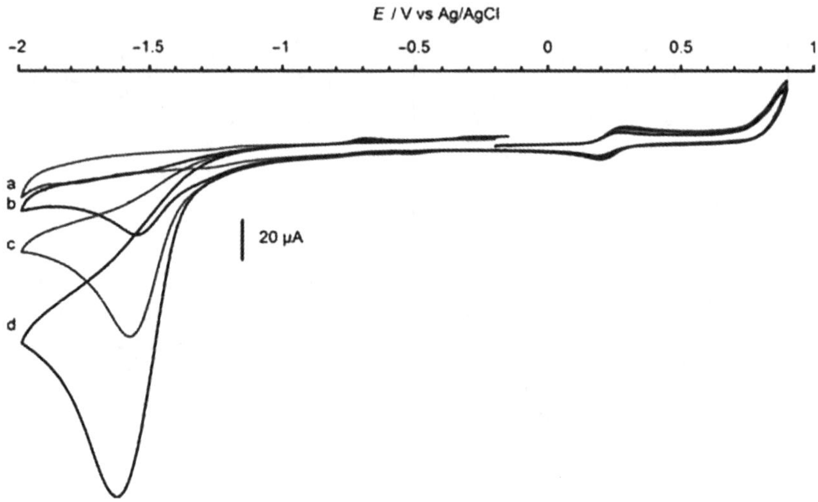

Figure 12.25 Cyclic voltammograms of complex $[Et_4N]$**3** in the presence of Et_3NHCl (a–d: 0; 1.5; 3; 10 equivalents) recorded in a DMF solution of $[^nBu_4N][BF_4]$ (0.1 M) on a glassy carbon electrode (3 mm diameter) at 100 mV s^{-1}.[147]

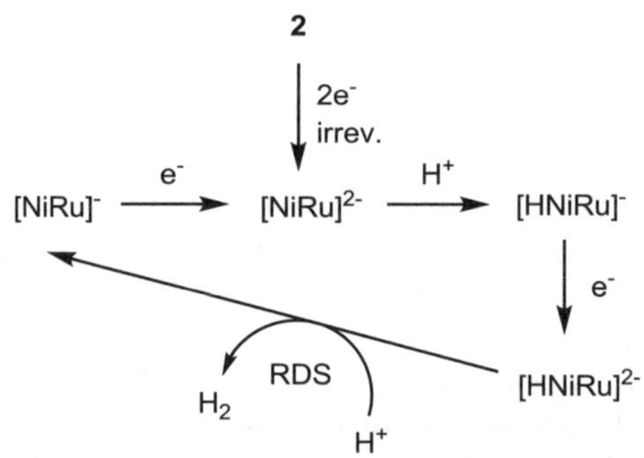

Scheme 12.3 The catalytic ECEC mechanism used for modelling the electrocatalytic behaviour of **2** (RDS = rate-determining step).[147]

A different mechanism was observed for complex $\mathbf{4}^{2-}$, which shows an irreversible reduction process at –1.54 V *vs.* Ag/AgCl. The electrocatalytic wave develops on this irreversible process and an ECCE (E = electron-transfer; C = chemical reaction) mechanism, confirmed by simulation of the cyclic voltammetric response, was proposed. In this case the two one-electron reduction steps occur before the corresponding protonation steps (Scheme 12.4).[147]

Complexes **1–4** proved to be stable under catalytic conditions and some trends may be identified.[147] An N_2S_2 environment at Ni enhances the catalytic rate, while the presence of two methyl groups adjacent to co-ordinated

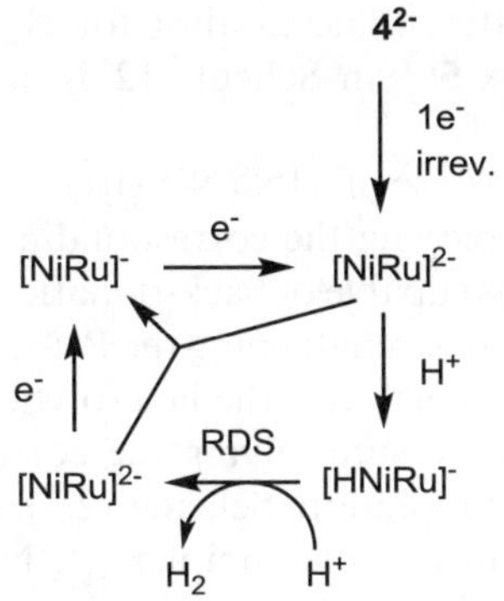

Scheme 12.4 The catalytic ECCE mechanism for the electrocatalytic behaviour of complex 4^{2-} (RDS: rate-determining step).[147]

S-donors inhibits hydrogen abstraction side-reactions that lead to decomposition of the complex.[106] The use of the [Ru(*p*-cymene)Cl]$^+$ unit enhances catalytic activity compared to a [Ru(CO)$_2$Cl$_2$] fragment suggesting that more electron-rich Ru centres are more effective. This implies that the rate-determining step involves protonation of a nucleophilic metal complex, a postulate supported by DFT calculations.[148] Interestingly, even though mononuclear Ni precursors may catalyse the production of H_2, NiRu complexes show enhanced rates of H_2 evolution suggesting a cooperative effect between the Ni and Ru centres, even if this is at the expense of the electrocatalytical potential becoming more cathodic.[147,149]

In summary, the complexes in Figure 12.24 are pre-catalysts that show irreversible redox processes, and the isolation of the catalytically competent species has yet to be achieved. The activation potential for these systems is highly cathodic,[147] although incorporation of more electron-rich groups such as [RuCp]$^+$ shifts the potential by +200 mV.[148] An important structural feature lacking in these compounds is the ability to modulate co-ordination at Ni between tetrahedral and square planar geometry, a feature believed to be crucial for the hydride acceptor function. Likewise, an in-built basic site close to the metal centre(s) to mediate proton transfer to the metal centre is a likely future development.[106]

12.5.2 Complexes with Co-ordinated Water

The development of water-stable catalysts is a major target in this area, since water is an environmentally benign solvent. However, in the case of modelling hydrogenase activity, the incorporation of a water ligand (that may also impart water-solubility) may be crucial in developing new catalysts for H_2 oxidation and proton reduction.[146,150] Water is a relatively labile ligand, and is able to act as a Lewis base or acid, to facilitate heterolytic cleavage of H_2 or to mediate proton transfer to the metal-hydride within a H_2 production cycle.[146] Ogo *et al.*

have prepared the first water-stable catalyst for H_2 cleavage using $[(L)NiRu(H_2O)(C_6Me_6)]^{2+}$ (complex **5**$^{2+}$ in Scheme 12.5) incorporating a bound H_2O ligand.[137]

It has been argued that the use of a $[Ni(N_2S_2)]$ core leads to complexes that are more stable than those employing the corresponding $[Ni(S_4)]$ system, while the hexamethylbenzene ligand is capable of back-donation, which increases the Lewis acidity of the Ru centre and so strengthening the Ru-thiolato bonds.[140,141,146] This complex can thus successfully catalyse the heterolytic cleavage of H_2 in water at ambient temperatures and pressures giving the corresponding hydride-bridged complex and liberating H^+ in solution (Scheme 12.5), with a consequent lowering of pH. It was possible to isolate the complex $[(L)NiRu(H)(C_6Me_6)]^+$, **6**$^+$, the structure of which was confirmed by single crystal X-ray diffraction.[137] Two possible mechanisms can be proposed for this reaction, one involving ring slippage of the hexamethylbenzene (Scheme 12.6), but both involving the active participation of co-ordinated H_2O.[146]

The bridging thiolato ligands in $[(L)NiRu(H)(C_6Me_6)]^+$ are regarded as very important in controlling the acute Ni–S–Ru angles to facilitate the binding of the hydride to both metal ions.[146] Between pH 4 and 6 $[(L)NiRu(H)(C_6Me_6)]^+$ does not react with H^+ to release H_2, because the bridging 'hydride' shows H^+ properties.[151] Under these conditions $[(L)NiRu(D)(C_6Me_6)]^+$ in D_2O catalyses the hydrogen isotope exchange reaction H^+/D^+ between gaseous H_2 and D_2O. Between pH 7 and 10 the bridging hydrogen acts as a hydride and $[(L)NiRu(H)(C_6Me_6)]^+$ is able to catalyse the hydrogenation of carbonyl compounds with a turnover frequency between 3 and 24 h^{-1} depending on the substrate,

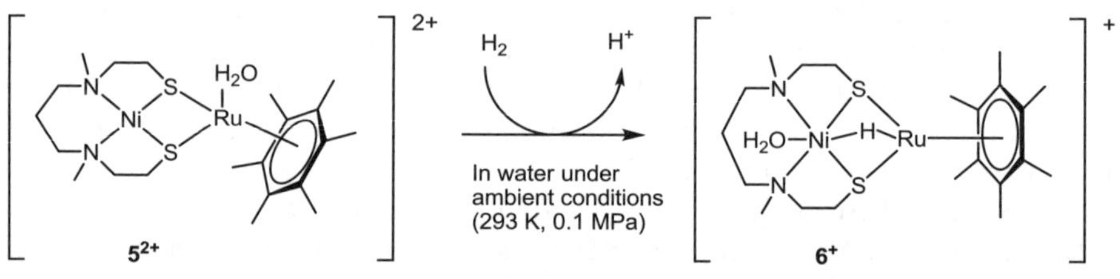

Scheme 12.5 Heterolytic cleavage of H_2 by compound **5**$^{2+}$ in water.[137]

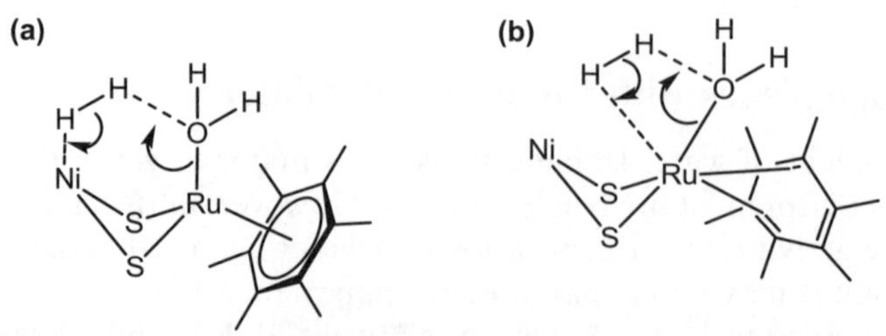

Scheme 12.6 Possible mechanisms for heterolytic cleavage of H_2 by complex **5**$^+$ (a) without or (b) with ring slippage.[146]

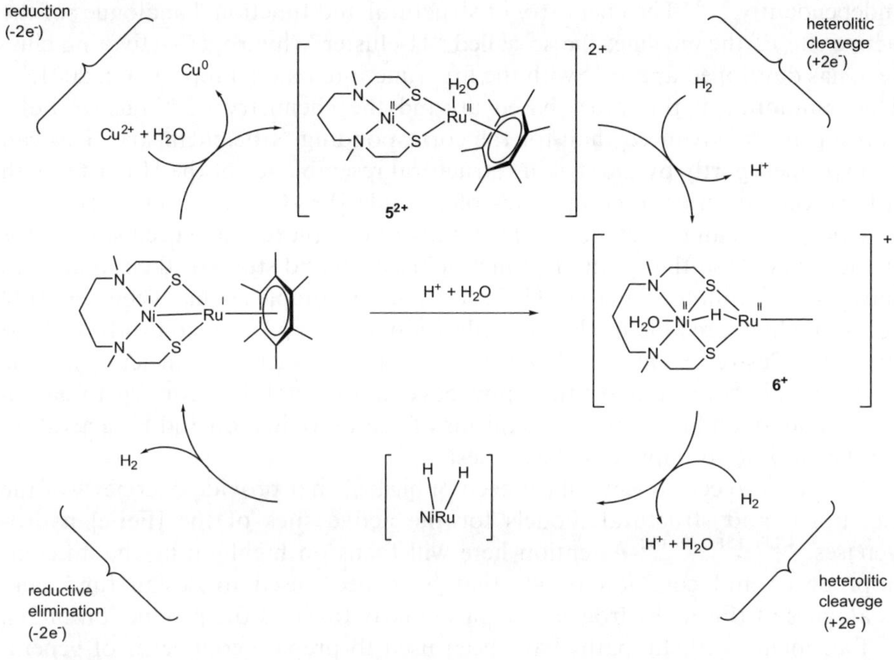

Scheme 12.7 A proposed mechanism for the catalytic reduction of Cu(II) to Cu(0) with $[(L)NiRu(H_2O)(C_6Me_6)]^{2+}$ in water at pH 4–6 at 0.1 MPa of H_2.[152]

and the complex can be cycled up to three times.[151] $[(L)NiRu(H_2O)(C_6Me_6)]^{2+}$ oxidises H_2, and this activation of H_2 can be coupled to the reduction of a Cu(II) to Cu(0) (Scheme 12.7).[152] The role of the water ligand was shown to be essential in the above cycle.[150,153] Thus, replacement of water with MeCN afforded $[(L)NiRu(MeCN)(C_6Me_6)]^{2+}$ which was unable to activate H_2 in MeCN solutions.

Overall, this is a promising methodology for applications in fuel cells, where H_2 needs to be activated to release electrons to fuel the cell. It is also possible to use this catalyst in conjunction with a natural [NiFe] hydrogenase, the two systems operating at different pH regimes, to activate H_2 over an extended pH range (2–10) in one pot under mild conditions at ambient pressure and temperature.[154] Future targets seek to improve the turnover frequency, to increase the reusability, and to exchange the Ru with a less expensive Fe centre.

12.6 [FeFe] Hydrogenase Model Complexes

12.6.1 Introduction

The crystal structure of [FeFe] hydrogenase was reported in 1998 (three years after reports of the structure of [NiFe] hydrogenase) by two groups working independently.[32,33] The chemistry of structural and functional analogues of the active site of the enzyme, the so called "H-cluster" (hydrogen-activating cluster), has developed apace[111] with the first functional model reported in 2001.[155] Understanding in this area, based around the chemistry of binuclear FeFe cores, is more advanced than for the corresponding NiFe chemistry. This can be explained partly by the striking structural resemblance of the H-cluster with a long known series of compounds of formula [$Fe_2(CO)_6$(μ-dithiolate)].[156,157] The properties and reactivity of this family of complexes have been studied for many years,[158] so the synthesis, functionalisation and study of their analogues have been facilitated greatly. Thus, a range of complexes has been reported most of which are able to electrocatalytically reduce protons to produce H_2 to greater or lesser extents.[159] However, they are all much less efficient than the enzyme and more fundamental knowledge and insight are required to gain a better understanding of the mechanisms of proton reduction and H_2 activation by the enzyme to improve catalyst design.

A series of recent reviews have been published that provide overviews of the functional and structural models for the active sites of the [FeFe] hydrogenases.[74,111,156,159–162] Attention here will focus on highlighting the different approaches and common trends that have been used to design functional analogues of [FeFe] hydrogenases, particularly towards the production of H_2.

Two main synthetic paths have been used to prepare complexes of general formula [$Fe_2(CO)_6$(μ-SR)$_2$] (Scheme 12.8).[163–167] These routes are flexible and are applicable to the preparation of a range of functionalised complexes. They allow not only the insertion of various functional groups on the dithiolate bridge, but also the exchange of CO groups with other molecules such as phosphines or cyanides.[163]

The general technique used to assess the catalytic activity of these compounds towards H_2 production is cyclic voltammetry and bulk electrolysis in the presence of variable concentrations of acid.[159] Parameters that are important to define how effective these catalysts include the potential at which

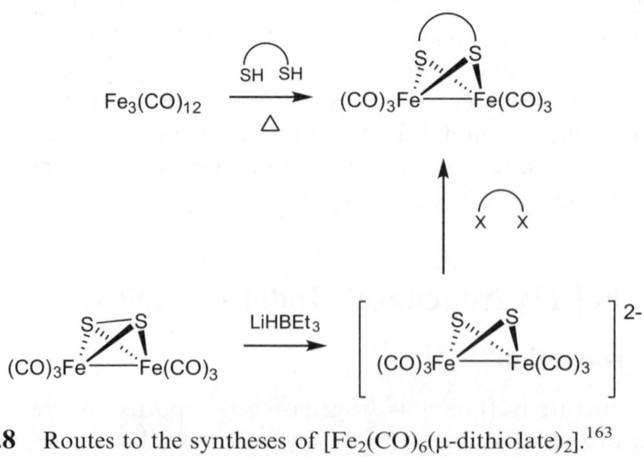

Scheme 12.8 Routes to the syntheses of [$Fe_2(CO)_6$(μ-dithiolate)$_2$].[163]

catalysis occurs (E_{cat}), and the turnover frequency and number. Ideally, the catalyst should possess a redox potential that is not too cathodic (less energy required to activate the complex), with a high turnover frequency and high turnover number. A range of acids have been used in the literature to test catalytic activity and direct comparisons between different catalysts can be problematic since different conditions have been used in each case. To facilitate comparisons, the use of the over-potential defined as the difference between E_{cat} and E^0_{HA} for HA/A^-, the H_2 half reaction, has been proposed.[159] However, it is not always easy to reliably estimate E^0_{HA} for weak acids, especially due to the limited data for $pK_{a,HA}$ in various organic solvents such as CH_2Cl_2 or THF. This notwithstanding, the observed catalytic peak current does allow general comparisons between different catalysts but only under identical conditions. Recently Felton *et al.* have introduced the term "catalytic-efficiency factor", to compare different catalysts and have correlated reduction potentials with catalytic activity for most of the analogues of the active sites of the [FeFe] hydrogenases published thus far.[159]

The conversion of protons to H_2 in the presence of a catalyst involves a 2-electron reduction and two protonation steps, the order of which varies according to the nature of the various ligands and the strength of the acid employed.[163] Two main approaches have been exploited to design improved catalysts, starting from compounds of general formula $[Fe_2(CO)_6(\mu\text{-}SR)_2]$:[111,163] (i) variation of the terminal carbonyl ligands and their exchange with phosphines or cyanides ligands; and (ii) the introduction of different bridging dithiolate groups with pendant functional groups including basic sites. $[Fe_2(CO)_6(\mu\text{-pdt})]$ (pdt^{2-} = propanedithiolate) is a well-studied complex and represents the prototypical compound.[156,168,169] $[Fe_2(CO)_6(\mu\text{-pdt})]$ catalyses H_2 production at *ca.* –1.7 V *vs.* Fc^+/Fc, and the observed mechanism is dependent upon the strength of the acid used. For example, in the presence of strong acid, an ECEC (E: one electron-transfer; C: chemical reaction, in this case protonation) mechanism is involved. After the first protonation, a bridging Fe–H–Fe hydride complex is formed and this is more readily reduced than the starting material.[156]

12.6.2 Exchanging the Terminal Carbonyl Ligands

The exchange of the carbonyl ligands for cyanide ligands more closely replicates the structure of the active site in terms of co-ordination of cyanide, thus controlling the electron density distribution, redox and protonation of the complex.[162] Increasing the electron density at the metal centres will shift the reduction potential cathodically, but will also facilitate the formation of a protonated complex, which is then more readily reduced. In this way, the first step of catalysis is favoured by protonation, which compensates for the cathodic shift in the reduction potential on changing the ligands about the Fe centres.

Although $[(\mu\text{-}S_2C_3H_6)Fe_2(CO)_4(CN)_2]^{2-}$ is stable in water, it yields only very small amounts of H_2 in organic solvents,[170–172] while $[(\mu\text{-}S_2C_3H_6)Fe_2(CO)_4(PMe_3)_2]$ shows greater catalytic activity than the hexacarbonyl

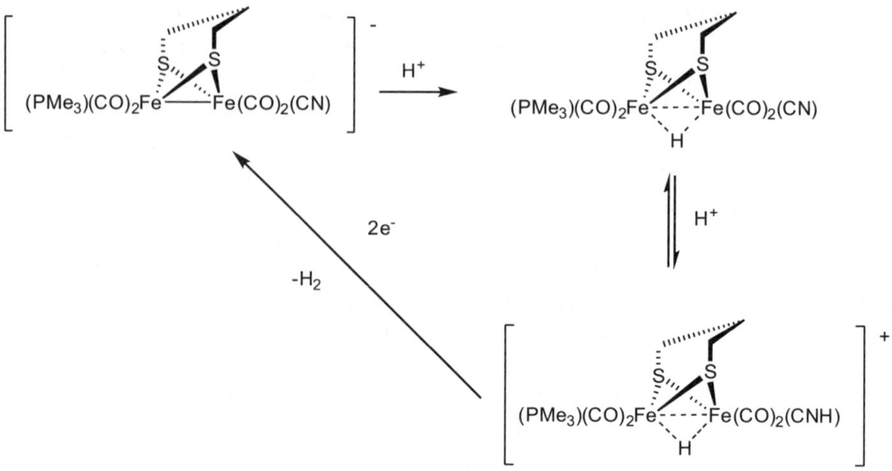

Scheme 12.9 The proposed mechanism for H_2 production by $[(\mu\text{-}S_2C_3H_6)Fe_2(CO)_4(CN)(PMe_3)]$ $[BF_4]$.[155]

analogue.[173] However, the combination of the two substitution patterns, as in $[(\mu\text{-}S_2C_3H_6)Fe_2(CO)_4(CN)(PMe_3)]^-$, affords an efficient catalyst for electrochemical H_2 evolution.[155] The first two steps in the mechanism are protonation at the Fe centres and at the CN^- ligand, and this shifts the reduction potential of the catalyst anodically (Scheme 12.9). The role of protonated CN^- is not fully understood, but CN^- may act as a proton relay, promoting intra-molecular coupling with the hydridic Fe–H–Fe bridge, or CN^- may weaken the bridging hydride bonds as it is a better π-acceptor ligand than, for example, PMe_3.[155]

To improve the catalytic activity of these systems, asymmetrically substituted analogues have been targeted (Scheme 12.10). These complexes should favour the 'rotated' structure (Figure 12.26) and are, according to theoretical calculations, more closely related to the enzyme. An asymmetric structure should facilitate the formation of a terminal rather than a bridging hydride,[156] since a terminal hydride is known to be more reactive than a bridging hydride towards H_2 formation in Fe(II)Fe(II) complexes.[161,174] This should enhance elimination of H_2 in the catalytic cycle, which is believed to be a relatively slow step.

A series of asymmetrical compounds have been reported that incorporate diphosphine ligands and some evidence for the formation of terminal hydrides, mostly at low temperatures, has been obtained (Scheme 12.11).[156] However, asymmetric substitution of ligands at the FeFe core is not the only factor in driving the formation of terminal hydrides; the basicity and overall electron density at the metal centres are also critical factors. For example, the tetra-substituted symmetrical complex $[(\mu\text{-}S_2C_3H_6)Fe_2(CO)_2(\kappa^2\text{-dppv})_2]$ can form a terminal hydride (Scheme 12.12) at 273 K and can catalyse the reduction of protons at a potential which is +200 mV more anodic than for the species incorporating a bridging hydride.[175]

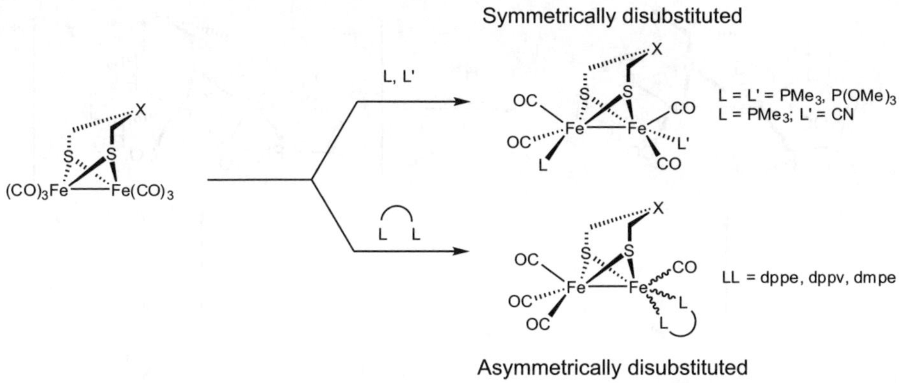

Scheme 12.10 Representation of symmetrically and asymmetrically substituted complexes.[156]

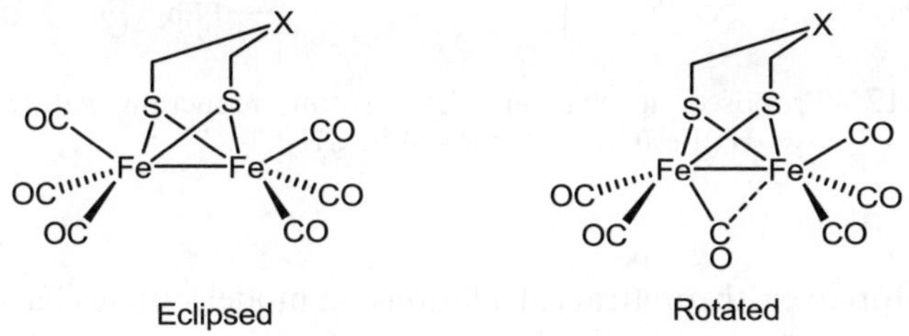

Figure 12.26 Rotated and eclipsed geometry in binuclear FeFe complexes.[161]

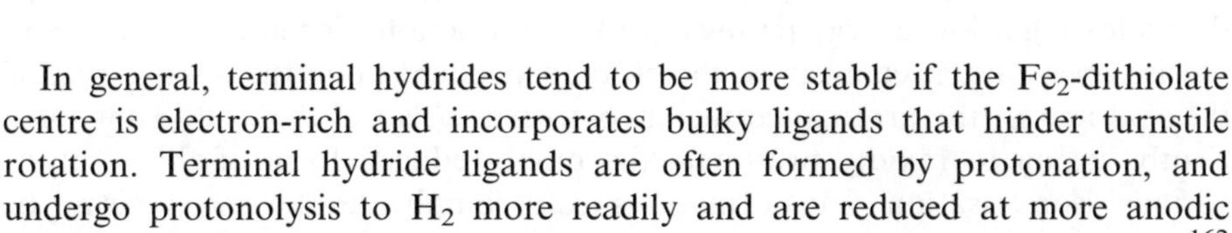

Scheme 12.11 Protonation of [(μ-$S_2C_3H_6$)$Fe_2(CO)_4$(κ^2-LL)] at low temperature.[156]

In general, terminal hydrides tend to be more stable if the Fe_2-dithiolate centre is electron-rich and incorporates bulky ligands that hinder turnstile rotation. Terminal hydride ligands are often formed by protonation, and undergo protonolysis to H_2 more readily and are reduced at more anodic potentials than analogous complexes incorporating bridging hydride ligands.[162]

12.6.3 Exchanging the Bridging Dithiolate

There has been much discussion about the precise nature of the dithiolate bridge in the enzyme; it has now been assigned as a di(thiomethyl)amine

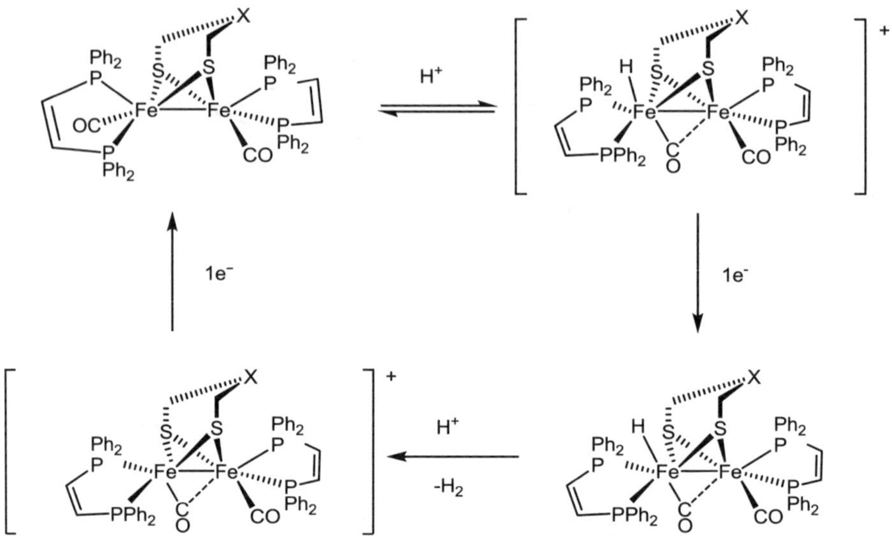

Scheme 12.12 Proposed mechanism for proton reduction catalysed by [(μ-$S_2C_3H_6$)$Fe_2(CO)_2$(κ^2-dppv)$_2$] in CH_2Cl_2.[175]

moiety.[111] Moreover, theoretical calculations on model systems have confirmed that the presence of an amine in the second co-ordination sphere of the metal centres provides a low energy pathway for the formation and cleavage of the H–H bond.[156,176–179] For this reason a series of complexes of general formula [$Fe_2(CO)_6$μ-$SCH_2NH(R)CH_2S$] has been reported.[111] The presence of an additional basic protonation site on the bridgehead of the dithiolate ligand that can deliver protons to the FeFe core imparts different protonation properties to this class of compound.[180,181] If the acid is not sufficiently strong to protonate the N-centre, proton reduction processes initiated by an electron-transfer are similar to those of propane dithiolate analogues. If the acid is strong enough to protonate the amine group, the first step in the catalytic cycle is protonation, leading to a shift to anodic potentials for the following electron-transfer step, thus allowing a low energy pathway to H_2 production (Scheme 12.13).[182] The precise operating mechanism is, therefore, dependent upon the relative pK_a of the acid and of the protonated metal complexes. Also, if the acid is not sufficiently strong to rapidly protonate the mono-reduced form of the catalyst, various isomerisation transformations can complicate the mechanism. In summary however, it is observed that catalysts based upon di(thiomethyl)-amine linkers are generally more efficient in terms of stability and activation over-potential compared to propane dithiolate analogues.[156,163,180,181]

The N-bridgehead centre has been exchanged for S donors and this substitution strongly affects the catalytic mechanism. The basicity of the S-centres in the thiolate bridge increases, and there is some evidence for protonation of the sulfur atom.[183] Thus, subtle changes in the second co-ordination sphere in

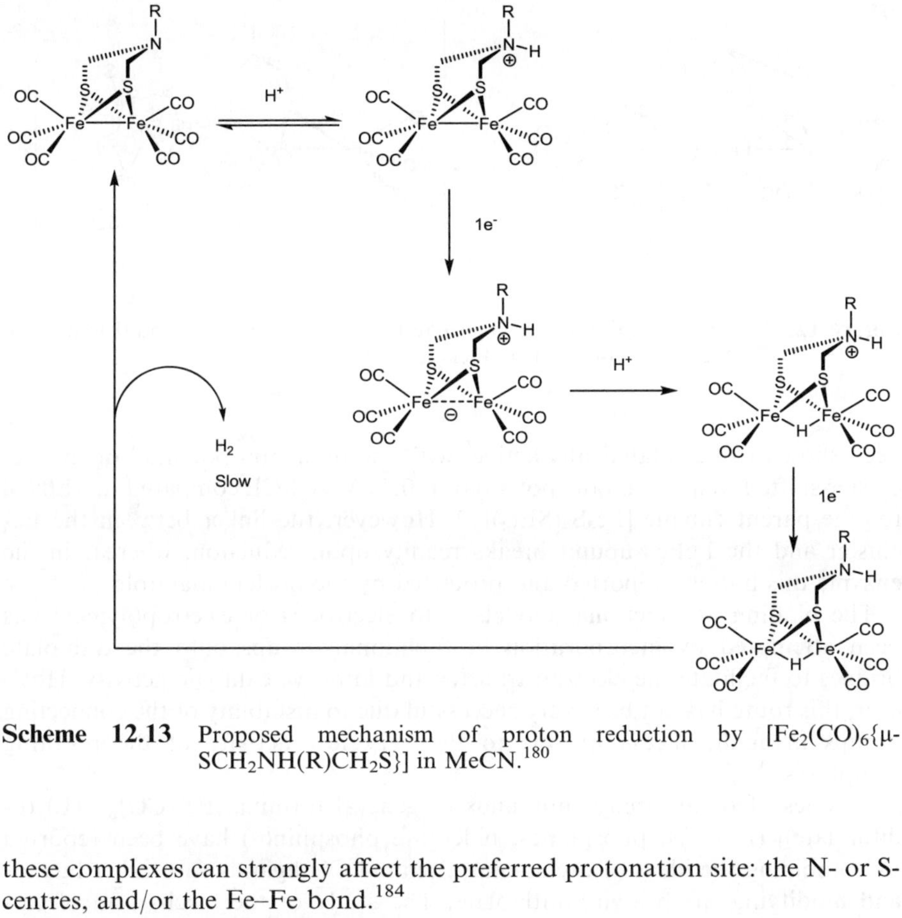

Scheme 12.13 Proposed mechanism of proton reduction by $[Fe_2(CO)_6\{\mu\text{-}SCH_2NH(R)CH_2S\}]$ in MeCN.[180]

these complexes can strongly affect the preferred protonation site: the N- or S-centres, and/or the Fe–Fe bond.[184]

The active site of [FeFe] hydrogenase performs proton reduction at the Fe(I)-Fe(II)/Fe(I)-Fe(I) level with limited exceptions; however, all *in vitro* models operate most effectively at the Fe(I)-Fe(I)/Fe(I)-Fe(0) level.[111] Pickett *et al.* have prepared a mixed valence Fe(I)-Fe(II)-Fe(II)-Fe(I) tetranuclear complex incorporating a tridentate facial trithiolate ligand (Figure 12.27a).[185] This species is kinetically efficient for electrocatalytic H_2 evolution, probably because, upon reduction by two electrons, the complex can readily rearrange to give the so-called "rotated-state" switching between terminal and bridging CO states during turnover. However, this is achieved at a thermodynamic cost with the complex operating at a potential (*ca.* –1.6 V *vs.* Fc^+/Fc, –1.07 V *vs.* Ag/AgCl) well removed from the thermodynamic equilibrium of the H^+/H_2 couple.

These bioinspired catalysts rely on the ability of electrodes to supply electrons. For this reason, the modelling of the –S(Cys)–$[Fe_4S_4]$ compound of the enzyme has not been considered as a priority. However, its incorporation into a model structure should provide insights into the chemistry underpinning the activity of the active site of the [FeFe] hydrogenases and into engineering biomimetic catalysts.[156] A model of the entire H-cluster (Figure 12.27b) has

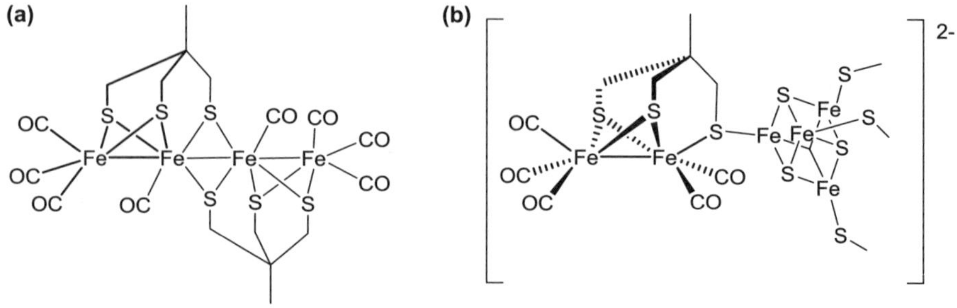

Figure 12.27 Views of (a) a mixed valence tetranuclear complex,[185] and (b) model of the entire H-cluster by Pickett *et al.*[186]

been shown to be catalytically active, with the reduction potential of the Fe_4 cluster shifted to more anodic potentials (–0.59 V *vs.* SCE compared to –1.09 V for the parent cubane $[Fe_4S_4(SEt_4)]^{2-}$). However, the linker between the Fe_4 cluster and the FeFe subunit breaks readily upon reduction, whereas in the enzyme this link is supported and protected by the protein manifold.[186,187]

The binding of functional models onto electrodes or electropolymers has been attempted by incorporation of functional groups onto the dithiolate bridges to facilitate the electron-transfer and improve catalytic activity. However, this route has not been very successful due to instability of the connecting groups in acidic media or due to the very low activity of the resulting complexes.[111]

A series of bioinspired compounds of general formula $[Fe_2(CO)_{6-x}(L)_x(\mu\text{-dithiolate})]$ (L = CN, phosphines, bidentate phosphines) have been reported by combining the ideas of exchanging terminal CO ligands with other ligands and modifying the bridging dithiolate. The study of the mechanism of catalysis becomes more complicated due to the increasing number of protonation sites. However, at least for some of them (Scheme 12.14 and Scheme 12.15 as examples) it has been possible to define the site(s) of protonation.[156,163] Kinetically, the first protonation site in these FeFe complexes is likely to be the N-donor bridgehead, because it requires a smaller structural rearrangement than protonation at the Fe sites (terminal or bridging). However, if the basicity of the metal centres is sufficiently high, for example in the case of tetra-substituted complexes, where 4 CO ligands are replaced by phosphines, equilibrium between the hydride and the ammonium derivatives can be observed. This tautomerisation results in a positive shift of the reduction potential, since it is easier to reduce the hydride (Scheme 12.14).[162,175]

The migration of the proton from the N-donor to the metal centres can be promoted by electron-transfer that increases basicity at the metal site.[180] The emerging picture from these studies is that protonation of the N-centre in the di(thiomethyl)amine bridge precedes and facilitates the formation of hydrides, and acts as a proton relay (Scheme 12.15).[156]

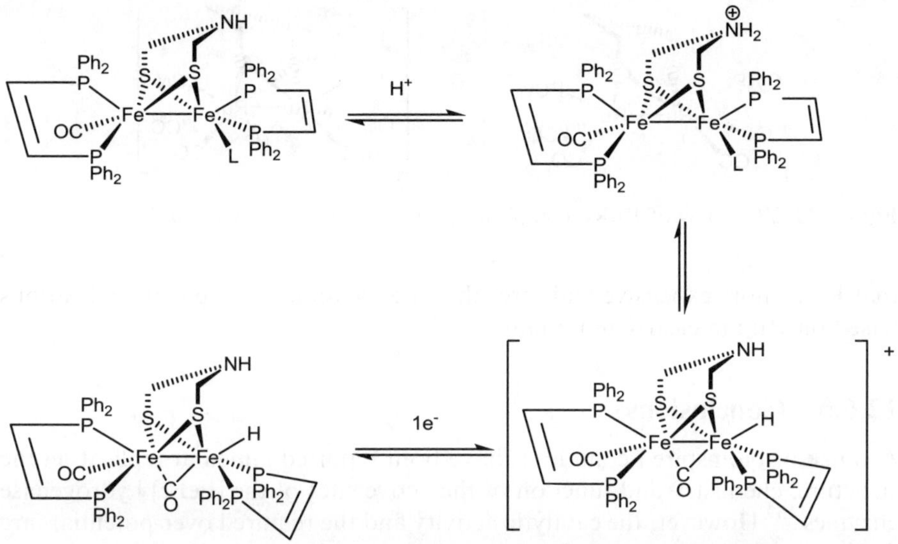

Scheme 12.14 Protonation of N- and Fe-centres in $[Fe_2(CO)_2(\kappa^2\text{-dppv})_2(\mu\text{-}SCH_2NHCH_2S)]$.[162,175]

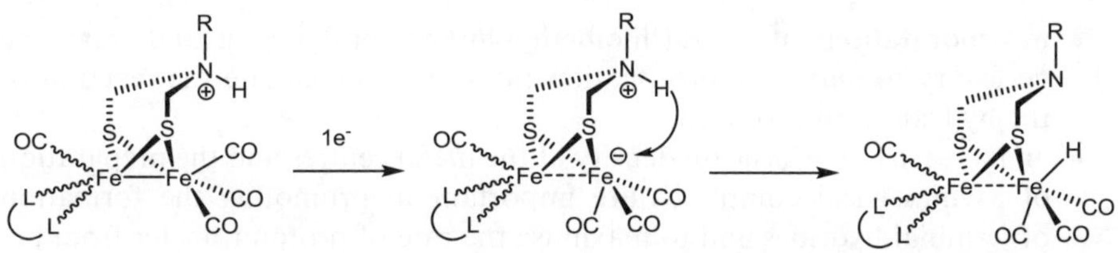

Scheme 12.15 Reductively induced tautomerisation and proton transfer in $[Fe_2(CO)_4(\kappa^2\text{-LL})\mu\text{-}SCH_2NH(R)CH_2S]^+$ (R = *μ*Pr, CH_2Ph, $CH_2C_6H_4X$, LL = dppe, dppv).[156,180]

12.6.4 Changing the Metal Centre

Exchanging the metal centres in FeFe complexes for Ru is of interest since complexes incorporating second-row transition metals are generally stronger Brønsted-bases than analogous complexes of first row metals.[111] Rauchfuss *et al.* reported the synthesis of $[Ru_2(\mu\text{-}S_2C_3H_6)(CO)_4(PCy_3)_2]$ (Figure 12.28) and showed that it can activate H_2 upon UV-photolysis to form bridging and terminal dihydride species. Upon addition of acid, the terminal hydride reacts to produce H_2, thus establishing that in this system a terminal H^- is more reactive than a bridging H^-.[188]

The synthesis of $[Ru_2(\mu\text{-}S_2C_3H_6)(\mu\text{-}H)(CO)_4(CN)_2]^-$ has also been reported and this complex catalyses proton reduction[139] but at a very cathodic potential (–1.8 V *vs.* Ag/AgCl) making $[Ru_2(\mu\text{-}S_2C_3H_6)(\mu\text{-}H)(CO)_4(CN)_2]^-$ a poorer catalyst than the analogous FeFe complex. Furthermore, taking into account

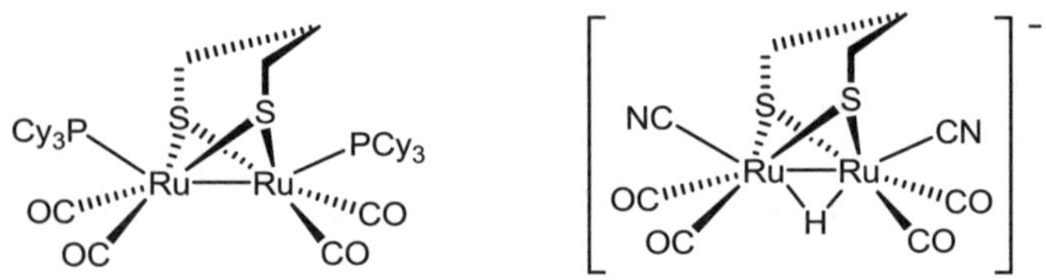

Figure 12.28 View of RuRu complexes prepared by Rauchfuss *et al.*[139,188]

that Ru is more expensive and rarer than Fe, possible commercial applications based on Ru are clearly more limited.[139,188]

12.6.5 Conclusions

A range of bioinspired complexes have been reported aimed at replicating the structure, chemistry and function of the active sites of the [FeFe] hydrogenase enzymes.[159] However, the catalytic activity and the required over-potentials are far from those of the natural systems. Many of the synthetic systems possess highly cathodic redox potentials (<-1.5 V *vs.* Fc^+/Fc) and lie significantly from *in vivo* capability. However, this research has highlighted some significant aspects of this chemistry:[156,160,161,163]

- incorporatation of a di(thiomethyl)amine bridging ligand may be necessary to maximise protonation rates and to facilitate the reaction of the hydride with a proton;
- an increase in the electron density at the metal centres and the preparation of asymmetrical complexes are important in promoting the formation of terminal hydrides and to maximise the rate of proton transfer from the N-centre bridgehead to the metal site;
- protonation of the Fe centres shifts the redox potential of the system to more anodic potentials thereby decreasing the energy required to produce H_2.

Current *in vitro* complexes generally fail to exhibit any reactivity towards H_2, while the enzyme is able to catalyse the activation reversibly.[111] This remains a critical challenge for the design of systems of potential technological relevance for replacing platinum in fuel cells. The following sections review recent advances in developing the above ideas to polyfunctional catalysts incorporating proton relays and photosensitive groups for photo-activation and photo-catalysis.

12.7 Proton Relays

A proton relay is a pendant base within a catalyst that may facilitate the transfer of protons between the solution and a catalyst.[189] It is known that, in comparison to an organic basic site, a metal centre can be slow to protonate.[190]

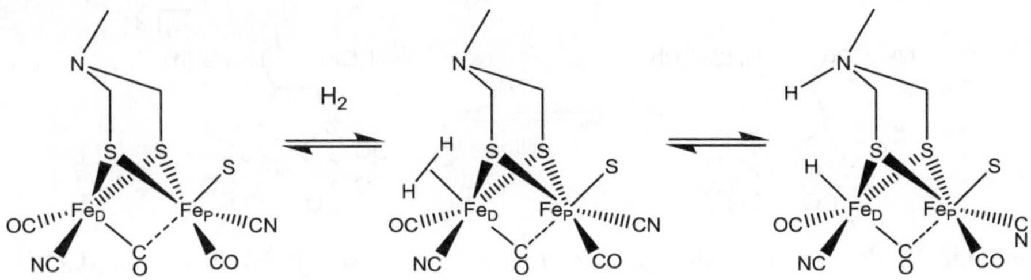

Scheme 12.16 A proposed mechanism of [FeFe] hydrogenase in reaction with H_2.[70,176,179,191]

Table 12.1 The thermodynamics of heterolytic cleavage of H_2 in the presence of a proton relay.[189]

	ΔG°/kcal mol⁻¹
$H_2 \rightarrow H^+ + H^-$	76^a
$ML_n^{2+} + H^- \rightarrow HML_n^+$	$-G^o_{H^-}$
$H^+ + Base \rightarrow HBase^+$	$-1.37pK_a$
$H_2 + ML_n^{2+} \rightarrow HML_n^+ + HBase^+$	$\Delta G^o = 76 - \Delta G^o_{H^-} - 1.37pK_a$

^{a}Heterolytic cleavage of H_2 requires 76 kcal mol^{-1} in CH_3CN.[194]

Thus, a basic site in close proximity to the catalytic metal centre may be ideally positioned to relay protons and promote the formation of the metal hydride intermediate required for the reversible oxidation of H_2.[174] Inspiration for the incorporation of a pendant base as a proton relay in hydrogenase models can be attributed to the presence of a di(thiomethyl)amine moiety at the active site of the [FeFe] hydrogenases.[70,176,179] The activation of H_2 by the [FeFe] hydrogenases is proposed to occur *via* the binding of H_2 to the vacant site of the distal Fe centre (Fe_D) (Scheme 12.16), followed by heterolytic cleavage to produce an Fe hydride and a protonated amine.

An understanding of how a basic site can influence the free energy of reaction for the heterolytic cleavage of H_2 can be obtained if we consider the individual reactions (Scheme 12.16) since the sum of their free energies equates to the free energy of the overall reaction (Table 12.1).[189] If the protonated base shows a pK_a that complements the hydride acceptor ability of the metal centre ($\Delta G^o_{H^-}$), the overall free energy of the reaction (ΔG^o) can then be close to zero (Table 12.1).[191–193] In the following sections we will discuss hydrogenase analogues that incorporate functional proton relays.

12.7.1 [FeFe] Hydrogenase Models Incorporating Protonation Sites

A number of studies have described dithiolate[155,168,181,186,188,195–202] or phosphido[203,204] bridged FeFe complexes, but only a limited number of systems

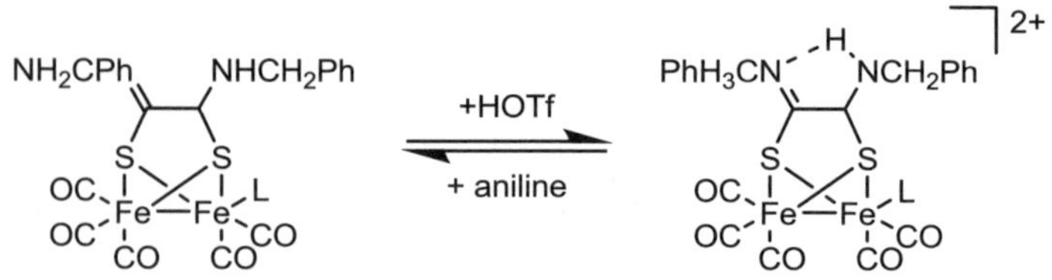

Scheme 12.17 The reversible protonation of [μ-SC(NBn)CH(NHBn)S-μ $Fe_2(CO)_5L$] [L = PPh_3, $P(Py)_3$] reported by Zhang *et al.*[209]

containing a N-donor functionality within di(thiomethyl)amine[165,181] or di(phosphidomethyl)amine[204] ligands have been reported. A range of octahedral Fe complexes that possess pendant amine groups in their co-ordination spheres are known,[205–208] and these bases promote the exchange of protons in solution with the hydride ligand, increase proton-coupled transfer reactions, and facilitate the heterolytic cleavage of H_2.

The construction of more elaborate functionalities is the subject of much current research. Recently, Zhang *et al.* have reported [FeFe] hydrogenase analogues [μ-SC(NBn)CH(NHBn)S-μ-$Fe_2(CO)_5L$] [L = PPh_3, $P(Py)_3$] which contain a functionalised C_2 bridge incorporating two basic sites (Scheme 12.17).[209] These complexes are readily protonated even with mild acids. The addition of one equivalent of trifluoromethanesulfonic acid (HOTf) to solutions of each complex results in the emergence of new peaks that are +610 to +650 mV more anodic, and are attributed to the reduction of the protonated species. Thus, protonation of the internal base of [(μ-SC(NBn)CH(NHBn)S-μ-$Fe_2(CO)_5L$] [L = PPh_3 or $P(Py)_3$] influences directly the reduction potential of the complex.

Lee *et al.* have substituted the di(thiomethyl)amine bridge with a di(thioethyl)amine ligand to form the tetranuclear complex [Fe_2(μ-$S(CH_2)_2$-$N^nPr(CH_2)_2S$)$(CO)_4(PMe_3)_2]_2$ (Figure 12.29).[210] The two amine sites within this complex can be protonated with 2 equivalents of HBF_4 to yield a doubly-protonated species. However, if [Fe_2(μ-$S(CH_2)_2N^nPr(CH_2)_2S$)$(CO)_4(PMe_3)_2]_2$ is treated with an excess of HBF_4, both Fe centres also protonate to form [Fe_2-(μ-H)(μ-$S(CH_2)_2N(H)^nPr(CH_2)_2S$)$(CO)_4(PMe_3)_2]_2(BF_4)_2$ which contains both NH protons and Fe hydrides. The protonation of the N- and Fe-centres shifts the reduction potential 1.2 V anodically and suggests that the ability of both the N and Fe sites to accept protons is essential for enzymatic H_2 production at mild potentials.[210]

Recently Li *et al.* designed pyridylphosphine complexes [(μ-pdt)$Fe_2(CO)_5L$] (L = Ph_2PCH_2py, Ph_2Ppy) and [$(CO)_2(PMe_3)$Fe(μ-pdt)Fe$(CO)_2L$] (Figure 12.30).[211] The pyridyl N-centre was targeted to parallel the N-heteroatom found in the bridgehead position at the active site of the [FeFe] hydrogenases.[181,201,204,212–215] The incorporation of pyridyl units leads to protonation in the presence of strong acid and a decrease of the over-potential by 360–490 mV for electrocatalytic proton reduction for these complexes.[211]

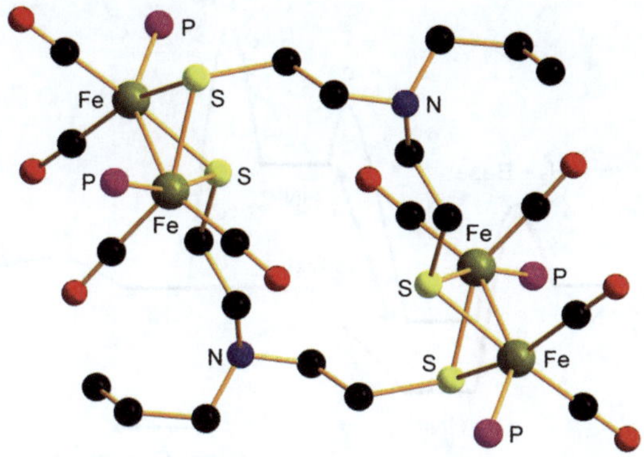

Figure 12.29 View of the single crystal X-ray structure of $[Fe_2(\mu\text{-}S(CH_2)_2N^nPr(CH_2)_2S)(CO)_4(PMe_3)_2]_2$ reported by Lee *et al.*[210] All methyl groups of PMe_3 are omitted for clarity.

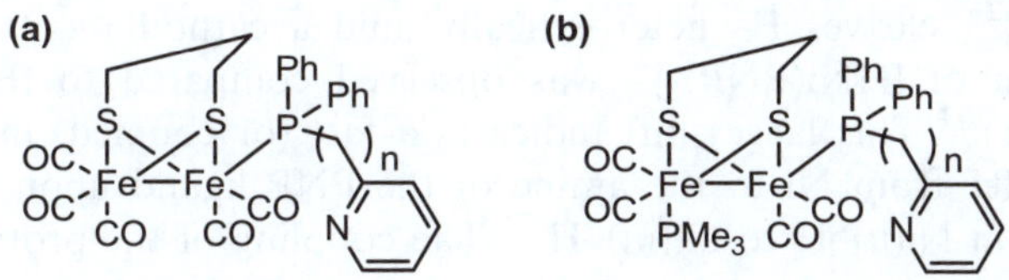

Figure 12.30 View of (a) $[(\mu\text{-pdt})Fe_2(CO)_5L]$ ($L = Ph_2PCH_2py$ or Ph_2Ppy) and (b) di-substituted complexes $[(CO)_2(PMe_3)Fe(\mu\text{-pdt})Fe(CO)_2L]$ ($L = Ph_2PCH_2py$ or Ph_2Ppy) reported by Li *et al.*[211]

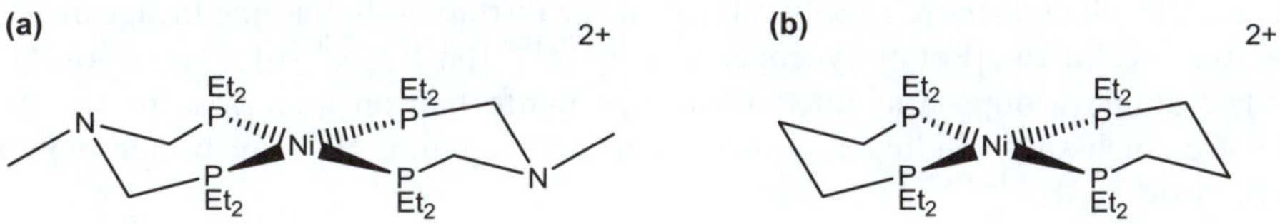

Figure 12.31 View of (a) $[Ni(PNP)_2]^{2+}$ containing two NMe basic groups and (b) the analogous complex $[Ni(depp)_2]^{2+}$ incorporating no amine sites.

12.7.2 Nickel Complexes Incorporating Protonation Sites

Curtis *et al.* have reported the complex $[Ni(PNP)_2]^{2+}$ ($PNP = Et_2PCH_2NMeCH_2PEt_2$) (Figure 12.31), which incorporates two basic amine groups within the ligand.[71] These amine groups are non-co-ordinating due to the two strained four-membered rings forming the backbone of the ligands, which also show some structural similarities to the bridging dithiolate ligand of the [FeFe] hydrogenases.

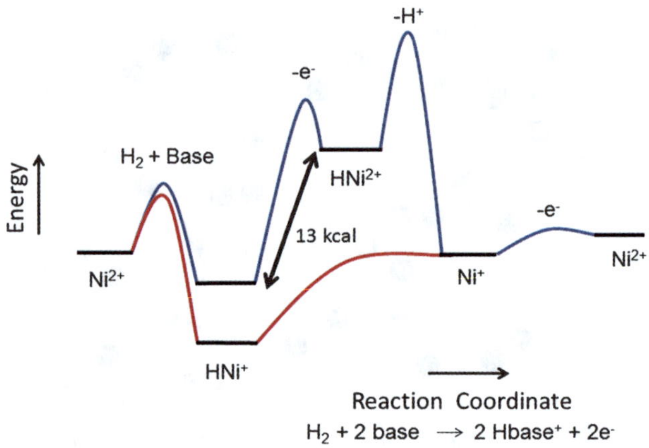

Figure 12.32 Comparison of the free energy reaction profiles for the electrocatalytic oxidation of H_2 using $[Ni(depp)_2]^{2+}$ (blue lines) and $[Ni(PNP)_2]^{2+}$ (red lines) under the same conditions.[189]

$[Ni(PNP)_2]^{2+}$ cleaves H_2 heterolytically, and a cathodic shift of 0.6 V in the oxidation of $[HNi(PNP)_2]^+$ was observed compared to $[HNi(depp)_2]^+$ (Figure 12.31).[71] This large shift indicates a fast (or coupled) intramolecular proton-transfer from Ni to the amine of the PNP ligand upon oxidation of the metal from Ni(II)–H to Ni(III)–H.[71] The coupling of the proton- and electron-transfer events overcomes the large energy barrier associated with the formation of a Ni(III)–H centre (Figure 12.32).[189] These results demonstrate that the pendant base not only facilitates intramolecular heterolytic cleavage of H_2, but also promotes intermolecular exchange between the hydride ligand and protons in solution.[189]

DuBois *et al.* targeted Ni(II) phosphine complexes incorporating a N-donor basic site which is more closely related to the di(thiomethyl)amine bridge at the active sites of the [FeFe] hydrogenases.[70,176,179] $[Ni(P^{R}{}_2N^{R'}{}_2)_2]^{2+}$ incorporates basic amine groups orientated in a boat conformation and close to the Ni centre, such that binding interactions to an incoming H_2 may be templated (Figure 12.33).[193,216,217]

These catalysts show relatively rapid electrocatalytic rates that can be tuned for either H_2 oxidation or production by changing the substituents of the diphosphine ligands.[189] More bulky substituents on the P-donors result in larger tetrahedral distortions and a greater affinity at Ni for hydride. Moreover, it was proposed that H_2 binding for the derivatives of $[Ni(P^{R}{}_2N^{R'}{}_2)_2]^{2+}$ was stabilised by an interaction with both pendant bases (Figure 12.33), which increases the observed rates of catalysis.[193,216,217]

More recently DuBois *et al.* have reported complexes $[Ni(P^{R}{}_2N^{R'}{}_2)_2]^{2+}$ (Figure 12.33) functionalised with thiophene derivatives in the R′ positions with a view to generating modified electrode surfaces.[218] This complex is very active towards H_2 production at a low potential of 280 mV. The introduction of *tert*-butyl groups into the R′ positions increases the basicity of the pendant

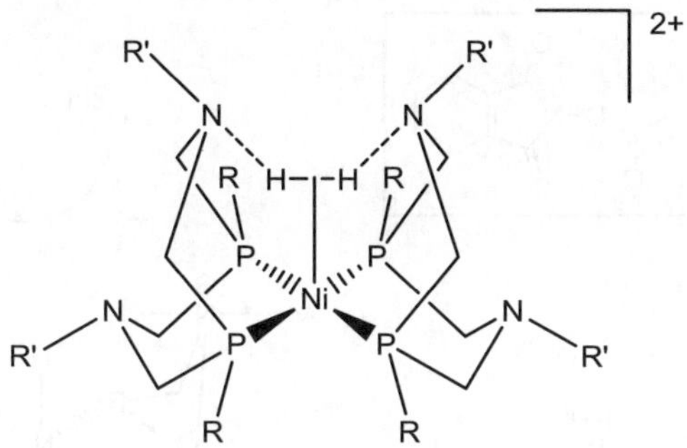

Figure 12.33 View of the complex $[Ni(P^{R}{}_{2}N^{R'}{}_{2})_{2}]^{2+}$ (R = cyclohexyl; R′ = benzyl) showing a potential interaction with H_2.[193,216,217]

amines in $[Ni(P^{R}{}_{2}N^{R'}{}_{2})_{2}]^{2+}$ (Figure 12.33), which results in a lowering of the free energy of H_2 addition or cleavage, the rate-determining step for catalysis. Recently, Yang *et al.* identified[219] the most active catalyst $[Ni(P^{R}{}_{2}N^{R'}{}_{2})_{2}]^{2+}$ (R′ = benzyl) for H_2 oxidation reported to date, with a turnover frequency of 50 s^{-1}. This result shows that the driving force for H_2 addition can be significantly tuned by variation of the ligand functionality.[219,220] Very recently a significant breakthrough has been reported in which the complex $[Ni(P^{Ph}{}_{2}N^{Ph})_{2}]^{2+}$ = ($P^{Ph}{}_{2}N^{Ph}$ = 1,3,6-tripehnyl-1-aza-3,6-diphosphacycloheptane) has been shown to act as an electrocatalyst for H_2 production with a ToF above 100 000 s^{-1}.[220a]

Fontecave *et al.* have reported the attachment of $[Ni(P^{R}{}_{2}N^{R'}{}_{2})_{2}]^{2+}$ (R = Ph) (Figure 12.33) in a new, efficient and robust electrode material.[221] The Ni-containing catalyst is both air- and moisture-stable and has been anchored to an amino-functionalised MWCNT-electrode material[222] *via* the formation of stable covalent amide linkages (Figure 12.34).[221] MWCNTs were chosen as the electrode material as they provide a large surface area for high catalyst loading, show high stability and high electrical conductivity,[223,224] and are versatile in terms of grafting molecular complexes to their surfaces.[224,225] The electro-catalytic properties of the Ni complex are retained in the grafted material and an estimated turnover frequency of 20 000 h^{-1} has been calculated for the material in the presence of [DMFH]OTf/DMF (1:1 mol/mol, 0.1 mol L^{-1}) in CH_3CN.[221] Practical applications of such molecular electrode materials require the use of aqueous electrolytes and, importantly, Fontecave *et al.* have demonstrated that a similar Ni-functionalised MWCNT electrode can produce H_2 catalytically in dilute (pH 0–1) aqueous sulfuric acid solutions.[221]

Yang *et al.* have recently reported a study probing the optimum number of positioned proton relays to maximise the catalytic rate of H_2 oxidation by a Ni complex. To this end, $[Ni(dppp)(NP^{Ph}{}_{2}N^{Bz}{}_{2})]^{2+}$ [$P^{Ph}{}_{2}N^{Bz}{}_{2}$ = 1,5-dibenzyl-3,7-diphenyl-1,5-diaza-3,7-diphosphacyclooctane; dppp = 1,3-bis(diphenyl-phosphino)propane] has been synthesised (Figure 12.35).[226]

$[Ni(dppp)(NP^{Ph}{}_{2}N^{Bz}{}_{2})]^{2+}$ is an electrocatalyst for H_2 oxidation, and for the first time, Ni–H and N–H intermediates have been identified

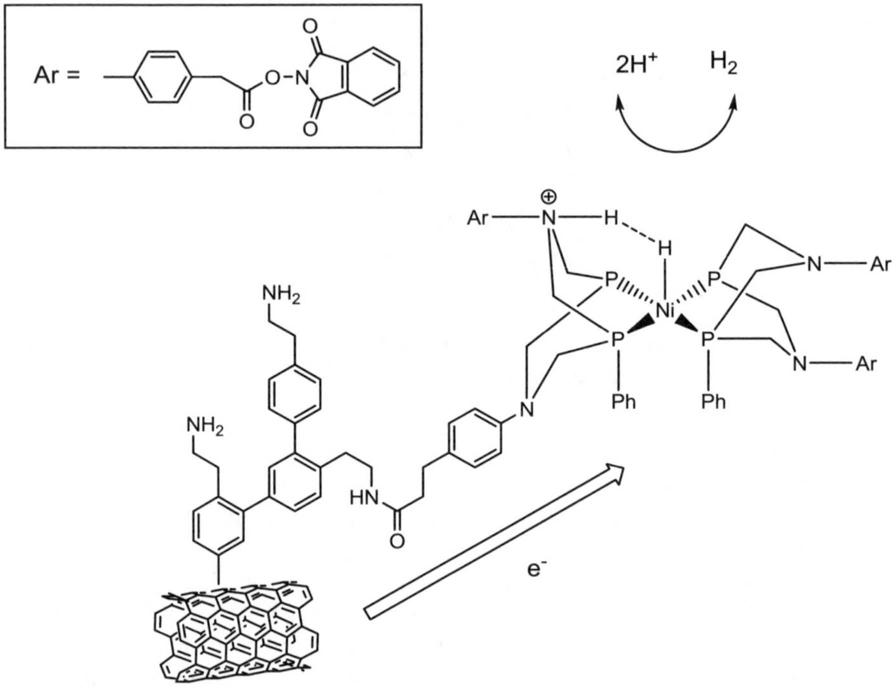

Figure 12.34 Representation of the $[Ni(P^{R}{}_2N^{R'}{}_2)_2]^{2+}$ derivative anchored to MWCNTs functionalised with 4-(2-aminoethyl)phenyl groups.[221]

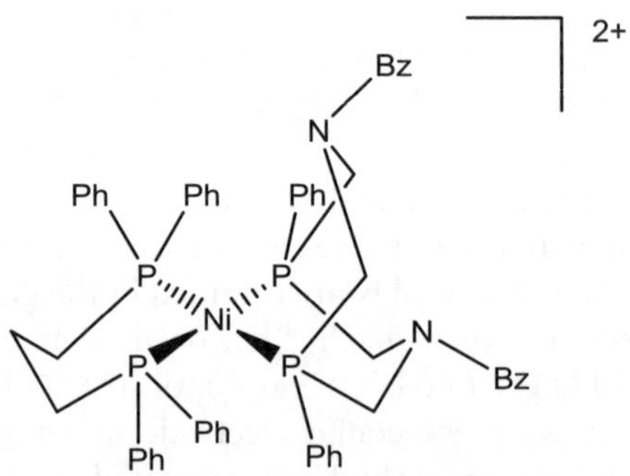

Figure 12.35 Complex $[Ni(dppp)(NP^{Ph}{}_2N^{Bz}{}_2)]^{2+}$ reported by Yang *et al.*[226]

spectroscopically.[226] However, $[Ni(dppp)(NP^{Ph}{}_2N^{Bz}{}_2)]^{2+}$ shows catalytic activity ~20 times lower than $[Ni(P^{R}{}_2N^{R'}{}_2)_2]^{2+}$ analogues,[226] this difference being attributed to the co-operative participation of two positioned pendant bases in the latter systems.[226] Thus, both the number and position of pendant bases play an important role in the design of this kind of molecular catalyst.[226]

It should be noted that Co analogues of $[Ni(P^{R}{}_2N^{R'}{}_2)_2]^{2+}$ and $[Ni(dppp)(NP^{Ph}{}_2N^{Bz}{}_2)]^{2+}$ (Figure 12.33 and Figure 12.35) have also been reported by DuBois *et al.*[220,227] $[Co(PPh_2NPh_2)_2(CH_3CN)]^{2+}$ is not catalytically active towards the formation of H_2.[227] However, a similar complex with a single cyclic diphosphine ligand, $[Co(PPh_2NPh_2)(CH_3CN)_3]^{2+}$, is an electrocatalyst for H_2 production with a catalytic turnover frequency of 90 s^{-1} at an estimated over-potential of 285 mV.[189,227] These values are generally comparable to those observed for the complexes $[Ni(P^R{}_2N^{R'}{}_2)_2]^{2+}$.[173,202,203] For $[Co(PPh_2NPh_2)(CH_3CN)_3]^{2+}$, high catalytic activity is obtained with only one pendant base rather than two bases as for Ni complexes.[189] Therefore, Co complexes containing a single diphosphine ligand with a positioned pendant base are a promising class of catalysts for H_2 production that will be investigated further and optimised in the future.[220]

12.7.3 [NiFe] Hydrogenase Models Incorporating Protonation Sites

There has been much less focus on work towards analogues of the active sites of the [NiFe] hydrogenases which contain protonation sites. Sun *et al.* have reported the trinuclear [NiFe$_2$] complexes $[Fe_2(CO)_6(\mu_3\text{-}S)_2Ni(Ph_2PCH_2)_2NR]$ (R = ^{n}Bu or Ph) (Figure 12.36).[228] Treatment of $[Fe_2(CO)_6(\mu_3\text{-}S)_2Ni(Ph_2PCH_2)_2N^nBu]$ with HOTf produces a stable protonated form of the complex, $[Fe_2(CO)_6(\mu_3\text{-}S)_2Ni(Ph_2PCH_2)_2NH(^nBu)][OTf]$, with the [2Fe-2S] units in these complexes adopting butterfly configurations with the Fe centres in square-pyramidal geometries. Six-membered rings adopting chair conformations are formed between the Ni(II) centres and the diphosphine ligands.[228] A comparison of $[Fe_2(CO)_6(\mu_3\text{-}S)_2Ni(Ph_2PCH_2)_2N(^nBu)]$ and its protonated form confirms that protonation of the basic site has almost no influence on the metal framework, but the orientation of the ^{n}Bu group on the N-centre of the NCPNiPC chair conformation changes from equatorial to axial.[228] In this axial conformation, the bridgehead N-centre is orientated such that it may play a role as a proton relay.

$[Fe_2(CO)_6(\mu_3\text{-}S)_2Ni(Ph_2PCH_2)_2N(^nBu)]$ and $[Fe_2(CO)_6(\mu_3\text{-}S)_2Ni(dppe)]$ show similar electrochemical processes (Table 12.2).[228] However, upon addition of two equivalents of HOTf to a solution of $[Fe_2(CO)_6(\mu_3\text{-}S)_2Ni(Ph_2PCH_2)_2N(^nBu)]$ and subsequent *in situ* formation of the protonated species, a new, more anodic reduction process is observed at $E_p{}^c = -1.40$ V *vs.*

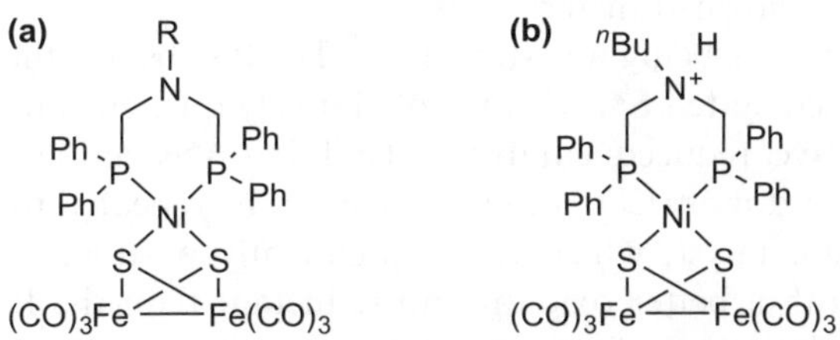

Figure 12.36 The structures of (a) $[Fe_2(CO)_6(\mu_3\text{-}S)_2Ni(Ph_2PCH_2)_2NR]$ (R = ^{n}Bu, Ph) and (b) $[Fe_2(CO)_6(\mu_3\text{-}S)_2Ni(Ph_2PCH_2)_2NH(^nBu)][OTf]$.[228]

Table 12.2 Redox potentials for $[Fe_2(CO)_6(\mu_3\text{-}S)_2Ni(Ph_2PCH_2)_2N(^nBu)]$, $[Fe_2(CO)_6(\mu_3\text{-}S)_2Ni(Ph_2PCH_2)_2NH(^nBu)][OTf]$ and $[Fe_2(CO)_6(\mu_3\text{-}S)_2Ni(dppe)]$ *vs.* Ag/Ag^+.[228]

Complex	E_p^a/ V Fe(II)Fe(I)/ Fe(II)Fe(II)	E_p^a/ V Fe(I)-Fe(I)/ Fe(II)Fe(I)	E_p^c/ V Fe(I)-Fe(I)/ Fe(0)Fe(I)
$[Fe_2(CO)_6(\mu_3\text{-}S)_2Ni(Ph_2PCH_2)_2N(^nBu)]$	+0.52	+0.18	–1.73
$[Fe_2(CO)_6(\mu_3\text{-}S)_2Ni(Ph_2PCH_2)_2NH(^nBu)][OTf]$		+0.36	–1.40
$[Fe_2(CO)_6(\mu_3\text{-}S)_2Ni(dppe)]$		+0.19	–1.68

Ag/Ag^+ (Table 12.2).[228] This new process quickly disappears on the addition of aniline, and the peaks associated with $[Fe_2(CO)_6(\mu_3\text{-}S)_2Ni(Ph_2PCH_2)_2N(^nBu)]$ are recovered quantitatively. Thus, the basic site in this heterotrinuclear complex can readily form a protonated species, which effects the redox potentials of the Fe(I)Fe(I)/Fe(0)Fe(I) couple. However, it is not reported how the protonation sites influence the catalytic activity of these complexes.[228]

12.8 Photo-activation and Photo-catalysis

12.8.1 Introduction

The efficient and clean production of H_2 from water *via* light-harvesting catalysts represents a major challenge in the development the hydrogen economy and its implementation to replace fossil fuels.[229,230] Thus, the development of low molecular weight complex models that mimic the function of hydrogenase biosites is also important in terms of synthesis of later generations of *in vitro* catalysts that incorporate light-harvesting capability. A major target is the coupling of a photosensitiser (PS), a group or compound which is able to absorb light, to a suitable catalyst, which is able to store the electrons necessary to convert protons into H_2.[229,231] The lifetime of the excited photosensitiser (PS*) should be sufficiently long to allow the electron-transfer to the catalyst, keeping in mind that quenching of PS* by the catalyst by energy transfer is a competing process. The redox properties of the separate components must be mutually compatible to allow electron-transfer from PS* to the catalyst with minimum back electron-transfer.

Two approaches are being investigated.[229] The PS absorbs one photon of light to form the excited-state PS*, which is oxidatively quenched by the pre-catalyst to form the active reduced catalyst. The PS is regenerated by an external sacrificial electron donor (ED) and the cycle can be repeated to accumulate two electrons on the catalyst, which can subsequently produce H_2. Alternatively, PS*, which is both a better oxidant and reductant than the PS, can be reductively quenched by ED to form PS^- which is a yet stronger reducing agent, which can donate an electron to the pre-catalyst, regenerating the PS. The cycle can then be repeated to accumulate two electrons on the catalyst.

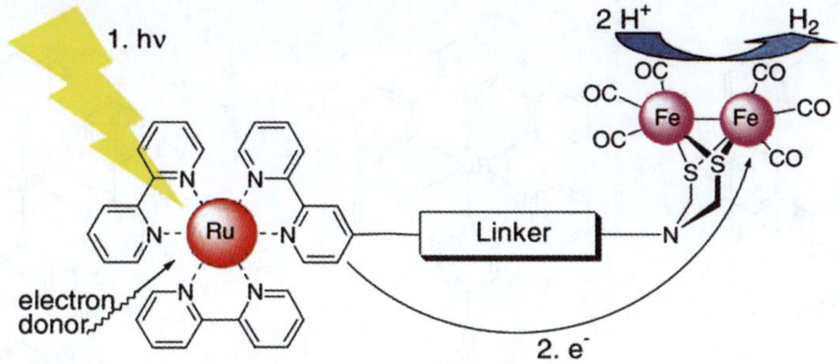

Scheme 12.18 Schematic representation of light-driven proton reduction.[229]

The transfer of the second electron to a mono-reduced catalyst is less favourable than the transfer of the first electron. However, protonation of the one-electron reduced catalyst and consequent neutralisation of charge will compensate to make the second electron-transfer more facile.[229] In recent years, progress has been made *via* covalently or co-ordinatively linking the PS directly onto the catalyst backbone through a linker, or by using all the required components together in homogeneous solution.[229–231] Complexes of Ru(II) with amine ligands, such as bipyridine, terpyrdine and their derivatives, are active photosensitisers[232,233] that can be targeted to harvest solar energy for H_2 production (Scheme 12.18).[229]

In the following sections a non-exhaustive overview of different approaches to photocatalysis using hydrogenase models is presented, with particular attention paid to the key points and the main achievements. Very recently, more comprehensive reviews have been published.[111,229–231,233]

12.8.2 Synthesis of a Photo-catalyst *via* Direct Linking and Assembly

$[Ru(bipy)_3]^{2+}$ complexes have been widely used as photo-sensitisers[234] and have been covalently-linked to FeFe cores *via* attachment to the dithiolate bridge.[230] Complexes **7**, **8**, **9** and **10** (Figure 12.37) have been prepared and characterised using various spectroscopic and electrochemical techniques, including cyclic voltammetry and IR, UV-Vis, fluorescence and time-resolved emission spectroscopies.[235–240]

Comparisons of data for complexes **7–10** with the corresponding precursors (the functionalised catalyst, PS and linker) were undertaken. No changes were observed in the UV-Vis and IR spectra of **7**, **8** and **9** with respect to the corresponding precursors, suggesting that there is no electronic interaction between the ground state of the FeFe complex and the Ru-centred PS.[230] In all cases, a decrease in the lifetime of the excited-state was reported for the complex, explained by dipole–dipole energy transfer quenching onto the FeFe core rather than invoking electronic interaction/coupling.[236–240] In general, it was found that the one-electron reduction potential of the catalytic centre was too cathodic,

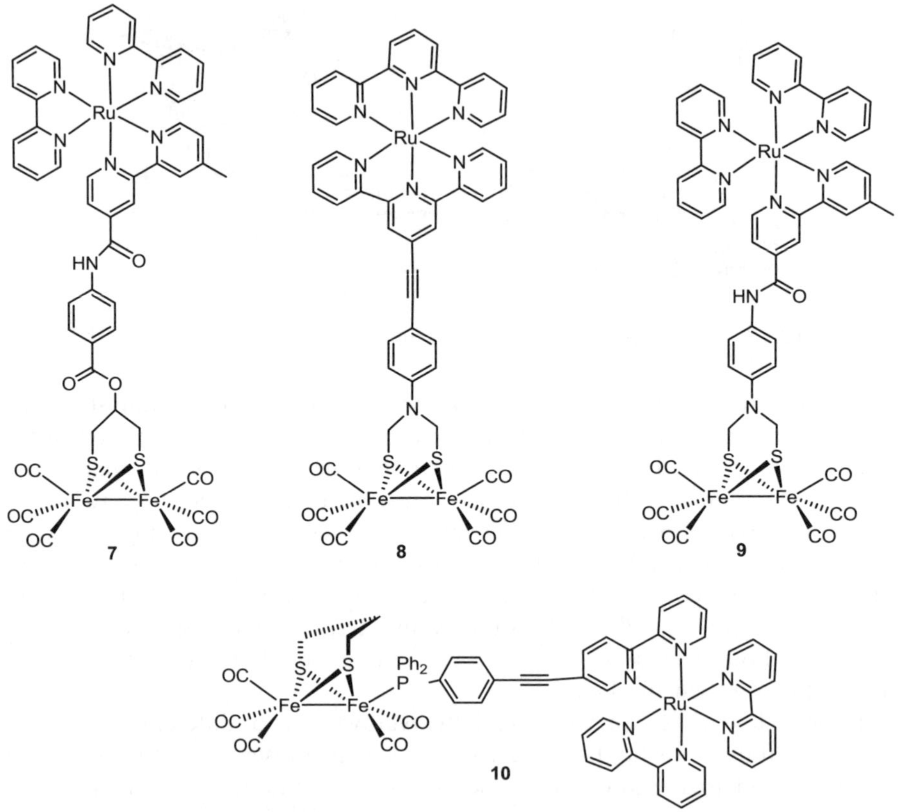

Figure 12.37 Some examples of complexes incorporating FeFe moieties attached covalently to a PS.[230]

and that electron-transfer from the PS* to the catalyst was not favoured.[230] Combining redox potentials with the excited state energies into the simplified Rehm–Weller equation, it was possible to calculate the driving force for electron-transfer from PS* to catalyst as uphill by 0.59 eV for **8**. However, at least for **8**, it is believed that protonation of the bridgehead N-donor could shift the reduction potential of the FeFe complex anodically, making the electron-transfer more favourable.[238] For **10**, the reduction potential of the FeFe centre was found to be too cathodic; however, in this case a non-negligible electronic coupling between the PS and FeFe moiety was observed in the UV-vis spectrum. Quenching of the PS* by the FeFe complex was observed *via* electron-transfer from metal to PS*, but this led to decomposition since the oxidised FeFe complex is known to be unstable.[235] To overcome these problems, shifting the reduction potential of the catalytic centre to more anodic potentials or the redox potential of the PS* cathodically was proposed. The second generation of these compounds focused on modification of the PS (Figure 12.38).[241–243]

A [Re(bipy)$(CO)_3$] centre was investigated as a PS since its excited state has a considerably more cathodic redox potential than $[Ru(bpy)_3]^{2+}$.[241,244,245] In this

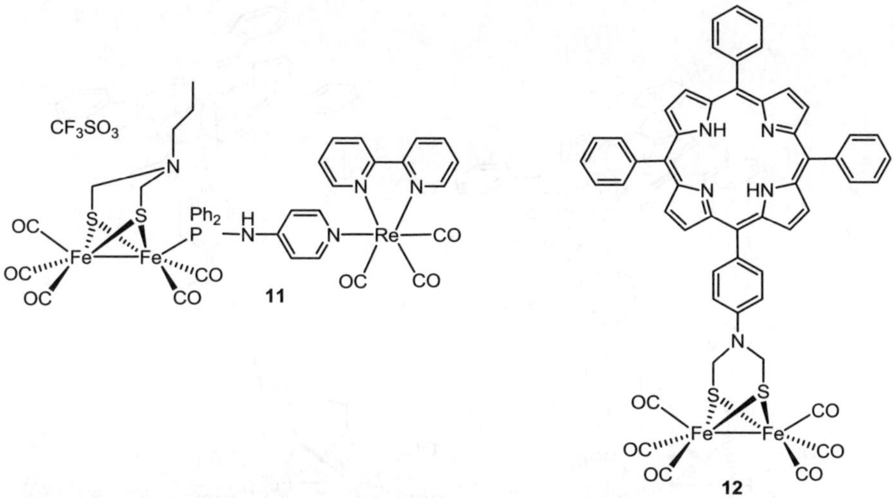

Figure 12.38 Examples of "second generation" complexes incorporating an FeFe core covalently-attached to a PS.[241–243]

case, the electron-transfer is thermodynamically feasible, as calculated for **11**.[241,246] Porphyrin derivatives of TPP (TPP = tetraphenylporphyrin) were selected due to their high light-harvesting efficiency (46% of light absorbed) and the lifetime of the excited state is longer than those for Ru(II) amine complexes. Also, porphyrin-based sensitisers are neutral, making electron-transfer potentially more facile.[243,247,248] In both cases, the FeFe catalyst incorporating an N-donor bridgehead was used. Both systems show some level of electronic interaction between the PS and the catalytic centre and the one-electron transfer from PS* to the FeFe core is thermodynamically feasible. Thus, upon irradiation of a solution of **12** in CH_2Cl_2 with EtSH (ED) and CF_3CO_2H (as proton source) with strong visible light, catalytic production of H_2 was observed. The turnover frequency is quite low (0.31 h^{-1}), probably due to the decomposition of the catalyst upon irradiation.[242] Nonetheless, **12** is a good starting point, and its photostability can be increased by substituting CO ligands of the FeFe unit and the photo and redox properties of the porphyrin PS can be tuned and optimised *via* coordination to a wide variety of metal ions.[242] A different approach has been successfully reported using non-covalent self-assembly of [Zn(TPP)] as a PS with pyridyl-functionalised FeFe cores (Figure 12.39).[249–251]

Thus, upon irradiation of solutions of **14**[249] and **15**[250] with strong visible light in the presence of an ED and a proton source, H_2 is produced. The turnover number is rather low: 0.16 ToN over two hours for **14**, and an estimated 1 ToN over 80 minutes for **15**, reflecting the instability of the catalyst upon irradiation and the limited concentration of protons available. Decreasing the pH leads to protonation of the pyridyl group and disruption of the supramolecular assembly, and thus low concentrations of protons have to be used in this system. It was also observed that photo-induced electron-transfer from PS* to the

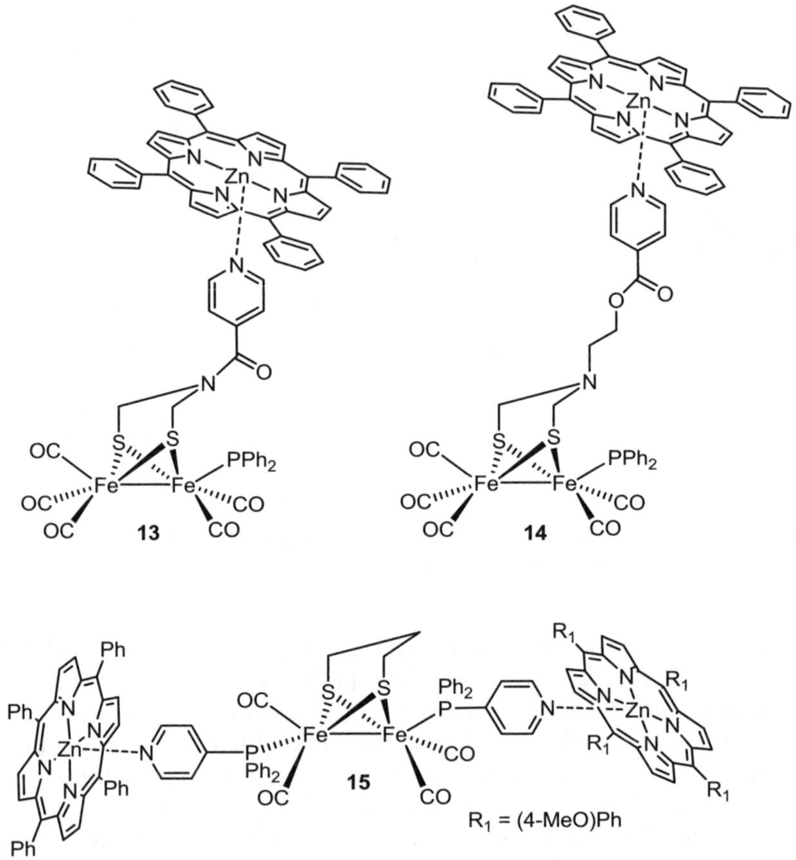

Figure 12.39 Examples of the assembly of FeFe catalysts with porphyrin centres as the PS.[249–251]

FeFe complex is faster and more efficient than for covalently-linked systems. In fact, in this kind of supramolecular assembly, the two units (PS and catalyst) are able to separate from each other after intramolecular electron-transfer, reducing both the back electron-transfer from reduced catalyst to the PS and the quenching by energy transfer compared to the covalently-linked PS system.[231] This approach is promising because of its wide applicability, and it also allows ready modification of all the subunits. Further studies are necessary to optimise all parameters and the components of these assemblies.

12.8.3 Mixing the Components

Light-driven H_2 production can also be achieved by the simple mixing of all of the necessary components (PS, catalyst, ED and proton source) without linking them chemically (Scheme 12.19).[252,253] The advantages of this approach are that synthesis is simplified dramatically and the rate of back electron-transfer from catalyst to PS is reduced. However, intermolecular electron-transfer is generally enthalpically less favorable than intramolecular electron-transfer due to the potentially greater interaction between the centres involved in the latter.[231]

A photo-induced intermolecular electron-transfer system was reported in 2007 by Sun *et al.*[253] combining $[Ru(bpy)_3]^{2+}$ as the PS with $[Fe_2(\mu\text{-}SCH_2XCH_2S)(CO)_6]$ ($X = CH_2$, $NCH_2C_6H_5$) as catalysts with diethyldithiocarbamate anion (dtc^-) as the ED (Scheme 12.20). Using laser-flash photolysis, the formation of the Fe(I)Fe(0) state, proposed to be a crucial intermediate for proton reduction, was observed. However, light-driven H_2 production was not observed, since the ED decomposes in the presence of proton donor.[253] The problem was solved using ascorbic acid as both proton source and ED, the ascorbate anion being able to reductively quench PS*.[252] Thus, upon irradiation of these solutions (Scheme 12.21), H_2 was successfully generated, even if the turnover frequency and number was still very low. The main problem was identified as instability of the catalytic centre under the conditions of

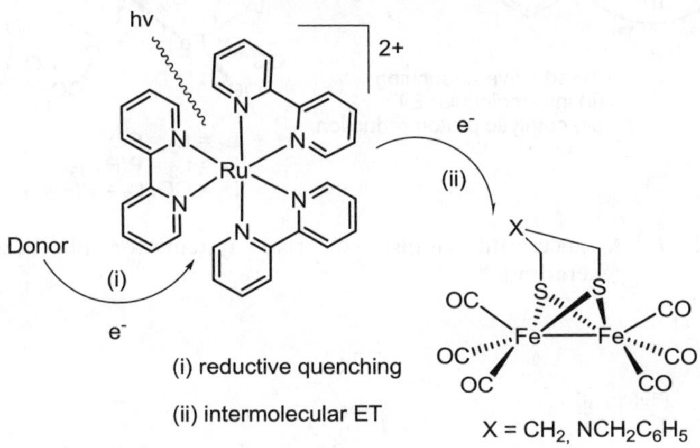

Scheme 12.19 Bioinspired triad system for photoinduced ET.[253]

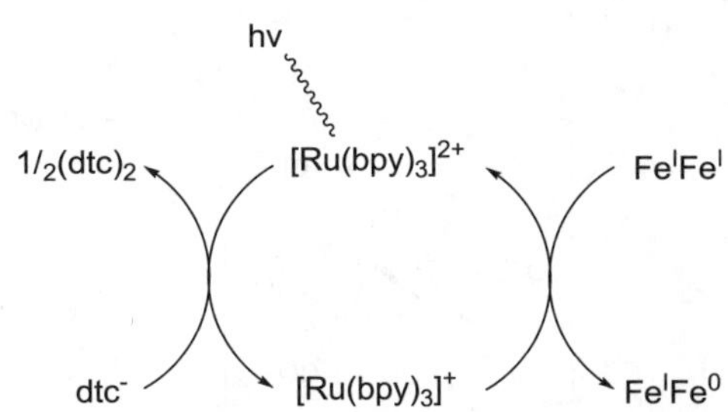

Scheme 12.20 Proposed light-driven electron-transfer process of the triad system composed of dtc^-, $[Ru(bpy)_3]^{2+}$ and $[Fe_2(\mu\text{-}SCH_2XCH_2S)(CO)_6]$ ($X = CH_2$, $NCH_2C_6H_5$).[253]

irradiation. However, the system is quite stable to water, which makes it potentially applicable to use in aqueous media as a proton source.[252]

The complex [Fe_2(μ-Cl_2bdt)$(CO)_6$] (μ-Cl_2bdt^{2-} = 3,6-dichlorobenzene-1,2-dithiolate)] is an active catalyst in water/DMF (1 : 1) when coupled to ascorbic acid as ED and proton source and $[Ru(bpy)_3]^{2+}$ as the PS (Scheme 12.22).[254] The relatively high turnover frequency (up to 2.7 min^{-1}) was explained by the increased stability of the reduced catalyst, [Fe_2(μ-Cl_2bdt)$(CO)_6$] being the first FeFe complex to exhibit reversible reductive

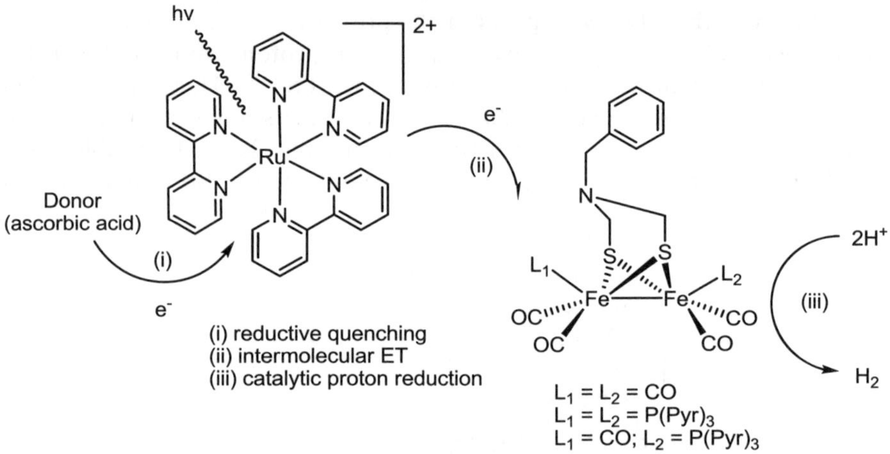

Scheme 12.21 A successful bioinspired triad system for photochemical H_2 generation.[252]

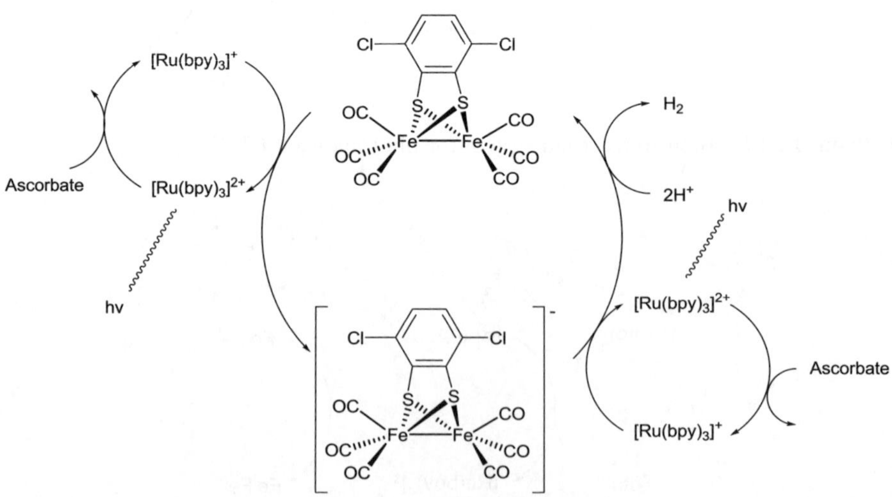

Scheme 12.22 Mechanism for catalysis proposed for the photochemical reduction of protons using [Fe_2(μ-Cl_2bdt)$(CO)_6$] (μ-Cl_2bdt^{2-} = 3,6-dichlorobenzene-1,2-dithiolate) in 1 : 1 H_2O/DMF.[254]

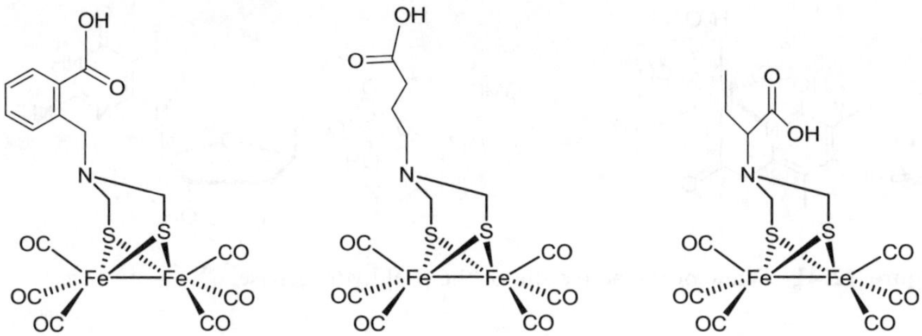

Figure 12.40 Complexes used as catalysts for light-driven H_2 production by Gao *et al.*[255]

electrochemistry in the absence of protons, with catalytic activity at mild potentials in the presence of protons.

Another water-stable system has been obtained using FeFe complexes with pendant carboxylic acid groups (Figure 12.40). However, the turnover frequency and number are not high, probably because these complexes do not exhibit reversible reductive electrochemistry.[255] Further studies are necessary to enhance the stability of catalytic centre and to improve its activity.

12.9 [Fe] Hydrogenase

The crystal structure of [Fe] hydrogenase was reported only in 2008,[256] and model and functional studies of this active site are at an early stage.[74,111] However, this area will undoubtedly develop rapidly in the coming years.[74,111]

The [Fe] hydrogenase (also known as Hmd, H_2-forming methylene-tetrahydromethanopterin dehydrogenase) is distinct from the other hydrogenases, since it is not redox active and performs a different chemistry.[257] It does not directly utilise H_2 and H^+ as substrates in hydrogen metabolism ($H_2 \rightarrow 2H^+ + 2e^-$), but catalyses the reduction of N^5,N^{10}-methenyl-tetrahydromethanopterin with H_2 to N^5,N^{10}-methylene-tetrahydromethanopterin and a proton. This is an intermediate step in the reduction of CO_2 to methane using H_2 by methanogens under Ni-deficient conditions.[257] The enzyme is specific to the pterin substrate and binding of H_2 is unfavourable in the absence of this substrate, the latter being directly involved in H_2 activation.[257] The active site is an Fe(II) complex bound by two carbonyls, a cysteinyl sulfur, an acylpyridyl group and a sixth oxygen-containing ligand (Figure 12.41).

The main target is to use low molecular weight analogues to understand the mechanism of [Fe] hydrogenase activity, so efforts have been directed to the preparation of structural models that resemble the active site as closely as possible and that show similar spectroscopic properties. No functional biomimetics have been reported; indeed, it is not clear whether small molecule mimics can function in the absence of the enzyme scaffold.[257] The first models were reported before the elucidation of the structure by X-ray crystallography and

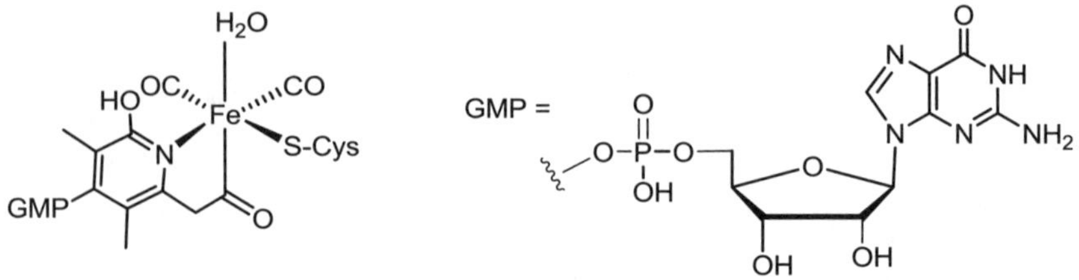

Figure 12.41 View of the active site of the [Fe] hydrogenase.[256]

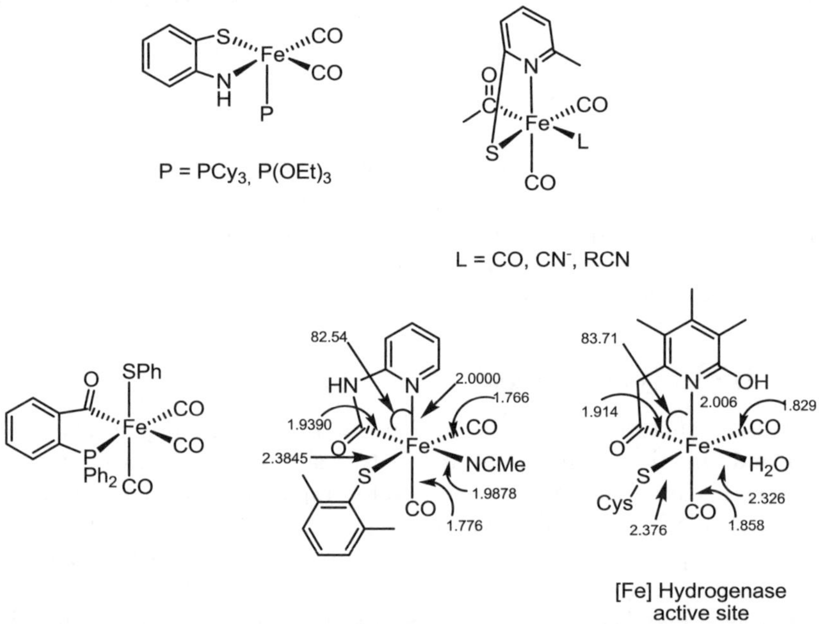

Figure 12.42 Examples of structural models of the active site of [Fe] hydrogenase.[258–261]

were based upon various spectroscopic data. A series of model complexes and targets have been prepared and these are summarised in Figure 12.42.[258–261]

Further studies are still required to understand better the mechanisms of action of the [Fe] hydrogenases and to improve the models including their functional catalytic properties.

12.10 Perspectives and Future Challenges

The future of the hydrogen economy hinges on the efficient and economic generation and utilisation of H_2 as an energy vector. Hydrogenase enzymes shed a sharp light on how Nature uses H_2 as an energy source, and our current ability to mimic such complex *in vivo* reactivity, control and efficiency using *in vitro* model complexes is developing apace, but in many ways remains embarrassingly limited. The generation of new catalysts and photo-catalysts for the light-driven production of H_2 is a major target world-wide, and new H_2 oxidation catalysts are urgently required for applications in fuel cells and to replace the currently-used precious and expensive Pt catalysts. Increasing the catalytic turnover numbers, frequency and rates of reaction *via* catalyst design, enhancing the rates and efficiency of electron-transfer, optimizing the interaction between catalyst, charge-transfer states, PS, ED, and substrates, and tuning the photo and redox properties of components all remain major challenges that need to be addressed. Chemical design, synthesis and functionalisation lie at the heart of the future development of *in vitro* catalysts. Hydrogenase enzymes are the masters in H_2 catalysis; we hope one day to catch up.

Acknowledgements

We thank EPSRC and the University of Nottingham for support. MS gratefully acknowledges receipt of an ERC Advanced Grant.

References

1. D. E. Fenton, *Biocoordination Chemistry*, Oxford University Press, Oxford, 1995.
2. P. C. Wilkins and R. G. Wilkins, *Inorganic Chemistry in Biology*, Oxford University Press, Oxford, 1997.
3. J. A. Cowan, *Inorganic Biochemistry*, VCH: Weinheim, 1993.
4. W. Kaim and B. Schwederski, *Bioinoganic Chemistry: Inorganic Elements in the Chemistry of Life*, Wiley, Chichester, 1994.
5. S. J. Lippard and J. M. Berg, *Principles of Bioinorganic Chemistry*, University Science Books, Mill Valley, CA, 1994.
6. J. R. R. Fraústo da Silva and R. J. P. Williams, *The Biological Chemistry of the Elements*, the Inorganic Chemsitry of Life, Clarendon Press, Oxford, 1991.
7. I. Bertini, H. B. Gray, J. S. Lippard and J. S. Valentine, *Bioinorganic Chemistry*, University Science Books, Sausalito, CA, 1994.
8. D. C. Rees and J. B. Howard, *Curr. Opin. Chem. Biol.*, 2000, **4**, 559–566.
9. G. C. Bond, *Platinum Met. Rev.*, 1994, **38**, 16–19.
10. D. Forster, F. G. A. Stone and R. West, *Advances Organometallic Chemistry*, Academic Press, 1979, vol. 17, pp. 255–267.
11. D. J. Evans, *Coord. Chem. Rev.*, 2005, **249**, 1582.
12. L. Yi, *Angew. Chem., Int. Ed.*, 2006, **45**, 5588.
13. J. A. Kent, *Riegel's Handbook of Industrial Chemistry*, Plenum Publishers, New York, 2003.

14. T. D. H. Bugg, *Introduction to Enzyme and Coenzyme Chemistry*, Blackwell Publishing Ltd, Oxford, UK, 2004.
15. R. H. Holm and E. I. Solomon, *Chem. Rev.*, 1996, **96**, 2237–2238.
16. F. Tian, O. B. Toon, A. A. Pavlov and H. De Sterck, *Science*, 2005, **308**, 1014–1017.
17. P. M. Vignais and B. Billoud, *Chem. Rev.*, 2007, **107**, 4206–4272.
18. M. Stephenson and L. H. Stickland, *Biochem. J.*, 1931, **25**, 205–214.
19. M. W. W. Adams, L. E. Mortenson and J. S. Chen, *Biochim. Biophys. Acta*, 1980, **594**, 105–176.
20. R. Cammack, M. Frey and R. Robson, *Hydrogen as a Fuel: Learning from Nature*, Taylor and Francis, London and New York, 2001.
21. P. M. Vignais, B. Billoud and J. Meyer, *Fems Microbiol. Rev.*, 2001, **25**, 455–501.
22. A. L. De Lacey, V. M. Fernandez, M. Rousset and R. Cammack, *Chem. Rev.*, 2007, **107**, 4304–4330.
23. A. Volbeda, M. H. Charon, C. Piras, E. C. Hatchikian, M. Frey and J. C. Fontecilla-Camps, *Nature*, 1995, **373**, 580–587.
24. Y. Higuchi, T. Yagi and N. Yasuoka, *Structure*, 1997, **5**, 1671–1680.
25. A. Volbeda, E. Garcin, C. Piras, A. L. De Lacey, V. M. Fernandez, E. C. Hatchikian, M. Frey and J. C. Fontecilla-Camps, *J. Am. Chem. Soc.*, 1996, **118**, 12989–12996.
26. E. Garcin, X. Vernede, E. C. Hatchikian, A. Volbeda, M. Frey and J. C. Fontecilla-Camps, *Structure*, 1999, **7**, 557–566.
27. Y. Higuchi, H. Ogata, K. Miki, N. Yasuoka and T. Yagi, *Structure*, 1999, **7**, 549–556.
28. H. Ogata, Y. Mizoguchi, N. Mizuno, K. Miki, S. Adachi, N. Yasuoka, T. Yagi, O. Yamauchi, S. Hirota and Y. Higuchi, *J. Am. Chem. Soc.*, 2002, **124**, 11628–11635.
29. P. M. Matias, C. M. Soares, L. M. Saraiva, R. Coelho, J. Morais, J. Le Gall and M. A. Carrondo, *J. Biol. Inorg. Chem.*, 2001, **6**, 63–81.
30. Y. Montet, P. Amara, A. Volbeda, X. Vernede, E. C. Hatchikian, M. J. Field, M. Frey and J. C. Fontecilla-Camps, *Nat. Struct. Biol.*, 1997, **4**, 523–526.
31. A. Volbeda, Y. Montet, X. Vernede, E. C. Hatchikian and J. C. Fontecilla-Camps, *Int. J. Hydr. Energy*, 2002, **27**, 1449–1461.
32. J. W. Peters, W. N. Lanzilotta, B. J. Lemon and L. C. Seefeldt, *Science*, 1998, **282**, 1853–1858.
33. Y. Nicolet, C. Piras, P. Legrand, C. E. Hatchikian and J. C. Fontecilla-Camps, *Struct. Fold. Des.*, 1999, **7**, 13–23.
34. S. E. Lamle, S. P. J. Albracht and F. A. Armstrong, *J. Am. Chem. Soc.*, 2005, **127**, 6595–6604.
35. C. Zirngibl, W. Van Dongen, B. Schworer, R. Von Bunau, M. Richter, A. Klein and R. K. Thauer, *Eur. J. Biochem.*, 1992, **208**, 511–520.
36. R. K. Thauer, A. R. Klein and G. C. Hartmann, *Chem. Rev.*, 1996, **96**, 3031–3042.
37. R. K. Thauer, *Microbiology*, 1998, **144**, 2377–2406.

38. S. Shima, E. J. Lyon, M. S. Sordel-Klippert, M. Kauss, J. Kahnt, R. K. Thauer, K. Steinbach, X. L. Xie, L. Verdier and C. Griesinger, *Angew. Chem., Int. Ed.*, 2004, **43**, 2547–2551.
39. E. J. Lyon, S. Shima, R. Boecher, R. K. Thauer, F. W. Grevels, E. Bill, W. Roseboom and S. P. J. Albracht, *J. Am. Chem. Soc.*, 2004, **126**, 14239–14248.
40. M. Korbas, S. Vogt, W. Meyer-Klaucke, E. Bill, E. J. Lyon, R. K. Thauer and S. Shima, *J. Biol. Chem.*, 2006, **281**, 30804–30813.
41. S. Shima and R. K. Thauer, *Chem. Rec.*, 2007, **7**, 37–46.
42. L. Margulis, *Proc. Natl. Acad. Sci. U. S. A.*, 1996, **93**, 1071–1076.
43. A. Volbeda and J. C. Fontecilla-Camps, *Coord. Chem. Rev.*, 2005, 1609–1619.
44. R. P. Happe, W. Roseboom, A. J. Pierik, S. P. J. Albracht and K. A. Bagley, *Nature*, 1997, **385**, 126.
45. A. J. Pierik, W. Roseboom, R. P. Happe, K. A. Bagley and S. P. J. Albracht, *J. Biol. Chem.*, 1999, **274**, 3331–3337.
46. Y. Higuchi, F. Toujou, K. Tsukamoto and T. Yagi, *J. Inorg. Biochem.*, 2000, **80**, 205–211.
47. R. P. Happe, W. Roseboom, G. Egert, C. G. Friedrich, C. Massanz, B. Friedrich and S. P. J. Albracht, *FEBS Lett.*, 2000, **466**, 259–263.
48. H. Ogata, S. Hirota, A. Nakahara, H. Komori, N. Shibata, T. Kato, K. Kano and Y. Higuchi, *Structure*, 2005, **13**, 1635–1642.
49. A. Volbeda, L. Martin, C. Cavazza, M. Matho, B. W. Faber, W. Roseboom, S. P. Albracht, E. Garcin, M. Rousset and J. C. Fontecilla-Camps, *J. Biol. Inorg. Chem.*, 2005, **10**, 239–249.
50. J. C. Fontecilla-Camps, A. Volbeda, C. Cavazza and Y. Nicolet, *Chem. Rev.*, 2007, **107**, 4273–4303.
51. J. C. Fontecilla-Camps, M. Frey, E. Garcin, C. Hatchikian, Y. Montet, C. Piras, X. Vernede and A. Volbeda, *Biochimie*, 1997, **79**, 661–666.
52. Y. Nicolet, C. Cavazza and J. C. Fontecilla-Camps, *J. Inorg. Biochem.*, 2002, **91**, 1–8.
53. F. M. C. Valente, A. S. Oliveira, N. Gnadt, I. Pacheco, A. V. Coelho, A. V. Xavier, M. Teixeira, C. M. Soares and I. A. C. Pereira, *J. Biol. Inorg. Chem.*, 2005, **10**, 667–682.
54. R. Bingemann and A. Klein, *Eur. J. Biochem.*, 2000, **267**, 6612–6618.
55. M. Teixeira, I. Moura, A. V. Xavier, J. J. G. Moura, J. Le Gall, D. V. DerVartanian and J. N. T. Peck, *J. Biol. Chem.*, 1989, **264**, 16435–16450.
56. C. C. Page, C. C. Moser and P. L. Dutton, *Curr. Opin. Chem. Biol.*, 2003, **7**, 551–556.
57. D. Leys and N. S. Scrutton, *Curr. Opin. Struct. Biol.*, 2004, **14**, 642–647.
58. P. Brzezinski and P. Adelroth, *Curr. Opin. Struct. Biol.*, 2006, **16**, 465–472.
59. J. W. Peters, *Curr. Opin. Struct. Biol.*, 1999, **9**, 670–676.
60. J. Meyer and J. Gagnon, *Biochemistry*, 1991, **30**, 9697–9704.
61. J. S. Chen and L. E. Mortenson, *Biochim. Biophys. Acta*, 1974, **371**, 283–298.
62. M. W. W. Adams, *Biochim. Biophys. Acta*, 1990, **1020**, 115–145.

63. M. Kaji, Y. Taniguchi, O. Matsushita, S. Katayama, S. Miyata, S. Morita and A. Okabe, *FEMS Microbiol. Lett.*, 1999, **181**, 329–336.
64. P. J. Van Dam, E. J. Reijerse and W. R. Hagen, *Eur. J. Biochem.*, 1997, **248**, 355–361.
65. T. M. Van der Spek, A. F. Arendsen, R. P. Happe, S. Yun, K. A. Bagley, D. J. Stufkens, W. R. Hagen and S. P. J. Albracht, *Eur. J. Biochem.*, 1996, **237**, 629–634.
66. R. L. Kerby, P. W. Ludden and G. P. Roberts, *J. Bacteriol.*, 1997, **179**, 2259–2266.
67. T. P. Chirpich, V. Zappia, R. N. Costilow and H. A. Barker, *J. Biol. Chem.*, 1970, **245**, 1778–1789.
68. Y. Nicolet, B. J. Lemon, J. C. Fontecilla-Camps and J. W. Peters, *Trends Biochem. Sci.*, 2000, **25**, 138–143.
69. A. J. Pierik, M. Hulstein, W. R. Hagen and S. P. J. Albracht, *Eur. J. Biochem.*, 1998, **258**, 572–578.
70. Y. Nicolet, A. L. De Lacey, X. Vernede, V. M. Fernandez, E. C. Hatchikian and J. C. Fontecilla-Camps, *J. Am. Chem. Soc.*, 2001, **123**, 1596–1601.
71. C. J. Curtis, A. Miedaner, R. Ciancanelli, W. W. Ellis, B. C. Noll, M. Rakowski DuBois and D. L. DuBois, *Inorg. Chem.*, 2003, **42**, 216–227.
72. L. Pieulle, X. Morelli, P. Gallice, E. Lojou, P. Barbier, M. Czjzek, P. Bianco, F. Guerlesquin and E. C. Hatchikian, *J. Mol. Biol.*, 2005, **354**, 73–90.
73. J. Cohen, K. Kim, P. King, M. Seibert and K. Schulten, *Structure*, 2005, **13**, 1321–1329.
74. D. M. Heinekey, *J. Organomet. Chem.*, 2009, **694**, 2671–2680.
75. T. Yamamura, H. Miyamae, Y. Katayama and Y. Sasaki, *Chem. Lett.*, 1985, **14**, 269–272.
76. T. Yamamura, *Chem. Lett.*, 1986, **15**, 801–804.
77. B. S. Snyder, C. P. Rao and R. H. Holm, *Aust. J. Chem.*, 1986, **39**, 963–974.
78. D. Sellmann, S. Funfgelder, G. Pohlmann, F. Knoch and M. Moll, *Inorg. Chem.*, 1990, **29**, 4772–4778.
79. S. Fox, Y. Wang, A. Silver and M. Millar, *J. Am. Chem. Soc.*, 1990, **112**, 3218–3220.
80. M. A. Halcrow and G. Christou, *Chem. Rev.*, 1994, **94**, 2421–2481.
81. S. P. J. Albracht, *Biochim. Biophys. Acta*, 1994, **1188**, 167–204.
82. M. A. Halcrow, *Angew. Chem., Int. Ed.*, 1995, **34**, 1193–1195.
83. S. G. Rosenfield, W. H. Armstrong and P. K. Mascharak, *Inorg. Chem.*, 1986, **25**, 3014–3018.
84. E. Bouwman and J. Reedijk, *Coord. Chem. Rev.*, 2005, **249**, 1555–1581.
85. J. R. Nicholson, G. Christou, J. C. Huffman and K. Folting, *Polyhedron*, 1987, **6**, 863–870.
86. N. Baidya, P. K. Mascharak, D. W. Stephan and C. F. Campagna, *Inorg. Chim. Acta*, 1990, **177**, 233–238.
87. T. Yamamura, H. Kurihara, N. Nakamura, R. Kuroda and K. Asakura, *Chem. Lett.*, 1990, **19**, 101–104.

88. A. Silver and M. Millar, *J. Chem. Soc., Chem. Commun.*, 1992, 948–949.
89. M. Cha, J. Sletten, S. Critchlow and J. A. Kovacs, *Inorg. Chim. Acta*, 1997, **263**, 153–159.
90. M. Y. Cha, C. L. Catlin, S. C. Critchlow and J. A. Kovacs, *Inorg. Chem.*, 1993, **32**, 5868–5877.
91. D. Sellmann, H. Binder, D. Haussinger, F. W. Heinemann and J. Sutter, *Inorg. Chim. Acta*, 2000, **300**, 829–836.
92. D. Sellmann, F. Geipel and F. W. Heinemann, *Eur. J. Inorg. Chem.*, 2000, 271–279.
93. D. Sellmann, F. Geipel and F. Heinemann, *Eur. J. Inorg. Chem.*, 2000, 59–63.
94. D. Sellmann, F. Geipel and M. Moll, *Angew. Chem. Int. Ed.*, 2000, **39**, 561–563.
95. D. Sellmann, T. Hofmann and F. Knoch, *Z. Naturforsch., B: Chem. Sci.*, 1994, **49**, 821–826.
96. F. Allen, *Acta Crystallogr. B*, 2002, **58**, 380–388.
97. A. J. Blake, M. A. Halcrow and M. Schröder, *J. Chem. Soc., Dalton Trans.*, 1994, 1463–1470.
98. A. J. Blake, R. O. Gould, M. A. Halcrow and M. Schröder, *J. Chem. Soc., Dalton Trans.*, 1993, 2909–2920.
99. A. J. Blake, R. O. Gould, M. A. Halcrow, A. J. Holder, T. I. Hyde and M. Schröder, *J. Chem. Soc., Dalton Trans.*, 1992, 3427–3431.
100. A. J. Blake, M. A. Halcrow and M. Schröder, *J. Chem. Soc., Dalton Trans.*, 1992, 2803–2808.
101. M. van Gastel, J. L. Shaw, A. J. Blake, M. Flores, M. Schröder, J. McMaster and W. Lubitz, *Inorg. Chem.*, 2008, **47**, 11688–11697.
102. W. Lubitz, E. Reijerse and M. van Gastel, *Chem. Rev.*, 2007, **107**, 4331–4365.
103. A. J. Atkins, A. J. Blake and M. Schröder, *J. Chem. Soc., Chem. Commun.*, 1993, 1662–1665.
104. A. J. Atkins, D. Black, A. J. Blake, A. Marin-Becerra, S. Parsons, L. Ruiz-Ramirez and M. Schröder, *Chem. Commun.*, 1996, 457–464.
105. H. J. Krüger and R. H. Holm, *J. Am. Chem. Soc.*, 1990, **112**, 2955–2963.
106. S. Canaguier, V. Artero and M. Fontecave, *Dalton Trans.*, 2008, 315–325.
107. A. C. Marr, D. J. E. Spencer and M. Schröder, *Coord. Chem. Rev.*, 2001, **219**, 1055–1074.
108. F. Osterloh, W. Saak, D. Haase and S. Pohl, *Chem. Commun.*, 1997, 979–980.
109. C. H. Lai, J. H. Reibenspies and M. Y. Darensbourg, *Angew. Chem., Int. Ed.*, 1996, **35**, 2390–2393.
110. W. F. Zhu, A. C. Marr, Q. Wang, F. Neese, D. J. E. Spencer, A. J. Blake, P. A. Cooke, C. Wilson and M. Schröder, *Proc. Natl. Acad. Sci. U. S. A.*, 2005, **102**, 18280–18285.
111. C. Tard and C. J. Pickett, *Chem. Rev.*, 2009, **109**, 2245–2274.
112. I. P. Georgakaki, L. M. Thomson, E. J. Lyon, M. B. Hall and M. Y. Darensbourg, *Coord. Chem. Rev.*, 2003, **238**, 255–266.

113. M. Y. Darensbourg, E. J. Lyon and J. J. Smee, *Coord. Chem. Rev.*, 2000, **206**, 533–561.
114. F. Osterloh, W. Saak and S. Pohl, *J. Am. Chem. Soc.*, 1997, **119**, 5648–5656.
115. P. A. Stenson, A. Marin-Becerra, C. Wilson, A. J. Blake, J. McMaster and M. Schröder, *Chem. Commun.*, 2006, 317–319.
116. S. C. Davies, D. J. Evans, D. L. Hughes, S. Longhurst and J. R. Sanders, *Chem. Commun.*, 1999, 1935–1936.
117. M. C. Smith, J. E. Barclay, S. P. Cramer, S. C. Davies, W. W. Gu, D. L. Hughes, S. Longhurst and D. J. Evans, *J. Chem. Soc., Dalton Trans.*, 2002, 2641–2647.
118. B. E. Barton, C. M. Whaley, T. B. Rauchfuss and D. L. Gray, *J. Am. Chem. Soc.*, 2009, **131**, 6942–6943.
119. R. Della Pergola, A. Fumagalli, L. Garlaschelli, C. Manassero, M. Manassero, M. Sansoni and A. Sironi, *Inorg. Chim. Acta*, 2008, **361**, 1763–1769.
120. B. E. Barton and T. B. Rauchfuss, *J. Am. Chem. Soc.*, 2010, **132**, 14877–14885.
121. C. Nunez, R. Bastida, A. Macias, L. Valencia, J. Ribas, J. L. Capelo and C. Lodeiro, *Dalton Trans.*, 2010, **39**, 7673–7683.
122. M. W. A. Steenland, W. Lippens, G. G. Herman and A. M. Goeminne, *Bull. Soc. Chim. Belg.*, 1993, **102**, 239–242.
123. L. C. Song, Y. L. Li, L. Li, Z. C. Gu and Q. M. Hu, *Inorg. Chem.*, 2010, **49**, 10174–10182.
124. D. Sellmann, F. Geipel, F. Lauderbach and F. W. Heinemann, *Angew. Chem., Int. Ed.*, 2002, **41**, 632–634.
125. A. Perra, Q. Wang, A. J. Blake, E. S. Davies, J. McMaster, C. Wilson and M. Schröder, *Dalton Trans.*, 2009, 925–931.
126. Z. L. Li, Y. Ohki and K. Tatsumi, *J. Am. Chem. Soc.*, 2005, **127**, 8950–8951.
127. A. Volbeda, M. H. Charon, C. Piras, E. C. Hatchikian, M. Frey and J. C. Fontecilla-Camps, *Nature*, 1995, **373**, 580–587.
128. C. V. Romao, I. A. C. Pereira, A. V. Xavier, J. Le Gall and M. Teixeira, *Biochem. Biophys. Res. Commun.*, 1997, **240**, 75–79.
129. S. Tanino, Z. L. Li, Y. Ohki and K. Tatsumi, *Inorg. Chem.*, 2009, **48**, 2358–2360.
130. C. Fichtner, C. Laurich, E. Bothe and W. Lubitz, *Biochemistry*, 2006, **45**, 9706–9716.
131. W. Roseboom, A. L. De Lacey, V. M. Fernandez, E. C. Hatchikian and S. P. J. Albracht, *J. Biol. Inorg. Chem.*, 2006, **11**, 102–118.
132. Y. Ohki, K. Yasumura, M. Ando, S. Shimokata and K. Tatsumi, *Proc. Natl. Acad. Sci. U. S. A.*, 2010, **107**, 3994–3997.
133. D. Sellmann, F. Lauderbach, F. Geipel, F. W. Heinemann and M. Moll, *Angew. Chem., Int. Ed.*, 2004, **43**, 3141–3144.
134. (a) F. Lauderbach, R. Prakash, A. W. Gotz, M. Munoz, F. W. Heinemann, U. Nickel, B. A. Hess and D. Sellmann, *Eur. J. Inorg. Chem.*, 2007, 3385–3393; (b) D. Seyferth, R. S. Henderson and L. C. Song,

Organometallics, 1982, **1**, 125–133; (c) A. A. Lozano, M. D. Santana, G. Garcia, J. E. Barclay, S. C. Davies and D. J. Evans, *Z. Anorg, Allg. Chem.*, 2005, **631**, 2062–2066.
135. Q. Wang, J. E. Barclay, A. J. Blake, E. S. Davies, D. J. Evans, A. C. Marr, E. J. L. McInnes, J. McMaster, C. Wilson and M. Schröder, *Chem. Eur. J.*, 2004, **10**, 3384–3396.
136. A. Perra, E. S. Davies, J. R. Hyde, Q. Wang, J. McMaster and M. Schröder, *Chem. Commun.*, 2006, 1103–1105.
137. S. Ogo, R. Kabe, K. Uehara, B. Kure, T. Nishimura, S. C. Menon, R. Harada, S. Fukuzumi, Y. Higuchi, T. Ohhara, T. Tamada and R. Kuroki, *Science*, 2007, **316**, 585–587.
138. T. Naota, H. Takaya and S. I. Murahashi, *Chem. Rev.*, 1998, **98**, 2599–2660.
139. A. K. Justice, R. C. Linck and T. B. Rauchfuss, *Inorg. Chem.*, 2006, **45**, 2406–2412.
140. T. B. Rauchfuss, *Science*, 2007, **316**, 553–554.
141. M. V. Rampersad, S. P. Jeffery, J. H. Reibenspies, C. G. Ortiz, D. J. Darensbourg and M. Y. Darensbourg, *Angew. Chem., Int. Ed.*, 2005, **44**, 1217–1220.
142. Y. Oudart, V. Artero, J. Pecaut and M. Fontecave, *Inorg. Chem.*, 2006, **45**, 4334–4336.
143. J. M. Goicoechea, M. F. Mahon, M. K. Whittlesey, P. G. A. Kumar and P. S. Pregosin, *Dalton Trans.*, 2005, 588–597.
144. F. Basolo, *New J. Chem.*, 1994, **18**, 19–24.
145. J. M. Oconnor and C. P. Casey, *Chem. Rev.*, 1987, **87**, 307–318.
146. S. Ogo, *Chem. Commun.*, 2009, 3317–3325.
147. Y. Oudart, V. Artero, J. Pecaut, C. Lebrun and M. Fontecave, *Eur. J. Inorg. Chem.*, 2007, 2613–2626.
148. S. Canaguier, L. Vaccaro, V. Artero, R. Ostermann, J. Pecaut, M. J. Field and M. Fontecave, *Chem. Eur. J.*, 2009, **15**, 9350–9364.
149. L. Vaccaro, V. Artero, S. Canaguier, M. Fontecave and M. J. Field, *Dalton Trans.*, 2010, **39**, 3043–3049.
150. K. Ichikawa, T. Matsumoto and S. Ogo, *Dalton Trans.*, 2009, 4304–4309.
151. B. Kure, T. Matsumoto, K. Ichikawa, S. Fukuzumi, Y. Higuchi, T. Yagi and S. Ogo, *Dalton Trans.*, 2008, 4747–4755.
152. T. Matsumoto, B. Kure and S. Ogo, *Chem. Lett.*, 2008, **37**, 970–971.
153. C. B. Zheng, K. Kim, T. Matsumoto and S. Ogo, *Dalton Trans.*, 2010, **39**, 2218–2225.
154. K. Ichikawa, K. Nonaka, T. Matsumoto, B. Kure, K. S. Yoon, Y. Higuchi, T. Yagi and S. Ogo, *Dalton Trans.*, 2010, **39**, 2993–2994.
155. F. Gloaguen, J. D. Lawrence and T. B. Rauchfuss, *J. Am. Chem. Soc.*, 2001, **123**, 9476–9477.
156. J. F. Capon, F. Gloaguen, F. Y. Petillon, P. Schollhammer and J. Talarmin, *Coord. Chem. Rev.*, 2009, **253**, 1476–1494.
157. L. C. Song, *Acc. Chem. Res.*, 2005, **38**, 21–28.
158. R. B. King and T. E. Bitterwolf, *Coord. Chem. Rev.*, 2000, **206**, 563–579.

159. G. A. N. Felton, C. A. Mebi, B. J. Petro, A. K. Vannucci, D. H. Evans, R. S. Glass and D. L. Lichtenberger, *J. Organomet. Chem.*, 2009, **694**, 2681–2699.
160. J. F. Capon, F. Gloaguen, F. Y. Petillon, P. Schollhammer and J. Talarmin, *Eur. J. Inorg. Chem.*, 2008, 4671–4681.
161. J. F. Capon, F. Gloaguen, F. Y. Petillon, P. Schollhammer and J. Talarmin, *C. R. Chimie*, 2008, **11**, 842–851.
162. F. Gloaguen and T. B. Rauchfuss, *Chem. Soc. Rev.*, 2009, **38**, 100–108.
163. J. F. Capon, F. Gloaguen, P. Schollhammer and J. Talarmin, *Coord. Chem. Rev.*, 2005, **249**, 1664–1676.
164. M. Razavet, S. C. Davies, D. L. Hughes, J. E. Barclay, D. J. Evans, S. A. Fairhurst, X. M. Liu and C. J. Pickett, *Dalton Trans.*, 2003, 586–595.
165. H. X. Li and T. B. Rauchfuss, *J. Am. Chem. Soc.*, 2002, **124**, 726–727.
166. F. Gloaguen, J. D. Lawrence, M. Schmidt, S. R. Wilson and T. B. Rauchfuss, *J. Am. Chem. Soc.*, 2001, **123**, 12518–12527.
167. J. D. Lawrence, H. X. Li and T. B. Rauchfuss, *Chem. Commun.*, 2001, 1482–1483.
168. S. J. Borg, T. Behrsing, S. P. Best, M. Razavet, X. M. Liu and C. J. Pickett, *J. Am. Chem. Soc.*, 2004, **126**, 16988–16999.
169. D. S. Chong, I. P. Georgakaki, R. Mejia-Rodriguez, J. Samabria-Chinchilla, M. P. Soriaga and M. Y. Darensbourg, *Dalton Trans.*, 2003, 4158–4163.
170. A. Le Cloirec, S. P. Best, S. Borg, S. C. Davies, D. J. Evans, D. L. Hughes and C. J. Pickett, *Chem. Commun.*, 1999, 2285–2286.
171. E. J. Lyon, I. P. Georgakaki, J. H. Reibenspies and M. Y. Darensbourg, *Angew. Chem., Int. Ed.*, 1999, **38**, 3178–3180.
172. M. Schmidt, S. M. Contakes and T. B. Rauchfuss, *J. Am. Chem. Soc.*, 1999, **121**, 9736–9737.
173. F. Gloaguen, J. D. Lawrence, T. B. Rauchfuss, M. Benard and M. M. Rohmer, *Inorg. Chem.*, 2002, **41**, 6573–6582.
174. J. I. van der Vlugt, T. B. Rauchfuss, C. M. Whaley and S. R. Wilson, *J. Am. Chem. Soc.*, 2005, **127**, 16012–16013.
175. B. E. Barton and T. B. Rauchfuss, *Inorg. Chem.*, 2008, **47**, 2261–2263.
176. C. Greco, M. Bruschi, L. De Gioia and U. Ryde, *Inorg. Chem.*, 2007, **46**, 5911–5921.
177. Z. P. Liu and P. Hu, *J. Am. Chem. Soc.*, 2002, **124**, 5175–5182.
178. Z. P. Liu and P. Hu, *J. Chem. Phys.*, 2002, **117**, 8177–8180.
179. H. J. Fan and M. B. Hall, *J. Am. Chem. Soc.*, 2001, **123**, 3828–3829.
180. J. F. Capon, S. Ezzaher, F. Gloaguen, F. Y. Petillon, P. Schollhammer and J. Talarmin, *Chem.–Eur. J.*, 2008, **14**, 1954–1964.
181. S. Ott, M. Kritikos, B. Akermark, L. C. Sun and R. Lomoth, *Angew. Chem., Int. Ed.*, 2004, **43**, 1006–1009.
182. J. L. Stanley, Z. M. Heiden, T. B. Rauchfuss, S. R. Wilson, L. De Gioia and G. Zampella, *Organometallics*, 2008, **27**, 119–125.

183. U. P. Apfel, D. Troegel, Y. Halpin, S. Tschierlei, U. Uhlemann, H. Gorls, M. Schmitt, J. Popp, P. Dunne, M. Venkatesan, M. Coey, M. Rudolph, J. G. Vos, R. Tacke and W. Weigand, *Inorg. Chem.*, 2010, **49**, 10117–10132.
184. S. Ezzaher, A. Gogoll, C. Bruhn and S. Ott, *Chem. Commun.*, 2010, **46**, 5775–5777.
185. C. Tard, X. M. Liu, D. L. Hughes and C. J. Pickett, *Chem. Commun.*, 2005, 133–135.
186. C. Tard, X. M. Liu, S. K. Ibrahim, M. Bruschi, L. De Gioia, S. C. Davies, X. Yang, L. S. Wang, G. Sawers and C. J. Pickett, *Nature*, 2005, **433**, 610–613.
187. M. H. Cheah, C. Tard, S. J. Borg, X. M. Liu, S. K. Ibrahim, C. J. Pickett and S. P. Best, *J. Am. Chem. Soc.*, 2007, **129**, 11085–11092.
188. A. K. Justice, R. C. Linck, T. B. Rauchfuss and S. R. Wilson, *J. Am. Chem. Soc.*, 2004, **126**, 13214–13215.
189. M. Rakowski DuBois and D. L. DuBois, *Chem. Soc. Rev.*, 2009, **38**, 62–72.
190. K. W. Kramarz and J. R. Norton, in *Progress in Inorganic Chemistry*, John Wiley & Sons Inc, New York, 1994, vol. 42, pp. 1–65.
191. D. E. Berning, A. Miedaner, C. J. Curtis, B. C. Noll, M. Rakowski DuBois and D. L. Dubois, *Organometallics*, 2001, **20**, 1832–1839.
192. C. J. Curtis, A. Miedaner, W. W. Ellis and D. L. DuBois, *J. Am. Chem. Soc.*, 2002, **124**, 1918–1925.
193. K. Fraze, A. D. Wilson, A. M. Appel, M. Rakowski DuBois and D. L. DuBois, *Organometallics*, 2007, **26**, 3918–3924.
194. W. W. Ellis, J. W. Raebiger, C. J. Curtis, J. W. Bruno and D. L. DuBois, *J. Am. Chem. Soc.*, 2004, **126**, 2738–2743.
195. J. W. Tye, J. Lee, H. W. Wang, R. Mejia-Rodriguez, J. H. Reibenspies, M. B. Hall and M. Y. Darensbourg, *Inorg. Chem.*, 2005, **44**, 5550–5552.
196. R. Mejia-Rodriguez, D. S. Chong, J. H. Reibenspies, M. P. Soriaga and M. Y. Darensbourg, *J. Am. Chem. Soc.*, 2004, **126**, 12004–12014.
197. C. A. Boyke, J. I. van der Vlugt, T. B. Rauchfuss, S. R. Wilson, G. Zampella and L. De Gioia, *J. Am. Chem. Soc.*, 2005, **127**, 11010–11018.
198. L. C. Song, Z. Y. Yang, H. Z. Bian and Q. M. Hu, *Organometallics*, 2004, **23**, 3082–3084.
199. E. J. Lyon, I. P. Georgakaki, J. H. Reibenspies and M. Y. Darensbourg, *J. Am. Chem. Soc.*, 2001, **123**, 3268–3278.
200. X. Zhao, I. P. Georgakaki, M. L. Miller, J. C. Yarbrough and M. Y. Darensbourg, *J. Am. Chem. Soc.*, 2001, **123**, 9710–9711.
201. J. D. Lawrence, H. X. Li, T. B. Rauchfuss, M. Benard and M. M. Rohmer, *Angew. Chem. Int. Ed.*, 2001, **40**, 1768–1771.
202. D. J. Evans and C. J. Pickett, *Chem. Soc. Rev.*, 2003, **32**, 268–275.
203. M. H. Cheah, S. J. Borg, M. I. Bondin and S. P. Best, *Inorg. Chem.*, 2004, **43**, 5635–5644.
204. P. Das, J. F. Capon, F. Gloaguen, F. Y. Petillon, P. Schollhammer, J. Talarmin and K. W. Muir, *Inorg. Chem.*, 2004, **43**, 8203–8205.
205. G. M. Jacobsen, R. K. Shoemaker, M. Rakowski DuBois and D. L. DuBois, *Organometallics*, 2007, **26**, 4964–4971.

206. G. M. Jacobsen, R. K. Shoemaker, M. J. McNevin, M. Rakowski DuBois and D. L. DuBois, *Organometallics*, 2007, **26**, 5003–5009.
207. R. M. Henry, R. K. Shoemaker, R. H. Newell, G. M. Jacobsen, D. L. DuBois and M. Rakowski DuBois, *Organometallics*, 2005, **24**, 2481–2491.
208. R. M. Henry, R. K. Shoemaker, D. L. DuBois and M. Rakowski DuBois, *J. Am. Chem. Soc.*, 2006, **128**, 3002–3010.
209. T. T. Zhang, M. Wang, N. Wang, P. Li, Z. Y. Li and L. C. Sun, *Polyhedron*, 2009, **28**, 1138–1144.
210. M. H. Chiang, Y. C. Liu, S. T. Yang and G. H. Lee, *Inorg. Chem.*, 2009, **48**, 7604–7612.
211. P. Li, M. Wang, L. Chen, J. H. Liu, Z. B. Zhaoa and L. C. Sun, *Dalton Trans.*, 2009, 1919–1926.
212. L. Schwartz, G. Eilers, L. Eriksson, A. Gogoll, R. Lomoth and S. Ott, *Chem. Commun.*, 2006, 520–522.
213. G. Eilers, L. Schwartz, M. Stein, G. Zampella, L. de Gioia, S. Ott and R. Lomoth, *Chem. Eur. J.*, 2007, **13**, 7075–7084.
214. B. E. Barton, M. T. Olsen and T. B. Rauchfuss, *J. Am. Chem. Soc.*, 2008, **130**, 16834–16835.
215. F. J. Wang, M. Wang, X. Y. Liu, K. Jin, W. B. Dong, G. H. Li, B. Akermark and L. C. Sun, *Chem. Commun.*, 2005, 3221–3223.
216. A. D. Wilson, R. K. Shoemaker, A. Miedaner, J. T. Muckerman, D. L. DuBois and M. Rakowski DuBois, *Proc. Natl. Acad. Sci. U. S. A.*, 2007, **104**, 6951–6956.
217. A. D. Wilson, R. H. Newell, M. J. McNevin, J. T. Muckerman, M. Rakowski DuBois and D. L. DuBois, *J. Am. Chem. Soc.*, 2006, **128**, 358–366.
218. D. H. Pool and D. L. DuBois, *J. Organomet. Chem.*, 2009, **694**, 2858–2865.
219. J. Y. Yang, S. T. Chen, W. G. Dougherty, W. S. Kassel, R. M. Bullock, D. L. DuBois, S. Raugei, R. Rousseau, M. Dupuis and M. Rakowski DuBois, *Chem. Commun.*, 2010, **46**, 8618–8620.
220. E. S. Wiedner, J. Y. Yang, W. G. Dougherty, W. S. Kassel, R. M. Bullock, M. Rakowski DuBois and D. L. DuBois, *Organometallics*, 2010, **29**, 5390–5401.
221. A. Le Goff, V. Artero, B. Jousselme, P. D. Tran, N. Guillet, R. Metaye, A. Fihri, S. Palacin and M. Fontecave, *Science*, 2009, **326**, 1384–1387.
222. S. Griveau, D. Mercier, C. Vautrin-Ul and A. Chausse, *Electrochem. Commun.*, 2007, **9**, 2768–2773.
223. V. Sgobba and D. M. Guldi, *Chem. Soc. Rev.*, 2009, **38**, 165–184.
224. B. Jousselme, G. Bidan, M. Billon, C. Goyer, Y. Kervella, S. Guillerez, E. Abou Hamad, C. Goze-Bac, J. Y. Mevellec and S. Lefrant, *J. Electroanal. Chem.*, 2008, **621**, 277–285.
225. J. Pinson and F. Podvorica, *Chem. Soc. Rev.*, 2005, **34**, 429–439.
226. J. Y. Yang, R. M. Bullock, W. J. Shaw, B. Twamley, K. Fraze, M. Rakowski DuBois and D. L. DuBois, *J. Am. Chem. Soc.*, 2009, **131**, 5935–5945.

227. G. M. Jacobsen, J. Y. Yang, B. Twamley, A. D. Wilson, R. M. Bullock, M. Rakowski DuBois and D. L. DuBois, *Energy Environ. Sci.*, 2008, **1**, 167–174.
228. L. L. Duan, M. Wang, P. Li, N. Wang, F. J. Wang and L. C. Sun, *Inorg. Chim. Acta*, 2009, **362**, 372–376.
229. L. C. Sun, B. Akermark and S. Ott, *Coord. Chem. Rev.*, 2005, **249**, 1653–1663.
230. M. Wang, Y. Na, M. Gorlov and L. C. Sun, *Dalton Trans.*, 2009, 6458–6467.
231. R. Lomoth and S. Ott, *Dalton Trans.*, 2009, 9952–9959.
232. A. Juris, V. Balzani, F. Barigelletti, S. Campagna, P. Belser and A. Vonzelewsky, *Coord. Chem. Rev.*, 1988, **84**, 85–277.
233. A. J. Esswein and D. G. Nocera, *Chem. Rev.*, 2007, **107**, 4022–4047.
234. V. Balzani, G. Bergamini, F. Marchioni and P. Ceroni, *Coord. Chem. Rev.*, 2006, **250**, 1254–1266.
235. J. Ekstrom, M. Abrahamsson, C. Olson, J. Bergquist, F. B. Kaynak, L. Eriksson, S. C. Licheng, H. C. Becker, B. Akermark, L. Hammarstrom and S. Ott, *Dalton Trans.*, 2006, 4599–4606.
236. S. Ott, M. Kritikos, B. Akermark and L. C. Sun, *Angew. Chem. Int. Ed.*, 2003, **42**, 3285–3288.
237. H. Wolpher, M. Borgstrom, L. Hammarstrom, J. Bergquist, V. Sundstrom, S. Stenbjorn, L. C. Sun and B. Akermark, *Inorg. Chem. Commun.*, 2003, **6**, 989–991.
238. S. Ott, M. Borgstrom, M. Kritikos, R. Lomoth, J. Bergquist, B. Akermark, L. Hammarstrom and L. C. Sun, *Inorg. Chem.*, 2004, **43**, 4683–4692.
239. S. Salyi, M. Kritikos, B. Akermark and L. C. Sun, *Chem.–Eur. J.*, 2003, **9**, 557–560.
240. H. G. Cui, W. Mei, L. L. Duan and L. C. Sun, *J. Coord. Chem.*, 2008, **61**, 1856–1861.
241. W. M. Gao, J. H. Liu, W. N. Jiang, M. Wang, L. H. Weng, B. Akermark and L. C. Sun, *C. R. Chimie*, 2008, **11**, 915–921.
242. L. C. Song, L. X. Wang, M. Y. Tang, C. G. Li, H. B. Song and Q. M. Hu, *Organometallics*, 2009, **28**, 3834–3841.
243. L. C. Song, M. Y. Tang, F. H. Su and Q. M. Hu, *Angew. Chem., Int. Ed.*, 2006, **45**, 1130–1133.
244. L. Yang, A. M. Ren, J. K. Feng, X. D. Liu, Y. G. Ma and H. X. Zhang, *Inorg. Chem.*, 2004, **43**, 5961–5972.
245. P. Y. Chen, R. Duesing, D. K. Graff and T. J. Meyer, *J. Phys. Chem.*, 1991, **95**, 5850–5858.
246. W. G. Wang, F. Wang, H. Y. Wang, G. Si, C. H. Tung and L. Z. Wu, *Chem.–Asian. J.*, 2010, **5**, 1796–1803.
247. I. Okura, *Coord. Chem. Rev.*, 1985, **68**, 53–99.
248. J. R. Darwent, P. Douglas, A. Harriman, G. Porter and M. C. Richoux, *Coord. Chem. Rev.*, 1982, **44**, 83–126.
249. X. Q. Li, M. Wang, S. P. Zhang, J. X. Pan, Y. Na, J. H. Liu, B. Akermark and L. C. Sun, *J. Phys. Chem. B*, 2008, **112**, 8198–8202.

250. A. M. Kluwer, R. Kapre, F. Hartl, M. Lutz, A. L. Spek, A. M. Brouwer, P. van Leeuwen and J. N. H. Reek, *Proc. Natl. Acad. Sci. U.S.A.*, 2009, **106**, 10460–10465.
251. L. C. Song, M. Y. Tang, S. Z. Mei, J. H. Huang and Q. M. Hu, *Organometallics*, 2007, **26**, 1575–1577.
252. Y. Na, M. Wang, J. X. Pan, P. Zhang, B. Akermark and L. C. Sun, *Inorg. Chem.*, 2008, **47**, 2805–2810.
253. Y. Na, J. X. Pan, M. Wang and L. C. Sun, *Inorg. Chem.*, 2007, **46**, 3813–3815.
254. D. Streich, Y. Astuti, M. Orlandi, L. Schwartz, R. Lomoth, L. Hammarstrom and S. Ott, *Chem.–Eur. J.*, 2010, **16**, 60–63.
255. W. M. Gao, J. L. Sun, T. Akermark, M. R. Li, L. Eriksson, L. C. Sun and B. Akermark, *Chem.–Eur. J.*, 2010, **16**, 2537–2546.
256. S. Shima, O. Pilak, S. Vogt, M. Schick, M. S. Stagni, W. Meyer-Klaucke, E. Warkentin, R. K. Thauer and U. Ermler, *Science*, 2008, **321**, 572–575.
257. J. A. Wright, P. J. Turrell and C. J. Pickett, *Organometallics*, 2010, **29**, 6146–6156.
258. (a) D. F. Chen, R. Scopelliti and X. L. Hu, *J. Am. Chem. Soc.*, 2010, **132**, 928–929; (b) X. Wang, Z. Li, X. Zeng, Q. Luo, D. J. Evans, C. J. Pickett and X. Liu, *Chem. Commun.*, 2008, 3555–3557.
259. T. B. Liu, B. Li, C. V. Popescu, A. Bilko, L. M. Perez, M. B. Hall and M. Y. Darensbourg, *Chem. Eur. J.*, 2010, **16**, 3083–3089.
260. P. J. Turrell, J. A. Wright, J. N. T. Peck, V. S. Oganesyan and C. J. Pickett, *Angew. Chem., Int. Ed.*, 2010, **49**, 7508–7511.
261. A. M. Royer, M. Salomone-Stagni, T. B. Rauchfuss and W. Meyer-Klaucke, *J. Am. Chem. Soc.*, 2010, **132**, 16997–17003.

CHAPTER 13

Theoretical Studies of O–O and H–H Bond Formation in Enzymes

PER E. M. SIEGBAHN

Department of Physics, ALBA NOVA and Department of Biochemistry and Biophysics, Arrhenius Laboratory, Stockholm University, SE-106 91, Stockholm, Sweden

13.1 Introduction

The greenhouse effect and the energy crisis are two connected major problems which have become more and more acute. An interesting possibility to address these problems is the production of hydrogen from water and sunlight. At present, a very large research effort is therefore spent to try to obtain biomimetic systems that can be functional and economic. Two main catalytic processes need to be developed. The first and most difficult one, is water oxidation with sunlight as energy source. The second one is where the protons and electrons, generated in the first step, should form hydrogen. In nature, photosystem II (PSII) is the only enzyme that can oxidize water with the help of sunlight. For this reason, and since dioxygen formation from water is one of the most fundamental reactions to the evolution of life on earth, it has been a high-priority scientific goal for decades to understand this process. Hydrogenases are the enzymes in nature that catalyze hydrogen formation from protons and electrons. Until rather recently, experimental techniques were the only means to study these processes. However, since water oxidation in particular is a very

RSC Energy and Environment Series No. 5
Molecular Solar Fuels
Edited by Thomas J. Wydrzynski and Warwick Hillier

Published by the Royal Society of Chemistry, www.rsc.org

complicated process involving a large number of transition states and at least a dozen intermediates, many of them short-lived, it has now been realized that it is not possible to understand these processes without detailed model calculations. In the present review, the current standing on theoretical research on these two processes will be briefly described.

13.2 O–O Bond Formation – Photosystem II

In photosystem II, an oxygen molecule is formed from two water molecules after four flashes of light, as illustrated by the Kok S-state cycle, see Figure 13.1. The first event in each step consists of a photon entering the chlorophylls and the energy being transferred to P_{680}, the reaction center. At P_{680} a charge separation occurs, with the electron being rapidly transferred to the quinone Q_A in 200 ps. In a proton-coupled step requiring 100 μs, the electron is then transferred further to a second quinone Q_B, which finishes the charge separation process. The ionized P^{+}_{680} has the highest redox potential known to biology with 1.25 V.[1,2] This potential is large enough to take an electron from a tyrosine, Tyr_z, located in between P_{680} and the oxygen evolving center (OEC). Tyr_z is simultaneously deprotonated to a nearby histidine, His190, creating a neutral tyrosyl radical. In the key water oxidation step, an electron is then transferred to the tyrosyl radical from the Mn_4Ca-complex of the OEC, where the substrate water molecules are bound. Coupled to this oxidation of the OEC, the substrate water molecule will in most cases lose a proton to bulk water. After repetition of these S-state transition steps four times, an oxygen molecule is produced. To understand water oxidation in PSII entails understanding all the S-state transitions involved as well as the mechanism for the O–O bond formation in the S_4-state. In the present section, the level of knowledge obtained to date of this process by theoretical studies will be briefly described.

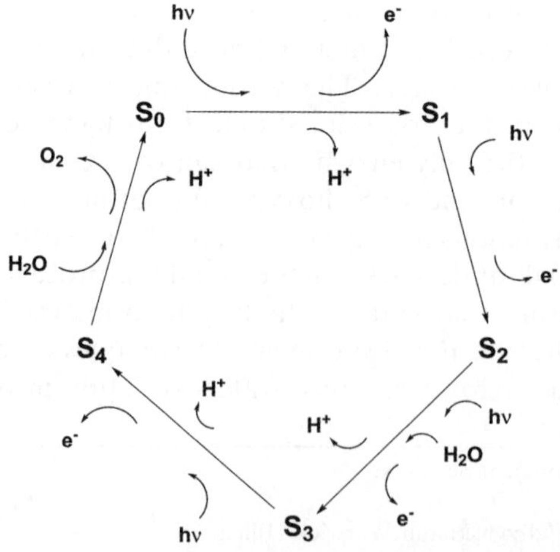

Figure 13.1 The Kok S-state cycle for water oxidation, updated by recent research.

The chemical models used for the theoretical studies of O–O bond formation have in recent years been built on X-ray structures,[3–5] schematically shown in Figure 13.2. These structures were obtained at a rather low resolution of 3.0–3.5 Å and the exact ligation to the metal atoms therefore had to be partly assumed. In the London X-ray structure,[3] the binding of the carboxylate amino acid ligands (aspartates and glutamates) was assumed to be mostly monodentate to one metal center. In contrast, in the Berlin X-ray structure,[4] most of these ligands were assumed to be bridging between two different metals. Possibilities in-between these two types of structures are also possible. This means that a ligation pattern has to be assumed as a starting point for the computational model studies. In the more recent studies, a ligation pattern close to the one from the Berlin structure was adopted as being more chemically reasonable. It should be added that, in principle the optimal ligation pattern could be obtained by direct minimization, but the number of possible conformations are so many that this is in practice not possible.

The first computational attempts to understand water oxidation were made before the first X-ray structures. The starting point therefore had to be rather crude models, where the goal was only to understand the general chemical requirements for forming an O–O bond.[6,7] The most important result from these studies was that the initial creation of an oxygen radical ligand appears to be necessary for a low-barrier O–O bond formation. The O–O bond was suggested to be formed by an attack of a second sphere water on the oxygen radical. This type of result was obtained essentially independently of the model used for the OEC. Attempting to obtain a picture of the S-state transitions was not meaningful at that early stage. Still, an important conclusion concerning electron and proton transfers could be drawn based on these studies and those of other enzymes. The general rule found is that for processes in enzymes where protons and electrons are removed or added, a metal complex tries to minimize changes of the total charge.[8,9] This means that two electron transfer steps can not occur after each other without a proton transfer in between.

Three different theoretical approaches have been used for trying to obtain mechanisms and better structures than are available from experiments. In the first approach,[10] a cluster model of the OEC was used with up to 200 atoms.

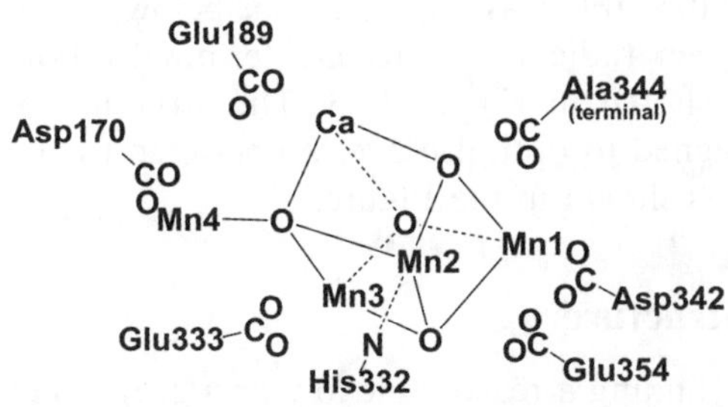

Figure 13.2 Simplified picture of the structure of the oxygen evolving complex, suggested by X-ray crystallography. This numbering of the Mn-atoms will be used throughout in the text.

In the second approach,[11] the QM/MM (Quantum Mechanics/Molecular Mechanics) methodology was employed. A small QM part was surrounded by a large MM part, together making up the entire protein. In the third approach,[12] different models for the OEC were constructed based on the core topology derived by polarized EXAFS (Extended X-ray Absorption Fine-Structure) spectra,[13] and with a ligand structure chosen to fit reasonably well into the X-ray structure. Based on the agreement with experiments for the computed spin-spectrum, the best candidates for the actual structure of the OEC were selected. A comparison of these three approaches have recently been made, and it was concluded that the results of the cluster model are at present the most reliable ones.[14] The present chapter will therefore focus on the results of that approach.

13.2.1 Mechanism for O–O Bond Formation

As mentioned above, the earliest attempts to model O–O bond formation all led to a mechanism where a preformed oxygen radical reacts with an external water in the S_4-state. However, the barrier for this mechanism was much too high compared to experiments. An important step toward a low-barrier mechanism was taken when essentially all possibilities to form the O–O bond with the oxygen radical were investigated for the best available S_4-state, which had four Mn(IV)-centers.[15] It should be noted that the oxygen radical in this state is terminally bound to a manganese. Rather surprisingly, by far the lowest barrier was found for a reaction between the terminal oxygen radical and a bridging oxo-group. What was even more surprising in the new mechanism was a spin-requirement for a low barrier, shown in Figure 13.3. This requirement means that the directions of the spins on the four most directly interacting atoms have to be alternating. The two oxygens have to have opposite spins to form a bond, and the manganese binding these atoms have to have opposite spins to the respective oxygens. The reasons for the requirements on the manganese spins are in one case a formation of the reduced Mn(III) (Mn4 in the Figure) in a high-spin state, and in the other case to allow O–O bond formation without crossing to another spin-surface. In the first mechanism of this type found, the preformed terminal oxygen radical was bound to the dangling manganese Mn4. As the structure of the OEC was improved, see below, it was found to be more optimal if the oxygen radical was instead terminally bound to Mn2 in the Mn_3Ca-cube,[16] as shown in Figure 13.3. The structure of the OEC can be rationalized as designed to optimize overlap between the interacting O–O and Mn–O σ-orbitals, as shown in the Figure.

13.2.2 OEC Structure

The next step after finding a reasonable low-barrier mechanism for O–O bond formation, was to try to find a better structure of the OEC than the one provided by X-ray crystallography and EXAFS. The type of TS discussed above should therefore first be fitted into the X-ray structures in some way.

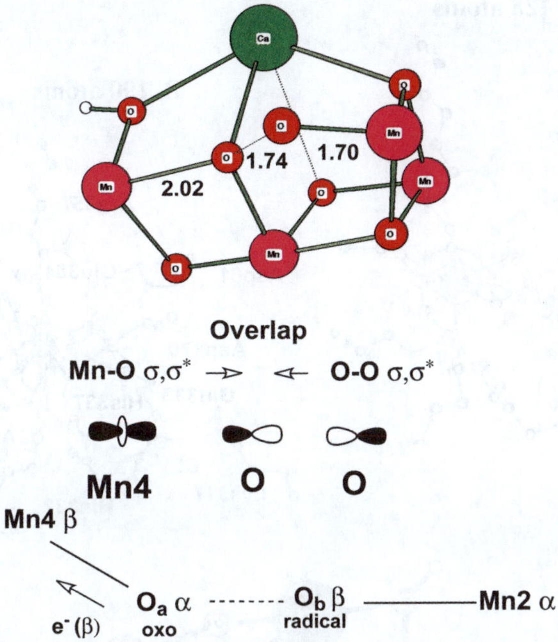

Figure 13.3 Optimized structure and schematic drawing of the low-barrier TS and mechanism for O–O bond formation discussed in the text. α and β denote spin-directions.

In order to proceed, two important assumptions were made. It was concluded that the most accurate positions determined by the X-ray analysis are the positions of the backbone atoms. Their positions are, for example, rather close in the two X-ray structures even though the OEC metal complexes are quite different.[3,4] The first important assumption made was therefore that these positions are sufficiently well described by the low resolution X-ray structures, so that a further optimization of the rest of the OEC-structure is meaningful. To ensure that the optimized models should stay as closely as possible to the X-ray structures, the backbone atoms (the α-carbons and their two backbone neighbors) were therefore fixed at the positions given by the X-ray structure.[3] This meant that 19 atoms, of a total of 128 for Model A in Figure 13.4, were fixed, while the rest of the atomic positions were energy-minimized. With this approach, a search for a better structure of the OEC was initiated. Still, there are a very large number of possibilities, so only the chemically most reasonable ones were fully investigated. The second important assumption made was based on experience during the past decade. It was assumed that the lowest energy structure obtained in this way, with the backbone constraints, is the one that is adopted by nature. Against this assumption, it can be argued that it is in principle possible that a higher energy structure is used in the enzyme, if the barriers for its decay are high enough, but this is considered as very unlikely, see further ref. 14.

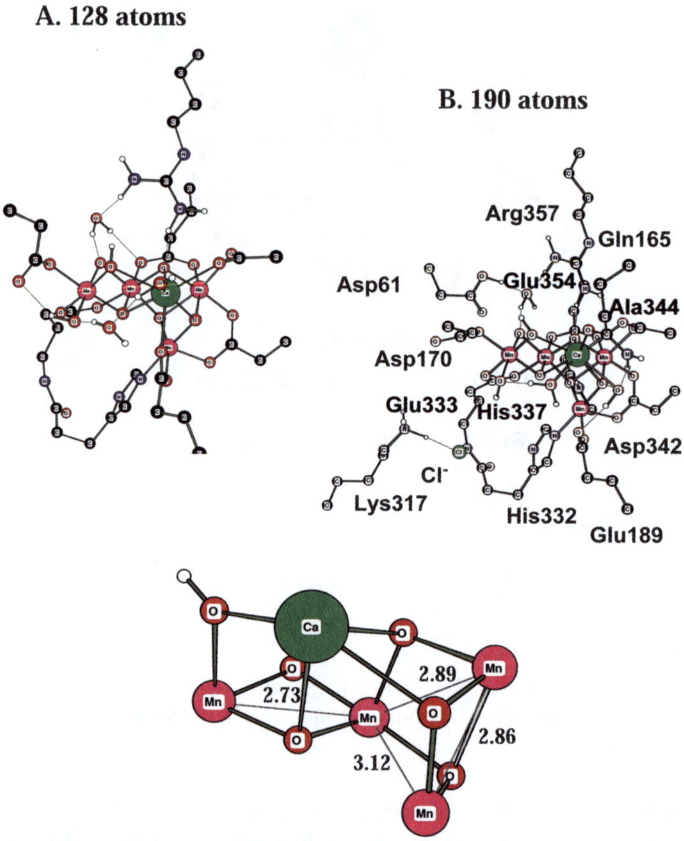

Figure 13.4 Two different models used to study water oxidation. The structures are for the resting state and have been fully optimized with backbone constraints. Most amino acid protons have been left out. A close-up of the metal cluster in Model B is also shown.

It turned out that it was relatively easy to fit the TS-structure, obtained in the earlier studies, into the back-bone structure given by the X-ray analysis.[10] In fact, this protein-fitted structure gave an even lower barrier than before for O–O bond formation. However, to obtain a more direct comparison to the measured densities and suggested X-ray structures, a structure for the resting state was needed. Electrons and protons were therefore added to the TS-structure and the geometries were energy minimized with the same backbone constraints until the resting state was reached. This structure, for the largest model used so far with about 190 atoms, is shown as Model B in Figure 13.4, illustrating the size of the model and the amino acids included in the model.

The optimized structure for the cluster model of the resting state has been placed into the electron density from the X-ray density measurement,[10] and it was shown that the structure fits very well into this density. It should again be emphasized that the structure was not fitted to the density but was optimized only with the constraints of fixed backbone atoms. There are quite significant differences between the optimized metal positions and those obtained from the X-ray analysis, with deviations of up to 1.8 Å. This illustrates the problem with the low resolution, since all three complexes fit well into the density.

The DFT structure agrees in many aspects with structures suggested by EXAFS studies,[13,17] where it had been emphasized that substantial discrepancies exist between EXAFS and the X-ray structures. One characteristic feature of the suggested EXAFS structures is that there should be a short Mn–Mn distance between the outer manganese and the nearest manganese in the Mn_3Ca-cube. These manganese atoms should therefore be connected by two μ-oxo bonds. This agrees with what is found in the optimized structure, see the close-up in Figure 13.4. The number of short Mn–Mn distances in the resting state has been a controversial issue with different suggestions. The best optimized structure at present has Mn–Mn distances of 2.73 Å, 2.86 Å, 2.89 Å and 3.11 Å. This is in as good agreement with EXAFS suggestions as one can hope for using DFT. In one EXAFS interpretation,[13] three distances of 2.7–2.8 Å and one of 3.3 Å are suggested, while in another one only two short 2.7 Å distances are suggested,[17] illustrating the remaining EXAFS uncertainties. Recently, a comparison has also been made between the calculated solution EXAFS spectrum for the cluster model, with the corresponding experimental spectrum showing good agreement.[18]

Even though a quite reasonable structure of the OEC has been obtained, where most aspects agree with observations, it is still clear that there can be minor errors in the structure. In particular, if a feature of the OEC does not affect the energetics of the S-state cycle significantly, that feature will not be very well defined by energy minimization. For example, it has been found that many of the amino acids can be mutated with only small effects on oxygen evolution.[19] Glu189 is a striking case, for which almost no mutation, even the ones that change the charge,[20] affect the rate of oxygen evolution. Therefore, it would not be surprising if Glu189 does not actually bind to the OEC. Investigations of structures where Glu189 is replaced by an hydroxide are presently being investigated. Gln165 is another case, where the calculations have shown that this residue can be replaced by a water molecule without significant energy effects.

For the above, and other reasons, the stability of the present structure of the OEC is continuously being investigated. For example, the present constraints on the backbone atoms have been somewhat released, by only holding the α-carbon fixed rather than three atoms as in the above structures. Only small effects were found, some of them slightly improving agreement with experiments, others leading to results in the other direction. The discrepancies between these results can be taken as a measure of the present uncertainties, which can probably only be improved by a better positioning of the backbone atoms by X-ray crystallography. Fixing the backbone atoms from the Berlin structure, rather than from the London structure, has also been investigated. In that case it was found that the backbone positions of Asp170 have to be modified to reach the same type of agreement as obtained with the London structure. Asp170 appears to be placed too far away from the center of the OEC in the Berlin structure. The backbone positions for the other amino acids are quite close in the two X-ray structures.

A way to circumvent the problem of energy insensitive features of the OEC discussed above, is to compare computed and measured (by Fourier Transform InfraRed, FTIR, spectroscopy) amino acid carboxylate C–O stretching frequencies.[21] However, the accuracy of this type of frequency calculations (in the harmonic approximation) is presently not clear, and is therefore still being investigated. For the present rather small models, using limited basis sets, the frequency shifts differ slightly from experiments. Still, for the *main conclusions* drawn from the FTIR experiments, the agreement is strikingly good, see further below.

13.2.3 S-state Transitions

By removing electrons and protons and adding substrate water, structures and energies for all the different S-states can be obtained. These structures have been optimized with the same backbone constraints as described above for the resting structure in Figure 13.4. This leads to the sequence of S-transitions as shown schematically in Figure 13.5. The structures are taken from the optimizations but with all amino acids removed from the figure for clarity. The starting point S'_0 is a state where O_2 has just been removed in the previous cycle. In this state, there is a relatively large opening in the middle of the OEC and three manganese atoms are only five coordinated (Mn2, Mn3 and Mn4) and these three will therefore be the ones that are Mn(III). The six-coordinated Mn1 will be Mn(IV). The first water substrate is easily bound in the large opening. It binds with a simultaneous loss of a proton to the bulk. The product is shown as S_0, where an arrow indicates the position of the substrate hydroxide. When the substrate binds in S_0, there will be a change of oxidation states. Mn4 will bind the substrate hydroxide rather tightly, changing its oxidation state from Mn(III) to Mn(IV), while Mn1 will now be Mn(III). It should be added that this order of oxidation states (IV,III,III,III) for S_0 is a quite new finding and occurs only for the largest models used. In previous calculations for slightly smaller models, Mn3 was the atom with oxidation state Mn(IV) in S_0. In the S_0 to S_1 transition, an electron is removed from Mn3 (a * marks the oxidation in S_1) leading to the set of oxidation states (IV,III,IV,III) for S_1. A proton taken from the substrate hydroxide is released to the bulk. In the S_1 to S_2 transition only an electron is removed in agreement with experiments.[22,23] The electron is taken from Mn4, leading to the set of oxidation states (IV,III,IV,IV) for S_2.

In the S_2 to S_3 transition, the second substrate water binds. However, the large opening present in S'_0 is no longer there. To make space for the second water substrate, the reconstruction illustrated in Figure 13.6 takes place. A proton is simultaneously removed from the substrate water as it binds, exactly as in the formation of S_0, and an electron is taken from Mn2. The substrate hydroxide becomes terminally bound to Mn2. In the S_3-state all manganese are

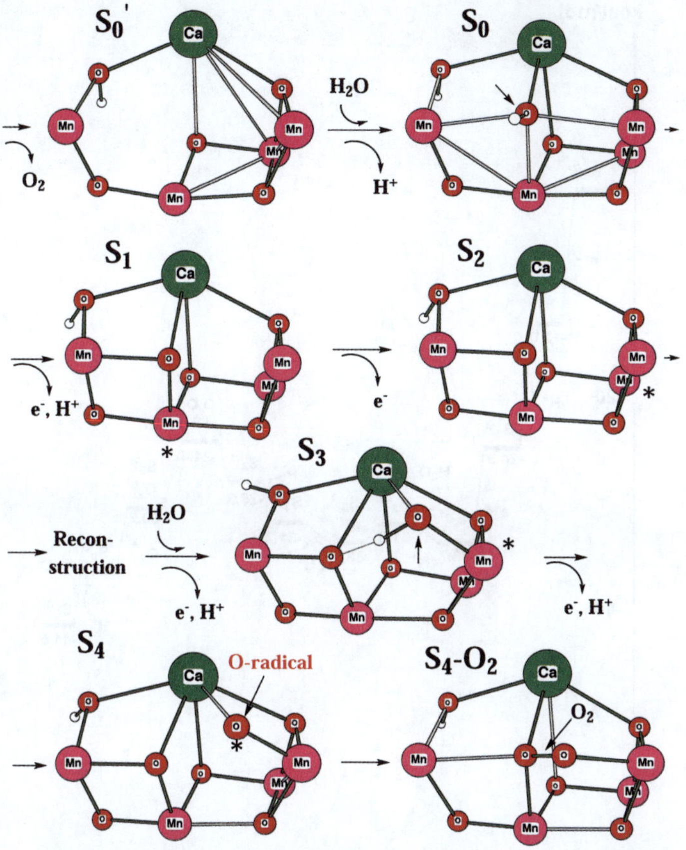

Figure 13.5 Schematic picture of the different S-transitions. The structures have been optimized but only the most important atoms are shown. A * marks the atom that has been oxidized in that transition.

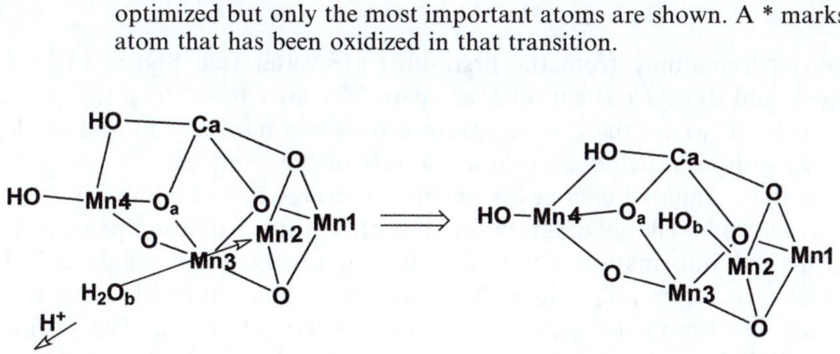

Figure 13.6 The mechanism for water insertion leading to the reconstruction in the S_2 to S_3 transition.

Mn(IV). In the final transition from S_3 to S_4, a proton is taken from the second substrate hydroxide and an electron from the resulting substrate oxygen. Finally, the O–O bond is formed between the terminal oxygen radical and the

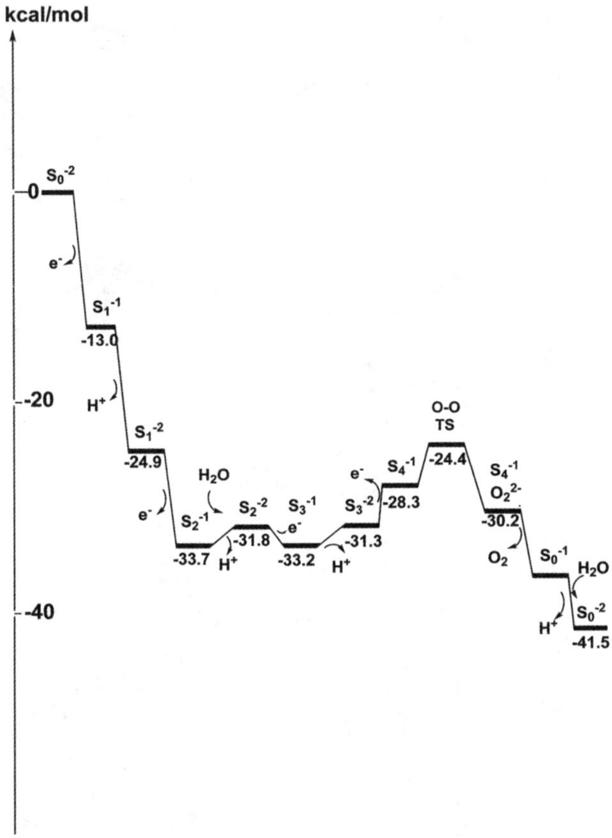

Figure 13.7 Energy diagram for dioxygen formation in PSII.

oxo-group remaining from the first substrate water (see Figure 13.3), O_2 is released, and the cycle starts all over again. The structures show that the OEC forms a basin where the two substrate waters can naturally fit and be deprotonated, without much occurring in the rest of the complex.

With the calculated energies from the structures described above, it is possible to set up an energy diagram for the entire water oxidation process. Using experimental information about the driving force, and a single adjustable parameter, accurate pK_a values and redox potentials can be obtained without explicitly describing the enzyme surrounding the active site. The results are essentially independent of the choice of dielectric constant.[9] The driving force for the catalytic cycle is obtained as 41.5 kcal mol^{-1}, by using the redox potential of 1.25 V for P^+_{680},[1,2] and 0.80 V for oxygen. The resulting diagram is shown in Figure 13.7. In S_n^m, n is the number of the S-state, and m is the charge of the complex (only including direct ligands to the OEC).

A general comment can first be made about the diagrams, and this is that the protons and electrons are removed in an alternating fashion. This preserves the charge of the catalyst as much as possible, which has been found to be an energetic advantage in enzyme mechanisms in general.[9] The general shape of the energy diagram can be described as follows. The first two S-transitions are quite exergonic by 24.9 and 8.8 kcal mol^{-1}, while the third one from S_2 to S_3 is actually calculated to be slightly endergonic by 0.5 kcal mol^{-1}. It is clear that the latter value is not entirely correct since all S-transitions must be exergonic. The reason for this is not completely clear at the present stage, but could be due to a remaining error in the structure. Calculations are in progress to investigate this problem and promising results are obtained by changing Glu189 to an hydroxide (see also above). The formation of the oxygen radical in S_4 is endergonic by 4.9 kcal mol^{-1}, which means that this S-state will not be observed. The local O–O formation barrier is only 3.9 kcal mol^{-1}. This means that the total barrier for O–O bond formation is $3.9+4.9=8.8$ kcal mol^{-1}, which is well below what is required for a process that takes milliseconds. The transition from S_3 to S_0 is found to be exergonic by 10.5 kcal mol^{-1}, which is not consistent with the experimental suggestion of a reversible O–O bond cleavage from S_0.[24] However, since this transition energy is likely to be less well determined by the present methods due to large structural and entropy changes in the release of oxygen, further work is needed to definitively make a conclusion of the energetics of this transition.

Many features described above, apart from the energetics, are also consistent with experimental observations. For example, the release pattern of the protons, where one striking feature is the release of only an electron and no proton in the S_1 to S_2 transition. Furthermore, the oxidation state assignments for the different Mn-atoms in the different S-transitions, are all consistent with the main conclusions drawn from FTIR.[21] Mn4 is the only atom with oxidation state IV in S_0. In S_0 to S_1, Mn3 is oxidized, while in S_2 to S_3, Mn2 is oxidized. The incorporation of the first water substrate in the S_4 to S_0 transition, and the second one in the S_2 to S_3 transition is also in essential agreement with experimental interpretations.[25,26] In the much debated question of whether an Mn-center is oxidized or not in the latter transition, the calculations support the suggestion of Mn-oxidation.

13.3 H–H Bond Formation – Hydrogenases

Hydrogenases are enzymes designed to reversibly cleave dihydrogen into protons and electrons,

$$H_2 \Leftrightarrow 2H^+ + 2e^- \tag{1}$$

They are classified by specifying the transition metal content of their active sites into three main classes, the NiFe,[27] the FeFe,[28] and the iron–sulfur cluster free hydrogenases.[29] Of the two main classes, the NiFe-enzymes are primarily used for hydrogen oxidation, and the FeFe-enzymes for proton reduction. There are two recent reviews on theoretical work on hydrogenases.[30,31] Most of this work considers H–H bond cleavage, but the reaction is commonly considered as easily reversible by external means, and the mechanism is therefore assumed to also be the same for the H–H bond formation. Only the pathway for H–H bond cleavage will be discussed here.

13.3.1 NiFe-Hydrogenase

The X-ray structure of the active site of NiFe-hydrogenase for *Desulfovibrio fructosovorans*[32] is shown in Figure 13.8 for a Ser499Ala mutant. There are four cysteinate ligands, two of them bridging between nickel and iron and two of them terminally bound to nickel. Iron has, in addition, three diatomic ligands, of which two are cyanides and one is a carbonyl, very unusual for a biological metal complex. Another unusual aspect of the active site is that so many of the nearest amino acids not bound to the metals, are charged. There is one positive arginine hydrogen bonding to two negative aspartates. There is also one nearby glutamic acid and one histidine, both of which could be charged. The electrons released (or accepted) are transferred over one or more FeS-clusters, one of which is shown in the Figure. A likely proton transfer pathway from the active complex, starting at Cys543 and going over Glu25, has also been assigned.[33] In many NiFe-hydrogenases, the sulfur in Cys543 has been replaced by a selenium.[34]

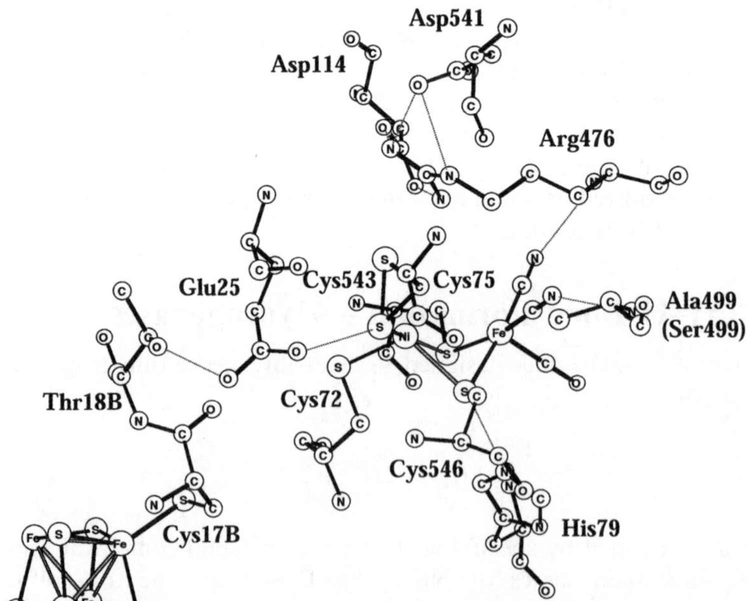

Figure 13.8 The X-ray structure of the active site for [NiFe] hydrogenase from *Desulfovibrio fructosovorans.*

The catalytic cycle of NiFe-hydrogenases is commonly illustrated by a scheme such as the one in Figure 13.9A.[35,36] It contains the three intermediate states observed experimentally, Ni_a-C*, Ni_a-S and Ni_a-SR. Ni_a-C* is the EPR-active resting state, which, based on its X-ray structure[32] and recent ENDOR results,[37–39] has been suggested to have a bridging hydride between the metals. The EPR-silent Ni_a-SR state is reached by reduction of Ni_a-C*, while Ni_a-S, which is also EPR-silent, is obtained by oxidation. To complete the cycle, Ni_a-SR can be reached by adding dihydrogen to Ni_a-S. Several other states have been observed under varying conditions but these will not be discussed in the present paper.

An interesting alternative scenario to the one in Figure 13.9A, shown as B in the same figure, has also been suggested.[40] The main difference between these mechanisms is that Ni_a-S does not participate in the active cycle. Instead, catalysis starts only after an initial reaction between Ni_a-S and dihydrogen where Ni_a-C* is created. Ni_a-C* is then only in equilibrium with Ni_a-SR during turnover.[41] The structural implications of this suggestion has not been completely clear, but it has been suggested that a bridging hydride may be present all the time during catalysis.[40,42]

Until quite recently, theoretical studies on NiFe-hydrogenases all led to a mechanism following the scheme in Figure 13.9A. A transition state (TS) with a heterolytic cleavage of the H–H bond was found. The TS structure from one of the most recent studies[43] is shown in Figure 13.10. There is essential agreement among all studies on most details of this type of mechanism.[30,31] The cleavage of the H–H bond is heterolytic with one hydrogen ending up as a bridging hydride and the other one as a proton on Cys543. Due to the character of the TS, it can be referred to as the heterolytic mechanism. A disagreement exists concerning the oxidation state of nickel, which is assigned as Ni(II) by most, but with one suggestion being Ni(III).[44] However, in the case of Ni(III) the reaction is unlikely to be reversible. This would imply a quite different mechanism for H–H bond formation, which can not be ruled out at the present stage. The oxidation state on iron in all studies is a low-spin Fe(II).

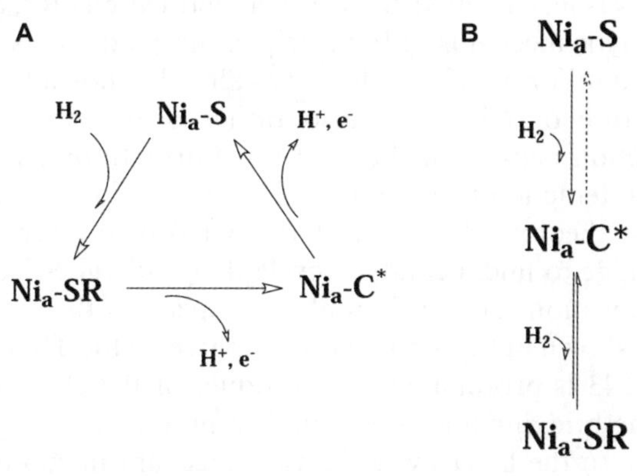

Figure 13.9 Two different types of catalytic cycles for NiFe-hydrogenases, suggested by experiments.

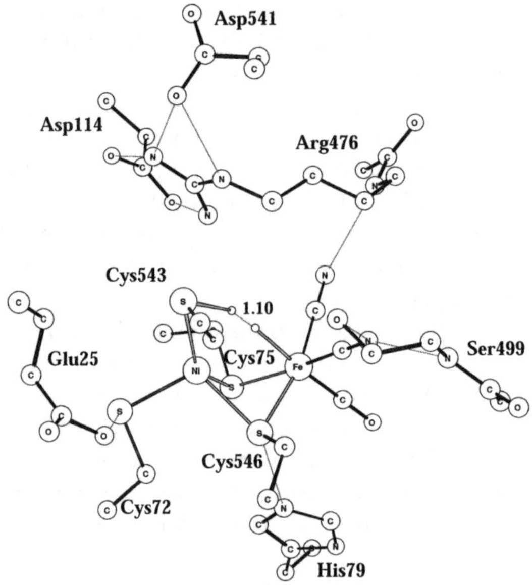

Figure 13.10 Model with 137 atoms showing the transition state for heterolytic cleavage of the H–H bond. Hydrogen atoms on the amino acids are omitted for clarity.

There were several reasons to continue searching for alternative mechanisms to the heterolytic cleavage discussed above. One reason was that the final barrier for the heterolytic mechanism was still rather high. Another reason came from indications from some experiments[40,41] that Ni_a-S and Ni_a-C* should not be in equilibrium during the catalytic cycle, see Figure 13.9B. This led to suggestions that a heterolytic cleavage should occur with a bridging hydride always present in the catalytic cycle. However, attempts to find a low energy pathway of this type all failed.[30] An interesting state in the context of a new mechanism was instead a Ni(I)-state, here termed Ni_a-R*, appearing at the end of the heterolytic mechanism. In principle, this oxidation state would allow oxidative addition to form Ni(III), which was already known to be an accessible state. The structure of Ni_a-R* is also rather open which should allow a favorable interaction between nickel and H_2. Furthermore, some experiments have favored an interaction between H_2 and nickel,[45,46] rather than with iron as obtained in most theoretical studies of the heterolytic mechanism. Attempts were therefore made to find a catalytic cycle that undergoes oxidative addition on nickel. The transition state for homolytic cleavage at the Ni-site, obtained at the DFT level, is shown in Figure 13.11. Compared to the TS for the heterolytic mechanism, Cys543 is protonated. The product of the cleavage (Ni_a-X*) has two hydrides, one bridging and one bound to nickel.

The alternative to the heterolytic cleavage mechanism shown in the scheme in Figure 13.9A, is thus to start out with only one turn of this heterolytic

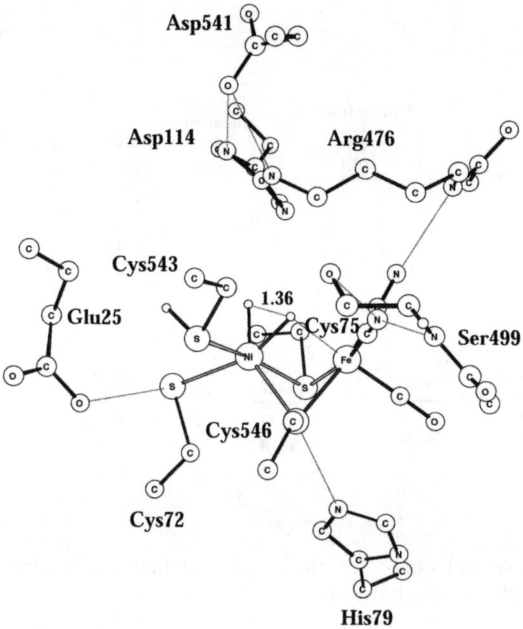

Figure 13.11 Model with 137 atoms showing the transition state for oxidative addition. Hydrogen atoms on the amino acids are omitted for clarity.

mechanism, but after that proceed with the oxidative addition mechanism. This would follow a scheme such as the one in Figure 13.9B. The computed energy diagram is given in Figure 13.12. This mechanism can be described in the following way. One turn of the heterolytic mechanism initiates the catalysis. First, H_2 binds at the Fe(II)-site in state **Ni$_a$S**. The heterolytic cleavage of the H–H bond has a barrier of 16.0 kcal mol^{-1} and leads to state **Ni$_a$SR**. Both these states are Ni(II) and are EPR silent. After release of an electron and a proton, the resting **Ni$_a$C***-state is reached, which is an EPR active Ni(III)-state. **Ni$_a$C*** has a bridging hydride and a protonated Cys543. After releasing an electron and the proton on Cys543, the hydride can be transferred to Cys543 (barrier 10.3 kcal mol^{-1}), leading to the critical **Ni$_a$R***-state, which is a Ni(I)-state. At this point, the cycle can either release another electron and proton to come back to the starting point **Ni$_a$S**, and then continue using heterolytic cleavage of H_2. Or, it can take the oxidative addition route, going over the **Ni$_a$X***-state. With release of an electron and a proton the **Ni$_a$SR**-state is reached from where the oxidative addition route can be continued. Energetically, the latter pathway is the preferred one, and a mechanism such as the one given in Figure 13.9B is therefore followed. The heterolytic barrier from **Ni$_a$R*** is 17.6 kcal mol^{-1}, but it is only 14.2 kcal mol^{-1} for the oxidative addition pathway. Heterolytic cleavage will therefore only be used in the first cycle, and there will be no equilibrium between **Ni$_a$S** and **Ni$_a$C*** in the rest of the cycles, as illustrated in Figure 13.9B.

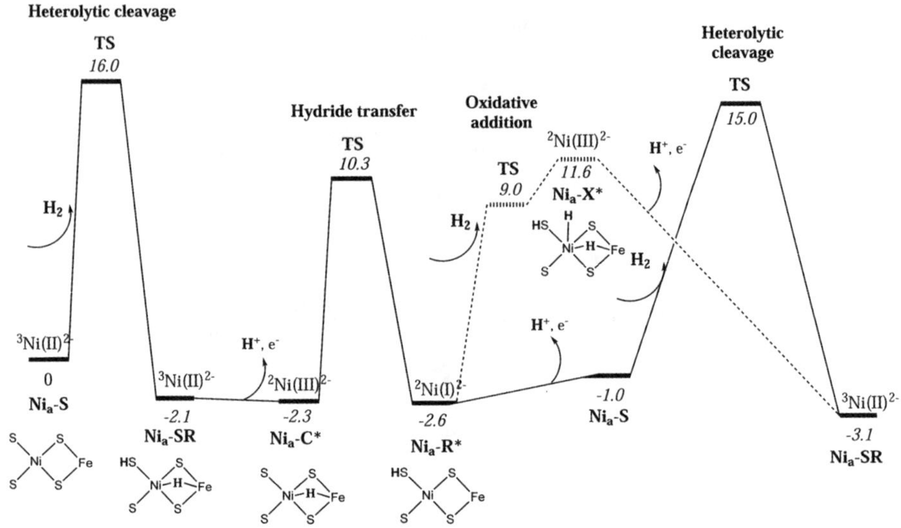

Figure 13.12 Potential curve for the combined heterolytic cleavage and oxidative addition mechanisms.

13.3.2 FeFe-Hydrogenase

The second major class of hydrogenases are FeFe-hydrogenases, which are generally found to catalyze H–H bond formation, in contrast to most of the NiFe-hydrogenases. The active site structure derived from X-ray analysis is shown in Figure 13.13.[47] The active site, known as the H-cluster, consists of an iron dimer bridged by a cysteine to an FeS-cluster. The iron dimer is the putative site of H^+ reduction and H_2 oxidation. Each iron atom of the dimer is coordinated by one terminally bound cyanide and one terminally bound carbonyl ligand. The two irons are bridged by a five-atom dithiolate, generally assigned as either 1,3-propanedithiolate (PDT, $-SCH_2CH_2CH_2S-$)[28] or 1,3-di(thiomethyl)amine (DTMA, $-SCH_2NHCH_2S-$).[48] A third carbonyl ligand is found either bridging the two iron atoms[49] or terminally bound to the distal iron atom,[47] apparently depending on the redox state of the iron dimer.

The work of Hall *et al.*[50] suggests a catalytic cycle for H_2 oxidation, such as the one given in Figure 13.14. This cycle begins with the inactive, over-oxidized form of FeFe-hydrogenase at an $Fe^{II}Fe^{II}$ redox level, with OH^- bound at the distal iron center. Protonation of the OH^- ligand and one-electron reduction leads to loss of H_2O yielding a catalytically active $Fe^{I}Fe^{II}$ form. Dihydrogen binds to the distal iron center and the complex undergoes a one-electron oxidation to yield an $(\eta_2\text{-}H_2)Fe^{II}Fe^{II}$ form. A proton is transferred from the bound η_2-H_2 to a nitrogen of the μ-SCH_2NHCH_2S, and then moved away from the iron dimer by the protein. The second proton is transferred from the hydride ligand to the nitrogen of the μ-SCH_2NHCH_2S, and then moved away from the iron dimer by the protein. The iron dimer undergoes one-electron oxidation to $Fe^{I}Fe^{II}$ to close the catalytic cycle.

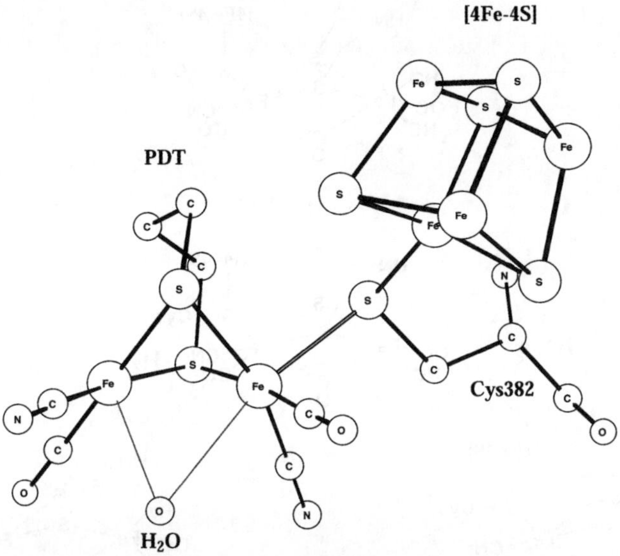

Figure 13.13 Structure of the active site of FeFe-hydrogenase given by Nicolet *et al.*[47]

As an alternative to the above mechanism, De Gioia and coworkers[51] suggested a catalytic cycle for H_2 oxidation where H_2 is activated in the area between the iron centers of the dimer. Their cycle also begins with the iron dimer in the $Fe^{II}Fe^{II}$ redox level with OH^- bound at the distal iron center. Protonation of the OH^- ligand and one-electron reduction leads to loss of H_2O, yielding the catalytically active $Fe^{I}Fe^{II}$ form. Up to this point, the two mechanisms are the same. In the De Gioia mechanism the μ-CO $Fe^{I}Fe^{II}$ form then rearranges to a form where the bridging carbonyl moves to a terminal position. Dihydrogen binds to the proximal iron center in between iron, where the bridging CO was originally bound. The complex undergoes a one-electron oxidation to yield an $Fe^{II}(\eta_2\text{-}H_2)Fe^{II}$ form. A proton is transferred from the bound $\eta_2\text{-}H_2$ to a sulfur atom of the bridging dithiolate, rather than to a nitrogen as in the Hall mechanism. The iron- and sulfur-bound protons are moved away from the iron dimer site by the protein. The di-iron active site undergoes one-electron oxidation to $Fe^{I}Fe^{II}$ to close the catalytic cycle.

There are some clear similarities between the mechanisms for NiFe-hydrogenase, shown in Figure 13.12, and for FeFe-hydrogenase, shown in Figure 13.14. In both mechanisms, there is an activation step that leads to a catalytically active metal dimer at the (I,II) redox level. In the oxidative addition part of the NiFe-mechanism and in the Hall version of the FeFe-mechanism, the activation of H_2 furthermore occurs at a terminal position of one of the metals. A major difference is that heterolytic cleavage is used throughout in the FeFe-case, while this type of cleavage is only used in the initial activation in the NiFe-case.

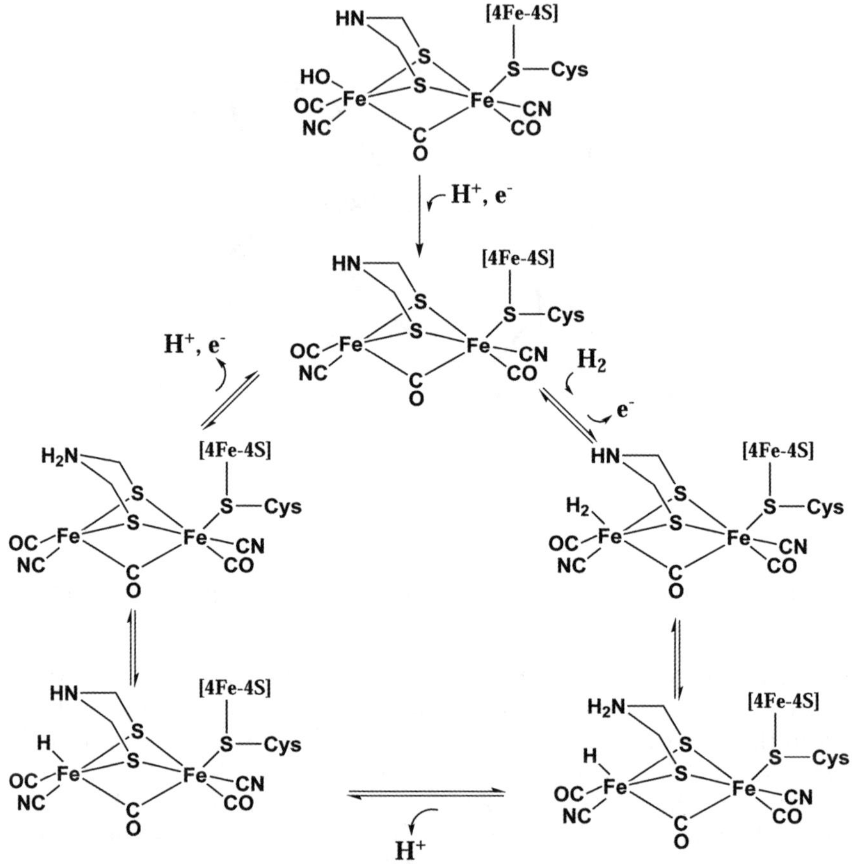

Figure 13.14 Proposed catalytic cycle for FeFe-hydrogenase using the central nitrogen of the μ-SCH_2 $NHCH_2$ S linker for H^+ transfer.[50] The oxidation states have been left purposely ambiguous.

13.4 Summary

Theoretical model calculations have been used to study the mechanisms of hydrogenase and photosystem II for about a decade. The present level of understanding obtained from theory of these processes has been briefly reviewed. The models have grown from a starting point with about 20–30 atom models of the active sites, to the present size of nearly 200 atoms. Hybrid DFT has been the main method used. Progress has been gradual in a constant comparison with new data from experiments.

For water oxidation by the OEC of PSII, the appearance of X-ray crystal structures meant a large improvement in the level of detailed modeling. Still, since the X-ray resolution is rather low, major modifications of the experimental structures have been needed to reach a reasonable water oxidation mechanism. The present mechanism for O–O bond formation, which is believed to be close to the actual one, involves a critical terminal oxygen radical reacting with a μ-oxo ligand. This mechanism would not be possible to find if the X-ray structures were used directly. In one of the X-ray structures,[3] the position of the oxygen radical is occupied by an oxo-ligand bound to three manganese centers, which would not allow radical formation. In the second X-ray structure,[4] the dangling manganese is too far away for providing the μ-oxo ligand that should react with the oxygen radical. Some of the details of the ligand binding have also been found to be critical for a low-barrier mechanism. The X-ray structures have led to two quite different suggestions for the ligand assignments, where the present model is closer to one of these suggestions.[4] Furthermore, by theoretical calculations it has been possible to investigate and obtain accurate information for the illusive S_4-state, where the O–O bond is formed. The lifetime of this state is too short for measurements at present.

For the mechanism of the reversible H–H bond formation in the hydrogenases, almost full consensus was reached for the mechanism at a rather early stage. However, for NiFe hydrogenase, the suggested heterolytic cleavage mechanism has recently been questioned. Instead, a mechanism where heterolytic cleavage is used only to generate the active Ni(I)Fe(II)-species has been suggested. This step is then followed by homolytic cleavage where nickel oscillates between the oxidation states (I) and (III). This general type of mechanism is in better agreement with some of the experiments. The involvement of the rather unusual Ni(I)-state is similar to the leading mechanism for FeFe-hydrogenase, where Fe(I)Fe(II) is the key species.

References

1. B. A. Diner, *Biochim. Biophys. Acta*, 2001, **1503**, 147–163.
2. F. Rappaport and J. Lavergne, *Biochim. Biophys. Acta*, 2001, **1503**, 246–259.
3. K. N. Ferreira, T. M. Iverson, K. Maghlaoui, J. Barber and S. Iwata, *Science*, 2004, **303**, 1831–1838.
4. B. Loll, J. Kern, W. Saenger, A. Zouni and J. Biesiadka, *Nature*, 2005, **438**, 1040–1044.
5. A. Guskov, J. Kern, A. Gabdulkhakov, M. Broser, A. Zouni and W. J. Saenger, *Nat. Struct. Biol.*, 2009, **16**, 334–341.
6. P. E. M. Siegbahn and R. H. Crabtree, *J. Am. Chem. Soc.*, 1999, **121**, 117–127.
7. P. E. M. Siegbahn, *Inorg. Chem.*, 2000, **39**, 2923–2935.
8. P. E. M. Siegbahn and M. Lundberg, *Photochem. Photobiol. Sciences*, 2005, **4**, 1035–1043.
9. P. E. M. Siegbahn and M. R. A. Blomberg, in *Computational Modeling for Homogeneous Catalysis and Biocatalysis*, ed. K. Morokuma and J. Musaev, Wiley-VCH, Germany, 2008, pp. 57–81.
10. P. E. M. Siegbahn, *Chem.–Eur. J.*, 2008, **27**, 8290–8302.
11. E. M. Sproviero, J. A. Gascon, J. P. McEvoy, G. W. Brudvig and V. S. Batista, *J. Am. Chem. Soc.*, 2008, **130**, 3428–3442.

12. D. A. Pantazis, M. Orio, T. Petrenko, S. Zein, W. Lubitz, J. Messinger and F. Neese, *Phys. Chem. Chem. Phys.*, 2009, **11**, 6788–6798.
13. J. Yano, J. Kern, K. Sauer, M. J. Latimer, Y. Pushkar J. Biesiadka, B. Loll, W. Saenger, J. Messinger A. Zouni and V. K. Yachandra, *Science*, 2006, **314** 821–825; J. Yano, J. Kern Y. Pushkar, K. Sauer, P. Glatzel, U. Bergmann, J. Messinger, A. Zouni and V. K. Yachandra, *Philos. Trans. R. Soc.*, B, 2008, **363**, 1139–1147.
14. P. E. M. Siegbahn, *J. Am. Chem. Soc.*, 2009, **131**, 18238–18239.
15. P. E. M. Siegbahn, *Chem.–Eur. J.*, 2006, **12**, 9217–9227.
16. P. E. M Siegbahn, *Acc. Chem. Res.*, 2010, in press.
17. H. Dau, A. Grundmeier, P. Loja and M. Haumann, *Philos. Trans. R. Soc., B*, 2008, **363**, 1237–1244.
18. U. Ryde and P. E. M. Siegbahn, in preparation.
19. P. J. Nixon, M. Sarcina and B. A. Diner, Photosystem II; The Light-Driven Water: Plastoquinone Oxidoreductase, in *Advances in Photosynthesis and Respiration*, ed. T. J. Wydrzynski and K. Satoh, Springer, The Netherlands, vol. 22, 2005, p. 71–94.
20. J. Clausen, S. Winkler, A. M. Hays, M. Hundelt, R. J. Debus and W. Junge, *Biochim. Biophys. Acta*, 2001, **1506**, 224–235.
21. R. J. Debus, *Coord. Chem. Rev.*, 2008, **252**, 244–258.
22. G. Renger, *Physiol. Plant.*, 1997, **100**, 828–841.
23. V. Förster and W. Junge, *Photochem. Photobiol.*, 1985, **41**, 183–190.
24. J. Clausen and W. Junge, *Nature*, 2004, **430**, 480–483.
25. W. Hillier and J. Messinger, in Photosystem II; The Light-Driven Water: Plastoquinone Oxidoreductase, *Advances in Photosynthesis and Respiration*, ed. T. J. Wydrzynski and K.Satoh, Springer, The Netherlands, vol. 22, 2005, p. 567–608.
26. T. Noguchi, *Philos. Trans. R. Soc., B*, 2008, **363**, 1189–1195.
27. A. E. Przbyla, J. Robbins, N. Menon and H. D. Peck Jr, *FEMS Microbiol. Rev.*, 1992, **88**, 109–113.
28. Y. Nicolet, B. J. Lemon, J. C. Fontecilla-Camps and J. W. Peters, *Trends Biochem. Sci.*, 2000, **25**, 138–143.
29. E. J. Lyon, S. Shima, G. Buurman, S. Chowdhuri, A. Batschauer, K. Steinbach and R. K. Thauer, *Eur. J. Biochem.*, 2004, **271**, 195–204.
30. P. E. M. Siegbahn, J. W. Tye and M. B. Hall, *Chem. Rev.*, 2007, **107**, 4414–4435.
31. M. Bruschi, G. Zampella, C. Greco, L. Bertini, P. Fantucci and L. De Gioia, in *Computational Inorganic and Bioinorganic Chemistry*, ed. E. I. Solomon, B. King and R. Scott, John Wiley and Sons, Ltd, Chichester, 2009.
32. A. Volbeda, L. Martin, C. Cavazza, M. Matho, B. Faber, W. Roseboom, S. Albracht, E. Garcin, M. Rousset and J. Fontecilla-Camps, *J. Biol. Inorg. Chem.*, 2005, **10**, 239–249.
33. M. Frey and J. C. Fontecilla-Camps, in *Handbook of Metalloproteins*; ed. A. Messerschmidt, R. Huber, T. Poulos and K. Wieghardt, John Wiley & Sons, Chichester, 2001, vol. 2, pp. 880–896.

34. E. Garcin, X. Vernede, E. C. Hatchikian, A. Volbeda, M. Frey and J. C. Fontecilla-Camps, *Structure*, 1999, **7**, 557–566.
35. A. L. De Lacey, V. M. Fernandez, M. Rousset and R. Cammack, *Chem. Rev.*, 2007, **107**, 4304–4330.
36. F. A. Armstrong and S. P. J. Albracht, *J. Philos. Trans. R. Soc., A*, 2005, **363**, 937–954.
37. M. Brecht, M. van Gastel, T. Buhrke, B. Friedrich and W. Lubitz, *J. Am. Chem. Soc.*, 2003, **125**, 13075–13083.
38. S. Foerster, M. Stein, M. Brecht, H. Ogata, Y. Higuchi and W. Lubitz, *J. Am. Chem. Soc.*, 2003, **125**, 83–93.
39. S. Foerster, M. van Gastel, M. Brecht and W. Lubitz, *J. Biol. Inorg. Chem.*, 2005, **10**, 51–62.
40. S. Kurkin, S. J. George, R. N. F. Thorneley and S. P. J. Albracht, *Biochemistry*, 2004, **43**, 6820–6831.
41. L. M. Roberts and P. A. Lindahl, *J. Am. Chem. Soc.*, 1995, **117**, 2565–2572.
42. P. Amara, A. Volbeda, J. C. Fontecilla-Camps and M. J. Field, *J. Am. Chem. Soc.*, 1999, **121**, 4468–4477.
43. S. O. Nilsson Lill and P. E. M. Siegbahn, *Biochemistry*, 2009, **48**, 1056–1066.
44. S. Q. Niu, L. M. Thomson and M. B. Hall, *J. Am. Chem. Soc.*, 1999, **121**, 4000–4007; A. Pardo, A. de Lacey, V. M. Fernandez, Y. Fan and M. B. Hall, *J. Biol. Inorg. Chem.*, 2006, **11**, 286–306.
45. Y. Montet, P. Amara, A. Volbeda, X. Vernede, E. C. Hatchikian, M. J. Field, M. Frey and J. C. Fontecilla-Camps, *Nat. Struct. Biol.*, 1997, **4**, 523–526.
46. A. Volbeda and J. C. Fontecilla-Camps, *Dalton. Trans.*, 2003, 4030–4038.
47. Y. Nicolet, C. Piras, P. Legrand, C. E. Hatchikian and J. C. Fontecilla-Camps, *Structure*, 1999, **7**, 13–23. (Protein Data Bank code 1HFE.)
48. Y. Nicolet, A. L. De Lacey, X. Vernede, V. M. Fernandez, E. C. Hatchikian and J. C. Fontecilla-Camps, *J. Am. Chem. Soc.*, 2001, **123**, 1596–1601.
49. J. W. Peters, W. N. Lanzilotta, B. J. Lemon and L. C. Seefedt, *Science*, 1998, **282**, 1853. (Errata: *Science*, **283**, 35; **283**, 2102.) (Protein Data Bank code 1FEH.)
50. H.-J. Fan and M. B. Hall, *J. Am. Chem. Soc.*, 2001, **123**, 3828–3829.
51. G. Zampella, C. Greco, P. Fantucci and L. De Gioia, *Inorg. Chem.*, 2006, **45**, 4109–4118.

CHAPTER 14

Redox Active Protein Maquettes: Multi-functional "Green Enzymes"

JAMES MURRAY

Division of Molecular Biosciences, Imperial College, Exhibition Road, London, SW7 2AZ, UK

14.1 Introduction

14.1.1 The Energy Problem

Most of humanity's energy needs are supplied, directly or indirectly by burning biomass, either directly, or as stored fossil fuels. However the supply of fossil fuels is finite, and their use has increased atmospheric levels of CO_2 to nearly 50% above pre-industrial levels,[1,2] which is believed to already be causing climate change due to an increased greenhouse effect. Therefore, humanity must develop new "carbon-neutral" energy sources. The largest potential "renewable" source of energy is sunlight, which also powers most of the biosphere. The power of sunlight on the Earth is $\sim$100 000 TW. Humanity uses around 15 TW of energy,[3] so there is a large amount available for use, we "merely" need to harvest it in an appropriate form.

One solution to using the energy of the sun is to create catalysts that will act in concert to generate fuels and chemical feedstocks from the same basic starting materials as biological systems – light, water and CO_2. Photons are used to release electrons from water, which are then used to reduce CO_2 to organic molecules, which are used structurally by the organism, and as "fuel"

RSC Energy and Environment Series No. 5
Molecular Solar Fuels
Edited by Thomas J. Wydrzynski and Warwick Hillier

Published by the Royal Society of Chemistry, www.rsc.org

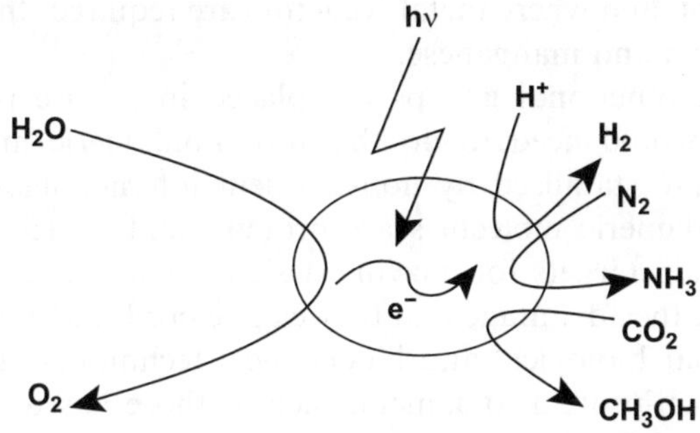

Figure 14.1 A "green energy" scheme. Electrons from the light-driven oxidation of water may be used to catalyse a variety of reactions, including hydrogen synthesis, nitrogen fixation and methane and methanol synthesis.

for respiration. Alternatively other valuable high energy chemicals may be generated by reduction of *e.g.* nitrogen gas or protons (Figure 14.1).

Sunlight can be used to generate electricity using photovoltaic cells, but these cannot be used when the sun is not shining. What is needed is a way to store the energy. Storing this energy as chemical bonds in fuels offers a very high energy density, so is most desirable. To achieve this involves mimicking the photochemistry of photosynthesis. In biology, these reactions are catalysed by a large number of enzymes, beginning with the light reactions of photosynthesis.

The first enzyme in oxygenic photosynthesis is photosystem II (PSII), a large membrane-bound protein–pigment complex, which uses four photons to oxidize water to oxygen, releasing four protons, and using the electrons to reduce plastoquinones.

Simple PSII analogues have been constructed using ruthenium pyridine photosensitizers, linked to a tyrosine group, and manganese ions.[4,5] These can perform light-driven manganese oxidation, but have not yet been shown to oxidise water using visible light. It is likely that the complexity of PSII function, requiring the management of high-energy intermediates, protons and electrons would require a catalyst with the complexity of a protein.

An enzyme is a biological catalyst, usually made of a protein, that catalyses a specific chemical reaction, or class of reactions. Many enzymes are unable to perform their function solely with the chemistry of the standard amino acids and so bind cofactors or coenzymes. These cofactors vary in size from a single metal ion to larger molecules such as chlorophyll. Most enzymes involved in bioenergetics are redox enzymes, involved in the movement of electrons. These enzymes are particularly rich in cofactors, as most biological redox reactions use non-protein cofactors. One third of all proteins are estimated to be metalloproteins.[6]

Enzymes have a degree of specificity and activity that it would be highly desirable to emulate. Industrial syntheses often involve high temperatures, expensive and toxic metal catalysts, and volatile solvents, and generate side-products. In contrast, most enzymatic reactions operate at room temperature, in aqueous solvent, and where metal cofactors are required, these are abundant metals such as iron and manganese.

In an enzyme, functional groups are placed in precise positions, and an artificial system would need to do this too. Foldameric molecules adopt a secondary structure stabilised by non-covalent interactions. The best understood sets of foldameric molecules are proteins and nucleic acids, and these have been the main targets for enzyme design and engineering. Proteins and nucleic acids have the advantage that they can be produced and manipulated by standard molecular biological and biochemical techniques, as well as laboratory synthesis. Non-protein foldamers, such as those based on β-amino acids are being developed for functions,[7,8] but have not been used yet in the development of metalloenzymes.

14.2 Approaches to Protein Design

The design of artificial proteins is of major interest as a source of novel catalysts, and for the understanding of protein structure and folding. Several approaches have been applied to this problem. The "protein folding problem" is to determine the tertiary structure to which a given sequence will fold. In the general case this is still unsolved, but for smaller protein domains, and where there are homologous structures, prediction structure prediction is becoming increasingly accurate.[9]

Protein design involves solutions to the "inverse folding problem", determining a sequence that will fold to a given conformation, with other conformations destabilised relative to the designed fold. Thus, protein design must involve both positive and negative design. Positive design ensures that the sequence folds to the desired conformation. Negative design changes to the sequence ensure that non-desired conformations are destabilised. Protein design is complicated by the fact that local folding is context dependent, and a given sequence can form an α-helix or β-strand depending on the environment.[10]

Two broad strategies for protein design are "top-down" and "bottom-up".[11] Top-down design involves the rational re-engineering of existing folded scaffolds, such as that to generate a specific site. Another "top-down" approach is the repacking of existing folds with new amino acids.[12] Bottom-up design is the construction of *de novo* structures, either by detailed energy calculations or simple binary patterning. A major success was the design and construction of the artificial Top7 protein, with a newly designed fold not found in biology.[13]

Many early designed protein structures adopted were very stable, but adopted a "molten globule"[14] structure. This is a state, thought also to be characteristic of intermediates in protein folding, in which the structure is compact, but the side-chains are mobile leading to an ensemble of interchanging structures. The hydrophobic effect drives the folding of soluble proteins, but the precise structure is determined by more specific interactions that are absent in a molten globule. Replacing the generic leucine interior with aromatic and β-branched amino acids helps to provide a more native-like specific structure.[15] Buried polar interactions[16,17] can improve the stability and specificity of the designed structure. Specific structures do not display the "molten globule" state, they show cooperative chemical and thermal denaturation, and NMR spectra corresponding to single states. A small number of artificial proteins have a sufficiently singular structure to have been crystallised[18] and their structures determined by X-ray crystallography.

14.2.1 Artificial and Engineered Enzymes

Many enzymes are used in industrial processes, and have been extensively modified from the wild-type protein. However, the creation of new enzyme activities is much harder. An approach taking advantage of the natural diversity of antibodies is the generation of catalytic antibodies or "abzymes".[19] Enzymes achieve a lot of their catalytic power by stabilising the transition state of a reaction. Therefore, antibodies raised against a compound that is a transition state analogue, could catalyse that reaction. Such catalytic antibodies have shown modest activity, and even have therapeutic potential.

Many classes of enzyme reaction are catalysed by multiple families of proteins, of independent origin. Yet in many of these families, the active site side chains are identical. An example are the serine proteases, chymotrypsin and subtilisin, which both have the "Ser–His–Asp" catalytic triad, but very different tertiary folds. Similarly, metal binding sites often show the same ligation patterns in different proteins. There are computer programs to "graft" a functional site onto a given scaffold protein,[20,21] which has led to the insertion of metal sites into proteins, and even to novel enzyme activities.[22,23]

14.3 Maquette Concept

Natural proteins are complex, having been subjected to billions of years of "tinkering" by evolution. Every amino acid in a protein may have several possible functions, such as contributing to protein stability, interacting with partner proteins and contributing to substrate binding. Each protein function is formed by a set of amino acids. The fitness of the protein is optimised over all of these functions. A given amino acid change may have a small positive or neutral effect on fitness. However, subsequent mutations become dependent on this first change, so that it cannot be reversed without a loss of fitness. Therefore, naturally evolved proteins are a complex interlocking network of functional dependencies, and it can be difficult to change their activity by successive point mutations (Figure 14.2).

A different philosophy, illustrated by the maquette principle, is to start with the simplest possible structure, and slowly introduce complexity.[24] A maquette is a small-scale model or rough draft of a sculpture or architectural work. The term was first used by the Dutton group to refer to peptide models of enzymes, and the term has been taken up by others. Here, I have referred to several designed peptides and proteins as "maquettes" even if not so-identified by the original authors.

The first maquettes were based on a four helix bundle design, a common motif in natural protein structures. The alpha helix has 3.6 amino acids per

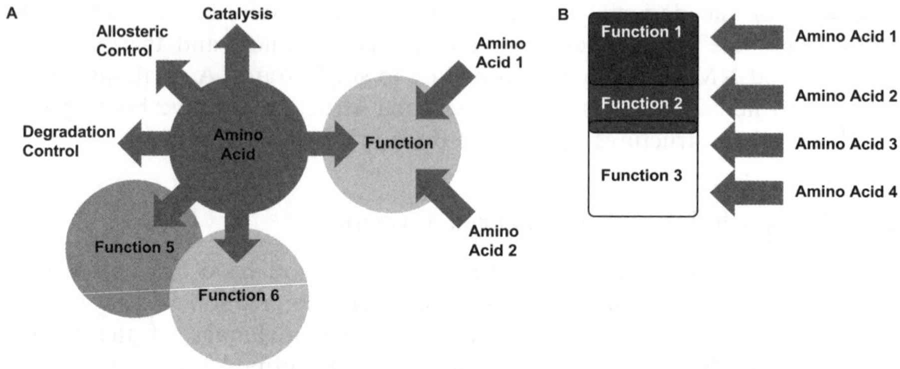

Figure 14.2 Interlocking amino acid functions. (A) In a natural protein, a single amino acid contributes to multiple functions, and a single function is determined by many amino acids. (B) In a maquette, the relationship between amino acids and the functions to which they contribute is determined from the beginning and can be manipulated in a tractable manner. Reproduced with permissions from reference 38.

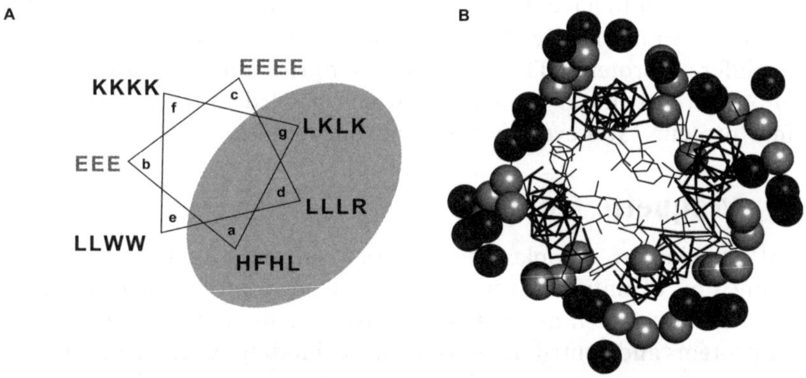

Figure 14.3 (A) Helical wheel diagram of the heptad-based repeat of the L31M maquette. Positively charged residues are black, negatively charged residues in grey and hydrophobic residues in outline. The interior of the bundle, shaded with a grey ellipse, is intended to drive tetramerization. The alternating stripes of oppositely charged side chains are intended to impose helix orientation. (B) Schematic of the crystal structure of the L31M maquette.[18] Positive residue side chains are represented by dark spheres, positively charges residues by light spheres. Hydrophobic side chains are drawn in outline. Opposite charges are spread over the protein surface. Reproduced with permission from reference 24. This and other molecular graphics figures produced with PyMol (The PyMOL Molecular Graphics System, Schrödinger, LLC).

turn, giving rise to a near-heptad repeat. Therefore, the faces of the helix may be given similar properties by placing similar residues down a "helical wheel" projection of the helix (Figure 14.3). A hydrophobic face on a helix will associate with other hydrophobic faces to form a tetrameric bundle, with a hydrophobic core (Figure 14.4). The burying of such hydrophobic cores is a near universal feature of protein folding. This simple sequence-patterning is present in natural helical bundles as well as in other regular protein structures. Naturally occurring β-solenoid proteins, which have highly regular sequences and structures,[25] show very strong sequence patterning. An α/β barrel has been redesigned using simple sequence patterning,[12] showing that simplified sequences are capable of folding into complex protein domains.

Natural proteins are "improved" and diversified by natural selection. A similar process of "directed evolution"[26] can be used to select or screen proteins for particular functions. In this process, a library of mutants is generated, which is then screened for improved function, in a manner analogous to Darwinian natural selection. This bypasses the need for "rational" design, and has been successful in many cases, *e.g*. the development of thermostable cellulases, with possible applications to the generation of biofuels.[27]

Although the maquette approach does not preclude the use of directed evolution techniques, the maquette "philosophy" as originally envisaged is to make simple iterative changes to incorporate a specific function.[28] For example, histidine residues can be introduced to allow heme binding. In this way, each amino acid change has a well-defined purpose.

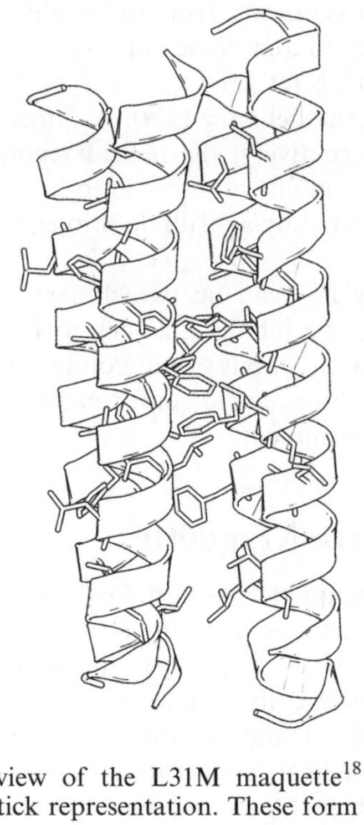

Figure 14.4 A cartoon view of the L31M maquette[18] showing the hydrophobic residues in stick representation. These form a hydrophobic core.

14.3.1 Synthesis and Assembly

Maquettes may be made both by chemical synthesis and *via* recombinant expression from synthetic genes. Chemical synthesis is usually by solid phase 9-fluorenylmethoxycarbonyl (Fmoc) protected chemistry.[29] Successive residues are added to the growing peptide at the C-terminus. Using this method, non-biological amino acids may be incorporated into the design. These have already proved useful in providing peptides with altered activity,[30,31] or allowing the ligation of cofactors to the modified amino acid.[32] A disadvantage is that sequence length is limited to around 50 amino acids due to declining yield. This may be circumvented by the ligation of two smaller synthesized peptides.[33]

Maquettes may also be produced recombinantly in bacteria using the standard techniques of molecular biology. A DNA gene corresponding to the protein is synthesized and incorporated into an expression vector, which is then used to transform an appropriate bacterial host strain. Production of the maquette is then induced in the transformed host. Recombinant expression can be scaled up at little cost. Additionally, all the tools of molecular biology are available. Of interest are random mutagenesis and library generation. These can be used to generate random diversity in maquettes, from which improved variants may be selected or screened. Increasingly, it is possible to use biological expression to incorporate unnatural amino acids into proteins.[34] Maquettes can bind a wide variety of redox cofactors, which can usually be added to the protein after synthesis. Iron–sulfur clusters are usually assembled chemically on the scaffold, in anaerobic conditions.

Two single helices can be tethered together *via* a disulfide bridge formed by cysteine at the end of the helix, (α-ss-α). A dimer of these dimers $(\alpha\text{-ss-}\alpha)_2$ forms the maquette. Alternatively, the helical segments may be tethered together with flexible loop regions, a "helix-loop-helix" motif. Two helix-loop-helix motifs may be connected by a disulfide bridge; the so-called "candelabra" arrangement.

The configuration may be *syn*: the two dimers parallel, or *anti*: with the dimers antiparallel. More specific design, such as by introducing buried polar interactions,[16] is required to stabilise one configuration over another. More specificity still is required for a single conformation to be stable enough for a crystal structure to be determined.[18]

14.4 Representative Maquettes

The original published maquette[35] was a four-helix bundle that bound four heme molecules with bis-histidine ligation. These hemes bound the protein in a well-defined way, and were able to perform long-range electron transfer.

Many enzymes are allosterically regulated, that is their activity is influenced by binding of a molecule at another (allo) site. As well as activation and repression, allostery can lead to cooperative substrate binding by multimeric enzymes. This key mechanism of enzyme regulation has been demonstrated in a maquette.[36] The binding of heme to an $(\alpha\text{-ss-}\alpha)_2$ dimer changed the orientation of the dimer from *syn* to *anti*.

This design philosophy reached a peak with a model O_2 transport protein.[37] This protein binds two hemes and oxygen. The protein binds O_2 more tightly than the toxic carbon monoxide, an improvement over the natural hemoglobin and myoglobin proteins. The restriction of water access[38] in this maquette is key to maintaining a stable oxyferrous heme state. Oxgyen bound to heme is the first step in many oxidative catalytic reactions.

14.4.1 Non Four-Helix Maquettes

The association of helices into bundles is largely driven by hydrophobic collapse. However the number of helices in the bundle is largely controlled by specific interactions. Some bundles have been constructed with different numbers of helices such as the trimeric coiled coil Coil-Ser maquette[39] and derivatives,[40,41] which have been used as a scaffold for a model of protein radicals.[42,43]

A heme peptide complex using only two helices tethered by disulfide bonds[44] shows a native-like solution structure.

As well as hydrophobic collapse, and disulfide tethering, folding and assembly can be driven by metal binding, as it is for the natural zinc finger motif.[45] Triple helical maquettes have been assembled with metals.[46,47] A six-coordinate metal ion tethers 3 2,2′-bipyridine ligands covalently attached to peptides, which then undergo a hydrophobic collapse to yield a folded structure.

Non-helical maquettes are rarer. Designing β-sheet proteins is in general much more challenging than helical ones, although there are an increasing number of examples. Rubredoxin is a largely β-sheet protein which binds a single iron site with four Cys side chains in a tetrahedral tetrathiolate geometry. A rubredoxin mimic has been constructed by techniques similar to those used for four helix bundles.[48] A three-strand motif corresponding to one side of the metal binding region of the protein was duplicated, for a C_2 symmetrical structure, then the two parts were linked with a rigid "Trypzip"[49] peptide. The simplified protein was monomeric and stable, bound metal ions stoichiometrically and could undergo reversible redox cycling.

Successive proline residues with *trans* peptide bonds form a left-handed polyproline II helix, a rigid element of secondary structure.[50] This has been used as a scaffold for assembling ruthenium bipyridine-based systems for photoconversion, with precisely-spaced electron donors and acceptors.[32,51,52]

14.5 Electron Transfer

A near-universal feature of electron transfer proteins is two catalytic sites, linked by redox chains.[53] Electrons are exchanged with substrates at the catalytic sites, which act as sources or sinks of electrons. The redox centres at the catalytic sites may be capable of multi-electron reactions. The redox chain contains a sequence of single-electron redox centres.

All of these components may be separate polypeptide chains, which may interact transiently or permanently. The rate of transfers between centres can be approximated by the following equation.

$$\log(k) = 13 - 0.6(R - 3.6) - 3.1(\Delta G + \lambda)^2/\lambda \quad (1)$$

where k is the rate in s^{-1}, R is the edge-to-edge distance in Å, ΔG is the driving force in eV and λ is the reorganization energy in eV.[54] The ΔG and λ terms give a Gaussian dependence of rate on the driving force that is characteristic of Marcus-like descriptions of tunnelling.[55] In this view, the intervening protein medium is of lesser importance, which would be a boon for the protein designer as extensive optimization is not necessary. In alternative view,[56,57] the metal–metal distance is more important than the edge-to-edge distance and the intervening medium has a greater effect. Maquettes are ideal tools with which to investigate protein electron transfer, as the protein environment, and edge-to-edge cofactor distance can be varied without affecting the overall structure.

14.6 Photosensitizers and Light Harvesting

Natural reaction centres use variants of chlorophyll to drive photochemistry. Chlorophyll is highly stabilised by the protein environment, which often contains carotenoids to quench triplet states.[58] In non-protein synthetic systems, chlorophyll is difficult to work with as it is easily oxidised and difficult to stabilise, and in general porphyrins are more stable than chlorophylls. However chlorophyll maquettes have been constructed with chlorophyll binding to small peptides,[59] or incorporated into four-helix bundles.[60] Natural proteins can be modified with non-natural chlorophylls; the water-soluble chlorophyll protein[61] can bind different chlorophyllides.[62] Some natural reaction centres use Zn-chlorophyll,[63] or this can be engineered into "normal" centres.[64] Zinc chlorins, which have a long-lived excited state have been used to bind to synthetic maquettes[65] and to a heme binding site in an engineered natural protein.[66] Rabanal *et al.*[67] describe a tetrameric four helix bundle maquette derivatized with two coproporphyrin I molecules to mimic the "special pair"[68] of photosynthetic reaction centres. The same group has constructed maquettes based on the light-harvesting proteins of purple bacteria that bind bacteriochlorophyll in detergent micelles.[69]

Flavins are used as photosensors[70] and can be cofactors in light-activated enzyme reactions, such as DNA photolyase.[71] Irradiation with blue photons generates a highly oxidizing triplet state. Flavins have been covalently incorporated into artificial maquettes for light-activated electron transfer,[72] despite these structures being entirely different from known flavin binding proteins.

Ruthenium pyridine[73] photosensitizers are very useful in artificial systems. These pigments have strong absorption in the visible region, and the excited state is long lived, and of sufficient potential to oxidise water. These compounds are useful in test systems, but are likely to be unsuitable for very large-scale deployment due to the high cost and scarcity of ruthenium.

Multi-electron reactions such as water oxidation require multiple photons, and so must have "storage states", such as the S-states of photosystem II,[74] and multiphoton reactions. Because the time between photons for a single pigment molecule is likely to be longer than the relaxation time, light-harvesting systems are required to ensure that photochemistry at each reaction centre occurs rapidly enough by coupling a large number of light-harvesting pigments to a single reaction centre. After a photon is absorbed, the exciton is "funnelled" by Förster resonance transfer to the reaction centre. All natural antennae have extremely high efficiency of transfer, approaching 100%, which is much better than any artificial system.[75] This high efficiency is thought to be due to the optimization of angles and distances between pigments, and their environment. Natural antennae also have a range of optimal absorptions so that the visible spectrum can be efficiently harvested. This highly efficient function has not yet been mimicked in a maquette system.

14.7 System Integration

Any scheme for using maquettes to perform a set of reactions will involve linking their functionalities together. In natural systems, enzymes activities are frequently assembled into larger complexes. Photosynthetic and respiratory membrane integral complexes are functionally coupled by the exchange of cofactors, and the proton-motive force maintained by energetic lipid bilayer membranes. Soluble protein enzymes are also linked *via* the exchange of cofactors such as quinones and NADPH. Electrons may be transferred within and between complexes over short distances *via* direct tunnelling. Synthetic maquettes could, in principle, use all of these methods. Although initially conceived as soluble proteins, maquettes may also be designed to be inserted into membranes by engineering a hydrophobic surface over part or all of their lengths.

14.7.1 Coupling to Electrodes

Organisms generate electrochemical gradients across membranes, and can transfer electrons along "wires" by quantum mechanical tunnelling. Artificial systems can use electrodes and wires coupled to catalytic units. There are many examples of redox enzymes and maquettes being functionally coupled to electrodes,[76] including photosystem I to form a photoelectrochemical cell.[77,78] The mechanism may be simple electrostatic interaction,[79] or covalent binding *via* cysteine residues.[80,81]

14.7.2 Membrane Integration

Many respiratory and photosynthetic complexes transport electrons or protons from one side of a membrane to another. In an artificial system, maquette-based proteins could transfer electrons from one active site to another across a membrane. In principle, there could even be a different solvent on each side of the membrane.

Natural integral membrane proteins fall into two main classes, helical bundles and beta-barrels. Proteins of both these classes satisfy internal hydrogen bonding requirements for maintaining a singular stable fold. However, the outside of the proteins contains largely hydrophobic residues projecting into the membrane region. The top and bottom of the transmembrane regions contain polar groups that interact with the polar lipid groups. The design of beta-barrel membrane proteins suffers from the same problems as soluble beta-sheet protein designs, however it is relatively straightforward to design helical peptides that will insert into biological membranes.

The maquette principle may be readily extended to membrane inserted maquettes. One approach,[82–84] is to engineer the surface regions of existing maquettes to be largely hydrophobic. Another is to begin with helical regions known to insert into membranes, and then engineer their association *via* the burying of polar interactions. For example, an asparagine residue on a transmembrane helix is sufficient to drive dimerization.[85] However, construction of membrane proteins in the general case is an even less well-solved problem than the soluble protein inverse folding problem.

14.8 Metal Clusters

In photosystem II, the water oxidation reaction is catalysed at the OEC, a metal-oxo complex of four manganese ions and a calcium ion ligated by protein side chains.[86,87] Chloride is also required for activity.[88] This site is precisely constructed, and so far is unique in biology. However, many other metal clusters are known to exist in redox enzymes. Examples include the [Ni-4Fe-5S] cluster of carbon monoxide dehydrogenase[89] and the many iron–sulfur clusters.[90] The di-iron cluster of ribonucleotide reductase (and other di-iron carboxylate proteins), has some similarity to the OEC of PSII. Both centres use a redox-active tyrosine as a link between the metal centre and the oxidant.

The "Due Ferri" (two-iron; DF) series of maquettes from the deGrado group,[91] consciously mimics the R2 di-metal centre environment, binding two metal ions with two EXXH sequence motifs. The di-Fe(III) active site is generated spontaneously on addition of Fe(II), by reaction with molecular oxygen to generate the oxo-bridged species. A DF maquette has even been shown to oxidise a phenol substrate.[92] This is similar to the recently discovered class of ribonucleotide reductase homologues,[93] the R2c-like proteins, which are organic ligand oxidases. Like the di-iron carboxylate proteins, the metal binding site is "pre-ordered" to accept the metal ions.[94] Despite the close similarity of these structures (Figure 14.5), the designed structure was produced before the discovery of the R2c-like proteins.

The only larger clusters that have been incorporated into maquettes are iron–sulfur clusters. These can be precisely specified by the incorporation of appropriately positioned cysteine residues. This has been done by mimicking biological sequences,[95] or by grafting a binding site into an existing protein[96] or maquette scaffold.[97,98] The cluster must be anaerobically assembled on the scaffold by chemical means.

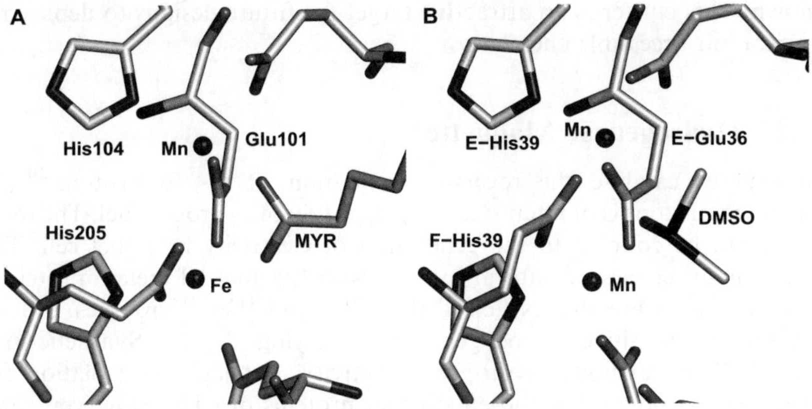

Figure 14.5 A comparison of the active sites of newly discovered class of R2c-like oxidases. (A) Drawn from PDB 3EE4[38] showing the myristate product. (B) A due-ferri-1 maquette, with an engineered di-metal binding site and substrate channel accessible to DMSO, drawn from PDB 1JM0.[116]

14.9 Future Directions and Challenges

Although a great deal has been accomplished with protein maquettes by way of proof-of-principle, the major challenges, such as the development of water oxidation catalysis, lie in the future.

14.9.1 Water Oxidation

The most obvious way to construct a water oxidation catalyst in a protein matrix is to mimic the oxygen evolving centre (OEC) of photosystem II, about which a great deal is known.

Other water oxidation catalysts require expensive metals such as ruthenium or iridium, and none have been shown to work using only visible light as a driver.

The manganese cluster of PSII contains higher ((III) and (IV)) oxidation states, unlike the diferric centres of ribonucleotide reductase or the DF maquettes. This manganese cluster is not assembled spontaneously on reaction with molecular oxygen. *In vivo*, the cluster is generated by the oxidative action of the light-activated enzyme itself, a complex process known as "photo-activation".[99] Mimicking this process will be another intermediate step in the construction of a water oxidation catalyst.

The recent publication of a cobalt-based catalyst by the Nocera group for the electrolysis of water with a low overpotential[100] has opened up new possibilities for water oxidation catalysis. The Nocera catalyst, like the OEC of PSII, is also believed to have a cubane-like structure,[101] but with cobalt rather than manganese. Cobalt is present in a few natural enzymes, but no enzyme cobalt cluster is known. This cluster is an attractive target for future designs to deploy non-biological but accessible chemistry.

14.9.2 Hydrogenase Maquettes

Hydrogenases catalyse the reversible oxidation of H_2 to protons.[102] The reduction of protons is of interest in the generation of hydrogen fuel. The reverse reaction can be coupled to the generation of electricity in a fuel cell. There are three main classes of natural hydrogenase enzyme, all metalloproteins.[102] The main classes are the [NiFe],[103] [FeFe][108] and [Fe][104] enzymes. All hydrogenases are sensitive to oxygen, to a varying degree. Synthetic metal clusters[105,106] have shown hydrogenase activity, but lack the isolation from solvent and second shell ligand effects that proteins provide. However, similar clusters have been incorporated into protein maquettes,[107] which will enable the effect of the protein matrix on the cluster structure and activity to be investigated.[108] A key aim of engineering natural and synthetic hydrogenases is to engineer an enzyme with reduced sensitivity to oxygen.

14.9.3 Proton Coupling

Multiple electron redox reactions are often coupled to proton exchange with the external solvent. This provides a mechanism for the protein to cope with the changes in electrostatic potential caused by the reaction. For some cofactors,

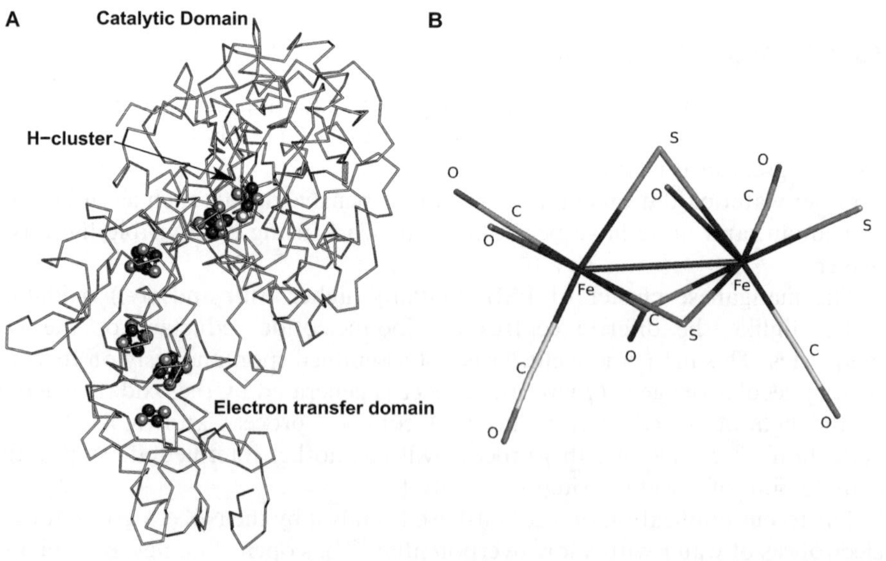

Figure 14.6 (A) A hydrogenase[108] showing the catalytic H-cluster and the chain of iron–sulfur clusters for electron transfer. (B) The catalytic cluster, a Fe–Fe centre with sulfur, oxygen and carbon monoxide ligands.

such as iron–sulfur clusters, the cofactor itself cannot bind protons, so the protein matrix must accommodate the change, usually *via* aspartate or glutamate residues.[109–111] These mechanisms have also been investigated using maquettes with heme and iron–sulfur cofactors[112] proton coupling.[113,114] However, more work is needed before the complex multi-electron and protein reactions of *e.g.* photosystem II and nitrogenase[115] can be replicated.

References

1. Dr. Pieter Tans, NOAA/ESRL (www.esrl.noaa.gov/gmd/ccgg/trends/) and Dr. Ralph Keeling, Scripps Institution of Oceanography (scrippsco2.ucsd.edu/).
2. T. Wigley, *Climatic Change*, 1983, **5**, 315.
3. N. Lewis and D. Nocera, *Proc. Natl. Acad. Sci. U. S. A.*, 2006, **103**, 15729.
4. R. Lomoth, A. Magnuson, M. Sjödin, P. Huang, S. Styring and L. Hammarström, *Photosynth. Res.*, 2006, **87**, 25.
5. L. Hammarström, *Curr. Opin. Chem. Biol.*, 2003, **7**, 666.
6. R. Holm, P. Kennepohl and E. Solomon, *Chem. Rev.*, 1996, **96**, 2239.
7. C. Goodman, S. Choi, S. Shandler and W. DeGrado, *Nat. Chem. Biol.*, 2007, **3**, 252.
8. D. Hill, M. Mio, R. Prince, T. Hughes and J. Moore, *Chem. Rev.*, 2001, **101**, 3893.
9. J. Moult, K. Fidelis, A. Kryshtafovych, B. Rost and A. Tramontano, *Proteins: Struct., Funct. Bioinf.*, 2009, **77**, 4.
10. D. Minor and P. Kim, *Nature*, 1996, **380**, 730.
11. P. Barker, *Curr. Opin. Struct. Biol.*, 2003, **13**, 490.
12. J. Silverman, R. Balakrishnan and P. Harbury, *Proc. Natl. Acad. Sci. U. S. A.*, 2001, **98**, 3092.
13. B. Kuhlman, G. Dantas, G. Ireton, G. Varani, B. Stoddard and D. Baker, *Science*, 2003, **302**, 1364.
14. M. Ohgushi and A. Wada, *FEBS Lett.*, 1983, **164**, 21.
15. B. Gibney, F. Rabanal, J. Skalicky, J. Wand and L. Dutton, *J. Am. Chem. Soc.*, 1997, **119**, 2323.
16. R. Koder, K. Valentine, J. Cerda, D. Noy, K. Smith, J. Wand and L. Dutton, *J. Am. Chem. Soc.*, 2006, **128**, 14450.
17. J. Schneider, J. Lear and W. DeGrado, *J. Am. Chem. Soc.*, 1997, **119**, 5742.
18. S. Huang, B. Gibney, S. Stayrook, P. Leslie Dutton and M. Lewis, *J. Mol. Biol.*, 2003, **326**, 1219.
19. C. V. Hanson, Y. Nishiyama and S. Paul, *Curr. Opin. Biotechnol.*, 2005, **16**, 631.
20. H. Hellinga and F. Richards, *J. Mol. Biol.*, 1991, **222**, 763.
21. N. Clarke and S. Yuan, *Proteins*, 1995, **23**, 256.
22. D. Rothlisberger, O. Khersonsky, A. Wollacott, L. Jiang, J. DeChancie, J. Betker, J. Gallaher, E. Althoff, A. Zanghellini, O. Dym, S. Albeck, K. Houk, D. Tawfik and D. Baker, *Nature*, 2008, **453**, 190.

23. L. Jiang, E. Althoff, F. Clemente, L. Doyle, D. Rothlisberger, A. Zanghellini, J. Gallaher, J. Betker, F. Tanaka, C. Barbas, D. Hilvert, K. Houk, B. Stoddard and D. Baker, *Science*, 2008, **319**, 1387.
24. R. Koder and P. Dutton, *Dalton Trans.*, 2006, 3045.
25. M. Vetting, S. Hegde, J. Fajardo, A. Fiser, S. Roderick, H. Takiff and J. Blanchard, *Biochemistry*, 2006, **45**, 1.
26. P. Romero and F. Arnold, *Nat. Rev. Mol. Cell Biol.*, 2009, **10**, 866.
27. P. Heinzelman, C. Snow, I. Wu, C. Nguyen, A. Villalobos, S. Govindarajan, J. Minshull and F. Arnold, *Proc. Natl. Acad. Sci. U. S. A.*, 2009, **106**, 5610.
28. B. Gibney, F. Rabanal, J. Skalicky, J. Wand and L. Dutton, *J. Am. Chem. Soc.*, 1999, **121**, 4952.
29. W. Chan and P. White, *Fmoc Solid Phase Peptide Synthesis: A Practical Approach*, Oxford University Press, USA, 2000.
30. A. Petros, S. Shaner, A. Costello, D. Tierney and B. Gibney, *Inorg. Chem.*, 2004, **43**, 4793.
31. H. Privett, C. Reedy, M. Kennedy and B. Gibney, *J. Am. Chem. Soc.*, 2002, **124**, 6828.
32. D. McCafferty, D. Friesen, E. Danielson, C. Wall, M. Saderholm, B. Erickson and T. Meyer, *Proc. Natl. Acad. Sci. U. S. A.*, 1996, **93**, 8200.
33. T. Hackeng, C. Mounier, C. Bon, P. Dawson, J. Griffin and S. Kent, *Proc. Natl. Acad. Sci. U. S. A.*, 1997, **94**, 7845.
34. Q. Wang, A. Parrish and L. Wang, *Chem. Biol.*, 2009, **16**, 323.
35. D. Robertson, R. Farid, C. Moser, J. Urbauer, S. Mulholland, R. Pidikiti, J. Lear, A. Wand, W. DeGrado and P. Dutton, *Nature*, 1994, **368**, 425.
36. A. Grosset, B. Gibney, F. Rabanal, C. Moser and P. Dutton, *Biochemistry*, 2001, **40**, 5474.
37. R. Koder, R. Anderson, L. Solomon, K. Reddy, C. Moser and L. Dutton, *Nature*, 2009, **458**, 305.
38. R. Anderson, R. Koder, C. Moser and L. Dutton, *Biochem. Soc. Trans.*, 2008, **36**, 1106.
39. B. Lovejoy, S. Choe, D. Cascio, D. McRorie, W. DeGrado and D. Eisenberg, *Science*, 1993, **259**, 1288.
40. N. Ogihara, M. Weiss, W. Degrado and D. Eisenberg, *Protein Sci.*, 1997, **6**, 80.
41. J. Johansson, B. Gibney, J. Skalicky, J. Wand and L. Dutton, *J. Am. Chem. Soc.*, 1998, **120**, 3881.
42. C. Tommos, J. Skalicky, D. Pilloud, A. Wand and P. Dutton, *Biochemistry*, 1999, **38**, 9495.
43. Q. Dai, C. Tommos, E. Fuentes, M. Blomberg, L. Dutton and J. Wand, *J. Am. Chem. Soc.*, 2002, **124**, 10952.
44. M. Rosenblatt, J. Wang and K. Suslick, *Proc. Natl. Acad. Sci. U. S. A.*, 2003, **100**, 13140.
45. E. Cox and G. McLendon, *Curr. Opin. Chem. Biol.*, 2000, **4**, 162.
46. A. Doerr, M. Case, I. Pelczer and G. McLendon, *J. Am. Chem. Soc.*, 2004, **126**, 4192.

47. A. Doerr and G. McLendon, *Inorg. Chem.*, 2004, **43**, 7916.
48. V. Nanda, M. Rosenblatt, A. Osyczka, H. Kono, Z. Getahun, L. Dutton, J. Saven and W. Degrado, *J. Am. Chem. Soc.*, 2005, **127**, 5804.
49. A. Cochran, N. Skelton and M. Starovasnik, *Proc. Natl. Acad. Sci. U. S. A.*, 2001, **98**, 5578.
50. A. Adzhubei and M. Sternberg, *Protein Sci.*, 1994, **3**, 2395.
51. S. Serron, S. Aldridge, C. Fleming, R. Danell, M. Baik, M. Sykora, D. Dattelbaum and T. Meyer, *J. Am. Chem. Soc.*, 2004, **126**, 14506.
52. S. Aldridge, B. Hornstein, S. Serron, D. Dattelbaum, J. Schoonover and T. Meyer, *J. Org. Chem.*, 2006, **71**, 5186.
53. C. Page, C. Moser and L. Dutton, *Curr. Opin. Chem. Biol.*, 2003, **7**, 551.
54. C. Moser, J. Keske, K. Warncke, R. Farid and P. Dutton, *Nature*, 1992, **355**, 796.
55. R. Marcus, *Biochim. Biophys. Acta, Rev. Bioenerg.*, 1985, **811**, 265.
56. H. Gray and J. Winkler, *Annu. Rev. Biochem.*, 1996, **65**, 537.
57. H. Gray and J. Winkler, *Q. Rev. Biophys.*, 2003, **36**, 341.
58. A. Young and H. Frank, *J. Photochem. Photobiol., B*, 1996, **36**, 3.
59. L. Eggink and J. Hoober, *J. Biol. Chem.*, 2000, **275**, 9087.
60. H. Rau, H. Snigula, A. Struck, B. Robert, H. Scheer and W. Haehnel, *Eur. J. Biochem.*, 2001, **268**, 3284.
61. D. Horigome, H. Satoh, N. Itoh, K. Mitsunaga, I. Oonishi, A. Nakagawa and A. Uchida, *J. Biol. Chem.*, 2007, **282**, 6525.
62. K. Schmidt, C. Fufezan, A. Krieger-Liszkay, H. Satoh and H. Paulsen, *Biochemistry*, 2003, **42**, 7427.
63. N. Wakao, N. Yokoi, N. Isoyama, A. Hiraishi, K. Shimada, M. Kobayashi, H. Kise, M. Iwaki, S. Itoh and S. Takaichi, *Plant Cell Physiol.*, 1996, **37**, 889.
64. P. Jaschke and T. Beatty, *Biochemistry*, 2007, **46**, 12491.
65. R. Razeghifard and T. Wydrzynski, *Biochemistry*, 2003, **42**, 1024.
66. B. Conlan, N. Cox, J. Su, W. Hillier, J. Messinger, W. Lubitz, L. Dutton and T. Wydrzynski, *Biochim. Biophys. Acta*, 2009, **1787**, 1112.
67. F. Rabanal, W. DeGrado and L. Dutton, *J. Am. Chem. Soc.*, 1996, **118**, 473.
68. J. Deisenhofer, O. Epp, K. Miki, R. Huber and H. Michel, *Nature*, 1985, **318**, 618.
69. D. Noy and L. Dutton, *Biochemistry*, 2006, **45**, 2103.
70. C. Lin, D. Robertson, M. Ahmad, A. Raibekas, M. Jorns, P. Dutton and A. Cashmore, *Science*, 1995, **269**, 968.
71. A. Sancar, *Science*, 1994, **266**, 1954.
72. R. Sharp, C. Moser, F. Rabanal and P. Dutton, *Proc. Natl. Acad. Sci. U. S. A.*, 1998, **95**, 10465.
73. A. Juris, *Coord. Chem. Rev.*, 1988, **84**, 85.
74. B. Kok, B. Forbush and M. McGloin, *Photochem. Photobiol.*, 1970, **11**, 457.
75. N. Adir, *Photosynth. Res.*, 2005, **85**, 15.
76. E. Reisner, J. Fontecilla-Camps and F. Armstrong, *Chem. Commun.*, 2009, 550.

77. P. Ciesielski, F. Hijazi, A. Scott, C. Faulkner, L. Beard, K. Emmett, S. Rosenthal, D. Cliffel and G. Kane Jennings, *Bioresource Technol.*, 2010, **101**, 3047.
78. N. Terasaki, N. Yamamoto, T. Hiraga, Y. Yamanoi, T. Yonezawa, H. Nishihara, T. Ohmori, M. Sakai, M. Fujii, A. Tohri, M. Iwai, Y. Inoue, S. Yoneyama, M. Minakata and I. Enami, *Angew. Chem., Int. Ed.*, 2009, **48**, 1585.
79. E. Topoglidis, B. Discher, C. Moser, L. Dutton and J. Durrant, *ChemBioChem*, 2003, **4**, 1332.
80. M. Case, G. McLendon, Y. Hu, K. Vanderlick and G. Scoles, *Nano Lett.*, 2003, **3**, 425.
81. J. Davis, C. Wrathmell, J. Zhao and J. Fletcher, *J. Mol. Recog.*, 2004, **17**, 167.
82. B. Discher, R. Koder, C. Moser and L. Dutton, *Curr. Opin. Chem. Biol.*, 2003, **7**, 741.
83. B. Discher, D. Noy, J. Strzalka, S. Ye, C. Moser, J. Lear, K. Blasie and L. Dutton, *Biochemistry*, 2005, **44**, 12329.
84. S. Ye, B. Discher, J. Strzalka, T. Xu, S. Wu, D. Noy, I. Kuzmenko, T. Gog, M. Therien, L. Dutton and K. Blasie, *Nano Lett.*, 2005, **5**, 1658.
85. C. Choma, H. Gratkowski, J. Lear and W. DeGrado, *Nat. Struct. Biol.*, 2000, **7**, 161.
86. K. Ferreira, T. Iverson, K. Maghlaoui, J. Barber and S. Iwata, *Science*, 2004, **303**, 1831.
87. A. Guskov, J. Kern, A. Gabdulkhakov, M. Broser, A. Zouni and W. Saenger, *Nat. Struct. Mol. Biol.*, 2009, **16**, 334.
88. C. F. Yocum, *Coord. Chem. Rev.*, 2008, **252**, 296.
89. H. Dobbek, V. Svetlitchnyi, L. Gremer, R. Huber and O. Meyer, *Science*, 2001, **293**, 1281.
90. D. Johnson, D. Dean, A. Smith and M. Johnson, *Annu. Rev. Biochem.*, 2005, **74**, 247.
91. A. Lombardi, C. Summa, S. Geremia, L. Randaccio, V. Pavone and W. DeGrado, *Proc. Natl. Acad. Sci., U. S. A.*, 2000, **97**, 6298.
92. M. Faiella, C. Andreozzi, R. T. M. de Rosales, V. Pavone, O. Maglio, F. Nastri, W. DeGrado and A. Lombardi, *Nat. Chem. Biol.*, 2009, **5**, 882.
93. C. Andersson and M. Högbom, *Proc. Natl. Acad. Sci. U. S. A.*, 2009, **106**, 5633.
94. O. Maglio, F. Nastri, V. Pavone, A. Lombardi and W. DeGrado, *Proc. Natl. Acad. Sci. U. S. A.*, 2003, **100**, 3772.
95. M. Antonkine, M. Koay, B. Epel, C. Breitenstein, O. Gopta, W. Gärtner, E. Bill and W. Lubitz, *Biochim. Biophys. Acta*, 2009, **1787**, 995.
96. C. Coldren, H. Hellinga and J. Caradonna, *Proc. Natl. Acad. Sci. U. S. A.*, 1997, **94**, 6635.
97. B. Gibney, S. Mulholland, F. Rabanal and P. Dutton, *Proc. Natl. Acad. Sci. U. S. A.*, 1996, **93**, 15041.
98. J. Grzyb, F. Xu, L. Weiner, E. Reijerse, W. Lubitz, V. Nanda and D. Noy, *Biochim. Biophys. Acta*, 2009.

99. J. Dasgupta, G. Ananyev and C. Dismukes, *Coord. Chem. Rev.*, 2008, **252**, 347.
100. M. Kanan and D. Nocera, *Science*, 2008, **321**, 1072.
101. M. Risch, V. Khare, I. Zaharieva, L. Gerencser, P. Chernev and H. Dau, *J. Am. Chem. Soc.*, 2009, **131**, 6936.
102. P. Vignais and B. Billoud, *Chem. Rev.*, 2007, **107**, 4206.
103. H. Ogata, Y. Mizoguchi, N. Mizuno, K. Miki, S. Adachi, N. Yasuoka, T. Yagi, O. Yamauchi, S. Hirota and Y. Higuchi, *J. Am. Chem. Soc.*, 2002, **124**, 11628.
104. O. Pilak, B. Mamat, S. Vogt, C. Hagemeier, R. Thauer, S. Shima, C. Vonrhein, E. Warkentin and U. Ermler, *J. Mol. Biol.*, 2006, **358**, 798.
105. C. Tard and C. Pickett, *Chem. Rev.*, 2009, **109**, 2245.
106. C. Tard, X. Liu, S. Ibrahim, M. Bruschi, L. De Gioia, S. Davies, X. Yang, L. Wang, G. Sawers and C. Pickett, *Nature*, 2005, **433**, 610.
107. A. Jones, B. Lichtenstein, A. Dutta, G. Gordon and P. Dutton, *J. Am. Chem. Soc.*, 2007, **129**, 14844.
108. J. Peters, W. Lanzilotta, B. Lemon and L. Seefeldt, *Science*, 1998, **282**, 1853.
109. Z. Zhang, L. Huang, V. Shulmeister, Y. Chi, K. Kim, L. Hung, A. Crofts, E. Berry and S. Kim, *Nature*, 1998, **392**, 677.
110. R. Hienerwadel, S. Grzybek, C. Fogel, W. Kreutz, M. Okamura, M. Paddock, J. Breton, E. Nabedryk and W. Mäntele, *Biochemistry*, 1995, **34**, 2832.
111. P. Hellwig, B. Rost, U. Kaiser, C. Ostermeier, H. Michel and W. Mäntele, *FEBS Lett.*, 1996, **385**, 53.
112. J. Shifman, C. Moser, W. Kalsbeck, D. Bocian and P. Dutton, *Biochemistry*, 1998, **37**, 16815.
113. A. Reddi, C. Reedy, S. Mui and B. Gibney, *Biochemistry*, 2007, **46**, 291.
114. M. Kennedy and B. Gibney, *J. Am. Chem. Soc.*, 2002, **124**, 6826.
115. L. Seefeldt, B. Hoffman and D. Dean, *Annu. Rev. Biochem.*, 2009, **78**, 701.
116. L. Di Costanzo, H. Wade, S. Geremia, L. Randaccio, V. Pavone, W. F. DeGrado and A. Lombardi, *J. Am. Chem. Soc.*, 2001, **123**, 12749–12757.

CHAPTER 15

Light-activated Bioconjugate Complexes

DAVID HVASANOV, DANIEL C GOLDSTEIN AND PALL THORDARSON*

School of Chemistry, The University of New South Wales, Sydney, NSW 2052, Australia

15.1 Introduction

Proteins play an important role in nature, responsible for the enzymatic conversions of nearly all processes in living organisms, including DNA replication and photosynthesis. They have structural roles, acting as scaffolds in cells to support and protect tissues and organisms. Furthermore, proteins are involved in cell signalling within and between cells which regulate the functions of cells *via* signal transduction. Commonly, protein functions depend on post-translational modifications after expression, such as glycosylation, which allows for proper folding and increased stability.

Scientists have long sought to understand and exploit the properties of proteins by incorporating synthetic molecules to introduce and enhance novel (unnatural) functionalities. This has led to the development of biotechnological, biomedical and pharmaceutical industry related devices such as biosensors,[1] bioelectronics,[2] biofuel cells[3] and bioconjugated therapeutic proteins/drugs.[4] In the field of *molecular solar fuels*, this is an attractive strategy as this will allow researchers to mix and match the best from synthetic and biological light-harvesters (chromophores) and electron donor/acceptor systems.

Bioconjugation can be defined as the linking of two or more components to form a novel complex with the combined properties and structural

RSC Energy and Environment Series No. 5
Molecular Solar Fuels
Edited by Thomas J. Wydrzynski and Warwick Hillier

Published by the Royal Society of Chemistry, www.rsc.org

characteristics of the individual component molecules.[5] In general, bioconjugation combines a biological component (protein, biomolecule) with a synthetic component creating a chimeric system resulting in a novel class of biomaterials to perform novel functions. Theoretically, the properties of the bioconjugate is the linear combination of the properties of the multi-component system.[6] This allows bioconjugates to overcome their intrinsic limitations and to possess properties and functions which otherwise would not exist in nature.

The advent of bioconjugation methods over the last few decades has led to the development of different chemoselective conjugation methods.[7,8] Bioconjugation chemistry has contributed to significant research progress and understanding in the field of life sciences. Through careful modification of proteins, scientists have been able to understand protein structure, functions and interactions.[9]

One of the most extensively studied areas of protein conjugates is PEGylation, as it has allowed for enhancement of peptides and proteins for pharmacological and therapeutic use *in vivo* and *in vitro* since it was first demonstrated in 1977.[10] Attachment of the hydrophilic poly(ethylene glycol) (PEG) polymer *via* PEGylation has led to improved targeting and tissue penetration for protein therapeutics in the biomedical industry.[11]

Biomolecules other than proteins, such as low molecular weight co-factors and ligands, are also common targets for bioconjugation and functionalisation. These often include porphyrins (heme),[12] $NAD(P)^+$,[13] FAD,[14] lipids,[15] sugars and oligosaccharides,[16] nucleic acids and nucleotides[17] and biotin.[18] However, these targets are beyond the scope of this review chapter. In addition, non-covalent bioconjugation techniques such as physisorption and layer-by-layer methods will not be dealt with here.[19]

The bulk of this book chapter will focus on giving the reader an overview of useful bioconjugation techniques for molecular solar fuels. Emphasis is placed on the methods for the covalent attachment of synthetic ligands to redox metalloproteins (biomolecule); including traditional methods such as targeting natural amino acids as well as recent conjugation techniques involving 'click' chemistry and non-natural amino acids. Studies of electron transfer, controlled biological activity, mimicry of biologic functions and use in bioelectrodes and biosensors will be used to illustrate the power of bioconjugate chemistry. The Chapter will conclude by highlighting the relatively few light-activated chromophore–electron acceptor bioconjugates that have appeared in the literature to date; but this class of bioconjugates is set to expand rapidly with the growing interest in merging biological and synthetic solutions to current challenges in the field of generating renewable energy.

15.2 Bioconjugation Methods

15.2.1 Classes for Specific Protein Functionalisation

Bioconjugation reactions involving chemical modifications and chemical cross-linking are often divided into classes of reactions. These can be separated into three generic types of functionalisation approaches; including direct

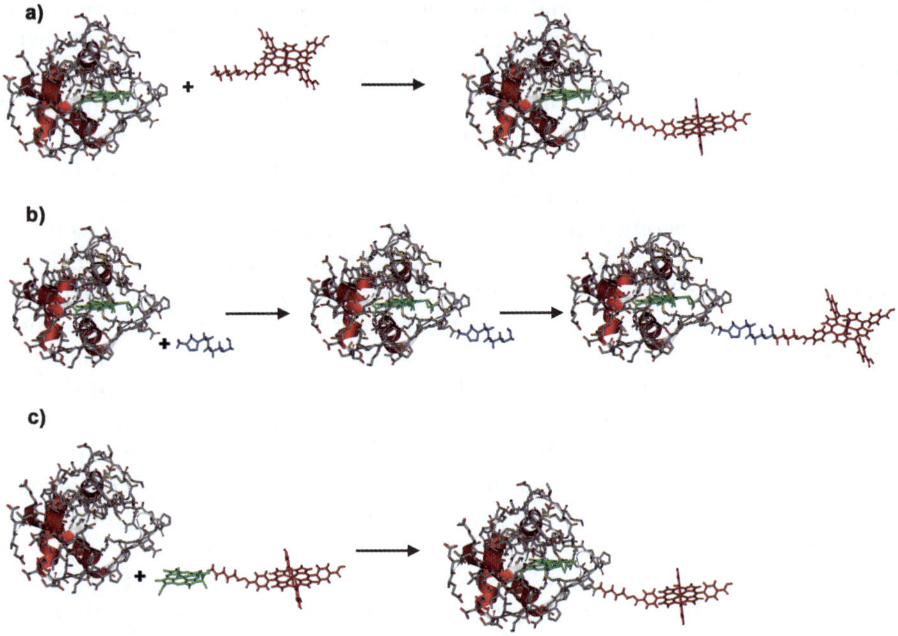

Figure 15.1 Schematic for the classes of protein functionalisation: (a) direct functionalisation; (b) indirect functionalisation; (c) indirect functionalisation with a cofactor/ligand.

functionalisation, indirect functionalisation and functionalisation with cofactors/ligands (Figure 15.1).[6]

Direct functionalisation involves chemically modifying a synthetic ligand possessing a specific functional group with a protein of interest. Indirect functionalisation requires bioconjugation between the synthetic/biomolecule components *via* a spacer containing heterobifunctionality. Finally, indirect fuctionalisation with ligands/cofactors is achieved by reaction of a synthetic ligand with a cofactor, which is then reconstituted into an apoprotein through self-association.

15.2.2 Classical Bioconjugation Reactions

Proteins and enzymes are most commonly modified by chemical modification between a synthetic ligand and a targeted amino acid residue. These residues often include lysine (Lys-NH_2), cysteine (Cys-SH), histidine (His), tyrosine (Tyr-PhOH) and glutamine (Gln).

15.2.2.1 Modification of Amines

The most common method for chemical modification in bioconjugate chemistry is the covalent linkage between a ligand and the amines of a protein.

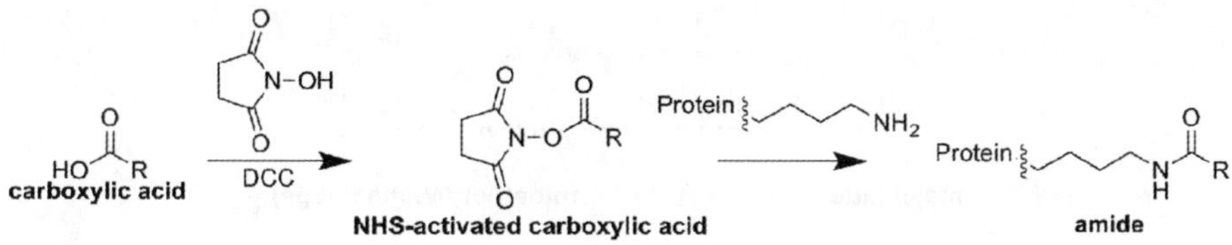

Scheme 15.1 Reactions of amines with *N*-hydroxysuccimide activated ester.

Coupling to the ε-amino group on a lysine residue is often targeted as they are abundant on the surface of proteins as well as the α-N terminus of a protein. They can be easily modified by reacting with an *N*-hydroxysuccimide (NHS) activated ester (Scheme 15.1).[6,20] It should be noted that chemical bond formation between NHS-esters and amines offer poor site-specificity, as the natural occurrence of lysine residues in mammalian proteins is estimated to be approximately 6%.[21]

Bioconjugation with amino groups can react efficiently due to their exposure to an aqueous environment based on their polar nature.[22] Although difficult, bioconjugation with the ε-amine group and α-N terminus can be selectively tailored based on their relative reactivity due to differing pK_a of 10 and 7.8, respectively.[23] For amine selective reactions to proceed, alkaline pH (above 8) is required due to the requirement of unprotonated amine groups.[9] Aldehyde-functionalised ligands can be used as an alternative to NHS-esters for chemical modifications; however, they are often more challenging to synthesise and involve a two-step reaction.[24]

An example of poor site-specific chemical modifications with amines was demonstrated by Pan *et al.*[25] who synthesised 10 singly labelled horse heart cytochrome *c* bioconjugates by reaction with an NHS-activated bipyridine ligand, as well as other products including unmodified protein.

15.2.2.2 Modification of Cysteine

Another popular target for chemical modification of natural amino acids is cysteine (cys). Bioconjugation with proteins are generally dependent on nucleophilic addition or displacement reaction mechanisms of the amino acid residue with an activated synthetic ligand.[9] Based on the theory of nucleophilicity proposed by Edwards and Pearson,[26] the thiol group of cysteine is one of the strongest nucleophiles in proteins.

Cysteine residues possess thiol groups with a pK_a of approximately 9.[5] Due to these properties, specific chemical modifications of cysteines can proceed selectively and rapidly under benign conditions.[6] Common modification reagents used in the literature include haloalkyl compounds, thiosulfonates and maleimides. Maleimides will be the focus in this section as they allow protein modification at neutral pH (6.8–7) coupled with control of specificity. They can be used to chemically modify cysteines as maleimide-functionalised ligands are Michael acceptors (Scheme 15.2).[27]

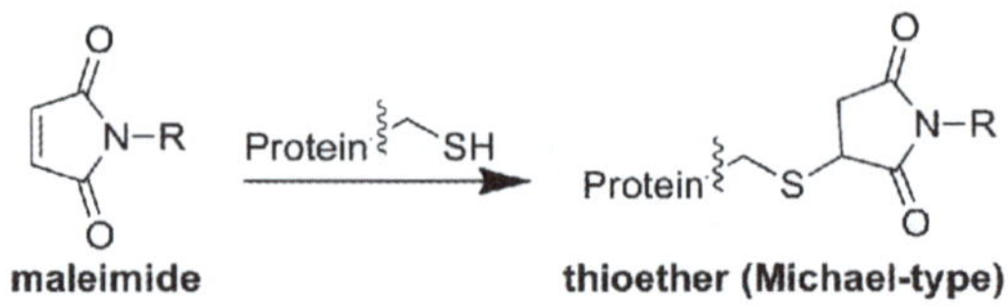

Scheme 15.2 Reactions of cysteine with a maleimide functionalised ligand.

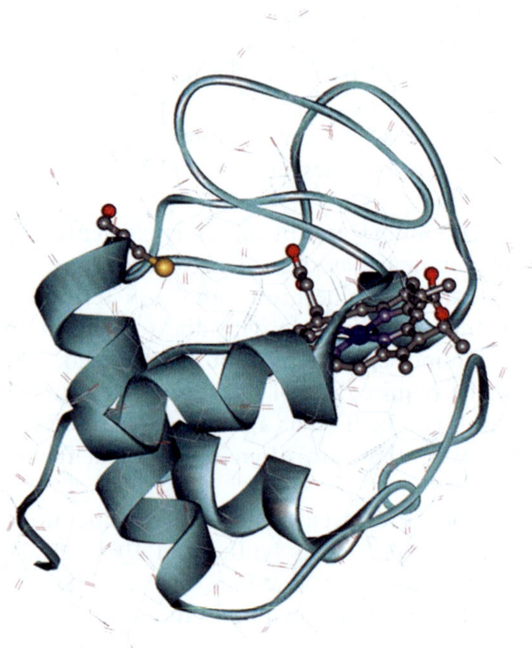

Figure 15.2 Structure of yeast iso-1 cytochrome *c*.[31]

Cysteine residues are rare in proteins and often only contain a single accessible residue available for modification. Alternatively, proteins which lack surface cysteine residues can be introduced *via* mutagenesis for site-specific modification.[28] This is a major advantage for synthesis of site-specific bioconjugates when targeting natural amino acids. Furthermore, maleimides are suitable for site-specific bioconjugation as they specifically react with cysteines at neutral pH and only side-react with other amino acids above pH 8.[6]

However, it should be noted that the reactivity of these residues are dependent on their local environment in the protein.[21] This effect is observed in iso-1 cytochrome *c* derived from *Saccharomyces cerevisiae* (yeast), where the CYS102 residue is buried in a hydrophobic pocket (Figure 15.2).[22,29] Peterson *et al.*[30] have demonstrated the reduced reactivity of cysteine in iso-1 cytochrome *c*[31] compared to bovine serum albumin (BSA), where BSA was fully functionalised in less than 3 minutes compared to at least 50 minutes for cytochrome *c* when reacting with the cysteine selective reagent, 4,4′-dithiodipyridine.

Additionally, chemical modification of cysteine may be hindered due to formation of disulfide bridges which are responsible for stability and folding in protein quarternary structures,[32] leading to loss of the free thiol group. Exposure of free thiol from disulfide bridges can be achieved easily by using mild chemical reducing agents such as mercaptoethanol or dithiothreitol. However, care must be taken to ensure destruction of the disulfide bridge does not inactivate or denature the protein.[33,34]

Peterson *et al.*[30,35] have demonstrated site-specific modification of iso-1 cytochrome *c* from the CYS102 residue of yeast and maleimide-functionalised ruthenium-terpyridine complexes to form light-harvesting donor-acceptor bioconjugates.

15.2.2.3 Modification of Histidine

Histidine residues in proteins play an important function in biology and protein chemistry. They are often essential for catalytic processes. As an example, it has been shown that histidine is essential in D-xylose isomerase enzymatic activity derived from *Streptomyces violaceoruber*.[36] As a result, chemical modification of histidine has been predominantly employed as a tool for determining catalytic mechanisms of enzymes as well as protein-protein or substrate/cofactor interactions.

The most common method for histidine modification is using diethylpyrocarbonate (Scheme 15.3).[37] However, less common alternative methods are available, such as photooxidation[38] (low specificity) and reactions with α-halo carboxylic acids and amides.[9]

Modification of histidine residues by diethylpyrocarbonate (DEP) results in the substitution of one of the nitrogen positions on the imidazole ring. Histidine has a pK_a of approximately 6.9 and reactions are generally specific to histidine between pH 5.5 to 7.5 during bioconjugation.[9,39] Reactions generally proceed under slightly acidic conditions to minimise regeneration of histidine during the reaction. Modification with DEP can be monitored by UV-Vis spectroscopy at 240 nm. The monosubstitution of histidine is a reversible reaction under neutral (and alkali pH) conditions and regeneration of histidine can be achieved by use

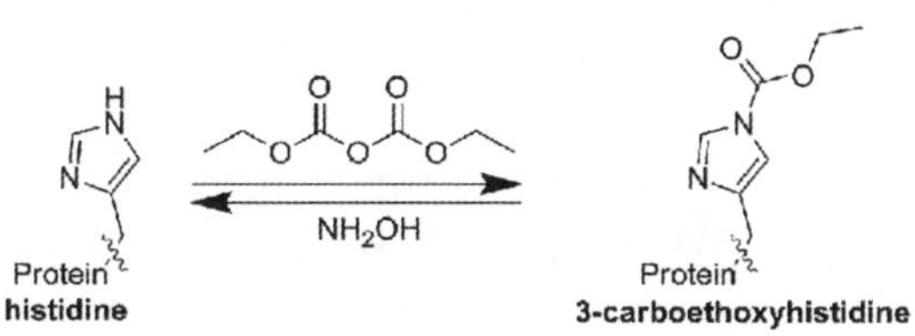

Scheme 15.3 Reversible reaction of histidine with diethylpyrocarbonate.

of hydroxylamine.[40] However, it should be noted that disubstitution reactions can also occur resulting in an irreversible product even after hydroxylamine treatment.[9]

Mattsson *et al.*[39] have shown that the cyclising activity of cyclomaltodextrin glucanotransferase (CGTase) derived from *Bacillus circulans* var. *alkalophilus* is dependent on the presence of unmodified histidine, confirmed by inhibition of activity when modified with DEP. Partial restoration of enzyme activity was achieved when incubated with hydroxylamine for 20 h.

An alternative method to modify and functionalise histidine residues is the formation of coordination complexes *via* coordinate covalent bonding. This is due to the lone pair electrons of the nitrogen histidine which can coordinate with transition metal ions. An elegant example of such a coordination complex was reported by Winkler *et al.*[41] who have synthesised a $Ru(NH_3)_5(His\text{-}33)^{3+}$-ferricytochrome *c* bioconjugate to study intramolecular electron transfer kinetics between the two redox centres.

15.2.2.4 Modification of Tyrosine

Tyrosine is a common target for modification as it can be functionalised with a variety of reagents and modification can be achieved with high specificity under mild conditions. Modification of tyrosine is generally carried out under mild alkaline conditions (pH $>$ 8) which increases reactivity and nucleophilicity due to ionisation of the hydroxyl group. In recent years, modification of tyrosine has generally been achieved using tetranitromethane (TNM).[42] However, iodination[43] and diazonium salts[44] can be used to functionalise tyrosine. Alkyation and acylation reactions have also been used but are less preferred due to cross-reactivity towards cysteine and lysine as well as harsh reaction conditions.[6]

Reactions with TNM under mild basic conditions nitrates tyrosyl residues, yielding the *o*-nitro derivative which lowers the pK_a from approximately 10 to 7 (Scheme 15.4).[9] The nitro functionality can also be further reduced to an amine which allows further modification using amine bioconjugation chemistry previously mentioned. Additionally, TNM can be used to directly cross-link tyrosyl residues in proteins to form a dityrosine.[45]

Iodination (Scheme 15.5) has allowed the modification of tyrosine residues in proteins to probe their reactivity and have been used in the study of tyrosine reactivity in cytochrome b_5.[46]

A novel and elegant method developed by Tilley and Francis[47] introduced a modification technique catalysed by palladium between a π-allyl species and

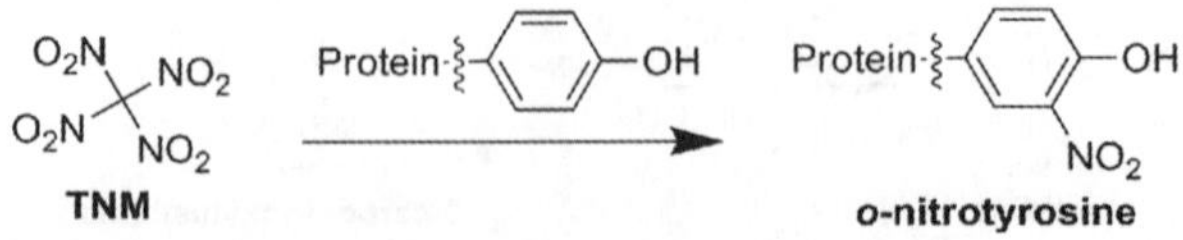

Scheme 15.4 Reactions of tyrosine with tetranitromethane (TNM).

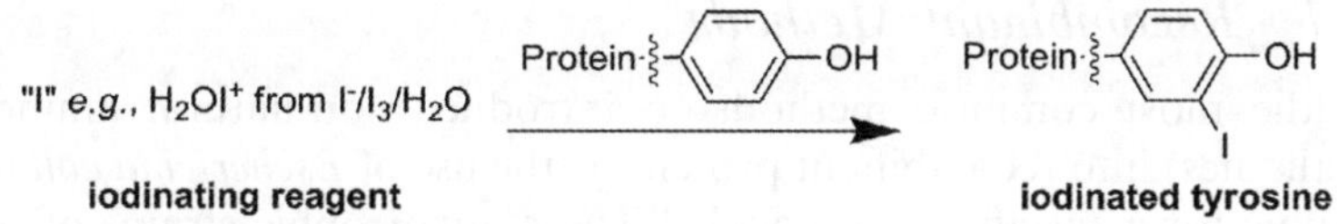

Scheme 15.5 Iodination reaction of tyrosine.

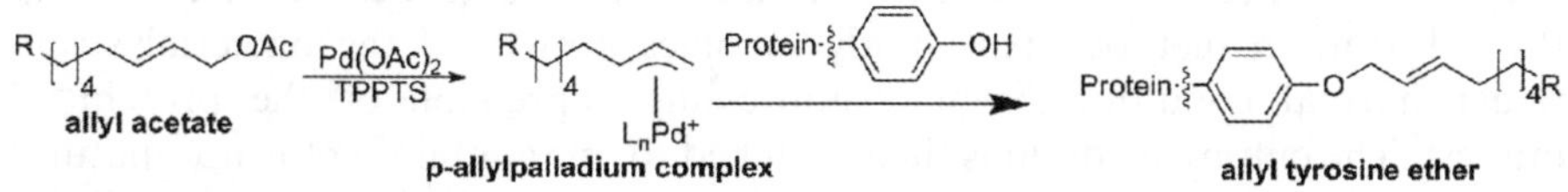

Scheme 15.6 Alkylation of tyrosine using π-allyl palladium complexes.

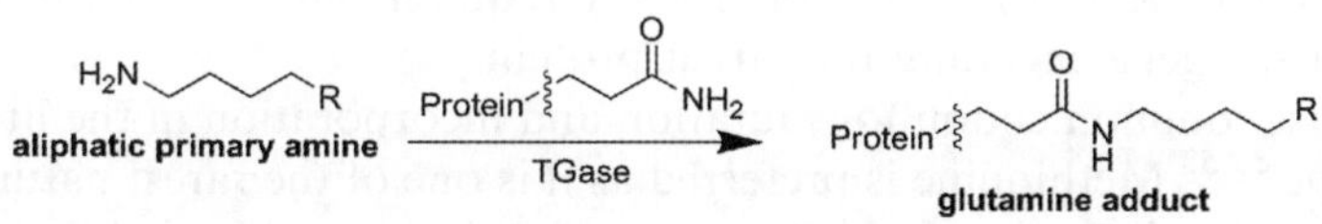

Scheme 15.7 Reaction of glutamine mediated by transglutaminase.

model protein chymotrypsinogen A, while requiring only a 5-fold excess of synthetic ligand (Scheme 15.6).

15.2.2.5 Modification of Glutamine

Glutamine is an important target for bioconjugation and is unique as the conjugation reaction is catalysed *via* the enzyme transglutaminase (TGase) rather than mediation by synthetic reagents. TGase catalyses the covalent cross-linking reaction between the γ-carboxamide of a glutamine residue and primary amines such as the ε-amino group of lysine (Scheme 15.7).[5] However, it should be noted that TGases are highly selective and modification of glutamine requires the exposure of the primary amine for the reaction to proceed.[6]

Glutamine modification reactions have been utilised for clinical diagnosis assays, as the over-expression of TGase is associated with cancer and bleeding disorders. Lee *et al.*[48] have used 5-(biotinamido)pentylamine to functionalise the glutamine residue of the protein *N*,*N*′-dimethylcasein to quantify TGase in plasma. Additionally, PEGylation has been achieved using protein-bound glutamine and amino-derivatives of poly(ethylene glycol) (PEG).[49]

15.2.3 Novel Non-classical Bioconjugation Reactions

As a result of the abundance of natural amino acids in proteins, synthetic biologists and chemists have developed techniques to introduce non-natural amino acids, allowing for chemoselective bioconjugation reactions. These non-natural amino acids include azides, alkynes, carbonyl and aldehydes which can be introduced selectively into proteins by semisynthetic or recombinant methods.[50]

15.2.3.1 Recombinant Methods

One of the most common methods to introduce non-natural amino acids (azides/alkynes) into recombinant proteins is the use of *Escherichia coli* that are auxotrophic for a specific amino acid.[51] These auxotrophic strains of bacteria are unable to synthesise a specific amino acid and are therefore dependent on amino acid uptake *via* growth media.

In general, incorporation of the non-natural amino acid can be achieved by the substitution method. This firstly involves growth of the bacterial strain under normal conditions, while suppressing expression of the protein of interest. The culture medium is then switched to one containing the non-natural amino acid analogue while incubating the bacteria to deplete the intracellular stock of the natural amino acid. Finally, the target protein is induced to express the mutant protein containing the non-natural amino acid residue at defined positions (Scheme 15.8).[50] One of the key requirements for this technique is control of the gene encoding the target protein.

The most common codon for mutation and incorporation in the literature is methionine.[52,53] Methionine is preferred as it is one of the rarest natural amino acids and due to its hydrophobic properties is often found buried in the protein core.[5,22] Due to the low frequency of methionine in natural proteins, incorporation of the non-natural amino acid has less effect on structure and function of the mutant. Other codons have also been selected for incorporation such as proline,[54] phenylalanine,[55] leucine,[56] isoleucine[57] and tryptophan.[58]

The incorporation efficiency can be increased by incubating the cells with large excess of appropriate aminoacyl tRNA transferase and reducing the synthetase substrate specificity and proofreading sites.[52] Finally, one of the issues to consider when incorporating non-natural amino acids such as azides/alkynes is that it may lead to toxic effects to the cell.[51]

An improved method to incorporate non-natural amino acids was developed by Schultz *et al.*[59] which translationally incorporates unnatural amino acids using *Xenopus* oocytes as the host organism (Scheme 15.9). This technique

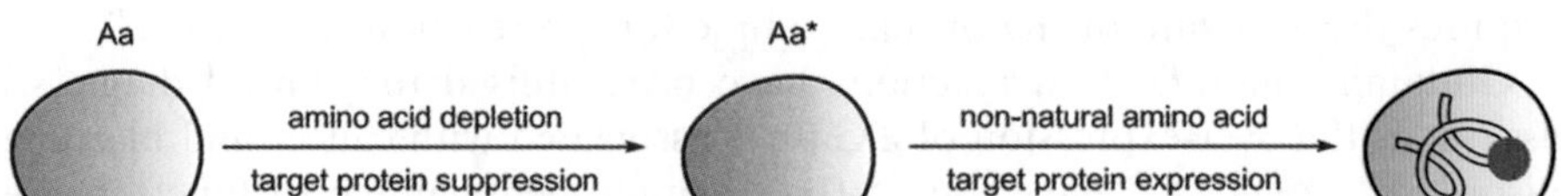

Scheme 15.8 Incorporation of non-natural amino acids *via* substitution method.

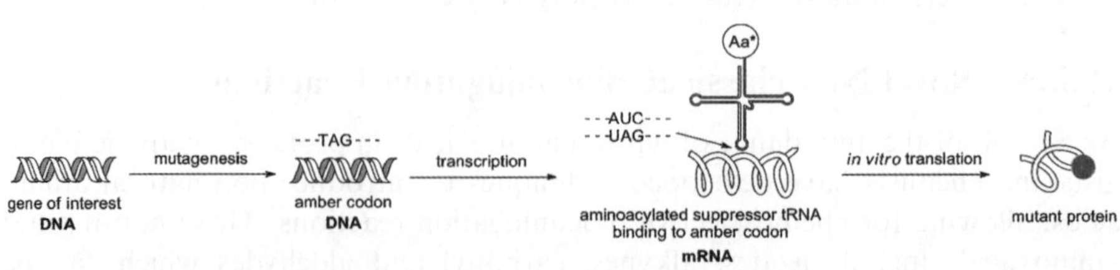

Scheme 15.9 Incorporation of non-natural amino acids *via* amber suppression method.

utilises an amber suppressor tRNA isolated from yeast due to its high suppression efficiency,[51] which recognises the amber (UAG) stop codon during translation and elongation by the ribosome. Incorporation and modification is achieved by chemical aminoacylation of the suppressor tRNA with an unnatural amino acid analogue and subsequent injection into the large *Xenopus* oocytes. Therefore, the binding of release factor I (RF-I) is prevented, which continues further elongation of the peptide (protein) while introducing the synthetic amino acid.[60] It is possible to use this method to incorporate an unnatural 21st amino acid in addition to the 20 natural amino acids.[50]

In recent years, alternative recombinant methods have been introduced to overcome the drawbacks of the substitution and amber suppression methods. The two most predominant methods are expanding the codon to four and five nucleotide bases to code for a non-natural amino acid[61] and introducing artificial nucleotides into a gene.[62] Incorporation of non-natural amino acids such as azides/alkynes using recombinant technologies has allowed these mutant proteins to be considered as targets for bioconjugation.

15.2.3.2 'Click' Reactions

One of the most common methods for chemical modification of proteins bearing unnatural amino acids has been through *click chemistry*. These include the aqueous version of the [3+2] Huisgen cycloaddition[63] and the (non) traceless Staudinger ligation.[64,65] The functionalities for *click* reactions can be introduced by chemical modification using classical bioconjugation methods or recombinant techniques.

The most popular approach in protein modification is the Cu(I)-catalysed [3+2] Huisgen cycloaddition (CCHC) of azides and terminal alkynes. The CCHC approach has advantages over other modification techniques, such as high coupling efficiency and high selectivity, which prevents side reactions with other functional groups in a reaction mixture.[66] Furthermore, due to the weak acid–base properties of azides and alkynes, CCHC is suitable for protein modification due to the resulting benign reaction conditions as well as its chemoselectivity (Scheme 15.10).[57]

In an elegant approach, Davis *et al.*[63] have incorporated unnatural amino acids (azidohomoalanine (azide) and homopropargylgylcine (alkyne)) and cysteine through recombinant methods. These chemical tags allow for orthogonal bioconjugation reactions through *click* chemistry and classical cysteine conjugation techniques. Using these techniques, they have post-translationally

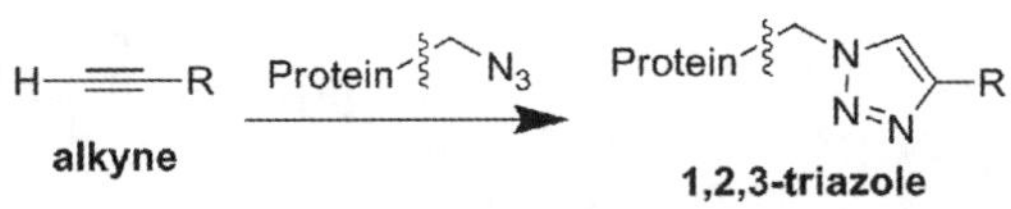

Scheme 15.10 Aqueous version of Cu(I)-catalysed [3+2] Huisgen cycloaddition (CCHC).

modified a LacZ reporter enzyme scaffold (SSβG) by glycosylation using *click* chemistry while simultaneously modifying cysteine with a sulphotyrosine mimic group. This has been used as a functional mimic of P-selectin-glycoprotein-ligand-1 which detects mammalian brain inflammation and disease.

van Hest *et al.*[66] have used CCHC *click* chemistry to functionalise alkyne modified green fluorescent protein (GFP) with an amphiphilic polystyrene-*block*-poly(acrylic acid) (PS-*b*-PAA) polymersome. The alkyne modified GFP was incorporated through semisynthetic means by lysine (amine) classical conjugation techniques and subsequently modified by *click* reaction (Scheme 15.11).

The less common *click* chemistry reactions for modifying proteins with synthetic ligands include the non-traceless Staudinger ligation.[67] This involves reaction between a phosphinothioester group of the ligand and an azide functionalised protein, resulting in the conjugated product and phosphine by-product (Scheme 15.12). The disadvantage of this reaction is that diphenyl phosphine oxide traces always remain in the product mixture.[50] A traceless Staudinger ligation has been developed to overcome this issue.[65,68]

15.2.3.3 Carbonyl Condensation

Proteins with carbonyl group functionalities are a novel class of reactions for modification and bioconjugation with synthetic ligands. Carbonyl groups are novel for protein modification as they are not found in natural proteins and can be modified selectively. These groups undergo keto–enol tautomerisation resulting in a slightly acidic α-hydrogen which can be attacked by nucleophiles.[50]

Keto/aldehyde functionalities can be introduced *via* recombinant methods,[69] synthetically with a bifunctional spacer[70] or by sodium periodate oxidation of

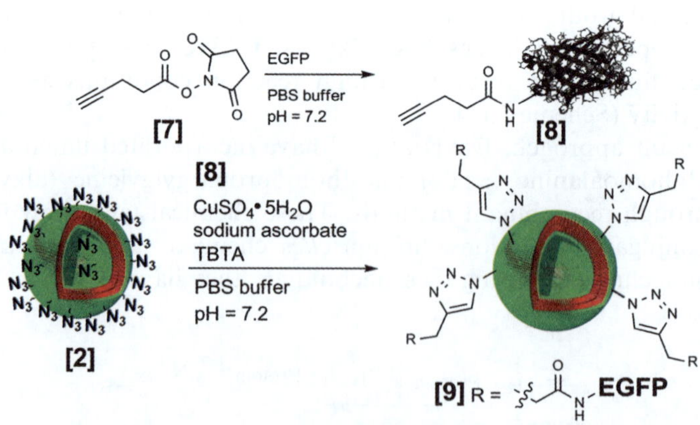

Scheme 15.11 Functionalisation of PS-*b*-PAA polymersomes with GFP using *click* chemistry.[66] Reproduced by permission of the Royal Society of Chemistry.

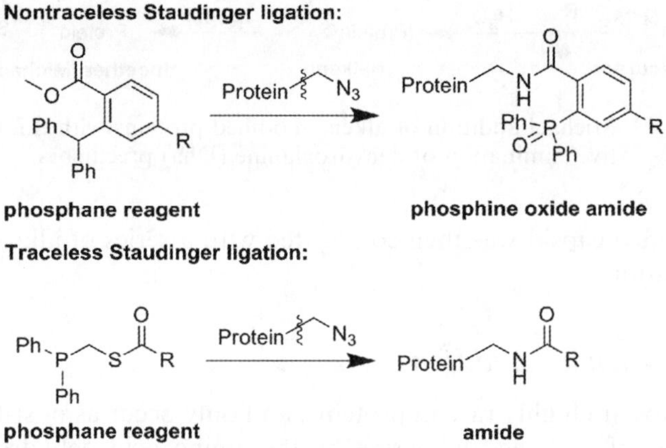

Scheme 15.12 Protein modification using the (non) traceless Staudinger ligation *click* reactions.

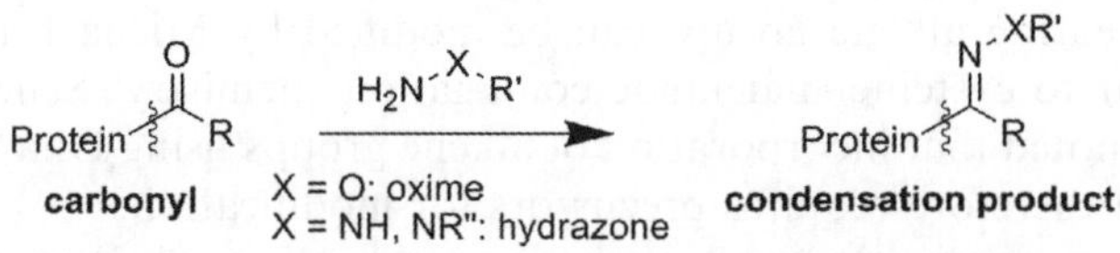

Scheme 15.13 Protein modification of carbonyl groups *via* oxime/hydrazone ligation methods.

an *N*-terminal threonine or serine.[71] It should be noted that carbonyl groups incorporated into proteins recombinantly form imines with ε-amino groups of lysine residues. However, this reaction is reversible and at physiological conditions the equilibrium lies towards the free amine and carbonyl groups.[72]

Carbonyl functionalised proteins can react readily and quantitatively with ligands containing hydrazide, alkoxyamine or semicarbazide functional groups, which form hydrazones, oximes and semicarbazones, respectively.[50] These reactions proceed under slightly acidic aqueous conditions and are stable under physiological conditions.

One of the most common methods for modification is the oxime ligation method which involves imine capture between the carbonyl group and an aminooxy nucelophile, forming the oxime product.[70] An alternative technique, the hydrazone ligation method is similar to the oxime method. However, a hydrazone product is formed.[73] These ligation methods are highly chemoselective due to the high reactivity between aminooxy/hydrazide groups and carbonyl groups (Scheme 15.13).

Francis *et al.*[70] have reported the modification of carbonyl functionalised tobacco mosaic virus capsid using the oxime ligation reaction. The ketone was incorporated on the protein surface by modification of the tyrosyl residue (TYR139) with a *p*-diazonium salt derived from *p*-aminoacetophenone. The

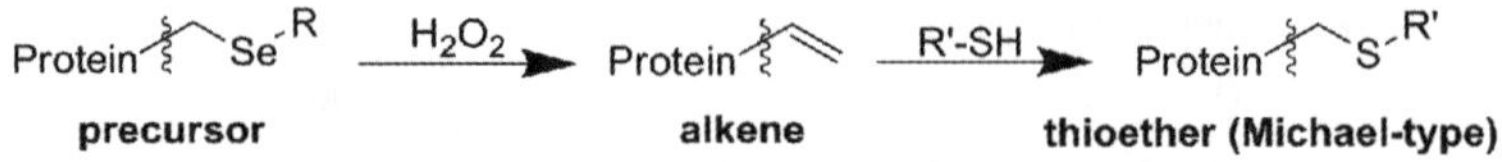

Scheme 15.14 Michael addition of alkene modified proteins with thiols after oxidative elimination of dehydroalanine (Dha) precursors.

ketone-modified capsid was then conjugated with a series of alkoxyamines *via* oxime formation.

15.2.3.4 Michael Addition

Alkene groups are highly rare in proteins, and only occur as post-translational modifications of cysteine and serine by the amino acid dehydroalanine.[74,75] Alkene modified proteins are a less common target for bioconjugation. In recent years, incorporation of alkenes has been achieved only through recombinant methods, such as phenylselenocysteine[75] and selenolysine[76] which are precursors of dehydroalanine (Dha).

Proteins bearing alkene groups can be modified by Michael addition with thiols, similar to cysteine–maleimide conjugation chemistry (Scheme 15.14).[77] It should be noted that incorporation of alkene groups using Dha in proteins is unstable and therefore requires precursors for modification.

Schultz *et al.*[75] have demonstrated the modification of Dha with thiols by incorporation of phenylselenocysteine precursor using the amber suppression method in *Escherichia coli*. The modification was initiated by oxidative elimination of the precursor with hydrogen peroxide to yield alkene functionality, which was subsequently reacted with desired thiol ligands to afford the thioether by Michael addition.

15.2.4 Surface Attachment Techniques

The immobilisation of biomolecules onto surfaces has been widely studied, and has proved invaluable for a number of applications, including biocatalysis for the conversion of chemical substrates,[78] and enzyme assays and biosensors for sensitive and selective detection of substrates.[1] A surface provides stability and reusability, and can even lead to higher enzyme activity, provided the properties of the enzyme and surface are taken into account.[78]

Surface attachment can generally be divided into two strategies: non-covalent (van der Waals/hydrophobic, hydrogen bonding, ionic interactions, encapsulation/entrapment, and bioaffinity), and covalent.[19] Covalent binding provides greatest stability across different media, as the enzyme is tightly fixed, minimising any leaching from the surface. In addition, covalent methods can potentially provide greater control over protein orientation, which allows for optimisation of the activity. The disadvantage is that the enzyme requires chemical modification, requiring careful consideration of the attachment method if enzyme function is to be maintained.

Traditional substrates for protein attachment include glass slides, polymer assay wells, and carbon, metal, semiconductor, and metal oxide electrodes. However, advances in material design has brought about an abundance of novel materials, including sol gels, hydrogels, micro- and nano-size particles, and porous and meso-porous materials.[19] It is often possible to tailor these materials to provide a large surface area that maximises interactions with proteins, allowing high loadings of protein material. In addition, porous materials can lead to higher stability by providing a suitable micro-environment for the enzyme.[79]

In applications where redox enzymes are immobilised on electrodes, good electrical communication between the redox centre of the enzyme and the conductive surface is desired. The protein matrix, however, generally acts as an insulator, leading to poor performance.[80] This barrier can be overcome either with the use of redox mediators (allowing indirect communication with the electrode surface) or through covalent immobilisation strategies.[2] For applications that require the transmission of activating light, glass and conductive tin oxide layers doped with fluorine or indium can be employed.[81]

If a surface already possesses a functional group, as in the case of a polymer, then direct conjugation with the protein is possible. For example, polymers possessing thioacetal groups can react with the primary amines of proteins to form isoindoles, and these have been used for the direct conjugation of a number of enzymes, including: bovine serum albumin (BSA), hemoglobin, cytochrome *c*, and horseradish peroxidase.[82,83] An additional advantage of this 'smart' polymer is that the resulting isoindole is fluorescent, allowing for quantification of protein attachment.

Direct attachment to gold surfaces is also possible for proteins possessing a single cysteine residue, through the formation of thiol–gold bonds. For example, reduced iso-1 cytochrome *c* exhibits electron transfer and retains its native functionality upon adsorption through its cysteine residue to a chemically untreated gold surface.[84]

In most cases, however, single or multiple hetero or homo bi-functional linker molecules are employed to chemically attach the protein to the surface. Silane coupling agents can be employed on a wide range of materials, including glass, metals/metal oxides, and semiconductors. These agents can be purchased or modified to incorporate a wide range of functional groups that can react with amino acid residues, including: isocyanate, azide, alkyne, maleimide, epoxy groups, or carboxy/NHS-esters.[5]

For gold and platinum surfaces, linkers possessing thiol groups provide a robust method of attachment. A recent example from the literature is the attachment of Photosystem I to gold and platinum nanoparticles.[85] The nanoparticles, functionalised with dithiol linkers, were bound directly to a cofactor variant of Photosystem I. Upon illumination of the bioconjugates, light absorbed by the photosystem resulted in electron transfer to the nanoparticles, catalysing the reduction of water to hydrogen.

On carbon surfaces, linkers functionalised with nitrene or diazonium groups can react to form stable covalent bonds with the surface.[86] Electrochemical reduction of diazonium groups is an effective method to produce functionalised surfaces. This method was recently used to assemble glucose oxidase on conductive diamond films.[87]

Hydrosilation is the reaction of Si–H bonds across the unsaturated bonds of an alkene or an alkyne, and can be used to functionalise silicon surfaces with well-defined monolayers.[88] This has been used for the attachment of peptides, and is a promising technique for the bioconjugation of redox proteins.

There are advantages and disadvantages to the different methods of immobilisation, and the needs of the system often dictate the method chosen. For example, the stability and activity requirements of an enzyme are very different for a biocatalysis application as compared to a disposable strip biosensor, and those requirements may mean a particular immobilisation approach is favoured.

15.3 Light-Activated Chromophore–Electron Acceptor Bioconjugates

In order to design molecular solar fuels using bio-active molecules such as metalloproteins, the bioconjugation to a light-harvesting electron donor, such as metal complexes, for photoinduced electron transfer is essential. Alternatively, biological light-harvesting systems can be linked to synthetic acceptors, *e.g.*, for photocatalytic formation of hydrogen. In both cases, specific linkages between the synthetic and biological components allow controlled electron transfer. The resulting bioconjugates could be applied in a photoelectrochemical cell based on biomolecules. This section will focus on previously reported work relating to the conjugation of light-harvesting chromophores to enzymes, involving intramolecular electron transfer of the bioconjugate, leading to potential applications in electricity generation (fuel cells)[89] or chemical transformations such as production of hydrogen.[85] Only a few of these photoactive donor–acceptor systems involving electron transfer has been reported in literature.

Bioconjugation of chromophores (biological/synthetic) to biomolecules have been reported, such as modification of fluorescent proteins with DNA oligonucleotides,[90] fluorescent dyes to proteins[91] and conjugated nanoparticles to enzymes.[92] However, these chromophore bioconjugates do not involve direct intramolecular electron transfer and are beyond the scope of this review.

Over the last 20 years, Millett *et al.*[93–95] have employed the use of ruthenium(II) bipyridine complexes (electron donor chromophore) conjugated to various cytochromes for electron transfer studies between redox proteins which are part of the electron transport chain. The main enzymes studied include cytochrome *c*, cytochrome bc_1, cytochrome *c* oxidase and cytochrome *c* peroxidase. In this section, only a few of these studies will be highlighted.

One of the approaches Pan and co-workers have adopted is the modification of lysine residues of horse heart cytochrome *c* with a Ru(II) bipyridine photoactive chromophore.[96] Modification was achieved by reaction of lysine residues with an *N*-hydroxysuccinimide functionalised bipyridine ligand, followed by treatment with a Ru(II) bipyridine complex (Scheme 15.15). Reaction conditions were 10 mM cytochrome *c* in Tris-HCl at pH 8 with a 4-fold molar excess of NHS-ligand for 2 h at 25 °C. The intermediate bioconjugate was then purified using an ion-exchange column and subsequently reacted with a Ru(II) bipyridine complex in sodium acetate at pH 4 with a 10-fold molar excess for 24 h at 25 °C. The final bioconjugate was further purified using an ion-exchange column, yielding seven singly labelled cytochrome derivatives. The luminescence decay rates of the resulting bioconjugates were found to be between (3.5–21) $\times$ 10^6 s^{-1}, indicating possible electron transfer from the excited triplet state of ruthenium to the ferric heme group of cytochrome.

Another approach undertaken by Geren and co-workers for the modification of yeast iso-1 cytochrome *c* with a Ru(II) bipyridine complex was achieved *via* conjugation of the CYS102 residue with a bromomethyl functionalised ligand (Scheme 15.16).[97] Resulting ligand attachment at CYS102 occurred after reaction of 1 mM iso-1 cytochrome *c* with 5-fold molar excess of ligand in 5% dry DMF, buffered in Tris-HCl at pH 8 for 16 h at room temperature and purification by cation exchange chromatography. Electron transfer studies of the Ru-102-cyt *c* bioconjugate was characterised by luminescence emission and was found to have a decay rate of 5.2 $\times$ 10^6 s^{-1}.

An analogous functionalisation technique of cytochrome *c* is the conjugation of photoactive ligand with histidine residues through a metal–ligand coordination complex.[98–100]

Gray *et al.*[98] have prepared histidine functionalised horse heart cytochrome *c* with a pentaammineruthenium(III) complex ($[Ru(NH_3)_5(H_2O)]^{3+}$). Histidine 33 was functionalised by reaction with 50-fold molar excess of ligand, buffered in

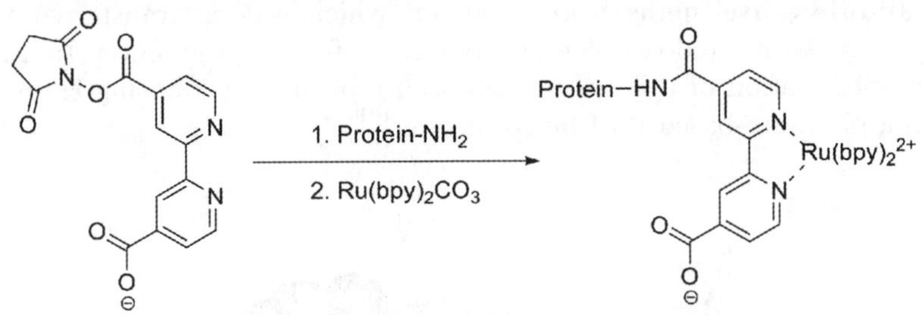

Scheme 15.15 Bioconjugation of cytochrome *c* lysine residues.[25]

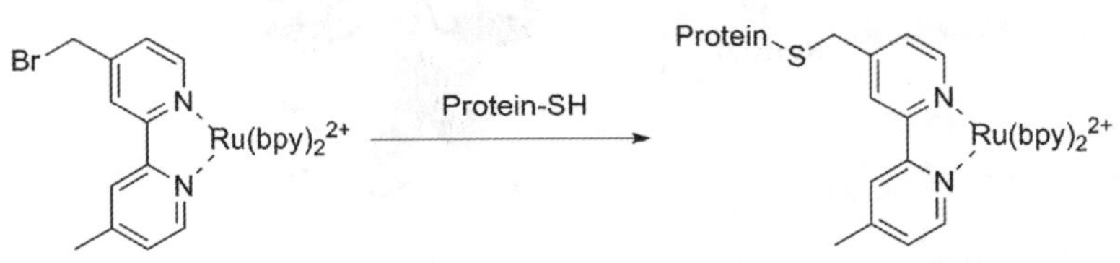

Scheme 15.16 Bioconjugation of cytochrome *c* cysteine 102 (CYS102) residue.[97]

0.2 mM phosphate buffer at pH 7 under argon at room temperature for up to 72 h. Winkler *et al.*[41] has determined the intramolecular electron transfer rate of the bioconjugate to be $(2 \pm 0.5) \times 10^1$ s^{-1}, independent of temperature between 0–37 °C.

Peterson *et al.*[30,35] have conjugated Ru(II) bisterpyridine chromophores with yeast iso-1 cytochrome *c* creating a light-activated donor–acceptor bioconjugate *via* a long and short chain spacer (Figure 15.3). Bioconjugates were modified using maleimide functionalised ligands with the CYS102 residue through a Michael addition reaction. Typically, bioconjugates were prepared by reacting 10 μm iso-1 cytochrome *c* with 5 to 10-fold molar excess of ligand in 5% acetonitrile, buffered in phosphate/EDTA buffer at pH 7 under physiological temperature (37 °C) for 18 h. Electron transfer rates of resulting short and long chain bioconjugates were 5.95×10^5 and 2.78×10^5 s^{-1}, respectively.

In an elegant approach, Grimme *et al.*[85] have developed a light-harvesting system resulting in chemical transformation producing H_2 (Scheme 15.17). They have utilised a biological electron donor, photosystem I (PS I) found in photosynthetic organisms and conjugated it to a platinum or gold terminal electron acceptor nanoparticle. Conjugation was initiated by formation of [4Fe–4S] clusters in solution by reacting sodium sulphide, ferrous ammonium sulphate and mercaptoethanol. The resulting cluster was then reconstituted into the PsaC subunit of PS I to yield holo-C13G/C33S variant PsaC. PS I was then modified by combining reconstituted PsaC and P700/F_x cores in the presence of PsaD. Finally, 1,6-hexanedithiol was introduced to displace the mercaptoethanol ligand from the [4Fe–4S] cluster and to covalently link PS I and Pt/Au nanoparticle. This resulting PS I photoactive bioconjugate evolves H_2 upon white light illumination using a Xenon arc lamp. The resulting chemical transformation is due to the formation of a charge-separated state when PS I absorbs wavelengths below 700 nm which is then transferred to the nanoparticle with a redox potential favourable for H_2 production. In a recent paper, optimisation of this system has been reported, maximising H_2 production in a platinum-based PS I bioconjugate.[101]

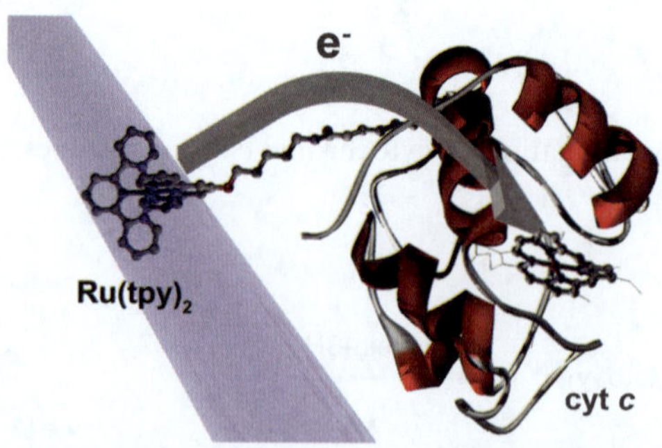

Figure 15.3 Photoinduced electron transfer of Ru(II) bisterpyridine-cytochrome *c* bioconjugate.[35]

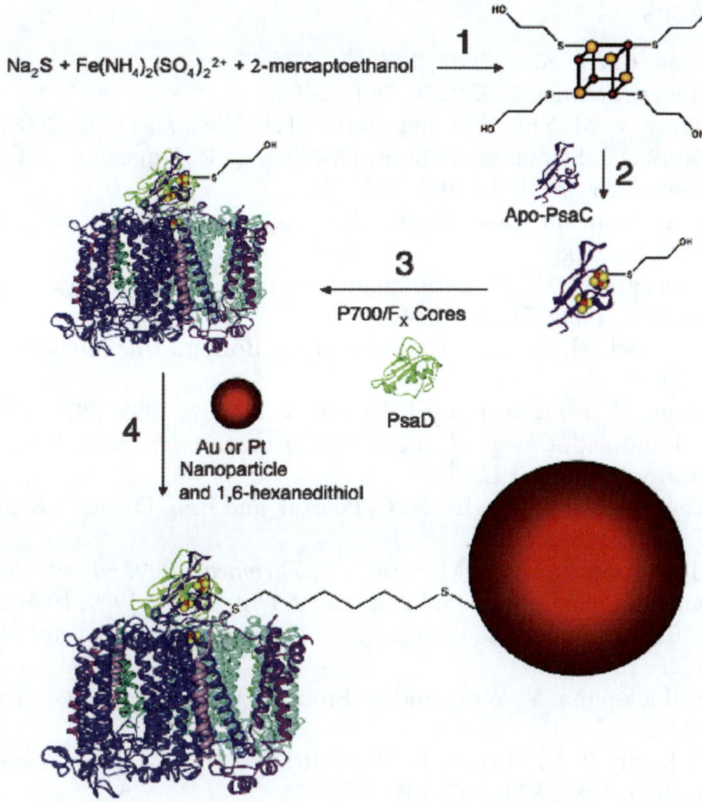

Scheme 15.17 Photosystem I/nanoparticle bioconjugate evolving H_2 upon light-activation. Reprinted with permission from ref. 85. Copyright 2007 American Chemical Society.

15.4 Conclusions

The advent of conjugation techniques developed over the last few decades through classical modification techniques, such as lysine and cysteine chemistry, as well as novel methods, such as introducing non-natural amino acids bearing azides/alkynes, has led to novel chimeric bioconjugates. These chimeric proteins have new properties and functions which otherwise were not present in nature.

As a result, bioconjugates in aqueous solutions or modified on surfaces such as electrodes has led to their diverse applications in biosensors, bio-arrays and bio-fuel cells. Chromophore functionalisation with redox metalloproteins that undergo electron transport, such as cytochrome *c*, allows new applications, and their resulting photoinduced electron transport (electrical generation or chemical transformation) properties makes them suitable candidates for molecular solar fuel cells.

References

1. J. Wang, *Chem. Rev.*, 2008, **108**, 814–825.
2. I. Willner, *Science*, 2002, **298**, 2407–2408.
3. I. Willner, Y.-M. Yan, B. Willner and R. Tel-Vered, *Fuel Cells*, 2009, **9**, 7–24.
4. K. Ulbrich, J. Strohalm, V. Subr, D. Plocova, R. Duncan and B. Rihova, *Macromol. Symp.*, 1996, **103**, 177–192.
5. G. T. Hermanson, *Bioconjugate Techniques*, 2nd edn, Elsevier Inc., San Diego, CA, 2008.
6. P. Thordarson, B. L. Droumaguet and K. Velonia, *Appl. Microbiol. Biotechnol.*, 2006, **73**, 243–254.
7. J. Hentschel, M. G. J. ten Cate and H. G. Börner, *Macromolecules*, 2007, **40**, 9224–9232.
8. C. Wang, R. J. Stewart and J. KopeCek, *Nature*, 1999, **397**, 417–420.
9. R. L. Lundblad, *Chemical reagents for protein modification*, 3rd edn, CRC Press, Boca Raton, FL, 2005.
10. A. Abuchowski, T. Van Es, N. C. Palczuk and F. F. Davis, *J. Biol. Chem.*, 1977, **252**, 3578–3581.
11. F. M. Veronese and M. Morpurgo, *Il Farmaco*, 1999, **54**, 497–516.
12. I. Hamachi, Se, ijlig and i. Shinkai, *Eur. J. Org. Chem.*, 1999, **1999**, 539–549.
13. A. F. Bückmann and G. Carrea, *Adv. Biochem. Eng. Biotechnol.*, 1989, **39**, 97–152.
14. A. F. Bückmann, V. Wray and A. Stocker, *Methods Enzymol.*, 1997, **280**, 360–374.
15. V. T. Kung, P. E. Maxim, R. W. Veltri and F. J. Martin, *Biochim. Biophys. Acta*, 1985, **839**, 105–109.
16. V. Pozsgay, N. E. Vieira and A. Yergey, *Org. Lett.*, 2002, **4**, 3191–3194.
17. C. M. Niemeyer, T. Sano, C. L. Smith and C. R. Cantor, *Nucleic Acids Res.*, 1994, **22**, 5530–5539.
18. D. S. Wilbur, M.-K. Chyan, P. M. Pathare, D. K. Hamlin, M. B. Frownfelter and B. B. Kegley, *Bioconjug. Chem.*, 2000, **11**, 569–583.
19. D. C. Goldstein, P. Thordarson and J. R. Peterson, *Aust. J. Chem.*, 2009, **62**, 1320–1327.
20. F. M. Veronese and G. Pasut, *Drug Discov. Today*, 2005, **10**, 1451–1458.
21. G. Wu, T. L. Ott, D. A. Knabe and F. W. Bazer, *J. Nutr.*, 1999, **129**, 1031–1038.
22. D. Bordo and P. Argos, *J. Mol. Biol.*, 1991, **217**, 721–729.
23. H.-A. Klok, *J. Polym. Sci., Part A: Polym. Chem.*, 2005, **43**, 1–17.
24. O. B. Kinstler, D. N. Brems, S. L. Lauren, A. G. Paige, J. B. Hamburger and M. J. Treuheit, *Pharm. Res.*, 1996, **13**, 996–1002.
25. L. P. Pan, B. Durham, J. Wolinska and F. Millett, *Biochemistry*, 1988, **27**, 7180–7184.
26. J. O. Edwards and R. G. Pearson, *J. Am. Chem. Soc.*, 1962, **84**, 16–24.
27. D. R. W. Hodgson and J. M. Sanderson, *Chem. Soc. Rev.*, 2004, **33**, 422–430.
28. S. Kanaya, S. Kimura, C. Katsuda and M. Ikehara, *Biochem. J.*, 1990, **271**, 59–66.

29. G. D. Brayer and M. E. P. Murphy, *Cytochrome c: A Multidisciplinary Approach*, 1st edn, University Science Books, Sausalito, California, **1996**.
30. J. R. Peterson, T. A. Smith and P. Thordarson, *Org. Biomol. Chem.*, 2010, **8**, 151–162.
31. G. V. Louie and G. D. Brayer, *J. Mol. Biol.*, 1990, **214**, 527–555.
32. C. S. Sevier and C. A. Kaiser, *Nat. Rev. Mol. Cell Biol.*, 2002, **3**, 836–847.
33. J. Futami, H. Tada, M. Seno, S. Ishikami and H. Yamada, *J. Biochem.*, 2000, **128**, 245–250.
34. K. L. Heredia, D. Bontempo, T. Ly, J. T. Byers, S. Halstenberg and H. D. Maynard, *J. Am. Chem. Soc.*, 2005, **127**, 16955–16960.
35. J. R. Peterson, T. A. Smith and P. Thordarson, *Chem. Commun.*, 2007, 1899–1901.
36. W. Vangrysperre, C. Ampe, H. Kerstershilderson and P. Tempst, *Biochem. J.*, 1989, **263**, 195–199.
37. C. J. Britten and M. I. Bird, *Biochim. Biophys. Acta*, 1997, **1334**, 57–64.
38. S. R. Fahnestock, *Biochemistry*, 1975, **14**, 5321–5327.
39. P. Mattsson, T. Pohjalainen and T. Korpela, *Biochim. Biophys. Acta.*, 1992, **1122**, 33–40.
40. D. K. Bhattacharyya, U. Bandyopadhyay and R. K. Banerjee, *J. Biol. Chem.*, 1993, **268**, 22292–22298.
41. J. R. Winkler, D. G. Nocera, K. M. Yocom, E. Bordignon and H. B. Gray, *J. Am. Chem. Soc.*, 1982, **104**, 5798–5800.
42. J. F. Riordan, M. Sokolovs and B. L. Vallee, *J. Am. Chem. Soc.*, 1966, **88**, 4104–4105.
43. O. A. Roholt and D. Pressman, *Methods Enzymol.*, 1972, **25**, 438–449.
44. M. G. D. Oliveira, E. Rogana, J. C. Rosa, B. B. Reinhold, M. H. Andrade, L. J. Greene and M. Maresguia, *J. Biol. Chem.*, 1993, **268**, 26893–26903.
45. G. R. Kunkel, M. Mehrabian and H. G. Martinson, *Mol. Cell. Biochem.*, 1981, **34**, 3–13.
46. T. E. Huntley and P. Strittmatter, *J. Biol. Chem.*, 1972, **247**, 4648–4653.
47. S. D. Tilley and M. B. Francis, *J. Am. Chem. Soc.*, 2006, **128**, 1080–1081.
48. K. N. Lee, P. J. Birckbichler and M. K. Patterson, *Clin. Chem.*, 1988, **34**, 906–910.
49. A. Fontana, B. Spolaore, A. Mero and F. M. Veronese, *PEGylated Protein Drugs: Basic Science and Clinical Applications*, Birkhäuser Basel, Basel, Switzerland, 2009.
50. A. J. de Graaf, M. Kooijman, W. E. Hennink and E. Mastrobattista, *Bioconjug. Chem.*, 2009, **20**, 1281–1295.
51. T. L. Hendrickson, V. de Crécy-Lagard and P. Schimmel, *Annu. Rev. Biochem.*, 2004, **73**, 147–176.
52. K. L. Kiick, R. Weberskirch and D. A. Tirrell, *FEBS Lett.*, 2001, **502**, 25–30.
53. K. L. Kiick and D. A. Tirrell, *Tetrahedron*, 2000, **56**, 9487–9493.
54. C. Renner, S. Alefelder, J. H. Bae, N. Budisa, R. Huber and L. Moroder, *Angew. Chem., Int. Ed.*, 2001, **40**, 923–925.

55. E. Yoshikawa, M. J. Fournier, T. L. Mason and D. A. Tirrell, *Macromolecules*, 1994, **27**, 5471–5475.
56. Y. Tang and D. A. Tirrell, *J. Am. Chem. Soc.*, 2001, **123**, 11089–11090.
57. Q. Wang, T. R. Chan, R. Hilgraf, V. V. Fokin, K. B. Sharpless and M. G. Finn, *J. Am. Chem. Soc.*, 2003, **125**, 3192–3193.
58. J. H. Bae, S. Alefelder, J. T. Kaiser, R. Friedrich, L. Moroder, R. Huber and N. Budisa, *J. Mol. Biol.*, 2001, **309**, 925–936.
59. C. J. Noren, S. J. Anthony-Cahill, M. C. Griffith and P. G. Schultz, *Science*, 1989, **244**, 182–188.
60. M. W. Nowak, P. C. Kearney, J. R. Sampson, M. E. Saks, C. G. Labarca, S. K. Silverman, W. Zhong, J. Thorson, J. N. Abelson, N. Davidson, P. G. Schultz, D. A. Dougherty and H. A. Lester, *Science*, 1995, **268**, 439–442.
61. M. Sisido and T. Hohsaka, *Appl. Microbiol. Biotechnol.*, 2001, **57**, 274–281.
62. C. Switzer, S. E. Moroney and S. A. Benner, *J. Am. Chem. Soc.*, 1989, **111**, 8322–8323.
63. S. I. van Kasteren, H. B. Kramer, H. H. Jensen, S. J. Campbell, J. Kirkpatrick, N. J. Oldham, D. C. Anthony and B. G. Davis, *Nature*, 2007, **446**, 1105–1109.
64. K. L. Kiick, E. Saxon, D. A. Tirrell and C. R. Bertozzi, *Proc. Natl. Acad. Sci. U. S. A.*, 2002, **99**, 19–24.
65. E. Saxon, J. I. Armstrong and C. R. Bertozzi, *Org. Lett.*, 2000, **2**, 2141–2143.
66. J. A. Opsteen, R. P. Brinkhuis, R. L. M. Teeuwen, D. Lowik and J. C. M. van Hest, *Chem. Commun.*, 2007, 3136–3138.
67. D. J. Vocadlo, H. C. Hang, E.-J. Kim, J. A. Hanover and C. R. Bertozzi, *Proc. Natl. Acad. Sci. U. S. A.*, 2003, **100**, 9116–9121.
68. M. L. Tsao, F. Tian and P. G. Schultz, *ChemBioChem*, 2005, **6**, 2147–2149.
69. L. Wang, Z. W. Zhang, A. Brock and P. G. Schultz, *Proc. Natl. Acad. Sci. U. S. A.*, 2003, **100**, 56–61.
70. T. L. Schlick, Z. Ding, E. W. Kovacs and M. B. Francis, *J. Am. Chem. Soc.*, 2005, **127**, 3718–3723.
71. K. F. Geoghegan and J. G. Stroh, *Bioconjugate Chem.*, 1992, **3**, 138–146.
72. C. Godoy-Alcántar, A. K. Yatsimirsky and J.-M. Lehn, *J. Phys. Org. Chem.*, 2005, **18**, 979–985.
73. J. Shao and J. P. Tam, *J. Am. Chem. Soc.*, 1995, **117**, 3893–3899.
74. R. Bar-Or, L. T. Rael and D. Bar-Or, *Rapid Commun. Mass Spectrom.*, 2008, **22**, 711–716.
75. J. Wang, S. M. Schiller and P. G. Schultz, *Angew. Chem., Int. Ed.*, 2007, **46**, 6849–6851.
76. F. P. Seebeck and J. W. Szostak, *J. Am. Chem. Soc.*, 2006, **128**, 7150–7151.
77. L. A. Marcaurelle and C. R. Bertozzi, *Chem.–Eur. J.*, 1999, **5**, 1384–1390.
78. U. Hanefeld, L. Gardossi and E. Magner, *Chem. Soc. Rev.*, 2009, **38**, 453–468.
79. H. Sarah, C. Jakki and M. Edmond, *Angew. Chem., Int. Ed.*, 2008, **47**, 8582–8594.

80. A. Heller, *Acc. Chem. Res.*, 1990, **23**, 128–134.
81. L. Deng, Y. Liu, G. Yang, L. Shang, D. Wen, F. Wang, Z. Xu and S. Dong, *Biomacromolecules*, 2007, **8**, 2063–2071.
82. D. Kyprianou, A. R. Guerreiro, I. Chianella, E. V. Piletska, S. A. Fowler, K. Karim, M. J. Whitcombe, A. P. F. Turner and S. A. Piletsky, *Biosens. Bioelectron.*, 2009, **24**, 1365–1371.
83. E. V. Piletska, S. A. Piletsky, S. Subrahmanyam, K. Karim and A. P. F. Turner, *Polymer*, 2001, **42**, 3603–3608.
84. H. A. Heering, F. G. M. Wiertz, C. Dekker and S. de Vries, *J. Am. Chem. Soc.*, 2004, **126**, 11103–11112.
85. R. A. Grimme, C. E. Lubner, D. A. Bryant and J. H. Golbeck, *J. Am. Chem. Soc.*, 2008, **130**, 6308–6309.
86. W. Yang, P. Thordarson, J. J. Gooding, S. P. Ringer and F. Braet, *Nanotechnology*, 2007, **18**.
87. J. Wang and J. A. Carlisle, *Diamond Relat. Mater.*, 2006, **15**, 279–284.
88. K. A. Kilian, T. Böcking, K. Gaus, M. Gal and J. J. Gooding, *ACS Nano*, 2007, **1**, 355–361.
89. N. Mano, F. Mao and A. Heller, *J. Am. Chem. Soc.*, 2003, **125**, 6588–6594.
90. V. Lapienc, F. Kukolka, K. Kiko, A. Arndt and C. M. Niemeyer, *Bioconjug. Chem.*, 2010, **21**, 921–927.
91. R. A. Miller, N. Stephanopoulos, J. M. McFarland, A. S. Rosko, P. L. Geissler and M. B. Francis, *J. Am. Chem. Soc.*, 2010, **132**, 6068–6074.
92. Q. Huo, *Colloids Surf., B*, 2007, **59**, 1–10.
93. F. Millett and B. Durham, *Biochemistry*, 2002, **41**, 11315–11324.
94. L. Geren, B. Durham and F. Millett, *Methods Enzymol.*, 2009, **456**, 507–520.
95. F. Millett and B. Durham, *Methods Enzymol.*, 2009, **456**, 95–109.
96. L. P. Pan, B. Durham, J. Wolinska and F. Millett, *Biochemistry*, 1988, **27**, 7180–7184.
97. L. Geren, S. Hahm, B. Durham and F. Millett, *Biochemistry*, 1991, **30**, 9450–9457.
98. K. M. Yocom, J. B. Shelton, J. R. Shelton, W. A. Schroeder, G. Worosila, S. S. Isied, E. Bordignon and H. B. Gray, *Proc. Natl. Acad. Sci. U. S. A.*, 1982, **79**, 7052–7055.
99. X.-J. Yang, F. Drepper, B. Wu, W.-H. Sun, W. Haehnel and C. Janiak, *Dalton Trans.*, 2005, 256–267.
100. S. S. Isied, G. Worosila and S. J. Atherton, *J. Am. Chem. Soc.*, 1982, **104**, 7659–7661.
101. R. A. Grimme, C. E. Lubner and J. H. Golbeck, *Dalton Trans.*, 2009, 10106–10113.

CHAPTER 16

Synthetic Photo-catalytic Proteins – a Model of Photosystem II

BRENDON CONLAN,* WARWICK HILLIER AND TOM WYDRZYNSKI

Research School of Biology, Australian National University, Canberra 0200, Australia

16.1 Introduction

In March, 2011, the world population reached 7 billion, whereas approximately 200 years ago the world population was only about 1 billion. This rapid increase in the world population is creating a high energy demand to power our energy-intensive lifestyles. Consequently, global energy consumption is expected to rise dramatically and more than double within the next 50 years. At present, approximately 80% of the global energy consumption comes from the combustion of fossil fuels. However, the extensive use of fossil fuels is beginning to have an impact on the environment. In 2008 alone, 10 880 million metric tons of CO_2 were released into the atmosphere by world petroleum consumption and another 13 300 million metric tons by world coal consumption.[1] This release of vast quantities of CO_2 has contributed to the highest atmospheric CO_2 levels since some 10 million years ago. In addition the combustion of fossil fuels also releases other pollutants into the atmosphere,

*Present address: Department of Biochemistry, La Trobe University, Melbourne, Australia

RSC Energy and Environment Series No. 5
Molecular Solar Fuels
Edited by Thomas J. Wydrzynski and Warwick Hillier

Published by the Royal Society of Chemistry, www.rsc.org

such as CO, NO_x, SO_x, C_xH_x, soot, ash, tar droplets, and various organic compounds all of which may have detrimental effects on the environment.[2]

In contrast, solar energy could provide a pollution-free, viable alternative to fossil fuels. Approximately 1361 W m^{-2} of light energy is incident on the planet surface,[3] which is more than enough to meet global needs. However, due to the diurnal cycle of the sun, an efficient storage mechanism is necessary in order to fully exploit solar energy. This can readily be accomplished by storing the energy in the chemical bonds of molecular solar fuels.[4]

Molecular hydrogen (H_2) is an ideal chemical fuel, since it is carbon neutral and is available in significant quantities. Although the earth's atmosphere is composed of only about 0.07% hydrogen and the surface about 0.14%,[4] hydrogen is virtually limitless in the form of water in the earth's oceans.

The splitting of water into O_2 and H_2 can readily be achieved through electrolysis. However, the electrolysis of water typically requires large activation or overpotential energies such that more energy is put in to the reaction than is stored chemically. Yet Nature has pointed the way to split water. About two and a half billion years ago primitive bacteria developed the ability to oxidize water into O_2, protons and electrons using the energy from sunlight.[5] This process, known as oxygenic photosynthesis, is driven by photosystem II (PSII), the light activated oxido-reductase found in green plants, algae and cyanobacteria (Figure 16.1).[6] This enzyme utilizes sunlight to carry out the most energetically demanding reaction in biology – the formation of the O–O bond. PSII has only evolved once in Nature and is one of the most elaborate nanoscale biological machines.

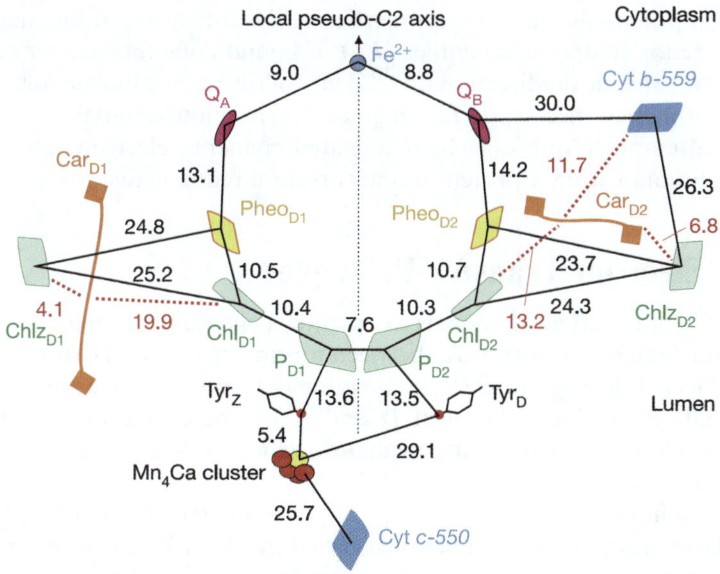

Figure 16.1 Photosystem II is a complex enzyme with many cofactors. Figure adapted from Loll *et al.*[62]

Understanding oxygenic photosynthesis and exploiting the ability to directly convert sunlight and water into clean H_2 fuel is a research field of growing significance around the world. The initial charge separation events in PSII have a remarkable efficiency of approximately 90%, making it extremely attractive to artificially mimic this process.[7–8] The first step towards this goal is to replicate PSII function in a well-defined, robust supporting matrix. The following sections outline the photosynthetic capture of energy and describe a reverse engineering approach for developing PSII model proteins.

16.2 Design of Photo-catalytic Proteins

The PSII enzyme in higher plants is a membrane-based complex made up of at least 27 protein subunits and numerous cofactors involved in electron transfer reactions and several hundred pigments for light harvesting.[6] The large size and complexity of this enzyme makes it difficult to work with and any mimic of this protein must by definition be simpler. Development of a minimalist photo-catalytic structural unit requires that only the essential redox cofactors are engineered into a simple framework. The essential cofactors for the function of PSII include: a bound plastoquinone electron acceptor; a high potential photo-oxidizable chlorophyll (Chl) complex; a redox-active tyrosine; and a tetra-nuclear manganese/calcium cluster (Mn_4O_xCa), *i.e.* quinone–Chl complex–tyrosine–manganese cluster.

The cofactors present in PSII are placed at specific positions in space by the protein framework. The amino acid residues responsible for binding these cofactors function not only as a scaffold to hold the cofactors in place, but to tune redox potentials and other properties of the cofactors. In the majority of enzymes, redox midpoint potentials (E_m) of bound cofactors are set such that electron transfer is unidirectional.[9–10] The following section details electron transfer principles necessary for engineering a photo-catalytic system and previous attempts to introduce light activated pigments, electron acceptors and electron donors within a protein framework in a redox active form.

16.3 Electron Transfer Principles

In order for light activated systems to operate, a charge separated state must be created through electron transfer between a primary donor (**D**) and a primary acceptor (**A**) molecule. In PSII, a functional reaction centre minimizes the recombination of charge between **D** and **A** by incorporation of additional secondary electron transfer intermediates, which separate the charges by a greater distance.

Intraprotein electron transfer is typically rate-limited by tunnelling through the insulating protein medium between the edges of the **D** and **A** redox centres. This electron tunnelling is reasonably accurately described by a simple exponential decay with distance.[9,11] The intervening protein medium also has some role to play in electron tunnelling rates[12–15] but more important than packing density is that of driving force. This is estimated from the difference in the redox midpoint potentials between donor and acceptor, assuming that redox centres are not too close. One other major factor that affects tunnelling rates is that of reorganization energy. If one or both donor or acceptor complexes is solvent exposed then the reorganization energy can become quite large (>1 eV).

Natural proteins with productive electron transfer reactions have been found to have cofactors separated by no more than ~14 Å.[9,16,17] This provides an upper distance for localising redox cofactors. It is also important that cofactors are not placed too close to each other, so as to prevent energy wasting back reactions. Therefore, when engineering a light activated electron transport chain it is wise to place all cofactors within 10–14 Å.

For the purposes of engineering photo-active proteins, aside from changing the cofactor distance, altering the driving force is probably the most tangible method of tuning electron transfer reactions in a protein. Unless engineering membrane associated proteins, the redox centres are likely to be partially solvent exposed resulting in large reorganization energies. As such a highly oxidizing pigment is needed to provide sufficient driving force for electron transfer. It is possible to increase or decrease the oxidation potential of **D** and **A** centres in proteins by varying the protein environment. Large changes in redox potentials for pigments and cofactors can be achieved by introducing or removing hydrogen bonds from the inner sphere ligands of the cofactors by changing single amino acids in a protein sequence.[18,19] Membrane based proteins are ideal for engineering redox reactions with low reorganization energies due to **D** and **A** molecules being buried far from the polar solvent, but these proteins are challenging to work with. The majority of light activated proteins engineered to date are small soluble four-helix bundles.[20–24] In principle, these systems are easier to synthesize and assemble, but carry the burden that engineered donor and/or acceptor groups would be solvent exposed and require high driving force to imitate electron transfer reactions.

16.4 Light Activated Pigments

Attempts at producing mimics of PSII have centred on the introduction of a pigment into a peptide matrix. Efforts to bind Chl, which is found in all natural reaction centres including PSII, have met with limited success. Chl is challenging to work with due to its long phytol tail, which makes the molecule very hydrophobic and not amenable for binding to smaller soluble proteins in solution. There are relatively few studies on the incorporation of Chl into non-photosynthetic proteins, although recently a *de novo* amphiphilic protein containing a lipophilic domain was found to bind a Ni-BChl molecule.[25]

A simple way to produce a light activated pigment bound to a protein matrix is to chemically modify a covalently attached heme, such as those found in the *b*- and *c*-type cytochromes. Removal of the heme iron atom using hydrofluoric acid and replacement with zinc produces the light active zinc protoporphyrin (ZnPP).[26] The *b*-type cytochromes have also been used successfully to bind light activated pigments. The heme group in these proteins is not covalently linked to the protein, enabling removal from the binding pocket[27] and replacement with a light active porphyrin.[28–30] Upon light excitation, ZnPP forms a triplet state with high efficiency which is long lived ($\tau \sim 14$ ms).[31,32] These electron transfer reactions are quite slow 10^3–10^5 s^{-1} despite donor/acceptor distances of 8–10Å.[29,33] A major drawback of utilizing ZnPP is that the long lived triplet states have a tendency to produce reactive singlet oxygen $^1O_2$ species. By utilizing *b*-type cytochromes, porphyrins that primarily donate an electron from the singlet state can be bound to the heme pocket. Zinc chlorin e_6 ($ZnCe_6$) has been used which is very similar to heme *b* except for changes in the length of the carbon chains terminating in carboxylates and also addition of an extra carboxylic acid. This makes the pigment more soluble and easier to handle whilst maintaining a high binding affinity for the heme site. $ZnCe_6$ is also utilized as a photoactive pigment because of its strong Q_y absorption band in the red ($\sim$638 nm), and high estimated oxidation potential of $\sim +1.1$ eV.[34]

Razeghifard and Wydrzynski found that $ZnCe_6$ could be bound in place of heme in a *de novo* 4-helix bundle protein with two histidine residues acting as axial ligands. Upon replacing one histidine with phenylalanine, only half the amount of $ZnCe_6$ was bound, and when both histidines were replaced with phenylalanine, $ZnCe_6$ binding was completely ablated. The bound $ZnCe_6$ could undergo some photochemistry after repetitive light flash illumination in the presence of an external quinone acceptor, though with low yield.[35] Light activated electron transfer has also been demonstrated with $ZnCe_6$ bound within a *de novo* protein to a quinone electron acceptor in solution.[36] Flavin molecules can be also be utilized to carry out photo-initiated electron transfer within proteins.[37–39]

16.5 Electron Acceptors

The use of electron acceptors present in solution have major drawbacks in that they must be present in high concentration and the reorganization energies associated with electron transfer are large producing low yields of electron transfer. In order to optimize electron transfer reactions between cofactors it is ideal to bury the cofactors within the protein matrix. Quinones have been utilized extensively as electron acceptors due to their efficiency in transferring charge in natural systems. Quinones in electron transport chains are ideally suited as electron acceptors with redox potentials ranging from -130 mV up to $+250$ mV.[40–42]

In PSII the first stable quinone electron acceptor is a non-covalently bound plastoquinone molecule (Q_A). Engineering a binding site for a plastoquinone molecule in a protein is difficult and in artificial reaction centres covalent attachment is more practical. A quinone will covalently link to a free sulfhydryl group (–SH), such as that found in cysteine, through a sulfur addition reaction.[43] Hay *et al.* were the first group to make use of this novel addition reaction, and by modifying cytochrome b_{562} placed a cysteine residue in the interior of the protein and covalently bound a quinone molecule within ~11 Å of a bound heme group. The heme was removed and replaced with the light active $ZnCe_6$. Light induced electron transfer from the bound pigment to the quinone was achieved with approximately 20% efficiency.[34]

16.6 Electron Donors

In order for PSII to function, a tyrosine molecule positioned near the P680 complex is oxidized donating an electron to $P680^+$. The tyrosine radical is then reduced by an electron taken from the Mn_4O_xCa complex. In producing a mimic of PSII it may be advantageous to involve a tyrosine residue to mediate the donation of electrons from the metal centre. The Gray laboratory, in their work on metal-modified redox proteins, have produced rhenium-based photosensitizers capable of generating tyrosyl radicals.[44–46]

It is difficult to engineer a protein that binds both a light activated group and an electron donating metal centre. When looking for examples of light activated oxidation of a metal centre the vast majority of studies involve the use of ruthenium and rhenium based photosensitizer's covalently attached to the surface of proteins.[47] These photosensitizers achieve extremely high driving forces and have been shown to oxidize a single bound metal ion.[48]

Utilizing larger, natural enzymes is a good way to introduce multiple cofactors and produce a light activated electron transport chain. The key advantage here is that the proteins selected can already contain existing features that facilitate the redesign of the protein. An example of reengineering natural systems is the modification of the purple bacterial reaction centre (RC) for PSII function. A reaction centre protein is a biological solution for devising physical charge separation between a donor–acceptor pair. The oxidation potential of the purple bacterial RC utilizes a bacteriochlorophyll dimer, P870, is quite low (~500 mV) and mutations to introduce hydrogen bonds to this centre increased the oxidation potential to a level approaching the potential of PSII.[49] An oxidizable tyrosine was added[50,51] and carboxylate residues were introduced to produce a manganese binding site in a position homologous to the manganese centre of PSII,[52] creating a monomeric manganese binding site. Upon illumination, the bound Mn was able to reduce $P870^+$ in a first order reaction.[53]

16.7 Engineering Bacterioferritin as a 'Reaction Centre'

The sheer complexity of the PSII enzyme has limited attempts at reverse engineering a synthetic water oxidase.[54] Previous efforts have concentrated on constructing biological motifs that mimic one aspect of the enzyme's mechanism, such as porphyrin excitation/oxidation[34] or metal binding/complex formation.[52] These attempts all faced limitations, either through use of an overly complex system which is difficult to work with, or due to use of small peptides which can't bind multiple cofactors. Thus, a new approach was needed to meet key design requirements in the production of an artificial photoactive protein with multiple cofactors. A naturally occurring protein scaffold was used to bind several introduced redox active cofactors.[55] The protein utilized contained existing useful architecture so that cofactors could be placed within appropriate electron transfer distances of each other. The physiological limitations of intra-protein electron transfer meant that all cofactors needed to be placed within <14 Å of each other.[9,16,55]

In order to mimic the reactions of PSII, a modified bacterioferritin (BFR) protein was used to bind several redox active cofactors and initiate light-driven electron transfer. The native BFR protein forms into a homodimer as the basic unit, binding a single heme group symmetrically at the interface between the two protein monomers.[56] BFR is a soluble 4-helix bundle protein ($\sim$18.5 kDa) which naturally contains a di-iron metal site and a heme *b* binding site.[57] *In vitro*, the BFR protein self-assembles as an oligomeric unit made up of 12 homodimeric units that form a large spherical shaped shell which encloses a hollow cavity with a diameter of $\sim$8 nm.[56] The cavity normally stores iron as ferric oxide, hydroxide or phosphate minerals and is connected to the outer protein surface by channels. The natural BFR protein shell binds 12 heme groups and 48 iron atoms.[56] However, functionally the heme group can be removed[27] and replaced with other similar structured but light active porphyrins.[58] Each monomer of bacterioferritin contains a binuclear metal binding site that links the four helices of the protein together and acts as a ferroxidase site, in which Fe^{II} is oxidized to Fe^{III} by O_2.[59] The organization of the di-iron binding site is very similar to other class II di-iron proteins. The crystal structure for BFR was initially produced with two Mn^{II} ions bound in the metal binding site (Figure 16.2).[56] Removal of the native cofactors and replacement with manganese in the metal centre and a light active chlorophyll analogue – zinc chlorin – within the heme binding site produce a protein with light dependent catalytic properties.

BFR presents an attractive starting point for engineering a multi-step, light activated protein as: (i) it is a highly stable and robust protein, (ii) the heme can be removed[27] and replaced with a photoactive chlorin;[34] (iii) the di-nuclear metal site has similar structural ligation as found in manganese catalase, where the di-manganese site can reach an oxidation state of $Mn_2^{III,IV}$;[60,61] and (iv) the cofactors when bound to BFR are separated by similar distances to that found in PSII.[62] Bacterioferritin was modified to produce a minimalist model of the PSII donor side in which analogues of PSII cofactors were bound. These included a photo-oxidizable chlorin, a redox active tyrosine, and a di-nuclear manganese centre (Figure 16.3).

Manganese was found to ligate to the di-nuclear metal binding site of bacterioferritin.[55] BFR bound $\sim$2 Mn ions per monomer with one site binding strongly and a second weaker binding site. It was also determined that 0.82 $ZnCe_6$ molecules bound per BFR dimer. The oxidation states of the Mn centre were characterized by EPR studies. Simulation of spectra for BFR in the dark strongly suggested that manganese ions bound to the BFR protein remained in

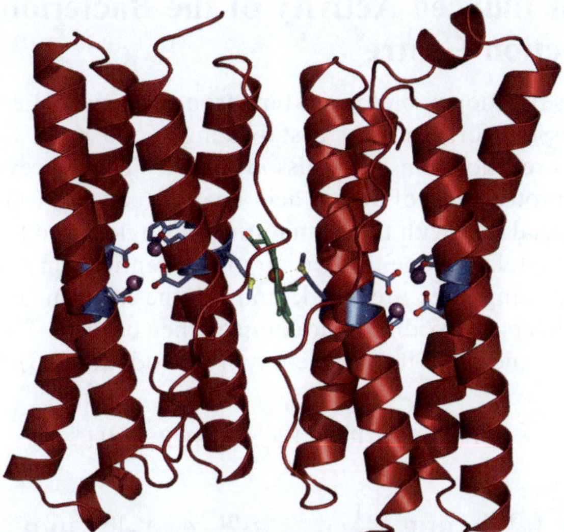

Figure 16.2 The BFR homodimer with two identical subunits each hosting dinuclear metal centres (purple spheres) and a heme (green) which is bound in a hydrophobic pocket at the subunit interface. Two manganese atoms bind per protein subunit. Adapted from the crystal structure PDB file 1BCF.

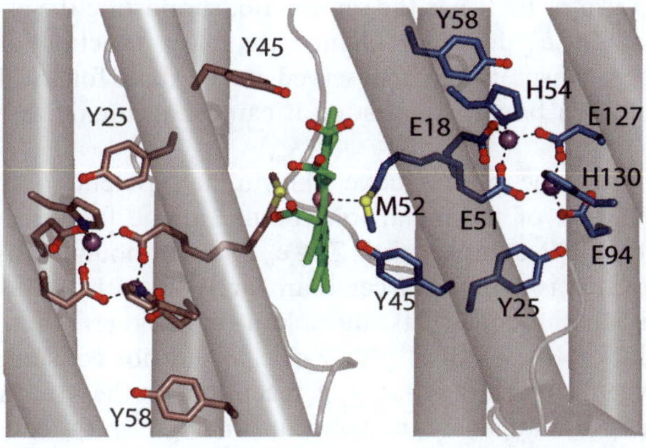

Figure 16.3 Diagram of the amino acid ligands to the dimanganese centre with $ZnCe_6$ bound at the interface of the BFR dimeric complex. Residues from the A subunit are in blue and the B subunit are in pink. The three tyrosines have been labelled on either side of the $ZnCe_6$ group to show the orientation of the dimer subunits relative to each other. The alpha helices are pictured as grey rods. The picture is modified from the PDB file 1BFR.

the 2+ oxidation state and were present as a weakly spin-coupled $Mn_2^{II,II}$ centre.[63] The estimated Mn–Mn distance from EPR simulations was ~4 Å, a distance agreeing with that obtained in the crystal structure.[56]

16.8 Light Induced Activity of the Bacterioferritin 'Reaction Centre'

Bacterioferritins function in biology to bind iron and oxidize Fe_2^{II} to Fe_2^{III}. This process involves the formation of a stabilizing μ-oxo bond between the two metal ions. This reaction consumes dissolved O_2.[64] The process involves metal binding and deprotonation of amino acid ligands (eqn (1)) followed by oxidation of bound metal ions with the formation of a single μ-oxo bond between the metal ions and uptake of one oxygen atom per centre (eqn (2)). Once all ferroxidase iron binding sites are filled, and all the 48 iron atoms bound per dodecamer of dimers are oxidized, the complex then begins to oxidize excess Fe^{II} within the sphere and store it as ferric-oxy-hydroxide (eqn (3)).[65]

$$2Fe^{II} + BFR \rightarrow Fe_2^{II,II} - BFR + 4H^+ \tag{1}$$

$$Fe_2^{II,II} - BFR + \frac{1}{2}O_2 \rightarrow Fe^{III} - O - Fe^{III} - BFR \tag{2}$$

$$4Fe^{II} + O_2 + 6H_20 \rightarrow 4Fe^{III}O(OH)(core) + 8H^+ \tag{3}$$

The BFR protein with iron present consumes 0.25 O_2 molecules per iron atom oxidized. This agrees with the literature values for wild type BFR. The $ZnCe_6$ bound form of BFR has the same ratio, supporting the assumption that the binding of $ZnCe_6$ does not change the basic function of the protein. Minimal oxygen consumption is observed in the dark for the BFR complex with $ZnCe_6$ and Mn bound, suggesting it cannot auto-oxidize Mn^{II} to Mn^{III} without light.

The addition of the photo-active chlorin had potential to induce light dependent oxidation of the metal ions bound at the ferroxidase site. Upon illumination, the BFR complex with $ZnCe_6$ and Mn bound consumed ~2 O_2 per Mn. This value is 4 times larger than for BFR with Fe bound and may suggest the di-Mn site constructs multiple μ-oxo bridges. Control samples – samples that lack Mn *i.e.*, BFR plus $ZnCe_6$ – did not consume a significant amount of oxygen, confirming that O_2 is reacting with the Mn. In samples with a fivefold excess of manganese, a significant increase in O_2 consumption was observed. This likely indicates that 'free' Mn^{2+} is oxidized within the multimeric sphere structure forming Mn^{III}-oxy-hydroxide.

The BFR complex produced a tyrosine radical upon illumination which was only present when Mn was bound to the protein. The most likely mechanism for tyrosine oxidation is electron hole migration from the oxidized $ZnCe_6^+$ radical to a nearby tyrosine residue. $ZnCe_6$ has a high estimated oxidation potential of ~ +1.1V,[34] which is sufficient to oxidize a tyrosine. It is unclear why tyrosine oxidation does not occur when no manganese ions are bound. When Mn (or any metal) binds to the BFR metal site, deprotonation of the metal ligands occurs along with structural changes to the ligands themselves.

These changes may alter the hydrogen bonding network around the tyrosine allowing proton coupled electron transfer.

There are three tyrosine residues in BFR (Y25, Y58, and Y45) that are close to the $ZnCe_6$ (10.6 Å, 11.0 Å, and 4.2 Å respectively) and could potentially act as electron donors (Figure 16.3). Tyrosine 45 rests within van der Waals bonding distance of the $ZnCe_6$ and makes up part of the heme binding pocket. It is unlikely that this tyrosine would be affected by Mn binding and as such is not likely to be the electron donating residue. Tyrosines 25 and 58, found at the metal binding site, make up part of the hydrogen bonding network of the site. Of these two residues, only Y25 is highly conserved in ferritins and bacterioferritins.[66] Tyrosine 25 is hydrogen bonded to glutamate 94, which is one of the metal ligands, and the oxygen atom of this tyrosine is placed 4.3 Å from the nearest Mn ion. Studies suggest that tyrosine (Y25) is likely to be the redox active residue in this system.[67]

Illumination of the BFR complex containing $ZnCe_6$ and Mn produced oxidation of the metal centre with the loss of Mn^{II} and the appearance of a Mn^{III} signal as measured by EPR. No intermediate state ($Mn_2^{II,III}$) was observed, so the second oxidation state may not require light excitation of the chlorin. The Mn^{III} signal was only observed in samples with excess Mn, suggesting the signal comes from a new Mn species unlikely to be bound to BFR. Comparing this to the native function of ferritins, it is tempting to assign the species to Mn^{III}-oxy-hydroxide [Mn^{III}(O)OH] which is accumulated in the core of the sphere structure.[68]

The bacterioferritin homodimer studies up to this point were limited in that they lacked a bound electron acceptor. Molecular engineering of BFR synthesised a new BFR heterodimer construct. The BFR gene was modified to include a 22 amino acid linker joining two monomeric subunits to produce a heterodimer (*i.e.*, BFR-L-BFR). The linker motif joined the C-terminus of one subunit to the N-terminus of the second subunit. This linker allowed one half of the dimer to be modified without changing the other half. The metal centre was intentionally removed in one half of the dimer and a quinone was covalently bound within this site. The heterodimer protein, when assembled, had a redox active metal centre and tyrosine on one side, a photo-oxidizable porphyrin centrally located, and a protein bound quinone electron acceptor on the other side. This linearized electron transfer pathway is shown in Figure 16.4. Overlaying the new cofactors onto the wild type BFR crystal structure and modelling the changes associated with binding produces a set of cofactors spaced as per Figure 16.5. This model system demonstrated porphyrin excitation, quinone reduction and metal centre oxidation. This protein was an improved mimic of PSII that produced electron transfer along a chain of redox cofactors culminating in the reduction of a quinone and oxidation of a Mn centre. The analysis of this complex was relatively simple as each individual cofactor could be introduced systematically, allowing the characterization of its interactions with the protein environment.

The heterodimer protein bound Mn as a weakly interacting $Mn_2^{II,II}$ dimer but with weaker binding than found in the BFR homodimer.[63] The BFR

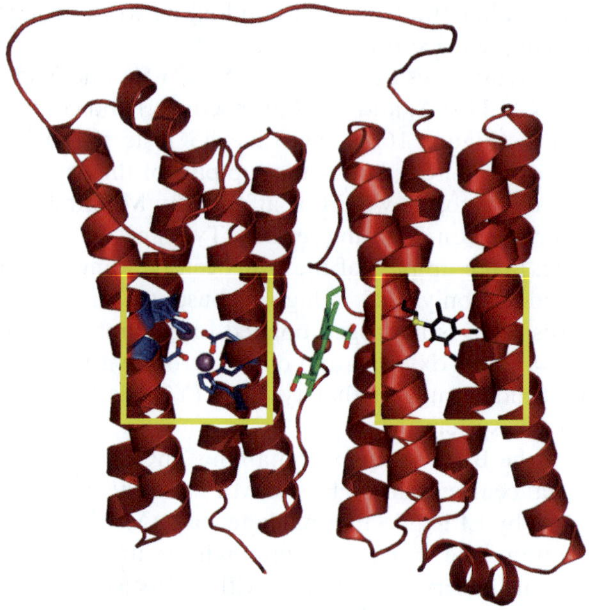

Figure 16.4 Bacterioferritin heterodimer (BFR-H2) with one subunit of the dimer changed relative to the other. This heterodimer contains a complete electron transport chain, with a light active $ZnCe_6$ molecule, a covalently attached quinone electron acceptor, and a di-manganese centre as electron donor. Molecular mechanics optimization of the structure was carried out using a Polak-Ribiere algorithm (conjugate gradient) on the original dimer structure with the 22 amino acid linker region incorporated.[76]

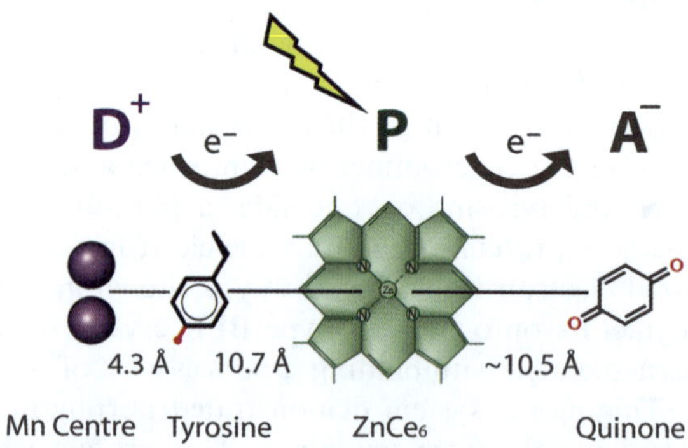

Figure 16.5 Diagram showing the relative distance of separation of all of the cofactors bound to the BFR heterodimer protein. Distances are given as edge to edge for the conjugated systems and metal ions.

heterodimer bound approximately one $ZnCe_6$ molecule and one 2,3-dimethoxy-5-methyl-1,4-benzoquinone (CoQ0) per dimer. The BFR heterodimer was found to behave similarly to BFR. However, the addition of a quinone electron acceptor was found to significantly increase uptake of oxygen. This increase is likely due to improvements in the efficiency of transferring the electron away from the $ZnCe_6$ pigment resulting in a longer lived charge-separated state.

Comparison of the steady-state fluorescence spectra of the BFR heterodimer with CoQ_0 bound indicates approximately 20% efficiency of charge transfer. The quenching of fluorescence as a result of electron transfer to an oxidized quinone is very similar to that reported by Hay *et al.* for the cyt b_{562} quino-protein with $ZnCe_6$ bound.[34] Measurements of the fluorescence lifetime quenching gave very similar efficiency values to that found for the steady state measurements (unpublished results).

16.9 Conclusions and Future Work

The goal for artificial photosynthetic complexes is to capture light energy and transiently store chemical oxidation potential before utilizing them for work. In the PSII system this photochemistry happens efficiently and is coupled to the oxidation of water, which itself is a remarkably thermodynamically efficient reaction. To this end, a homodimer of bacterioferritin was engineered to emulate the PSII reaction centre light induced electron transfer from a di-manganese complex to a photo-oxidized pigment. The protein complex was capable of using light energy to drive photochemical charge separation and couple the oxidizing potential to multiple oxidation events at the bound Mn_2^{II} centre. This work produced the first bioengineered protein capable of photo-catalytic, multi-electron oxidation of a di-manganese centre.

To improve upon the design of the bacterioferritin system, a bacterioferritin heterodimer construct was designed which incorporated both donor and acceptor groups. PSII is made up of a heterodimeric protein core that binds the multiple cofactors necessary to create a light dependent charge separated state. The modified bacterioferritin is also a heterodimer with multiple cofactors capable of forming such a state. The heterodimer bound a light active pigment which could transfer an electron to a quinone electron acceptor bound to the protein complex and subsequently oxidize a Mn centre. The heterodimer protein is one of the first successful attempts in engineering a linear electron transport chain containing a number of the cofactors found in PSII. The quinone-protein form of the heterodimer appears to carry out light induced electron transfer from the chlorin to the quinone with approximately 20% efficiency.

Limited attempts have been made in the generation of a protein-bound manganese centre that acts as an electron donor species. However numerous synthetic di-manganese compounds have been produced as models of the OEC, of which several are able to catalyze the oxidation of water to molecular oxygen.[69–72] The use of a di-manganese centre in our protein based system could therefore potentially facilitate the oxidation of water. Theoretically, introduction of mutations to the pigment interface of the BFR system could increase hydrogen bonds surrounding the $ZnCe_6$ and further increase the oxidation potential.[73] This could promote oxidation of the di-manganese centre to the higher oxidation states that are necessary for water splitting.[74]

The unique spherical dodecamer of dimers of the BFR model system has a large internal cavity of ~8 Å, which could potentially be used to compartmentalize reactions within the same molecular assembly. This could allow separation of oxidative and reductive processes across a protein interface. It is proposed that the internal cavity of the BFR sphere structure is anaerobic,[75] in which case the addition of a hydrogenase site within the sphere would limit O_2 poisoning of the hydrogenase catalyst. If, for instance, the Mn centre was capable of photo-catalytic oxidation of water this oxidative reaction could be coupled to reduction of protons to hydrogen gas on the inside of the complex.

The ultimate aim of this ongoing research is to produce an artificial catalyst whereby water is efficiently split into molecular oxygen, protons and electrons. This work represents one of the first steps in developing synthetic photocatalytic, 'green enzymes' that utilize light energy to catalyze oxygen evolution and hydrogen production.

Abbreviations

A	electron acceptor
BFR	bacterioferritin
BFR-M	BFR with two Mn ions bound
BFR-Z	BFR with ZnCe6 bound
BFR-ZM	BFR with stoichiometric amounts of ZnCe6 and Mn bound
D	electron donor
EPR	electron paramagnetic resonance
RC	reaction centre
PSII	Photosystem II
Q	quinone
ZnCe6	zinc chlorin e6.

References

1. US energy information administration (www.eia.gov)
2. J. S. Gaffney and N. A. Marley, *Atmos. Environ.*, 2009, **43**, 23.
3. NASA Goddard Space Flight Center: Solar Radiation (http://atmospheres.gsfc.nasa.gov/climate/index.php?section=136)
4. T. R. Cook, D. K. Dogutan, S. Y. Reece, Y. Surendranath, T. S. Teets and D. G. Nocera, *Chem. Rev.*, 2010, **110**, 6474.
5. R. E. Blankenship, *Molecular Mechanisms of Photosynthesis*, Blackwell Science Ltd, Oxford, 2002, pp. 321.
6. T. Wydrzynski and K. Satoh, Photosystem II : the light-driven water: plastoquinone oxidoreductase in *Advances in Photosynthesis and Respiration*, Springer, Dordrecht, 2005, vol. 22, pp. 786.

7. O. Kruse, J. Rupprecht, J. R. Mussgnug, G. C. Dismukes and B. Hankamer, *Photochem. Photobiol. Sci.*, 2005, **4**, 957.
8. N. S. Lewis and D. G. Nocera, *Proc. Natl. Acad. Sci. U. S. A.*, 2006, **103**, 15729.
9. C. C. Page, C. C. Moser, X. X. Chen and P. L. Dutton, *Nature*, 1999, **402**, 47.
10. F. Rabanal, B. R. Gibney, W. F. DeGrado, C. C. Moser and P. L. Dutton, *Inorg. Chim. Acta*, 1996, **243**, 213.
11. C. Moser, J. M. Keske, K. Warncke, R. S. Farid and P. L. Dutton, *Nature*, 1992, **355**, 796.
12. H. B. Gray and J. R. Winkler, *Annu. Rev. Biochem.*, 1996, **65**, 537.
13. J. R. Winkler, A. J. Di Bilio, N. A. Farrow, J. H. Richards and H. B. Gray, *Pure Appl. Chem.*, 1999, **71**, 1753.
14. J. R. Winkler and H. B. Gray, *J. Biol. Inorg. Chem.*, 1997, **2**, 399.
15. H. B. Gray and J. R. Winkler, *Q. Rev. Biophys.*, 2003, **36**, 341.
16. D. Noy, C. C. Moser and P. L. Dutton, *Biochim. Biophys. Acta.*, 2006, **1757**, 90.
17. C. C. Page, C. C. Moser and P. L. Dutton, *Curr. Opin. Chem. Biol.*, 2003, **7**, 551.
18. J. P. Allen and J. C. Williams, *J. Bioenerg. Biomembr.*, 1995, **27**, 275.
19. M. R. Gunner, *Curr. Top. Bioenerg.*, 1991, **16**, 319.
20. M. Chen, L. L. Eggink, J. K. Hoober and A. W. D. Larkum, *J. Am. Chem. Soc.*, 2005, **127**, 2052.
21. L. L. Eggink and J. K. Hoober, *J. Biol. Chem.*, 2000, **275**, 9087.
22. K. A. Meadows, P. S. Parkes-Loach, J. W. Kehoe and P. A. Loach, *Biochemistry*, 1998, **37**, 3411.
23. D. Noy and P. L. Dutton, *Biochemistry*, 2006, **45**, 2103.
24. H. K. Rau, H. Snigula, A. Struck, B. Robert, H. Scheer and W. Haehnel, *Eur. J. Biochem.*, 2001, **268**, 3284.
25. D. Noy, B. M. Discher, I. V. Rubtsov, R. A. Hochstrasser and P. L. Dutton, *Biochemistry*, 2005, **44**, 12344.
26. J. M. Vanderkooi, F. Adar and M. Erecinska, *Eur. J. Biochem.*, 1976, **64**, 381.
27. F. W. J. Teale, *Biochim. Biophys. Acta*, 1959, **35**, 543.
28. S. E. Petersonkennedy, J. L. McGourty and B. M. Hoffman, *J. Am. Chem. Soc.*, 1984, **106**, 5010.
29. K. P. Simolo, G. L. McLendon, M. R. Mauk and A. G. Mauk, *J. Am. Chem. Soc.*, 1984, **106**, 5012.
30. J. L. McGourty, S. E. Petersonkennedy, W. Y. Ruo and B. M. Hoffman, *Biochemistry*, 1987, **26**, 8302.
31. B. P. Sudha, N. Dixit, V. T. Moy and J. M. Vanderkooi, *Biochemistry*, 1984, **23**, 2103.
32. H. Zemel and B. M. Hoffman, *J. Am. Chem. Soc.*, 1981, **103**, 1192.
33. G. L. McLendon, J. R. Winkler, D. G. Nocera, M. R. Mauk, A. G. Mauk and H. B. Gray, *J. Am. Chem. Soc.*, 1985, **107**, 739.
34. S. Hay, B. B. Wallace, T. A. Smith, K. P. Ghiggino and T. Wydrzynski, *Proc. Natl. Acad. Sci. U. S. A.*, 2004, **101**, 17675.

35. A. R. Razeghifard and T. Wydrzynski, *Biochemistry*, 2003, **42**, 1024.
36. A. Mennenga, W. Gartner, W. Lubitz and H. Gorner, *Phys. Chem. Chem. Phys.*, 2006, **8**, 5444.
37. I. Ahmad, M. A. Cusanovich and G. Tollin, *Proc. Natl. Acad. Sci. U. S. A.*, 1981, **78**, 6724.
38. M. B. Twitchett, J. C. Ferrer, P. Siddarth and A. G. Mauk, *J. Am. Chem. Soc.*, 1997, **119**, 435.
39. R. E. Sharp, C. C. Moser, F. Rabanal and P. L. Dutton, *Proc. Natl. Acad. Sci. U. S. A.*, 1998, **95**, 10465.
40. F. Rappaport and B. A. Diner, *Coord. Chem. Rev.*, 2008, **252**, 259.
41. A. Krieger-Liszkay and A. W. Rutherford, *Biochemistry*, 1998, **37**, 17339.
42. W. W. Li, P. Hellwig, M. Ritter and W. Haehnel, *Chem–Eur. J.*, 2006, **12**, 7236.
43. E. Redfearn, *Biochemistry of Quinones*, Academic Press, New York, 1965, pp. 149–181.
44. J. R. Winkler, D. G. Nocera, K. M. Yocom, E. Bordignon and H. B. Gray, *J. Am. Chem. Soc.*, 1982, **104**, 5798.
45. A. J. Di Bilio, B. R. Crane, W. A. Wehbi, C. N. Kiser, M. M. Abu-Omar, R. M. Carlos, J. H. Richards, J. R. Winkler and H. B. Gray, *J. Am. Chem. Soc.*, 2001, **123**, 3181.
46. A. M. Blanco-Rodriguez, M. Busby, C. Gradinaru, B. R. Crane, A. J. Di Bilio, P. Matousek, M. Towrie, B. S. Leigh, J. H. Richards, A. Vlcek and H. B. Gray, *J. Am. Chem. Soc.*, 2006, **128**, 4365.
47. D. G. Whitten, *Acc. Chem. Res.*, 1980, **13**, 83.
48. J. E. Miller, A. J. Di Bilio, W. A. Wehbi, M. T. Green, A. K. Museth, J. R. Richards, J. R. Winkler and H. B. Gray, *Biochim. Biophys. Acta*, 2004, **1655**, 59.
49. J. C. Williams, A. L. M. Haffa, J. L. McCulley, N. W. Woodbury and J. P. Allen, *Biochemistry*, 2001, **40**, 15403.
50. B. A. Diner and R. D. Britt, in *Photosystem II-The Light Driven Water:-Plastoquinone Oxidoreductase*, ed. T. Wydrzynski and K. Satoh, Springer, Dordrecht, 2005, vol. 22, pp. 207–233.
51. L. Kalman, R. LoBrutto, J. P. Allen and J. C. Williams, *Nature*, 1999, **402**, 696.
52. M. Thielges, G. Uyeda, A. Camara-Artigas, L. Kalman, J. C. Williams and J. P. Allen, *Biochemistry*, 2005, **44**, 7389.
53. L. Kalman, R. LoBrutto, J. C. Williams and J. P. Allen, *Biochemistry*, 2006, **45**, 13869.
54. T. Wydrzynski, W. Hillier and B. Conlan, *Photosynth. Res.*, 2007, **94**, 225.
55. B. Conlan, *Photosynth. Res.*, 2008, **98**, 687.
56. F. Frolow, A. J. Kalb and J. Yariv, *Nat. Struct. Biol.*, 1994, **1**, 453.
57. S. C. Andrews, J. M. A. Smith, C. Hawkins, J. M. Williams, P. M. Harrison and J. R. Guest, *Eur. J. Biochem.*, 1993, **213**, 329.
58. B. Conlan, T. Wydrzynski and W. Hillier, *Photosynth. Res.*, 2007, **91**, 281.
59. N. E. LeBrun, G. R. Moore and A. J. Thomson, *Mol. Phys.*, 1995, **85**, 1061.

60. V. V. Barynin, M. M. Whittaker, S. V. Antonyuk, V. S. Lamzin, P. M. Harrison, P. J. Artymiuk and J. W. Whittaker, *Structure*, 2001, **9**, 725.
61. K. O. Schafer, R. Bittl, F. Lendzian, V. Barynin, T. Weyhermuller, K. Wieghardt and W. Lubitz, *J. Phys. Chem. B*, 2003, **107**, 1242.
62. B. Loll, J. Kern, W. Saenger, A. Zouni and J. Biesiadka, *Nature*, 2005, **438**, 1040.
63. B. Conlan, N. Cox, J. H. Su, W. Hillier, J. Messinger, W. Lubitz, P. L. Dutton and T. Wydrzynski, *Biochim. Biophys. Acta*, 2009, **1787**, 1112.
64. H. L. Liu, H. N. Zhou, W. M. Xing, H. F. Zhao, S. X. Li, J. F. Huang and R. C. Bi, *FEBS Lett.*, 2004, **573**, 93.
65. X. Yang, N. E. Le Brun, A. J. Thomson, G. R. Moore and N. D. Chasteen, *Biochemistry*, 2000, **39**, 4915.
66. M. A. Carrondo, *EMBO J.*, 2003, **22**, 1959.
67. K. Hingorani, B. Conlan, W. Hillier and T. Wydrzynski, *Aust. J. Chem.*, 2009, **62**, 1351.
68. B. Zhang, J. N. Harb, R. C. Davis, J. W. Kim, S. H. Chu, S. Choi, T. Miller and G. D. Watt, *Inorg. Chem.*, 2005, **44**, 3738.
69. R. Tagore, R. H. Crabtree and G. W. Brudvig, *Inorg. Chem.*, 2008, **47**, 1815.
70. J. Limburg, J. S. Vrettos, L. M. Liable-Sands, A. L. Rheingold, R. H. Crabtree and G. W. Brudvig, *Science*, 1999, **283**, 1524.
71. J. Limburg, J. S. Vrettos, H. Y. Chen, J. C. de Paula, R. H. Crabtree and G. W. Brudvig, *J. Am. Chem. Soc.*, 2001, **123**, 423.
72. A. K. Poulsen, A. Rompel and C. J. McKenzie, *Angew. Chem., Int. Ed.*, 2005, **44**, 6916.
73. L. Kalman, J. Williams and J. Allen, in *Photosystem II: The light-driven water:plastoquinone oxidoreductase*, ed. T. Wydrzynski and K. Satoh, Springer, Dordrecht, 2005, vol. 22, pp. 715–727.
74. P. Kurz, G. Berggren, M. F. Anderlund and S. Styring, *Dalton Trans.*, 2007, 4258.
75. S. Baaghil, A. Lewin, G. R. Moore and N. E. Le Brun, *Biochemistry*, 2003, **42**, 14047.
76. E. Polak and G. Ribiere, *Rev. Fr. Inform. Rech. Operation.*, 1969, **3**, 35.

CHAPTER 17

Wired Reaction Centers

CAROLYN E. LUBNER,[a] DONALD A. BRYANT[b] AND JOHN H. GOLBECK*[a,b]

[a] Department of Chemistry, The Pennsylvania State University, University Park, PA 16802, USA; [b] Department of Biochemistry and Molecular Biology, The Pennsylvania State University, University Park, PA 16802, USA

17.1 Introduction

Our highly interconnected global economy remains critically dependent on the availability of inexpensive portable (gas or liquid) transportation fuels. These fuels are typically short or medium chain hydrocarbons derived exclusively from geological sources. They have been efficiently exploited in the 20th Century using a well-developed infrastructure that spans exploration, extraction, processing, and distribution. Their availability is essential to the transportation sector, yet shortages are expected in the 21st Century due to demand that will at some point exceed supply, political instabilities that result in artificial shortages, and, not least, failure to take serious measures to conserve this limited resource. Further aggravating the problem is the buildup of greenhouse gases. The potential energy contained within the chemical bonds of a hydrocarbon molecule can only be released by oxidizing it to carbon dioxide and water, with the predictable consequence that the increase of carbon dioxide in the atmosphere will continue to exacerbate the problem of global climate change. Either consideration gives adequate reason to search for alternative molecules that either contain no carbon or contain carbon but can be continuously recycled so as to be carbon-neutral.

RSC Energy and Environment Series No. 5
Molecular Solar Fuels
Edited by Thomas J. Wydrzynski and Warwick Hillier

Published by the Royal Society of Chemistry, www.rsc.org

17.1.1 Hydrogen as an Alternative Fuel

Hydrogen is one alternative to hydrocarbon fuels. Its use depends on releasing the 143 kJ g^{-1} of energy available in the heat of combustion of the reaction: $H_2 + \frac{1}{2}O_2 \rightarrow H_2O$. However, hydrogen is not a fuel, *i.e.* it does not exist in mineable quantities on Earth, and the production of hydrogen by the opposite reaction, $H_2O \rightarrow H_2 + \frac{1}{2}O_2$, requires an energy input of at least 143 kJ g^{-1} of energy. The current methods of generating hydrogen, including the electrolysis of water and steam reforming of methane, are neither inexpensive nor are they (necessarily) carbon-neutral. The generation of hydrogen (and oxygen) from the electrolysis of water takes place in an electrolytic cell, wherein an electromotive force (emf) is induced between the cathode and the anode, each of which is located in a separate vessel connected *via* a salt bridge. The negatively charged electrode combines electrons with protons, thereby producing hydrogen. The positively charged electrode extracts electrons from water, thereby producing oxygen. The wire carries the electrons from the anode to the cathode, and the salt bridge maintains the necessary electrical neutrality. To maintain a finite driving force and overcome resistive losses, electrolysis of water requires an overpotential, which means that a larger input of energy than that attained by the thermodynamic midpoint potential is needed. Thus, not all of the electrical power put into the cell is converted into the chemical bond energy of hydrogen.[1]

The heating of water and methane to high temperatures in the presence of a nickel catalyst results in the generation of hydrogen in the highly endothermic reaction $CH_4 + H_2O \rightarrow CO + 3H_2$. This so-called steam reforming process is less expensive than the electrolysis of water and is the preferred industrial method of hydrogen production. Additional hydrogen can be recovered using the resulting carbon monoxide as a feedstock in the water-gas shift reaction $CO + H_2O \leftrightarrow CO_2 + H_2$. The by-product of both reactions, carbon dioxide, is a greenhouse gas that exacerbates the problems of global warming and climate change. Furthermore, because energy is lost in the conversion of methane into hydrogen, and because hydrogen is more difficult to handle than is methane, little is gained by converting one usable transportation fuel into another.

Given the drawbacks associated with currently utilized methods of generating hydrogen, renewable sources are being sought to supply the necessary energy input. In principle, the energy source can be tidal, geothermal, wind or solar. There are good reasons for choosing solar energy: sunlight is widely, albeit unevenly, distributed over the Earth's surface; it is plentiful (the solar constant is 1.37 kW m^{-2}); it is not subject to overt political interference; and it is inexhaustible. The amount of solar energy that strikes the surface of the Earth in one hour is more than sufficient to satisfy the total global energy demand for an entire year.[2] Hence, direct photocatalytic methods for generating hydrogen and oxygen from water remains a much sought-after goal.

17.1.2 The Design Strategy of Natural Photosynthesis

One place to seek design rules for a direct photocatalytic scheme for hydrogen production is to study the process of oxygenic photosynthesis that occurs naturally in cyanobacteria, algae and plants. Over the course of 3.5×10^9 years, cells have evolved two types of light-dependent enzymes that carry out the conversion of solar energy into chemical bond energy.[3] These two enzymes work together in series with one another in a similar fashion to an electrochemical cell; the reaction is broken down into two half-reactions wherein the oxidation of water, which occurs at the anode, is separated spatially from the reduction of $NADP^+$, which occurs at the cathode. Electrons are carried from the anode to the cathode through the use of redox-active proteins in the biological system. The Type II reaction center Photosystem II (PS II) produces a strong oxidant and a weak reductant while performing the half-cell reaction $H_2O + 2h\nu \rightarrow \frac{1}{2}O_2 + 2H^+ + 2e^-$. The Type I reaction center Photosystem I (PS I) produces a weak oxidant and a strong reductant while performing the half-cell reaction $NADP^+ + H^+ + 2e^- + 2h\nu \rightarrow NADPH$. Both of these half-cell reactions are thermodynamically unfavorable and are driven to completion by utilizing the energy of a visible photon. PS II and PS I operate in series so that the following net reaction is carried out: $2H_2O + 2NADP^+ + 4h\nu \rightarrow O_2 + 2H^+ + 2NADPH$. As in a chemical cell, electrons flow from PS II to PS I *via* a series of organic molecules (quinones) and proteins (plastocyanin or cytochrome c_6) and ionic balance is maintained by conducting these reactions in an aqueous solution.

It is interesting that the standard biochemical midpoint potential of the $NADP^+/NADPH$ half-cell couple ($E^{o\prime} = -320$ mV) is only 90 mV more oxidizing than that of the H^+/H_2 half-cell couple ($E^{o\prime} = -430$ mV). Although NADPH (along with NADH) is an indispensable reductant of living cells, it is not particularly useful as a source of stored energy for industry. However, were it possible to re-engineer PS I to generate hydrogen instead of NADPH, a method to generate a transportation fuel directly from sunlight might become feasible. Minimally, a demonstration device would become available that would provide a 'gold standard' to evaluate artificial photosynthesis and synthetic catalysts. Two important design features have emerged from the study of natural photosynthesis: the need for two light reactions to carry out the independent processes of water splitting and generation of stored chemical energy (NADPH), and the need to separate the light reactions from the catalytic reactions.

17.1.2.1 Design Features of Photosystem I

PS I is the reaction center complex located in the photosynthetic membranes of plants and cyanobacteria that carries out the light-driven, immediate reaction $Pc^{red} + Fd^{ox} + h\nu \rightarrow Pc^{ox} + Fd^{red}$, where Pc is the soluble protein plastocyanin, and Fd is the soluble protein ferredoxin. It carries out this otherwise thermodynamically unfavorable reaction by using light energy to transfer electrons across the thylakoid membrane.[4] PS I is a membrane-bound, multi-subunit, protein-pigment complex. In the case of cyanobacteria, PS I consists of 11 or 12 polypeptides and approximately 100 chlorophyll and 22 carotenoid molecules that are active in light harvesting.[5] The protein subunits PsaA and PsaB comprise the core of the reaction center and, along with PsaC, bind the electron

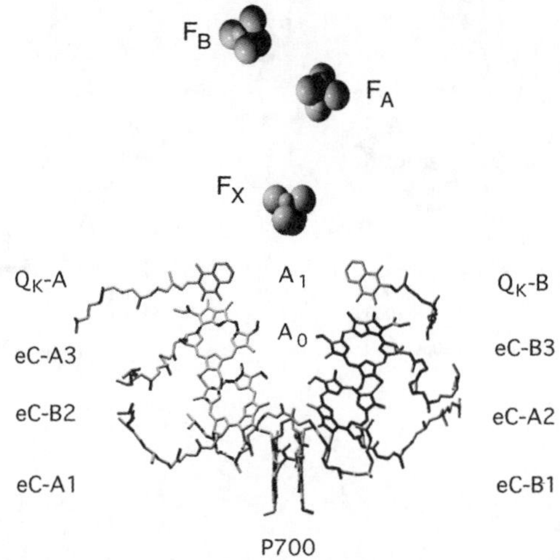

Figure 17.1 Arrangement of the electron transfer cofactors in PS I from *T. elongatus*. The C2 axis of symmetry occurs between the center of P_{700} and the F_X cluster. The spectroscopic names of the cofactors are depicted in the center (the crystallographic names are on the sides). (PDB entry 1JB0.)

transfer cofactors. PsaC acts as an interface to shuttle electrons from within the membrane to soluble electron accepters, such as Fd.

The principal antenna pigments in cyanobacterial PS I are the chlorophyll *a* (Chl *a*) molecules that are capable of absorbing photons at wavelengths shorter than 700 nm. When a Chl *a* absorbs a photon, the resulting exciton is transferred on the fs to ps time scale by excitonic or resonance energy transfer processes to the primary electron donor of PS I, a special pair of Chl *a*/Chl *a′* molecules termed P_{700} (where Chl *a′* is the 3′-epimer of Chl *a*). The arrangement of the core electron transfer cofactors is shown in Figure 17.1. Charge separation occurs between the eC-A1/eC-B1 special pair and either eC-A3 or eC-B3, which are more commonly identified by their spectroscopic names, P_{700} and A_0. The electron is transferred within 30 ps to Q_K-A or Q_K-B,[6] which are bound phylloquinones identified by their spectroscopic names A_{1A} and A_{1B}, to a PsaA/PsaB interpolypeptide [4Fe-4S] cluster, F_X. The electron is then transferred serially through the two [4Fe-4S] clusters, F_A and F_B, which are associated with PsaC, to a soluble, oxygen-stable [2Fe-2S] ferredoxin. The F_B cluster has a midpoint potential of –580 mV, which is more than sufficient to reduce protons to H_2. Electron transfer on the PS I acceptor side is thermodynamically favorable, as the standard biochemical midpoint potential of the A_0/A_0^- couple is considerably more negative than that of F_B/F_B^- couple. Figure 17.2 depicts the redox potentials of the electron transfer cofactors as a function of their distance from P_{700}, as well as forward electron transfer and charge-recombination times. Electron transfer from P_{700} to F_B is very rapid, occurring within 20 to 200 ns, and the lifetime of the charge-separated state, $P_{700}^+ F_B^-$, is rather long, ~65 ms. Provided the electron is transferred from the

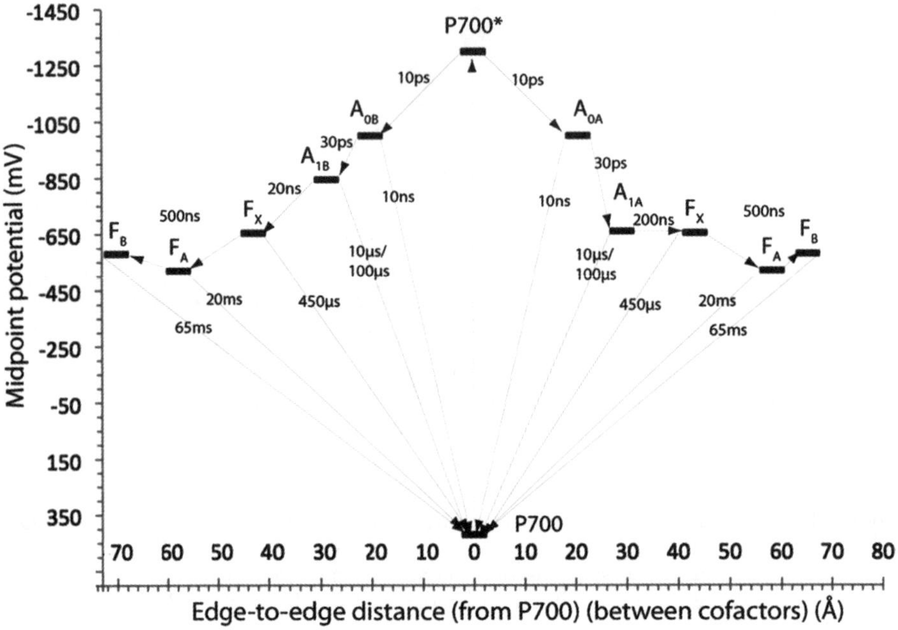

Figure 17.2 Forward and backward electron transfer kinetics, as well as the redox potentials (ordinate) and distance in the membrane normal (abscissa) on the A branch (right) and B branch (left) of the electron cofactors in PS I. Since the electron transfer pathways converge at F_X, the left and right sides represent identical clusters, F_X, F_A, and F_B. (Reproduced with permission from ref. 146. Copyright 2010 American Chemical Society.)

F_B cluster within this lifetime, charge recombination is precluded, and the electron can be harnessed for useful work.[7,8] In the case of PS I in living cells, this work is the reduction of ferredoxin or flavodoxin and the subsequent production of NADPH, but if the electron could be removed from F_B^- and delivered to a hydrogen-producing catalyst, it could be used to reduce protons directly to hydrogen.

17.1.2.2 Properties of Photosystem I Appropriate for Hydrogen Generation

Over the course of *ca.* 3.5 Ga, the PS I reaction center has evolved a number of favorable characteristics that provide important design clues for artificial photosynthetic systems. (i) The pigments absorb all visible wavelengths shorter than 700 nm. This spectral response represents 43 to 46% of the total solar emission that reaches the surface of the Earth under normal weather conditions.[9,10] The absorption in the blue and red regions by the ~100 chlorophyll *a* molecules and 22 carotenoids is especially high, although at high pigment concentrations, total absorption for $\lambda < 700$ nm can be readily approached. The typical leaf or cyanobacterial cell can therefore be considered a black body, from which the surface reflected light only makes it appear green. (ii) The antenna chlorophylls in PS I provide a high optical cross section. Given the area of one chlorophyll *a* molecule and a noontime solar flux at the latitude of Washington, DC, in summer, a photon falls on the surface area of each chlorophyll *a* in PS I about 10 times per second. Because there are about 100 chlorophyll *a* molecules per P_{700}, up to 1000 photons are absorbed per second per reaction center, yielding a rate constant of 1 ms. Thus, in full sunlight, 1000 electrons can be processed per second. Provided no rate limitations occur on the electron donor or acceptor sides of PS I, a maximum of 500 hydrogen molecules per reaction center per second could be generated at this photon flux. (In reality, not all photons are absorbed by a given chlorophyll because the extinction coefficient of 80 000 M^{-1} cm^{-1} is not constant throughout its spectral response, limiting the rate constant to less than 1/3 of the ideal (David Kramer, personal communication). Of course, were it possible to concentrate sunlight, higher rates should be possible, provided that electron donation to $P_{700}{}^{+}$ does not become the rate-limiting step. (iii) The quantum yield of PS I approaches 1.0. Regardless of wavelength, nearly each and every photon that is absorbed by PS I is processed into the charge separated state $P_{700}{}^{+}$ $F_B{}^{-}$. (iv) The thermodynamic efficiency of photochemical charge separation is high. The 1.01 V in the charge-separated state $P_{700}{}^{+}$ $F_B{}^{-}$ represents a remarkable 59% conversion efficiency for a red photon, and a still impressive 38% conversion efficiency for a blue photon.[11,12] The charge-separated state is stable for 65 ms, which is sufficient time to transfer the electron and to perform useful work. (v) The PS I reaction center is able to deliver electrons at a redox potential sufficient to drive H_2 evolution. The F_B cluster has a pH-independent standard midpoint potential of –580 mV,[11,12] which is 140 mV more reducing than the half-cell potential of the H^+/H_2 couple at pH 7.0. Thus, assuming standard conditions, the equilibrium constants would differ by a factor of 100, which means that sufficient electromotive force exists to drive the reaction (nearly) to completion. (vi) The PS I reaction center is robust on a biochemical timescale. There are no light-sensitive or overly heat-sensitive components in PS I. The only known form of damage is a chilling-induced (4 °C) photoinhibition described in leaves[13] and membranes[14] of the cold-sensitive cucumber plant that results in destruction of the Fe/S clusters. The causative agent is the •OH radical, formed in the presence of H_2O_2 by the reduced Fe/S clusters of PS I.[15] (vii) The cofactors and proteins that constitute PS I can be readily manipulated. Over the last 20 years, the following manipulations have been developed: (i) the biochemical removal[16–18] and reconstitution[18–20] of the PsaC, PsaD and PsaE polypeptides;[21–23] (ii) the chemical denaturation and renaturation of the F_X, F_A and F_B clusters;[24–27] (iii) the genetic replacement of phylloquinone with a variety of non-native quinines;[28–32] and (iv) the selective removal[33,34] and the directed mutagenesis of many of the proteins that constitute PS I.[35] PS I is therefore a highly promising demonstration platform for using solar energy to generate the reducing power required for hydrogen evolution.

17.1.3 Traditional Catalysts for Generation of Hydrogen

Hydrogen catalysis has typically been achieved through the use of noble metals, such as platinum and gold. These metals are able to catalyze the reaction $2H^+ + 2e^- \rightarrow H_2$ following a straightforward mechanism. First, H^+ ions adsorb to the metal surface and combine with an electron to form an H atom. A covalent bond is catalytically generated between two adsorbed H atoms to yield H_2, which then desorbs from the Pt surface. This process occurs at the standard biochemical midpoint potential of –420 mV at pH 7.0, *i.e.,* there is little or no overpotential. Non-noble metals are also able to catalyze the reduction of protons; however large overpotentials are generally required, which leads to an unacceptable loss of efficiency.

By combining these metal catalysts with semiconductors, such as TiO_2 or Al_2O_3, the photocatalytic generation of hydrogen has been attained.[36–38] This has been studied with Au, Pt and Pd nanoparticles supported on these semiconductor materials in aqueous ethanol solution. H_2 production occurs by means of alcohol reforming on the surface of the nanoparticles following a known reaction mechanism in which ethanol decomposes to produce stoichiometric amounts of H_2 and CO. Upon absorption of a photon, an electron-hole pair is created within the semiconductor material. Electrons and holes rapidly migrate to opposite poles of the semiconductor surface into conduction and valence bands, respectively. The electrons are then used by the nanoparticles to catalyze hydrogen production. An electron–hole pair may only be created when photons possess enough energy to span the band gap of the semiconductor material. For the case of TiO_2, the band gap is *ca.* 3.2 eV, which corresponds to light with wavelengths shorter than 350 nm. On the basis of this requirement, only a very small fraction of incident solar radiation has sufficient energy to produce a charge-separated state. In an attempt to overcome the band-gap limitation, organic dyes have been introduced into these systems to increase the absorption of solar radiation.[36–38] To complement the limited cross section of the semiconductor materials, these dyes absorb photons in the visible region and inject electrons directly into the conduction band. While these dye-sensitized systems increase the range of wavelengths that are capable of producing the charge-separated state necessary for hydrogen production, organic dyes are degraded by illumination, moisture, and oxidation. Further hindering the widespread utility of these devices, noble metals are both expensive and rare.

17.1.4 Properties of Hydrogenases

Microorganisms have evolved enzymes, namely hydrogenases (H_2ases), which can also perform the reversible reaction: $2H^+ + 2e^- \leftrightarrow H_2$ under thermodynamically favorable conditions, *i.e.*, without a large overpotential and with remarkably high catalytic rates (for reviews, see ref. 39–43. Most importantly, the largely inorganic active sites of these enzymes employ inexpensive and abundant metal ions, iron and nickel, as cofactors. The reaction catalyzed by H_2ases is deceptively simple, and its directionality depends on the properties of redox conditions and partner proteins. In the context of cellular metabolism, H_2ases function in H_2 uptake in the presence of suitable electron acceptors, but produce H_2 when cells are grown in the presence of appropriate electron donors. Because they obligatorily produce H_2 as a by-product of the reduction of N_2 gas to ammonia, it should also be noted that all nitrogenases are also H_2-producing H_2ases.[44,45] However, because of the very high ATP requirements of nitrogenases during N_2 reduction and H_2 production, as well as the complexity of their biogenesis, these enzymes may not be suitable for direct solar energy conversion schemes with wired reaction centers and will not be further considered here.

Three distinctive classes of H_2ases are known: [NiFe]-H_2ases, [FeFe]-H_2ases, and [Hmd]-H_2ases.[41–43,46] Sequence comparisons, as well as structural comparisons based on X-ray crystallography, show that these three enzyme classes clearly arose independently during evolution.[39,41] In general, [NiFe]-H_2ases are active in H_2 uptake/oxidation reactions and are often more stable and oxygen tolerant; conversely, [FeFe]-H_2ases are frequently involved in H_2 evolution and are usually much more readily inactivated by exposure to oxygen. The [Hmd]-H_2ases were discovered by Thauer and coworkers in methanogens,[47] in which they function as methylenetetrahydromethanopterin dehydrogenases. Although the [Hmd]-H_2ases were initially believed to be metal-free, they are now known to have a single Fe atom in their active site.[48–51] Because the [Hmd]-H_2ases lack Fe/S clusters[47,51] and are readily inactivated by exposure to UV-A and blue light,[49] they are not suitable for use as the catalytic module for H_2 production using wired reaction centers, and they also will thus not be considered further here.

17.1.4.1 [NiFe] Hydrogenases

The largest and best-characterized family of H_2ases are the [NiFe]-H_2ases, which are widely distributed among members of the domains *Bacteria* and *Archaea*.[39,41,42,52] The catalytic modules of [NiFe]-H_2ases are heterodimers formed from subunits with masses of approximately 60 and 30 kDa. Structures of several [NiFe]-H_2ases from sulfate-reducing *Deltaproteobacteria* have been determined by X-ray crystallography (Figure 17.3). These include the [NiFe]-H_2ases of *Desulfovibrio gigas*,[53] *D. vulgaris* Miyazaki,[54–56] *D. fructosovorans*[57,58] and *D. desulfuricans* ATCC 27774.[59] In some enzymes, *e.g.*, that of *Desulfomicrobium baculatum*, a selenide atom replaces one of the sulfur ligands to the [NiFe] moiety at the active site.[60] The bimetallic [NiFe] center at the active site is deeply buried and is associated with the large subunit near its interface with the small subunit. This subunit interface is nearly flat and is extensive (~3500 $Å^2$).[53] The Ni atom is ligated by the sulfur atoms of four Cys residues, two of which act as bridging ligands to the Fe atom, and the Fe atom additionally carries two CN^- ligands and one CO ligand.[42,46,61]

The bimetallic active site of [NiFe]-H_2ases is connected to electron donors and acceptors at the external surface of the heterodimer by a chain of three Fe/S clusters, which are bound to the small subunit (Figure 17.3).[42,46,61]

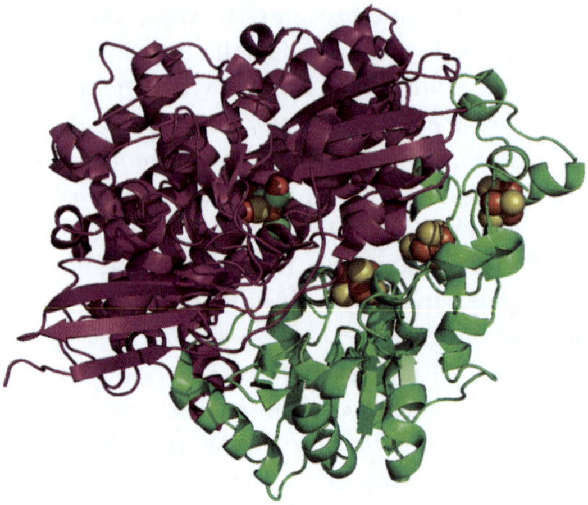

Figure 17.3 X-Ray crystal structure of the [NiFe]-H_2ase from *Desulfovibrio vulgaris* Miyazaki F. (PDB entry 1H2R.)

In most cases the central cluster is a [3Fe-4S] cluster with a relatively high midpoint potential (+65 mV); the flanking Fe/S clusters are [4Fe-4S] clusters with midpoint potentials of approximately –340 mV.[58,61] However, the small subunit of the *D. baculatum* enzyme harbors three [4Fe-4S] clusters.[60] A site-specific variant of the *D. fructosovorans* small subunit, in which the [3Fe-4S] cluster was converted into a [4Fe-4S] cluster, produced some interesting effects on catalysis.[58] This mutation shifted the redox potential of the central Fe/S cluster by ~300 mV (from +65 to –250 mV); this was associated with a 50% decrease in the H_2 oxidation rate and a 50% increase in the H^+ reduction rate.[58] Somewhat larger effects were observed when similar experiments were performed with *Methanococcus voltae*.[62] Evidently, the principal consequence of the high-potential [3Fe-4S] cluster in the electron transport chain is a slowing of the rate of electron transfer to the active-site proximal [4Fe-4S] cluster due to the thermodynamically uphill electron transfer step.

Four major structural variations on the central theme of the [NiFe]-H_2ase heterodimer are known.[39,41,42,52] Group 1 contains uptake [NiFe]-H_2ases, which are membrane-bound, periplasmic enzymes that allow cells to use H_2 as an energy source. These enzymes are typically translocated to the periplasmic space by the "hitchhiker" mechanism of the twin-arginine translocon pathway after the enzyme is completely assembled in the cytosol. A single signal peptide occurs at the amino terminus of the small subunit of these enzymes, which is sufficient to direct the translocation of both subunits across the cytoplasmic membrane. Cytoplasmic H_2-sensing H_2ases and the uptake [NiFe]-H_2ases of cyanobacteria are cytoplasmic enzymes that comprise Group 2. In Group 3 enzymes, the core heterodimer associates with additional subunits to produce larger, hetero-multimeric complexes that remain in the cytoplasm and that characteristically bind 8-hydroxy-5-deazaflavin, FMN, NAD^+ and/or $NADP^+$. These so-called "bidirectional H_2ases" reversibly transfer electrons from H_2 to soluble cofactors, or they re-oxidize reduced coenzymes under anoxic conditions using protons as electron acceptors. The Group 4 [NiFe]-H_2ases are complex, membrane-associated, multimeric enzymes with six or more subunits, which reduce protons in an energy conserving process under anoxic conditions.

17.1.4.2 Biogenesis of [NiFe]-Hydrogenases

Successful protein engineering of H_2ases as catalytic modules for biotechnological applications requires that one can produce site-specifically modified variant proteins with purposefully altered properties. The ability to generate variant enzymes requires a genetic system in the source organism that produces the H_2ase of interest, or alternatively requires the ability to express the protein heterologously in a foreign host such as *Escherichia coli*. However, the metal insertion and maturation processes for [NiFe]-H_2ases are complex, and in the case of the HycEG H_2ase-3 of *E. coli,* the participation of at least eight accessory proteins is required (for reviews, see ref. 63–65). These assembly factors often exhibit specificity for individual H_2ase subunits, and for this reason, efficient heterologous expression of [NiFe]-H_2ases in *E. coli* has not yet been reported. For example, studies in *Thiocapsa roseopersicina* have shown that some accessory proteins function selectively while others act pleiotropically.[66] These observations were recently exploited in the expression of *hupL* and *hyaD* from a marine *Alteromonas* sp. in *T. roseopersicina,* which led to the production of an active H_2ase.[67] In this case, *T. roseopersicina* presumably supplied the required *hypABCDEF* products, and they apparently functioned to produce an active H_2ase only when the HyaD endoprotease from *Alteromonas* sp. was additionally provided. Expression of a 13-gene operon, which encodes the soluble, oxygen-tolerant, bidirectional [NiFe]-H_2ase and all of its associated maturation factors from *Ralstonia eutropha,*[68,69] in a H_2ase-free null mutant of the cyanobacterium *Synechococcus* sp. PCC 7002 has also recently been achieved.[70] Although the structural genes for the endogenous H_2ase had been deleted, it is possible that some of the remaining accessory Hyp proteins participated in the biogenesis of the heterologous enzyme.

Based upon detailed studies of H_2ase-3 (HycGE) from *E. coli*, the current working model for the assembly of the large subunit of [NiFe]-H_2ase is as follows (for recent reviews, see ref. 41, 46, 71 and 72). HypD, an Fe/S protein, and HypC form a heterodimeric complex; the structures of these proteins as well as HypE were recently determined.[73] Together, these proteins are involved in the assembly of the Fe atom, which will become a part of the bimetallic [NiFe] center.[64,74] HypE and HypF also form a complex, which utilizes ATP and carbamoyl phosphate produced from bicarbonate and glutamine to generate the two CN^- ligands for the Fe atom of the bimetallic [NiFe] center. Although the pathway for CO production has not yet been defined, exogenous CO can be incorporated, and CO insertion is thought to occur prior to transfer of the completed $Fe(CN^-)_2CO$ unit into pre-HycE. The $Fe(CN^-)_2CO$ unit is then transferred from HypC into the H_2ase large subunit precursor, pre-HycE, to which HypC remains associated. Insertion of Ni requires the actions of three proteins: HypA, HypB, and SlyD.[72] HypA is believed to be the Ni carrier, and HypB is thought to be a GTP-dependent Ni insertase, whose affinity for Ni is modulated by SlyD binding.[75] After the Ni has been inserted, HypC is released, and the HycI endopeptidase cleaves 33 amino acid residues from the C-terminus of pre-HycE. Once the C-terminal residues of HycE have been proteolytically removed, the HycG small subunit with its previously assembled Fe/S clusters binds to holo-HycE to produce mature heterodimeric H_2ase-3. Mutants lacking the small subunit can still produce mature, soluble large subunits, and mature, soluble small subunits can be produced in mutants lacking the large subunit.[76] Stable subcomplexes that permit the *in vitro* reconstitution of the soluble NAD^+-reducing bidirectional hydrogenase of *R. eutropha* have also been reported.[77]

17.1.4.3 [FeFe]-Hydrogenases

[FeFe]-H_2ases are widely distributed in anaerobic members of the domain *Bacteria* and are commonly associated with clostridia and sulfate-reducers; and enzymes of this class are the only H_2ases that have been reported to date in *Eucarya* (for recent reviews see ref. 39, 41, 42, 46, 78 and 79). Although some [FeFe]-H_2ases function in H_2 oxidation, the majority of these typically monomeric enzymes produce H_2 by catalyzing proton reduction. These versatile catalysts can accept electrons from a variety of physiological electron donors, and they exhibit astounding turnover rates. For example, the [FeFe]-H_2ase of *Clostridium pasteurianum* can produce up to 6000 molecules of H_2 per second.[80] Green algae, *e.g.*, *Chlamydomonas reinhardtii*, have the simplest [FeFe]-H_2ases known.[79] These enzymes are 45–48 kDa proteins that contain an H-cluster binding site but lack auxiliary Fe/S clusters.[78] The X-ray structure of an assembly intermediate, which contains the [4Fe-4S] cluster but not the diatomic unit of the H-cluster, of the *C. reinhardtii* H_2ase-1 was recently determined (see below).[81] Consistent with the NAD(P)H dependency of these enzymes, heterotrimeric and heterotetrameric [FeFe]-H_2ases, which contain subunits related to peripheral subunits of the NADH-ubiquinone oxidoreductase and that harbor additional Fe/S clusters, are also known.[78]

Structures are known for the [FeFe]-H_2ases of *C. pasteurianum*[82] and *D. desulfuricans*.[83] The [FeFe]-H_2ase of the former organism is a mushroom-shaped, monomeric enzyme in which the H-cluster active site is deeply buried within the protein (Figure 17.4). The H-cluster is formed from a binuclear [FeFe] center, which is connected to a [4Fe-4S] cluster by the sulfur atom of a bridging cysteine residue; three additional cysteine residues complete the ligation of the [4Fe-4S] cluster. Each Fe of the diatomic subcluster is currently believed to bind one CO and one CN^- ligand, and an additional bridging CO ligand is also currently believed to be present.[46] Sulfur atoms that bridge the two Fe atoms of the diatomic subcluster were initially modeled as 1,3-propanedithiolate; however, the identity of this dithiolate bridging molecule has remained controversial. Mechanistic and spectroscopic considerations

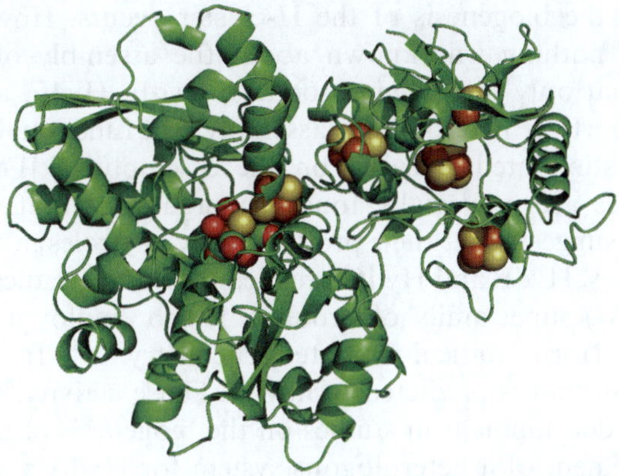

Figure 17.4 X-Ray crystal structure of the [FeFe]-H_2ase from *C. pasteurianum*. (PDB entry 1FEH.)

concerning proton extraction and delivery processes to the active site have led to the suggestion that the bridgehead atom is nitrogen, which would imply that the small bridging organic molecule would be di(thiomethyl)amine (see ref. 42, 46 and 84). However, further refinement of the X-ray structure and density functional theory optimization calculations have also suggested that the bridging atom could be oxygen rather than nitrogen, which would mean that the bridging diatomic ligand would be dithiomethylether.[85]

In addition to the H-cluster, the cytoplasmic [FeFe]-H_2ase of *C. pasteurianum* carries three [4Fe-4S] clusters and one [2Fe-2S] cluster per 63 kDa monomer (see Figure 17.4).[82] These cofactors could allow the enzyme to interact with different electron donors for H_2 production under different physiological conditions. In contrast, the periplasmic [FeFe]-H_2ase of *D. desulfuricans* is a heterodimer of 42 kDa and 11 kDa subunits that only carries two additional [4Fe-4S] clusters that are associated with a bacterial ferredoxin-like domain.[83] In this enzyme the smaller subunit encircles the larger subunit like a belt, and its sequence corresponds to the C-terminal tail of monomeric, cytoplasmic [FeFe]-H_2ases like that of *C. pasteurianum*. Translocation of such periplasmic [FeFe]-H_2ases across the cytoplasmic membrane also occurs by the hitchhiker mechanism as described above for [NiFe]-H_2ases. The small subunit is initially translated with a Sec-independent, TAT translocon signal peptide at its N-terminus, and this signal peptide is proteolytically removed during translocation of the holo-enzyme heterodimer to the periplasm. Although the small subunit does not participate in cofactor ligation, the N-terminus of the small subunit and the C-terminus of the large subunit interact through a bound cysteine molecule that presumably stabilizes the overall structure in the *D. desulfuricans* enzyme.[42]

17.1.4.4 Biogenesis of [FeFe]-Hydrogenases

In order to use [FeFe]-H_2ases as catalytic modules for biotechnological purposes, it is essential to have the ability to produce variant enzymes and to understand how the biogenesis of the H-cluster occurs. However, until very recently, almost nothing was known about the assembly of [FeFe]-H_2ases. The discovery that only three gene products, HydE, HydF, and HydG, were sufficient to support the heterologous assembly of a functional HydA H_2ase in *E. coli,* greatly stimulated research on the assembly of [FeFe]-H_2ases.[86–89] These three genes are universally found in organisms that produce [FeFe]-H_2ases, and this suggests that their products play key roles in the biogenesis of this class of H_2ases. HydE and HydG are members of the radical *S*-adenosyl-*L*-methionine (SAM) superfamily of proteins, which employ a [4Fe-4S] cluster and SAM to perform radical-mediated chemistry.[86,90] In contrast, HydF includes a domain that is predicted to have GTPase activity.[87,91]

An important development in studies on the biogenesis of the H-cluster has been the development of a heterologous system for HydA assembly based on overproduction of biogenesis proteins in *E. coli*[88] (for a review see ref. 89). When *hydA* from *C. reinhardtii* and the *hydE, hydF,* and *hydG* genes from *C. acetobutylicum* were heterologously expressed in *E. coli*, functional holo-HydA H_2ase was assembled in *E. coli.*[92] This heterologous system was recently exploited to show that, in the absence of HydE, HydF, and HydG, HydA is first synthesized with an inserted [4Fe-4S] cluster,[93] and the structure of this intermediate has been determined by X-ray crystallography.[81] This HydA assembly intermediate can subsequently be converted into active holo-[FeFe]-H_2ase by incubating HydA with an *E. coli* extract derived from cells expressing *hydE, hydF,* and *hydG.*[92–94] Furthermore, when HydF was purified from the extracts of cells expressing all three accessory proteins and then incubated with the [4Fe-4S]-containing HydA, addition of the di-iron unit to form an active H_2ase also occurred. These results and others strongly imply that HydF acts as a scaffold protein for assembly of the diatomic subcluster of the H-cluster.[94] Furthermore, these studies clearly establish that the roles of the HydE, HydF, and HydG accessory proteins are to synthesize and insert the di-iron unit of the H-cluster into an assembly intermediate form of HydA that harbors a previously assembled [4Fe-4S] cluster.

Current results suggest that HydE and HydG synthesize the inorganic ligands required for synthesis of the diatomic unit of the H-cluster. Although the structure of HydE from *Thermotoga maritima* has been determined,[95] the substrate for this enzyme has not yet been identified.[96,97] Based upon its similarity to biotin and lipoate synthases, HydE is currently suspected to synthesize the dithiolate bridging ligand. HydG exhibits significant sequence similarity to ThiH, a radical SAM enzyme that catalyzes the cleavage of tyrosine,[98] and conflicting reports concerning its enzymatic activity have recently appeared. It was initially reported that HydG cleaves tyrosine to produce *p*-cresol and dehydroglycine; the latter compound was suggested to be the precursor for the synthesis of the dithiolate bridging ligand in the H-cluster.[99] However, a subsequent study has found that HydG catalyzes tyrosine cleavage to produce *p*-cresol and CN^- in equal amounts.[97] Although the substrate was not glycine as first suggested, this finding is consistent with the proposal that HydG functions in the production of the diatomic ligands of the H-cluster.[100]

Finally, Mulder *et al.*[93] have proposed that HydE and HydG modify a [2Fe-2S] cluster on HydF to introduce the CO and CN^- ligands while utilizing the bridging sulfides of that cluster as the source of the sulfur atoms for the formation of the dithiolate bridging molecule, whose precise origin and identity still remains to be unequivocally established (see ref. 85). In the absence of HydE and HydG, HydF has EPR properties consistent with the presence of both [4Fe-4S] and [2Fe-2S] clusters.[101] However, the EPR signal associated with the [2Fe-2S] cluster is no longer observed when HydE and HydG are coexpressed with HydF. Fourier-transform infrared spectroscopy of the latter protein shows that this form of HydF harbors a cluster intermediate that is spectroscopically similar to the diatomic subcluster of the H-cluster.[102] Although the biogenesis of the H-cluster in [FeFe]-H_2ases is currently not as well understood as the biogenesis of the diatomic centers of [NiFe]-H_2ases, very rapid progress on this problem is now being made and further insights are likely to emerge very quickly.

17.2 Connecting Photosystem I with Hydrogen Catalysts

A realistic photochemical half-cell for generating fuel from sunlight consists of three components: a module that converts sunlight into an emf, a module that catalyzes the reduction of protons to hydrogen, and a linker that facilitates the transfer of electrons from the light module to the catalytic module. The challenge is to deliver electrons from the photochemical module to the catalytic module at high quantum yields, high thermodynamic efficiencies, and high rates. The design put forth below is based on the biological requirement of utilizing two independent photochemical half-cells to carry out the reaction. We focus here exclusively on the design and optimization of a photochemical module that is able to carry out the cathodic half-cell reaction: $2H^+ + 2e^- + 2h\nu \rightarrow H_2$. To this end, we discuss below systems that utilize PS I, hydrogen generating catalysts, and a variety of linker schemes for the photocatalytic production of H_2.

17.2.1 *In Vivo* Hydrogen Generating Systems

It has long been established that a mixture of crude extracts of photosynthetic reaction centers and hydrogen-generating enzymes can generate hydrogen in the light.[103–107] When these mixtures include whole chloroplasts or thylakoid membrane extracts, PS II serves as the electron donor to PS I through the oxidation of water. Methyl viologen or ferredoxin serves as the electron carrier between PS I and H_2ase. These systems generally have very limited long term hydrogen production abilities due to the instability of PS II under continuous light as well as the inactivation of the H_2ase enzyme in the presence of oxygen evolved from PS II.

This process also occurs naturally in living cells under certain conditions. In these *in vivo* systems, the low potential electrons produced by PS I are coupled *via* soluble ferredoxins to either hydrogenase or nitrogenase. For the case of nitrogenase, hydrogen is formed as a byproduct of nitrogen fixation. Currently, there are efforts underway to manipulate the biochemical pathways within cyanobacteria to favor electron transfer from PS I to H_2ase enzymes.[108] The major advantages of these systems are that they are self-repairing and sustainable, and they require very little input once assembled. Ultimately, hydrogen production depends upon slow diffusion chemistry, making it difficult to envision a system that could operate with a high enough efficiency to produce hydrogen in relevant quantities.

Because a large body of literature is already devoted to hydrogen generation by microorganisms, we will not discuss the topic further.

17.2.2 Platinized Photosystem I Reaction Centers

In principle, metal catalysts can be used for the photo-production of hydrogen in the same way that hydrogen is evolved when PS I and H_2ase enzymes are in close proximity. However, since both PS I and metal nanoparticle catalysts are both relatively large in size, a diffusion limited process is likely to be slow and inefficient. These limitations are further confounded by the requirement that electrons must be transferred from the F_B cluster of PS I to the catalyst within 65 ms. Ideally, the two photochemical and catalytic modules should be in very close proximity either by direct contact or through the use of an electron-transferring linker molecule.

An approach pioneered by Greenbaum and co-workers successfully met the requirement of the direct deposition of platinum onto thylakoid membranes and isolated PS I complexes.[109–111] Their strategy took advantage of the Hill reaction of photosynthesis, in which ferric ions are reduced to ferrous ions. A platinum salt (hexachloroplatinate, $[PtCl_6]^{2-}$) is combined with spinach chloroplasts and illuminated, thereby causing the Pt^{4+} ion to be reduced to Pt(s) and deposited onto the chloroplast surface. Continued illumination leads to the production of hydrogen, which is produced at a maximal rate when the deposits contain between 50 and 500 Pt atoms.[110] This method additionally works with hexachloroosmiate ($[OsCl_6]^{2-}$)[111] and hexachlororuthenate ($[RuCl_6]^{2-}$).[112] The highest rates of hydrogen production (0.113 μmol H_2 mg Chl^{-1} h^{-1}) occurred with $OsCl_6^{2-}$ precipitates, and somewhat lower rates were obtained with platinum and ruthenium precipitates.

Photoreduction and deposition of platinum can also be carried out on isolated PS I reaction centers.[113,114] When the sacrificial donor, sodium ascorbate, and the mediator, plastocyanin, the native electron donor to spinach PS I, are used to reduce oxidized P_{700}, hydrogen was evolved at a rate of 0.025 μmol mg Chl^{-1} h^{-1} in the light.[113] A 3-fold higher rate was achieved by eliminating diffusion through the cross-linking of plastocyanin to the reaction center ($\sim$0.080 μmol H_2 mg Chl^{-1} h^{-1}).[114]

More recently, the rate of light-induced H_2 production by platinized PS I complexes has been optimized to yield 5.5 μmol H_2 mg Chl^{-1} h^{-1}.[115] This increase is largely due to the use of PS I isolated from a thermophilic cyanobacterium, which allowed the platinized system to operate at higher temperatures (55 °C). This rate was maintained for 85 days when tested intermittently.

A major drawback to this approach is the large amount of Pt (or Os or Ru) needed for either chloroplasts or isolated PS I particles. It is possible that many of the deposited metal atoms are never actively involved in catalysis because they are not properly located on the external surface of the PS I. These atoms may not receive electrons or they may be inaccessible to the hydrogen ions available in solution. The resulting inefficiencies lead to a higher device cost in the long run.

17.2.3 Photosystem I-Hydrogenase Fusion Systems

To bypass the need for large amounts of noble metals, a system that employs Earth-abundant metals is highly desirable. Such systems may utilize a H_2ase enzyme as the catalyst, which make use of iron and/or nickel cofactors to carry out the reduction of protons.

By fusing parts of the photochemical and catalytic components together, a system that is less dependent upon diffusion chemistry can be realized. One strategy is to engineer genetically a semi-rigid complex that brings the terminal electron acceptor of PS I, F_B, into close contact with the proximal electron acceptor of a H_2ase enzyme.[116] As deduced from the X-ray crystal structures, this distance can be as short as ~14 Å if the small subunit of a [NiFe]-H_2ase from *Ralstonia eutropha* H16 is fused to the stromal PS I protein, PsaE from *Thermosynechococcus elongatus*. The expectation was that the distance would be sufficiently short that electrons would be efficiently transferred from F_B to a [4Fe-4S] cluster on the H_2ase within the available time scale (*i.e.*, 65 ms). A His_6-tagged variant of PsaE was fused to a variant of the H_2ase small subunit, HoxK, in which the membrane anchor domain was substituted with a linker peptide (Ser–Gly–Gly). The resulting HoxK-PsaE-His_6 fusion was able to assemble with the H_2ase large subunit, HoxG, and to generate hydrogen with the use of reduced methyl viologen, albeit at a rate 16% of the wild-type enzyme. When this H_2ase-PsaE fusion protein was rebound to trimeric PS I complexes derived from a *psaE* null mutant, light-induced hydrogen was produced at a rate of 0.58 μmol H_2 mg Chl^{-1} h^{-1} (Figure 17.5). To address the question of whether the donation of electrons from the sacrificial source to PS I is rate-limiting, the fusion complex was immobilized onto a gold electrode.[117] When a potential of –90 mV was applied to the illuminated complex, hydrogen was evolved at an initial rate of 2812 μmol H_2 mg Chl^{-1} h^{-1}. Unfortunately, this PS I-H_2ase fusion complex is only active for a matter of minutes. This lack of robustness limits the possibility of integration into a photochemical hydrogen device.

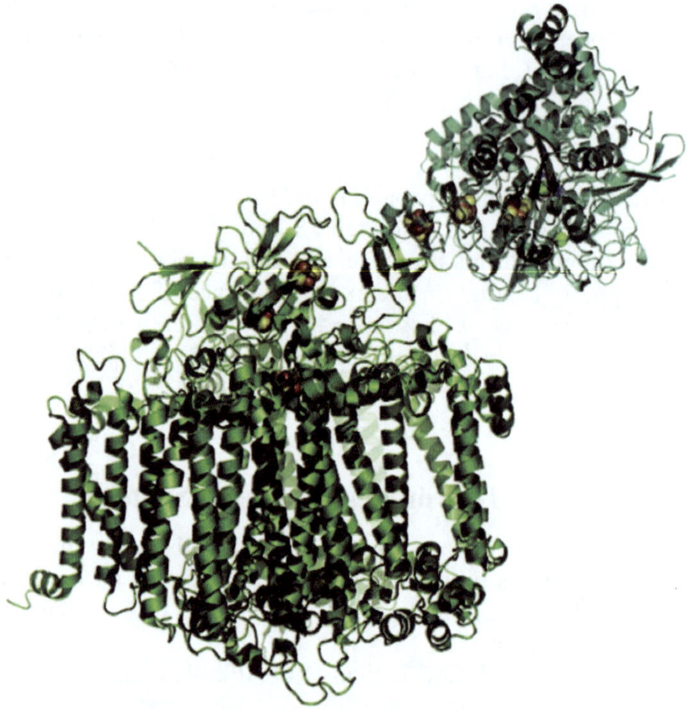

Figure 17.5 Cartoon of the PS I-H_2ase fusion construct. The H_2ase enzyme is attached to the PsaE subunit of PS I. (Reproduced with permission from ref. 117. Copyright 2010 American Chemical Society.)

A similar fusion complex was constructed using PS I from *Synechocystis* sp. PCC 6803, [NiFe]-H_2ase from *Desulfovibrio vulgaris* Miyazaki F, and cytochrome (Cyt) c_3 also from *D. vulgaris*.[118] In this complex, the Cyt c_3 was cross-linked to PsaE and the fusion protein was rebound to a PsaE-less PS I. H_2ase binds to the fusion protein through interaction with the Cyt c_3, which is the native electron donor in *D. vulgaris*, and light-induced hydrogen evolution occurred at a rate of 0.3 μmol H_2 mg Chl^{-1} h^{-1}.

While these systems aimed to bring the photochemical and catalytic components in close proximity, the rates in solution and the stability are lower than expected when compared to H_2ase activities in solution. It appears that electron transfer between the two components is not optimal and thus may still be diffusion-based. Depending on the mutual orientation of H_2ase relative to PS I, the distance between the Fe/S clusters can be as short as 14 Å and as long as 25 Å. Both the distance and the orientation of the Fe/S clusters would affect the rate of electron transfer. Because of the low probability that both of these factors are optimal, the rate of hydrogen production is diminished. However, provided that the long-term stability problem can be addressed, the fact that the rate can be so much improved after immobilization on a gold electrode is extremely promising.

17.2.4 Molecular Wire Technology: Eliminating Diffusion Chemistry

The technologies discussed up to this point for generating light-induced hydrogen do not employ a direct (*i.e.*, covalent) connection between the photochemical and catalytic components. This may account for the lowered efficiency of electron transfer between the two modules and may be the missing design feature. The benefit of using a molecular wire to connect the reaction modules is twofold: it tethers the two components at a fixed distance that is advantageous for rapid electron transfer, and it covalently attaches two Fe/S clusters, thereby extending the electron transfer pathway from the light module and thus providing a direct pathway for an electron from PS I to the catalytic component. In theory, a molecular wire could be attached to any of the electron transfer cofactors of PS I, although in practice this has only been successfully done at the A_1 quinone binding site,[119,120] and with the F_B cluster of the PsaC subunit.[121]

Due to the limited lifetime for forward electron transfer from PS I, the length of the wire and thus the distance between the two cofactors becomes important. The distance between cofactors and the rate of electron transfer can be described by Marcus theory, which relates the rate to distance, the Gibbs free energy change, reorganization energy, and temperature.[122] To achieve a high quantum yield and optimal rates, an electron should be transferred forward with a rate constant about two orders of magnitude faster than the competing charge recombination in PS I. This limitation results in a required forward electron transfer rate on the order of nanoseconds for the A_1 site, and microseconds for the F_B site. Therefore the distance limitations of a molecular wire are especially short for the A_1 site, and considerably longer for the F_B sites.

17.2.4.1 Attachment of a Molecular Wire to the A_{1A} and A_{1B} Phylloquinone Sites

The native phylloquinone can be extracted from PS I with the use of organic solvents, namely 50% diethyl ether in water, which leaves empty A_{1A} and A_{1B} sites.[123] It is possible to reconstitute foreign quinones back into the binding pocket with high efficiency. Using a synthetic naphthoquinone with an alkyl chain extension at the phytyl tail of the quinone to a 4,4′-bipyridinium derivative, the electron transfer chain of PS I has been extended out of the protein moiety (Figure 17.6).[124] The 4,4′-bipyridinium derivative acts as an electron transfer cofactor, while the naphthoquinone head group connects this wire to the electron transfer chain of PS I. To show that this approach is viable, transient absorption measurements at 600 nm were carried out and, upon excitation at 440 nm, a broad absorbance peak was observed that is not seen for wild-type PS I.

Not only has this approach been used to connect the A_{1A} and A_{1B} sites to a redox-active dye, but this approach has also been used to attach PS I to a gold nanoparticle as well as to an electrode surface.[119] Photooxidation of P_{700} was observed when the viologen unit was replaced with a gold nanoparticle,

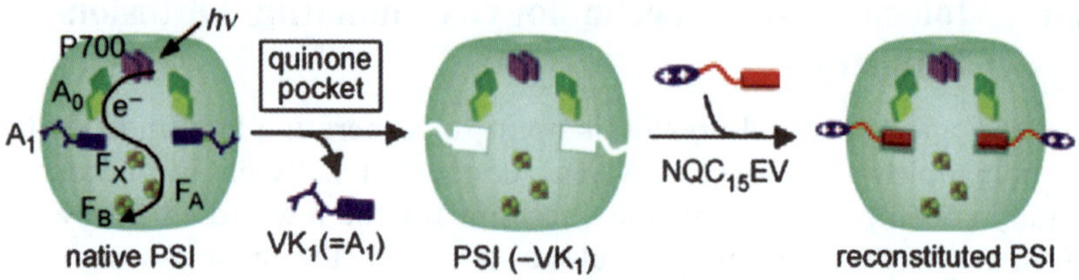

Figure 17.6 Reconstitution of the A_{1A}/A_{1B} sites with a modified external quinone.[124] Copyright Wiley-VCH Verlag GmbH & Co. KGaA. Reproduced with permission.

suggesting that the PS I complex is fully active with reconstituted quinones in the A_{1A} and A_{1B} sites. Furthermore, when this construct was adsorbed to a gold electrode through Au–S bonds, photoactivity was retained. Both the A_{1A} and A_{1B} sites can be reconstituted as visualized by TEM. In this manner, two nanoparticles can be attached to a single PS I complex, indicating that it should be possible to extract electrons from both the A- and B-branches of electron transfer cofactors.

It is possible that this tailorable technology could be adapted for attachment of PS I to a catalytic module to generate hydrogen, although this has not yet been attempted. This strategy would have the advantage that electrons would be extracted at much lower redox potentials ($E^{o\prime}$ = –800 mV for A_{1B} and –620 mV for A_{1A}) than at the F_B cluster ($E^{o\prime}$ = –580 mV), thereby providing a larger driving force for hydrogen generation. However, this approach has the disadvantage that, due to the much faster charge recombination with P_{700}^+ (lifetimes of 10 μs and 100 μs), the electron must be extracted from A_{1A} and A_{1B} in much shorter time periods (∼100 ns to 1 μs) so as to achieve a high quantum yield. Because the electron must quantum mechanically tunnel from the A_{1A} and A_{1B} sites to the catalyst, this necessarily limits the length of the molecular wire.

17.2.4.2 Basis for the Chemical Rescue of a [4Fe-4S] Cluster

Affixing a molecular wire to a [4Fe-4S] cluster involves the direct attachment of the wire to the most surface exposed iron atom. For this to be possible, control over the incorporation of the iron and sulfur atoms that make up the cluster is required. The technique involves reconstituting the Fe/S clusters, F_A and F_B, into the PsaC apoprotein *in vitro*.[125–130] The *psaC* gene is overexpressed in *E. coli*, which results in the accumulation of the PsaC apoprotein as an inclusion body. Inorganic Fe/S clusters are pre-formed in solution by using a ferrous salt, sodium sulfide and 2-mercaptoethanol; the 2-mercaptoethanol and sulfide ligate the iron atoms in solution resulting in cubane [4Fe-4S] clusters. The pre-formed cluster is then offered to the solubilized PsaC apoprotein and incorporated through ligand-exchange reactions. Because the [4Fe-4S] clusters are bound through coordination bonds with thiolates, the 2-mercaptoethanol ligands in the pre-formed cluster and the Cys residues in the holoprotein are of similar

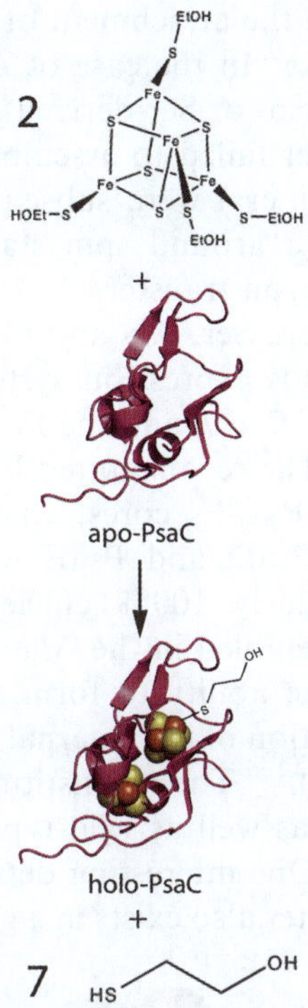

Figure 17.7 Proposed method of Fe/S cluster insertion into the $Cys_{13}Gly$ variant of unbound PsaC.

strength. Thus, any enthalpic contribution to the free energy of a ligand exchange reaction is likely to be negligible. Instead, the cluster insertion is driven by the entropic contribution to the free energy change because of the statistics of converting one PsaC apoprotein and two pre-formed Fe/S clusters into one PsaC holoprotein and eight 2-mercaptoethanol molecules (Figure 17.7).

17.2.4.3 Attachment of a Molecular Wire to the Terminal F_B Iron–Sulfur Cluster

The idea of wiring the F_B iron–sulfur cluster was ultimately conceived in a comparative analysis of *in vivo* and *in vitro* variants of PsaC, which were initially introduced for the purpose of identifying the specific cysteine ligands to the F_A and F_B clusters.[131,132]

In the *in vivo* study, the second Cys in the Cys-xx-Cys-xx-Cys-xxx-Cys-Pro motif harboring the F_B cluster was changed to Asp, Ser, Ala and Gly in *Synechocystis* sp. PCC 6803.[131,132] This position was chosen because there are examples of both naturally occurring and mutant dicluster ferredoxins in which the second Cys is replaced with an oxygen-containing amino acid, usually Asp. Additionally, the second Cys in this motif is usually located at the surface of the protein, which could facilitate the attachment of a molecular wire to the most exposed Fe atom of the cluster. In the case of Asp or Ser, a [4Fe-4S] cluster assembled with 3 Cys and 1 Asp- or Ser-derived oxygen ligand, but in the case of Ala or Gly, the Fe/S cluster failed to assemble and the PsaC protein (and most of the PS I reaction centers) were subsequently degraded. The mixed-ligand [4Fe-4S] clusters had a ground spin state of $S = 3/2$, but they were nonetheless competent in electron transfer.

In the *in vitro* study, the Asp, Ser, Ala and Gly variations were introduced into PsaC using a heterologous expression system.[23,127,133–135] Variant PsaC apoproteins were expressed in *E. coli* and the Fe/S clusters were reconstituted, as described in section 2.4.2. The reconstituted PsaC holoprotein was rebound along with PsaD ($\pm$ PsaE) to P_{700}-F_X cores, which were prepared by removing the stromal proteins PsaC, PsaD and PsaE with chaotrophic agents. The reassembly occurred with nearly 100% efficiency.[21,22] Unlike the *in vivo* mutants, [4Fe-4S] clusters assembled in the Ala and Gly variants as well as in the Asp and Ser variants. The ability to form a [4Fe-4S] cluster was subsequently explained by the retention of the external thiolate at the otherwise open coordination site (Figure 17.8).[136] The reconstituted PS I complexes were fully active and functioned nearly as well as wild-type PS I in electron transfer to ferredoxin and flavodoxin.[35] One interesting detail is that the Cys_3-2-mercaptoethanol ligated [4Fe-4S] cluster also exists in a ground spin state of $S = 3/2$.[137]

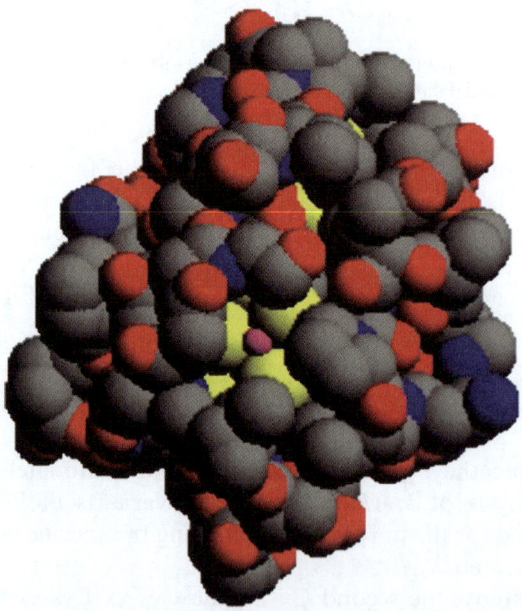

Figure 17.8 Depiction of exposed iron atom in the Cys_{13}Gly variant of unbound PsaC in the absence of 2-mercaptoethanol. The surface accessibility of one of the cubane iron atoms is clearly seen in this CPK depiction (PDB entry 1JB0). (Reproduced with permission from ref. 146. Copyright 2010 American Chemical Society.)

The similar spectroscopic properties of the Ala, Gly, Ser and Asp variants of PsaC suggest that the 2-mercaptoethanol ligand is not displaced by the hydroxyl group of Ser or the carboxyl group of Asp. This finding is consistent with studies of inorganic complexes, which have shown that sulfur is a better ligand than oxygen to [4Fe-4S] clusters.[135] The cluster exchange mechanism ultimately allows an external rescue ligand to be attached to the protein *in vitro*. In contrast, the mechanism of biological insertion must be substantially different because the *in vivo* Gly and Ala variants do not yield a functional PsaC. Assuming that the Fe/S clusters are exchanged into PsaC *in vivo* using a scaffold protein such as NfuA, the Cys ligand at the mixed coordination site would not be transferred into PsaC. Hence, the iron atom associated with the open coordination site is likely to be unstable. In summary, the *in vivo* and *in vitro* studies showed that: (i) a thiol-containing external ligand can chemically rescue PsaC by substituting for a missing Cys residue in the second position of a traditional [4Fe-4S] cysteine motif (Cys-xx-<u>C</u>ys-xx-Cys-xxx-Cys-Pro); and (ii) a sulfur ligand is preferred over an oxygen ligand to a [4Fe-4S] cluster.

When *p*-F^{19}-benzenethiol was used to pre-form the clusters instead of mercaptoethanol, the presence of the external ligand could be detected by F^{19}-NMR.[121] The chemical shift and paramagnetic broadening of the F^{19} proved that an external thiolate had chemically rescued the F_B cluster of PsaC. Because there is no amino acid side chain in the vicinity of the Gly-proximal iron atom, molecular wires with a variety of functionalities can be attached without steric hindrance. It is possible to introduce other thiolate-containing ligands into this site by one of two methods. In the first, Fe/S clusters are pre-formed using thiol-containing molecular wires and are subsequently incorporated into the protein by a mass-action-driven ligand exchange reaction. All but the thiol-containing wire at the Gly site of the PsaC variant are displaced. In the second, a facile thiol displacement reaction is carried out on the reconstituted PsaC variant in which a thiol-containing molecular wire displaces the single 2-mercaptoethanol rescue ligand. These exchange reactions have been best studied in non-aqueous media,[138] with substitution affinities that roughly parallel the pKa values of the thiolate down to a pKa of ≤ 6.5 without any significant degradation of the cubane core. Similar exchange reactions likely occur in aqueous medium with the Gly variant of PsaC. Both methods have successfully been employed to incorporate an external rescue ligand at the open coordination site on the F_B cluster of PsaC.

The ability of electrons to pass through the molecular wire has been assessed by attaching a viologen derivative, 1-(3-thiopropyl)-1′-(3-(acetylthio)propyl)-4,4′-bipyridinium, to the open coordination site of the PsaC variant using the thiol displacement reaction described above. The bipyridinium is redox-active, has a midpoint potential similar to that of paraquat ($E'^0 = -440$ mV), and has a strong absorbance at 600 nm when reduced by a single electron.[139] Electron transfer is thermodynamically favorable from F_B ($E'^0 = -580$ mV) to this dye molecule. Low-temperature EPR studies after reduction with sodium hydrosulfite, show a broad, axial set of resonances characteristic of an $S = 1/2$ [4Fe-4S] in the F_A site, and a sharp, derivative resonance due to the $S = 1/2$ bipyridinium radical, indicating that the bipyridinium wire is integral with the PsaC protein. This PsaC variant can be rebound to P_{700}-F_X cores in the presence of PsaD to construct a wired PS I construct that is photochemically active.

When cyt c_6 is used as the mediator and ascorbate as the sacrificial donor to $P_{700}{}^+$, the bipyridinium radical is reduced at a rate of 58.3 μmol mg Chl^{-1} h^{-1} in the light.[140] With continued illumination, the absorbance at 600 nm reaches a peak and then decreases as the reduction peaks and is replaced by the doubly reduced bipyridinium. This construct establishes proof-of-concept that a photochemical module such as PS I that can be directly wired to a catalytic component with the use of a molecular wire is achievable.

17.2.5 Catalytic H_2 Production with PS I–Molecular Wire–Nanoparticle Bioconjugates

Gold and platinum nanoparticles can be linked to the wired PS I construct through the use of alkyl or aryl dithiol-containing molecular wires.[141,142] One end of the dithiol wire displaces the 2-mercaptoethanol retained at the F_B cluster of the Gly variant of PsaC, while the other end binds to the nanoparticle surface. When cyanobacterial PS I is rebuilt with the Gly variant of PsaC containing the 2-mercaptoethanol ligand, and combined with 12 nm Au or 3 nm Pt nanoparticles and a 1,6-hexanedithiol wire, PS I-1,6-hexanedithiol-nanoparticle bioconjugates form spontaneously (Figure 17.9). When sodium ascorbate and 2,6-dichlorophenolindophenol (DPIP) are used as the sacrificial electron donor and mediator, respectively, light-induced hydrogen is generated

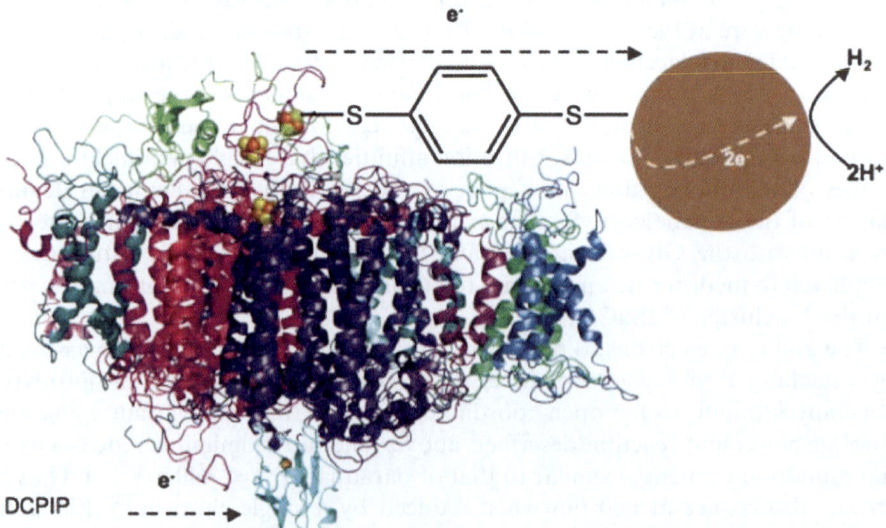

Figure 17.9 Cartoon of the PS I-molecular wire-nanoparticle bioconjugate.[142] (Reproduced by permission of The Royal Society of Chemistry.)

at a rate of 3.4 μmol H_2 mg Chl^{-1} h^{-1} for Au nanoparticle bioconjugates and 9.6 μmol H_2 mg Chl^{-1} h^{-1} for Pt nanoparticle bioconjugates. When Cyt c_6 was used as the mediator instead of DPIP (ascorbate still functions as the sacrificial donor), the rate of light-induced hydrogen increased to 49.3 μmol H_2 mg Chl^{-1} h^{-1}.[141] This observation suggests that the limiting factor for light-induced hydrogen production is the rate of $P_{700}{}^{+}$ reduction. Light-induced H_2 was not observed in controls that lack one or more of the components, or when PS I was reconstructed with wild-type PsaC.

17.2.5.1 Optimizing Light-induced H_2 Generation

The rate limitation of PS I-molecular wire-nanoparticle bioconjugates can occur at three sites: (i) an inefficiency at the PS I donor side, (ii) an inefficiency in electron transfer between the photochemical and catalytic components, and (iii) global factors effecting the rate of H_2 production.

Native electron donors such as Cyt c_6 and Pc can be cross-linked to the PS I reaction center near the P_{700} site to minimize donor side diffusion. Pc is more efficiently cross-linked to spinach PS I than Cyt c_6 is to cyanobacterial PS I. This is largely due to an extension in a lumenal loop of PsaF in spinach PS I that is not present in cyanobacterial PS I.[143,144] Another advantage to using spinach PS I is the incorporation of the light harvesting complex I (LHCI) to the reaction center core, resulting in a higher chlorophyll to P_{700} ratio for these LHCI-PS I complexes, which effectively increases the optical cross section and allows more photons to be absorbed per unit time under non-saturating light conditions. The method for generating Pc-crosslinked, wired-PS I complexes using spinach is the same as for the cyanobacteria. Spinach PS I–molecular wire–Pt nanoparticle bioconjugates with cross-linked Pc produced H_2 at a rate twice that for bioconjugates using the same components but using soluble Pc as the electron donor to $P_{700}{}^{+}$. This result indicates that the rate is donor-side limited, but at least some of this limitation can be overcome by cross-linking the mediator to PsaF.

Electron transfer between the photochemical and catalytic components has been shown to be a function of the length and aromaticity of the molecular wire. When different dithiol wires were incorporated into the bioconjugate, varying amounts of H_2 were produced (Table 17.1).[142] For the 1,6-hexanedithiol and 1,4-benzenedithiol wires, the average sulfur-to-sulfur distance was 11.5 Å and 8.2 Å, respectively, as determined by energy-minimized DFT calculations. A general trend is observed for the aliphatic wires: the best rate was found using 1,6-hexanedithiol, with shorter and longer wires producing lower rates. One possible explanation is that forward electron transfer through the molecular wire may be sufficiently slow at long wire lengths, so that charge recombination between $F_B{}^{-}$ and $P_{700}{}^{+}$ outcompetes forward electron transfer. An exception to this trend is observed for 1,3-propanedithiol, which produced the lowest rate of hydrogen production. This may be due to protein denaturation as a result of the close proximity to the nanoparticle surface. Another possibility for the low rate is that 1,3-propanedithiol may be too short to tether

Table 17.1 Molecular wire length and bond saturation affect the rate of H_2 production by plastocyanin cross-linked rebuilt spinach PS I-dithiol molecular wire-Pt nanoparticle bioconjugates.[142] (Reproduced by permission of The Royal Society of Chemistry).

Molecular wire	*Rate of H_2 production/ µmol H_2 mg Chl^{-1} h^{-1}*	*Rate of H_2 production/ µmol H_2 mol PS I^{-1} s^{-1}*
 1,3-propanedithiol	2.5	0.09
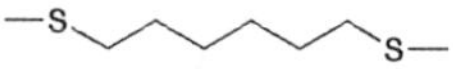 1,6-hexanedithiol	98.6	3.52
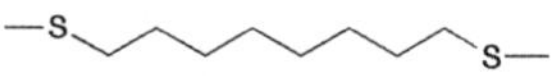 1,8-octanedithiol	49.0	1.75
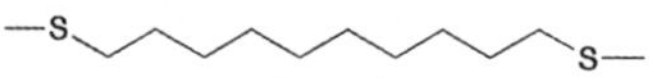 1,10-decanedithiol	16.1	0.57
 1,4-benzenedithiol	150.5	5.37
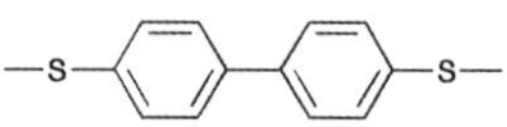 4,4'-biphenyldithiol	92.5	3.31

the two modules effectively. Aromatic wires lead to higher rates of light-driven H_2 evolution than aliphatic wires. Differences in the efficiency of bioconjugate formation for the various wires may explain these effects.

Solution pH and ionic strength are factors that affect the system globally. These two solution conditions can alter the binding affinities of proteins, but because Pc is chemically cross-linked to PS I, any effect on the donor side would be minimal. However, a major effect is likely to occur on the acceptor side, because altering the pH will change the concentration of H^+ ions in solution. This would thus alter the concentration available for adsorption, and ultimately for H_2 catalysis, at the surface of the Pt nanoparticle. As expected, the highest values of H_2 production were achieved at pH 6, below which the nanoparticles irreversibly aggregate. Nanoparticle aggregation also occurs at moderate ionic strengths.

Maximum light-induced H_2 production occurs when each of the above parameters are optimized. These conditions were: spinach LHCI-PS I complexes with cross-linked Pc as the electron donor, 1,4-benzenedithiol as the molecular wire, Pt nanoparticles, MES buffer at pH 6.0, 0.04 ionic strength, and illumination as a thin film at a light intensity of 996 µE m^{-2} s^{-1}. Using these bioconjugates and conditions resulted in a maximum rate of 312 µmol H_2 mg Chl^{-1} h^{-1},[142] which reflects an approximate six-fold increase in rate over that observed for bioconjugates without cross-linked Pc that had been made with 1,6-hexanedithiol and buffered at pH 8.3. By extrapolating to saturating light intensities by determining the V_{max} from the fit of the Michaelis–Menten kinetic equation, the theoretical maximal rate of hydrogen evolution should be 638 µmol H_2 mg Chl^{-1} h^{-1}, which is close to the theoretical maximal rate at one sun.

These experiments represent a proof-of-concept, demonstrating that the electron transfer chain of PS I can be extended outside the protein moiety and into an external catalytic unit for the efficient generation of light-induced hydrogen. One outstanding problem is that these bioconjugates employ platinum, an expensive and rare metal, to carry out the catalysis. Nevertheless, these experiments lay the groundwork for wiring PS I to biological catalysts such as H_2ase, or to Fe- or Ni-based inorganic catalysts based on abundant transition metals.

17.2.6 Light-induced Hydrogen Production with PS I–Molecular Wire–H_2ase Constructs

H_2ase can be wired to PS I in a similar manner to the Pt bioconjugates described above. The HydA protein of [FeFe]-H_2ase from *C. acetobutylicum* can be tethered to PS I by employing the same approach that was used to generate the Gly variant of PsaC (Figure 17.10). The surface-located, distal [4Fe-4S] cluster is altered so that the second Cys residue in the cluster-binding motif (Cys 98) is changed by site-specific mutagenesis to Gly, thus creating an open coordination site. The variant *hydA* gene is coexpressed in *E. coli* together with the H_2ase maturation genes from *C. acetobutylicum*, *hydEFG*, producing

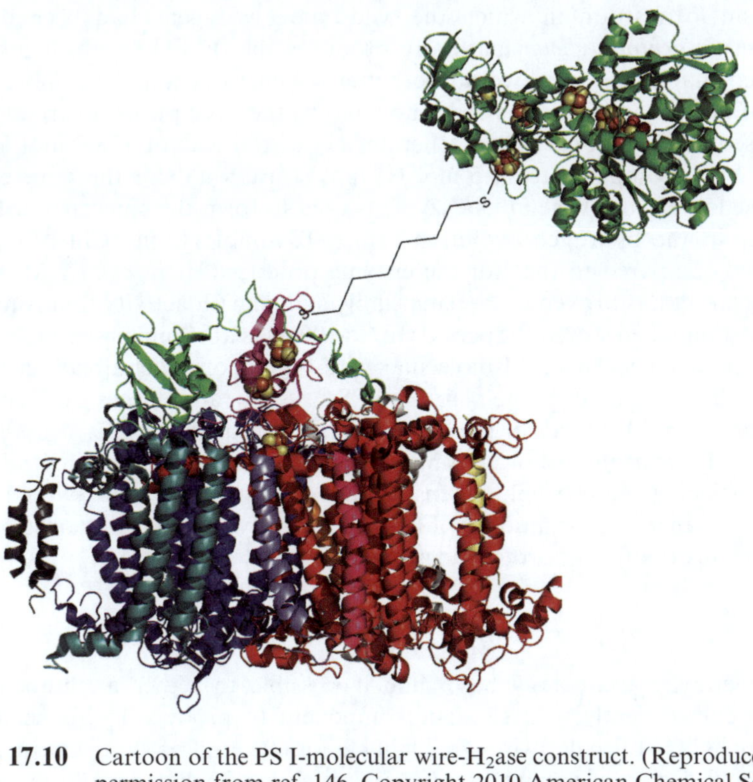

Figure 17.10 Cartoon of the PS I-molecular wire-H_2ase construct. (Reproduced with permission from ref. 146. Copyright 2010 American Chemical Society.)

the variant HydA holoprotein.[145] Cyclic voltammetry shows that electron transfer to and from the active site does not appear to have been altered significantly in the variant HydA enzyme. The variant enzyme is active in H_2 evolution activity (11.5 mol H_2 mol H_2ase^{-1} s^{-1}) using reduced methyl viologen as the electron donor, but it has a significantly lower activity compared to the wild-type enzyme (80.6 mol H_2 mol H_2ase^{-1} s^{-1}).

The F_B cluster of PsaC can be connected to the distal [4Fe-4S] cluster of the Gly variant of the HydA H_2ase enzyme using 1,6-hexanedithiol as the molecular wire. The PS I–molecular wire–H_2ase construct must be kept under strictly anaerobic conditions due to the oxygen sensitivity of the [FeFe]-H_2ase. A light-induced H_2 evolution rate of 3.9 μmol H_2 mg Chl^{-1} h^{-1} was measured using Cyt c_6 as the mediator and sodium ascorbate as the sacrificial donor.[146] This rate is approximately an order of magnitude higher than achieved previously using PS I in combination with H_2ase in solution.

While this construct produces hydrogen, the rate can probably be increased by employing a variant H_2ase enzyme with higher activity, by cross-linking an electron donor to PS I, by utilizing more efficient wires, and/or by altering the solution conditions (see Section 2.5.1 above). The expression of H_2ase in an *E. coli* background may not be ideal for obtaining a highly active enzyme since some of the machinery to assemble the enzyme properly may be suboptimal. The variant *C. acetobutylicum hydA* gene can also be expressed in the native organism in a strain in which the wild-type *hydA* gene had been deleted. Because *C. acetobutylicum* naturally produces the [FeFe]-H_2ase, it might be expected that this system would be better suited to synthesize a more active form of the enzyme. This has been shown to be the case: purified variant HydA expressed in *C. acetobutylicum* generates H_2 with a rate of 138.7 mol H_2 mol H_2ase^{-1} s^{-1}, compared to 11.5 mol H_2 mol H_2ase^{-1} s^{-1} for the same enzyme expressed in *E. coli*. When this enzyme is used to form the construct, a 10-fold increase in the hydrogen evolution rate (30.3 μmol H_2 mg Chl^{-1} h^{-1}) was observed compared to that for the enzyme produced in *E. coli*.[145] Moreover, this construct has proven to be remarkably stable, with activity that is more or less constant when stored for periods up to 100 days at room temperature under anoxic conditions. This PS I-molecular wire-H_2ase construct already generates more hydrogen than does the platinized PS I system, and it does so without the need for any noble metals or the excess energy to heat the device. Cross-linking Cyt c_6 to PS I, using a shorter conjugated molecular wire, lowering the pH to 6.0, and increasing the light intensity above 1000 μE m^{-2} s^{-1} would all be expected to increase the amount of light-induced hydrogen. This construct also outperforms the fusion proteins described above.

17.2.6.1 Optimization of a Molecular Wire

Molecular wire technology has made it possible to tether a photochemical component efficiently to a catalytic component to produce hydrogen directly from sunlight. Molecular wires are highly tailorable, and properties such as their length and degree of conjugation have been shown to significantly modify the rate of hydrogen production. It is assumed that the electron quantum mechanically tunnels between the F_B cluster and the catalytic component through the σ and π bonds of the molecular wire. For wires with lengths between 6.6 and 28 Å, electron transfer occurs on the nanosecond and picosecond timescale for either aliphatic or completely conjugated wires, respectively.[147] Because the wires employed in the PS I–molecular wire–catalyst systems described here (Table 17.1) are significantly shorter than those reported by Sikes *et al.*, rates of electron transfer through molecular wires are almost certain to outcompete the inherent back-reaction kinetics between P_{700}^{+} and F_B^{-}.

Alternatively, the electron may quantum mechanically tunnel through space (or more precisely, the aqueous medium) between the F_B cluster and catalyst, owing only to the proximity of the two redox cofactors to the molecular wire. The effect of aromatic wires on H_2 production rates might be less significant if this mechanism is dominant. Due to the flexibility of the molecular wire, the through-space distance between the two components is likely to change as a function of time. Therefore, it is probably realistic to assume that some combination of through-bond and through-space electron tunneling occurs in these tethered systems.

17.3 Functionalizing Electrode Surfaces

The incorporation of any of the aforementioned technologies into a sustainable solar energy device requires a connection with an electron source, *i.e.*, an anode. Ideally, the source of electrons should also be sustainable with the source of energy based on solar, instead of fossil fuel-based energy. By immobilizing both the cathode and anode onto an electrode, the two half-reactions can be coupled, and a complete electrolytic cell could be constructed.

This section will focus only on the cathode and will discuss the immobilization of PS I onto gold electrodes. The orientation and coverage (local concentration) are generally fixed when the photochemical component is immobilized on a surface, a configuration that may allow for more specific interactions between the photochemical component and the catalytic component. Much research has been devoted to the adsorption of proteins onto electrode surfaces; however, we choose to focus here on those approaches that would orient PS I with the P_{700} (donor) side adjacent to the electrode and the F_B site in a position favorable for interaction with the catalytic component. This can be achieved through a variety of methods, which include the direct attachment of unmodified and modified PS I reaction centers, through covalent attachment, and though docking that involves an intervening layer of Cyt c_6.

17.3.1 Adsorbing Unmodified Reaction Centers to Electrodes

The electrochemical and photoelectrochemical properties of photosynthetic reaction centers have been studied after their adsorption onto gold electrode surfaces. The attachment can be performed in several ways: for example, by direct adsorption of the reaction center on a surface[120,148–154] as well as by modifications to the reaction center itself.[155–158] Direct adsorption of PS I onto gold electrodes has been achieved using a mixture of self-assembled monolayers (SAMs) composed of alkanethiols terminated with functionalities that either attract PS I, through hydrophilic OH- or COOH-groups, or repel protein moieties, with hydrophobic CH_3- and CF_3-groups. While the surface coverage can be controlled easily by altering the ratio of attractive versus repulsive functionalities, the orientation of the reaction centers is random, with some complexes directing the F_B cluster away from the electrode in a manner that is receptive to interaction with the catalytic module, and with others oriented with F_B facing the electrode surface. When OH-terminated SAMs are used, 30% of the reaction center complexes orient randomly, while 70% orient with P_{700} proximal to the electrode.[149] The latter orientation is preferable because it ensures that electrons can be efficiently transferred between the electrode and the photochemical component, which requires short distances.

Additionally, PS I can be covalently immobilized on electrode surfaces through a reaction of any exposed Lys residues on the protein surface with an aldehyde group on the chemical terephthaldialdehyde (TPDA), forming an imine bond.[153,154] The TPDA group can be adhered to an electrode surface through attachment at the free end of a SAM. Nanoporous gold leaf electrodes functionalized with TPDA-capped SAMs result in a large content of bound PS I reaction centers, due to the larger surface area of the electrode. Unfortunately, since there are Lys resides on each side of PS I, this method results in a mixture of orientations.

Recently, a PS II complex has been immobilized at an electrode surface by employing an osmium functionalized redox polymer (poly(vinyl)imidazole $Os(bipy)_2Cl$-polymer).[159] The redox-active polymer effectively holds the photosynthetic reaction center close to the electrode surface, while providing a series of cofactors (–$Os(bipy)_2Cl$) for electron transfer between the reaction center and electrode. Additionally, this matrix is a hydrogel, which allows small molecules to diffuse almost freely through the polymer immobilization matrix (Figure 17.11). Because of the additional electron transfer cofactors between reaction center and electrode, and the permeability of the matrix, the exact orientation of the reaction center becomes less important. If this immobilization procedure can be adapted for PS I, a PS I functionalized electrode with high surface coverage and efficiency may result.

17.3.2 Attaching Variant Reaction Centers to Electrodes

To overcome the orientation heterogeneity problem, photosynthetic reaction centers have been physically modified to include residues that increase their binding affinity to a surface, either through genetically engineered Cys residues or poly-histidine tags. With both attachment strategies, the residues or tag should position the reaction center with the primary donor proximal to the electrode surface. Additionally, if the electrode is the source of electrons for the photochemical component, then the primary donor should be in very close proximity to the surface to ensure high rates and efficiencies of electron transfer to $P_{700}{}^{+}$.

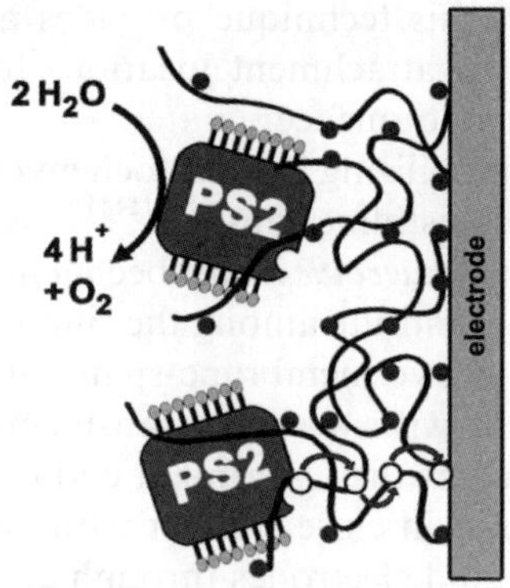

Figure 17.11 PS 2 entrapped within a mediator modified redox polymer. Arrows depict the electron transfer pathway by a hopping mechanism. (— polymer chains, * osmium complexes).[159] (Copyright Wiley-VCH Verlag GmbH & Co. KGaA. Reproduced with permission.)

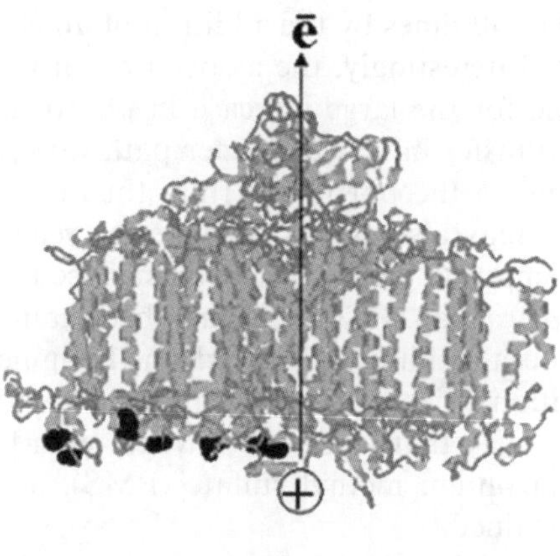

Figure 17.12 Covalent attachment of PS I to an electrode surface through introduced Cys residues.[155] Copyright Wiley-VCH Verlag GmbH & Co. KGaA. Reproduced with permission.

Using PS I from the cyanobacterium *Synechocystis sp.* PCC 6803, several residues in the extra membrane loops closest to P_{700} have been changed to cysteines (Figure 17.12).[155] To ensure that these positions are surface-located, the PS I complexes were reacted with a surface-active reagent, biotin-maleimide, which has the ability to modify thiol compounds. Four separate mutants ($D_{479}C$, $S_{499}C$, $S_{599}C$, and $Y_{634}C$) were shown to contain surface-exposed thiols. Each of these variants spontaneously assembled onto the surface of a clean gold electrode. Monolayers of the individual mutant proteins were assessed for their photochemical ability using Kelvin probe force microscopy. It was shown that bound PS I complexes were functionally photoactive and that they oriented in the same direction. Because all mutants in the study demonstrated this ability, this technique provides a broad approach, which does not depend on specific attachment locations for placing photochemical components onto an electrode surface.

Another method for immobilizing the photochemical component to a surface is through the use of poly-histidine tags.[157,158,160] A suitably tagged reaction center from *Rhodobacter sphaeroides* has been characterized extensively.[158] Bacterial reaction centers constitute among the simplest type of reaction center, consisting of three subunits: two membrane-spanning proteins L and M, and a largely extrinsic H protein with a single-transmembrane α-helix. The poly-histidine tag was integrated at the C-terminal end of the M-subunit, and thus can place the primary donor in close contact with an electrode surface. Reaction centers are bound to gold electrodes through an interaction of the His-tag and Ni^{2+} chelated nitrillotriacetic acid terminated SAMs. The oxidation of the primary donor was analyzed by cyclic voltammetry, and the reaction centers were found to be electrochemically active. This could only occur if the primary donor were positioned sufficiently close to the electrode surface, allowing electron transfer to occur between the electrode and the special pair.

Illumination of these devices generated low photocurrent levels that were shown to increase 20 to 40 times by the addition of an electron-donating Cyt *c* (from horse heart).[160] Interestingly, the association of Cyt *c* with the reaction center was responsible for the large increase in photocurrent. Cyt *c* not only divided the electron transfer into two shorter pathways, but it also allows for electron delocalization on the porphyrin ring, thus decreasing the total tunneling distance. Cyt *c* may also act to stabilize the reaction center in a more fixed and optimal orientation on the electrode surface (Figure 17.13).

PS I from *Synechocystis* sp. PCC 6803 has been immobilized onto a gold electrode through a decahistidine tag placed on the lumenal side of the PsaF subunit.[117] This position orients the reaction center with P_{700} closest to the electrode surface, in a way that is able to interact with the sacrificial electron donor *N*-methylphenazonium methyl sulfate (PMS), and upon illumination, photocurrents are sustained.

Stabilization of bacterial and plant reaction centers on electrode surfaces often requires peptide surfactants, which may ultimately interfere with the interaction of the catalytic module. Therefore, cyanobacterial and algal rather than higher plant reaction centers might prove more useful in the construction of a photochemical device. By employing Cys residues or poly-histidine tags, monolayers of photosynthetic reaction centers can be formed on electrodes in the proper orientation for contacting a catalytic module.

17.3.3 Optimizing Electron Transfer Between Electrode and Photoactive Component

While photocurrents have been measured for all of the aforementioned PS I/electrode systems, the efficiency of electron transfer between the electrode and the primary donor of PS I, P_{700}, has not been reported. It is likely that the

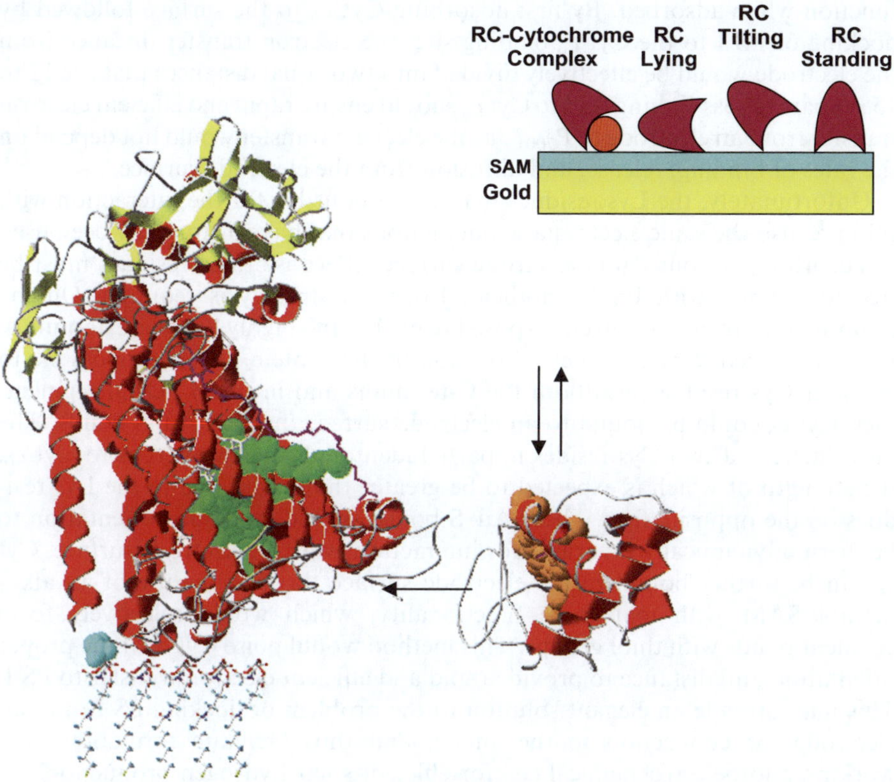

Figure 17.13 Bacterial Reaction Center adsorbed to electrode and stabilized with cytochrome *c*. Reproduced with permission from ref. 160. Copyright 2010 American Chemical Society.

distance between the electrode surface and P_{700} is rather long, especially for systems that rely on the use of highly functionalized surfaces employing several different SAMs of varying length and composition. This is usually the case for PS I given that it is a large membrane bound protein complex that, in its isolated form, contains a layer of detergent on its lateral surface. One possibility to maintain the appropriate distance of PS I from the surface (30 Å), while allowing for high rates of electron transfer, is to incorporate the natural electron donor Cyt c_6 between the electrode and PS I.

Cyt c_6 is a small, ~10 kDa, soluble redox protein that contains a single heme group. A patch of Lys residues are present on the solvent exposed surface of the heme group, which is specifically involved in binding to PS I. Electrostatic and hydrophobic interactions presumably position Cyt c_6 and PS I with the correct cofactor orientation and distance to allow for rapid electron transfer (15 μs).[161] In addition, Cyt c_6 can be chemically cross-linked to PS I in this orientation. Cyt c_6 is such a small protein that it interacts minimally with the electrode surface, meaning that it should be able to maintain both its structure and its function while adsorbed. By first adsorbing Cyt c_6 to the surface followed by docking of PS I to the Cyt c_6 binding site, the electron transfer distance from the electrode would be effectively divided into two equal distances that are 12 to 15 Å apart. Cross-linking PS I to Cyt c_6 should ensure rapid and efficient electron transfer from an electrode to $P_{700}{}^{+}$, as the electron transfer would not depend on the rates of binding, release, and diffusion from the electrode surface.

Unfortunately, the Lys residues that are responsible for the interaction with PS I are also the same electrostatic interactions that have historically been used to adsorb Cyt c_6 onto gold electrode surfaces. Because these residues must be free to interact with PS I, another adsorption strategy is required. On the opposite face from the solvent exposed heme lies the highly flexible C-terminus, which is preceded by an α-helical portion of the protein. If it were possible to engineer Cys residues into both the C-terminus and in this α-helical segment, then Cyt c_6 could be bound to an electrode surface through Au–S bonds. The introduction of two Cys resides imparts bidentate ligand properties to Cyt c_6, the strength of which is expected to be greater than those due to the Lys residues on the opposite face. These Au–S bonds should cause this orientation to be thermodynamically favorable for interactions with an electrode surface. Cyt c_6 can be further bound to the electrode surface through the use of an alkanethiol SAM with maleimide functionality, which would selectively form covalent bonds with thiol groups. This method would poise Cyt c_6 at the proper orientation and distance to provide rapid and efficient electron transfer to PS I. This may provide an elegant solution to the problem of docking PS I onto an electrode for connection to the anode, and thus facilitate formation of a working photoelectrochemical cell for efficient solar hydrogen production.

17.4 Summary

17.4.1 Future Prospects

While there is currently not a working photoelectrochemical device composed mainly of biological machinery, several fields from surface chemistry to biochemistry and molecular biology are making advances applicable to the creation of a device that efficiently couples the energy of the sun with hydrogen gas production. Ultimately, the implementation of this type of device depends on several factors. (i) Each component requires exhaustive optimization that, as shown from the technologies described above, should decrease the effect of diffusional parameters and increase stability. This step relies on understanding the reaction mechanisms and the underlying chemistry of electron transfer and H_2 catalysis in the system. Specifically, the optical cross section of PS I may be increased by employing a larger antenna system, such as that found when the chlorophyll binding protein, IsiA, forms rings around PS I complexes. Should this become available, an inherent mismatch in electron throughput between PS I and H_2ase may be addressed. Additionally, much research is focused on creating H_2ases with higher oxygen tolerances. Both of these advances would increase the value of a device, allowing it to work under a wider range of conditions. (ii) Perhaps most importantly is the integration of these biological components into an actual device. Drawing from the design of a classical electrochemical cell, the most promising way to do this is through the use of electrodes. Stability of these biological components on electrode surfaces remains one of the greater challenges for this technology. The efficacy of this device will increase exponentially as new methods are developed for immobilizing PS I and/or Cyt c_6 onto surfaces. (iii) As inorganic hydrogen catalysts utilizing small amounts of abundant metals become available, they can be substituted for the H_2ase enzyme in an existing device. These may be able to offer higher stability and a lower device cost depending on the particular catalyst. This same logic can be applied to advances in synthetic photosynthesis or photochemical charge separation. Because a variety of component substitutions are feasible, as well as the ability to work in combination with new technologies, the versatility of such a device may allow for its application in a number of diverse environments.

17.4.2 A Reality Check

Biologically inspired systems that employ PS I as a means of photon capture and as a source of strongly reducing electrons are yielding promising results in solar hydrogen production. Nevertheless, there are challenges to the use of hydrogen as a viable fuel. In spite of its advantages, one formidable problem is that the widespread use of H_2 would require large changes in energy/fueling infrastructure, starting from generation through storage to distribution and utilization. Another problem is that although H_2 has nearly three times the energy content of gasoline on a weight basis (120 MJ kg^{-1} *vs.* 44 MJ kg^{-1}), the situation is reversed on a volume basis ((3 MJ L^{-1}) at 5000 psi or 8 MJ L^{-1} as a liquid *vs.* 32 MJ L^{-1} for gasoline), which translates to an immensely larger storage, distribution and utilization system. An additional problem is that any H_2 generated would be at atmospheric pressure, and it would need to be compressed, liquefied or stored as a hydride for transportation from the site of generation to the site of utilization. The production of liquid H_2 involves using 30–35% of the available energy in liquefaction alone, and because of its low heat of vaporization, the best Dewar storage vessels have boil-off rates of 15% per day. The compression of H_2 involves using 10–15% of the energy available to achieve 10 000 psi. Metal hydrides are expensive and heavy; the 'holy grail' would be the discovery of a material based on a light metal that would be capable of storing large amounts of H_2. In spite of these technological challenges, initial studies have provided solid evidence that a PS I-based biohybrid system has significant potential to be used in a H_2-producing device. To establish a functional device, these drawbacks will have to be eliminated or diminished.

17.4.3 Final Thoughts on a Bioinspired Device

What has been presented here is an overview of the current status of solar hydrogen production utilizing PS I (Table 17.2). PS I is able to convert solar

Table 17.2 Comparative table of H_2 production rates among different technologies discussed here.

Technology	*H_2 evolution rate/μmol H_2 mg Chl^{-1} h^{-1}*
Osmium coated Chloroplasts	0.113
Platinized PS I	0.025
Platinized PS I, cross-linked Pc	0.080
Platinized PS I, cross-linked Pc, 55°C	5.5
PS I-wire-Au NP	3.4
PS I-wire-Pt NP	9.6
PS I-wire-Pt NP, cross-linked Cyt c_6	49.3
PS I-wire-Pt NP, optimized	312
PS I-Cyt c_3-H_2ase Fusion	0.3
PS I-H_2ase Fusion	0.58
PS I-H_2ase Fusion, on electrode	2812 (unstable)
PS I-wire-H_2ase (H_2ase *E. coli* expression)	3.9
PS I-wire-H_2ase (H_2ase *C. acetobutylicum* expression)	30.3

energy into useful chemical reducing power with high quantum yields and efficiencies. It is also very robust and easily manipulated. Utilizing Pt for the catalytic component is an unfortunately impractical design, owing to its low natural abundance and high cost. Another major issue will be its competitive use in other products, such as batteries and hydrogen fuel cells, which will likely be in higher demand as the world moves towards an economy based less on fossil fuels and more on renewable sources. H_2ase enzymes can catalyze the formation of hydrogen at very high rates, although due to their oxygen sensitivity, they are not as robust as the PS I reaction center. The long-term value of such a device will rely on the ability to improve the oxygen sensitivity of H_2ase enzymes or the implementation of inorganic catalysts that are constructed from cheap, Earth-abundant elements. Of course, there can be no device if this cathodic system cannot be integrated onto an electrode that connects it with an operating anode that would split water from sunlight. Ideally, both the anode and cathode need to be cost effective: the components should be inexpensive and abundant, the construct should be easy to assembly or preferably self-assemble, and if at all possible it should be sustainable and self-repairing. To this end, there are no devices that can promise all of these features, however the building blocks are present and they are waiting to be assembled.

References

1. M. A. Rosen and D. S. Scott, *Int. J. Hydrogen Energy*, 1998, **23**, 653–659.
2. U. N. D. Program, *World Energy Assesment Report: Energy and the Challenge of Sustainability*, New York, 2003.
3. J. Golbeck H., *Proc. Nat. Acad. Sci. U. S. A.*, 1993, **90**, 1642–1646.
4. P. Heathcote, M. R. Jones and P. K. Fyfe, *Philos. Trans. R. Soc. London, Ser. B*, 2003, **358**, 231–243.

5. P. Jordan, P. Fromme, H. T. Witt, O. Klukas, W. Saenger and N. Krauß, *Nature*, 2001, **411**, 909–917.
6. K. Brettel and M. H. Vos, *FEBS Lett.*, 1999, **447**, 315–317.
7. J. H. Golbeck, *Photosynthetic Reaction Centers*, Biophysical Society, Bethesda, MD, 2004.
8. C. C. Moser, C. C. Page, R. J. Cogdell, J. Barber, C. A. Wraight and P. L. Dutton, *Adv. Protein Chem.*, 2003, **63**, 71–109.
9. S. Lien and A. San Pietro, *An Inquiry into Biophotolysis of Water to Produce Hydrogen*, 1971.
10. M. Gibbs, A. Hollaender, B. Kok, L. O. Krampitz and A. San Pietro, *Proceedings of the Workshop on Bio-Solar Conversion*, 1973.
11. B. Ke, R. E. Hansen and H. Beinert, *Proc. Nat. Acad. Sci. U. S. A.*, 1973, **70**, 2941–2945.
12. P. Heathcote, D. L. Williams-Smith, C. K. Sihra and M. C. W. Evans, *Biochim. Biophys. Acta*, 1978, **503**, 333–342.
13. I. Terashima, S. Funayama and K. Sonoike, *Planta*, 1994, **193**, 300–306.
14. K. Sonoike, I. Terashima, M. Iwaki and S. Itoh, *FEBS Lett.*, 1995, **362**, 235–238.
15. I. Terashima, K. Noguchi, T. ItohNemoto, Y. M. Park, A. Kubo and K. Tanaka, *Physiol. Plant*, 1998, **103**, 295–303.
16. K. G. Parrett, T. Mehari, P. G. Warren and J. H. Golbeck, *Biochim. Biophys. Acta*, 1989, **973**, 324–332.
17. J. H. Golbeck, K. G. Parrett, T. Mehari, K. L. Jones and J. J. Brand, *FEBS Lett.*, 1988, **228**, 268–272.
18. N. Li, P. Warren, J. Golbeck, G. Frank, H. Zuber and D. Bryant, *Biochim. Biophys. Acta*, 1991, **1059**, 215–225.
19. J. H. Golbeck, T. Mehari, K. Parrett and I. Ikegami, *FEBS Lett.*, 1988, **240**, 9–14.
20. K. G. Parrett, T. Mehari and J. H. Golbeck, *Biochim. Biophys. Acta*, 1990, **1015**, 341–352.
21. J. Zhao, P. V. Warren, N. Li, D. A. Bryant and J. H. Golbeck, *FEBS Lett.*, 1990, **276**, 175–180.
22. N. Li, J. Zhao, P. Warren, J. Warden, D. Bryant and J. Golbeck, *Biochem.*, 1991, **30**, 7863–7872.
23. J. Zhao, N. Li, P. Warren, J. Golbeck and D. Bryant, *Biochem.*, 1992, **31**, 5093–5099.
24. J. H. Golbeck and J. M. Cornelius, *Biochim. Biophys. Acta*, 1986, **849**, 16–24.
25. J. T. Warden and J. H. Golbeck, *Biochim. Biophys. Acta*, 1986, **849**, 25–31.
26. J. H. Golbeck, K. G. Parrett and A. E. McDermott, *Biochim. Biophys. Acta*, 1987, **893**, 149–160.
27. J. H. Golbeck, K. G. Parrett and L. L. Root, in *Prog. Photosynth. Res., Proc. Int. Congr. Photosynth. 7th Meeting*, ed. J. Biggins, 1986, vol. 1, 253–256.
28. T. W. Johnson, G. Shen, B. Zybailov, D. Kolling, R. Reategui, S. Beauparlant, I. R. Vassiliev, D. A. Bryant, A. D. Jones, J. H. Golbeck and P. R. Chitnis, *J. Biol. Chem.*, 2000, **275**, 8523–8530.

29. A. Y. Semenov, I. R. Vassiliev, A. van der Est, M. D. Mamedov, B. Zybailov, G. Shen, D. Stehlik, B. A. Diner, P. R. Chitnis and J. H. Golbeck, *J. Biol. Chem.*, 2000, **275**, 23429–23438.
30. B. Zybailov, A. van der Est, S. G. Zech, C. Teutloff, T. W. Johnson, G. Shen, R. Bittl, D. Stehlik, P. R. Chitnis and J. H. Golbeck, *J. Biol. Chem.*, 2000, **275**, 8531–8539.
31. T. W. Johnson, B. Zybailov, A. D. Jones, R. Bittl, S. Zech, D. Stehlik, J. H. Golbeck and P. Chitnis, *J. Biol. Chem.*, 2001, **276**, 31512–31521.
32. Y. Sakuragi, B. Zybailov, G. Shen, A. D. Jones, P. R. Chitnis, A. van der Est, R. Bittl, S. Zech, D. Stehlik, J. H. Golbeck and D. A. Bryant, *Biochemistry*, 2002, **41**, 394–405.
33. A. Van Der Est, A. I. Valieva, Y. E. Kandrashkin, G. Shen, D. A. Bryant and J. H. Golbeck, *Biochemistry*, 2004, **43**, 1264–1275.
34. F. Yang, G. Shen, W. M. Schluchter, B. Zybailov, A. O. Ganago, I. R. Vassiliev, D. A. Bryant and J. H. Golbeck, *J. Phys. Chem.*, 1998, **102**, 8288–8299.
35. J. H. Golbeck, *Photosynth. Res.*, 1999, **61**, 107–149.
36. Y. Mizukoshi, Y. Makise, T. Shuto, J. Hu, A. Tominaga, S. Shironita and S. Tanabe, *Ultrason. Sonochem.*, 2007, **14**, 387–392.
37. K. Gurunathan, P. Maruthamuthu and M. V. C. Sastri, *Int. J. Hydrogen Energy*, 1997, **22**, 57–62.
38. K. B. Dhanalakshmi, S. Latha, S. Anandan and P. Maruthamuthu, *Int. J. Hydrogen Energy*, 2001, **26**, 669–674.
39. P. M. Vignais and A. Colbeau, *Curr. Issues Mol. Biol.*, 2004, **6**, 159–188.
40. P. M. Vignais, *Results Probl. Cell Differ.*, 2008, **45**, 223–252.
41. P. M. Vignais and B. Billoud, *Chem. Rev.*, 2007, **107**, 4206–4272.
42. J. C. Fontecilla-Camps, A. Volbeda, C. Cavazza and Y. Nicolet, *Chem. Rev.*, 2007, **107**, 4273–4303.
43. J. C. Fontecilla-Camps, *Amino Acids*, 2009, **37**, 37–46.
44. L. C. Seefeldt, B. M. Hoffman and D. R. Dean, *Annu. Rev. Biochem.*, 2009, **78**, 701–722.
45. M. L. Ghirardi, A. Dubini, J. Yu and P. C. Maness, *Chem. Soc. Rev.*, 2009, **38**, 52–61.
46. J. C. Fontecilla-Camps, P. Amara, C. Cavazza, Y. Nicolet and A. Volbeda, *Nature*, 2009, **460**, 814–822.
47. C. Zirngibl, W. Vandongen, B. Schworer, R. Vonbunau, M. Richter, A. Klein and R. K. Thauer, *Eur. J. Biochem.*, 1992, **208**, 511–520.
48. E. J. Lyon, S. Shima, R. Boecher, R. K. Thauer, F. W. Grevels, E. Bill, W. Roseboom and S. P. Albracht, *J. Am. Chem. Soc.*, 2004, **126**, 14239–14248.
49. E. J. Lyon, S. Shima, G. Buurman, S. Chowdhuri, A. Batschauer, K. Steinbach and R. K. Thauer, *Eur. J. Biochem.*, 2004, **271**, 195–204.
50. S. Shima, E. J. Lyon, M. Sordel-Klippert, M. Kauss, J. Kahnt, R. K. Thauer, K. Steinbach, X. Xie, L. Verdier and C. Griesinger, *Angew. Chem., Int. Ed.*, 2004, **43**, 2547–2551.

51. S. Shima, O. Pilak, S. Vogt, M. Schick, M. S. Stagni, W. Meyer-Klaucke, E. Warkentin, R. K. Thauer and U. Ermler, *Science*, 2008, **321**, 572–575.
52. P. Tamagnini, E. Leitão, P. Oliveira, D. Ferreira, F. Pinto, D. J. Harris, T. Heidorn and P. Lindblad, *FEMS Microbiol. Rev.*, 2007, **31**, 692–720.
53. A. Volbeda, M. H. Charon, C. Piras, E. C. Hatchikian, M. Frey and J. C. Fontecilla-Camps, *Nature*, 1995, **373**, 580–587.
54. Y. Higuchi, H. Ogata, K. Miki, N. Yasuoka and T. Yagi, *Structure*, 1999, **7**, 549–556.
55. Y. Higuchi, F. Toujou, K. Tsukamoto and T. Yagi, *J. Inorg. Biochem.*, 2000, **80**, 205–211.
56. Y. Higuchi, T. Yagi and N. Yasuoka, *Structure*, 1997, **5**, 1671–1680.
57. Y. Montet, P. Amara, A. Volbeda, X. Vernede, E. C. Hatchikian, M. J. Field, M. Frey and J. C. Fontecilla-Camps, *Nature Struct. Biol.*, 1997, **4**, 523–526.
58. M. Rousset, Y. Montet, B. Guigliarelli, N. Forget, M. Asso, P. Bertrand, J. C. Fontecilla-Camps and E. C. Hatchikian, *Proc. Natl. Acad. Sci. U. S. A.*, 1998, **95**, 11625–11630.
59. P. M. Matias, C. M. Soares, L. M. Saraiva, R. Coelho, J. Morais, J. Le Gall and M. A. Carrondo, *J. Biol. Inorg. Chem.*, 2001, **6**, 63–81.
60. E. Garcin, X. Vernede, E. C. Hatchikian, A. Volbeda, M. Frey and J. C. Fontecilla-Camps, *Structure*, 1999, **7**, 557–566.
61. H. Ogata, W. Lubitz and Y. Higuchi, *Dalton Trans.*, 2009, 7577–7587.
62. R. Bingemann and A. Klein, *Eur. J. Biochem.*, 2000, **267**, 6612–6618.
63. L. Casalot and M. Rousset, *Trends Microbiol.*, 2001, **9**, 228–237.
64. M. Blokesch and A. Bock, *J. Mol. Biol.*, 2002, **324**, 287–296.
65. M. Blokesch, A. Paschos, E. Theodoratou, A. Bauer, M. Hube, S. Huth and A. Böck, *Biochem. Soc. Trans.*, 2002, **30**, 674–680.
66. G. Maroti, B. D. Fodor, G. Rakhely, A. T. Kovacs, S. Arvani and K. L. Kovacs, *Eur. J. Biochem.*, 2003, **270**, 2218–2227.
67. G. Maroti, Y. Tong, S. Yooseph, H. Baden-Tillson, H. O. Smith, A. T. Kovacs, M. Frazier, J. C. Venter and Q. Xu, *Appl. Environ. Microbiol.*, 2009, **75**, 5821–5830.
68. T. Burgdorf, O. Lenz, T. Buhrke, E. van der Linden, A. K. Jones, S. P. Albracht and B. Friedrich, *J. Mol. Microbiol. Biotechnol.*, 2005, **10**, 181–196.
69. R. Cramm, *J. Mol. Microbiol. Biotechnol.*, 2009, **16**, 38–52.
70. Y. Xu, The Pennsylvania State University, 2010.
71. L. Forzi and R. G. Sawers, *Biometals*, 2007, **20**, 565–587.
72. M. R. Leach and D. B. Zamble, *Curr. Opin. Chem. Biol.*, 2007, **11**, 159–165.
73. S. Watanabe, R. Matsumi, T. Arai, H. Atomi, T. Imanaka and K. Miki, *Mol. Cell*, 2007, **27**, 29–40.
74. A. Böck, P. W. King, M. Blokesch and M. C. Posewitz, *Adv. Microb. Physiol.*, 2006, **51**, 1–71.
75. J. W. Zhang, G. Butland, J. F. Greenblatt, A. Emili and D. B. Zamble, *J. Biol. Chem.*, 2005, **280**, 4360–4366.

76. A. Magalon and A. Bock, *FEBS Lett.*, 2000, **473**, 254–258.
77. C. Massanz, S. Schmidt and B. Friedrich, *J. Bacteriol.*, 1998, **180**, 1023–1029.
78. J. Meyer, *Cell Mol. Life Sci.*, 2007, **64**, 1063–1084.
79. S. T. Stripp and T. Happe, *Dalton Trans.*, 2009, 9960–9969.
80. Z. Chen, B. J. Lemon, S. Huang, D. J. Swartz, J. W. Peters and K. A. Bagley, *Biochemistry*, 2002, **41**, 2036–2043.
81. D. W. Mulder, E. S. Boyd, R. Sarma, R. K. Lange, J. A. Endrizzi, J. B. Broderick and J. W. Peters, *Nature*, 2010, **465**, 248–252.
82. J. W. Peters, W. N. Lanzilotta, B. J. Lemon and L. C. Seefeldt, *Science*, 1998, **282**, 1853–1858.
83. Y. Nicolet, C. Piras, P. Legrand, E. C. Hatchikian and J. C. Fontecilla-Camps, *Structure*, 1999, **7**, 13–23.
84. Y. Nicolet, B. J. Lemon, J. C. Fontecilla-Camps and J. W. Peters, *Trends Biochem. Sci.*, 2000, **25**, 138–143.
85. A. S. Pandey, L. J. Giles, J. W. Peters and R. K. Szilagyi, *J. Am. Chem. Soc.*, 2008, **130**, 4533–4540.
86. M. C. Posewitz, P. W. King, S. L. Smolinski, R. D. Smith, A. R. Ginley, M. L. Ghirardi and M. Seibert, *Biochem. Soc. Trans.*, 2005, **33**, 102–104.
87. M. C. Posewitz, P. W. King, S. L. Smolinski, L. P. Zhang, M. Seibert and M. L. Ghirardi, *J. Biol. Chem.*, 2004, **279**, 25711–25720.
88. P. W. King, M. C. Posewitz, M. L. Ghirardi and M. Seibert, *J. Bacteriol.*, 2006, **188**, 2163–2172.
89. C. M. English, C. Eckert, K. Brown, M. Seibert and P. W. King, *Dalton Trans.*, 2009, 9970–9978.
90. J. K. Rubach, X. Brazzolotto, J. Gaillard and M. Fontecave, *FEBS Lett.*, 2005, **579**, 5055–5060.
91. X. Brazzolotto, J. K. Rubach, J. Gaillard, S. Gambarelli, M. Atta and M. Fontecave, *J. Biol. Chem.*, 2006, **281**, 769–774.
92. S. E. McGlynn, S. S. Ruebush, A. V. Naumov, L. E. Nagy, A. Dubini, P. W. King, J. B. Broderick, M. C. Posewitz and J. W. Peters, *J. Biol. Inorg. Chem.*, 2007, **12**, 443–447.
93. D. W. Mulder, D. O. Ortillo, D. J. Gardenghi, A. V. Naumov, S. S. Ruebush, R. K. Szilagyi, B. Huynh, J. B. Broderick and J. W. Peters, *Biochemistry*, 2009, **48**, 6240–6248.
94. S. E. McGlynn, E. M. Shepard, M. A. Winslow, A. V. Naumov, K. S. Duschene, M. C. Posewitz, W. E. Broderick, J. B. Broderick and J. W. Peters, *FEBS Lett.*, 2008, **582**, 2183–2187.
95. Y. Nicolet, J. K. Rubach, M. C. Posewitz, P. Amara, C. Mathevon, M. Atta, M. Fontecave and J. C. Fontecilla-Camps, *J. Biol. Chem.*, 2008, **283**, 18861–18872.
96. S. E. McGlynn, D. W. Mulder, E. M. Shepard, J. B. Broderick and J. W. Peters, *Dalton Trans.*, 2009, 22.
97. R. C. Driesener, M. R. Challand, S. E. McGlynn, E. M. Shepard, E. S. Boyd, J. B. Broderick, J. W. Peters and P. L. Roach, *Angew. Chem., Int. Ed.*, 2010, **49**, 1–5.

98. M. R. Challand, F. T. Martins and P. L. Roach, *J. Biol. Chem.*, 2010, **285**, 5240–5248.
99. E. Pilet, Y. Nicolet, C. Mathevon, T. Douki, J. C. Fontecilla-Camps and M. Fontecave, *FEBS Lett.*, 2009, **583**, 506–511.
100. J. W. Peters, R. K. Szilagyi, A. V. Naumov and T. Douglas, *FEBS Lett.*, 2006, **580**, 363–367.
101. E. M. Shepard, S. E. McGlynn, A. L. Bueling, C. S. Grady-Smith, S. J. George, M. A. Winslow, S. P. Cramer, J. W. Peters and J. B. Broderick, *Proc. Natl. Acad. Sci. U. S. A.*, 2010, **107**, 10448–10453.
102. I. Czech, A. Silakov, W. Lubitz and T. Happe, *FEBS Lett.*, 2009, **584**, 638–642.
103. D. I. Arnon, A. Paneque and A. Mitsui, *Science*, 1961, **134**, 1425–1437.
104. J. R. Benemann, J. A. Berenson, N. O. Kaplan and M. D. Kamen, *Proc. Natl. Acad. Sci. U. S. A.*, 1973, **70**, 2317–2320.
105. A. Ben-Amotz and M. Gibbs, *Biochem. Biophys. Res. Commun.*, 1975, **64**, 355–359.
106. A. A. Krasnovsky, C. Van Ni, V. V. Nikandrov and G. P. Brin, *Plant Physiol.*, 1980, **66**, 925–930.
107. H. McTavish, *J. Biochem.*, 1998, **123**, 644–649.
108. N. Waschewski, G. Bernat and M. Rogner, in *Biomass to Biofuels*, ed. A. Vertes, N. Qureshi, H. Yukawa and H. P. Blaschek, John Wiley & Sons, Chichester, UK, 2009.
109. E. Greenbaum, *Science*, 1985, **230**, 1373–1375.
110. E. Greenbaum, *J. Phys. Chem.*, 1988, **92**, 4571–4574.
111. J. W. Lee, C. V. Tevault, S. L. Blankinship, R. T. Collins and E. Greenbaum, *Energy Fuels*, 1994, **8**, 770–773.
112. J. W. Lee, I. Lee and E. Greenbaum, *J. Phys. Chem. B*, 2005, **109**, 5409–5413.
113. J. F. Millsaps, B. D. Bruce, J. W. Lee and E. Greenbaum, *Photochem. Photobiol.*, 2001, **73**, 630–635.
114. B. R. Evans, H. M. O'Neill, S. A. Hutchens, B. D. Bruce and E. Greenbaum, *Nano Lett.*, 2004, **4**, 1815–1819.
115. I. J. Iwuchukwu, M. Vaughn, N. Myers, H. O'Neill, P. Frymier and B. D. Bruce, *Nature Nanotech.*, 2009, **5**, 73–79.
116. M. Ihara, H. Nishihara, K. S. Yoon, O. Lenz, B. Friedrich, H. Nakamoto, K. Kojima, D. Honma, T. Kamachi and I. Okura, *Photochem. Photobiol.*, 2006, **82**, 676–682.
117. H. Krassen, A. Schwarze, B. Friedrich, K. Ataka, O. Lenz and J. Heberle, *ACS Nano*, 2009, **3**, 4055–4061.
118. M. Ihara, H. Nakamoto, T. Kamachi, I. Okura and M. Maeda, *Photochem. Photobiol.*, 2006, **82**, 1677–1685.
119. N. Terasaki, N. Yamamoto, K. Tamada, M. Hattori, T. Hiraga, A. Tohri, I. Sato, M. Iwai, S. Taguchi, I. Enami, Y. Inoue, Y. Yamanoi, T. Yonezawa, K. Mizuno, M. Murata, H. Nishihara, S. Yoneyama, M. Minakata, T. Ohmori, M. Sakai and M. Fujii, *Biochim. Biophys. Acta, Bioenerg.*, 2007, **1767**, 653–659.

120. N. Terasaki, N. Yamamoto, T. Higara, I. Sato, Y. Inoue and S. Yamada, *Thin Solid Films*, 2006, **499**, 153–156.
121. M. L. Antonkine, E. M. Maes, R. S. Czernuszewicz, C. Breitenstein, E. Bill, C. J. Falzone, R. Balasubramanian, C. Lubner, D. A. Bryant and J. H. Golbeck, *Biochim. Biophys. Acta, Bioenerg.*, 2007, **1267**, 712–724.
122. R. A. Marcus and N. Sutin, *Biochim. Biophys. Acta*, 1985, **811**, 265–322.
123. S. Itoh, M. Iwaki and I. Ikegami, *Biochim. Biophys. Acta*, 2001, **1507**, 115–138.
124. N. Terasaki, N. Yamamoto, T. Hiraga, Y. Yamanoi, T. Yonezawa, H. Nishihara, T. Ohmori, M. Sakai, M. Fujii, A. Tohri, M. Iwai, Y. Inoue, S. Yoneyama, M. Minakata and I. Enami, *Angew. Chem., Int. Ed.*, 2009, **48**, 1585–1587.
125. J. D. Zhao, N. Li, P. V. Warren, J. H. Golbeck and D. A. Bryant, *Biochemistry*, 1992, **31**, 5093–5099.
126. Y. S. Jung, I. R. Vassiliev, F. Y. Qiao, F. Yang, D. A. Bryant and J. H. Golbeck, *J. Biol. Chem.*, 1996, **271**, 31135–31144.
127. L. Yu, J. D. Zhao, W. P. Lu, D. A. Bryant and J. H. Golbeck, *Biochemistry*, 1993, **32**, 8251–8258.
128. T. Mehari, F. Y. Qiao, M. P. Scott, D. F. Nellis, J. D. Zhao, D. A. Bryant and J. H. Golbeck, *J. Biol. Chem.*, 1995, **270**, 28108–28117.
129. L. Yu, D. A. Bryant and J. H. Golbeck, *Biochemistry*, 1995, **34**, 7861–7868.
130. L. Yu, I. R. Vassiliev, Y. S. Jung, D. A. Bryant and J. Golbeck, H., *J. Biol. Chem.*, 1995, **270**, 28118–28125.
131. Y. S. Jung, I. R. Vassiliev, J. P. Yu, L. McIntosh and J. H. Golbeck, *J. Biol. Chem.*, 1997, **272**, 8040–8049.
132. J. P. Yu, I. R. Vassiliev, Y. S. Jung, J. H. Golbeck and L. McIntosh, *J. Biol. Chem.*, 1997, **272**, 8032–8039.
133. L. Yu, D. A. Bryant and J. H. Golbeck, *Biochemistry*, 1995, **34**, 7861–7868.
134. T. Mehari, F. Y. Qiao, M. P. Scott, D. F. Nellis, J. D. Zhao, D. A. Bryant and J. H. Golbeck, *J. Biol. Chem.*, 1995, **270**, 28108–28117.
135. Y. S. Jung, I. R. Vassiliev, F. Y. Qiao, F. Yang, D. A. Bryant and J. H. Golbeck, *J. Biol. Chem.*, 1996, **271**, 31135–31144.
136. M. Antonkine and J. H. Golbeck, Molecular Interactions of the Stromal Subunit PsaC with the PsaA/PsaB Heterodimer, in *Photosystem I: The Light- The Light-Driven Plastocyanin: Ferredoxin Oxidoreductase*, ed. J. H. Golbeck, Springer, Dordrecht, 2006, pp. 79–98.
137. J. H. Golbeck, in *Photosynthesis: Molecular Bio*, ed. L. Mcintosh, Academic Press Inc, San Diego, CA, 1998, vol. 297, pp. 95–123.
138. L. Que, M. A. Bobrick, J. A. Ibers and R. Holm, *J. Am. Chem. Soc.*, 1974, **96**, 4168–4177.
139. L. A. Summers, *The Bipyridinium Herbicides*, Academic Press Inc., New York, NY, 1980.
140. C. E. Lubner, Ph. D. dissertation, The Pennsylvania State University, 2011.

141. R. A. Grimme, C. E. Lubner, D. A. Bryant and J. H. Golbeck, *J. Am. Chem. Soc.*, 2008, **130**, 6308–6309.
142. R. A. Grimme, C. E. Lubner and J. Golbeck, *Dalton Trans.*, 2009, 10106–10113.
143. J. Farah, F. Rappaport, Y. Choquet, P. Joliot and J. D. Rochaix, *EMBO J.*, 1995, **14**, 4976–4984.
144. M. Hippler, F. Drepper, W. Haehnel and J. D. Rochaix, *Proc. Natl. Acad. Sci. U. S. A.*, 1998, **95**, 7339–7344.
145. C. E. Lubner, P. Knorzer, P. J. N. Silva, K. A. Vincent, T. Happe, D. A. Bryant and J. H. Golbeck, *Biochemistry*, 2010, **49**, 10264–10266.
146. C. E. Lubner, R. A. Grimme, D. A. Bryant and J. H. Golbeck, *Biochemistry*, 2010, **49**, 404–414.
147. H. D. Sikes, J. F. Smalley, S. P. Dudek, A. R. Cook, M. D. Newton, C. E. D. Chidsey and S. W. Feldberg, *Science*, 2001, **291**, 1519–1523.
148. B. S. Ko, B. Babcock, G. K. Jennings, S. G. Tilden, R. R. Peterson, D. Cliffel and E. Greenbaum, *Langmuir*, 2004, **20**, 4033–4038.
149. I. Lee, J. W. Lee and E. Greenbaum, *Phys. Rev. Lett.*, 1997, **79**, 3294–3297.
150. M. Ciobanu, H. A. Kincaid, V. Lo, A. D. Dukes, G. K. Jennings and D. Cliffel, *J. Electroanal. Chem.*, 2007, **599**, 72–78.
151. I. Lee, J. W. Lee, A. Stubna and E. Greenbaum, *J. Phys. Chem. B*, 2000, **104**, 2439–2443.
152. H. A. Kincaid, T. Niedringhaus, M. Ciobanu, D. E. Cliffel and G. K. Jennings, *Langmuir*, 2006, **22**, 8114–8120.
153. C. J. Faulkner, S. Lees, P. N. Ciesielski, D. Cliffel and G. K. Jennings, *Langmuir*, 2008, **24**, 8409–8412.
154. P. N. Ciesielski, A. M. Scott, C. J. Faulkner, B. J. Berron, D. E. Cliffel and G. K. Jennings, *ACS Nano*, 2008, **2**, 2465–2472.
155. L. Frolov, Y. Rosenwaks, C. Carmeli and I. Carmeli, *Adv. Mater.*, 2005, **17**, 2434–2437.
156. I. Carmeli, C. Carmeli, L. Frolov and S. Richter, *J. Am. Chem. Soc.*, 2007, **129**, 12352–12353.
157. R. Das, P. J. Kiley, M. Segal, J. Norville, A. A. Yu, L. Y. Wang, S. A. Trammell, L. E. Reddick, R. Kumar, F. Stellacci, N. Lebedev, J. Schnur, B. D. Bruce, S. G. Zhang and M. Baldo, *Nano Lett.*, 2004, **4**, 1079–1083.
158. S. A. Trammell, L. Wang, J. M. Zullo, R. Shashidar and N. Lebedev, *Biosens. Bioelectron.*, 2004, **19**, 1649–1655.
159. A. Badura, D. Guschin, B. Esper, T. Kothe, S. Neugebauer, W. Schuhmann and M. Rogner, *Electroanalysis*, 2008, **20**, 1043–1047.
160. N. Lebedev, S. A. Trammell, A. Spano, E. Lukashev, I. Griva and J. Schnur, *J. Am. Chem. Soc.*, 2006, **128**, 12044–12045.
161. A. Diaz, F. Navarro, M. Hervas, J. A. Navarro, S. Chavez, F. J. Florencio and M. A. de la Rosa, *FEBS Lett.*, 1994, **347**, 173–177.

CHAPTER 18

Future Perspectives on Solar Fuels

T. A. FAUNCE*[a]

[a]Australian National University, College of Medicine, Biology and the Environment and College of Law (joint appointment), Australian Research Council Future Fellow, College of Law, Fellows Road, Acton, Canberra ACT, 0200, Australia

18.1 Introduction

In Zebra River Canyon, Western Namibia it is possible to view evidence of three vastly separated ages of life on earth, each with poignant implications for this book and the theme of this chapter in particular. First, the region possesses some of the best-preserved Proterozoic stromatolites on the planet. These fossils derived from the cementing and accreting of sedimentary grains bound by the biofilms of cyanobacteria about 1.5 billion years ago. Cyanobacteria were amongst the earliest forms of life on earth, transducing sunlight, water and carbon dioxide into their food, in the process filling our atmosphere with oxygen and providing sustenance for more complex organisms.

Second, just a short walk from the stromatolites, is a field littered with stone tools made about 1.5 million years ago when such human technology was state-of-the-art and set to usher in then unimaginable eras of social and environmental change.

Third, a hundred years ago, this same desert country was brutally cleared of its Herero and Namaqua inhabitants with the assistance of military technology (the construction of which was powered by fossil fuels). To encounter the remnants of these displaced people, you have to drive an hour from Zebra River Canyon to a town where they survive in a shanty suburb, desperately

RSC Energy and Environment Series No. 5
Molecular Solar Fuels
Edited by Thomas J. Wydrzynski and Warwick Hillier

Published by the Royal Society of Chemistry, www.rsc.org

short of food and fuel for light (and thus education), as well as cooking and heat. Such conjunctions of plant fossils, early technology artifacts and poverty poignantly highlight just how our existence and moral worth as a species is dependent not only on photosynthesis, but the decisions we make about how to use technology directly or indirectly related to it.

The earlier chapters in this book have demonstrated how humanity is rapidly approaching the point where artificially improved photosynthesis has a reasonable prospect of becoming the cutting edge technological development that defines an age of humanity, just as the production of stone tools did all those years ago. This chapter opens by examining whether the research and development of molecular solar fuels will be characterized in future by its promotion of fundamental societal virtues such as equality and environmental sustainability. As a thought experiment, it presents a vision of some important elements of such a future world—one where energy is primarily not only a matter of global artificial photosynthesis (GAP), but of such virtues.

Central to the future perspective presented here is nanotechnological construction with enhanced efficiency of each aspect of the natural photosynthetic process into units capable of inexpensive mass production for domestic use. This involves a vision of artificial photosynthesis research that incorporates the utilization of genetically-engineered (or even wholly synthetic) organisms undertaking photosynthesis, but emphasizes worldwide household utilization of structures that are chiefly a matter of engineered nanochemistry.

The challenge of how to promptly develop nanotechnologically-based artificial photosynthesis is explored through a future perspective on the scientific challenges involved in three key areas covered by sections of this book: *Light Capture* (including light-harvesting complexes, synthetic pigment arrays and dye-sensitive solar cells), *Photochemical Conversion* (including structure of the photosynthetic reaction centres, optimizing photochemical quantum yield) and *Energy Storage* (including optimization of photosynthetic water oxidation and catalysis).

As the world's human population rises, GAP may take the pressure off natural photosynthesis as our primary (and water-intensive) source of food and fuel. The chapter concludes by considering how inequalities and injustices in the use of solar fuel technology can best be avoided or minimized. It extends the analysis presented in Chapter 1 to scope the governance, legal and regulatory obstacles likely to confront the future research and development of solar fuels. In particular, it presents the case for a macro-science GAP Project designed not only to accelerate the pace of such research to meet the urgency of climate change and human energy and food needs, but to do so in a manner coherent with equity and environmental sustainability.

18.2 What Will a World Powered by Solar Fuels Look Like?

At the Copenhagen Climate Conference in December 2009, the world's nation states created the *Copenhagen Accord*. This non-binding political agreement recognized the critical impacts of population growth and fossil fuel-driven climate change as well as the need to establish a comprehensive adaptation program including international support for those countries most vulnerable to its adverse effects.[1] For the first time, all major CO_2-emitting countries agreed to a target of keeping global warming to less than 2°C above pre-industrial levels. The Copenhagen Accord also contained important undertakings concerning mitigation (including the Copenhagen Green Climate Fund and establishing a mechanism to accelerate renewable energy technology development and transfer).[2] That agreement must be viewed in the context of internationally agreed targets to reduce poverty and lack of necessary fuel and food as expressed in the United Nations *Millennium Development Goals*.[3] These critical survival issues for the poor will be exacerbated as global population grows towards 10 billion by 2050 and energy consumption rises over 600 EJ/yr.

Many observers have derided such so-called 'soft-norm' (unenforceable and non-legally binding) agreements as facilitating a model for business-as-usual by the fossil fuel-related industries that are so central to energy supply and anthropogenic climate change. What if, however, science could provide in the next twenty years a means by which such goals and principles could be satisfied without compromising the capacity of people to obtain sufficient energy for survival and flourishing?

It is a salient fact, often reproduced in this context, that more solar energy strikes the Earth's surface in one hour of each day than the energy used by all human activities in one year.[4,5] At present daily power consumption for a citizen of a developed nation is about 125 kWh/day (~250 kWh/day for a citizen of the United States); much of this power being devoted to transport (~40 kWh/day), heating (~40 kWh/day) and domestic electrical appliances (~18 kWh/day), with the remainder lost in electricity conversion and distribution.[6] Table 18.1 sets out energy/power conversion units. As Table 18.2 demonstrates, world energy consumption (which varies greatly on estimates, for example, by the United National Development Program or the International Energy Agency) is currently in the region of 400 EJ/yr.

The solar energy potentially usable at ~1.0 kilowatts per square metre of the earth is 3.9×10^6 EJ/yr. As Table 18.2 reveals this is vastly more than our current or projected global levels of energy consumption. Raw sunshine at midday on a cloudless day, for example, can deliver ~1000 W m^{-2} in mid-latitude regions and ~1200 W m^{-2} in low-latitude dry desert areas.[7] If we take into account, however, the earth's tilt, diurnal and atmospheric influences on solar intensity, then this figure becomes approximately ~110 W m^{-2}.[8]

Photosynthesis, the ultimate source of our oxygen, food and fossil fuels, has been operating on earth for 2.5 GYr, since a time known in geological circles as the great oxidation event (GOE).[9] Photosynthesis accounts for a global annual CO_2 flux of 124 PgC/yr[10] and an annual O_2 flux of ~10^{11} t/yr.[11] In its present technologically-unenhanced form, photosynthesis globally already traps around 4000 EJ/yr solar energy in the form of biomass.[12] The global biomass energy potential for human use from photosynthesis as it currently operates globally (focusing, for example, on crop and forestry residues, energy crops,

Table 18.1 Energy and power unit conversion.

Energy and power units Power is the rate at which energy is used. 1 Joule of energy = work done when a force of 1 Newton moves its point of application 1 metre. 1 Newton of force = force which will give a mass of 1 kg an acceleration of 1 metre per second per second. 1 Watt of power = 1 Joule per sec (1 kilowatt = 1000 Watts). 1 kWh = 3.6 x 10^6 J = 3.6 MJ The kilowatt-hour is a standard unit of electricity production and consumption. One 40 W light bulb constantly on uses 1 kWh/day. The average citizen in a developed nation uses 125 kWh/d.
Exajoule (EJ): (exa (E) = 10^{18}) is a measure of global energy supply and consumption. 1 EJ = 10^{18} J The EJ/yr is a measure of power conveniently fitting with yearly estimates of global power consumption.
Terawatt (TW): (tera (T) = 10^{12}) is a measure of global power supply and consumption. 1 TW/yr = 8.76 × 10^{12} kWh = 31.54 EJ The terawatt is a common unit of global power consumption, taking into account time in seconds and so making TW/yr a more convenient measure of global power.
Gigawatt (GW): (giga (G) = 10^9) is a measure of industrial power supply and consumption 1 GW/yr = 8.76 × 10^9 kWh. Large individual plants have capacities in the neighborhood of 1 GW and the gigawatt-year (GW/yr) is a common unit in discussions of global electricity power production.
Megatonne of oil equivalent (Mtoe10^6): 1 Mtoe = 4.1868 × 10^{16} J = 4.1868 × 10^4 TJ = 11630 GWh Gigatonne of oil equivalent (Gtoe10^9): 1 Gtoe = 41.868 EJ The International Energy Agency uses Mtoe in its calculations of the world energy statistics

and animal and municipal wastes) has been variously estimated to be 33 to 1135 EJ/year[13], 104 EJ/year[14] or 91 to 675 EJ/year.[15]

Working on the basis that genetically-unenhanced plants are less than 2% efficient in converting solar energy into carbohydrates (producing say 0.5 W/m^2), then even if 3000 m^2 per person is devoted to it, biomass will indirectly (*via* intermediate energy carriers) contribute only 36 kWh/day per person.[16] Current non-life-based solar energy technology is even less efficient. If houses in the future, for instance, have 10 m^2 of 20%-efficient solar panels per person, then only ~5 kWh/day per person will be produced. Even large solar farms (for example taking up 200 m^2 per person with 10%-efficient solar panels) could produce but ~50kWh/day per person.[17] This is still a long way short of ~125 kWh/day average developed nation consumption and still leaves the problem of power storage for night use and transport. Clearly, if we are to make better use of the ratio of available solar energy to world annual energy consumption then a better way to convert solar energy into fuel needs to be developed.

As the earlier chapters of this book demonstrate, researchers now are actively redesigning photosynthesis to achieve, for example, low cost, localised, direct

Table 18.2 Annual global solar energy received compared to energy consumption.[a]

Parameter	*Energy/Power*
Annual solar energy intercepted by the Earth at ~1.37 kW m^{-2}	5.5×10^6 EJ/yr
Annual solar energy reflected by the atmosphere back to space at ~0.3 kW m^{-2}	1.6×10^6 EJ/yr
Annual solar energy potentially usable at ~1 kW m^{-2}	3.9×10^6 EJ/yr
Annual solar energy practically usable (with existing solar technology) at ~110W m^{-2}	1575 (min) EJ/yr 49 837 (max) EJ/yr
2004 primary energy consumption *UNDP estimate*	402 EJ (12.76 TW/yr)
2008 primary energy consumption *IEA estimate*	352.8 EJ (8428 Mtoe) (11.18 TW/yr)
2050 primary energy consumption *UNDP estimate*	590–1050 EJ
2100 primary energy consumption *UNDP estimate*	880–1900 EJ
Ratio of potentially usable annual solar energy to current primary energy consumption/yr	~9000 to 1
Ratio of practically usable annual solar energy to current primary energy consumption/yr (with existing solar technology)	~3.9 (min) to 1 or ~124 (max) to 1

[a]Sources: H-H Rogner in United Nations Development World Energy Assessment, United Nations, 2004, ch. 5, p. 162 and International Energy Agency, Key World Energy Statistics, IEA, Paris, 2010, p. 30.

(without intermediate energy carriers) conversion of sunlight, water and carbon dioxide into fuel for heating and cooking.[18] One such approach is to genetically manipulate or even synthetically reproduce photosynthetic plants and bacteria to maximize their light capture and carbon reduction activities.[19] Another, more focused on here, is to use nanotechnology to create enhanced non-life-based models of the photosynthetic process.

Artificially enhanced photosynthesis, if applied equitably, could assist crop production on marginal lands, reduce atmospheric CO_2 levels, lower geopolitical and military tensions over fossil fuel, food and water scarcity and create carbon-neutral hydrogen fuel for domestic, community and industrial storage.[20] What if the global economy's energy dependence on fossil fuels could be reduced as buildings, cars, planes and ships became producers of their own fuel? Economies would restructure to emphasize smaller locally-powered and controlled units, minimizing energy use in transportation (much present-day energy consumption is dedicated to gathering raw materials and low-cost labor to make and transport goods for use in other countries). Nanotechnology-based artificial photosynthesis systems might be programmed to remove carbon dioxide from the atmosphere in proportion as it was used in the burning of the produced ethanol, or help coastal industrial plants split sea water using sunlight to produce carbon-neutral hydrogen-based fuels.

Such thought experiments involve a future perspective in which solar fuels created though humanity's capacity to fully understand the principles and enhance (particularly through nanotechnology) the operation of photosynthesis, have become the predominant form of energy generation on the planet. Such micro or local generation of food and fuel will challenge the present paradigm of centralized fossil-fuel oriented power generation controlled by multinational corporations. It may not be an easy transition for them to manage and they may resist or try to delay the change to a GAP-fuelled world.

A proposition advanced in this chapter is that establishing the ethical and legal principles for dissemination of GAP technology will be equally important with facilitating the scientific collaborations that will allow it to take place in time to address the major societal and environmental challenges that the expanding human population and its dependence on fossil fuels are currently creating.

Societies, just as individuals, acquire virtues or character traits through the consistent application of principles in the face of obstacles.[21] In the past, those principles were predominantly a matter of philosophic and religious, as well as legislative and judicial debate. Presently, and certainly into the future, it will be statements on bioethics and human rights by international organizations such as the United Nations or UNESCO that will consensually shape the principles that should be applied in globally rolling out new technologies such as artificial photosynthesis. The United Nations *Millennium Development Goals*, for instance, are particularly focused on issues of energy storage, production and conversion, agricultural productivity enhancement, water treatment and remediation and experts have encouraged nanotechnology to systematically contribute to their achievement.[22] Principles supporting similar goals (and directed to individuals, communities and private corporations and well as States (article 1)) appear in the UNESCO *Universal Declaration on Bioethics and Human Rights* (particularly the social responsibility principle in article 14(b)—'access to adequate nutrition and water', 14(c)—'improvement in living conditions and the environment' and 14(e)—'reduction in poverty and illiteracy').[23]

It may be that GAP is destined to be the moral culmination of nanotechnology. Individuals and communities seeking virtue through supporting environmental sustainability as a core part of their implicit social contract will not only strive to minimize energy consumption but embrace artificial photosynthesis as a means of powering their dwellings and transportation. If such a future perspective is to be realized, changes in scientific progress must be made in the three key areas previously presented as sections in this book. The following paragraphs aim to provide a future perspective on these areas as a basis for discussing how such breakthroughs can be used for societal and environmental good.

18.3 Future Challenges for Global Dissemination of Solar Fuels

18.3.1 Light Capture

One future perspective on GAP light capture involves nanostructured materials or synthetic organisms absorbing photons from a much wider region of the solar spectrum (photon absorption by antenna chlorophyll molecules in thylakoid membranes of chloroplasts, for example, is currently restricted primarily to ~430–700 nm).[24]

The technology behind this improved light capture system may include mesoporous thin film dye-sensitive solar cells of semiconductor nanoparticles[25] and carbon nanotubes harvesting and conducting the resultant electricity.[26] Nanomaterials and hybrid organic–inorganic nanostructures will have improved the conversion efficiency of existing PV units and have pushed the boundary of the Shockley–Queisser limit, possibly by extrapolation from the technique of multiple stacking of solar cells with increasing band gap energies to exploit the solar spectrum more profitably.[27,28] In addition, the mid-infrared (IR) range (7–14 μm) in the solar spectrum, under-utilized by current technology, may have become a routinely utilized source of electricity, even at night when the radiation persists. Such IR radiation could be collected by Nanoantenna Electromagnetic Collectors, the electrons flowing freely in the antenna generating an alternating current[29] which is converted into a direct current (DC) by high frequency rectification.[30] In the case of microwave-to-DC conversion under ideal conditions, the efficiency could be as high as 84%.[31]

Whichever method is ultimately used, light capture will separate charge efficiently over macroscopic distances using inexpensive materials.[32,33] Full understanding will have been obtained of how the molecular dynamics of bridge molecules in a donor–bridge–acceptor system control long distance electron transport, allowing the preparation and characterization of new self-assembling antenna building blocks that elicit formation of a reaction centre over multiple cycles of photon absorption and charge separation.[34] This will probably involve new knowledge of how self-assembly is obtained, for example, by entropy-driven 'depletion–attraction'[35] as well as physical forces such as hydrogen bonding.[36]

One of the most exciting prospects in the area of light capture is that the excitation energies of each nanoscale arranged pigment chromophore (excitons) in the antenna array will have been mapped in terms of protein spatial, orientational and energetic factors, facilitating the construction of artificial photosynthetic electron pathways to the reaction centre that perform a single quantum computation, sensing many states simultaneously and so enhancing the efficiency of the energy capture and transfer at physiological temperatures.[37–39] The premise that this ideally would allow a future nanotechnological artificial photosynthesis antenna system to sample all the potential electron pathways instantaneously and 'choose' the most efficient has been popularized through Ian McEwan's novel *Solar*.[40]

18.3.2 Photochemical Conversion

The future successful nanostructured light harvesting antenna system will interface at high energy transfer efficiency with fully characterised artificial versions of the PS II and PSI reaction centres.[41,42] A nanotechnological version of a fully characterized PS II may not be limited to using four successive charge–separation reactions in four manganese ions and a calcium ion to oxidize two water molecules (stored by the oxygen-evolving complex (OEC)) to yield a molecule of diatomic oxygen, four hydrogen ions and electrons (then siphoned off for storage in chemical bonds).

PS II in plants is a complex protein with 27 subunits and 32 cofactors involved in electron transfer and light harvesting. Any synthetic mimic of this protein (maquette) must be simpler.[43] The ultimately successful GAP model of PS II is likely to involve significantly more efficient versions of the chlorophyll complex, redox-active tyrosine and tetra-nuclear manganese/calcium cluster (Mn_4Ca).[44] Higher structural resolution and computation analysis will have clarified the water substrate binding sites.[45] Research in photochemical conversion will have discovered ways to decrease entropy by imposing constraints on the spatial relations amongst the pigments, donors and acceptors.[46]

Household artificial photosynthesis systems will incorporate designer molecules that prolong charge separation.[47] These will probably apply one or either of two strategies: chemical bonds in supramolecular structures, or supports such as polymers, zeolites, sol-gel glasses, lipid membranes and self-assembled films.[48]

18.3.3 Energy Storage

The perfected GAP catalysts for oxidation of water and for the reduction of CO_2 will be based on detailed understanding of multi-electron and multi-step catalytic events.[49] The quantum yield will have increased to near unity and the turnover number to $\sim 10^9$ perhaps by exploiting the capacity of the built-in electric field of nanoscale and microscale semiconductor p–n junctions to add photovoltages in series.[50] The most globally widespread water catalytic system will involve inexpensive and self-repairing components that operate under ambient conditions at neutral pH with non-pure (salty or bacterially and chemically contaminated) water.[51] Such a novel water oxidation centre will also be stable to a variety of exposure conditions in air, water and heat.[52]

The water-splitting sections of GAP devices in widespread use may involve metal complexes and molecular assemblies on surfaces and in rigid media.[53] A variety of other approaches may also have proven successful.[54] A major challenge overcome will have been to optimize the free energy (ΔG) required for the overall water splitting process.[55,56] New generation semiconducting oxide photoelectrodes may have been perfected.[57] Multiwalled carbon nanotubes[58] and singlewalled carbon nanotubes might have produced the critical breakthrough.[59] Water oxidation systems utilizing photosensitive components grafted by core-shell nanowires to a genetically engineered virus may have proven important.[60] So may the role of carotenoid polyenes in providing protection from singlet oxygen damage.[61]

New catalysts for H_2 production and methods for efficient H_2 storage will have been built. Nanotechnology is likely to have made smaller and more efficient components that are less susceptible to oxygen induced free-radical damage than H_2-evolving bacteria genetically modified with less oxygen-sensitive hydrogenases, or even wholly synthetic photosynthetic bacteria with electron flow directed towards the H_2-producing enzymes and away from competing pathways.[62]

In the GAP 'dark reaction', ATP and NADPH as well as carbon dioxide (CO_2) will be used in an enhanced version of the Calvin–Benson cycle to make locally usable food or fuel (for domestic, heating, cooking, light and transport) in the form of carbohydrate *via* the enzyme RuBisCO (ribulose-1,5-bisphosphate carboxylase oxygenase).[63] Redox-active CO_2 carriers will recover the necessary CO_2 directly from the atmosphere and pump it over significant concentration ranges.[64] CO_2 reduction and H_2O oxidation half-reactions may have been coupled to remove the need for a sacrificial electron donor.[65] Photochemical methods coupled with time-resolved spectroscopy will have identified and monitored intermediates and rate-limiting processes in the reaction cycle leading to H_2 evolution, aiding the development of photosensitizers and electron relays that are not inhibited by acids.[66] Multiple photoinduced charge separation pathways will accumulate several redox equivalents at a single redox site to improve the efficiency of CO_2 reduction, and bio-inspired self-repair strategies will have ensured this aspect of GAP survives damage from repeated cycles of thermodynamically demanding reactions.[67]

One GAP model involves bio-mimetic polymer photovoltaic generators plugged in to the national electricity grid to power hydrogen fuel and waterless agriculture, chemical feedstocks and polymers for fibre production.[68] This model has the advantage of the 'light' and 'dark' reactions being uncoupled in relation not only to energy/material flow balance, but also to the requirement to be co-located in space. Such an uncoupling will vastly extend the area for capturing light over otherwise barren land, and also allow the elimination or reduction of molecular oxygen in GAP reactions, enhancing longevity of the components.

The model presented here, however, emphasizes the greater potential for individual and community economic autonomy implicit in micro or local generation of fuel and food through GAP units installed as a policy priority on domestic dwellings and vehicles. Large GAP facilities providing fuel for industry or backup supply can still be preferentially located under such a model near large sources of seawater, CO_2, waste heat, high solar irradiation and proximity to end use facilities.

18.4 Governance Models for Global Artificial Photosynthesis

For a future perspective in which artificial photosynthesis (utilizing the breakthroughs described above) emerges as a dominant form of local (domestic and household), as well as industrial, power production, a crucial issue will be whether this contributes meaningfully to the issues of climate change or global poverty before environmental and social 'tipping points' are reached. As can been seen from the discussion in Section 18.3, advances in light capture, energy conversion and storage related to GAP will require the development of novel structures and systems over which, in the normal course of events, intellectual monopoly privileges (IMPs), such as patents, will be claimed.

At one level such privileges provide incentives for innovation; at another they may inhibit collaboration and prompt, effective development of the field.[69] Already, for example, there are major 'dual-use' and intellectual monopoly privilege concerns about synthetic organisms and proteins, such as those that may be used in the creation in enhanced GAP versions of PSII and PSI.[70]

Potential governance models for GAP to get around these issues could either involve full access to patents, open-source sharing or variations of governance techniques designed to protect global public goods. Such models and ways to effectively promote global artificial photosynthesis will be discussed in the following sections. As we will see, one way to address them may be as part of the governance structure for a macro-science GAP Project.

18.4.1 Fostering International Collaborations

Numerous competitively funded research teams have dedicated GAP-related projects already underway in many developed nations, as is shown in Table 18.3.[71] A dozen European research partners, for example, form the Solar-H network, supported by the European Union. These include Uppsala University, Uppsala, Sweden; Commissariat a l'Energie Atomique (CEA), Paris, France; Université Paris-Sud, Orsay, France; Max Planck Gesellschaft, Germany; Ruhr-Universität Bochum, Germany; Biological Research Center of the Hungarian Academy of Sciences; Wageningen Universiteit, The Netherlands; and the Universite de Geneve, Switzerland.[72] The US Dept of Energy (DOE) *Joint Center for Artificial Photosynthesis* (JCAP) led by the California Institute of Technology (Caltech) and Lawrence Berkeley National Laboratory has US$122m over 5 years to build a solar fuel system. Caltech and the Massachusetts Institute of Technology have a $20 million National Science Foundation (NSF) grant to improve photon capture and catalyst efficiency, while several Energy Frontier Research Centers funded by the US DOE are focused on GAP-related endeavours.[73] In Japan, Nobel Prize winner Eiichi Negishi has established an Artificial Photosynthesis Group, based on the Catalysis Research Centre, Hokkaido University, and including researchers at University of Tokyo, Tokyo Institute of Technology, University of Tsukuba, Kyoto University and Kyushu University.[74]

Advancement in global artificial photosynthesis could involve gradual evolution of structured collaboration amongst large regional and national research projects like those set out in Table 18.3. Catalysts for such collaboration might include active promotion by a scientific body, such as the International Society of Photosynthesis Research (ISPR), or a series of conferences or workshops on Global Artificial Photosynthesis (GAP) governance involving research leaders as well as senior policy officials from bodies such as UNESCO. One such conference (GAP I) took place on Lord Howe Island, Australia on 14–18 August 2011.[75] Coordination with established international renewable energy organizations such the International Renewable Energy Agency (IRENA) and the World Council for Renewable Energy (WCRE), as well as EUROSOLAR,

Table 18.3 Representative artificial photosynthesis researchers and groups by region.

Region	*Representative AP Centres, Core Researchers & Focus*
United States	**Joint Center on Artificial Photosynthesis (JCAP)** Caltech Pasadena, California and Lawrence Berkeley National Laboratory Berkeley, California *(Lewis, Alivisatos, Yang, Brunschwig, Chandler, Frei, McFarland, Gray, Peters, Atwater, Norskov)* Demonstrate a scalable and cost-effective solar fuels generator that without use of rare materials or wires, robustly produces fuel from the sun ten times more efficiently than typical current crops. **Nocera Research Group**, MIT. *(Nocera).* Reactions of multielectron, proton-coupled transformations of energy poor substrates. **Yale Solar Group**, Yale University. (*Schmuttenmaer, Brudvig, Crabtree, Batista*) Modelling water oxidising chemistry with synthetic catalysts. **Center for Bioenergy and Photosynthesis** Arizona University *(Moore, Moore, Gust)* Carotenoid pigments and polymer polyenes in chlorophyll-based artificial reaction centers **Photosynthetic Antenna Research Center (PARC)** Pennsylvania University and Washington University in St Louis. *(Biswas, Bocian, Dutton, Lindsey, Moser, Shreve, Blankenship)* Designer peptide maquettes for scaffoldings to house tetrapyrrole chromophores with electronic properties under synthetic control, understanding photosynthesis **Argonne-Northwestern Solar Energy Research (ANSER) Center** Northwestern University, Illinois *(Wasielewski)* Self-assembly strategies for integrating light harvesting and charge separation
Europe	**Solar H Network**
Sweden	**Uppsala University**, Uppsala, Sweden *(Styring, Hammarström, Lindblad, Magnuson, Ho, Streich)* EPR spectroscopy of the natural water splitting complex in Photosystem II as well as on synthetic Ru-Mn compounds, H2 production from genetically modified cyanobacteria, Photochemical electron transfer and electrochemistry in Ru-Mn complexes and other supra molecular systems
France	**Commissariat a l'Energie Atomique** (CEA), Paris, France *(Rutherford, Leibl, Peltier, Cournac, Fontecave)* Spectroscopy of photosynthesis, hydrogenases and synthetic systems, with close attention to EPR spectroscopy of the water splitting complex in Photosystem II, bio-energetic and molecular characterization of novel electron transfer pathways in chloroplasts. **Université Paris-Sud**, Orsay, France *(Girerd, Aukauloo)* Metal-organic synthesis, spectroscopic and magnetic characterization of Mn- and Fe-containing multinuclear metal complexes

Table 18.3 (*Continued*).

Region	*Representative AP Centres, Core Researchers & Focus*
Germany	**Max Planck Gesellschaft**, Germany *(Lubitz, Wieghardt)* Synthesis and spectroscopic characterization of catalytically active coordination compounds including Fe- and Mn-systems, Ru-Mn compounds and metal/radical systems, studies of photosynthetic reaction centers **Ruhr-Universität Bochum**, Bochum, Germany *(Rögner, Happe)* Identified, cloned and characterized novel genes encoding unique Fe-hydrogenases in Chlamydomonas and in other green algae, water oxidizing photosynthesis in cyanobacteria and molecular structural studies of photosynthetic reaction centers, mass culture of cyanobacteria in photo-biofermentors
Hungary	**Biological Research Center** Hungarian Academy of Sciences, Szeged, Hungary *(Vass, Kovacs)* Hydrogen metabolism in micro-organisms; metabolic and genetic regulation of bio-hydrogen formation in both photo-synthetic and non-photosynthetic bacteria, studies of photosynthetic electron transfer in plants, cyanobacteria and algae and DNA array development to study the regulation of the photosynthetic systems upon environmental stress and variation
Netherl.	**Wageningen Universiteit**, Wageningen, The Netherlands *(Wijffels)* Photo-bioreactors
Switz.	**Universite de Geneve**, Geneva, Swizerland *(Rochaix)* Regulation of photosynthetic genes
	Other European AP Groups
Germany	**UniCat Cluster of Excellence (Unifying Concepts in Catalysis)** Freie Universitat Berlin *(Dau and Zaharieva)* Water oxidation
Germany	**Jülich Institute of Solid State Research**, Germany *(Kögerler, Botar)* Ruthenium inorganic metal oxide catalyst
UK	**Centre for Photomolecular Sciences**, Imperial College London (*Barber, Durrant*) Photosystems I and II.
Switz.	**Laboratory of Photonics and Interfaces**, Ecole Polytechnique de Lausanne (*Grätzel*) Dye-sensitized solar cells
Australia	**Water Oxidation Catalysts for Bio-Inspired Photoelectrochemical Cells** Monash University *(Spicca, MacFarlane, Swiegers)* Water splitting with manganese and nafion polymer coating **Australian National University AP Research Groups,** Australian National University *(Hillier, Wydrzynski, Chow)* Synthetic photo-catalytic proteins *(Pace, Stranger, Krausz)* **Intelligent Polymer Research Institute** University of Wollongong *(Officer)* Inorganic/organic hybrid solar cells

Table 18.3 (*Continued*).

Region	*Representative AP Centres, Core Researchers & Focus*
China	**State Key Lab of Matrix Composites**, Shanghai Jiaotong University. *(Fan)*
	Artificial leaf with titanium dioxide **State Key Laboratory of Catalysis**, Dalian (*Ho*)
	Organic Solids Laboratory, Center for Molecular Science, Institute of Chemistry, Chinese Academy of Sciences, Beijing *(Jing, Zhu)*
	Fullerene–fluorescein–anthracene hybrids
Japan	**Negishi Artificial Photosynthesis Group**, Catalysis Research Centre, Hokkaido University, University of Tokyo, Tokyo Institute of Technology, University of Tsukuba, Kyoto University and Kyushu University. *(Negishi, Takahashi, Ueda, Ohtani, Tsukuda, Ishitani, Ikeuchi)*
	Metal catalysts **Kyoto University** (*Aratani*)
	Osaka University (*Kamiya, Fukuzumi*)
Singapore	**Solar Fuels Laboratory**, Nanyang Technological University, Singapore *(Andersson)*
	Solar Fuels

Table 18.4 Global Artificial Photosynthesis (GAP) compared to other macroscience projects.[a]

Macroscience project	*Projected cost ($ billion)*	*Target completion date*	*Est. life/ years*
Apollo Space Program	40.0	1970's	10
Space Station Freedom	30.0	1999	30
Earth Observing System	17.0	2000	15
Superconducting Super Collider	11.0	1999	30
Human Genome Project	3.0	2005	Perpetual
Hubble Space Telescope	1.5	1990	15 to 20
Global Artificial Photosynthesis	15(?)	2025(?)	Perpetual

[a]Modified from: F. S. Collins, M. Morgan and A. Patrinos, The Human Genome Project: Lessons from Large-Scale Biology Science, 2003, 300, 286–290.

the non-profit European Association for Renewable Energy, will be important for GAP Project regulatory stability and credibility.[76]

A successful GAP Project must not only overcome a range of scientific challenges but organizational difficulties exemplified by similar macro-science projects (see Table 18.4). The scientific challenge for the Human Genome Project (HGP) was perhaps more clearly defined than is likely to be the case with GAP.[77] The HGP set aside a dedicated amount of money to deal with related ethical and social issues, though little of this went to perhaps the most successful related governance outcome: the UNESCO *Universal Declaration on the Human Genome and Human Rights.*[78] As with the HGP, GAP Project work is likely to be distributed across a variety of laboratories in different nations, rather than being focused in one place like the European Organization for Nuclear Research

(CERN) or the international project on fusion energy (ITER). CERN's scientific popularity and public profile have been enhanced by allowing many nations to fund new equipment (such as the Large Hadron Collider) open to use by independently-funded physicists from around the world. The Hubble Space Telescope (funded by NASA in collaboration with the European Space Agency) similarly allows any qualified scientist to submit a research proposal, successful applicants having a year after observation before their data is released to the entire scientific community.[79] A large scale GAP model might offer similar opportunities if part of its 'ramping-up' process involved providing internationally-accessible large-scale centralized research facilities. The ITER success highlights the benefits of signatories to a GAP collaboration agreeing to share scientific data, procurements, finance and staffing.[80]

18.4.2 Open-access Model

An open-access model for a GAP Project is a future perspective involving funding rules requiring public good licensing, technology transfer, ethical and social implications research, as well as rapid and free access to data. A public–private GAP partnership model might facilitate members' access to non-exclusive licenses over IMPs. Many of the debates that will impact on GAP here are already being played out in relation to synthetic biology.[81] IMPS claimed over GAP components (such as antenna systems, reaction centers and water catalysts) will be hard to identify, fragmented across many owners and sometimes overly broad.[82] These factors may make it harder for would-be GAP innovators to get licences and patents as GAP research advances.[83]

As GAP research progresses towards the mass-production and distribution phase, the need for common standards may promote a 'tipping dynamic' in which one solution owned by a single corporation quickly comes to dominate the field. GAP will probably benefit not from following the 'dog-eat-dog' patent wars of pharmaceutical research and development, but from examples such as the mobile telephone industry where no single manufacturer owns every patent that covers its product—forcing sharing arrangements. The number of licensing transactions that GAP firms face could be reduced by facilitating limited or unlimited open-access to the core, widely used parts of GAP technology.[84] A variety of strategies are available in this context under domestic legislation and international law, as will be discussed in Section 18.4.3.

Patent pools between the public–private sectors with zero royalties could be used to promote GAP collaboration, but contributors will have to ensure the terms they agree on don't violate anti-monopoly laws by excluding competitors. The complex technologies involved in GAP may cause researchers competing in this rapidly advancing area to unintentionally infringe IMPs, especially when patents are allowed to be broad and numerous or vaguely written. Companies called 'patent trolls' may infest the GAP area as they have in that of biotechnology. Such firms will acquire patents (for example from bankrupt GAP-related firms) simply to parasitically profit from the evolving research needs of GAP researchers. In another field (the software industry), the Open Innovation Network gets around this by buying up Linux-related patents.[85]

A further issue is that GAP firms and their researchers will tend to become locked into researching parts of artificial photosynthesis with which they have become familiar, have ready legal access, or are already widely used across the industry and less likely to be the subject of IMP challenges. The important point is that if the most widely used GAP components parts are subject to some form of open-access permission then the research will proceed faster. This effect can be enhanced if GAP researchers are encouraged to deliberately choose "open" rather than closed (patented) parts. GAP-involved firms and universities (given incentives such as raised profile, obtaining reciprocal access and building a user base) could donate part of their data to open source projects.[86] Additionally, companies involved in public–private linkage projects with GAP researchers might be required to specify as part of their grant application the nature and length of the IMP protection they will need. Competition from other applicants would then provide a powerful incentive for voluntarily limiting the patent duration claimed in such a submission.[87]

Many specific open-source models exist that the governance structures of a GAP Project might draw upon. Cambia's Bioforge initiative *BiOS* (Biological Open Source), for instance, is a legally enforceable framework to enable the sharing of the capability to use patented and non-patented technology materials and methods, within a dynamically expanding group whose entry was conditional upon consent to the same principles of responsible sharing, in a 'protected commons'. Those who join a BiOS 'concordance' agree not to assert IMP rights against other members' use of the technology to do research or develop products (whether for profit or public good).[88] Similarly, the *Initiative for Open Innovation* (IOI) is a new global facility, funded through grants from the Bill & Melinda Gates Foundation and The Lemelson Foundation.[89] The *BioBricks Foundation* (BBF), likewise, is a not-for-profit organization founded by engineers and scientists from MIT, Harvard, and UCSF with significant experience in both non-profit and commercial biotechnology research. BBF encourages the development and responsible use of technologies based on BioBrick™ standard DNA parts that encode basic biological functions.[90]

Members who join an open-source-structured GAP Project collaboration might be contractually granted brief periods of exclusive ownership (much less than 20y patent terms) in return for a promise to afterwards share data and receive access to a confidential database governed by trade secrets and copyright laws which are less expensive or restrictive than patents. Such researchers could publish information they supplied at any time—so blocking third parties from obtaining patents—as happens with the *Merck Gene Index* (MGIP). The MGIP promotes the unrestricted exchange of human genomic data, and facilitates progress in biomedical research by reducing duplication of efforts, speeding the identification of disease-related genes.[91]

One example of how open-source governance may assist to speed up that pace of GAP research derives from how multinational pharmaceutical companies have supported the *SNP Consortium*.[92] The *SNP Consortium* was established in 1999 as a collaboration of several companies and institutions to produce a public resource of single nucleotide polymorphisms (SNPs) in the human genome. The initial goal was to discover 300 000 SNPs in two years, but by 2001, 1.4 million SNPs had been released to the public domain.[93]

18.4.3 International Law: Photosynthesis as Planetary Common Heritage

A GAP Project governance structure emphasizing international law might protect photosynthesis from excessive patents promoting inequitable or unsustainable use within the class of United Nations treaties involved with protecting the common heritage of humanity (such provisions cover, for instance, outer space,[94] the moon,[95] deep sea bed,[96] Antarctica[97] and world natural heritage sites[98]). Five core components are generally regarded as encompassing the common heritage of humanity concept under public international law. First, there can be no private or public appropriation; no one legally owns common heritage spaces or materials. Second, representatives from all nations must manage such resources on behalf of all (this often necessitating a special agency to coordinate shared management). Third, all nations must actively share with each other the benefits acquired from exploitation of the resources from the commons heritage region, this requiring restraint on the profit-making activities of private corporate entities and linking the concept to that of global public good. Fourth, there can be no weaponry developed using common heritage materials. Fifth, the commons should be preserved for the benefit of future generations.[99,100]

The claim for GAP and its core components to common heritage status would likely be at an inchoate stage initially. Probably the closest analogies involve claims that genetic diversity of agricultural crops,[101] plant genetic resources in general,[102] biodiversity[103] or the atmosphere[104] should be treated as not just areas of common concern but subject to common heritage requirements under international law. The non-binding UNESCO *Universal Declaration on the Human Genome and Human Rights*, for example, only goes so far as to declare in Article 1 that: "The human genome underlies the fundamental unity of all members of the human family, as well as the recognition of their inherent dignity and diversity. In a symbolic sense, it is the heritage of humanity." Article 4 states: "The human genome in its natural state shall not give rise to financial gains".[105] Other international law concepts that could be influential are those that may declare GAP a global public good,[106] an aspect of technology sharing obligations,[107] or those arising under the international right to health (set out for example in article 12 of the United Nations *International Covenant on Civil and Political Rights*).[108] Some basic principles for GAP as common heritage are set out in Table 18.5. The UNESCO *Declaration on the Responsibilities of the Present Generations Towards Future Generations* expresses a concept of planetary common heritage that could encompass GAP in article 4:

Table 18.5 Photosynthesis as common heritage of humanity-basic principles.

1. Photosynthesis in its natural form is the common heritage of humanity and a global public and environmental good.
2. Photosynthesis in its artificial (for example genetically and technologically modified) forms, being derived from an understanding of natural photosynthesis, is also the common heritage of humanity and a global public and environmental good.
3. States, individuals, groups, communities, institutions and corporations, public and private shall:
- **(a)** Ensure equitable sharing (particularly with people living in poverty) of access to scientific and technological developments in artificial photosynthesis and its benefits as a source of fuel and food.
- **(b)** Ensure prompt and open-sharing amongst research communities of scientific theories and data involving artificial photosynthesis, including by encouraging and requiring limitations on intellectual monopoly rights and privileges.
- **(c)** Have due regard to the sustainability of the environment, biosphere and biodiversity as well as the interests of future generations, in the development of artificial photosynthesis and protection of natural photosynthesis.
- **(d)** Safeguard and promote equitable access to natural photosynthesis including by placing public and environmental interest restrictions on the commercial exploitation of research and development of genetically and nanotechnologically-modified photosynthetic processes.
- **(e)** Take active steps towards the establishment of an *International Global Artificial Photosynthesis Authority* (IGAPA) to collectively manage research and development of artificial photosynthesis according to the above principles.

"The present generations have the responsibility to bequeath to future generations an Earth which will not one day be irreversibly damaged by human activity. Each generation inheriting the Earth temporarily should take care to use natural resources reasonably and ensure that life is not prejudiced by harmful modifications of the ecosystems and that scientific and technological progress in all fields does not harm life on Earth."[109]

18.4.4 The Symbolic Angle: Planetary Nanomedicine

Planetary medicine is now a growing field in which the expertise of medical professionals is directed towards issues of global health and environmental protection, particularly including climate change.[110] A GAP Project could well be promoted through domestic and international media as a defining symbolic endeavour of planetary nanomedicine.[111,112] One significance of this for artificial photosynthesis researchers is that funding agencies respond indirectly to public and governmental national interest concerns, and nanotechnology, despite its great promise, still has a problematic place in the popular imagination owing to safety issues. A GAP Project therefore represents an excellent opportunity to create a high profile awareness of nanotechnology as a positive contributor to overcoming major contemporary public health and environmental problems.

18.4.5 The Corporatised, Fully Patented Model

As mentioned previously, many of the nanotechnological techniques and structures, as well as the artificial proteins involved in GAP, will be the subject of patent and other IMP claims. In most jurisdictions the relevant patent offices will require that their inventors claim such contributions to be novel, inventive (non-obvious in the USA) and useful, with a specification complete enough to allow others to make the device without undue experimentation.

The process of photosynthesis is as central to life on earth as DNA; thus there are likely to be major debates over whether patents should be allowed over any part of the photosynthetic process. Such a debate will be unlikely to inhibit patents being taken out over many aspects of GAP. The US Supreme Court, for example, has ruled that genes (despite the symbolic importance of DNA to human heritage) can be patentable if they are isolated and purified.[113]

GAP research and development will also face major issues about whether patents should cover GAP products as well as processes and functions.[114] It is likely that in the US the 'utility' for a GAP patent (as is the case for DNA) will be that it must be specific, substantial and credible.[115] If GAP IMP ownership becomes fragmented, researchers in the field may find their 'follow-on' research hampered by the high cost and difficulty in negotiating contracts with large numbers of GAP IMP owners. Each individual GAP patent owner, for example, without some prior licensing and sharing arrangement, will have an incentive to overcharge other researchers requiring access.[116]

Using IMPS to foster appropriate industry involvement (either as suppliers of equipment or resources or customers of outputs) in a GAP project will be a major issue given the tensions between public and private rights exhibited in the final stages of the HGP. Lessons from the SEMATECH (SEmiconductor MAnufacturing TECHnology) non-profit consortium may be that while large scale national funding and industry partners are necessary for initial momentum, global impact requires inclusion of industry from multiple nations and division into pure research and manufacturing subsidiaries. The *Center for Revolutionary Solar Photoconversion* (CRSP) involves public funding from two separate sources (US DOE and NSF) carefully demarcated from involvement by multinational corporate members (including DuPont, General Motors, Konarka, Lockheed Martin, Sharp and Toyota).[117]

18.5 Conclusion

The future perspective presented here is that capturing, converting and storing secure, carbon-neutral, sustainable energy from its most abundant source, the sun, will be recognized at national and international governance levels as the most important scientific and technical challenge facing humanity in the 21st century. A high profile multidisciplinary GAP Project involving some measure of open-source data sharing is likely to be a feature of such a GAP initiative if it is to satisfy societal virtues of equity and environmental sustainability. It is possible that international law may become involved in such an initiative through efforts to declare aspects of photosynthesis common heritage of humanity and so open to restricted claims of intellectual monopoly privilege.

Acknowledgements

The author is in receipt of an Australian Research Council Future Fellowship on nanotechnology and global health. He wishes to particularly thank his colleagues Warwick Hillier, Ron Pace and Fred Chow for their comments on early stages of this chapter.

References

1. United Nations. Framework Convention on Climate Change. Draft decision -/CP.15 CONFERENCE OF THE PARTIES Fifteenth session Copenhagen, 7-18 December 2009 FCCC/CP/2009/L.7 18 December 2009.
2. United Nations. Framework Convention on Climate Change. Draft decision -/CP.15 CONFERENCE OF THE PARTIES Fifteenth session Copenhagen, 7-18 December 2009 FCCC/CP/2009/L.7 18 December 2009.
3. United Nations, Millennium Development Goals http://www.un.org/millenniumgoals/.
4. D. G. Nocera, *Daedalus*, 2006, **135**, 112.
5. L. Hammarström and S. Hammes-Schiffer, *Acc. Chem. Res.*, 2009, **42**, 1859.
6. D. J. C. MacKay, *Sustainable Energy-Without the Hot Air. UIT*, Cambridge, 2009, p. 204.
7. A. B. Pittock, *Climate Change. The Science Impacts, and Solutions*, CSIRO Publishing, Collingwood, 2nd edn, 2009, p 177.
8. D. J. C. MacKay, *Sustainable Energy-Without the Hot Air*, UIT, Cambridge, 2009, p. 38.
9. M. Leslie, *Science*, 2009, **323**, 1286.
10. C. Beer, M. Reichstein, E. Tomelleri, P. Ciais, M. Jung, N. Carvalhais, C. Rödenbeck, M. A. Arain, D. Baldocchi, G. B. Bonan, A. Bondeau, A. l. Cescatti, G. Lasslop, A. Lindroth, M. Lomas, S. Luyssaert, H. Margolis, K. W. Oleson, O. Roupsard, E. Veenendaal, N. Viovy, C. Williams, F. I. Woodward 11 and D. Papale, *Science*, 2010, **329**, 834.
11. W. Hillier and T. Wydrzynski, *Coord. Chem. Rev.*, 2008, **252**, 306.
12. A. Kumar, D. D. Jones and M. A. Hann, *Energies*, 2009, **2**, 556.
13. M. Hoogwijk, A. Faaij, R. van den Broek, G Berndes, D. Gielen and W Turkenburg, *Biomass Bioenergy*, 2003, **25**, 119.
14. M. Parikka, *Biomass Bioenergy*, 2004, **27**, 613.
15. G. Fischer and L. Schrattenholzer, *Biomass Bioenergy*, 2001, **20**, 151.
16. D. J. C. MacKay, *Sustainable Energy-Without the Hot Air*, UIT, Cambridge, 2009, pp. 43–44.
17. D. J. C. MacKay, *Sustainable Energy-Without the Hot Air*, UIT, Cambridge, 2009, p. 41.
18. J. K. Hurst, *Science*, 2010, **328**, 315.

19. A. J. Ragauskas, C. K. Williams, B. H. Davison, G. Britovsek, J. Cairney, C. A. Eckert, W. J. Frederick Jr, J. P. Hallett, D. J. Leak, C. L. Liotta, J. R. Mielenz, R. Murphy, R. Templer and T. Tschaplinski, *Science*, 2006, **484**, 311.
20. R. Pace, in *Artificial Photosynthesis: from basic biology to industrial application*, ed. A. Collings and C. Critchley, Wiley-VCH Verlag, Weinheim, 2005, p. 13.
21. I. Kant, Principles of the Doctrine of Virtue, in *Collected Works*, ed. M. J. Gregor, Harper & Row, San Francisco, 1996, p. 502.
22. F. Salamanca-Buentello, D. W. Persad, E. B. Court, D. K. Martin and A. S. Daar, *PloS Med.*, 2005, **2**, e97.
23. T. A. Faunce and H. Nasu., *J. Med. Philos.*, 2009, **34**, 296.
24. R. E. Blankenship, *Molecular Mechanisms of Photosynthesis*. Blackwell Science, Oxford/Malden, 2002.
25. K. Kalyanasundaram and M. Graëtzel, *Curr. Opin. Biotechnol.*, 2010, **21**, 298.
26. V. Sgobba and D. M. Guldi, *Chem. Soc. Rev.*, 2009, **38**, 165.
27. W. Guter, J. S. Chone, S. P. Philipps, M. Steiner, G. Siefer, A. Wekkell, E. Welser, E. Oliva, A. W. Bett and F. Dimroth, *Appl. Phys. Lett.*, 2009, **94**, 223504–1.
28. G. Konstantatos and E. H. Sargent, *Nat. Nanotechnol.*, 2010, **5**, 391.
29. D. K. Kotter, S. D. Novack, W. D. Slater and P. J. Pinhero, *J. Solar Energy Eng.*, 2010, **132**, 011014.
30. I. Iñiguez-de-laTorre, J. Mateos, D. Pardo, A. M. Song and T. González, *Appl. Phys. Lett.*, 2009, **94**, 093512.
31. W. C. Brown, *IEEE Trans. Microwave Theory Tech.*, 1988, **32**, 1230.
32. N. S. Lewis and D. G. Nocera, *Proc. Natl. Acad. Sci. U. S. A.*, 2006, **103**(43), 15729.
33. M. T. Colvin, A. B. Ricks, A. M. Scott, A. L. Smeigh, R. Carmieli, T. Miura and M. R. Wasielewski, *J. Am. Chem. Soc.*, 2011, **133**, 1240.
34. S. Karlsson, J. Boixel, Y. Pellegrin, E. Blart, H.-C. Becker, F. Odobel and L. Hammarstrom, *J. Am. Chem. Soc.*, 2010, **132**, 17977.
35. W.S. Chow, *Aust. J. Plant Physiol.*, 1999, **26**, 641.
36. M. R. Wasielewski, *J. Org. Chem.*, 2006, **71**, 5051.
37. H. Lee, Y.-C. Cheng and G. R. Fleming, *Science*, 2007, **316**, 1462.
38. G. S. Engel, T. R. Calhoun, E. L. Read, T.-K. Ahn, T. Mancal, Y.-C. Cheng, R. E. Blankenship and G. R. Fleming, *Nature*, 2007, **446**, 782.
39. P. Ball, *Nat. Mater.*, 2010, **9**, 614.
40. I. McEwan, *Solar*, Jonathan Cape, 2010.
41. D. Gust, T. A. Moore and A. L. Moore, *Acc. Chem. Res.*, 2009, **42**, 1890.
42. H. N. Chapman, P. Fromme, A. Barty, T. A. White, R. A. Kirian, A. Acquila and M. S. Hunter, *et al.*, *Nature*, 2011, **470**, 73.
43. R. L. Koder, J. L. R. Anderson, L. A. Solomon, K. S. Reddy, C. C. Moser and P. L. Dutton, *Nature*, 2009, **458**, 305–309.
44. B. Conlan, *Photosynth. Res.*, 2008, **98**, 687.
45. W. Hillier and T. Wydrzynski, *Coord. Chem. Rev.*, 2008, **252**, 306.

46. D. Gust and T. A. Moore, *Science*, 1989, **244**, 35.
47. R. Carmieli, Q. Mi, A. B. Ricks, E. M. Giacobbe, S. M. Mickley and M. R. Wasielewski, *J. Am. Chem. Soc.*, 2009, **131**, 8372.
48. M. Sykora, K. A. Maxwell, J. M. DeSimone and T. J. Meyer, *Proc. Natl. Acad. Sci. U. S. A.*, 2000, **97**, 7687.
49. P. Alivisatos, *Nature*, 2011, **469**, 24.
50. W. J. Youngblood, A. L. Seung-Hyun, K. Maeda and T. E. Mallouk, *Acc. Chem. Res.*, 2009, **42**, 1966.
51. M. W. Kanan and D. G Nocera, *Science*, 2008, **321**, 1072.
52. Q. Yin, J. M. Tan, C. Besson, Y. V. Geletti, D. G. Musaev, A. E. Kuznetsov, Z. Luo, K. I. Hardcastle and C. L. Hill, *Science*, 2010, **328**, 342.
53. J. H. Alstrum-Acevedo, M. K. Brennaman and T. J. Meyer, *Inorg. Chem.*, 2005, **44**, 6802.
54. T. M. McCormick, B. D. Calitree, A. Orchard, N. D. Kraut, F. V. Bright, M. R. Detty and R. Eisenberg, *J. Am. Chem. Soc.*, 2010, **132**, 15480.
55. V. Balzani, A. Credi and M. Venturi, *ChemSusChem*, 2008, **1**, 26.
56. H. B. Gray, *Nature Chem.*, 2009, **1**, 7.
57. M. Woodhouse and B. A. Parkinson, *Chem. Soc. Rev.*, 2009, **38**, 197.
58. F. M. Toma, A. Sartorel, M. Iurlo, M. Carraro, P. Parisse, C. Maccato, S. Rapino, B. R. Gonzalez, H. Amenitsch, T. D. Ros, L. Casalis, A. Goldoni, M. Marcaccio, G. Scorrano, G. Scoles, F. Paolucci, M. Prato and M. Bonchio, *Nat. Chem.*, 2010, **2**, 826.
59. V. Sgobba and D. M. Guidi, *Chem. Soc. Rev.*, 2009, **38**, 165.
60. Y. S. Nam, A. P. Magyar, D. Lee, J. W. Kim, D. S. Yun, H. Park, T. S. Pollom, D. A Weitz and A. M Belcher, *Nat. Nanotechnol.*, 2010, **5**, 340.
61. D. Gust, T. A. Moore and A. L. Moore, *Acc. Chem. Res.*, 2009, **42**, 1890.
62. A. Magnuson, M. Anderlund, O. Johansson, P. Lindblad, R. Lomoth, T. Polivka, S. Ott, K. Stensjö, S. Styring, V. Sundström and L. Hammarström, *Acc. Chem. Res.*, 2009, **42**(12), 1899.
63. D. Gust and T. A. Moore, *Science*, 1989, **244**, 35.
64. M. R. Dubois and D. L. Dubois, *Acc. Chem. Res.*, 2009, **42**, 1974.
65. A. J. Morris, G. J. Meyer and E. Fujita, *Acc. Chem. Res.*, 2009, **42**, 1983.
66. J. L. Dempsey, B. S. Brunschwig, J. R. Winkler and H. B. Gray, *Acc. Chem. Res.*, 2009, **42**, 1995.
67. M. R. Wasielewski, *J. Org. Chem.*, 2006, **71**, 5051.
68. R. Pace, in *Artificial Photosynthesis: from basic biology to industrial application*, ed. A. Collings and C. Critchley, Wiley-VCH Verlag, Weinheim, 2005, p. 13.
69. M. Blakeney, in *Intellectual Property in the New Millennium*, ed. D. Vaver and L. Bently, Cambridge University Press, Cambridge, 2004, p. 3.
70. M. K. Cho and D. A. Relman, *Science*, 2010, **329**, 38.
71. K. Sanderson, *Nature*, 2008, **452**, 400.
72. Solar H network. http://www.fotomol.uu.se/Forskning/Biomimetics/solarh/index.shtm.
73. J. Tollefson, *Nature*, 2010, **466**, 541.

74. Nikkei.com, accessed Jan 19 2011, http://e.nikkei.com/e/fr/tnks/Nni20110118D18JFA13.htm.
75. Towards Global Artificial Photosynthesis: Energy, Nanochemistry and Governance. http://law.anu.edu.au/coast/tgap/conf.htm.
76. International Renewable Energy Agency, http://www.irena.org/; American Council on Renewable Energy, http://www.acore.org/front; European Association for Renewable Energy, http://www.eurosolar.de/en/.
77. R. H. Waterston, E. S. Lander and J. E. Sulston, *Proc. Natl. Acad. Sci. U. S. A.*, 2002, **99**, 3712.
78. T. A. Faunce and H. Nasu, *J. Med. Philos.*, 2009, **34**, 296.
79. NASA, Hubble Telescope, http://www.nasa.gov/mission_pages/hubble/main/index.html.
80. R. Hiwatari, K. Okano, Y. Asaoka, K. Shinya and Y. Ogawa, *Nucl. Fusion*, 2005, **45**, 96.
81. J. A. Goldstein, *Gene Patents and Collaborative Licensing Models*, ed. G. van Overwalle, Cambridge University Press, Cambridge, 2009, p. 50.
82. G. van Overwalle, E. van Zimmeren, B. Verbeure and G. Matthijs, *Nat. Rev. Genet.*, 2006, **7**, 143.
83. Intellectual Property Office, *UK Innovation Nanotechnology Patent Landscape Analysis*, 2009, http://www.ipo.gov.uk/types/patent/p-informatic/p-informatic-report.htm.
84. J. Henkel and S. M. Maurer, *Nat. Biotechnol.*, 2009, **12**, 1095.
85. B. Verbeure, *Gene Patents and Collaborative Licensing Models*, ed. G. van Overwalle, Cambridge University Press, Cambridge, 2009, p. 3.
86. M. Spence, *Gene Patents and Collaborative Licensing Models*, ed. G. van Overwalle, Cambridge University Press, Cambridge, 2009, p. 161.
87. J. Henkel and S. M. Maurer, *Nat. Biotechnol.*, 2009, **12**, 1095.
88. BIOS, http://www.bios.net/daisy/bios/mta.html.
89. Initiative for Open Innovation, http://www.openinnovation.org/daisy/ioi/home.html.
90. Biobrick Foundation, http://bbf.openwetware.org/.
91. A. R. Williamson, *Drug Discov. Today*, 1999, **4**, 115.
92. J. Henkel and S. M. Maurer, *Nat. Biotechnol.*, 2009, **12**, 1095.
93. G. A. Thorisson and L. D. Stein, *Nucl. Acids Res.*, 2003, **31**, 124.
94. United Nations, *Treaty on Principles Governing the Activities of States in the Exploration and Use of Outer Space, Including the Moon and Other Celestial Bodies*, art 1, Jan. 27, 1967, 18 U.S.T. 2410, 610 U.N.T.S. 205.
95. United Nations, *Agreement Governing Activities of States on the Moon and Other Celestial Bodies* art. 1, Dec. 17, 1979, 18 I.L.M. 1434.
96. United Nations, *Convention on the Law of the Sea* art. 1, para. 1, Dec. 10, 1982, 1833 U.N.T.S. 397.
97. United Nations, *Antarctic Treaty*, art. VI., Dec. 1, 1959, 12 U.S.T. 794, 402 U.N.T.S. 72.
98. UNESCO, World Heritage Convention, http://whc.unesco.org/en/conventiontext/.
99. J. Frakes, *Wisconsin International Law Journal*, 2003, **21**, 409.

100. A. Pardo, *Proceedings of the American Society of International Law*, 1968, **62**, 216.
101. C. Fowler, *Green Globe Yearbook*, Oxford University Press, Oxford, 1993, p. 33.
102. United Nations Food and Agricultural Organisation (FAO), *International Undertaking on Plant Genetic Resources*. Art 1. Res 8/83 (1983).
103. M. Bowman and C. Redgwell, *International Law and Conservation of Biological Diversity*, Kluwer Law International, The Hague, 1996, pp. 39–40.
104. United Nations, *Legal Status of the Atmosphere*. Para 1. UN Res 43/53 6 Dec 1988.
105. UNESCO, Universal Declaration on the Human Genome and Human Rights, http://portal.unesco.org/shs/en/ev.php-URL_ID=1881&URL_DO=DO_TOPIC&URL_SECTION=201.htm.
106. I. Kaul, in *Limits to Privatisation*, ed. E. U. von Weisacker, O. R. Young and M. Finger, Earthscan, London, 2006, p. 311.
107. T. A. Faunce and H. Nasu, *Public Health Ethics*, 2008, **1**, 146.
108. United Nations, *International Covenant on Economic, Social and Cultural Rights*. Adopted and opened for signature, ratification and accession by General Assembly resolution 2200A (XXI) of 16 December 1966. http://www2.ohchr.org/english/law/cescr.htm; entry into force 3 January 1976.
109. UNESCO, *Declaration on the Responsibilities of the Present Generations Towards Future Generations*, November 12, 1997 http://portal.unesco.org/en/ev.php-URL_ID=13178&URL_DO=DO_TOPIC&URL_SECTION=201.html.
110. T. McMichael, *Science.*, 2002, **297**, 1093.
111. T. A. Faunce, *Nanotechnology for Sustainable Energy Conference*, sponsored by the European Science Foundation, July 2010, Obergurgl, Austria.
112. T. A. Faunce, *15th International Congress of Photosynthesis*, August 2010, Beijing.
113. L. J. Demaine and A. X. Fellmeth, *Stanford Law Review*, 2002, **55**, 303.
114. J. Calvert, *Science as Culture*, 2007, **16**, 207.
115. USPTO (2001), Utility examination guidelines, Federal register Friday 5 Jan 2001 66 (4) http://www.ustpo.gov/web/offices/com/sol/notices/utilexmguide.pdf.
116. J. Henkel and S. M. Maurer, *Nat. Biotechnol.*, 2009, **12**, 1095.
117. Center for Revolutionary Solar Photoconversion, http://www.colorado-collaboratory.org/crsp/, accessed 13 July 2010.

Subject Index